DIFFERENTIAL AND INTEGRAL CALCULUS

DIFFERENTIAL AND INTEGRAL CALCULUS

BY

R. COURANT

Professor of Mathematics in New York University

TRANSLATED BY

E. J. McSHANE

Professor of Mathematics in the University of Virginia

VOLUME II

Wiley Classics Library Edition Published 1988

WILEY

WILEY - INTERSCIENCE

A DIVISION OF JOHN WILEY & SONS, INC.

First Published 1936
Reprinted 1937, 1940, 1941, 1942 (*twice*), 1943, 1944, 1945, 1947, 1948, 1949, 1950, 1951, 1952, 1953, 1954, 1956 (*twice*), 1957 (*twice*), 1958, 1959, 1961, 1962, 1964 (*twice*), 1966, 1967, 1968

4 5 6 7 8 9 10

ISBN: 0-471-17853-5

ISBN 0-471-60840-8 (pbk.).

Printed in the United States of America

PREFACE

The present volume contains the more advanced parts of the differential and integral calculus, dealing mainly with functions of several variables. As in Volume I, I have sought to make definitions and methods follow naturally from intuitive ideas and to emphasize their physical interpretations—aims which are not at all incompatible with rigour.

I would impress on readers new to the subject, even more than I did in the preface to Volume I, that they are not expected to read a book like this consecutively. Those who wish to get a rapid grip of the most essential matters should begin with Chapter II, and next pass on to Chapter IV; only then should they fill in the gaps by reading Chapter III and the appendices to the various chapters. It is by no means necessary that they should study Chapter I systematically in advance.

The English edition differs from the German in many details, and contains a good deal of additional matter. In particular, the chapter on differential equations has been greatly extended. Chapters on the calculus of variations and on functions of a complex variable have been added, as well as a supplement on real numbers.

I have again to express my very cordial thanks to my German publisher, Julius Springer, for his generous attitude in consenting to the publication of the English edition. I have also to thank Blackie & Son, Ltd., and their staff, especially Miss W. M. Deans, for co-operating with me and my assistants and relieving me of a considerable amount of proof reading. Finally

I must express my gratitude to the friends and colleagues who have assisted me in preparing the manuscript for the press, reading the proofs, and collecting the examples; in the first place to Dr. Fritz John, now of the University of Kentucky, and to Miss Margaret Kennedy, Newnham College, Cambridge, and also to Dr. Schönberg, Swarthmore College, Swarthmore, Pa.

R. COURANT.

NEW ROCHELLE, NEW YORK.
March, 1936.

CONTENTS

CHAPTER I

PRELIMINARY REMARKS ON ANALYTICAL GEOMETRY AND VECTOR ANALYSIS

CHAPTER II

FUNCTIONS OF SEVERAL VARIABLES AND THEIR DERIVATIVES

APPENDIX

CHAPTER III

DEVELOPMENTS AND APPLICATIONS OF THE DIFFERENTIAL CALCULUS

APPENDIX

CHAPTER IV

MULTIPLE INTEGRALS

APPENDIX

CHAPTER V

INTEGRATION OVER REGIONS IN SEVERAL DIMENSIONS

APPENDIX

CHAPTER VI

DIFFERENTIAL EQUATIONS

CHAPTER VII

CALCULUS OF VARIATIONS

Chapter VIII

FUNCTIONS OF A COMPLEX VARIABLE

SUPPLEMENT

CHAPTER I

Preliminary Remarks on Analytical Geometry and Vector Analysis

In the interpretation and application of the mathematical facts which form the main subject of this second volume it is often convenient to use the simple fundamental concepts of analytical geometry and vectoı analysis. Hence, even though many readers will already have a certain knowledge of these subjects, it seems advisable to summarize their elements in a brief introductory chapter. This chapter, however, need not be studied before the rest of the book is read; the reader is advised to refer to the facts collected here only when he finds the need of them in studying the later parts of the book.

1. Rectangular Co-ordinates and Vectors

1. Co-ordinate Axes.

To fix a point in a plane or in space, as is well known, we generally make use of a rectangular co-ordinate system. In the plane we take two perpendicular lines, the x-axis and the y-axis; in space we take three mutually perpendicular lines, the x-axis, the y-axis, and the z-axis. Taking the same unit of length on each axis, we assign to each point of the plane an x-co-ordinate and a y-co-ordinate in the usual way, or to each point in space an x-co-ordinate, a y-co-ordinate, and a z-co-ordinate (fig. 1). Conversely, to every set of values (x, y) or (x, y, z) there corresponds just one point of the plane, or of space, as the case may be; a point is completely determined by its co-ordinates.

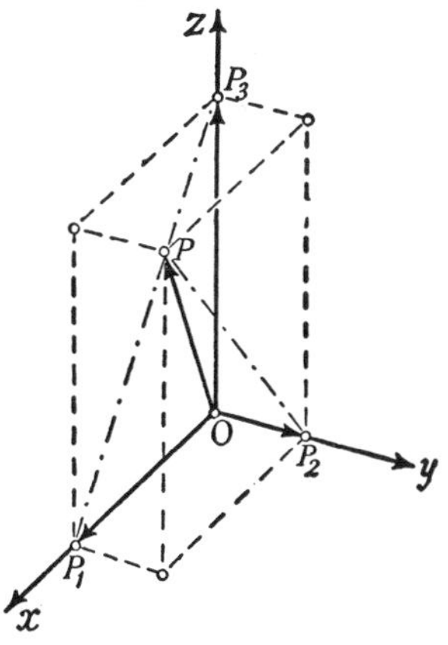

Fig. 1.—Co-ordinate axes in space

Using the theorem of Pythagoras we find that *the distance between two points* (x_1, y_1) and (x_2, y_2) is given by

$$r = \sqrt{(x_1 - x_2)^2 + (y_1 - y_2)^2},$$

while the distance between the points with co-ordinates (x_1, y_1, z_1) and (x_2, y_2, z_2) is

$$r = \sqrt{(x_1 - x_2)^2 + (y_1 - y_2)^2 + (z_1 - z_2)^2}.$$

In setting up a system of rectangular axes we must pay attention to the *orientation* of the co-ordinate system.

In Vol. I, Chap. V, § 2 (p. 268) we distinguished between positive and

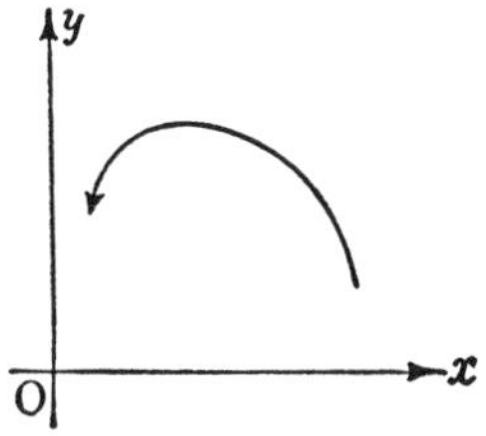

Fig. 2.—Right-handed system of axes

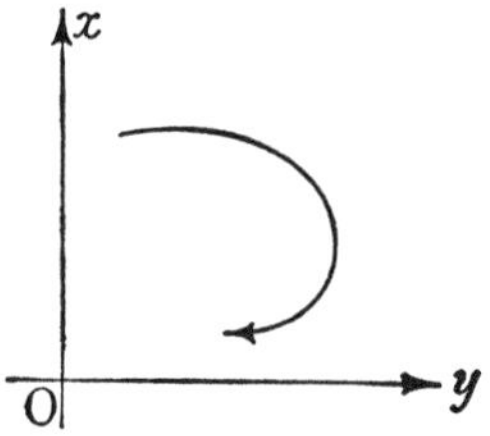

Fig. 3.—Left-handed system of axes

negative senses of rotation in the plane. The rotation through 90° which brings the *positive* x-axis of a plane co-ordinate system into the position of the *positive* y-axis in the shortest way defines a sense of rotation. According as this sense of rotation is positive or negative, we say that the system of axes is *right-handed* or *left-handed* (cf. figs. 2 and 3). It is impossible to change a right-handed system into a left-handed system by a rigid motion confined to the plane. A similar distinction occurs with co-ordinate systems

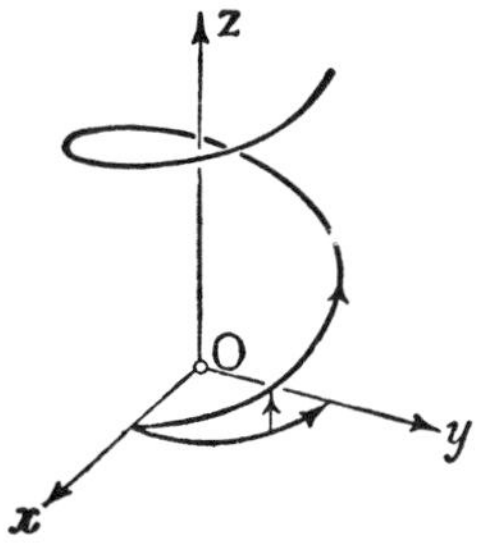

Fig. 4.—Right-handed screw

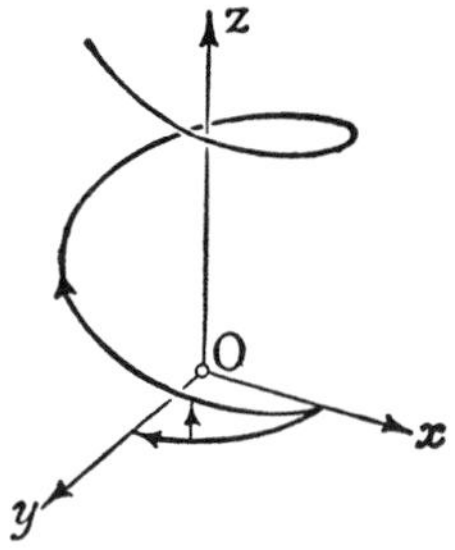

Fig. 5.—Left-handed screw

in space. For if one imagines oneself standing on the xy-plane with one's head in the direction of the positive z-axis, it is possible to distinguish two types of co-ordinate system by means of the apparent orientation of the co-ordinate system in the xy-plane. If this system is right-handed the system in space is also said to be right-handed, otherwise left-handed (cf. figs. 4 and 5). A right-handed system corresponds to an ordinary right-handed screw; for if we make the xy-plane rotate about the z-axis (in the sense prescribed by its orientation) and simultaneously give it a motion of translation along the positive z-axis, the combined motion is obviously that of a right-handed screw. Similarly, a left-handed system corresponds to a left-handed screw. No rigid motion in three dimensions can transform a left-handed system into a right-handed system.

In what follows we shall always use right-handed systems of axes.

We may also assign an orientation to a system of three arbitrary axes passing through one point, provided these axes do not all lie in one plane, just as we have done here for a system of rectangular axes.

2. Directions and Vectors. Formulæ for Transforming Axes.

An oriented line l in space or in a plane, that is, a line traversed in a definite sense, represents a *direction*; every oriented line that can be made to coincide with the line l in position and sense by displacement parallel to itself represents the same direction. It is customary to specify a direction relative to a co-ordinate system by drawing an oriented half-line in the given direction, starting from the origin of the co-ordinate system, and on this half-line taking the point with co-ordinates (α, β, γ) which is at unit distance from the origin. The numbers α, β, γ are called the *direction cosines* of the direction. They are the cosines of the three angles δ_1, δ_2, δ_3 which the oriented line l makes with the positive x-axis, y-axis, and z-axis * (cf. fig. 6); by the distance formula, they satisfy the relation

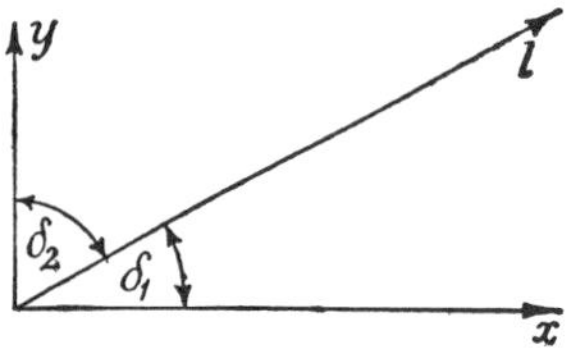

Fig. 6.—The angles which a straight line makes with the axes

$$\alpha^2 + \beta^2 + \gamma^2 = 1.$$

If we restrict ourselves to the xy-plane, a direction can be specified by the angles δ_1, δ_2 which the oriented line l having this direction and passing through the origin forms with the positive x-axis and y-axis; or by the direction cosines $\alpha = \cos\delta_1$, $\beta = \cos\delta_2$, which satisfy the equation

$$\alpha^2 + \beta^2 = 1.$$

A line-segment of given length and given direction we shall call a *vector*; more specifically, a *bound vector* if the initial point is fixed in space, and a *free vector* if the position of the initial point is immaterial. In the following pages, and indeed throughout most of the book, we shall omit the adjectives free and bound, and if nothing is said to the contrary we shall always take the vectors to be free vectors. We denote vectors by heavy type, e.g. $\boldsymbol{a}$, $\boldsymbol{b}$, $\boldsymbol{c}$, $\boldsymbol{x}$, $\boldsymbol{A}$. Two free vectors are said to be equal if one of them can be made to coincide with the other by displacement parallel to itself. We sometimes call the length of a vector its absolute value and denote it by $|\boldsymbol{a}|$.

* The angle which one oriented line forms with another may always be taken as being between 0 and π, for in what follows only the cosines of such angles will be considered.

If from the initial and final points of a vector v we drop perpendiculars on an oriented line l, we obtain an oriented segment on l corresponding to the vector. If the orientation of this segment is the same as that of l, we call its length the *component of v in the direction of l*; if the orientations are opposite, we call the negative of the length of the segment the *component of v in the direction of l*. The component of v in the direction of l we denote by v_l. If δ is the angle between the direction of v and that of l (cf. fig. 7), we always have

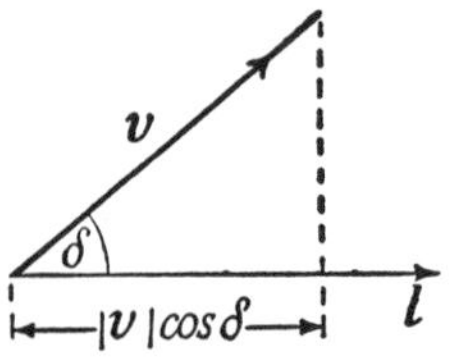

Fig. 7.—Projection of a vector

$$v_l = |v| \cos\delta.$$

A vector v of length 1 is called a *unit vector*. Its component in a direction l is equal to the cosine of the angle between l and v. The components of a vector v in the directions of the three axes of a co-ordinate system are denoted by v_1, v_2, v_3. If we transfer the initial point of v to the origin, we see that

$$|v| = \sqrt{v_1^2 + v_2^2 + v_3^2}.$$

If α, β, γ are the direction cosines of the direction of v, then

$$v_1 = |v|\alpha, \quad v_2 = |v|\beta, \quad v_3 = |v|\gamma.$$

A free vector is completely determined by its components v_1, v_2, v_3.

An equation

$$v = w$$

between two vectors is therefore equivalent to the three ordinary equations

$$v_1 = w_1,$$
$$v_2 = w_2,$$
$$v_3 = w_3.$$

There are two different reasons why the use of vectors is natural and

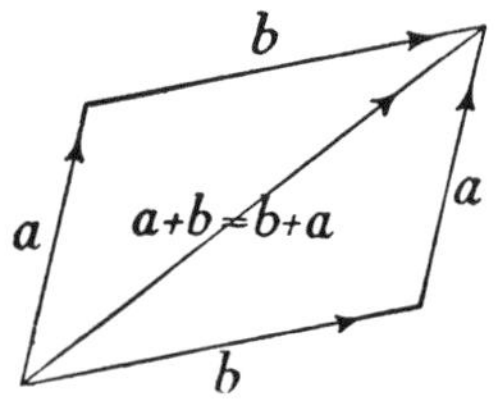

Fig. 8.—Commutative law of vector addition

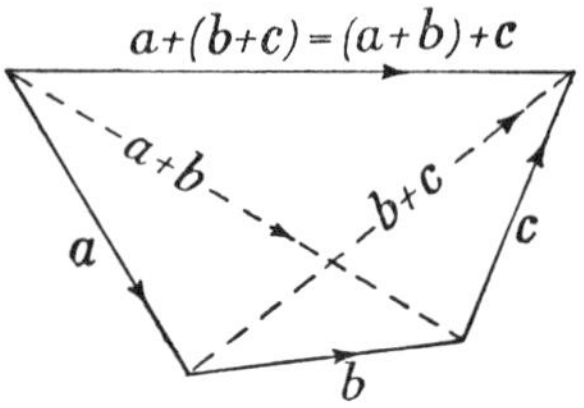

Fig. 9.—Associative law of vector addition

advantageous. Firstly, many geometrical concepts, and a still greater number of physical concepts, such as force, velocity, acceleration, &c., immediately reveal themselves as vectors independent of the particular co-ordinate system. Secondly, we can set up simple rules for calculating with vectors analogous to the rules for calculating with ordinary numbers; by means of these many arguments can be developed in a simple way, independently of the particular co-ordinate system chosen.

We begin by defining the *sum* of the two vectors $\boldsymbol{a}$ and $\boldsymbol{b}$. For this purpose we displace the vector $\boldsymbol{b}$ parallel to itself until its initial point coincides with the final point of $\boldsymbol{a}$. Then the initial point of $\boldsymbol{a}$ and the final point of $\boldsymbol{b}$ determine a new vector $\boldsymbol{c}$ (see fig. 8) whose initial point is the initial point of $\boldsymbol{a}$ and whose final point is the final point of $\boldsymbol{b}$. We call $\boldsymbol{c}$ the sum of $\boldsymbol{a}$ and $\boldsymbol{b}$ and write

$$\boldsymbol{a} + \boldsymbol{b} = \boldsymbol{c}.$$

For this additive process the *commutative law*

$$\boldsymbol{a} + \boldsymbol{b} = \boldsymbol{b} + \boldsymbol{a}$$

and the *associative law*

$$\boldsymbol{a} + (\boldsymbol{b} + \boldsymbol{c}) = (\boldsymbol{a} + \boldsymbol{b}) + \boldsymbol{c} = \boldsymbol{a} + \boldsymbol{b} + \boldsymbol{c}$$

obviously hold, as a glance at figs. 8 and 9 shows.

From the definition of vector addition we at once obtain the "projection theorem": *the component of the sum of two or more vectors in a direction l is equal to the sum of the components of the individual vectors in that direction*, that is,

$$(\boldsymbol{a} + \boldsymbol{b})_l = a_l + b_l.$$

In particular, the components of $\boldsymbol{a} + \boldsymbol{b}$ in the directions of the co-ordinate axes are $a_1 + b_1$, $a_2 + b_2$, $a_3 + b_3$.

To form the sum of two vectors we accordingly have the following simple rule. *The components of the sum are equal to the sums of the corresponding components of the summands.*

Every point P with co-ordinates (x, y, z) may be determined by the *position vector* from the origin to P, whose components in the directions of the axes are just the co-ordinates of the point P. We take three unit vectors in the directions of the three axes, $\boldsymbol{e}_1$ in the x-direction, $\boldsymbol{e}_2$ in the y-direction, $\boldsymbol{e}_3$ in the z-direction. If the vector $\boldsymbol{v}$ has the components v_1, v_2, v_3, then

$$\boldsymbol{v} = v_1\boldsymbol{e}_1 + v_2\boldsymbol{e}_2 + v_3\boldsymbol{e}_3.$$

We call $\boldsymbol{v}_1 = v_1\boldsymbol{e}_1$, $\boldsymbol{v}_2 = v_2\boldsymbol{e}_2$, $\boldsymbol{v}_3 = v_3\boldsymbol{e}_3$ the *vector components of* $\boldsymbol{v}$.

Using the projection theorem stated above, we easily obtain the *transformation formulæ* which determine (x', y', z'), the co-ordinates of a given point P with respect to the axes Ox', Oy', Oz', in terms of (x, y, z), its co-ordinates with respect to another set * of axes Ox, Oy, Oz which has the same origin as the first set and may be obtained from it by rotation. The three new axes form angles with the three old axes whose cosines may be

* It is to be noted that in accordance with the convention adopted on p. 3 both systems of axes are to be right-handed.

expressed by the following scheme, where for example γ_1 is the cosine of the angle between the x'-axis and the z-axis:

	x	y	z
x'	α_1	β_1	γ_1
y'	α_2	β_2	γ_2
z'	α_3	β_3	γ_3

From P we drop perpendiculars to the axes Ox, Oy, Oz, their feet being P_1, P_2, P_3 (cf. fig. 1, p. 1). The vector from O to P is then equal to the sum of the vectors from O to P_1, from O to P_2, and from O to P_3. The direction cosines of the x'-axis relative to the axes Ox, Oy, Oz are α_1, β_1, γ_1, those of the y'-axis α_2, β_2, γ_2, and those of the z'-axis α_3, β_3, γ_3. By the projection theorem we know that x', which is the component of the vector $\overrightarrow{OP}$ in the direction of the x'-axis, must be equal to the sum of the components of $\overrightarrow{OP_1}$, $\overrightarrow{OP_2}$, $\overrightarrow{OP_3}$ in the direction of the x'-axis, so that

$$x' = \alpha_1 x + \beta_1 y + \gamma_1 z,$$

for $\alpha_1 x$ is the component of x in the direction of the x'-axis, and so on. Carrying out similar arguments for y' and z', we obtain the *transformation formulæ*

$$\begin{aligned} x' &= \alpha_1 x + \beta_1 y + \gamma_1 z \\ y' &= \alpha_2 x + \beta_2 y + \gamma_2 z \\ z' &= \alpha_3 x + \beta_3 y + \gamma_3 z, \end{aligned}$$

and conversely

$$\begin{aligned} x &= \alpha_1 x' + \alpha_2 y' + \alpha_3 z' \\ y &= \beta_1 x' + \beta_2 y' + \beta_3 z' \\ z &= \gamma_1 x' + \gamma_2 y' + \gamma_3 z'. \end{aligned}$$

Since the components of a bound vector $\boldsymbol{v}$ in the directions of the axes are expressed by the formulæ

$$\begin{aligned} v_1 &= x_2 - x_1 \\ v_2 &= y_2 - y_1 \\ v_3 &= z_2 - z_1, \end{aligned}$$

in which (x_1, y_1, z_1) are the co-ordinates of the initial point and (x_2, y_2, z_2) the co-ordinates of the final point of $\boldsymbol{v}$, it follows that the *same* transformation formulæ hold for the components of the *vector* as for the co-ordinates:

$$\begin{aligned} v_1' &= \alpha_1 v_1 + \beta_1 v_2 + \gamma_1 v_3 \\ v_2' &= \alpha_2 v_1 + \beta_2 v_2 + \gamma_2 v_3 \\ v_3' &= \alpha_3 v_1 + \beta_3 v_2 + \gamma_3 v_3. \end{aligned}$$

3. Scalar Multiplication of Vectors.

Following conventions like those for the addition of vectors, we now define the product of a vector $\boldsymbol{v}$ by a number c: if $\boldsymbol{v}$ has the components

v_1, v_2, v_3, then cv is the vector with components cv_1, cv_2, cv_3. This definition agrees with that of vector addition, for $v + v = 2v$, $v + v + v = 3v$, and so on. If $c > 0$, cv has the same direction as v, and is of length $c\,|\,v\,|$; if $c < 0$, the direction of cv is opposite to the direction of v, and its length is $(-c)\,|\,v\,|$. If $c = 0$, we see that cv is the *zero vector* with the components 0, 0, 0.

We can also define the *product* of two vectors u and v, where this "multiplication" of vectors satisfies rules of calculation which are in part similar to those of ordinary multiplication. There are two different kinds of vector multiplication. We begin with *scalar multiplication*, which is the simpler and the more important for our purposes.

By the scalar product * uv *of the vectors* u *and* v *we mean the product of their absolute values and the cosine of the angle* δ *between their directions*:

$$uv = |\,u\,|\,|\,v\,|\cos\delta.$$

The scalar product, therefore, is simply the component of one of the vectors in the direction of the other multiplied by the length of the second vector.

From the projection theorem the *distributive law* for multiplication,

$$(u + v)w = uw + vw,$$

follows at once, while the *commutative law*,

$$uv = vu,$$

is an immediate consequence of the definition.

On the other hand, there is an essential difference between the scalar product of two vectors and the ordinary product of two numbers, for *the product can vanish although neither factor vanishes.*

If the lengths of u *and* v *are not zero, the product* uv *vanishes if, and only if, the two vectors* u *and* v *are perpendicular to one another.*

In order to express the scalar product in terms of the components of the two vectors, we take both the vectors u and v with initial points at the origin. We denote their vector components by u_1, u_2, u_3 and v_1, v_2, v_3 respectively, so that $u = u_1 + u_2 + u_3$ and $v = v_1 + v_2 + v_3$. In the equation $uv = (u_1 + u_2 + u_3)(v_1 + v_2 + v_3)$ we can expand the product on the right in accordance with the rules of calculation which we have just established; if we notice that the products u_1v_2, u_1v_3, u_2v_1, u_2v_3, u_3v_1, and u_3v_2 vanish because the factors are perpendicular to one another, we obtain $uv = u_1v_1 + u_2v_2 + u_3v_3$. Now the factors on the right have the same direction, so that by definition $u_1v_1 = u_1v_1$, &c., where u_1, u_2, u_3 and v_1, v_2, v_3 are the components of u and v respectively. Hence

$$uv = u_1v_1 + u_2v_2 + u_3v_3.$$

This equation could have been taken as the definition of the scalar product, and is an important rule for calculating the scalar product of two vectors

* Often called the *inner product*.

given in terms of their components. In particular, if we take $\boldsymbol{u}$ and $\boldsymbol{v}$ as unit vectors with direction cosines α_1, α_2, α_3 and β_1, β_2, β_3 respectively, the scalar product is equal to the cosine of the angle between $\boldsymbol{u}$ and $\boldsymbol{v}$, which is accordingly given by the formula

$$\cos\delta = \alpha_1\beta_1 + \alpha_2\beta_2 + \alpha_3\beta_3.$$

The *physical* meaning of the scalar product is exemplified by the fact, proved in elementary physics, that a force $\boldsymbol{f}$ which moves a particle of unit mass through the directed distance $\boldsymbol{v}$ does work amounting to $\boldsymbol{fv}$.

4. The Equations of the Straight Line and of the Plane.

Let a straight line in the xy-plane or a plane in xyz-space be given. In order to find their equations we erect a perpendicular to the line (or

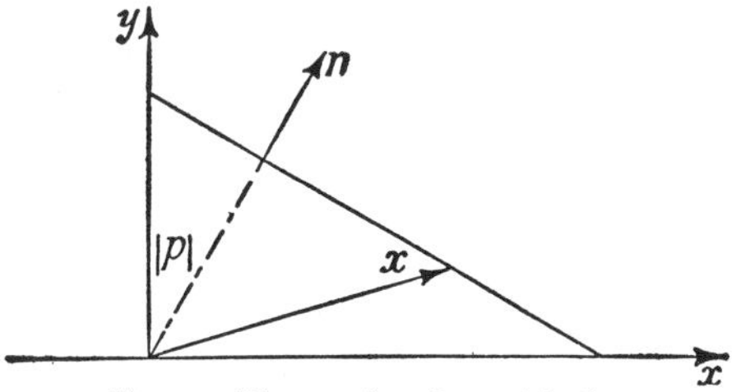

Fig. 10.—The equation of a straight line

the plane) and specify a definite " positive direction along the normal ", perpendicular to the line (or plane); it does not matter which of the two possible directions is taken as positive (cf. fig. 10). The vector with unit length and the direction of the positive normal we denote by $\boldsymbol{n}$. The points of the line (or plane) are characterized by the property that the position vector $\boldsymbol{x}$ from the origin to them has a constant projection p on the direction of the normal; in other words, the scalar product of this position vector and the normal vector $\boldsymbol{n}$ is constant. If α, β (or α, β, γ) are the direction cosines of the positive direction of the normal, that is, the components of $\boldsymbol{n}$, then

$$\alpha x + \beta y - p = 0$$

$$(\text{or} \quad \alpha x + \beta y + \gamma z - p = 0)$$

is the required equation of the line (or plane). Here p has the following meaning: the absolute value $|p|$ of p is the distance of the line (or plane) from the origin. Moreover, p is positive if the line (or plane) does not pass through the origin and $\boldsymbol{n}$ is in the direction of the perpendicular *from* the origin *to* the line (or plane); p is negative if the line (or plane) does not pass through the origin and $\boldsymbol{n}$ has the opposite direction; p is zero if the line (or plane) passes through the origin. Conversely, if α, β (or α, β, γ) are direction cosines, this equation represents a line (or plane) at a distance p from the origin, whose normal has these direction cosines.

The expression $\alpha x + \beta y - p$ (or $\alpha x + \beta y + \gamma z - p$) on the left-hand side of this so-called *normal* or *canonical form* of the equation of the straight line (or plane) also has a geometrical meaning for any point P (x, y) not lying on the line (or plane). Since $\alpha x + \beta y$ (or $\alpha x + \beta y + \gamma z$) is the projection of the position vector from O to P on the normal, we see at once that *the expression* $\alpha x + \beta y - p$ (*or* $\alpha x + \beta y + \gamma z - p$) *is the perpendicular distance of the point* P *from the line* (*or plane*) *and is positive for points on one side of the line or plane* (*namely, that on which the normal is positive*) *and negative for points on the other side.*

From the canonical form of the equation we obtain other forms of equation for the straight line (or plane) by multiplying by an arbitrary non-vanishing factor. Conversely, an arbitrary linear equation

$$Ax + By + D = 0 \text{ (or } Ax + By + Cz + D = 0)$$

represents a straight line (or plane) provided the coefficients A, B (or A, B, C) are not all zero.* In the second of these equations, for example, we may divide by $\sqrt{A^2 + B^2 + C^2}$ and put

$$\alpha = \frac{A}{\sqrt{A^2 + B^2 + C^2}}, \quad \beta = \frac{B}{\sqrt{A^2 + B^2 + C^2}},$$

$$\gamma = \frac{C}{\sqrt{A^2 + B^2 + C^2}}, \quad p = -\frac{D}{\sqrt{A^2 + B^2 + C^2}}.$$

In this way we obtain an equation which is seen to represent a plane at a distance p from the origin, whose normal has the direction cosines α, β, γ. Corresponding remarks hold for the equation of the straight line.

A straight line in space may be determined by any two planes passing through the line. For a line in space we thus obtain two linear equations

$$A_1x + B_1y + C_1z + D_1 = 0,$$
$$A_2x + B_2y + C_2z + D_2 = 0,$$

which are satisfied by (x, y, z), the co-ordinates of any point on the line. Since an infinite number of planes pass through a given line, this representation of a line in space is not unique.

Frequently it is more convenient to represent a line analytically in *parametric form* by means of a parameter t. If we consider three linear functions of t,

$$x = a_1 + b_1t,$$
$$y = a_2 + b_2t,$$
$$z = a_3 + b_3t,$$

where the b's are not all zero, then as t traverses the number axis the point (x, y, z) describes a straight line. This we see at once by eliminating t between each pair of equations, whereby we obtain two linear equations for x, y, z.

* If $A = B = 0$ (or $A = B = C = 0$), D must also be zero, and any point of the plane (or of space) satisfies the equation.

The direction cosines α, β, γ of the line in its parametric form are proportional to the coefficients b_1, b_2, b_3. For these direction cosines are

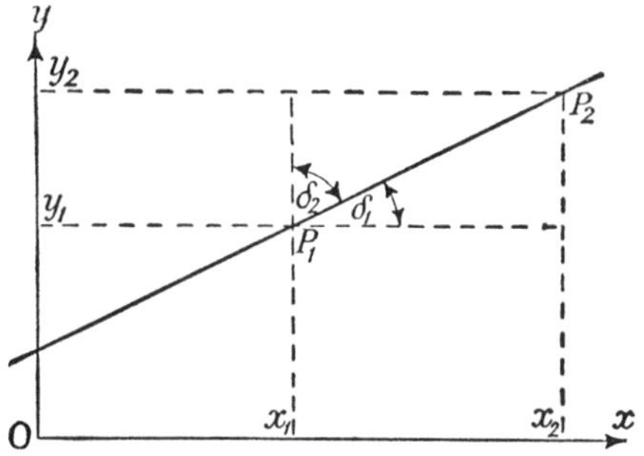

Fig. 11.—Parametric representation of a straight line passing through two points

proportional (cf. fig. 11) to $x_1 - x_2$, $y_1 - y_2$, $z_1 - z_2$, the differences of the co-ordinates of two points P_1, P_2 with co-ordinates

$$x_1 = a_1 + b_1 t_1, \quad y_1 = a_2 + b_2 t_1, \quad z_1 = a_3 + b_3 t_1$$

and

$$x_2 = a_1 + b_1 t_2, \quad y_2 = a_2 + b_2 t_2, \quad z_2 = a_3 + b_3 t_2.$$

Hence

$$\overline{P_1P_2} \cos \delta_1 = x_2 - x_1 = b_1(t_2 - t_1),$$
$$\overline{P_1P_2} \cos \delta_2 = y_2 - y_1 = b_2(t_2 - t_1),$$
$$\overline{P_1P_2} \cos \delta_3 = z_2 - z_1 = b_3(t_2 - t_1),$$

where $\overline{P_1P_2}$ denotes the length of the segment P_1P_2. Consequently

$$\alpha = \rho b_1, \quad \beta = \rho b_2, \quad \gamma = \rho b_3 \quad \left(\text{where } \rho = \frac{t_2 - t_1}{\overline{P_1P_2}}\right).$$

Since the sum of the squares of the direction cosines is unity, it follows that

$$\alpha = \frac{b_1}{\pm\sqrt{b_1^2 + b_2^2 + b_3^2}}, \quad \beta = \frac{b_2}{\pm\sqrt{b_1^2 + b_2^2 + b_3^2}}, \quad \gamma = \frac{b_3}{\pm\sqrt{b_1^2 + b_2^2 + b_3^2}},$$

where the double sign of the square root corresponds to the fact that we can choose either of the two possible senses on the line.

By means of the direction cosines we can easily bring the parametric representation of the line into the form

$$x = x_0 + \alpha\tau,$$
$$y = y_0 + \beta\tau,$$
$$z = z_0 + \gamma\tau,$$

where (x_0, y_0, z_0) is a fixed point on the line; the new parameter τ is connected with the previous parameter t by the equation

$$x_0 + \alpha\tau = a_1 + b_1 t.$$

From the fact that $\alpha^2 + \beta^2 + \gamma^2 = 1$ it follows that

$$\tau^2 = (x - x_0)^2 + (y - y_0)^2 + (z - z_0)^2.$$

Hence the absolute value of τ is the distance between (x_0, y_0, z_0) and (x, y, z). The sign of τ indicates whether the direction of the line is from the point (x_0, y_0, z_0) to the point (x, y, z), or vice versa; in the first case τ is positive, in the second negative.

From this we obtain a useful expression for (x, y, z), the co-ordinates of a point P on the segment joining the points $P_0(x_0, y_0, z_0)$ and $P_1(x_1, y_1, z_1)$, namely,

$$x = \lambda_0 x_0 + \lambda_1 x_1, \quad y = \lambda_0 y_0 + \lambda_1 y_1, \quad z = \lambda_0 z_0 + \lambda_1 z_1,$$

where λ_0 and λ_1 are positive and $\lambda_0 + \lambda_1 = 1$. If τ and τ_1 denote the distances from P_0 of the points P and P_1 respectively, we find that $\lambda_0 = 1 - \dfrac{\tau}{\tau_1}$ and $\lambda_1 = \dfrac{\tau}{\tau_1}$. For if we calculate α, say, from $x_1 = x_0 + \alpha\tau_1$, and substitute this value, $\alpha = (x_1 - x_0)/\tau_1$, in the equation $x = x_0 + \alpha\tau$, we obtain the expression given above.

Let a straight line be given by

$$\begin{aligned} x &= x_0 + \alpha\tau, \\ y &= y_0 + \beta\tau, \\ z &= z_0 + \gamma\tau. \end{aligned}$$

We now seek to find the equation of the plane which passes through the point (x_0, y_0, z_0) and is perpendicular to this line. Since the direction cosines of the normal to this plane are α, β, γ, the canonical form of the required equation is

$$\alpha x + \beta y + \gamma z - p = 0,$$

and since the point (x_0, y_0, z_0) lies on the plane

$$p = \alpha x_0 + \beta y_0 + \gamma z_0.$$

The equation of the plane through (x_0, y_0, z_0) perpendicular to the line with direction cosines α, β, γ is therefore

$$\alpha(x - x_0) + \beta(y - y_0) + \gamma(z - z_0) = 0.$$

In the same way, the equation of a straight line in the xy-plane which passes through the point (x_0, y_0) and is perpendicular to the line with direction cosines α, β is

$$\alpha(x - x_0) + \beta(y - y_0) = 0.$$

Later we shall need a formula for δ, the *angle between two planes* given by the equations

$$\begin{aligned} \alpha x + \beta y + \gamma z - p &= 0, \\ \alpha' x + \beta' y + \gamma' z - p' &= 0. \end{aligned}$$

Since the angle between the planes is equal to the angle between their normal vectors, the scalar product of these vectors is $\cos\delta$, so that

$$\cos\delta = \alpha\alpha' + \beta\beta' + \gamma\gamma'.$$

In the same way, for the angle δ between the two straight lines

$$\alpha x + \beta y - p = 0 \text{ and } \alpha' x + \beta' y - p' = 0$$

in the xy-plane we have

$$\cos\delta = \alpha\alpha' + \beta\beta'.$$

EXAMPLES

1. Prove that the quantities $\alpha_1, \alpha_2, \ldots, \gamma_3$ (p. 6), defining a rotation of axes, satisfy the relations

$$\begin{aligned} \alpha_1\alpha_2 + \beta_1\beta_2 + \gamma_1\gamma_2 &= 0, & \alpha_1^2 + \beta_1^2 + \gamma_1^2 &= 1,\\ \alpha_2\alpha_3 + \beta_2\beta_3 + \gamma_2\gamma_3 &= 0, & \alpha_2^2 + \beta_2^2 + \gamma_2^2 &= 1,\\ \alpha_3\alpha_1 + \beta_3\beta_1 + \gamma_3\gamma_1 &= 0, & \alpha_3^2 + \beta_3^2 + \gamma_3^2 &= 1. \end{aligned}$$

2. If $\boldsymbol{a}$ and $\boldsymbol{b}$ are two vectors with initial point O and final points A and B, then the vector with O as initial point and the point dividing AB in the ratio $\theta : 1 - \theta$ as final point is given by

$$(1 - \theta)\boldsymbol{a} + \theta\boldsymbol{b}.$$

3. The centre of mass of the vertices of a tetrahedron $PQRS$ may be defined as the point dividing MS in the ratio $1 : 3$, where M is the centre of mass of the triangle PQR. Show that this definition is independent of the order in which the vertices are taken and that it agrees with the general definition of the centre of mass (Vol. I, p. 283).

4. If in the tetrahedron $PQRS$ the centres of the edges PQ, RS, PR, QS, PS, QR are denoted by A, A', B, B', C, C' respectively, then the lines AA', BB', CC' all pass through the centre of mass and bisect one another there.

5. Let $P_1, \ldots, P_n$ be n arbitrary particles in space, with masses $m_1, m_2, \ldots, m_n$ respectively. Let G be their centre of mass and let $\boldsymbol{p}_1, \ldots, \boldsymbol{p}_n$ denote the vectors with initial point G and final points $P_1, \ldots, P_n$. Prove that

$$m_1\boldsymbol{p}_1 + m_2\boldsymbol{p}_2 + \ldots + m_n\boldsymbol{p}_n = 0.$$

2. THE AREA OF A TRIANGLE, THE VOLUME OF A TETRAHEDRON, THE VECTOR MULTIPLICATION OF VECTORS

1. The Area of a Triangle.

In order to calculate the area of a triangle in the xy-plane we imagine it moved parallel to itself until one of its vertices is at the origin; let the other two vertices be $P_1(x_1, y_1)$ and $P_2(x_2, y_2)$ (cf. fig. 12). We write down the equation of the line joining P_1 to the origin in its canonical form

$$\frac{-y_1}{\sqrt{x_1^2 + y_1^2}}\,x + \frac{x_1}{\sqrt{x_1^2 + y_1^2}}\,y = 0;$$

hence for the distance h of the point P_2 from this line we have (except perhaps for sign) the expression

$$\pm h = \frac{-y_1 x_2}{\sqrt{x_1^2 + y_1^2}} + \frac{x_1 y_2}{\sqrt{x_1^2 + y_1^2}}.$$

Since the length of the segment OP_1 is $\sqrt{x_1^2 + y_1^2}$, we find that twice the

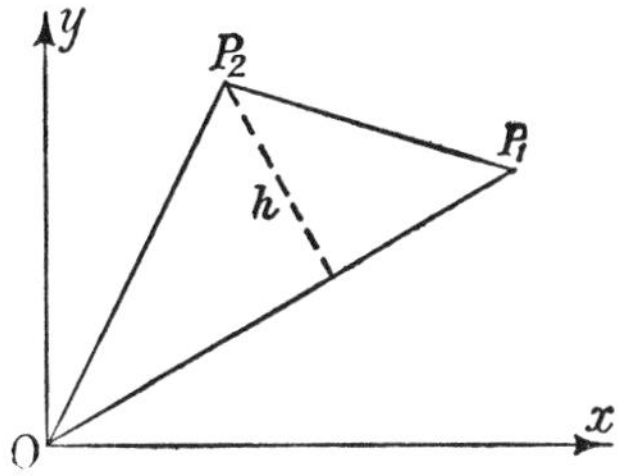

Fig. 12.—To illustrate the method for finding the area of a triangle

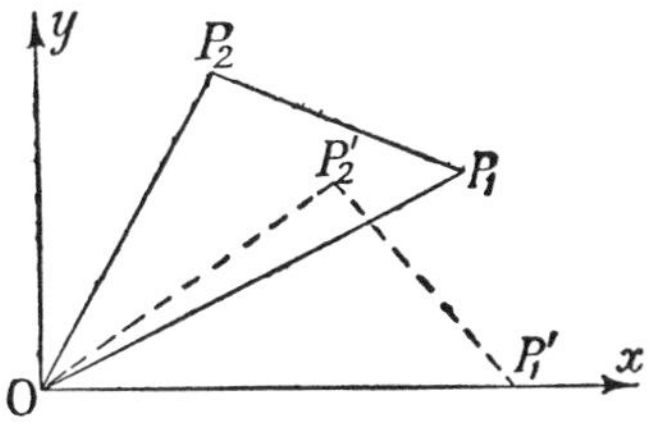

Fig. 13.—Determination of the sign of the area of a triangle

area of the triangle, which is the product of the "base" OP_1 and the altitude h, is given (except perhaps for sign) by the expression

$$2A = x_1 y_2 - x_2 y_1.$$

This expression can be either positive or negative; it changes sign if we interchange P_1 and P_2. We now make the following assertion. *The expression A has a positive or negative value according as the sense in which the vertices OP_1P_2 are traversed is the same as the sense of the rotation associated with the co-ordinate axes, or not.* Instead of proving the fact by more detailed investigation of the argument given above, which is quite feasible, we prefer to prove it by the following method. We rotate the triangle OP_1P_2 about the origin O until P_1 lies on the positive x-axis. (The case in which O, P_1, P_2 lie on a line, so that $A = \frac{1}{2}(x_1y_2 - x_2y_1) = 0$, can be omitted.) This rotation leaves the value of A unaltered. After the rotation P_1 has the co-ordinates $x_1' > 0$, $y_1' = 0$, and the co-ordinates of the new P_2 are x_2' and y_2'. The area of the triangle is now

$$A = \frac{1}{2} x_1' y_2',$$

and therefore has the same sign as y_2'. The sign of y_2', however, is the same as the sign of the sense in which the vertices OP_1P_2 are traversed (cf. fig. 13). Our statement is thus proved.

For the expression $x_1y_2 - x_2y_1$, which gives twice the area with its proper sign, it is customary to introduce the symbolic notation

$$x_1y_2 - x_2y_1 = \begin{vmatrix} x_1 & x_2 \\ y_1 & y_2 \end{vmatrix},$$

which we call a *two-rowed determinant*, or *determinant of the second order.*

If no vertex of the triangle is at the origin of the co-ordinate system, e.g. if the three vertices are (x_0, y_0), (x_1, y_1), (x_2, y_2), by moving the axes parallel to themselves we obtain the formula

$$A = \frac{1}{2}\begin{vmatrix} x_1 - x_0 & x_2 - x_0 \\ y_1 - y_0 & y_2 - y_0 \end{vmatrix}$$

for the area of the triangle.

2. Vector Multiplication of two Vectors.

In addition to the scalar product of two vectors we have the important concept of the *vector product.** The vector product $[\boldsymbol{ab}]$ of the vectors $\boldsymbol{a}$ and $\boldsymbol{b}$ is defined as follows (cf. fig. 14):

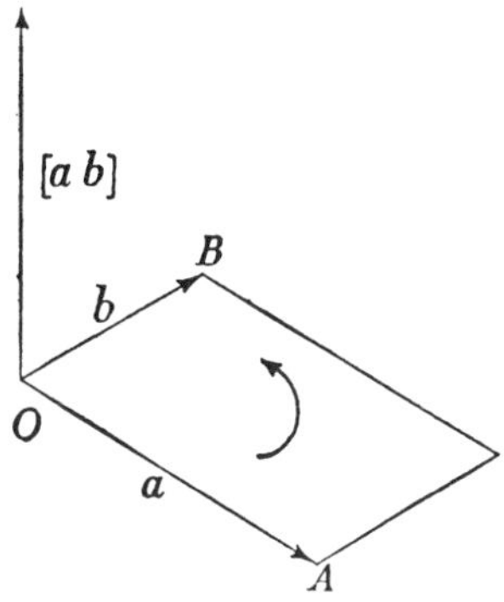

Fig. 14.—Vector product of two vectors a and b

We measure off $\boldsymbol{a}$ and $\boldsymbol{b}$ from a point O. Then $\boldsymbol{a}$ and $\boldsymbol{b}$ are two sides of a parallelogram in space. The vector product $[\boldsymbol{ab}] = \boldsymbol{c}$ is a vector whose length is numerically equal to the area of the parallelogram and whose direction is perpendicular to the plane of the parallelogram, the sense of direction being such that the rotation from $\boldsymbol{a}$ to $\boldsymbol{b}$ and $\boldsymbol{c} = [\boldsymbol{ab}]$ is right-handed. (That is, if we look at the plane from the final point of the vector $\boldsymbol{c}$, we see the shortest rotation from the direction of $\boldsymbol{a}$ to that of $\boldsymbol{b}$ as a positive rotation.) If $\boldsymbol{a}$ and $\boldsymbol{b}$ lie in the same straight line, we must have $[\boldsymbol{ab}] = 0$, since the area of the parallelogram is zero.

Rules of Calculation for the Vector Product.

(1) If $\boldsymbol{a} \neq 0$ and $\boldsymbol{b} \neq 0$, then $[\boldsymbol{ab}] = 0$ if, and only if, $\boldsymbol{a}$ and $\boldsymbol{b}$ have the same direction or opposite directions.

For then, and only then, the area of the parallelogram with $\boldsymbol{a}$ and $\boldsymbol{b}$ as sides is equal to zero.

(2) The equation

$$[\boldsymbol{ab}] = -[\boldsymbol{ba}]$$

holds.

* Often called the *outer product*; other notations in use for it are $\boldsymbol{a} \times \boldsymbol{b}$, $\boldsymbol{a} \wedge \boldsymbol{b}$.

This follows at once from the definition of $[\boldsymbol{ab}]$.

(3) If a and b are real numbers, then

$$[a\boldsymbol{a}\, b\boldsymbol{b}] = ab\,[\boldsymbol{ab}].$$

For the parallelogram with sides $a\boldsymbol{a}$ and $b\boldsymbol{b}$ has an area ab times as great as that of the parallelogram with sides $\boldsymbol{a}$ and $\boldsymbol{b}$ and lies in the same plane as the latter.

(4) The distributive law holds:

$$[\boldsymbol{a}(\boldsymbol{b}+\boldsymbol{c})] = [\boldsymbol{ab}] + [\boldsymbol{ac}], \quad [(\boldsymbol{b}+\boldsymbol{c})\boldsymbol{a}] = [\boldsymbol{ba}] + [\boldsymbol{ca}].$$

We shall prove the first of these formulæ; the second follows from it when rule (2) is applied.

We shall now give a geometrical construction for the vector product $[\boldsymbol{ab}]$ which will demonstrate the truth of the distributive law directly.

Let E be the plane perpendicular to $\boldsymbol{a}$ through the point O. We project

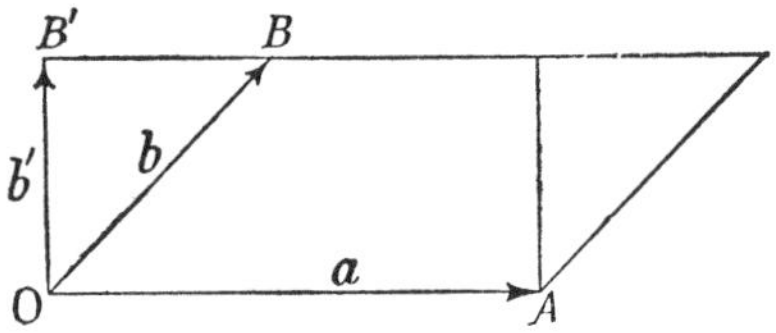

Fig. 15.—To show that $[\boldsymbol{ab}] = [\boldsymbol{ab}']$

$\boldsymbol{b}$ orthogonally on E, thus obtaining a vector $\boldsymbol{b}'$ (cf. fig. 15). Then $[\boldsymbol{ab}'] = [\boldsymbol{ab}]$, for in the first place the parallelogram with sides $\boldsymbol{a}$ and $\boldsymbol{b}$ has the same base and the same altitude as the parallelogram with sides $\boldsymbol{a}$ and $\boldsymbol{b}'$; and in the second place the directions of $[\boldsymbol{ab}']$ and $[\boldsymbol{ab}]$ are the same, since $\boldsymbol{a}$, $\boldsymbol{b}$, $\boldsymbol{b}'$ lie in one plane and the sense of rotation from $\boldsymbol{a}$ to $\boldsymbol{b}'$ is the same as that from $\boldsymbol{a}$ to $\boldsymbol{b}$. Since the vectors $\boldsymbol{a}$ and $\boldsymbol{b}$ are sides of a rectangle, the length of $[\boldsymbol{ab}'] = [\boldsymbol{ab}]$ is the product $|\boldsymbol{a}|\,|\boldsymbol{b}'|$. If, therefore, we increase the length of $\boldsymbol{b}'$ in the ratio $|\boldsymbol{a}| : 1$, we obtain a vector $\boldsymbol{b}''$ which has the same length as $[\boldsymbol{ab}']$. But $[\boldsymbol{ab}] = [\boldsymbol{ab}']$ is perpendicular to both $\boldsymbol{a}$ and $\boldsymbol{b}$, so that we obtain $[\boldsymbol{ab}] = [\boldsymbol{ab}']$ from $\boldsymbol{b}''$ by a rotation through 90° about the line $\boldsymbol{a}$. The sense of this rotation must be positive when looked at from the final point of $\boldsymbol{a}$. Such a rotation we shall call a *positive rotation about the vector* $\boldsymbol{a}$ *as axis*.

We can therefore form $[\boldsymbol{ab}]$ in the following way: project $\boldsymbol{b}$ orthogonally on the plane E, lengthen it in the ratio $|\boldsymbol{a}| : 1$, and rotate it positively through 90° about the vector $\boldsymbol{a}$.

To prove that $[\boldsymbol{a}(\boldsymbol{b}+\boldsymbol{c})] = [\boldsymbol{ab}] + [\boldsymbol{ac}]$ we proceed as follows: $\boldsymbol{b}$ and $\boldsymbol{c}$ are the sides OB, OC of a parallelogram $OBDC$, whose diagonal OD is the sum $\boldsymbol{b}+\boldsymbol{c}$. We now perform the three operations of projection, lengthening, and rotation on the whole parallelogram $OBDC$ instead of on the individual vectors $\boldsymbol{b}$, $\boldsymbol{c}$, $\boldsymbol{b}+\boldsymbol{c}$; we thus obtain a parallelogram $OB_1D_1C_1$ whose sides OB_1, OC_1 are the vectors $[\boldsymbol{ab}]$ and $[\boldsymbol{ac}]$ and whose diagonal is the product $[a(b + c)]$. From this the equation $[ab] + [ac] = [a(b + c)]$ clearly follows (cf. fig. 16).

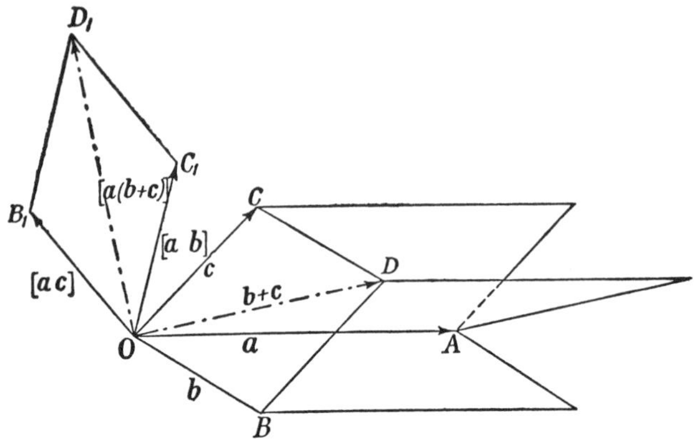

Fig. 16.—Distributive law for the vector product of two vectors a and b

(5) Let a and b be given by their components along the axes, a_1, a_2, a_3 and b_1, b_2, b_3 respectively. What is the expression for the vector product $[ab]$ in terms of the vector components?

We express a by the sum of its vector components in the directions of the axes. If e_1, e_2, e_3 are the unit vectors in the directions of the axes, then

$$a = a_1e_1 + a_2e_2 + a_3e_3,$$

and similarly

$$b = b_1e_1 + b_2e_2 + b_3e_3.$$

By the distributive law we obtain

$$\begin{aligned}[ab] = {} & [(a_1e_1)(b_1e_1)] + [(a_1e_1)(b_2e_2)] + [(a_1e_1)(b_3e_3)] \\ & + [(a_2e_2)(b_1e_1)] + [(a_2e_2)(b_2e_2)] + [(a_2e_2)(b_3e_3)] \\ & + [(a_3e_3)(b_1e_1)] + [(a_3e_3)(b_2e_2)] + [(a_3e_3)(b_3e_3)],\end{aligned}$$

which by rules (1) and (3) may also be written

$$\begin{aligned}[ab] = {} & a_1b_2[e_1e_2] + a_1b_3[e_1e_3] + a_2b_1[e_2e_1] \\ & + a_2b_3[e_2e_3] + a_3b_1[e_3e_1] + a_3b_2[e_3e_2].\end{aligned}$$

Now from the definition of vector product it follows that

$$e_1 = [e_2e_3] = -[e_3e_2], \quad e_2 = [e_3e_1] = -[e_1e_3], \quad e_3 = [e_1e_2] = -[e_2e_1].$$

Hence

$$[ab] = (a_2b_3 - a_3b_2)e_1 + (a_3b_1 - a_1b_3)e_2 + (a_1b_2 - a_2b_1)e_3.$$

The components of the vector product $[ab] = c$ are therefore

$$c_1 = \begin{vmatrix} a_2 & a_3 \\ b_2 & b_3 \end{vmatrix}, \qquad c_2 = \begin{vmatrix} a_3 & a_1 \\ b_3 & b_1 \end{vmatrix}, \qquad c_3 = \begin{vmatrix} a_1 & a_2 \\ b_1 & b_2 \end{vmatrix}.$$

In physics we use the vector product of two vectors to represent a *moment.* A force f acting at the final point of the position vector x has the moment $[fx]$ about the origin.

3. The Volume of a Tetrahedron.

We consider a tetrahedron (cf. fig. 17) whose vertices are the origin and three other points P_1, P_2, P_3 with co-ordinates (x_1, y_1, z_1), (x_2, y_2, z_2),

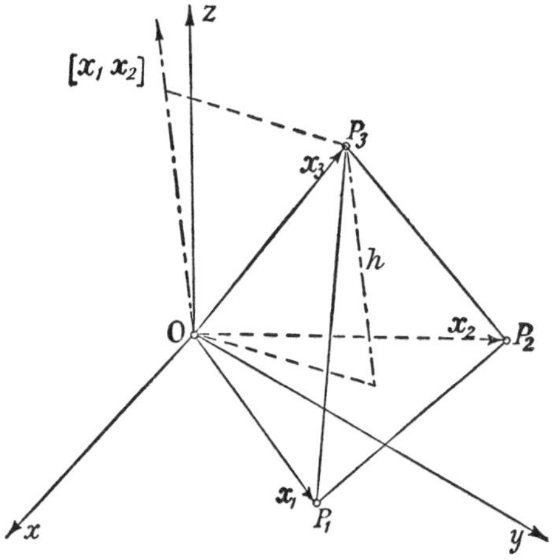

Fig. 17.—Determination of the volume of a tetrahedron

(x_3, y_3, z_3) respectively. To express the volume of this tetrahedron in terms of the co-ordinates of its vertices we proceed as follows. The vectors $x_1 = OP_1$ and $x_2 = OP_2$ are sides of a triangle whose area is half the length of the vector product $[x_1 x_2]$. This vector product has the direction of the perpendicular from P_3 to the plane of the triangle OP_1P_2; h, the length of this perpendicular (the altitude of the tetrahedron), is therefore given by the scalar product of the vector $x_3 = OP_3$ and the unit vector in the direction of $[x_1 x_2]$; for h is equal to the component of OP_3 in the direction of $[x_1 x_2]$. Since the absolute value of $[x_1 x_2]$ is twice the area A of the triangle OP_1P_2, and since the volume V of the tetrahedron is equal to $\frac{1}{3}Ah$, we have

$$V = \tfrac{1}{6}([x_1 x_2] x_3).$$

Or, since the components of $[x_1 x_2]$ are given by

$$\begin{vmatrix} y_1 & z_1 \\ y_2 & z_2 \end{vmatrix}, \qquad \begin{vmatrix} z_1 & x_1 \\ z_2 & x_2 \end{vmatrix}, \qquad \begin{vmatrix} x_1 & y_1 \\ x_2 & y_2 \end{vmatrix},$$

we can write

$$V = \frac{1}{6}\left\{x_3\begin{vmatrix} y_1 & z_1 \\ y_2 & z_2 \end{vmatrix} + y_3\begin{vmatrix} z_1 & x_1 \\ z_2 & x_2 \end{vmatrix} + z_3\begin{vmatrix} x_1 & y_1 \\ x_2 & y_2 \end{vmatrix}\right\}.$$

This also holds for the case in which O, P_1, P_2 lie on a straight line; in this case, it is true, the direction of $[x_1x_2]$ is indeterminate, so that h can no longer be regarded as the component of OP_3 in the direction of $[x_1x_2]$, but nevertheless $A = 0$, so that $V = 0$, and this follows also from the above expression for V, since in this case all the components of $[x_1x_2]$ vanish.

Here again the volume of the tetrahedron is given with a definite sign, as the area of the triangle was on p. 13; and we can show that the sign is positive if the three axes OP_1, OP_2, OP_3 taken in that order form a system of the same type (right-handed or left-handed, as the case may be) as the co-ordinate axes, and negative if the two systems are of opposite type. For in the first case the angle δ between $[x_1x_2]$ and x_3 lies in the interval $0 \leq \delta \leq \frac{\pi}{2}$, and in the second case in the interval $\frac{\pi}{2} \leq \delta \leq \pi$, as follows immediately from the definition of $[x_1x_2]$, and V is equal to

$$|[x_1x_2]|\ |x_3| \cos\delta.$$

The expression

$$x_3\begin{vmatrix} y_1 & z_1 \\ y_2 & z_2 \end{vmatrix} + y_3\begin{vmatrix} z_1 & x_1 \\ z_2 & x_2 \end{vmatrix} + z_3\begin{vmatrix} x_1 & y_1 \\ x_2 & y_2 \end{vmatrix}$$

occurring in our formulæ may be expressed more briefly by the symbol

$$\begin{vmatrix} x_1 & y_1 & z_1 \\ x_2 & y_2 & z_2 \\ x_3 & y_3 & z_3 \end{vmatrix},$$

which we call a *three-rowed determinant*, or *determinant of the third order*. Writing out the two-rowed determinants in full, we see that

$$\begin{vmatrix} x_1 & y_1 & z_1 \\ x_2 & y_2 & z_2 \\ x_3 & y_3 & z_3 \end{vmatrix} = x_3y_1z_2 - x_3y_2z_1 + x_2y_3z_1 - x_1y_3z_2 + x_1y_2z_3 - x_2y_1z_3.$$

Just as in the case of the triangle, we find that the volume of the tetrahedron with vertices (x_0, y_0, z_0), (x_1, y_1, z_1), (x_2, y_2, z_2), (x_3, y_3, z_3) is

$$V = \frac{1}{6}\begin{vmatrix} x_1 - x_0 & y_1 - y_0 & z_1 - z_0 \\ x_2 - x_0 & y_2 - y_0 & z_2 - z_0 \\ x_3 - x_0 & y_3 - y_0 & z_3 - z_0 \end{vmatrix}.$$

EXAMPLES *

1. What is the distance of the point $P(x_0, y_0, z_0)$ from the straight line l given by

$$x = at + b, \quad y = ct + d, \quad z = et + f\,?$$

* The more difficult examples are indicated by an asterisk.

2*. Find the shortest distance between two straight lines l and l' in space, given by the equations

$$\begin{aligned} x &= at + b \\ y &= ct + d \\ z &= et + f \end{aligned} \qquad \text{and} \qquad \begin{aligned} x &= a't + b' \\ y &= c't + d' \\ z &= e't + f'. \end{aligned}$$

3. Show that the plane through the three points (x_1, y_1, z_1), (x_2, y_2, z_2), (x_3, y_3, z_3) is given by

$$\begin{vmatrix} x_1 - x & y_1 - y & z_1 - z \\ x_2 - x & y_2 - y & z_2 - z \\ x_3 - x & y_3 - y & z_3 - z \end{vmatrix} = 0.$$

4. In a uniform rotation let (α, β, γ) be the direction cosines of the axis of rotation, which passes through the origin, and ω the angular velocity. Find the velocity of the point (x, y, z).

5. Prove Lagrange's identity

$$[\boldsymbol{x}\boldsymbol{y}]^2 = |\,\boldsymbol{x}\,|^2 |\,\boldsymbol{y}\,|^2 - (\boldsymbol{x}\boldsymbol{y})^2.$$

6. The area of a convex polygon with the vertices $P_1(x_1, y_1)$, $P_2(x_2, y_2)$, . . . , $P_n(x_n, y_n)$ is given by half the absolute value of

$$\begin{vmatrix} x_1 & x_2 \\ y_1 & y_2 \end{vmatrix} + \begin{vmatrix} x_2 & x_3 \\ y_2 & y_3 \end{vmatrix} + \dots + \begin{vmatrix} x_{n-1} & x_n \\ y_{n-1} & y_n \end{vmatrix} + \begin{vmatrix} x_n & x_1 \\ y_n & y_1 \end{vmatrix}.$$

3. Simple Theorems on Determinants of the Second and Third Order

1. Laws of Formation and Principal Properties.

The determinants of the second and third order occurring in the calculation of the area of a triangle and the volume of a tetrahedron, together with their generalization, the *determinant of the* n*th order*, or n*-rowed determinant*, are very important in that they enable formal calculations in all branches of mathematics to be expressed in a compact form. Here we shall develop the properties of determinants of the second and third order; those of higher order we shall need but seldom. It may, however, be pointed out that all the principal theorems may be generalized at once for determinants with any number of rows. For the theory of these we must refer the reader to books on algebra and determinants.*

By their definitions (pp. 14, 18) the determinants

$$\begin{vmatrix} a & b \\ c & d \end{vmatrix} \quad \text{and} \quad \begin{vmatrix} a & b & c \\ d & e & f \\ g & h & k \end{vmatrix}$$

* Cf. e.g. H. W. Turnbull, *The Theory of Determinants, Matrices, and Invariants* (Blackie & Son, Ltd., 1929).

are expressions formed in a definite way from their elements a, b, c, d and $a, b, c, d, e, f, g, h, k$ respectively. The horizontal lines of elements (such as d, e, f in our example) are called *rows* and the vertical lines (such as c, f, k) are called *columns.*

We need not spend any time in discussing the formation of the two-rowed determinant

$$\begin{vmatrix} a & b \\ c & d \end{vmatrix} = ad - bc.$$

For the three-rowed determinant we give the "diagonal rule" which exhibits the symmetrical way in which the determinant is formed:

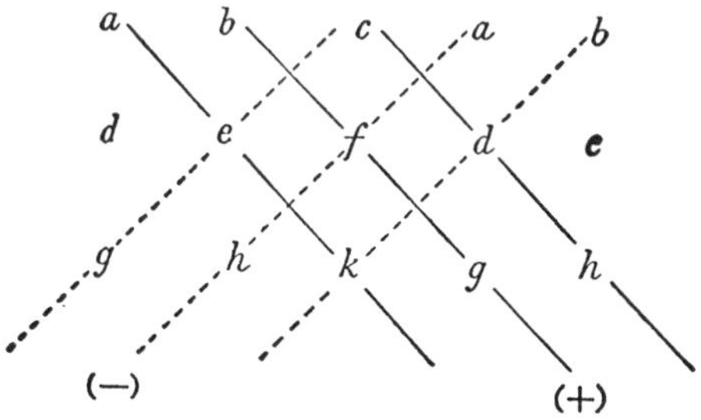

We repeat the first two columns after the third and then form the product of each triad of numbers in the diagonal lines, multiply the products associated with lines slanting downwards and to the right by $+1$, the others by -1, and add. In this way we obtain

$$\begin{vmatrix} a & b & c \\ d & e & f \\ g & h & k \end{vmatrix} = \begin{matrix} aek + bfg + cdh \\ -ceg - afh - bdk. \end{matrix}$$

We shall now prove several theorems on determinants:

(1) *If the rows and columns of a determinant are interchanged, the value of the determinant is unaltered.* That is,

$$\begin{vmatrix} a & b \\ c & d \end{vmatrix} = \begin{vmatrix} a & c \\ b & d \end{vmatrix},$$

$$\begin{vmatrix} a & b & c \\ d & e & f \\ g & h & k \end{vmatrix} = \begin{vmatrix} a & d & g \\ b & e & h \\ c & f & k \end{vmatrix}.$$

This follows immediately from the above expressions for the determinants.

(2) *If two rows (or two columns) of a determinant are interchanged, the sign of the determinant is altered,* that is, the determinant is multiplied by -1.

In virtue of (1) this need only be proved for the columns, and it can be verified at once by the law of formation of the determinant given above.

(3) In section 2 (p. 18) we introduced three-rowed determinants by the equations

$$\begin{vmatrix} x_1 & y_1 & z_1 \\ x_2 & y_2 & z_2 \\ x_3 & y_3 & z_3 \end{vmatrix} = x_3 \begin{vmatrix} y_1 & z_1 \\ y_2 & z_2 \end{vmatrix} + y_3 \begin{vmatrix} z_1 & x_1 \\ z_2 & x_2 \end{vmatrix} + z_3 \begin{vmatrix} x_1 & y_1 \\ x_2 & y_2 \end{vmatrix}.$$

Using (2), we write this in the form

$$\begin{vmatrix} x_1 & y_1 & z_1 \\ x_2 & y_2 & z_2 \\ x_3 & y_3 & z_3 \end{vmatrix} = x_3 \begin{vmatrix} y_1 & z_1 \\ y_2 & z_2 \end{vmatrix} - y_3 \begin{vmatrix} x_1 & z_1 \\ x_2 & z_2 \end{vmatrix} + z_3 \begin{vmatrix} x_1 & y_1 \\ x_2 & y_2 \end{vmatrix};$$

then in the determinants on the right the elements are in the same order as on the left. If we interchange the last two rows and then write down the same equation, using (2), we obtain:

$$\begin{vmatrix} x_1 & y_1 & z_1 \\ x_2 & y_2 & z_2 \\ x_3 & y_3 & z_3 \end{vmatrix} = -x_2 \begin{vmatrix} y_1 & z_1 \\ y_3 & z_3 \end{vmatrix} + y_2 \begin{vmatrix} x_1 & z_1 \\ x_3 & z_3 \end{vmatrix} - z_2 \begin{vmatrix} x_1 & y_1 \\ x_3 & y_3 \end{vmatrix},$$

and similarly

$$\begin{vmatrix} x_1 & y_1 & z_1 \\ x_2 & y_2 & z_2 \\ x_3 & y_3 & z_3 \end{vmatrix} = x_1 \begin{vmatrix} y_2 & z_2 \\ y_3 & z_3 \end{vmatrix} - y_1 \begin{vmatrix} x_2 & z_2 \\ x_3 & z_3 \end{vmatrix} + z_1 \begin{vmatrix} x_2 & y_2 \\ x_3 & y_3 \end{vmatrix}.$$

We call these three equations the *expansion in terms of the elements* of the third row, the second row, and the first row respectively. By interchanging columns and rows, which according to (1) does not alter the value of the determinant, we obtain the expansion by columns,

$$\begin{vmatrix} x_1 & y_1 & z_1 \\ x_2 & y_2 & z_2 \\ x_3 & y_3 & z_3 \end{vmatrix} = x_1 \begin{vmatrix} y_2 & z_2 \\ y_3 & z_3 \end{vmatrix} - x_2 \begin{vmatrix} y_1 & z_1 \\ y_3 & z_3 \end{vmatrix} + x_3 \begin{vmatrix} y_1 & z_1 \\ y_2 & z_2 \end{vmatrix},$$

$$\begin{vmatrix} x_1 & y_1 & z_1 \\ x_2 & y_2 & z_2 \\ x_3 & y_3 & z_3 \end{vmatrix} = -y_1 \begin{vmatrix} x_2 & z_2 \\ x_3 & z_3 \end{vmatrix} + y_2 \begin{vmatrix} x_1 & z_1 \\ x_3 & z_3 \end{vmatrix} - y_3 \begin{vmatrix} x_1 & z_1 \\ x_2 & z_2 \end{vmatrix},$$

$$\begin{vmatrix} x_1 & y_1 & z_1 \\ x_2 & y_2 & z_2 \\ x_3 & y_3 & z_3 \end{vmatrix} = z_1 \begin{vmatrix} x_2 & y_2 \\ x_3 & y_3 \end{vmatrix} - z_2 \begin{vmatrix} x_1 & y_1 \\ x_3 & y_3 \end{vmatrix} + z_3 \begin{vmatrix} x_1 & y_1 \\ x_2 & y_2 \end{vmatrix}.$$

An immediate consequence of this is the following theorem:

(4) *If all the elements of one row (or column) are multiplied by a number* ρ, *the value of the determinant is multiplied by* ρ.

From (2) and (4) we deduce the following:

(5) *If the elements of two rows (or two columns) are proportional*, that is, if every element of one row (or column) is the product of the corresponding

element in the other row (or column) and the same factor ρ, *then the determinant is equal to zero.*

For according to (4) we can write the factor outside the determinant. If we then interchange the equal rows, the value of the determinant is unchanged, but by (2) it should change sign. Hence its value is zero.

In particular, a determinant in which one row or column consists entirely of zeros has the value zero, as also follows from the definition of a determinant.

(6) *The sum of two determinants, having the same number of rows, which differ only in the elements of one row (or column) is equal to the determinant which coincides with them in the rows (or columns) common to the two determinants and in the one remaining row (or column) has the sums of the corresponding elements of the two non-identical rows (or columns).*

For example:

$$\begin{vmatrix} a & b & c \\ d & e & f \\ g & h & k \end{vmatrix} + \begin{vmatrix} a & m & c \\ d & n & f \\ g & p & k \end{vmatrix} = \begin{vmatrix} a & b+m & c \\ d & e+n & f \\ g & h+p & k \end{vmatrix}.$$

For if we expand in terms of the rows (or columns) in question, which in our example consist of the elements b, e, h and m, n, p respectively, and add, we obtain the expression

$$(-b-m)\begin{vmatrix} d & f \\ g & k \end{vmatrix} + (e+n)\begin{vmatrix} a & c \\ g & k \end{vmatrix} + (-h-p)\begin{vmatrix} a & c \\ d & f \end{vmatrix},$$

which clearly is just the expansion of the determinant

$$\begin{vmatrix} a & b+m & c \\ d & e+n & f \\ g & h+p & k \end{vmatrix}$$

in terms of the column $b+m$, $e+n$, $h+p$. This proves the statement.

Similar statements hold for two-rowed determinants.

(7) *If to each element of a row (or column) of a determinant we add the same multiple of the corresponding element of another row (or column), the value of the determinant is unchanged.*

By (6) the new determinant is the sum of the original determinant and a determinant which has two proportional rows (or columns); by (5) this second determinant is zero.*

* The rule for expansion in terms of rows or columns may be extended to define determinants of the fourth and higher order. Given a system of sixteen numbers

$$\begin{pmatrix} a_1 & b_1 & c_1 & d_1 \\ a_2 & b_2 & c_2 & d_2 \\ a_3 & b_3 & c_3 & d_3 \\ a_4 & b_4 & c_4 & d_4 \end{pmatrix},$$

for example, we define a determinant of the fourth order by the expression

The following examples illustrate how the above theorems are applied to the evaluation of determinants. We have

$$\begin{vmatrix} a & 0 & 0 \\ 0 & e & 0 \\ 0 & 0 & k \end{vmatrix} = aek,$$

as we can prove by the diagonal rule. *A determinant in which the elements in the so-called principal diagonal alone differ from zero is equal to the product of these elements.*

Evaluation of a determinant:

$$\begin{vmatrix} 1 & 1 & -1 \\ 1 & -1 & 1 \\ -1 & 1 & 1 \end{vmatrix} = \begin{vmatrix} 2 & 0 & 0 \\ 1 & -1 & 1 \\ -1 & 1 & 1 \end{vmatrix} \text{ (second row added to the first),}$$

$$\begin{vmatrix} 2 & 0 & 0 \\ 1 & -1 & 1 \\ -1 & 1 & 1 \end{vmatrix} = 2 \begin{vmatrix} -1 & 1 \\ 1 & 1 \end{vmatrix} \text{ (expansion in terms of the first row).}$$

Hence

$$\begin{vmatrix} 1 & 1 & -1 \\ 1 & -1 & 1 \\ -1 & 1 & 1 \end{vmatrix} = -4.$$

Another example is

$$\begin{vmatrix} 1 & x & x^2 \\ 1 & y & y^2 \\ 1 & z & z^2 \end{vmatrix} = \begin{vmatrix} 1 & x & x^2 \\ 0 & y-x & y^2-x^2 \\ 1 & z & z^2 \end{vmatrix} = \begin{vmatrix} 1 & x & x^2 \\ 0 & y-x & y^2-x^2 \\ 0 & z-x & z^2-x^2 \end{vmatrix}.$$

If we now expand this in terms of the first column we obtain

$$\begin{vmatrix} y-x & y^2-x^2 \\ z-x & z^2-x^2 \end{vmatrix} = (y-x)\ (z-x) \begin{vmatrix} 1 & y+x \\ 1 & z+x \end{vmatrix} = (y-x)(z-x)(z-y).$$

2. Application to Linear Equations.

Determinants are of fundamental importance in the theory of linear equations. In order to solve the two equations

$$ax + by = A,$$
$$cx + dy = B,$$

for x and y, we multiply the first equation by c and the second by a, and

---

$$a_1 \begin{vmatrix} b_2 & c_2 & d_2 \\ b_3 & c_3 & d_3 \\ b_4 & c_4 & d_4 \end{vmatrix} - b_1 \begin{vmatrix} a_2 & c_2 & d_2 \\ a_3 & c_3 & d_3 \\ a_4 & c_4 & d_4 \end{vmatrix} + c_1 \begin{vmatrix} a_2 & b_2 & d_2 \\ a_3 & b_3 & d_3 \\ a_4 & b_4 & d_4 \end{vmatrix} - d_1 \begin{vmatrix} a_2 & b_2 & c_2 \\ a_3 & b_3 & c_3 \\ a_4 & b_4 & c_4 \end{vmatrix};$$

and similarly we can introduce determinants of the fifth, sixth, . . . , nth order in succession. It turns out that in all essential properties these agree with the determinants of two or three rows. Determinants of more than three rows, however, cannot be expanded by the "diagonal rule". We shall not consider further details here.

subtract the second from the first; then we multiply the first equation by d and the second by b and subtract. We thus obtain

$$(bc - ad)y = Ac - Ba$$
$$(ad - bc)x = Ad - Bb,$$

or

$$x\begin{vmatrix} a & b \\ c & d \end{vmatrix} = \begin{vmatrix} A & b \\ B & d \end{vmatrix}, \quad y\begin{vmatrix} a & b \\ c & d \end{vmatrix} = \begin{vmatrix} a & A \\ c & B \end{vmatrix}.$$

If we assume that the determinant

$$\begin{vmatrix} a & b \\ c & d \end{vmatrix}$$

is different from zero, these equations at once give the solution

$$x = \frac{\begin{vmatrix} A & b \\ B & d \end{vmatrix}}{\begin{vmatrix} a & b \\ c & d \end{vmatrix}}, \quad y = \frac{\begin{vmatrix} a & A \\ c & B \end{vmatrix}}{\begin{vmatrix} a & b \\ c & d \end{vmatrix}},$$

which can be verified by substitution. If, however, the determinant $\begin{vmatrix} a & b \\ c & d \end{vmatrix}$ vanishes, the equations

$$x\begin{vmatrix} a & b \\ c & d \end{vmatrix} = \begin{vmatrix} A & b \\ B & d \end{vmatrix}, \quad y\begin{vmatrix} a & b \\ c & d \end{vmatrix} = \begin{vmatrix} a & A \\ c & B \end{vmatrix}$$

would lead to a contradiction if either of the determinants $\begin{vmatrix} a & A \\ c & B \end{vmatrix}$ and $\begin{vmatrix} A & b \\ B & d \end{vmatrix}$ were different from zero. If, however,

$$\begin{vmatrix} A & b \\ B & d \end{vmatrix} = \begin{vmatrix} a & A \\ c & B \end{vmatrix} = 0,$$

our formulæ tell us nothing about the solution.

We therefore obtain the fact, which is particularly important for our purposes, that *a system of equations of the above form, whose determinant is different from zero, always has a unique solution.*

If our system of equations is *homogeneous*, that is, if $A = B = 0$, our calculations lead to the solution $x = 0$, $y = 0$, provided that $\begin{vmatrix} a & b \\ c & d \end{vmatrix} \neq 0$.

For three equations with three unknowns,

$$ax + by + cz = A,$$
$$dx + ey + fz = B,$$
$$gx + hy + kz = C,$$

a similar discussion leads to a similar conclusion. We multiply the first equation by $\begin{vmatrix} e & f \\ h & k \end{vmatrix}$, the second by $-\begin{vmatrix} b & c \\ h & k \end{vmatrix}$, the third by $\begin{vmatrix} b & c \\ e & f \end{vmatrix}$, and add, thus obtaining

$$x\left\{a\begin{vmatrix} e & f \\ h & k \end{vmatrix} - d\begin{vmatrix} b & c \\ h & k \end{vmatrix} + g\begin{vmatrix} b & c \\ e & f \end{vmatrix}\right\}$$

$$+ y\left\{b\begin{vmatrix} e & f \\ h & k \end{vmatrix} - e\begin{vmatrix} b & c \\ h & k \end{vmatrix} + h\begin{vmatrix} b & c \\ e & f \end{vmatrix}\right\}$$

$$+ z\left\{c\begin{vmatrix} e & f \\ h & k \end{vmatrix} - f\begin{vmatrix} b & c \\ h & k \end{vmatrix} + k\begin{vmatrix} b & c \\ e & f \end{vmatrix}\right\} = A\begin{vmatrix} e & f \\ h & k \end{vmatrix} - B\begin{vmatrix} b & c \\ h & k \end{vmatrix} + C\begin{vmatrix} b & c \\ e & f \end{vmatrix}.$$

But by our formulæ for the expansion of a determinant in terms of the elements of a column, this equation can be written in the form

$$x\begin{vmatrix} a & b & c \\ d & e & f \\ g & h & k \end{vmatrix} + y\begin{vmatrix} b & b & c \\ e & e & f \\ h & h & k \end{vmatrix} + z\begin{vmatrix} c & b & c \\ f & e & f \\ k & h & k \end{vmatrix} = \begin{vmatrix} A & b & c \\ B & e & f \\ C & h & k \end{vmatrix}.$$

By rule (4) the coefficients of y and z vanish, so that

$$x\begin{vmatrix} a & b & c \\ d & e & f \\ g & h & k \end{vmatrix} = \begin{vmatrix} A & b & c \\ B & e & f \\ C & h & k \end{vmatrix}.$$

In the same way we derive the equations

$$y\begin{vmatrix} a & b & c \\ d & e & f \\ g & h & k \end{vmatrix} = \begin{vmatrix} a & A & c \\ d & B & f \\ g & C & k \end{vmatrix},$$

$$z\begin{vmatrix} a & b & c \\ d & e & f \\ g & h & k \end{vmatrix} = \begin{vmatrix} a & b & A \\ d & e & B \\ g & h & C \end{vmatrix}.$$

If the determinant

$$\begin{vmatrix} a & b & c \\ d & e & f \\ g & h & k \end{vmatrix}$$

is not zero, the last three equations give us the value of the unknowns. Provided that this determinant is not zero, the equations can be solved uniquely for x, y, z. If the determinant is zero, it follows that the right-hand sides of the above equations must also be zero, and the equations therefore cannot be solved unless A, B, C satisfy the special conditions which are expressed by the vanishing of every determinant on the right.

If, in particular, the system of equations is homogeneous, so that $A = B = C = 0$, and if its determinant is different from zero, it again follows that $x = y = z = 0$.

In addition to the cases above, in which the number of equations is equal to the number of unknowns, we shall occasionally meet with systems of two (homogeneous) equations with three unknowns, e.g.

$$ax + by + cz = 0,$$
$$dx + ey + fz = 0.$$

If the three determinants

$$D_1 = \begin{vmatrix} b & c \\ e & f \end{vmatrix}, \quad D_2 = \begin{vmatrix} c & a \\ f & d \end{vmatrix}, \quad D_3 = \begin{vmatrix} a & b \\ d & e \end{vmatrix}$$

are not all zero, if, for example, $D_3 \neq 0$, our equations can first be solved for x and y; this gives

$$x = \frac{zD_1}{D_3}, \quad y = \frac{zD_2}{D_3},$$

or

$$x : y : z = D_1 : D_2 : D_3.$$

Geometrically this has the following meaning: we are given two vectors $\mathfrak{u}$ and $\mathfrak{v}$ with the components a, b, c and d, e, f respectively. We seek a vector $\mathfrak{x}$ which is perpendicular to $\mathfrak{u}$ and $\mathfrak{v}$, that is, which satisfies the equations

$$\mathfrak{u}\mathfrak{x} = 0, \quad \mathfrak{v}\mathfrak{x} = 0.$$

Thus $\mathfrak{x}$ is in the direction of $[\mathfrak{u}\mathfrak{v}]$.

EXAMPLES

1. Show that the determinant

$$\begin{vmatrix} a & b & c \\ d & e & f \\ g & h & k \end{vmatrix}$$

can always be reduced to the form

$$\begin{vmatrix} \alpha & 0 & 0 \\ 0 & \beta & 0 \\ 0 & 0 & \gamma \end{vmatrix}$$

merely by repeated application of the following processes: (1) interchanging two rows or two columns, (2) adding a multiple of one row (or column) to another row (or column).

2. If the three determinants

$$\begin{vmatrix} a_1 & a_2 \\ b_1 & b_2 \end{vmatrix}, \quad \begin{vmatrix} a_1 & a_2 \\ c_1 & c_2 \end{vmatrix}, \quad \begin{vmatrix} b_1 & b_2 \\ c_1 & c_2 \end{vmatrix}$$

do not all vanish, then the necessary and sufficient condition for the existence of a solution of the three equations

$$a_1x + a_2y = d$$
$$b_1x + b_2y = e$$
$$c_1x + c_2y = f$$

is

$$D = \begin{vmatrix} a_1 & a_2 & d \\ b_1 & b_2 & e \\ c_1 & c_2 & f \end{vmatrix} = 0.$$

3. State the condition that the two straight lines

$$\begin{aligned} x &= a_1 t + b_1 \\ y &= a_2 t + b_2 \\ z &= a_3 t + b_3 \end{aligned} \qquad \text{and} \qquad \begin{aligned} x &= c_1 t + d_1 \\ y &= c_2 t + d_2 \\ z &= c_3 t + d_3 \end{aligned}$$

either intersect or are parallel.

4*. Prove the properties (1) to (7), given on pp. 20–22, for determinants of the fourth order (defined on p. 22 (footnote)).

5. Prove that the volume of a tetrahedron with vertices (x_1, y_1, z_1), (x_2, y_2, z_2), (x_3, y_3, z_3), (x_4, y_4, z_4) is given by

$$\frac{1}{6}\begin{vmatrix} x_1 & y_1 & z_1 & 1 \\ x_2 & y_2 & z_2 & 1 \\ x_3 & y_3 & z_3 & 1 \\ x_4 & y_4 & z_4 & 1 \end{vmatrix}.$$

4. Affine Transformations and the Multiplication of Determinants

We shall conclude these preliminary remarks by discussing the simplest facts relating to the so-called *affine transformations*; at the same time we shall obtain an important theorem on determinants.

1. Affine Transformations of the Plane and of Space.

By a *mapping* or *transformation* of a portion of space (or of a plane) we mean a law by which each point has assigned to it another point of space (or point of the plane) as *image point*; the point itself we call the *original point*, or sometimes the *model* (in antithesis to the *image*). We obtain a physical expression of the concept of mapping by imagining that the portion of space (or of the plane) in question is occupied by some deformable substance and that our transformation represents a deformation in which every point of the substance moves from its original position to a certain final position.

Using a rectangular system of co-ordinates, we take (x, y, z) as the co-ordinates of the original point and (x', y', z') as those of the corresponding image point.

The transformations which are not only the simplest and most easily understood, but are also of fundamental importance for the general case, are the *affine transformations*. An affine transformation is one in which the co-ordinates (x', y', z') (or in the plane (x', y')) of the image point are expressed linearly in terms of those of the original point. Such a transformation is therefore given by the three equations

$$\begin{aligned} x' &= ax + by + cz + m \\ y' &= dx + ey + fz + n \\ z' &= gx + hy + kz + p, \end{aligned}$$

or in the plane by the two equations

$$\begin{aligned} x' &= ax + by + m \\ y' &= cx + dy + n, \end{aligned}$$

with constant coefficients $a, b, \ldots$ These assign an image point to every point of space (or of the plane). The question at once arises whether we can interchange the relation of image point and original point, that is, whether every point of space (or of the plane) has an original point corresponding to it. The necessary and sufficient condition for this is that the equations

$$\begin{aligned} ax + by + cz &= x' - m \\ dx + ey + fz &= y' - n \\ gx + hy + kz &= z' - p \end{aligned} \quad \text{or} \quad \begin{aligned} ax + by &= x' - m \\ cx + dy &= y' - n \end{aligned}$$

shall be capable of being solved for the unknowns x, y, z (or x, y), no matter what the values of x', y', z' are. By section 3 (p. 24) an affine transformation *has* an inverse, and in fact a unique inverse,* provided that its determinant

$$\Delta = \begin{vmatrix} a & b & c \\ d & e & f \\ g & h & k \end{vmatrix}, \quad \text{or} \quad \Delta = \begin{vmatrix} a & b \\ c & d \end{vmatrix},$$

is different from zero. We shall confine our attention to affine transformations of this type, and shall not discuss what happens when $\Delta = 0$.

By introducing an intermediate point (x'', y'', z'') we can resolve the general affine transformation into the transformations

$$\begin{aligned} x'' &= ax + by + cz \\ y'' &= dx + ey + fz \\ z'' &= gx + hy + kz \end{aligned} \quad \text{or} \quad \begin{aligned} x'' &= ax + by \\ y'' &= cx + dy \end{aligned}$$

and

$$\begin{aligned} x' &= x'' + m \\ y' &= y'' + n \\ z' &= z'' + p \end{aligned} \quad \text{or} \quad \begin{aligned} x' &= x'' + m \\ y' &= y'' + n. \end{aligned}$$

Here (x, y, z) is mapped first on (x'', y'', z'') and then (x'', y'', z'') is mapped on (x', y', z'). Since the second transformation is merely a parallel translation of the space (or of the plane) as a whole and is therefore quite easily under-

* That is, every image point has one and only one original point.

stood, we may restrict ourselves to the study of the first. We shall therefore only consider affine transformations of the form

$$\begin{aligned} x' &= ax + by + cz \\ y' &= dx + ey + fz \\ z' &= gx + hy + kz \end{aligned} \quad \text{or} \quad \begin{aligned} x' &= ax + by \\ y' &= cx + dy \end{aligned}$$

with non-vanishing determinants.

The results of section 3 (p. 25) for linear equations enable us to express the inverse transformation by the formulæ

$$\begin{aligned} x &= a'x' + b'y' + c'z' \\ y &= d'x' + e'y' + f'z' \\ z &= g'x' + h'y' + k'z' \end{aligned} \quad \text{or} \quad \begin{aligned} x &= a'x' + b'y' \\ y &= c'x' + d'y', \end{aligned}$$

in which a', b', . . . are certain expressions formed from the coefficients a, b, . . . Because of the uniqueness of the solution, the original equations also follow from these latter. In particular, from $x = y = z = 0$ it follows that $x' = y' = z' = 0$, and conversely.

The characteristic geometrical properties of affine transformations are stated in the following theorems.

(1) *In space the image of a plane is a plane; and in the plane the image of a straight line is a straight line.*

For by section 1 (p. 9) we can write the equation of the plane (or the line) in the form

$$Ax + By + Cz + D = 0$$

$$(\text{or} \qquad Ax + By + D = 0).$$

The numbers A, B, C (or A, B) are not all zero. The co-ordinates of the image points of the plane (or of the line) satisfy the equation

$$A(a'x' + b'y' + c'z') + B(d'x' + e'y' + f'z') + C(g'x' + h'y' + k'z') + D = 0$$

$$(\text{or} \qquad A(a'x' + b'y') + B(c'x' + d'y') + D = 0).$$

Hence the image points themselves lie on a plane (or a line), for the coefficients

$$\begin{aligned} A' &= a'A + d'B + g'C \\ B' &= b'A + e'B + h'C \\ C' &= c'A + f'B + k'C \end{aligned} \qquad \left(\text{or} \quad \begin{aligned} A' &= a'A + c'B \\ B' &= b'A + d'B \end{aligned}\right)$$

of the co-ordinates x', y', z' (or x', y') cannot all be zero; otherwise the equations

$$\begin{aligned} a'A + d'B + g'C &= 0 \\ b'A + e'B + h'C &= 0 \\ c'A + f'B + k'C &= 0 \end{aligned} \qquad \left(\text{or} \quad \begin{aligned} a'A + c'B &= 0 \\ b'A + d'B &= 0 \end{aligned}\right)$$

would hold, and these we may regard as equations in the unknowns A, B, C

(or A, B). But we have shown above that from these equations it follows that $A = B = C = 0$ (or $A = B = 0$).

(2) *The image of a straight line in space is a straight line.*

This follows immediately from the fact that a straight line may be regarded as the intersection of two planes; by (1) its image is also the intersection of two planes and is therefore a straight line.

(3) *The images of two parallel planes of space (or of two parallel lines of the plane) are parallel.*

For if the images had points of intersection the originals would have to intersect in the original points of these intersections.

(4) *The images of two parallel lines in space are two parallel lines.*

For as the two lines lie in a plane and do not intersect one another, the same is true for their images, by (1) and (2). The images are therefore parallel.

The image of a vector $\mathbf{v}$ is of course a vector $\mathbf{v}'$ leading from the image of the initial point of $\mathbf{v}$ to the image of the final point of $\mathbf{v}$. Since the components of the vector are the differences of the corresponding coordinates of the initial and final points, under the most general affine transformation they are transformed according to the equations

$$\begin{aligned} v_1' &= av_1 + bv_2 + cv_3 \\ v_2' &= dv_1 + ev_2 + fv_3 \\ v_3' &= gv_1 + hv_2 + kv_3. \end{aligned}$$

2. The Combination of Affine Transformations and the Resolution of the General Affine Transformation.

If we map a point (x, y, z) on an image point (x', y', z') by means of the transformation

$$\begin{aligned} x' &= ax + by + cz \\ y' &= dx + ey + fz \\ z' &= gx + hy + kz \end{aligned}$$

and then map (x', y', z') on a point (x'', y'', z'') by means of a second affine transformation

$$\begin{aligned} x'' &= a_1x' + b_1y' + c_1z' \\ y'' &= d_1x' + e_1y' + f_1z' \\ z'' &= g_1x' + h_1y' + k_1z', \end{aligned}$$

then we readily see that (x, y, z) and (x'', y'', z'') are also related by an affine transformation: and in fact

$$\begin{aligned} x'' &= a_2x + b_2y + c_2z \\ y'' &= d_2x + e_2y + f_2z \\ z'' &= g_2x + h_2y + k_2z \end{aligned}$$

where the coefficients are given by the equations

$$
\begin{aligned}
a_2 &= a_1a + b_1d + c_1g, & b_2 &= a_1b + b_1e + c_1h, & c_2 &= a_1c + b_1f + c_1k,\\
d_2 &= d_1a + e_1d + f_1g, & e_2 &= d_1b + e_1e + f_1h, & f_2 &= d_1c + e_1f + f_1k,\\
g_2 &= g_1a + h_1d + k_1g, & h_2 &= g_1b + h_1e + k_1h, & k_2 &= g_1c + h_1f + k_1k.
\end{aligned}
$$

We say that this last transformation is the *combination* or *resultant* of the first two. If the determinants of the first two transformations are different from zero, their inverses can be formed; hence the compound transformation also has an inverse. The coefficients of the compound transformation are obtained from those of the original transformation by multiplying corresponding elements of a column of the first transformation and of a row of the second, adding the three products thus obtained, and using this "product" of column and row as the coefficient which stands in the column with the same number as the column used and in the row with the same number as the row used.

In the same way, combination of the transformations

$$
\begin{aligned}
x' &= ax + by \\
y' &= cx + dy
\end{aligned}
\quad \text{and} \quad
\begin{aligned}
x'' &= a_1x' + b_1y' \\
y'' &= c_1x' + d_1y'
\end{aligned}
$$

gives the new transformation

$$
\begin{aligned}
x'' &= (a_1a + b_1c)x + (a_1b + b_1d)y \\
y'' &= (c_1a + d_1c)x + (c_1b + d_1d)y.
\end{aligned}
$$

By a *primitive* transformation we mean one in which two (or one) of the three (or two) co-ordinates of the image are the same as the corresponding co-ordinates of the original points. Physically we may think of a primitive transformation as one in which the space (or plane) undergoes a stretching in one direction only (the stretching of course varying from place to place) so that all the points are simply moved along a family of parallel lines. A primitive affine transformation in which the motion takes place parallel to the x-axis is analytically represented by formulæ of the type

$$
\begin{aligned}
x' &= ax + by + cz \\
y' &= y \\
z' &= z
\end{aligned}
\quad \text{or} \quad
\begin{aligned}
x' &= ax + by \\
y' &= y.
\end{aligned}
$$

The general affine transformation in the plane,

$$
\begin{aligned}
x' &= ax + by \\
y' &= cx + dy,
\end{aligned}
$$

with a non-vanishing determinant, can be obtained by a combination of primitive transformations.

In the proof we may assume * that $a \neq 0$. We introduce an intermediate

* If $a = 0$, then $b \neq 0$, and we can return to the case $a \neq 0$ by interchanging x and y. Such an interchange, represented by the transformation $X = y$, $Y = x$, is itself effected by the three successive primitive transformations

$$
\begin{aligned}
\xi_1 &= x - y \\
\eta_1 &= y
\end{aligned}
;\quad
\begin{aligned}
\xi_2 &= \xi_1 \\
\eta_2 &= \xi_1 + \eta_1 = x
\end{aligned}
;\quad
\begin{aligned}
X &= -\xi_2 + \eta_2 = y \\
Y &= \eta_2 \qquad\;\; = x.
\end{aligned}
$$

point (ξ, η) by the primitive transformation

$$\xi = ax + by, \quad \eta = y,$$

whose determinant a is different from zero. From ξ, η we obtain x', y' by a second primitive transformation

$$x' = \xi, \quad y' = \frac{c}{a}\xi + \frac{ad - bc}{a}\eta$$

with the determinant

$$\frac{ad - bc}{a} = \frac{1}{a}\begin{vmatrix} a & b \\ c & d \end{vmatrix}.$$

This gives the required resolution into primitive transformations.

In a similar way *the affine transformation in space*

$$\begin{aligned} x' &= ax + by + cz \\ y' &= dx + ey + fz \\ z' &= gx + hy + kz, \end{aligned}$$

with a non-vanishing determinant, can be resolved into primitive transformations.

Of the three determinants

$$\begin{vmatrix} a & b \\ d & e \end{vmatrix}, \quad \begin{vmatrix} a & c \\ d & f \end{vmatrix}, \quad \begin{vmatrix} b & c \\ e & f \end{vmatrix}$$

at least one must be different from zero; otherwise, as the expansion in terms of the elements of the last row shows, we should have

$$\begin{vmatrix} a & b & c \\ d & e & f \\ g & h & k \end{vmatrix} = 0.$$

As in the previous case, we can then assume without loss of generality (1) that $\begin{vmatrix} a & b \\ d & e \end{vmatrix} \neq 0$, and (2) that $a \neq 0$. The first intermediate point (ξ, η, ζ) is given by means of the equations

$$\begin{aligned} \xi &= ax + by + cz \\ \eta &= \phantom{ax + {}} y \\ \zeta &= \phantom{ax + by + {}} z. \end{aligned}$$

The determinant of this primitive transformation is a, which is not zero. For the second transformation to ξ', η', ζ' we wish to put $\xi' = \xi$, $\zeta' = \zeta$, and also to have $\eta' = y'$. One primitive transformation then remains. If in the equation $\eta' = y' = dx + ey + fz$ we introduce the quantities ξ, η, ζ instead of x, y, z, we obtain the second primitive transformation in the form

$$\xi' = \xi$$
$$\eta' = \frac{d}{a}\xi + \frac{1}{a}\begin{vmatrix} a & b \\ d & e \end{vmatrix}\eta + \frac{1}{a}\begin{vmatrix} a & c \\ d & f \end{vmatrix}\zeta.$$
$$\zeta' = \zeta.$$

The determinant of this transformation is $\frac{1}{a}\begin{vmatrix} a & b \\ d & e \end{vmatrix} \neq 0$. The third transformation must then be

$$x' = \xi'$$
$$y' = \eta'$$
$$z' = -\frac{\begin{vmatrix} d & e \\ g & h \end{vmatrix}}{\begin{vmatrix} a & b \\ d & e \end{vmatrix}}\xi' + \frac{\begin{vmatrix} a & b \\ g & h \end{vmatrix}}{\begin{vmatrix} a & b \\ d & e \end{vmatrix}}\eta' + \frac{\begin{vmatrix} a & b & c \\ d & e & f \\ g & h & k \end{vmatrix}}{\begin{vmatrix} a & b \\ d & e \end{vmatrix}}\zeta'.$$

3. The Geometrical Meaning of the Determinant of Transformation, and the Multiplication Theorem.

From the considerations of the previous section we can find the geometrical meaning of the determinant of an affine transformation and at the same time an algebraic theorem on the multiplication of determinants.

We consider a plane triangle with vertices (0, 0), (x_1, y_1), (x_2, y_2), whose area is given (section 2, p. 14) by the formula

$$A = \frac{1}{2}\begin{vmatrix} x_1 & x_2 \\ y_1 & y_2 \end{vmatrix}.$$

We shall investigate the relation between A and the area A' of its image obtained by means of a primitive affine transformation

$$x' = ax + by$$
$$y' = y.$$

The vertices of the image triangle have the co-ordinates (0, 0), $(ax_1 + by_1, y_1)$, $(ax_2 + by_2, y_2)$, and therefore

$$A' = \frac{1}{2}\begin{vmatrix} x_1' & x_2' \\ y_1' & y_2' \end{vmatrix} = \frac{1}{2}\begin{vmatrix} ax_1 + by_1 & ax_2 + by_2 \\ y_1 & y_2 \end{vmatrix}.$$

This determinant, however, can be transformed by the theorems of section 3 (p. 22) in the following way:

$$A' = \frac{1}{2}\begin{vmatrix} ax_1 + by_1 & ax_2 + by_2 \\ y_1 & y_2 \end{vmatrix} = \frac{1}{2}\begin{vmatrix} ax_1 & ax_2 \\ y_1 & y_2 \end{vmatrix} = \frac{a}{2}\begin{vmatrix} x_1 & x_2 \\ y_1 & y_2 \end{vmatrix},$$

that is,

$$A' = aA.$$

If we had taken the primitive transformation

$$x' = x$$
$$y' = cx + dy,$$

we should have found in the same way that

$$A' = dA.$$

We see, therefore, that a primitive affine transformation has the effect of multiplying the area of a triangle by a constant independent of the triangle.* Since the general affine transformation can be formed by combining primitive transformations, the statement remains true for any affine transformation. *In the case of an affine transformation the ratio of the area of an image triangle to the area of the original triangle is constant and independent of the choice of triangle, depending only on the coefficients of the transformation.* In order to find this constant ratio we consider in particular the triangle with vertices (0, 0), (1, 0) and (0, 1), whose area A is $\frac{1}{2}$. Since the image of this triangle according to the transformation

$$x' = ax + by$$
$$y' = cx + dy$$

has the vertices (0, 0), (a, c), (b, d) its area is

$$\frac{1}{2}\begin{vmatrix} a & b \\ c & d \end{vmatrix} = A \begin{vmatrix} a & b \\ c & d \end{vmatrix},$$

and we thus see that *the constant ratio of area* A'/A *for an affine transformation is the determinant of the transformation.*

For transformations in space we can proceed in exactly the same way. If we consider the tetrahedron with the vertices (0, 0, 0), (x_1, y_1, z_1), (x_2, y_2, z_2), (x_3, y_3, z_3) and subject it to the primitive transformation

$$x' = ax + by + cz$$
$$y' = \qquad y$$
$$z' = \qquad\qquad z,$$

the image tetrahedron has the vertices (0, 0, 0), $(ax_1 + by_1 + cz_1, y_1, z_1)$, $(ax_2 + by_2 + cz_2, y_2, z_2)$, $(ax_3 + by_3 + cz_3, y_3, z_3)$, so that its volume V' is

$$V' = \frac{1}{6}\begin{vmatrix} ax_1 + by_1 + cz_1 & ax_2 + by_2 + cz_2 & ax_3 + by_3 + cz_3 \\ y_1 & y_2 & y_3 \\ z_1 & z_2 & z_3 \end{vmatrix}$$

$$= \frac{a}{6}\begin{vmatrix} x_1 & x_2 & x_3 \\ y_1 & y_2 & y_3 \\ z_1 & z_2 & z_3 \end{vmatrix}.$$

* If no vertex of the triangle lies at the origin, the same fact holds, in virtue of the general formula for the area given on p. 14.

Hence
$$V' = aV,$$

where V is the volume of the original tetrahedron. For the volume of the image given by the primitive transformation

$$\begin{aligned} x' &= x \\ y' &= dx + ey + fz \\ z' &= z \end{aligned}$$

we similarly find that

$$V' = eV,$$

and for the primitive transformation

$$\begin{aligned} x' &= x \\ y' &= y \\ z' &= gx + hy + kz \end{aligned}$$

we find that

$$V' = kV$$

From this it follows that an arbitrary affine transformation has the effect of multiplying the volume of a tetrahedron by a constant.* In order to find this constant for the transformation

$$\begin{aligned} x' &= ax + by + cz \\ y' &= dx + ey + fz \\ z' &= gx + hy + kz \end{aligned}$$

we consider the tetrahedron with the vertices (0, 0, 0), (1, 0, 0), (0, 1, 0), (0, 0, 1), whose image has the vertices (0, 0, 0), (a, d, g), (b, e, h), (c, f, k). For the volumes V' and V of the image and the original we therefore have

$$V' = \frac{1}{6}\begin{vmatrix} a & b & c \\ d & e & f \\ g & h & k \end{vmatrix}, \qquad V = \frac{1}{6};$$

hence the determinant $\begin{vmatrix} a & b & c \\ d & e & f \\ g & h & k \end{vmatrix}$ is the constant sought.

The sign of the determinant also has a geometrical meaning. For from what we have seen in section 2 (p. 18) on the connexion between the sense of rotation and the volume of the tetrahedron or area of the triangle, it follows at once that *a transformation with a positive determinant preserves the sense of rotation, while a transformation with a negative determinant reverses it.*

* If no vertex of the tetrahedron coincides with the origin, this theorem follows from the general formula for the volume of a tetrahedron (p. 18).

We now consider the combination of two transformations

$$\begin{aligned} x' &= ax + by + cz \qquad & x'' &= a_1x' + b_1y' + c_1z' \\ y' &= dx + ey + fz & y'' &= d_1x' + e_1y' + f_1z' \\ z' &= gx + hy + kz & z'' &= g_1x' + h_1y' + k_1z', \end{aligned}$$

$$\begin{aligned} x'' &= (a_1a + b_1d + c_1g)x + (a_1b + b_1e + c_1h)y + (a_1c + b_1f + c_1k)z \\ y'' &= (d_1a + e_1d + f_1g)x + (d_1b + e_1e + f_1h)y + (d_1c + e_1f + f_1k)z \\ z'' &= (g_1a + h_1d + k_1g)x + (g_1b + h_1e + k_1h)y + (g_1c + h_1f + k_1k)z. \end{aligned}$$

As we pass from x, y, z to x', y', z' the volume of a tetrahedron is multiplied by

$$\begin{vmatrix} a & b & c \\ d & e & f \\ g & h & k \end{vmatrix},$$

as we pass from x', y', z' to x'', y'', z'' by

$$\begin{vmatrix} a_1 & b_1 & c_1 \\ d_1 & e_1 & f_1 \\ g_1 & h_1 & k_1 \end{vmatrix},$$

and by direct change from x, y, z to x'', y'', z'' it is multiplied by

$$\begin{vmatrix} a_1a + b_1d + c_1g & a_1b + b_1e + c_1h & a_1c + b_1f + c_1k \\ d_1a + e_1d + f_1g & d_1b + e_1e + f_1h & d_1c + e_1f + f_1k \\ g_1a + h_1d + k_1g & g_1b + h_1e + k_1h & g_1c + h_1f + k_1k \end{vmatrix}.$$

This gives us the following relation, known as the theorem for the multiplication of determinants:

$$\begin{vmatrix} a_1 & b_1 & c_1 \\ d_1 & e_1 & f_1 \\ g_1 & h_1 & k_1 \end{vmatrix} \begin{vmatrix} a_2 & b_2 & c_2 \\ d_2 & e_2 & f_2 \\ g_2 & h_2 & k_2 \end{vmatrix}$$

$$= \begin{vmatrix} a_1a_2 + b_1d_2 + c_1g_2 & a_1b_2 + b_1e_2 + c_1h_2 & a_1c_2 + b_1f_2 + c_1k_2 \\ d_1a_2 + e_1d_2 + f_1g_2 & d_1b_2 + e_1e_2 + f_1h_2 & d_1c_2 + e_1f_2 + f_1k_2 \\ g_1a_2 + h_1d_2 + k_1g_2 & g_1b_2 + h_1e_2 + k_1h_2 & g_1c_2 + h_1f_2 + k_1k_2 \end{vmatrix}.$$

As before, we call the elements of the determinant on the right the "products" of the rows of $\begin{vmatrix} a_1 & b_1 & c_1 \\ d_1 & e_1 & f_1 \\ g_1 & h_1 & k_1 \end{vmatrix}$ and the columns of $\begin{vmatrix} a_2 & b_2 & c_2 \\ d_2 & e_2 & f_2 \\ g_2 & h_2 & k_2 \end{vmatrix}$; at the intersection of the i-th row and the k-th column of the product of the determinants there stands the expression formed from the i-th row of $\begin{vmatrix} a_1 & b_1 & c_1 \\ d_1 & e_1 & f_1 \\ g_1 & h_1 & k_1 \end{vmatrix}$ and the k-th column of $\begin{vmatrix} a_2 & b_2 & c_2 \\ d_2 & e_2 & f_2 \\ g_2 & h_2 & k_2 \end{vmatrix}$. Since rows and columns are interchangeable, the product of the determinants can also be obtained by combining columns and rows, columns and columns, or rows and rows.

For two-rowed determinants the corresponding theorem of course holds, namely

$$\begin{vmatrix} a_1 & b_1 \\ c_1 & d_1 \end{vmatrix} \begin{vmatrix} a_2 & b_2 \\ c_2 & d_2 \end{vmatrix} = \begin{vmatrix} a_1a_2 + b_1b_2 & a_1c_2 + b_1d_2 \\ c_1a_2 + d_1b_2 & c_1c_2 + d_1d_2 \end{vmatrix}$$

(combining rows and rows, &c.).

Examples

1. Evaluate the following determinants:

$$(a) \begin{vmatrix} 3 & 4 & 5 \\ 4 & 5 & 6 \\ 5 & 6 & 7 \end{vmatrix}, \quad (b) \begin{vmatrix} 1 & 1 & 1 \\ 1 & 2 & 4 \\ 1 & 3 & 9 \end{vmatrix}, \quad (c) \begin{vmatrix} 1 & 1 & 1 \\ 2 & 3 & 4 \\ 3 & -1 & 7 \end{vmatrix}, \quad (d) \begin{vmatrix} 1 & x & x^3 \\ 1 & y & y^3 \\ 1 & z & z^3 \end{vmatrix}.$$

2. Find the relation which must exist between a, b, c in order that the system of equations

$$\begin{aligned} 3x + 4y + 5z &= a \\ 4x + 5y + 6z &= b \\ 5x + 6y + 7z &= c \end{aligned}$$

may have a solution.

3*. (*a*) Prove the inequality

$$D = \begin{vmatrix} a & b & c \\ a' & b' & c' \\ a'' & b'' & c'' \end{vmatrix} \leqq \sqrt{(a^2 + b^2 + c^2)(a'^2 + b'^2 + c'^2)(a''^2 + b''^2 + c''^2)}.$$

(*b*) When does the equality sign hold?

4. What conditions must be satisfied in order that the affine transformation

$$x' = ax + by, \quad y' = cx + dy$$

may leave the distance between any two points unchanged?

5. Prove that in an affine transformation the image of a quadric

$$ax^2 + by^2 + cz^2 + dxy + exz + fyz + gx + hy + iz + j = 0$$

is another quadric.

6*. Prove that the affine transformation

$$\begin{aligned} x' &= ax + by + cz \\ y' &= dx + ey + fz \\ z' &= gx + hy + kz \end{aligned}$$

leaves at least one direction unaltered.

7. Give the formulæ for a rotation through the angle φ about the axis

$x : y : z = 1 : 0 : -1$ such that the rotation of the plane $x = z$ is positive when looked at from the point $(-1, 0, 1)$.

8. Prove that an affine transformation transforms the centre of mass of a system of particles into the centre of mass of the image particles.

9. If $\alpha_1, \ldots, \gamma_3$ denote the quantities on p. 6, defining a rotation of axes, then

$$\begin{vmatrix} \alpha_1 & \beta_1 & \gamma_1 \\ \alpha_2 & \beta_2 & \gamma_2 \\ \alpha_3 & \beta_3 & \gamma_3 \end{vmatrix} = \pm 1.$$

CHAPTER II

Functions of Several Variables and their Derivatives

We have already become acquainted with functions of several variables in Chapter X of Vol. I, and there learned enough to appreciate their importance and usefulness. We are now about to enter on a more thorough study of these functions, discussing properties which were not touched upon in the previous volume and proving theorems which there were merely made plausible. No proof in this volume will involve previous knowledge of any proof developed in Chapter X of Vol. I. Yet the student is recommended to read that chapter, as the intuitive discussion given there will assist him in forming mental images of matters which are perhaps somewhat abstract.

As a rule a theorem which can be proved for functions of two variables can be extended to functions of more than two variables without any essential change in the argument. In what follows, therefore, we shall usually confine ourselves to functions of two variables, and shall only discuss functions of three or more variables when some special point is involved.

1. The Concept of Function in the Case of Several Variables

1. Functions and their Ranges of Definition.

Equations of the form

$$u = x + y, \quad u = x^2y^2, \quad \text{or} \quad u = \log(1 - x^2 - y^2)$$

assign a *functional value* u to a pair of values (x, y). In the first two of these examples a value of u is assigned to *every* pair of values (x, y), while in the third the correspondence has a meaning only for those pairs of values (x, y) for which the inequality $x^2 + y^2 < 1$ is true.

In these cases we say that u is a *function* of the *independent variables* x and y. This expression we use in general whenever some law assigns a value of u as *dependent variable*, corresponding to each pair of values (x, y) belonging to a certain specified set. Similarly, we say that u is a function of the n variables $x_1, x_2, \ldots, x_n$ if for every set of values $(x_1, x_2, \ldots, x_n)$ belonging to a certain specified set there exists a corresponding value of u.

Thus, for example, the volume $u = xyz$ of a rectangular parallelepiped is a function of the lengths of the three sides x, y, z; the magnetic declination is a function of the latitude, the longitude, and the time; the sum $x_1 + x_2 + \ldots + x_n$ is a function of the n terms $x_1, x_2, \ldots, x_n$.

In the case of functions of two variables we represent the pair of values (x, y) by a *point* in a two-dimensional rectangular co-ordinate system with the co-ordinates x and y, and we occasionally call this point the *argument point* of the function. In the case of the functions $u = x + y$ and $u = x^2y^2$ this argument point can range over the whole of the xy-plane, and we say that these functions are defined in the whole xy-plane. In the case of the function $u = \log(1 - x^2 - y^2)$, the point must remain within the circle $x^2 + y^2 < 1$, and the function is defined only for points inside this circle.

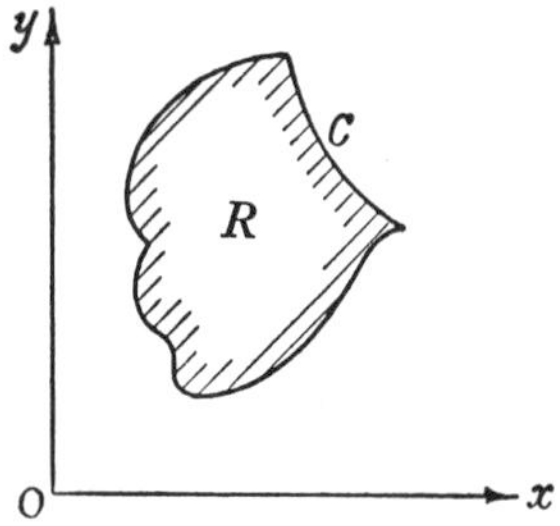

Fig. 1.—A simply-connected region

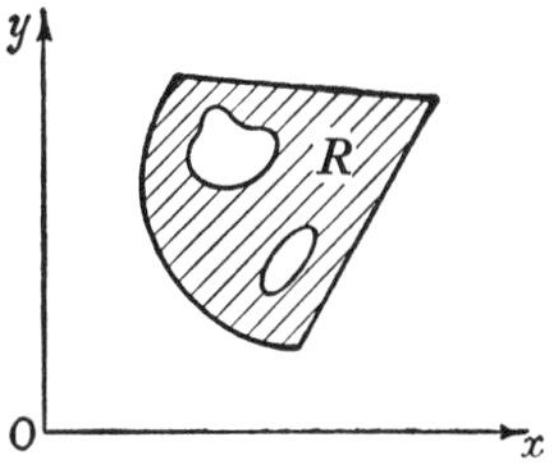

Fig. 2.—A triply-connected region

As in the case of functions of a single variable, the arguments in the case of functions of several variables may be either " discontinuous " or " continuous ". Thus the average population per state of the United States depends on the number of states and on the number of inhabitants, both of which are integers. On the other hand, lengths, weights, &c., are examples of continuous variables. In the future we shall deal almost exclusively with pairs of continuously variable arguments; the point (x, y) will be allowed to vary in a definite "region" (or "domain") of the xy-plane, corresponding to the "interval" in the case of functions of one variable. This region may consist of the whole xy-plane; or it may consist of a portion of the plane bounded by a single closed curve C which does not intersect itself (a "simply-connected region"; cf. fig. 1); or it may be bounded by several closed curves. In the last case it is said to be a "multiply-connected region", the number of the boundary curves giving the so-called "connectivity"; fig. 2, for example, shows a *triply-connected region.*

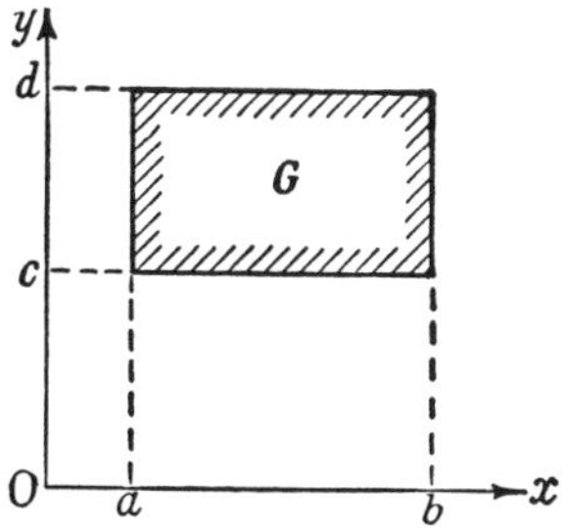

Fig. 3.—A rectangular region

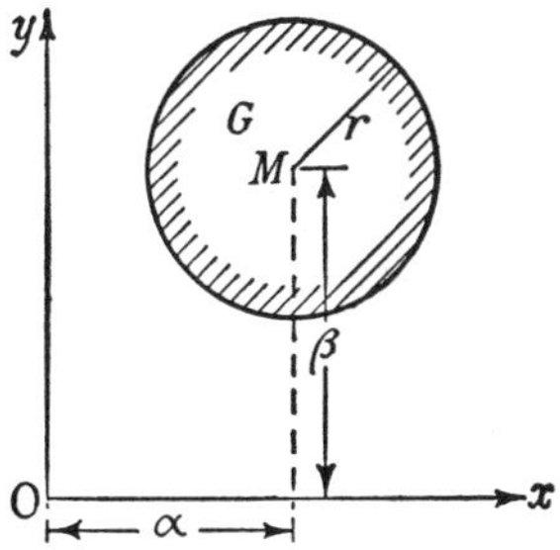

Fig. 4.—A circular region

The boundary curves, and in fact every curve considered in this volume, will be assumed to be *sectionally smooth.** That is, we assume once and for all that every such curve consists of a finite number of arcs, each one of which has a continuously-turning tangent at each of its points up to and including the end points. Such curves, therefore, can at most have a finite number of corners or cusps.

The most important types of region, which recur over and over again in the study of functions of several variables, are (1) the *rectangular region* (fig. 3), defined by inequalities of the form

$$a \leqq x \leqq b$$
$$c \leqq y \leqq d,$$

in which each of the independent variables is restricted to a definite interval, and the argument point varies in a rectangle;

* Ger. *stückweise glatt.*

and (2) the *circular region* (fig. 4), defined by an inequality of the form

$$(x - \alpha)^2 + (y - \beta)^2 \leqq r^2,$$

in which the argument point varies in a circle with radius r and centre (α, β).

A point P which belongs to a region R is said to be an *interior point* of R if we can find a circle with its centre at P lying entirely within R. If P is an interior point of R, we also say that R is a *neighbourhood* of P. Thus any neighbourhood of P will contain a sufficiently small circle with P as centre.

We may briefly remark that corresponding statements hold in the case of functions of more than two independent variables, e.g. of three variables x, y, z. In this case the argument point varies in a three-dimensional region instead of in a plane region. In particular, this region may be a *rectangular region*, defined by inequalities of the form

$$a \leqq x \leqq b, \quad c \leqq y \leqq d, \quad e \leqq z \leqq f,$$

or a *spherical region*, defined by an inequality of the form

$$(x - \alpha)^2 + (y - \beta)^2 + (z - \gamma)^2 \leqq r^2.$$

In conclusion, we shall mention a finer distinction, which, while scarcely essential for the purposes of this book, is nevertheless of importance in more advanced study. We sometimes have to consider regions which do not contain their boundary points, that is, the points of the curves bounding them. Such regions are called *open regions* (cf. the Appendix to this chapter, p. 98). Thus, for example, the region $x^2 + y^2 < 1$ is bounded by the circle $x^2 + y^2 = 1$, which does not belong to the region; the region is therefore open. If, on the other hand, the boundary points do belong to the region, as will be the case in most of the examples which we shall discuss, we say that the region is *closed*.

When we are dealing with more than three independent variables, say x, y, z, w, our intuition fails to provide a geometrical interpretation of the set of independent variables. Still, we shall occasionally make use of geometrical terminology, speaking of a system of n numbers as a point in n-dimensional space. By rectangular regions and spherical regions in such a space we naturally mean systems of points whose co-ordinates satisfy inequalities of the form

$$a_1 \leqq x \leqq a_2, \quad b_1 \leqq y \leqq b_2, \quad c_1 \leqq z \leqq c_2, \quad d_1 \leqq w \leqq d_2, \quad \ldots$$

or $(x-\alpha)^2+(y-\beta)^2+(z-\gamma)^2+(w-\delta)^2+\ldots \leqq r^2$

respectively.

We can now give precise expression to our definition of the concept of function in the following words. *If* R *is a region in which the independent variables* x, y, . . . *may vary, and if a definite value* u *is assigned to each point* (x, y, . . .) *of this region according to some law, then* u = f(x, y, . . .) *is said to be a function of the continuous independent variables* x, y, ...

It is to be noted that, just as in the case of functions of one variable, a functional correspondence associates a *unique* value of u with the system of independent variables $x, y, \ldots$. Thus if the functional value is assigned by an analytical expression which is multiple-valued, such as $\arctan \frac{y}{x}$, this expression does not determine the function completely. On the contrary, we have still to specify which of the several possible values of the expression is to be used; in the case mentioned, we have still to state that we are to take the value of $\arctan \frac{y}{x}$ which lies between $-\frac{\pi}{2}$ and $+\frac{\pi}{2}$, or the value between 0 and π, or we must make some other similar specification. In such a case we say that the expression defines several different single-valued *branches* of the function (cf. Vol. I, p. 17). If we wish to consider all these branches at once, without giving any one of them preference, we may regard them as together forming a *multiple-valued* function. In this book, however, we shall make use of this idea in Chap. VIII only.

2. The Simplest Types of Functions.

Just as in the case of functions of one variable, the simplest functions are the *rational integral* functions or *polynomials*. The most general polynomial of the first degree (*linear* function) is of the form

$$u = ax + by + c,$$

where a, b, and c are constants. The general polynomial of the second degree has the form

$$u = ax^2 + bxy + cy^2 + dx + ey + f.$$

The general polynomial of any degree is a sum of terms of the form $a_{mn}x^m y^n$, where the constants a_{mn} are arbitrary.

Rational fractional functions are quotients of polynomials; to this class belongs e.g. the *linear fractional* function

$$u = \frac{ax + by + c}{a'x + b'y + c'}.$$

By extraction of roots we pass from the rational functions to certain *algebraic* functions,* e.g.

$$u = \sqrt{\frac{x - y}{x + y}} + \sqrt[3]{\frac{(x + y)^2}{x^3 + xy}}.$$

In the construction of more complicated functions of several variables we almost always fall back on the well-known functions of one variable,† e.g.

$$u = \sin(x \operatorname{arc} \cos y) \quad \text{or} \quad u = \log_x y.$$

3. Geometrical Representation of Functions.

In Chapter X of Vol. I we discussed the two principal methods for representing a function of two independent variables, namely (1) by means of the surface $u = f(x, y)$ in xyu-space, described by the point with co-ordinates (x, y, u) as (x, y) ranges over the region of definition of the function u, and (2) by means of the curves (contour lines) in the xy-plane along which u has a definite fixed value k. We shall not repeat this discussion here. If the student is not already perfectly familiar with these methods of geometrical representation, he would be well advised to turn to the previous volume and read the discussion given there (p. 460 *et seq.*).

2. CONTINUITY

1. Definition.

The reader who is acquainted with the theory of functions of a single variable and has seen what an important part is played in it by the concept of continuity will naturally expect that a corresponding concept will figure prominently in the theory of functions of more than one variable. Moreover, he will know in advance that the statement that the function $u = f(x, y)$ is continuous at the point (x, y) will mean, roughly speaking, that *for all points* (ξ, η) *near* (x, y) *the value of the function* $f(\xi, \eta)$ *will differ but little from* $f(x, y)$. This idea we shall express more precisely as follows.

The function $f(x, y)$, *defined in the region* R, *is continuous at the point* (ξ, η) *of* R, *provided that for every positive number* ϵ *it is possible to find a positive number* $\delta = \delta(\epsilon)$ (*in general depending on*

* For an accurate definition of the term "algebraic function" see p. 119.

† Cf. also the section on compound functions (p. 69).

ϵ and tending to 0 with ϵ) such that for all points of the region whose distance from (ξ, η) is less than δ (that is, for which*

$$(x - \xi)^2 + (y - \eta)^2 \leqq \delta^2),$$

$$|f(x, y) - f(\xi, \eta)| \leqq \epsilon.$$

Or, in other words, the relation

$$|f(\xi + h, \eta + k) - f(\xi, \eta)| \leqq \epsilon$$

is to hold for all pairs of values (h, k) such that $h^2 + k^2 \leqq \delta^2$ and the point $(\xi + h, \eta + k)$ belongs to the region R.

If a function is continuous at every point of a region R, we say that it is *continuous in* R.

In the definition of continuity we can replace the distance condition $h^2 + k^2 \leqq \delta^2$ by the following equivalent condition:

To every $\epsilon > 0$ there shall correspond two positive numbers δ_1 and δ_2 such that

$$|f(\xi + h, \eta + k) - f(\xi, \eta)| \leqq \epsilon$$

whenever $|h| \leqq \delta_1$ and $|k| \leqq \delta_2$.

The two conditions are equivalent. For if the original condition is fulfilled, so is the second if we take $\delta_1 = \delta_2 = \delta/\sqrt{2}$; and conversely, if the second condition is fulfilled, so is the first if for δ we take the smaller of the two numbers δ_1 and δ_2.

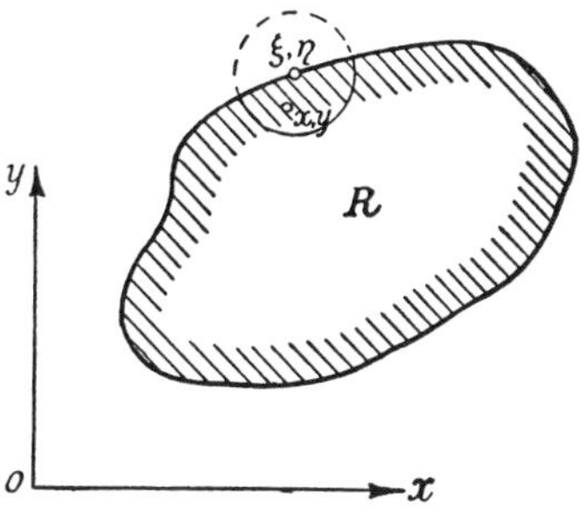

Fig. 5.—Boundary point

The following facts are almost obvious:

The sum, difference, and product of continuous functions are also continuous. The quotient of continuous functions is continuous except where the denominator vanishes. Continuous functions of continuous functions are themselves continuous (cf. section 5, No. 1, p. 70). In particular, *all polynomials are continuous, and all rational fractional functions are also continuous except where the denominator vanishes.*†

* Fig. 5 illustrates the case where (ξ, η) lies on the boundary of R.

† Another obvious fact, which, however, is worth stating, is as follows: *if a function* f(x, y) *is continuous in a region* R *and is different from zero at an*

A function of several variables may have discontinuities of a much more complicated type than a function of a single variable. For example, discontinuities may occur along whole arcs of curves, as in the case of the function $u = y/x$, which is discontinuous along the whole line $x = 0$. Moreover, a function $f(x, y)$ may be continuous in x for each fixed value of y and continuous in y for each fixed value of x and yet be a discontinuous function of x and y. This is exemplified by the function $f(x, y) = \frac{2xy}{x^2 + y^2}$, $f(0, 0) = 0$. If we take any fixed non-zero value of y, this function is obviously continuous as a function of x, as the denominator cannot vanish. If $y = 0$, we have $f(x, 0) = 0$, which is a continuous function of x. Similarly, $f(x, y)$ is continuous in y for each fixed value of x. But at every point on the line $y = x$, except the point $x = y = 0$, we have $f(x, y) = 1$; and there are points of this line arbitrarily close to the origin. Hence the function is discontinuous at the point $x = y = 0$.

Other examples of discontinuous functions will be found in Vol. I (p. 464).

2. The Concept of Limit in the Case of Several Variables.

The concept of the limit of a function of two variables is closely related to the concept of continuity. Let us suppose that the function $f(x, y)$ is defined in a region R, and that (ξ, η) is a point either within R or on its boundary. Then the statement that the limit of $f(x, y)$ as x tends to ξ and y tends to η is l is to be understood as having the following meaning: *for every* $\epsilon > 0$ *there is a* $\delta > 0$ *such that*

$$|f(x, y) - l| < \epsilon$$

for all points (x, y) *in* R *for which the inequality*

$$0 < (x - \xi)^2 + (y - \eta)^2 \leqq \delta^2$$

holds. It is to be noted that, just as in the case of functions of one variable, the point (x, y) is required to be distinct from the point (ξ, η).

We symbolize the existence of the limit l by writing

$$\lim_{\substack{x \to \xi \\ y \to \eta}} f(x, y) = l, \text{ or } f(x, y) \to l \text{ as } (x, y) \to (\xi, \eta).$$

interior point P *of the region, it is possible to mark off about* P *a neighbourhood, say a circle, belonging entirely to* R, *in which* f(x, y) *is nowhere equal to zero.* For if the value of the function at P is a, we can mark off about P a circle so small that the value of the function in the circle differs from a by less than $a/2$ and therefore is certainly not zero.

For emphasis this is sometimes read "the double limit as x tends to ξ and y tends to η of $f(x, y)$ is l".

Using the language of limits, we can say that a function $f(x, y)$ is continuous at a point (ξ, η) if, and only if,

$$\lim_{\substack{x \to \xi \\ y \to \eta}} f(x, y) = f(\xi, \eta).$$

We can see the matter in a new light if we consider sequences of points. We shall say that a sequence of points (x_1, y_1), (x_2, y_2), . . . , (x_n, y_n), . . . tends to a limit point (ξ, η) if the distance $\sqrt{\{(x_n - \xi)^2 + (y_n - \eta)^2\}}$ tends to 0 as n increases. We can then show at once (cf. Vol. I, p. 47) that if $f(x, y) \to l$ as $(x, y) \to (\xi, \eta)$, then $\lim_{n \to \infty} f(x_n, y_n) = l$ for every sequence of points (x_n, y_n) in R which tends to (ξ, η). The converse is also true; if $\lim_{n \to \infty} f(x_n, y_n)$ exists and is equal to l for every sequence (x_n, y_n) of points in R tending to (ξ, η), then the double limit of $f(x, y)$ as $x \to \xi$ and $y \to \eta$ exists and is equal to l. We omit the proof of this.

In our definition of limit we have allowed the point (x, y) to vary in the region R. If we so desire, however, we can impose restrictions on the point (x, y). For example, we may require it to lie in a sub-region R' of R, or on a curve C, or in a set of points M in R. In this case we say that $f(x, y)$ tends to l as (x, y) tends to (ξ, η) in R' (or on C, or in M). It is of course understood that R' (or C, or M) must contain points arbitrarily close to (ξ, η) in order that the definition may be applicable.

Our definition of continuity then implies the two following requirements: (1) *as* (x, y) *tends to* (ξ, η) *in* R *the function* f(x, y) *must possess a limit l; and* (2) *this limit l must coincide with the value of the function at the point* (ξ, η).

It is obvious that we could define continuity of a function, not only in a region R but also, for example, along a curve C, in the same way.

3. The Order to which a Function Vanishes.*

If the function $f(x, y)$ is continuous at the point (ξ, η), the difference $f(x, y) - f(\xi, \eta)$ tends to zero as x tends to ξ and y tends to η. By introducing the new variables $h = x - \xi$ and

* This sub-section may be omitted on a first reading.

$k = y - \eta$ we can express this as follows: the function $\phi(h, k) = f(\xi + h, \eta + k) - f(\xi, \eta)$ of the variables h and k tends to 0 as h and k tend to 0.

We shall frequently meet with functions such as $\phi(h, k)$ which tend to zero * as h and k do. As in the case of one independent variable, for many purposes it is useful to describe the behaviour of $\phi(h, k)$ as $h \to 0$ and $k \to 0$ more precisely by distinguishing between different "orders of vanishing" or "orders of magnitude" of $\phi(h, k)$. For this purpose we base our comparisons on the distance

$$\rho = \sqrt{h^2 + k^2} = \sqrt{(x - \xi)^2 + (y - \eta)^2}$$

of the point with co-ordinates $x = \xi + h$ and $y = \eta + k$ from the point with co-ordinates ξ and η, and make the following statement:

A function $\phi(\mathrm{h}, \mathrm{k})$ *vanishes as* $\rho \to 0$ *to the same order as* $\rho = \sqrt{\mathrm{h}^2 + \mathrm{k}^2}$, *or, more precisely, to at least the same order as* ρ, *provided that there is a constant* C, *independent of* h *and* k, *such that the inequality*

$$\left|\frac{\phi(h, k)}{\rho}\right| \leqq C$$

holds for all sufficiently small values of ρ; or, more precisely, when there is a $\delta > 0$ such that the inequality holds for all values of h and k such that $0 < \sqrt{h^2 + k^2} < \delta$. Further, we say that $\phi(h, k)$ *vanishes to a higher order* † *than* ρ *if the quotient* $\dfrac{\phi(\mathrm{h}, \mathrm{k})}{\rho}$ *tends to* 0 *as* $\rho \to 0$. This is sometimes expressed by the symbolical notation ‡ $\phi(h, k) = o(\rho)$.

* In the older literature the expressions "$\phi(h, k)$ becomes infinitely small as h and k do" or "$\phi(h, k)$ is infinitesimal" are also found. These statements have a perfectly definite meaning if we regard them simply as another way of saying "$\phi(h, k)$ tends to 0 as h and k do". We nevertheless prefer to avoid the misleading expression "infinitely small" entirely.

† In order to avoid confusion, we would expressly point out that a higher order of vanishing for $\rho \to 0$ implies smaller values in the neighbourhood of $\rho = 0$; for example, ρ^2 vanishes to a higher order than ρ, and ρ^2 is smaller than ρ, when ρ is nearly zero.

‡ The letter o is of course chosen because it is the first letter of the word *order*. If we wish to express the statement that $\phi(h, k)$ vanishes to at least the same order as ρ, but not necessarily to a higher order, we use the letter O instead of o, writing $\phi(h, k) = O(\rho)$. In this book, however, we shall not use this symbol.

Let us now consider a few examples. Since

$$\frac{|h|}{\sqrt{(h^2+k^2)}} \leqq 1 \quad\text{and}\quad \frac{|k|}{\sqrt{(h^2+k^2)}} \leqq 1,$$

the components h and k of the distance ρ in the directions of the x- and y-axes vanish to at least the same order as the distance itself. The same is true for a linear homogeneous function $ah+bk$ with constants a and b, or for the function $\rho \sin\frac{1}{\rho}$. For fixed values of α greater than 1 the power ρ^α of the distance vanishes to a higher order than ρ; symbolically, $\rho^\alpha = o(\rho)$ for $\alpha > 1$. Similarly, a homogeneous quadratic polynomial $ah^2 + bhk + ck^2$ in the variables h and k vanishes to a higher order than ρ as $\rho \to 0$:

$$ah^2 + bhk + ck^2 = o(\rho).$$

More generally, the following definition is used. If the comparison function $\omega(h, k)$ is defined for all non-zero values of (h, k) in a sufficiently small circle about the origin, and is not equal to zero, then $\phi(h, k)$ *vanishes to at least the same order as* $\omega(h, k)$ as $\rho \to 0$ if for some suitably chosen constant C the relation

$$\left|\frac{\phi(h, k)}{\omega(h, k)}\right| \leqq C$$

holds; and similarly, $\phi(h, k)$ *vanishes to a higher order than* $\omega(h, k)$, or $\phi(h, k) = o(\omega(h, k))$, if $\dfrac{\phi(h, k)}{\omega(h, k)} \to 0$ when $\rho \to 0$.

For example, the homogeneous polynomial $ah^2 + bhk + ck^2$ is at least of the same order as ρ^2, since

$$|ah^2 + bhk + ck^2| \leqq (|a| + \tfrac{1}{2}|b| + |c|)(h^2 + k^2).$$

Also $\rho = o\left(\dfrac{1}{|\log \rho|}\right)$, since $\lim\limits_{\rho\to 0} (\rho \log \rho) = 0$ (Vol. I, p. 195).

EXAMPLES

1. Discuss the behaviour of the functions

(*a*) $x^3 - 3xy^2$,
(*b*) $x^4 - 6x^2y^2 + y^4$

in the neighbourhood of the origin.

2. How many constants does the general form of a polynomial $P(x, y)$ of degree n contain?

3. Prove that the expression

$$ax^3 + bxy^2 + cx^2y + dy^3$$

vanishes at $x = y = 0$ to at least the same order as $\rho^3 = (x^2 + y^2)^{\frac{3}{2}}$.

4. Find the condition that the polynomial

$$P = ax^2 + 2bxy + cy^2$$

is of exactly the same order as ρ^2 in the neighbourhood of $x = 0$, $y = 0$ (i.e. both $\frac{P}{\rho^2}$ and $\frac{\rho^2}{P}$ are bounded).

5. Are the following functions continuous at $x = y = 0$?

(a) $\dfrac{x^2 - y^2}{x^2 + y^2}$. (b) $\dfrac{x^2 + 2xy + y^2}{x^2 + y^2}$. (c) $\dfrac{x^2 + 3xy + y^2}{x^2 + 4xy + y^2}$.

(d) $-\dfrac{|x - y|}{x^2 - 2xy + y^2}$. (e) $e^{-\frac{|x-y|}{x^2-2xy+y^2}}$.

(f) $|x|^y$. (g) $|x|^{1/|y|}$. (h)* $|y|^{|x|}\dfrac{\sqrt{(x^2 + y^2)}}{\sqrt{(x^2 + y^2)} + \left|\dfrac{y}{x}\right|}$.

6. Find a $\delta(\varepsilon)$ (p. 44) for those functions of Ex. 5 which are continuous.

3. THE DERIVATIVES OF A FUNCTION

1. Definition. Geometrical Representation.

If in a function of several variables we assign definite numerical values to all but one of the variables and allow only that one

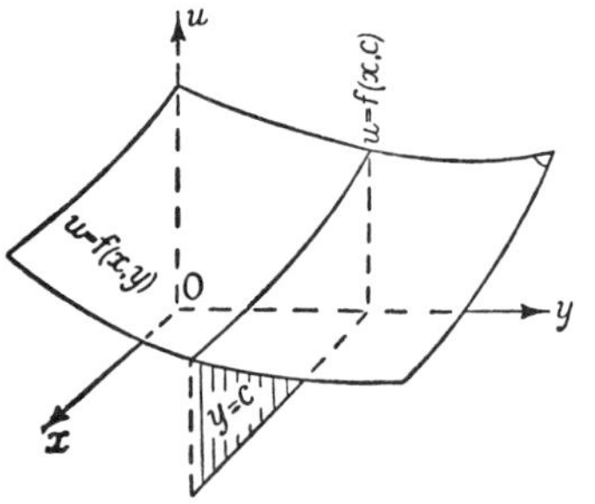

Fig. 6

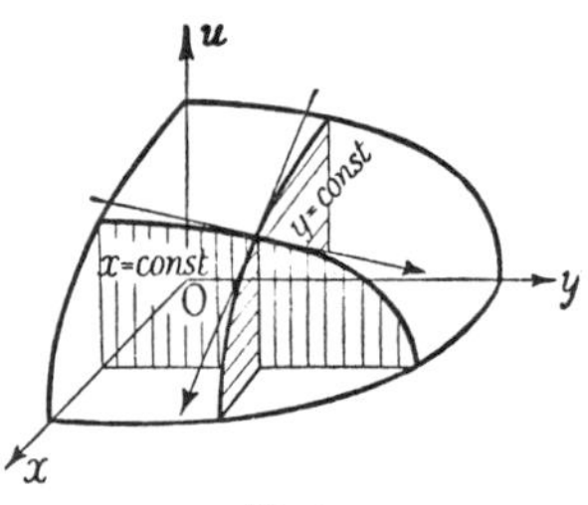

Fig. 7

Sections of $u = f(x, y)$

variable, say x, to vary, the function becomes a function of one variable. We consider a function $u = f(x, y)$ of the two variables x and y and assign to y a definite fixed value $y = y_0 = c$. The function $u = f(x, y_0)$ of the single variable x which is thus formed may be represented geometrically by letting the plane $y = y_0$ cut the surface $u = f(x, y)$ (cf. figs. 6 and 7). The curve of intersection thus formed in the plane is represented by the equation $u = f(x, y_0)$. If we differentiate this function in the usual way at the point $x = x_0$ (we assume that the derivative exists), we obtain the *partial derivative of* f(x, y) *with respect to* x at the point (x_0, y_0). According to the usual definition of the derivative, this is the limit *

$$\lim_{h \to 0} \frac{f(x_0 + h, y_0) - f(x_0, y_0)}{h}.$$

Geometrically this partial derivative denotes the tangent of the angle between a parallel to the x-axis and the tangent line to the curve $u = f(x, y_0)$. It is therefore the *slope of the surface* u = f(x, y) *in the direction of the* x*-axis.*

To represent these partial derivatives several different notations are used, of which we mention the following:

$$\lim_{h \to 0} \frac{f(x_0 + h, y_0) - f(x_0, y_0)}{h} = f_x(x_0, y_0) = u_x(x_0, y_0).$$

If we wish to emphasize that the partial derivative is the limit of a difference quotient, we denote it by

$$\frac{\partial f}{\partial x} \quad \text{or} \quad \frac{\partial}{\partial x} f.$$

Here we use a special round letter ∂, instead of the ordinary d used in the differentiation of functions of one variable, in order to show that we are dealing with a function of several variables and differentiating with respect to one of them.

It is sometimes convenient to use Cauchy's symbol D, mentioned on p. 90 of Vol. I, and write

$$\frac{\partial f}{\partial x} = D_x f,$$

but we shall seldom use this symbol.

In exactly the same way we define the partial derivative of

* If (x_0, y_0) is a point on the boundary of the region of definition, we make the restriction that in the passage to the limit the point $(x + h, y_0)$ must always remain in the region.

$f(x, y)$ with respect to y at the point (x_0, y_0) by the relation

$$\lim_{k \to 0} \frac{f(x_0, y_0 + k) - f(x_0, y_0)}{k} = f_y(x_0, y_0) = D_y f(x_0, y_0).$$

This represents the slope of the curve of intersection of the surface $u = f(x, y)$ with the plane $x = x_0$ perpendicular to the x-axis.

Let us now think of the point (x_0, y_0), hitherto considered fixed, as variable, and accordingly omit the suffixes 0. In other words, we think of the differentiation as carried out at any point (x, y) of the region of definition of $f(x, y)$. Then the two derivatives are themselves functions of x and y,

$$u_x(x, y) = f_x(x, y) = \frac{\partial f(x, y)}{\partial x} \text{ and } u_y(x, y) = f_y(x, y) = \frac{\partial f(x, y)}{\partial y}.$$

For example, the function $u = x^2 + y^2$ has the partial derivatives $u_x = 2x$ (in differentiation with respect to x the term y^2 is regarded as a constant and so has the derivative 0) and $u_y = 2y$. The partial derivatives of $u = x^3y$ are $u_x = 3x^2y$ and $u_y = x^3$.

We similarly make the following definition for any number n of independent variables,

$$\frac{\partial f(x_1, x_2, \ldots, x_n)}{\partial x_1} = \lim_{h \to 0} \frac{f(x_1 + h, x_2, \ldots, x_n) - f(x_1, x_2, \ldots, x_n)}{h}$$
$$= f_{x_1}(x_1, x_2, \ldots, x_n) = D_{x_1} f(x_1, x_2, \ldots, x_n),$$

it being assumed that the limit exists.

Of course we can also form *higher partial derivatives* of $f(x, y)$ by again differentiating the partial derivatives of the "first order", $f_x(x, y)$ and $f_y(x, y)$, with respect to one of the variables, and repeating this process. We indicate the order of differentiation by the order of the suffixes or by the order of the symbols ∂x and ∂y in the "denominator", from right to left,* and use the following symbols for the second derivatives:

$$\frac{\partial}{\partial x}\left(\frac{\partial f}{\partial x}\right) = \frac{\partial^2 f}{\partial x^2} = f_{xx} = D^2{}_{xx} f,$$

$$\frac{\partial}{\partial x}\left(\frac{\partial f}{\partial y}\right) = \frac{\partial^2 f}{\partial x \partial y} = f_{xy} = D^2{}_{xy} f,$$

* In Continental usage, on the other hand, $\frac{\partial}{\partial x}\left(\frac{\partial f}{\partial y}\right)$ is written $\frac{\partial^2 f}{\partial y \partial x}$.

$$\frac{\partial}{\partial y}\left(\frac{\partial f}{\partial x}\right)=\frac{\partial^2 f}{\partial y\,\partial x}=f_{yx}=D^2{}_{yx}f,$$

$$\frac{\partial}{\partial y}\left(\frac{\partial f}{\partial y}\right)=\frac{\partial^2 f}{\partial y^2}=f_{yy}=D^2{}_{yy}f.$$

We likewise denote the third partial derivatives by

$$\frac{\partial}{\partial x}\left(\frac{\partial^2 f}{\partial x^2}\right)=\frac{\partial^3 f}{\partial x^3}=f_{xxx},$$

$$\frac{\partial}{\partial y}\left(\frac{\partial^2 f}{\partial x^2}\right)=\frac{\partial^3 f}{\partial y\,\partial x^2}=f_{yxx},$$

$$\frac{\partial}{\partial x}\left(\frac{\partial^2 f}{\partial x\,\partial y}\right)=\frac{\partial^3 f}{\partial x^2\,\partial y}=f_{xxy},\ \text{\&c.};$$

and in general the n-th derivatives by

$$\frac{\partial}{\partial x}\left(\frac{\partial^{n-1} f}{\partial x^{n-1}}\right)=\frac{\partial^n f}{\partial x^n}=f_{x^n},$$

$$\frac{\partial}{\partial y}\left(\frac{\partial^{n-1} f}{\partial x^{n-1}}\right)=\frac{\partial^n f}{\partial y\,\partial x^{n-1}}=f_{yx^{n-1}},\ \text{\&c.}$$

In practice the performance of partial differentiations involves nothing that the student has not met with already. For according to the definition all the independent variables are to be kept constant except the one with respect to which we are differentiating. We therefore have merely to regard the other variables as constants and carry out the differentiation according to the rules by which we differentiate functions of a single independent variable. The student may nevertheless find it helpful to study the examples of partial differentiation given in Chapter X of Vol. I (p. 469 *et seq.*).

Just as in the case of one independent variable, the possession of derivatives is a special property of a function, not enjoyed even by all continuous functions.* All the same, this property is possessed by all functions of practical importance, except perhaps at isolated exceptional points.

* For an explanation of the term "differentiable", which implies *more* than that the partial derivatives with respect to x and y exist, see p. 60 *et seq.*

2. Continuity and the Existence of Partial Derivatives with respect to x and y.

In the case of functions of a single variable, we know that the existence of the derivative of a function at a point implies the continuity of the function at that point (cf. Vol. I, p. 97). In contrast with this, the possession of partial derivatives does *not* imply the continuity of a function of two variables: e.g. the function $u(x, y) = \frac{2xy}{x^2+y^2}$, with $u(0, 0) = 0$, has partial derivatives everywhere, and yet we have already seen (p. 46) that it is discontinuous at the origin. Geometrically speaking, the existence of partial derivatives restricts the behaviour of the function in the directions of the x- and y-axes only, and not in other directions. Nevertheless the possession of *bounded* partial derivatives does imply continuity, as is stated by the following theorem:

If a function $f(x, y)$ *has partial derivatives* f_x *and* f_y *everywhere in a region* R, *and these derivatives everywhere satisfy the inequalities*

$$|f_x(x, y)| < M, \quad |f_y(x, y)| < M,$$

where M *is independent of* x *and* y, *then* $f(x, y)$ *is continuous everywhere in* R.

To prove this we consider two points with co-ordinates (x, y) and $(x + h, y + k)$ respectively, both lying in the region R. We further assume that the two line-segments joining these points to the point $(x + h, y)$ both lie entirely in R; this is certainly true if (x, y) is a point interior to R and the point $(x + h, y + k)$ lies sufficiently close to (x, y). We then have

$$f(x + h, y + k) - f(x, y) = \{f(x + h, y + k) - f(x + h, y)\} + \{f(x + h, y) - f(x, y)\}.$$

The two terms in the first bracket on the right differ only in y, those in the second bracket only in x. We can therefore transform both the brackets on the right-hand side by means of the ordinary mean value theorem of the differential calculus (Vol. I, p. 103), regarding the first bracket as a function of y alone and the second as a function of x alone. We thus obtain the relation

$$f(x + h, y + k) - f(x, y) = kf_y(x + h, y + \theta_1 k) + hf_x(x + \theta_2 h, y),$$

where θ_1 and θ_2 are two numbers between 0 and 1. In other words, the derivative with respect to y is to be formed for a point of the vertical line joining $(x+h, y)$ to $(x+h, y+k)$, and the derivative with respect to x is to be formed for a point of the horizontal line joining (x, y) and $(x+h, y)$. Since by hypothesis both derivatives are less in absolute value than M, it follows that

$$|f(x+h, y+k) - f(x, y)| \leqq M(|h| + |k|).$$

For sufficiently small values of h and k the right-hand side is itself arbitrarily small, and the continuity of $f(x, y)$ is proved.

3. Change of the Order of Differentiation.

In the examples of partial differentiation given in Vol. I it will be found that $f_{yx} = f_{xy}$; in other words, it makes no difference whether we differentiate first with respect to x and then with respect to y, or first with respect to y and then with respect to x. This observation depends on the following important theorem:

If the "mixed" partial derivatives f_{xy} *and* f_{yx} of *a function* $f(x, y)$ *are continuous in a region* R, *then the equation*

$$f_{yx} = f_{xy}$$

holds throughout the interior of that region; that is, the order of differentiation with respect to x *and to* y *is immaterial.*

The proof, like that of the previous sub-section, is based on the mean value theorem of the differential calculus. We consider the four points (x, y), $(x+h, y)$, $(x, y+k)$, and $(x+h, y+k)$, where $h \neq 0$ and $k \neq 0$. If (x, y) is an interior point of the region R, and h and k are small enough, all four of these points belong to R. We now form the expression

$$A = f(x+h, y+k) - f(x+h, y) - f(x, y+k) + f(x, y).$$

By introducing the function

$$\phi(x) = f(x, y+k) - f(x, y)$$

of the variable x and regarding the variable y merely as a "parameter", we can write this expression in the form

$$A = \phi(x+h) - \phi(x).$$

Transforming the right-hand side by means of the ordinary

mean value theorem of the differential calculus, we obtain

$$A = h\phi'(x + \theta h),$$

where θ lies between 0 and 1. From the definition of $\phi(x)$, however, we have

$$\phi'(x) = f_x(x, y + k) - f_x(x, y);$$

and since we have assumed that the "mixed" second partial derivative f_{yx} does exist, we can again apply the mean value theorem and find that

$$A = hkf_{yx}(x + \theta h, y + \theta' k),$$

where θ and θ' denote two numbers between 0 and 1.

In exactly the same way we can start with the function

$$\psi(y) = f(x + h, y) - f(x, y)$$

and represent A by means of the equation

$$A = \psi(y + k) - \psi(y).$$

We thus arrive at the equation

$$A = hkf_{xy}(x + \theta_1 h, y + \theta_1' k), \text{ where } 0 < \theta_1 < 1 \text{ and } 0 < \theta_1' < 1,$$

and if we equate the two expressions for A we obtain the equation

$$f_{yx}(x + \theta h, y + \theta' k) = f_{xy}(x + \theta_1 h, y + \theta_1' k).$$

If here we let h and k tend simultaneously to 0 and recall that the derivatives $f_{xy}(x, y)$ and $f_{yx}(x, y)$ are continuous at the point (x, y), we immediately obtain

$$f_{yx}(x, y) = f_{xy}(x, y),$$

which was to be proved.*

* For more refined investigations it is often useful to know that the theorem on the reversibility of the order of differentiation can be proved with weaker hypotheses. It is, in fact, sufficient to assume that, in addition to the first partial derivatives f_x and f_y, only *one mixed partial derivative, say* f_{xy}, *exists, and that this derivative is continuous at the point in question.* To prove this, we return to the above equation

$$A = f(x + h, y + k) - f(x + h, y) - f(x, y + k) + f(x, y),$$

divide by hk, and then let k alone tend to 0. Then the right-hand side has a limit, and therefore the left-hand side also has a limit, and

$$\lim_{k \to 0} \frac{A}{kh} = \frac{f_y(x + h, y) - f_y(x, y)}{h}.$$

The theorem on the reversibility of the order of differentiation has far-reaching consequences. In particular, we see that the number of distinct derivatives of the second order and of higher orders of functions of several variables is decidedly smaller than we might at first have expected. If we assume that all the derivatives which we are about to form are continuous functions of the independent variables in the region under consideration, and if we apply our theorem to the functions $f_x(x, y), f_y(x, y), f_{xy}(x, y)$, &c., instead of to the function $f(x, y)$, we arrive at the equations

$$\begin{aligned} f_{xxy} &= f_{xyx} = f_{yxx}, \\ f_{xyy} &= f_{yxy} = f_{yyx}, \\ f_{xxyy} &= f_{xyxy} = f_{xyyx} = f_{yxxy} = f_{yxyx} = f_{yyxx}, \end{aligned}$$

and in general we have the following result:

*In the repeated differentiation of a function of two independent variables the order of the differentiations may be changed at will, provided only that the derivatives in question are continuous functions.**

Further, it was proved above with the sole assumption that f_{yx} exists that

$$\frac{A}{hk} = f_{yx}(x + \theta h, y + \theta' k).$$

In virtue of the assumed continuity of f_{yx}, we find that for arbitrary $\epsilon > 0$ and for all sufficiently small values of h and k

$$f_{yx}(x, y) - \epsilon < f_{yx}(x + \theta h, y + \theta' k) < f_{yx}(x, y) + \epsilon,$$

whence it follows that

$$f_{yx}(x, y) - \epsilon \leqq \frac{f_y(x + h, y) - f_y(x, y)}{h} \leqq f_{yx}(x, y) + \epsilon$$

or

$$\lim_{h \to 0} \frac{f_y(x + h, y) - f_y(x, y)}{h} = f_{yx}(x, y),$$

that is,

$$f_{xy}(x, y) = f_{yx}(x, y).$$

* It is of fundamental interest to show by means of an example that in the absence of the assumption of the continuity of the second derivative f_{xy} or f_{yx} the theorem need not be true, and that on the contrary f_{xy} can differ from f_{yx}. This is exemplified by the function $f(x, y) = xy\dfrac{x^2 - y^2}{x^2 + y^2}$, $f(0, 0) = 0$, for which all the partial derivatives of second order exist, but are not continuous. We find that

$$f_x(0, y) = \lim_{x \to 0} \frac{f(x, y) - f(0, y)}{x} = \lim_{x \to 0} y\frac{x^2 - y^2}{x^2 + y^2} = -y,$$

$$f_y(x, 0) = \lim_{y \to 0} \frac{f(x, y) - f(x, 0)}{y} = \lim_{y \to 0} x\frac{x^2 - y^2}{x^2 + y^2} = x,$$

and consequently

$$f_{yx}(0, 0) = -1 \text{ and } f_{xy}(0, 0) = +1.$$

These two expressions are different, which by the above theorem can only be due to the discontinuity of f_{xy} at the origin.

With our assumptions about continuity a function of two variables has *three* partial derivatives of the second order,

$$f_{xx},\ f_{xy},\ f_{yy},$$

four partial derivatives of the third order,

$$f_{xxx},\ f_{xxy},\ f_{xyy},\ f_{yyy},$$

and in general $(n+1)$ partial derivatives of the n-th order,

$$f_{x^n},\ f_{x^{n-1}y},\ f_{x^{n-2}y},\ \dots,\ f_{xy^{n-1}},\ f_{y^n}.$$

It is obvious that similar statements also hold for functions of more than two independent variables. For we can apply our proof equally well to the interchange of differentiations with respect to x and z or with respect to y and z, &c., for each interchange of two successive differentiations involves only two independent variables at a time.

EXAMPLES

1. How many n-th derivatives has a function of three variables?

2. Prove that the function

$$f(x_1, x_2, \dots, x_n) = \frac{1}{(x_1^2 + x_2^2 + \dots + x_n^2)^{(n-2)/2}}$$

satisfies the equation

$$f_{x_1x_1} + f_{x_2x_2} + \dots + f_{x_nx_n} = 0.$$

3. Calculate

$$\frac{\partial^2}{\partial x^2}\begin{vmatrix} a+x & b & c \\ d & e+x & f \\ g & h & k+x \end{vmatrix}.$$

4. Prove that

$$\frac{\partial^3}{\partial x\,\partial y\,\partial z}\begin{vmatrix} f_1(x) & f_2(x) & f_3(x) \\ g_1(y) & g_2(y) & g_3(y) \\ h_1(z) & h_2(z) & h_3(z) \end{vmatrix} = \begin{vmatrix} f_1'(x) & f_2'(x) & f_3'(x) \\ g_1'(y) & g_2'(y) & g_3'(y) \\ h_1'(z) & h_2'(z) & h_3'(z) \end{vmatrix}.$$

5. Considering

$$D = \begin{vmatrix} a & b & c \\ d & e & f \\ g & h & k \end{vmatrix}$$

as a function of the nine variables $a, b, \dots, k$, prove that

(*a*) $aD_a + bD_b + cD_c = D,$

(b)* $$\begin{vmatrix} D_a & D_b & D_c \\ D_d & D_e & D_f \\ D_g & D_h & D_k \end{vmatrix} = D^2.$$

4. The Total Differential of a Function and its Geometrical Meaning

1. The Concept of Differentiability.

In the case of functions of one variable the existence of a derivative is intimately connected with the possibility of approximating to the function $\eta = f(\xi)$ in the neighbourhood of the point x by means of a linear function $\eta = \phi(\xi)$. This linear function is defined by the equation

$$\phi(\xi) = f(x) + (\xi - x)f'(x).$$

Geometrically (ξ and η being current co-ordinates), this represents the tangent to the curve $\eta = f(\xi)$ at the point P with the co-ordinates $\xi = x$ and $\eta = f(x)$; analytically, its characteristic feature is that it differs from the function $f(\xi)$ in the neighbourhood of P by a quantity $o(h)$ of higher order than the abscissa $h = \xi - x$ (cf. p. 48). Hence

$$f(\xi) - \phi(\xi) = f(\xi) - f(x) - (\xi - x)f'(x) = o(h)$$

or, otherwise,

$$f(x + h) - f(x) - hf'(x) = o(h) = \epsilon h,$$

where ϵ denotes a quantity which tends to zero as h does. The term $hf'(x)$, the "linear part" of the increment of $f(x)$ corresponding to an increment of h in the independent variable, we have already (Vol. I, p. 107) called the *differential* of the function $f(x)$ and have denoted it by

$$dy = df(x) = hy' = hf'(x)$$

(or also by $dy = y'dx$, since for the function $y = x$ the differential has the value $dy = dx = 1 \times h$). We can now say that this differential is a function of the two independent variables x and h, and we need not restrict the variable h in any way. Of course this concept of differential is as a rule only used when h is small, so that the differential $hf'(x)$ forms an approximation to the difference $f(x+h)-f(x)$ which is accurate enough for the particular purpose.

Conversely, instead of beginning with the notion of the derivative, we could have laid the emphasis on the requirement that it should be possible to approximate to the function $\eta = f(\xi)$ in the neighbourhood of the point P by a linear function such that the difference between the function and the linear approximation function vanishes to a higher order than the increment h of the independent variable. In other words, we should require that for the function $f(\xi)$ at the point $\xi = x$ there should exist a quantity A, depending on x but not on h, such that

$$f(x+h)-f(x) = Ah + o(h) = Ah + \epsilon h,$$

where ϵ tends to 0 with h. This condition is equivalent to the requirement that $f(x)$ shall be differentiable at the point x; the quantity A must then be taken as the derivative $f'(x)$ at the point x. We see this immediately if we rewrite our condition in the form

$$\frac{f(x+h)-f(x)}{h} = A + \epsilon$$

and then let h tend to 0. Differentiability of a function with respect to a variable and the possibility of approximating to a function by means of a linear function in this way are therefore equivalent properties.

If we notice that $A + \epsilon = \alpha(x, h)$ is a function of h which tends to $A(x)$ as $h \to 0$, we arrive at the equivalent definition: $f(x)$ is said to be differentiable at the point x if $f(x+h)-f(x) = h\,\alpha(x, h)$, where the quantity $\alpha(x, h)$ is continuous, as a function of h, at $h = 0$.

These ideas can be extended in a perfectly natural way to functions of two and more variables.

We say that the function $u = f(x, y)$ is *differentiable* at the point (x, y) if it can be approximated to in the neighbourhood of this point by a linear function, that is, if it can be represented in the form

$$f(x+h, y+k) = f(x, y) + Ah + Bk + \epsilon_1 h + \epsilon_2 k,$$

where A and B are independent of the variables h and k and where ϵ_1 and ϵ_2 tend to 0 as h and k do. In other words, the difference between the function $f(x+h, y+k)$ at the point $(x+h, y+k)$ and the function $f(x, y)+Ah+Bk$ which is linear in h and k must be of the order of magnitude* $o(\rho)$, that is, must vanish as $\rho \to 0$ to a higher order than the distance $\rho=\sqrt{(h^2+k^2)}$ of the point $(x+h, y+k)$ from the point (x, y).

If such an approximate representation is possible, it follows at once that the function $f(x, y)$ can be partially differentiated with respect to x and to y at the point (x, y) and that

$$f_x = A \text{ and } f_y = B.$$

For if we put $k=0$ and divide by h we obtain the relation

$$\frac{f(x+h, y)-f(x, y)}{h} = A+\epsilon_1.$$

Since ϵ_1 tends to zero with h, as we pass to the limit $h \to 0$ the left-hand side has a limit, and that limit is A. Similarly, we obtain the equation $f_y(x, y)=B$.

Conversely, we shall prove that a function $u=f(x, y)$ is differentiable in the sense just defined, that is, it can be approximated to by a linear function, if it possesses *continuous* derivatives of the first order at the point in question. In fact, we can write the increment

$$\Delta u = f(x+h, y+k)-f(x, y)$$

of the function in the form

$$\Delta u = \{f(x+h, y+k)-f(x, y+k)\}+\{f(x, y+k)-f(x, y)\}.$$

As before (p. 54), the two brackets can be expressed in the form

$$\Delta u = hf_x(x+\theta_1 h, y+k)+kf_y(x, y+\theta_2 k),$$

using the ordinary mean value theorem of the differential calculus. Since by hypothesis the partial derivatives f_x and f_y are continuous at the point (x, y), we can write

$$f_x(x+\theta_1 h, y+k) = f_x(x, y)+\epsilon_1$$

* The equivalence of these two definitions follows from the following remark: the inequality $|\epsilon_1 h+\epsilon_2 k| \leqq |\epsilon| \sqrt{(h^2+k^2)}$ always holds, where $\epsilon = |\epsilon_1| + |\epsilon_2|$ and tends to 0 as ϵ_1 and ϵ_2 do. Hence the second definition of differentiability follows from the first. Again, since $|\epsilon\sqrt{(h^2+k^2)}| \leqq |\epsilon|(|h|+|k|)$, if the second condition is fulfilled the difference between the function and the linear approximation is of the form $\theta\epsilon(|h|+|k|)$, where $-1 \leqq \theta \leqq +1$, whence it follows that the requirements of the first definition are also fulfilled.

and
$$f_y(x, y + \theta_2 k) = f_y(x, y) + \epsilon_2,$$

where the numbers ϵ_1 and ϵ_2 tend to zero as h and k do. We thus obtain

$$\begin{aligned}\Delta u &= hf_x(x, y) + kf_y(x, y) + \epsilon_1 h + \epsilon_2 k \\ &= hf_x(x, y) + kf_y(x, y) + o(\sqrt{h^2 + k^2}),\end{aligned}$$

and this equation is the expression of the above statement.* We shall occasionally refer to a function with continuous first partial derivatives as a *continuously differentiable* function. If in addition all the second-order partial derivatives are continuous, we say that the function is twice continuously differentiable, and so on.

As in the case of functions of one variable, the definition of differentiability can be replaced by the following equivalent definition: the function $f(x, y)$ is said to be differentiable at the point (x, y) if

$$f(x + h, y + k) - f(x, y) = \alpha h + \beta k,$$

where α and β depend on h and k as well as on x and y, and are continuous as functions of h and k for $h = 0$, $k = 0$.

No further discussion is required to show how these considerations can be extended to functions of three and more variables.

2. Differentiation in a Given Direction.

An important property of differentiable functions is that they not only possess partial derivatives with respect to x and y, or, as we also say, in the x- and y-directions, but they also have partial derivatives in any other direction. By the *derivative in the direction* α we mean the following:

We let the point $(x + h, y + k)$ approach the point (x, y) in such a way that it is always on the straight line through (x, y) which makes the constant angle α with the positive x-axis. In other words, h and k do not tend to 0 independently of one another, but satisfy the relations

$$h = \rho \cos \alpha \quad \text{and} \quad k = \rho \sin \alpha,$$

where ρ is the distance $\sqrt{(h^2 + k^2)}$ of the point $(x + h, y + k)$

* If we assume the existence only, and not the continuity, of the derivatives f_x and f_y, the function is not necessarily differentiable (cf. p. 65 *et seq.*).

from the point (x, y) and tends to 0 as h and k do. If as usual we then form the difference $f(x+h, y+k) - f(x, y)$ and divide by ρ, we call the limit of the fraction

$$D_{(\alpha)}f(x, y) = \lim_{\rho \to 0} \frac{f(x + \rho\cos\alpha, y + \rho\sin\alpha) - f(x, y)}{\rho}$$

the derivative of the function $f(x, y)$ at the point (x, y) in the direction α, provided that the limit exists. In particular, when $\alpha = 0$ we have $k = 0$ and $h = \rho$, and we obtain the partial derivative with respect to x; when $\alpha = \pi/2$ we have $h = 0$ and $k = \rho$, and we obtain the partial derivative with respect to y.

If the function $f(x, y)$ is differentiable, we have

$$\begin{aligned} f(x+h, y+k) - f(x, y) &= hf_x + kf_y + \epsilon\rho \\ &= \rho(f_x\cos\alpha + f_y\sin\alpha + \epsilon). \end{aligned}$$

As ρ tends to 0, so does ϵ, and *for the derivative in the direction* α *we obtain the expression*

$$D_{(\alpha)}f(x, y) = f_x\cos\alpha + f_y\sin\alpha;$$

it is therefore a linear function of the derivatives f_x *and* f_y *in the* x- *and* y-*directions, with the coefficients* $\cos\alpha$ *and* $\sin\alpha$. This result always holds good, provided that the derivatives f_x and f_y exist and are continuous at the point in question.

Thus for the radius vector $r = \sqrt{(x^2 + y^2)}$ from the origin to the point (x, y) we have the partial derivatives

$$r_x = \frac{x}{\sqrt{x^2+y^2}} = \frac{x}{r} = \cos\theta \quad \text{and} \quad r_y = \frac{y}{\sqrt{x^2+y^2}} = \frac{y}{r} = \sin\theta,$$

where θ denotes the angle which the radius vector makes with the x-axis. Consequently, in the direction α the function r has the derivative

$$D_{(\alpha)}r = r_x\cos\alpha + r_y\sin\alpha = \cos\theta\cos\alpha + \sin\theta\sin\alpha = \cos(\theta - \alpha);$$

in particular, in the direction of the radius vector itself this derivative has the value 1, while in the direction perpendicular to the radius vector it has the value 0.

In the direction of the radius vector the function x has the derivative $D_{(\theta)}(x) = \cos\theta$ and the function y has the derivative $D_{(\theta)}(y) = \sin\theta$; in the direction perpendicular to the radius vector they have the derivatives $D_{(\theta+\pi/2)}x = -\sin\theta$ and $D_{(\theta+\pi/2)}y = \cos\theta$ respectively.

The derivative of a function $f(x, y)$ in the direction of the radius vector is in general denoted by $\frac{\partial f(x, y)}{\partial r}$. Thus we have the convenient relation

$$\frac{\partial}{\partial r} = \cos\theta \frac{\partial}{\partial x} + \sin\theta \frac{\partial}{\partial y},$$

where any differentiable function can be written after the symbols $\frac{\partial}{\partial r}, \frac{\partial}{\partial x}, \frac{\partial}{\partial y}$.

It is also worth noting that we obtain the derivative of the function $f(x, y)$ in the direction α if, instead of allowing the point Q with co-ordinates $(x + h, y + k)$ to approach the point P with co-ordinates (x, y) along a straight line with the direction α, we let Q approach P along an arbitrary curve whose tangent at P has the direction α. For then if the line $\overline{PQ}$ has the direction β, we can write $h = \rho\cos\beta$, $k = \rho\sin\beta$, and in the formulæ used in the above proof we have to replace α by β. But since by hypothesis β tends to α as $\rho \to 0$, we obtain the same expression for $D_{(\alpha)}f(x, y)$.

In the same way, a differentiable function $f(x, y, z)$ of three independent variables can be differentiated in a given direction. We suppose that the direction is specified by the cosines of the three angles which it forms with the co-ordinate axes. If we call these three angles α, β, γ, and if we consider two points (x, y, z) and $(x + h, y + k, z + l)$, where

$$\begin{aligned} h &= \rho\cos\alpha, \\ k &= \rho\cos\beta, \\ l &= \rho\cos\gamma, \end{aligned}$$

then just as above we obtain the expression

$$f_x \cos\alpha + f_y \cos\beta + f_z \cos\gamma$$

for the derivative in the direction given by the angles (α, β, γ).

3. Geometrical Interpretation. The Tangent Plane.

For a function $u = f(x, y)$ all these matters can easily be illustrated geometrically. We recall that the partial derivative with respect to x is the slope of the tangent to the curve in which the surface is intersected by a plane perpendicular to the xy-plane and parallel to the xu-plane. In the same way, the derivative in the direction a gives the slope of the tangent to the curve in which the surface is intersected by a plane perpendicular to the xy-plane and making the angle a with the x-axis. The formula $D_{(a)}f(x, y) = f_x \cos a + f_y \sin a$ now enables us to calculate the slopes of the tangents to all such curves, that is, of all tangents to the surface at a given point, from the slopes of two such tangents.

We approximated to the differentiable function $\zeta = f(\xi, \eta)$ in the neighbourhood of the point (x, y) by the linear function

$$f(\xi, \eta) = f(x, y) + (\xi - x)f_x + (\eta - y)f_y,$$

where ξ and η are the current co-ordinates. Geometrically this linear function represents a plane, which by analogy with the tangent line to a curve we shall call the *tangent plane* to the surface. The difference between this linear function and the function $f(\xi, \eta)$ tends to zero as $\xi - x = h$ and $\eta - y = k$ do, and in fact vanishes to a higher order than $\sqrt{(h^2 + k^2)}$. By the definition of the tangent to a plane curve, however, this states that the intersection of the tangent plane with any plane perpendicular to the xy-plane is the tangent to the corresponding curve of intersection. We thus see that *all these tangent lines to the surface at the point* (x, y, u) *lie in one plane, the tangent plane.*

This property is the geometrical expression of the differentiability of the function at the point $(x, y, u = f(x, y))$. If (ξ, η, ζ) are current co-ordinates, the equation of the tangent plane at the point $(x, y, u = f(x, y))$ is

$$\zeta - u = (\xi - x)f_x + (\eta - y)f_y.$$

As has already been shown on p. 61, the function is differentiable at a given point provided that the partial derivatives are continuous there. In contrast with the case where there is only one independent variable, the mere *existence* of the partial derivatives f_x and f_y is *not* sufficient to ensure the differentiability of the function. If the derivatives are not continuous at the point in question, the tangent plane to the surface at this point may fail to exist, or, analytically speaking, the difference between $f(x + h, y + k)$ and the function $f(x, y) + hf_x(x, y) + kf_y(x, y)$ which is linear in h and k may fail to vanish to a higher order than $\sqrt{(h^2 + k^2)}$.

This is clearly shown by a simple example. We write

$$f(x, y) = 0 \quad \text{if} \quad x = 0 \quad \text{or} \quad y = 0,$$
$$f(x, y) = |x| \quad \text{if} \quad x - y = 0 \quad \text{or} \quad x + y = 0.$$

Between these lines we define the function in such a way that it is represented geometrically by planes. The surface $u = f(x, y)$ therefore consists of eight triangular pieces of planes, meeting in roof-like edges above the lines $x = 0$, $y = 0$, $y = x$ and $y = -x$. This surface obviously has no tangent plane at the origin, although the derivatives $f_x(0, 0)$ and $f_y(0, 0)$ both exist and have the value 0. The derivatives are not continuous at the origin, however; in fact, as we readily see, they do not even exist on the edges.*

4. The Total Differential of a Function.

As in the case of functions of one variable, it is often convenient to have a special name and symbol for the linear part of the increment of a differentiable function $u = f(x, y)$. We call this linear part the *differential* of the function, and write

$$du = df(x, y) = \frac{\partial f}{\partial x} h + \frac{\partial f}{\partial y} k = \frac{\partial f}{\partial x} dx + \frac{\partial f}{\partial y} dy.$$

The differential, sometimes called the *total differential,* is a function of *four* independent variables, namely the co-ordinates x and y of the point under consideration and the increments h

* Another example of a similar type is given by the function

$$u = f(x, y) = \frac{xy}{\sqrt{x^2 + y^2}} \quad \text{if} \quad x^2 + y^2 \neq 0,$$
$$u = 0 \quad \text{if} \quad x = 0,\ y = 0.$$

If we introduce polar co-ordinates this becomes

$$u = \frac{r}{2} \sin 2\theta.$$

The first derivatives with respect to x and to y exist everywhere in the neighbourhood of the origin and have the value 0 at the origin itself. These derivatives, however, are not continuous at the origin, for

$$u_x = y\left(\frac{1}{\sqrt{x^2 + y^2}} - \frac{x^2}{\sqrt{(x^2 + y^2)^3}}\right) = \frac{y^3}{\sqrt{(x^2 + y^2)^3}}.$$

If we approach the origin along the x-axis, u_x tends to 0, while if we approach along the y-axis, u_x tends to 1. This function is not differentiable at the origin; at that point no tangent plane to the surface $u = f(x, y)$ exists. For the equations $f_x(0, 0) = f_y(0, 0) = 0$ show that the tangent plane would have to coincide with the plane $u = 0$. But at the points of the line $\theta = \pi/4$ we have $\sin 2\theta = 1$ and $u = r/2$; thus the distance u of the point of the surface from the point of the plane does not, as must be the case with a tangent plane, vanish to a higher order than r.

and k, which are the differentials of the independent variables or *independent differentials*. We need scarcely emphasize once more that this has nothing to do with the vague concept of "infinitely small quantities". It simply means that du approximates to $\Delta u = f(x+h, y+k) - f(x, y)$, the increment of the function, with an error which is an arbitrarily small fraction of $\sqrt{(h^2+k^2)}$ (itself arbitrarily small), provided that h and k are sufficiently small quantities. Incidentally, we thus collect the expressions for the different partial derivatives in one formula. For example, from the total differential we obtain the partial derivative $\dfrac{\partial f}{\partial x}$ by putting $dy = 0$ and $dx = 1$.

We again emphasize that to speak of the total differential of a function $f(x, y)$ has no meaning unless the function is differentiable in the sense defined above (for which the continuity, but not the mere existence, of the two partial derivatives suffices).

If the function $f(x, y)$ also possesses continuous partial derivatives of higher order, we can form the differential of the differential $df(x, y)$, that is, we can multiply its partial derivatives with respect to x and y by $h = dx$ and $k = dy$ respectively and then add these products. In this differentiation we must regard h and k as constants, corresponding to the fact that the differential $df = hf_x + kf_y$ is a function of the four independent variables x, y, h, and k. We thus obtain the *second differential* * of the function,

$$d^2f = d(df) = \frac{\partial}{\partial x}\left(\frac{\partial f}{\partial x}h + \frac{\partial f}{\partial y}k\right)h + \frac{\partial}{\partial y}\left(\frac{\partial f}{\partial x}h + \frac{\partial f}{\partial y}k\right)k$$

$$= \frac{\partial^2 f}{\partial x^2}h^2 + 2\frac{\partial^2 f}{\partial x \partial y}hk + \frac{\partial^2 f}{\partial y^2}k^2$$

$$= \frac{\partial^2 f}{\partial x^2}dx^2 + 2\frac{\partial^2 f}{\partial x \partial y}dx\,dy + \frac{\partial^2 f}{\partial y^2}dy^2.$$

Similarly, we can form the *higher differentials*

$$d^3f = d(d^2f) = \frac{\partial^3 f}{\partial x^3}dx^3 + 3\frac{\partial^3 f}{\partial x^2 \partial y}dx^2 dy + 3\frac{\partial^3 f}{\partial x \partial y^2}dx\,dy^2 + \frac{\partial^3 f}{\partial y^3}dy^3,$$

* We shall later see (p. 80 *et seq.*) that the differentials of higher order introduced formally here correspond exactly to the terms of the corresponding order in the increment of the function.

$$d^4f = \frac{\partial^4 f}{\partial x^4}\,dx^4 + 4\,\frac{\partial^4 f}{\partial x^3 \partial y}\,dx^3 dy + 6\,\frac{\partial^4 f}{\partial x^2 \partial y^2}\,dx^2 dy^2$$
$$+ 4\,\frac{\partial^4 f}{\partial x \partial y^3}\,dx\,dy^3 + \frac{\partial^4 f}{\partial y^4}\,dy^4,$$

and, as we can easily show by induction, in general

$$d^n f = \frac{\partial^n f}{\partial x^n}\,dx^n + \binom{n}{1}\frac{\partial^n f}{\partial x^{n-1} dy}\,dx^{n-1} dy + \dots$$
$$+ \binom{n}{n-1}\frac{\partial^n f}{\partial x\, dy^{n-1}}\,dx\,dy^{n-1} + \frac{\partial^n f}{\partial y^n}\,dy^n.$$

The last expression can be expressed symbolically by the equation

$$d^n f = \left(\frac{\partial f}{\partial x}\,dx + \frac{\partial f}{\partial y}\,dy\right)^{(n)} = (f_x dx + f_y dy)^{(n)}$$

where the expression on the right is first to be expanded formally by the binomial theorem, and then the expressions

$$\frac{\partial^n f}{\partial x^n}\,dx^n,\quad \frac{\partial^n f}{\partial x^{n-1} dy}\,dx^{n-1} dy,\ \dots,\ \frac{\partial^n f}{\partial y^n}\,dy^n$$

are to be substituted for the products and powers of the quantities $f_x dx$ and $f_y dy$.

For calculations with differentials the rule

$$d(fg) = f\,dg + g\,df$$

holds good; this follows immediately from the rule for the differentiation of a product.

In conclusion, we remark that the discussion in this sub-section can immediately be extended to functions of more than two independent variables.

5. Application to the Calculus of Errors.

The practical advantage of having the differential $df = hf_x + kf_y$ as a convenient approximation to the increment of the function $f(x, y)$, $\Delta u = f(x+h, y+k) - f(x, y)$, as we pass from (x, y) to $(x+h, y+k)$, is exhibited particularly well in the so-called " calculus of errors " (cf. Vol. I, p. 349). Suppose, for example, that we wish to find the possible error in the determination of the density of a solid body by the method of displacement. If m is the weight of the body in air and $\bar{m}$ its weight in water, by Archimedes' principle the loss of weight $(m - \bar{m})$ is the weight of the water displaced. If we are using the c.g.s. system of units, the weight of the water displaced is numerically equal to its volume, and hence to the volume of the solid. The density s is thus given in terms of the independent variables m and $\bar{m}$ by the formula $s = m/(m - \bar{m})$. The error in the measurement of the density s caused by an error dm in the measurement of m and an error $d\bar{m}$ in the measurement of $\bar{m}$ is given approximately by the total differential

$$ds = \frac{\partial s}{\partial m} dm + \frac{\partial s}{\partial \bar{m}} d\bar{m}.$$

By the quotient rule the partial derivatives are

$$\frac{\partial s}{\partial m} = -\frac{\bar{m}}{(m - \bar{m})^2} \quad \text{and} \quad \frac{\partial s}{\partial \bar{m}} = \frac{m}{(m - \bar{m})^2};$$

hence the differential is

$$ds = \frac{-\bar{m}\,dm + m\,d\bar{m}}{(m - \bar{m})^2}.$$

Thus the error in s is greatest if, say, dm is negative and $d\bar{m}$ is positive; that is, if instead of m we measure too small an amount $m + dm$ and instead of $\bar{m}$ too large an amount $\bar{m} + d\bar{m}$. For example, if a piece of brass weighs about 100 gm. in air, with a possible error of 5 mg., and in water weighs about 88 gm., with a possible error of 8 mg., the density is given by our formula to within an error of about

$$\frac{88 \,.\, 5 \,.\, 10^{-3} + 100 \,.\, 8 \,.\, 10^{-3}}{12^2} \approx 9 \,.\, 10^{-3},$$

or about one per cent.

5. Functions of Functions (Compound Functions) and the Introduction of New Independent Variables

1. General Remarks. The Chain Rule.

It often happens that the function u of the independent variables x, y is stated in the form of a compound function

$$u = f(\xi, \eta, \ldots)$$

where the arguments $\xi, \eta, \ldots$ of the function f are themselves functions of x and y:

$$\xi = \phi(x, y), \quad \eta = \psi(x, y), \ldots.$$

We then say that

$$u = f(\xi, \eta, \ldots) = f(\phi(x, y), \psi(x, y), \ldots) = F(x, y)$$

is given as a compound function of x and y.

For example, the function

$$u = e^{xy}\sin(x+y) = F(x, y)$$

may be written as a compound function by means of the relations

$$u = e^{\xi}\sin\eta = f(\xi, \eta); \quad \xi = xy, \ \eta = x + y.$$

Similarly, the function

$$u = \log(x^4 + y^4) \, . \, \text{arc}\sin\sqrt{1 - x^2 - y^2} = F(x, y)$$

can be expressed in the form

$$u = \eta \, \text{arc}\sin\xi = f(\xi, \eta);$$
$$\xi = \sqrt{1 - x^2 - y^2}, \quad \eta = \log(x^4 + y^4).$$

In order to make this concept more precise, we adopt the following assumption to begin with: the functions $\xi = \phi(x, y)$, $\eta = \psi(x, y)$, . . . are defined in a certain region R of the independent variables x, y. As the argument point (x, y) varies within this region, the point with the co-ordinates $(\xi, \eta, \ldots)$ always lies in a certain region S of $\xi\eta\ldots$-space, in which the function $u = f(\xi, \eta, \ldots)$ is defined. The compound function

$$u = f(\phi(x, y), \psi(x, y), \ldots) = F(x, y)$$

is then defined in the region R.

In many cases detailed examination of the regions R and S will be quite unnecessary, e.g. in the first example given above, in which the argument point (x, y) can traverse the whole of the xy-plane and the function $u = e^{\xi}\sin\eta$ is defined throughout the $\xi\eta$-plane. On the other hand, the second example shows the need for considering the regions R and S in the definition of compound functions. For the functions

$$\xi = \sqrt{1 - x^2 - y^2} \quad \text{and} \quad \eta = \log(x^4 + y^4)$$

are defined only in the region R consisting of the points $0 < x^2 + y^2 \leqq 1$, that is, the region consisting of the circle with unit radius and centre the origin, the centre being removed. Within this region $|\xi| < 1$, while η can have all negative values and the value 0. For the region S of points (ξ, η) defined by these relations the function η arc $\sin\xi$ is defined.

A continuous function of continuous functions is itself continuous. More precisely:

If the function $u = f(\xi, \eta, \ldots)$ *is continuous in the region* S, *and the functions* $\xi = \phi(x, y)$, $\eta = \psi(x, y)$, . . . *are continuous in the region* R, *then the compound function* $u = F(x, y)$ *is continuous in* R.

The proof follows immediately from the definition of continuity. Let (x_0, y_0) be a point of R, and let $\xi_0, \eta_0, \ldots$ be the corresponding values of $\xi, \eta, \ldots$. Then for any positive ϵ the difference

$$f(\xi, \eta, \ldots) - f(\xi_0, \eta_0, \ldots)$$

is numerically less than ϵ, provided only that the inequalities

$$|\xi - \xi_0| < \delta, \quad |\eta - \eta_0| < \delta, \ldots$$

are all satisfied, where δ is a sufficiently small positive number. But by the continuity of $\phi(x, y), \psi(x, y), \ldots$ these last inequalities are all satisfied if

$$|x - x_0| < \gamma, \quad |y - y_0| < \gamma,$$

where γ is a sufficiently small positive quantity. This establishes the continuity of the compound function.

Further, we shall prove that a differentiable function of differentiable functions is itself differentiable. This statement is formulated more precisely in the following theorem, which at the same time gives the rule for the differentiation of compound functions, or so-called *chain rule*:

If $\xi = \phi(x, y)$, $\eta = \psi(x, y)$, ... *are differentiable functions of* x *and* y *in the region* R, *and* $f(\xi, \eta, \ldots)$ *is a differentiable function of* $\xi, \eta, \ldots$ *in the region* S, *then the compound function*

$$u = f(\phi(x, y), \psi(x, y), \ldots) = F(x, y)$$

is also a differentiable function of x *and* y, *and its partial derivatives are given by the formulæ*

$$F_x = f_\xi \phi_x + f_\eta \psi_x + \ldots,$$
$$F_y = f_\xi \phi_y + f_\eta \psi_y + \ldots,$$

or, briefly, by

$$u_x = u_\xi \xi_x + u_\eta \eta_x + \ldots,$$
$$u_y = u_\xi \xi_y + u_\eta \eta_y + \ldots.$$

Thus in order to form the partial derivative with respect to x we must first differentiate the compound function with respect to all the functions $\xi, \eta, \ldots$ which depend on x, multiply each of these derivatives by the derivative of the corresponding function with respect to x, and then add all the products thus formed. This is the generalization of the chain rule for functions of one variable discussed in Vol. I, Chapter III (p. 153).

Our statement can be written in a particularly simple and suggestive form if we use the notation of differentials, namely

$$\begin{aligned} du &= u_x dx + u_y dy = u_\xi d\xi + u_\eta d\eta + \ldots \\ &= u_\xi(\xi_x dx + \xi_y dy) + u_\eta(\eta_x dx + \eta_y dy) + \ldots \\ &= (u_\xi \xi_x + u_\eta \eta_x + \ldots) dx + (u_\xi \xi_y + u_\eta \eta_y + \ldots) dy. \end{aligned}$$

This equation means that the linear part of the increment of the compound function $u = f(\xi, \eta, \ldots) = F(x, y)$ can be found by first writing down this linear part as if $\xi, \eta, \ldots$ were the independent variables and subsequently replacing $d\xi, d\eta, \ldots$ by the linear parts of the increments of the functions $\xi = \phi(x, y)$, $\eta = \psi(x, y), \ldots$. This fact exhibits the convenience and flexibility of the differential notation.

In order to prove our statement we have merely to make use of the assumption that the functions concerned are differentiable. From this it follows that if we denote the increments of the independent variables x and y by Δx and Δy, the quantities $\xi, \eta, \ldots$ change by the amounts

$$\begin{aligned} \Delta\xi &= \phi_x \Delta x + \phi_y \Delta y + \epsilon_1 \Delta x + \gamma_1 \Delta y, \\ \Delta\eta &= \psi_x \Delta x + \psi_y \Delta y + \epsilon_2 \Delta x + \gamma_2 \Delta y, \\ &\ldots\ldots\ldots\ldots\ldots \end{aligned}$$

where the numbers $\epsilon_1, \epsilon_2, \ldots, \gamma_1, \gamma_2, \ldots$ tend to 0 as Δx and Δy do, or as $\sqrt{(\Delta x^2 + \Delta y^2)}$ does. Moreover, if the quantities $\xi, \eta, \ldots$ undergo changes $\Delta\xi, \Delta\eta, \ldots$, the function $u = f(\xi, \eta, \ldots)$ is subject to an increment of the form

$$\Delta u = f_\xi \Delta\xi + f_\eta \Delta\eta + \ldots + \delta_1 \Delta\xi + \delta_2 \Delta\eta + \ldots,$$

where the quantities $\delta_1, \delta_2, \ldots$ tend to 0 as $\Delta\xi, \Delta\eta, \ldots$ do, or as $\sqrt{(\Delta\xi^2 + \Delta\eta^2 + \ldots)}$ does (and may be taken as exactly zero when the corresponding increments $\Delta\xi, \Delta\eta$ vanish).

If in the last expression we take the increments $\Delta\xi, \Delta\eta, \ldots$ as those due to a change of Δx in the value of x and a change of Δy in the value of y, as given above, we obtain

$$\begin{aligned} \Delta u = {} & (f_\xi \phi_x + f_\eta \psi_x + \ldots) \Delta x \\ & + (f_\xi \phi_y + f_\eta \psi_y + \ldots) \Delta y + \epsilon \Delta x + \gamma \Delta y. \end{aligned}$$

Here the quantities ϵ and γ have the values

$$\epsilon = f_\xi \epsilon_1 + f_\eta \epsilon_2 + \dots + \phi_x \delta_1 + \psi_x \delta_2 + \epsilon_1 \delta_1 + \epsilon_2 \delta_2 + \dots,$$
$$\gamma = f_\xi \gamma_1 + f_\eta \gamma_2 + \dots + \phi_y \delta_1 + \psi_y \delta_2 + \gamma_1 \delta_1 + \gamma_2 \delta_2 + \dots.$$

On the right we have a sum of products, each of which contains at least one of the quantities $\epsilon_1, \epsilon_2, \dots, \gamma_1, \gamma_2, \dots, \delta_1, \delta_2, \dots$. From this we see that ϵ and γ also tend to 0 as Δx and Δy do. By the results of the preceding section, however, this expresses the statement asserted in our theorem.

It is obvious that this result is quite independent of the number of independent variables $x, y, \dots$, and remains valid e.g. if the quantities $\xi, \eta, \dots$ depend on only one independent variable x, so that the quantity u is a compound function of the single independent variable x.

If we wish to calculate the higher partial derivatives, we have only to differentiate the right-hand sides of our equations with respect to x and y, treating $f_\xi, f_\eta, \dots$ as compound functions. Confining ourselves for the sake of simplicity to the case of three functions ξ, η, and ζ, we thus obtain

$$\begin{aligned} u_{xx} &= f_{\xi\xi}\xi_x^2 + f_{\eta\eta}\eta_x^2 + f_{\zeta\zeta}\zeta_x^2 + 2f_{\xi\eta}\xi_x\eta_x + 2f_{\eta\zeta}\eta_x\zeta_x + 2f_{\xi\zeta}\xi_x\zeta_x \\ &\quad + f_\xi\xi_{xx} + f_\eta\eta_{xx} + f_\zeta\zeta_{xx}, \end{aligned}$$

$$\begin{aligned} u_{xy} &= f_{\xi\xi}\xi_x\xi_y + f_{\eta\eta}\eta_x\eta_y + f_{\zeta\zeta}\zeta_x\zeta_y + f_{\xi\eta}(\xi_x\eta_y + \xi_y\eta_x) + f_{\eta\zeta}(\eta_x\zeta_y + \eta_y\zeta_x) \\ &\quad + f_{\xi\zeta}(\xi_x\zeta_y + \xi_y\zeta_x) + f_\xi\xi_{xy} + f_\eta\eta_{xy} + f_\zeta\zeta_{xy}, \end{aligned}$$

$$\begin{aligned} u_{yy} &= f_{\xi\xi}\xi_y^2 + f_{\eta\eta}\eta_y^2 + f_{\zeta\zeta}\zeta_y^2 + 2f_{\xi\eta}\xi_y\eta_y + 2f_{\eta\zeta}\eta_y\zeta_y + 2f_{\xi\zeta}\xi_y\zeta_y \\ &\quad + f_\xi\xi_{yy} + f_\eta\eta_{yy} + f_\zeta\zeta_{yy}. \end{aligned}$$

2. Examples.*

1. Let us consider the function

$$u = e^{x^2 \sin^2 y + 2xy \sin x \sin y + y^2}.$$

We put

$$\xi = x^2 \sin^2 y, \quad \eta = 2xy \sin x \sin y, \quad \zeta = y^2$$

and obtain

$$\xi_x = 2x \sin^2 y, \quad \eta_x = 2y \sin x \sin y + 2xy \cos x \sin y, \quad \zeta_x = 0;$$
$$\xi_y = 2x^2 \sin y \cos y, \quad \eta_y = 2x \sin x \sin y + 2xy \sin x \cos y, \quad \zeta_y = 2y;$$
$$u_\xi = u_\eta = u_\zeta = e^{\xi+\eta+\zeta}.$$

* We would emphasize that the following differentiations can also be carried out directly, without using the chain rule.

Hence

$$u_x = 2e^{x^2 \sin^2 y + 2xy \sin x \sin y + y^2}(x \sin^2 y + y \sin x \sin y + xy \cos x \sin y)$$

and

$$u_y = 2e^{x^2 \sin^2 y + 2xy \sin x \sin y + y^2}(x^2 \sin y \cos y + x \sin x \sin y + xy \sin x \cos y + y).$$

2. In the case of the function

$$u = \sin(x^2 + y^2)$$

we put $\xi = x^2 + y^2$, and obtain

$$u_x = 2x \cos(x^2 + y^2), \quad u_y = 2y \cos(x^2 + y^2),$$
$$u_{xx} = -4x^2 \sin(x^2 + y^2) + 2 \cos(x^2 + y^2), \quad u_{xy} = -4xy \sin(x^2 + y^2),$$
$$u_{yy} = -4y^2 \sin(x^2 + y^2) + 2 \cos(x^2 + y^2).$$

3. In the case of the function

$$u = \arctan(x^2 + xy + y^2)$$

the substitution

$$\xi = x^2, \quad \eta = xy, \quad \zeta = y^2$$

leads to

$$u_x = \frac{2x + y}{1 + (x^2 + xy + y^2)^2},$$

$$u_y = \frac{x + 2y}{1 + (x^2 + xy + y^2)^2}.$$

3. Change of the Independent Variables.

A particularly important application of the facts developed on pp. 69–74 occurs in the process of changing the independent variables. For example, let $u = f(\xi, \eta)$ be a function of the two independent variables ξ, η, which we interpret as rectangular co-ordinates in the $\xi\eta$-plane. If we introduce new rectangular co-ordinates x, y in that plane (cf. p. 6) by the transformation

$$\xi = \alpha_1 x + \beta_1 y, \quad x = \alpha_1 \xi + \alpha_2 \eta,$$
$$\eta = \alpha_2 x + \beta_2 y, \quad y = \beta_1 \xi + \beta_2 \eta,$$

the function $u = f(\xi, \eta)$ is transformed into a new function of x and y,

$$u = f(\xi, \eta) = F(x, y),$$

and this new function is formed from $f(\xi, \eta)$ by a process of compounding such as was described on p. 69. We then say that new independent variables x and y have been introduced into the relation $u = f(\xi, \eta)$ between the independent variables ξ and η and the dependent variable u.

The rules of differentiation given on p. 71 at once yield

$$u_x = u_\xi \alpha_1 + u_\eta \alpha_2,$$
$$u_y = u_\xi \beta_1 + u_\eta \beta_2,$$

where the symbols u_x, u_y denote the partial derivatives of the function $F(x, y)$, and the symbols u_ξ, u_η denote the partial derivatives of the function $f(\xi, \eta)$.

Thus the partial derivatives of any function are transformed according to the same law as the independent variables when the co-ordinate axes are rotated. This is true for rotation of the axes in space also.

Another important type of change of the independent variables is the change from rectangular co-ordinates (x, y) to polar co-ordinates (r, θ) which are connected with the rectangular co-ordinates by the equations

$$x = r\cos\theta, \quad r = \sqrt{(x^2 + y^2)},$$
$$y = r\sin\theta, \quad \theta = \operatorname{arc\,cos}\frac{x}{\sqrt{(x^2+y^2)}} = \operatorname{arc\,sin}\frac{y}{\sqrt{(x^2+y^2)}}.$$

On introducing the polar co-ordinates we have

$$u = f(x, y) = f(r\cos\theta, r\sin\theta) = F(r, \theta),$$

and the quantity u appears as a compound function of the independent variables r and θ. Hence by the chain rule we obtain

$$u_x = u_r r_x + u_\theta \theta_x = u_r \frac{x}{r} - u_\theta \frac{y}{r^2} = u_r \cos\theta - u_\theta \frac{\sin\theta}{r},$$

$$u_y = u_r r_y + u_\theta \theta_y = u_r \frac{y}{r} + u_\theta \frac{x}{r^2} = u_r \sin\theta + u_\theta \frac{\cos\theta}{r}.$$

These yield the equation

$$u_x{}^2 + u_y{}^2 = u_r{}^2 + \frac{1}{r^2} u_\theta{}^2,$$

which is frequently of use. By the chain rule the higher derivatives are given by

$$u_{xx} = u_{rr}\cos^2\theta + u_{\theta\theta}\frac{\sin^2\theta}{r^2} - 2u_{r\theta}\frac{\cos\theta\sin\theta}{r} + u_r\frac{\sin^2\theta}{r}$$
$$+ 2u_\theta \frac{\cos\theta\sin\theta}{r^2},$$

$$u_{xy} = u_{yx} = u_{rr} \cos\theta \sin\theta - u_{\theta\theta} \frac{\cos\theta \sin\theta}{r^2} + u_{r\theta} \frac{\cos^2\theta - \sin^2\theta}{r}$$

$$+ u_\theta \frac{\sin^2\theta - \cos^2\theta}{r^2} - u_r \frac{\sin\theta \cos\theta}{r},$$

$$u_{yy} = u_{rr} \sin^2\theta + u_{\theta\theta} \frac{\cos^2\theta}{r^2} + 2u_{r\theta} \frac{\cos\theta \sin\theta}{r} + u_r \frac{\cos^2\theta}{r}$$

$$- 2u_\theta \frac{\cos\theta \sin\theta}{r^2}.$$

This leads us to the following formula, giving the expression appearing in the well-known "Laplace's" or "potential" equation $\Delta u = 0$ in terms of polar co-ordinates:

$$\Delta u = u_{xx} + u_{yy} = u_{rr} + u_{\theta\theta} \frac{1}{r^2} + u_r \frac{1}{r} = \frac{1}{r^2} \left\{ r \frac{\partial}{\partial r} \left(r \frac{\partial u}{\partial r} \right) + \frac{\partial^2 u}{\partial \theta^2} \right\}.$$

Of the formulæ

$$u_r = u_x \frac{x}{r} + u_y \frac{y}{r} = u_x \cos\theta + u_y \sin\theta,$$

$$u_\theta = -u_x y + u_y x = -u_x r \sin\theta + u_y r \cos\theta,$$

which express the rules for the differentiation of a function $f(x, y)$ with respect to r and θ, the first is the expression for the derivative of $f(x, y)$ in the direction of the radius vector r which we previously met with on p. 64.

In general, whenever we are given a series of relations defining a compound function,

$$u = f(\xi, \eta, \ldots),$$

$$\xi = \phi(x, y), \qquad \eta = \psi(x, y), \ldots$$

we may regard it as an introduction of new independent variables x, y instead of $\xi, \eta, \ldots$. Corresponding sets of values of the independent variables assign the same value to u, whether it is regarded as a function of $\xi, \eta, \ldots$ or of x, y.

In all cases involving the differentiation of compound functions

$$u = f(\xi, \eta, \ldots)$$

the following point must carefully be noted. We must distinguish clearly between the dependent variable u and the function $f(\xi, \eta, \ldots)$ which connects u with the independent variables

$\xi, \eta, \ldots$. The symbols of differentiation $u_\xi, u_\eta, \ldots$ have no meaning until the functional connexion between u and the independent variables is specified. When dealing with compound functions $u = f(\xi, \eta, \ldots) = F(x, y)$, therefore, we really should not write u_ξ, u_η or u_x, u_y, but should instead write f_ξ, f_η or F_x, F_y respectively. Yet for the sake of brevity the simpler symbols u_ξ, u_η, u_x, u_y are often used when there is no risk that confusion will arise.

The following example will serve to show that the result of differentiating a quantity depends on the nature of the functional connexion between it and the independent variables, that is, it depends on which of the independent variables are kept fixed during the differentiation. With the "identical" transformation $\xi = x$, $\eta = y$ the function $u = 2\xi + \eta$ becomes $u = 2x + y$, and we have $u_x = 2$, $u_y = 1$. If, however, we introduce the new independent variables $\xi = x$ (as before) and $\xi + \eta = v$, we find that $u = x + v$, so that $u_x = 1$, $u_v = 1$. That is, differentiation with respect to the same independent variable x gives different results in the two different cases.

Examples

1. Prove that the tangent plane to the quadric

$$ax^2 + by^2 + cz^2 = 1$$

at the point (x_0, y_0, z_0) is

$$axx_0 + byy_0 + czz_0 = 1.$$

2. If $u = u(x, y)$ is the equation of a cone, then

$$u_{xx}u_{yy} - u_{xy}^2 = 0.$$

3. Prove that if a function $f(x)$ is continuous and has a continuous derivative, then the derivative of the function

$$g(x) = \begin{vmatrix} f(x) & x & 1 \\ f(x_1) & x_1 & 1 \\ f(x_2) & x_2 & 1 \end{vmatrix}$$

vanishes for a certain value between x_1 and x_2.

4. Let $f(x, y, z)$ be a function depending only on $r = \sqrt{(x^2 + y^2 + z^2)}$, i.e. let $f(x, y, z) = g(r)$.

(a) Calculate $f_{xx} + f_{yy} + f_{zz}$.

(b) Prove that if $f_{xx} + f_{yy} + f_{zz} = 0$, it follows that $f = \dfrac{a}{r} + b$ (where a and b are constants).

5. If $f(x_1, x_2, \ldots, x_n) = g(r) = g(\sqrt{(x_1^2 + x_2^2 + \ldots + x_n^2)})$, calculate

$$f_{x_1x_1} + f_{x_2x_2} + \ldots + f_{x_nx_n}$$

(cf. Ex. 2, p. 58).

6*. Find the expression for $f_{xx} + f_{yy} + f_{zz}$ in three-dimensional polar co-ordinates, i.e. transform to the variables r, θ, φ defined by

$$\begin{aligned} x &= r \sin\theta \cos\varphi \\ y &= r \sin\theta \sin\varphi \\ z &= r \cos\theta. \end{aligned}$$

Compare with example 4(*a*).

7. Prove that the expression

$$f_{xx} + f_{yy}$$

is unchanged by rotation of the co-ordinate system.

8. Prove that with the linear transformation

$$\begin{aligned} x &= \alpha\xi + \beta\eta \\ y &= \gamma\xi + \delta\eta, \end{aligned}$$

$f_{xx}(x, y)$, $f_{xy}(x, y)$, $f_{yy}(x, y)$ are respectively transformed by the same law as the coefficients a, b, c of the polynomial

$$ax^2 + 2bxy + cy^2.$$

6. THE MEAN VALUE THEOREM AND TAYLOR'S THEOREM FOR FUNCTIONS OF SEVERAL VARIABLES

1. Statement of the Problem. Preliminary Remarks.

We have already seen in Vol. I (Chapter VI, p. 320 *et seq.*) how a function of a single variable can be approximated to in the neighbourhood of a given point with an accuracy of order higher than the n-th, by means of a polynomial of degree n, the Taylor series, provided that the function possesses derivatives up to the $(n + 1)$-th order. The approximation by means of the linear part of the function, as given by the differential, is only the first step towards this closer approximation. In the case of functions of several variables, e.g. of two independent variables, we may also seek for an approximate representation in the neighbourhood of a given point by means of a polynomial of degree n. In other words, we wish to approximate to $f(x + h, y + k)$ by means of a " Taylor expansion " in terms of the differences h and k.

By a very simple device this problem can be reduced to what we already know from the theory of functions of one variable. Instead of considering the function $f(x + h, y + k)$, we introduce yet another variable t and regard the expression

$$F(t) = f(x + ht, y + kt)$$

as a function of t, keeping x, y, h, and k fixed for the moment. As t varies between 0 and 1. the point with co-ordinates $(x + ht,\ y + kt)$ traverses the line-segment joining (x, y) and $(x + h,\ y + k)$.

We begin by calculating the derivatives of $F(t)$. If we assume that all the derivatives of the function $f(x, y)$ which we are about to write down are continuous in a region entirely containing the line-segment, the chain rule (section 5, p. 71) at once gives

$$F'(t) = hf_x + kf_y,$$
$$F''(t) = h^2 f_{xx} + 2hkf_{xy} + k^2 f_{yy},$$
$$\cdots\cdots\cdots\cdots$$

and, in general, we find by mathematical induction that the n-th derivative is given by the expression

$$F^{(n)}(t) = h^n f_{x^n} + \binom{n}{1} h^{n-1} k f_{x^{n-1}y} + \binom{n}{2} h^{n-2} k^2 f_{x^{n-2}y^2} + \ldots + k^n f_{y^n}.$$

which, as on p. 68, can be written symbolically in the form

$$F^{(n)}(t) = (hf_x + kf_y)^{(n)}.$$

In this last formula the bracket on the right is to be expanded by the binomial theorem and then the powers and products of the quantities $\dfrac{\partial f}{\partial x}$ and $\dfrac{\partial f}{\partial y}$ are to be replaced by the corresponding n-th derivatives $\dfrac{\partial^n f}{\partial x^n}, \dfrac{\partial^n f}{\partial x^{n-1} \partial y}, \ldots$. In all these derivatives the arguments $x + ht$ and $y + kt$ are to be written in place of x and y.

2. The Mean Value Theorem.

In forming our polynomial of approximation we start from a *mean value theorem* analogous to that which we already know for functions of one variable. This theorem gives a relation between the difference $f(x + h,\ y + k) - f(x, y)$ and the partial derivatives f_x and f_y. We expressly assume that these derivatives are continuous. On applying the ordinary mean value theorem to the function $F(t)$ we obtain

$$\frac{F(t) - F(0)}{t} = F'(\theta t),$$

where θ is a number between 0 and 1, and from this it follows that

$$\frac{f(x+ht,\, y+kt)-f(x,\, y)}{t} = hf_x(x+\theta ht,\, y+\theta kt) + kf_y(x+\theta ht,\, y+\theta kt).$$

If we put $t=1$ in this, we obtain the required *mean value theorem for functions of two variables* in the form

$$f(x+h,\, y+k)-f(x,\, y) = hf_x(x+\theta h,\, y+\theta k)+kf_y(x+\theta h,\, y+\theta k) = hf_x(\xi,\, \eta)+kf_y(\xi,\, \eta).$$

That is, *the difference between the values of the function at the points* (x + h, y + k) *and* (x, y) *is equal to the differential at an intermediate point* (ξ, η) *on the line-segment joining the two points.* It is worth noting that the *same* value of θ occurs in both f_x and f_y.

The following fact, the proof of which we leave to the reader, is a simple consequence of the mean value theorem. A function $f(x, y)$ whose partial derivatives f_x and f_y exist and have the value 0 at every point of a region is a constant.

3. Taylor's Theorem for Several Independent Variables.

If we apply Taylor's formula with Lagrange's form of the remainder (cf. Vol. I, Chapter VI, p. 324) to the function $F(t)$ and finally put $t=1$, we obtain *Taylor's theorem* for functions of two independent variables,

$$\begin{aligned} f(x+h,\, y+k) &= f(x,\, y)+\{hf_x(x,\, y)+kf_y(x,\, y)\} \\ &+\frac{1}{2!}\{h^2 f_{xx}(x,\, y)+2hkf_{xy}(x,\, y)+k^2 f_{yy}(x,\, y)\}+\dots \\ &+\frac{1}{n!}\left\{h^n f_{x^n}(x,\, y)+\binom{n}{1}h^{n-1}kf_{x^{n-1}y}(x,\, y)+\dots+k^n f_{y^n}(x,\, y)\right\} \\ &+R_n, \end{aligned}$$

where R_n symbolizes the remainder term

$$R_n = \frac{1}{(n+1)!}\{hf_x(x+\theta h,\;\, y+\theta k)+kf_y(x+\theta h,\;\, y+\theta k)\}^{(n+1)},$$
$$0<\theta<1.$$

The homogeneous polynomials of degree 1, 2, . . . , n, $n+1$, into which the increment $f(x+h, y+k) - f(x, y)$ is thus split up, apart from the factors

$$\frac{1}{1!}, \frac{1}{2!}, \ldots, \frac{1}{n!}, \frac{1}{(n+1)!},$$

are respectively the first, second, . . . , n-th differentials

$$df = hf_x + kf_y,$$
$$d^2f = (hf_x + kf_y)^{(2)} = h^2 f_{xx} + 2hkf_{xy} + k^2 f_{yy},$$
$$\cdots\cdots\cdots\cdots\cdots\cdots$$
$$d^n f = (hf_x + kf_y)^{(n)} = h^n f_{x^n} + \binom{n}{1} h^{n-1} k f_{x^{n-1}y} + \ldots + k^n f_{y^n}$$

of $f(x, y)$ at the point (x, y) and the $(n+1)$-th differential $d^{n+1}f$ at an intermediate point on the line-segment joining (x, y) and $(x+h, y+k)$. Hence Taylor's theorem can be written more compactly as

$$f(x+h, y+k) = f(x, y) + df(x, y) + \frac{1}{2!} d^2 f(x, y) + \ldots$$
$$+ \frac{1}{n!} d^n f(x, y) + R_n,$$

where

$$R_n = \frac{1}{(n+1)!} d^{n+1} f(x + \theta h, y + \theta k), \qquad 0 < \theta < 1.$$

In general the remainder R_n vanishes to a *higher* order than the term $d^n f$ just before it; that is, as $h \to 0$ and $k \to 0$ we have $R_n = o\{\sqrt{(h^2 + k^2)^n}\}$.

In the case of Taylor's theorem for functions of one variable the passage ($n \to \infty$) to *infinite Taylor series* played an important part, leading us to the expansions of many functions in power series. With functions of several variables such a process is in general too complicated. Here to an even greater degree than in the case of functions of one variable we lay the stress rather on the fact that by means of Taylor's theorem the increment $f(x+h, y+k) - f(x, y)$ of a function is split up into increments df, d^2f, . . . of different orders.

Examples

1. Find the polynomial of the second degree which best approximates to the function $\sin x \sin y$ in the neighbourhood of the origin.

2. If $f(x, y)$ is a continuous function with continuous first and second derivatives, then

$$f_{xx}(0, 0) = \lim_{h \to +0} \frac{f(2h, e^{-1/2h}) - 2f(h, e^{-1/h}) + f(0, 0)}{h^2}.$$

3. Prove that the function e^{-y^2+2xy} can be expanded in a series of the form

$$\sum_{n=0}^{\infty} \frac{H_n(x)}{n!} y^n$$

which converges for all values of x and y and that

(a) $H_n(x)$ is a polynomial of degree n (so-called Hermite polynomials).

(b) $H'_n(x) = 2nH_{n-1}(x)$.

(c) $H_{n+1} - 2xH_n + 2nH_{n-1} = 0$.

(d) $H''_n - 2xH'_n + 2nH_n = 0$.

4. Find the Taylor series for the following functions and indicate their range of validity:

$$(a)\ \frac{1}{1 - x - y}; \quad (b)\ e^{x+y}.$$

7. The Application of Vector Methods

Many facts and relationships in the differential and integral calculus of several independent variables take a decidedly clearer and simpler form if we apply the ideas and notation of vector analysis. We shall accordingly conclude this chapter with some discussion of the matter.

1. Vector Fields and Families of Vectors.

The step which connects vector analysis with the subjects just discussed is as follows. Instead of considering a single vector or a finite number of vectors, as in Chapter I (p. 3), we investigate a *vector manifold* depending on one or more continuously varying parameters.

If, for example, we consider a solid body occupying a portion of space and in a state of motion, then at a given instant each point of the solid will have a definite velocity, represented by a vector $\mathbf{u}$. We say that these vectors form a *vector field* in the region in question. The three components of the field vector then appear as three functions

$$u_1(x_1, x_2, x_3), \quad u_2(x_1, x_2, x_3), \quad u_3(x_1, x_2, x_3)$$

of the three co-ordinates of position, which we here denote by (x_1, x_2, x_3) instead of (x, y, z).

A case of a velocity field is represented in fig. 8, which shows

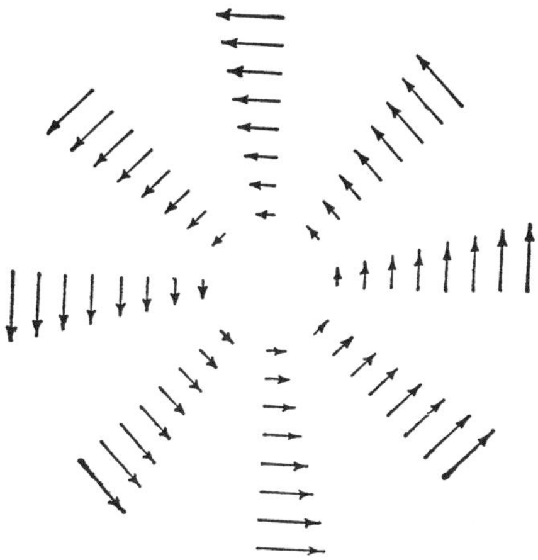

Fig. 8.—The velocity field in a rotation

the velocity field of a solid body rotating about an axis with constant angular velocity.

The forces acting on the points of a moving solid body likewise form a vector field. As an example of a force field we consider the attractive force per unit mass exerted by a heavy particle, according to Newton's law of gravitation. By Newton's law all the vectors of this field of force are directed towards the attracting particle, and their lengths are inversely proportional to the square of the distance from the particle.

If we pass to a new rectangular co-ordinate system by rotation of axes, all the vectors of the field will have new components with respect to the new system of axes. If the two co-ordinate systems are connected by equations of the form (Chapter I, section 1, p. 6)

$$\begin{aligned}\xi_1 &= \alpha_1 x_1 + \beta_1 x_2 + \gamma_1 x_3\\ \xi_2 &= \alpha_2 x_1 + \beta_2 x_2 + \gamma_2 x_3\\ \xi_3 &= \alpha_3 x_1 + \beta_3 x_2 + \gamma_3 x_3\end{aligned}$$

or

$$\begin{aligned}x_1 &= \alpha_1 \xi_1 + \alpha_2 \xi_2 + \alpha_3 \xi_3\\ x_2 &= \beta_1 \xi_1 + \beta_2 \xi_2 + \beta_3 \xi_3\\ x_3 &= \gamma_1 \xi_1 + \gamma_2 \xi_2 + \gamma_3 \xi_3\end{aligned}$$

respectively, then the relations between the components u_1, u_2, u_3 with respect to the x-system and the components $\omega_1(\xi_1, \xi_2, \xi_3)$, $\omega_2(\xi_1, \xi_2, \xi_3)$, $\omega_3(\xi_1, \xi_2, \xi_3)$ with respect to the new ξ-system are given by the equations of transformation

$$\omega_1 = \alpha_1 u_1 + \beta_1 u_2 + \gamma_1 u_3$$
$$\omega_2 = \alpha_2 u_1 + \beta_2 u_2 + \gamma_2 u_3$$
$$\omega_3 = \alpha_3 u_1 + \beta_3 u_2 + \gamma_3 u_3$$

and

$$u_1 = \alpha_1 \omega_1 + \alpha_2 \omega_2 + \alpha_3 \omega_3$$
$$u_2 = \beta_1 \omega_1 + \beta_2 \omega_2 + \beta_3 \omega_3$$
$$u_3 = \gamma_1 \omega_1 + \gamma_2 \omega_2 + \gamma_3 \omega_3$$

respectively. (Cf. Chap. I, p. 6.) The components $\omega_1, \omega_2, \omega_3$ in the new system thus arise from the introduction of the new variables and the simultaneous transformation of the functions representing the components in the old system.

When in physical applications each point of a portion of space has assigned to it a definite value of a function $u = f(x_1, x_2, x_3)$, such as the density at the point, and we wish to emphasize that the property is not a component of a vector, but on the contrary is a property which retains the same value although the co-ordinate system is altered, we say that the function is a *scalar function* or *scalar*; or, if we wish to emphasize the association between the values of the function and the points of the portion of space, we speak of a *scalar field*. Thus for every vector field $\boldsymbol{u}$ the quantity $|\boldsymbol{u}|^2 = u_1^2 + u_2^2 + u_3^2$ is a scalar; for it represents the square of the length of the vector and therefore retains the same value independently of the co-ordinate system to which the components of the vector are referred.

In the examples above the vector field $\boldsymbol{u}$ is given us to begin with, and its components with respect to any system of rectangular co-ordinates are therefore determined. If, conversely, in a definite co-ordinate system, say an x-system, there are given three functions $u_1(x_1, x_2, x_3)$, $u_2(x_1, x_2, x_3)$, $u_3(x_1, x_2, x_3)$, these three functions define a vector field with respect to that system, the components of the field being given by the three functions. To obtain the expressions for the components $\omega_1, \omega_2, \omega_3$ in any other system we have only to apply the equations of transformation deduced above.

In addition to vector fields, we also consider manifolds of vectors called *families of vectors*, which do not correspond to each point of a region in space, but are functions of a parameter t. We express this by writing $\boldsymbol{u} = \boldsymbol{u}(t)$. If we think of $\boldsymbol{u}$ as a position vector measured from the origin of co-ordinates in $u_1u_2u_3$-space, then as t varies the final point of this vector describes a curve in space given by three parametric equations,

$$u_1 = \phi(t), \quad u_2 = \psi(t), \quad u_3 = \chi(t).$$

Vectors which depend on a parameter t in this way can be differentiated with respect to t. By the derivative of a vector $\boldsymbol{u}(t)$ we mean the vector $\boldsymbol{u}'(t)$ which is obtained by the passage to the limit

$$\lim_{h \to 0} \frac{\boldsymbol{u}(t+h) - \boldsymbol{u}(t)}{h}$$

and which accordingly has the components

$$u_1' = \frac{du_1}{dt}, \quad u_2' = \frac{du_2}{dt}, \quad u_3' = \frac{du_3}{dt}.$$

We see at once that *the fundamental rules of differentiation hold for vectors.* Firstly, it is obvious that if

$$\boldsymbol{w} = \boldsymbol{u} + \boldsymbol{v}$$

then

$$\boldsymbol{w}' = \boldsymbol{u}' + \boldsymbol{v}'.$$

Further, the product rule applied to the *scalar* product of two vectors $\boldsymbol{u}$ and $\boldsymbol{v}$, $\boldsymbol{uv} = u_1v_1 + u_2v_2 + u_3v_3$ (cf. p. 7), gives

$$d\frac{(\boldsymbol{uv})}{dt} = \boldsymbol{uv}' + \boldsymbol{u}'\boldsymbol{v}.$$

In the same way we obtain the rule

$$d\frac{[\boldsymbol{uv}]}{dt} = [\boldsymbol{uv}'] + [\boldsymbol{u}'\boldsymbol{v}]$$

for the vector product.

2. Application to the Theory of Curves in Space. Resolution of a Motion into Tangential and Normal Components.

We shall now make some simple applications of these ideas. If $\mathbf{x}(t)$ is a position vector in $x_1x_2x_3$-space which depends on a parameter t, and therefore defines a curve in space, the vector $\mathbf{x}'(t)$ will be in the direction of the *tangent* to the curve at the point corresponding to t. For the vector $\mathbf{x}(t+h)-\mathbf{x}(t)$ is in the direction of the line-segment joining the points (t) and $(t+h)$ (cf. fig. 9); therefore so is the vector $\dfrac{\mathbf{x}(t+h)-\mathbf{x}(t)}{h}$, which differs from it only in the factor $1/h$. As $h \to 0$ the direction of this chord approaches the direction of the tangent. If instead of t we introduce as parameter the *length of the arc* of the curve measured from a definite starting-point, and denote differentiation with respect to s by means of a dot, we can prove that

$$\dot{x}_1^2 + \dot{x}_2^2 + \dot{x}_3^2 = 1;$$

this may also be written in the form

$$\dot{\mathbf{x}}\dot{\mathbf{x}} = \dot{\mathbf{x}}^2 = 1.$$

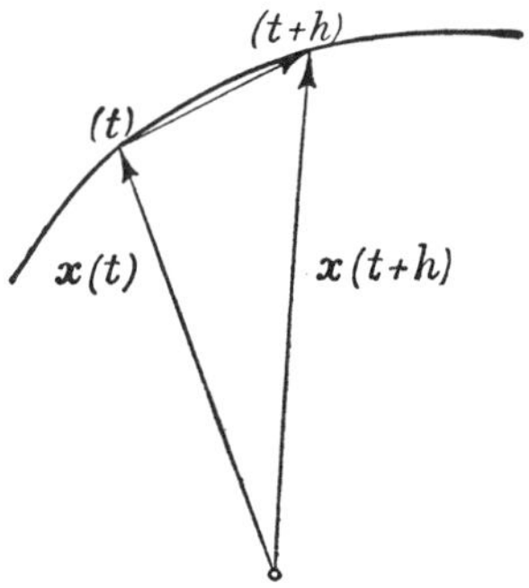

Fig. 9.—Differentiation of the position vector of a curve

The proof follows exactly the same lines as the corresponding proof for plane curves (cf. Vol. I, Chap. V, p. 280). The vector $\dot{\mathbf{x}}$ is therefore of unit length. If we again differentiate both sides of the equation $\dot{\mathbf{x}}\dot{\mathbf{x}} = 1$ with respect to s, we obtain

$$\dot{\mathbf{x}}\ddot{\mathbf{x}} = 0.$$

This equation states that the vector $\ddot{\mathbf{x}}$ with components $\ddot{x}_1(s)$, $\ddot{x}_2(s)$, $\ddot{x}_3(s)$ is *perpendicular to the tangent*. This vector we call the *curvature vector* or *principal normal vector*, and its absolute value, that is, its length

$$k = \frac{1}{\rho} = \sqrt{(\ddot{x}_1^2 + \ddot{x}_2^2 + \ddot{x}_3^2)},$$

we call the *curvature* of the curve at the corresponding point.

The reciprocal $\rho = 1/k$ of the curvature we call the *radius of curvature*, as before. The point obtained by measuring from the point on the curve a length ρ in the direction of the principal normal vector is called the *centre of curvature.*

We shall show that this definition of the curvature agrees with that given in Vol. I, Chap. V (pp. 280–3). For $\dot{\mathbf{x}}$ is a vector of unit length. If we think of the vectors $\dot{\mathbf{x}}(s+h)$ and $\dot{\mathbf{x}}(s)$ as measured from a fixed origin, then the difference $\dot{\mathbf{x}}(s+h) - \dot{\mathbf{x}}(s)$ will be represented, as in fig. 9, by the vector joining the final points of the vectors $\dot{\mathbf{x}}(s)$ and $\dot{\mathbf{x}}(s+h)$. If α is the angle between the vectors $\dot{\mathbf{x}}(s)$ and $\dot{\mathbf{x}}(s+h)$, the length of the vector joining their final points is $2 \sin \alpha/2$, since $\dot{\mathbf{x}}(s)$ and $\dot{\mathbf{x}}(s+h)$ are both of unit length. Hence if we divide the length of this vector by α and let $h \to 0$, the quotient tends to the limit 1. Consequently

$$\lim_{h\to 0} \frac{\alpha}{h} = \lim_{h \to 0} \frac{1}{h} \sqrt{\{(\dot{x}_1(s+h) - \dot{x}_1(s))^2 + (\dot{x}_2(s+h) - \dot{x}_2(s))^2 + (\dot{x}_3(s+h) - \dot{x}_3(s))^2\}}.$$

Here the limit on the right is exactly $\sqrt{(\ddot{x}_1{}^2 + \ddot{x}_2{}^2 + \ddot{x}_3{}^2)}$. But α/h is the ratio of the angle between the directions of the tangents at two points of the curve and the length of arc between those points, and the limit of that ratio is what we have previously defined as the curvature of the curve.

The curvature vector plays an important part in mechanics. We suppose that a particle of unit mass moves along a curve $\mathbf{x}(t)$, where t is the time. The velocity of the motion is then given both in magnitude and in direction by the vector $\mathbf{x}'(t)$, where the dash denotes differentiation with respect to t. Similarly, the acceleration is given by the vector $\mathbf{x}''(t)$. By the chain rule we have

$$\mathbf{x}' = \dot{\mathbf{x}} \frac{ds}{dt}$$

(where the dot denotes differentiation with respect to s), and also

$$\mathbf{x}'' = \dot{\mathbf{x}} \frac{d^2s}{dt^2} + \ddot{\mathbf{x}} \left(\frac{ds}{dt}\right)^2.$$

In view of what we already know about the lengths of $\dot{\mathbf{x}}$ and $\ddot{\mathbf{x}}$, this equation expresses the following facts:

The "acceleration vector" of the motion is the sum of two

vectors. One of these is directed along the tangent to the curve, and its length is equal to $\frac{d^2s}{dt^2}$, that is, to the acceleration of the point in its path (the *tangential acceleration*). The other is directed towards the centre of curvature, and its length is equal to the square of the velocity multiplied by the curvature (the *normal acceleration*).

3. The Gradient of a Scalar.

We now return to the consideration of vector fields and shall give a brief discussion of certain concepts which frequently arise in connexion with them.

Let $u = f(x_1, x_2, x_3)$ be any function defined in a region of $x_1x_2x_3$-space; that is, according to the terminology previously adopted, u is a scalar quantity. We may now regard the three partial derivatives

$$u_1 = f_{x_1}, \quad u_2 = f_{x_2}, \quad u_3 = f_{x_3}$$

in the x-system as forming the three components of a vector $\boldsymbol{u}$. If we now pass to a new system of rectangular co-ordinates, the ξ-system, by rotation of axes, the new components of the vector $\boldsymbol{u}$ are given according to the formulæ of p. 6 by the equations

$$\begin{aligned} \omega_1 &= \alpha_1 u_1 + \beta_1 u_2 + \gamma_1 u_3 \\ \omega_2 &= \alpha_2 u_1 + \beta_2 u_2 + \gamma_2 u_3 \\ \omega_3 &= \alpha_3 u_1 + \beta_3 u_2 + \gamma_3 u_3. \end{aligned}$$

On the other hand, if we introduce the rectangular co-ordinates ξ_1, ξ_2, ξ_3 as new independent variables in the function $f(x_1, x_2, x_3)$, the chain rule gives

$$\begin{aligned} f_{\xi_1} &= f_{x_1}\alpha_1 + f_{x_2}\beta_1 + f_{x_3}\gamma_1 \\ f_{\xi_2} &= f_{x_1}\alpha_2 + f_{x_2}\beta_2 + f_{x_3}\gamma_2 \\ f_{\xi_3} &= f_{x_1}\alpha_3 + f_{x_2}\beta_3 + f_{x_3}\gamma_3. \end{aligned}$$

Hence

$$\omega_1 = f_{\xi_1}, \quad \omega_2 = f_{\xi_2}, \quad \omega_3 = f_{\xi_3},$$

and we thus see that in the new co-ordinate system also the components of the vector $\boldsymbol{u}$ are given by the partial derivatives of the function f with respect to the three co-ordinates. Thus to every function f in three-dimensional space there corresponds a definite vector, whose components in *any* rectangular co-ordinate system are given by the three partial derivatives with respect to the co-ordinates. We call this vector the *gradient* of the function, and denote it by

$$\boldsymbol{u} = \operatorname{grad} f.$$

For a function of three variables the gradient is an analogue of the derivative for functions of one variable.

In order to form a graphical idea of the meaning of the gradient, we shall form the derivative of the function in the direction $(\alpha_1, \alpha_2, \alpha_3)$, where $\alpha_1, \alpha_2, \alpha_3$ are the three angles which this direction makes with the axes, so that $\cos^2\alpha_1 + \cos^2\alpha_2 + \cos^2\alpha_3 = 1$. For this derivative we have already obtained the expression

$$D^{(\alpha)}f = f_{x_1}\cos\alpha_1 + f_{x_2}\cos\alpha_2 + f_{x_3}\cos\alpha_3.$$

If we think of a vector $\boldsymbol{e}$ of unit length in the direction $(\alpha_1, \alpha_2, \alpha_3)$, this vector will have components $e_1 = \cos\alpha_1$, $e_2 = \cos\alpha_2$, $e_3 = \cos\alpha_3$. Thus for the derivative of the function in the direction $(\alpha_1, \alpha_2, \alpha_3)$ we obtain the expression

$$D^{(\alpha)}f = \boldsymbol{e} \operatorname{grad} f,$$

the scalar product of the gradient and the unit vector in the direction $(\alpha_1, \alpha_2, \alpha_3)$, i.e. the projection of the gradient on that vector (cf. Chap. I, p. 7).

It is this fact that accounts for the importance of the concept of gradient. If, for example, we wish to find the direction in which the value of the function increases or decreases most rapidly, we must choose the direction in which the above expression has the greatest or least value. This clearly occurs when the direction of $\boldsymbol{e}$ is the same as that of the gradient or is exactly opposite to it.

Thus the direction of the gradient is the direction in which the function increases most rapidly, while the direction opposite to that of the gradient is that in which the function decreases most rapidly; the magnitude of the gradient gives the rate of increase or decrease.

We shall return to the geometrical interpretation of the gradient in Chapter III (p. 124). We can, however, immediately give an intuitive idea of the direction of the gradient. If in the first instance we confine ourselves to vectors in two dimensions, we have to consider the gradient of a function $f(x, y)$. We shall suppose that this function is represented by its contour lines (or level lines)

$$f(x, y) = c$$

in the xy-plane. Then the derivative of the function $f(x, y)$ in the direction (cf. p. 62) of these level lines is obviously zero. For if P and Q are two points on the same level line, the equation $f(P) - f(Q) = 0$ holds (the meaning of the symbols is obvious), and the equation will still hold if we divide both sides by h, the distance between P and Q, and then let h tend to 0. The projection of the gradient in the direction of the tangent to the level line is therefore zero, and hence *at every point the gradient is perpendicular to the level line through that point.* An exactly analogous statement holds for the gradient in three dimensions. If we represent the function $f(x_1, x_2, x_3)$ by its level surfaces

$$f(x_1, x_2, x_3) = c,$$

the gradient has the component zero in every direction tangent to a level surface, and is therefore perpendicular to the level surface.

In applications we frequently meet with vector fields which represent the gradient of a scalar function. The gravitational field of force may be taken as an example.

If we denote the co-ordinates of the attracting particle by (ξ_1, ξ_2, ξ_3), those of the attracted particle by (x_1, x_2, x_3), and their masses by m and M, the components of the force of attraction are given by the expressions

$$C\frac{\xi_1 - x_1}{\sqrt{\{(\xi_1 - x_1)^2 + (\xi_2 - x_2)^2 + (\xi_3 - x_3)^2\}^3}}$$

$$C\frac{\xi_2 - x_2}{\sqrt{\{(\xi_1 - x_1)^2 + (\xi_2 - x_2)^2 + (\xi_3 - x_3)^2\}^3}}$$

$$C\frac{\xi_3 - x_3}{\sqrt{\{(\xi_1 - x_1)^2 + (\xi_2 - x_2)^2 + (\xi_3 - x_3)^2\}^3}}.$$

Here C is a constant with the value γmM, where γ is the "gravitational constant". (The factors

$$\frac{\xi_1 - x_1}{\sqrt{\{(\xi_1 - x_1)^2 + (\xi_2 - x_2)^2 + (\xi_3 - x_3)^2\}}}, \text{ \&c.,}$$

are the cosines of the angles which the line through the two points makes with the axes.) By differentiation we see at once that these components are the derivatives of the function

$$\frac{c}{\sqrt{\{(\xi_1 - x_1)^2 + (\xi_2 - x_2)^2 + (\xi_3 - x_3)^2\}}}$$

with respect to the co-ordinates x_1, x_2, x_3 respectively. The force vector apart from a constant factor is therefore the gradient of the function

$$\frac{1}{r} = \frac{1}{\sqrt{\{(\xi_1 - x_1)^2 + (\xi_2 - x_2)^2 + (\xi_3 - x_3)^2\}}}.$$

If a field of force is obtained from a scalar function by forming the gradient, this scalar function is often called the *potential function* of the field. We shall consider this concept from a more general point of view in the study of work and energy (Chapter V, p. 350, and Chapter VI, pp. 415, 468–81).

4. The Divergence and Curl of a Vector Field.

By differentiation we have assigned to every function or scalar a vector field, the gradient. Similarly, by differentiation we can assign to every vector field a certain scalar, known as the *divergence* of the vector field. Given a specific co-ordinate system, the x-system, we define the divergence of the vector $\boldsymbol{u}$ as the function

$$\operatorname{div} \boldsymbol{u} = \frac{\partial u_1}{\partial x_1} + \frac{\partial u_2}{\partial x_2} + \frac{\partial u_3}{\partial x_3},$$

i.e. the sum of the partial derivatives of the three components with respect to the corresponding co-ordinates. Suppose now that we change the co-ordinate system to the ξ-system. If the divergence is really to be a scalar function associated with the vector field and independent of the particular co-ordinate system, we must have

$$\operatorname{div} \boldsymbol{u} = \frac{\partial \omega_1}{\partial \xi_1} + \frac{\partial \omega_2}{\partial \xi_2} + \frac{\partial \omega_3}{\partial \xi_3},$$

where ω_1, ω_2, ω_3 are the components of $\boldsymbol{u}$ in the ξ-system. In fact, the truth of the equation

$$\frac{\partial u_1}{\partial x_1} + \frac{\partial u_2}{\partial x_2} + \frac{\partial u_3}{\partial x_3} = \frac{\partial \omega_1}{\partial \xi_1} + \frac{\partial \omega_2}{\partial \xi_2} + \frac{\partial \omega_3}{\partial \xi_3}$$

can be verified immediately by applying the chain rule and the transformation formulæ of p. 84.

Here we content ourselves with the formal definition of the divergence; its physico-geometrical interpretation will be discussed later (Chapter V, section 5, p. 388).

We shall adopt the same procedure for the so-called *curl* * of a vector field. The curl is itself a vector

$$\boldsymbol{r} = \text{curl}\, \boldsymbol{u}$$

whose components r_1, r_2, r_3 are defined by the equations

$$r_1 = \frac{\partial u_3}{\partial x_2} - \frac{\partial u_2}{\partial x_3}, \quad r_2 = \frac{\partial u_1}{\partial x_3} - \frac{\partial u_3}{\partial x_1}, \quad r_3 = \frac{\partial u_2}{\partial x_1} - \frac{\partial u_1}{\partial x_2}.$$

In order to show that our definition actually gives a vector independent of the particular co-ordinate system, we could verify by direct differentiation that the quantities

$$\rho_1 = \frac{\partial \omega_3}{\partial \xi_2} - \frac{\partial \omega_2}{\partial \xi_3}, \quad \rho_2 = \frac{\partial \omega_1}{\partial \xi_3} - \frac{\partial \omega_3}{\partial \xi_1}, \quad \rho_3 = \frac{\partial \omega_2}{\partial \xi_1} - \frac{\partial \omega_1}{\partial \xi_2},$$

which define the curl in terms of the new co-ordinates, are connected with the quantities r_1, r_2, r_3 by the equations of transformation for vector components. Here, however, we shall omit these computations, since in Chapter VI, section 6 (p. 396) we shall give a physical interpretation of the curl which clearly brings out its vectorial character.

The three concepts of gradient, divergence, and curl can all be related to one another if we use a symbolic vector with the components $\dfrac{\partial}{\partial x_1}, \dfrac{\partial}{\partial x_2}, \dfrac{\partial}{\partial x_3}$. This symbolic vector is often called *nabla* † and is denoted by the symbol ∇. The gradient of a scalar field $f(x_1, x_2, x_3)$, grad f, is the product ∇f of the scalar quantity f and the symbolic vector ∇, that is, it is a vector with the components

$$\frac{\partial f}{\partial x_1}, \quad \frac{\partial f}{\partial x_2}, \quad \frac{\partial f}{\partial x_3}.$$

The curl of a vector field $u(x_1, x_2, x_3)$, curl $\boldsymbol{u}$, is the vector product $[\nabla \boldsymbol{u}]$ of the vector $\boldsymbol{u}$ and the symbolic vector ∇; finally, the divergence is the scalar product

$$\text{div}\, \boldsymbol{u} = \nabla \boldsymbol{u} = \frac{\partial u_1}{\partial x_1} + \frac{\partial u_2}{\partial x_2} + \frac{\partial u_3}{\partial x_3}.$$

* Often called *rotation* (with the abbreviation rot).

† After a Hebrew stringed instrument of similar shape.

In conclusion we mention a few relations which constantly recur. *The curl of a gradient is zero*; in symbols

$$\operatorname{curl} \operatorname{grad} f = 0.$$

As we easily see, this relation follows from the reversibility of the order of differentiation.

The divergence of a curl is zero; in symbols

$$\operatorname{div} \operatorname{curl} \boldsymbol{u} = 0.$$

This also follows directly from the reversibility of the order of differentiation.

The divergence of a gradient is an extremely important expression frequently occurring in analysis, notably in the well-known " Laplace's " or " potential equation ". It is the sum of the three " principal " second-order partial derivatives of a function; in symbols

$$\operatorname{div} \operatorname{grad} f = \Delta f = \frac{\partial^2 f}{\partial x_1^2} + \frac{\partial^2 f}{\partial x_2^2} + \frac{\partial^2 f}{\partial x_3^2},$$

where Δf is written as an abbreviation for the expression on the right.* The symbol

$$\Delta = \frac{\partial^2}{\partial x_1^2} + \frac{\partial^2}{\partial x_2^2} + \frac{\partial^2}{\partial x_3^2}$$

is called the *Laplacian operator.*

Finally, we may mention that the terminology of vector analysis is often used in connexion with more than three independent variables; thus a system of n functions of n independent variables is sometimes called a vector field in n-dimensional space. The concepts of scalar multiplication and of the gradient then retain their meanings, but in other respects the state of affairs is more complicated than in the case of three dimensions.

Examples

1. Find the equation of the so-called *osculating plane* of a curve $x = f(t)$, $y = g(t)$, $z = h(t)$ at the point t_0, i.e. the limit of the planes passing through three points of the curve as these points approach the point with parameter t_0.

2. Show that the curvature vector and the tangent vector both lie in the osculating plane.

* The notation $\nabla^2 f$ is also used.

3*. Let $x = x(s)$ be an arbitrary curve in space, such that the vector $x(s)$ is three times continuously differentiable (s is the length of arc). Find the centre of the sphere of closest contact with the curve at the point s.

4. If C is a continuously differentiable closed curve and A a point not on C, there is a point B on C which has a shorter distance from A than any other point on C. Prove that the line AB is normal to the curve.

5. If $x = x(s)$ is a curve on a sphere of unit radius, the equation

$$\ddot{x}^2(\dddot{x}^2 - \ddot{x}^4) = ([\dot{x}\ddot{x}]\dddot{x})^2$$

holds.

6. If $x = x(t)$ is any parametric representation of a curve, then the vector $\dfrac{d^2x}{dt^2}$ with initial point x lies in the osculating plane at x.

7. The limit of the ratio of the angle between the osculating planes at two neighbouring points of a curve and the length of arc between these two points, i.e. the derivative of the unit normal vector with respect to the arc (s), is called the *torsion* of the curve. Let $\xi_1(s)$, $\xi_2(s)$ denote the unit vectors along the tangent and the curvature vector of the curve $x(s)$; by $\xi_3(s)$ we mean the unit vector orthogonal to ξ_1 and ξ_2 (the so-called binormal vector), which is given by $[\xi_1\xi_2]$. Prove Frenet's formulae

$$\begin{aligned}\dot{\xi}_1 &= \xi_2/\rho,\\ \dot{\xi}_2 &= -\xi_1/\rho + \xi_3/\tau,\\ \dot{\xi}_3 &= -\xi_2/\tau,\end{aligned}$$

where $1/\rho = k$ is the curvature and $1/\tau$ the torsion of $x(s)$.

8. Using the vectors ξ_1, ξ_2, ξ_3 of Ex. 7 as co-ordinate vectors, find expressions for (a) the vector $\dddot{x}$, (b) the vector from the point x to the centre of the sphere of closest contact at x.

9. Show that a curve of zero torsion is a plane curve.

10*. Prove that if $z = u(x, y)$ represents the surface formed by the tangents of an arbitrary curve, then (a) every osculating plane of the curve is a tangent plane to the surface; (b) $u(x, y)$ satisfies the equation

$$u_{xx}u_{yy} - u_{xy}^2 = 0.$$

11. Prove that

$$\text{curl curl } \mathbf{u} = \text{grad div } \mathbf{u} - \Delta\mathbf{u}.$$

Appendix to Chapter II

1. The Principle of the Point of Accumulation in Several Dimensions and its Applications

If we wish to refine the concepts of the theory of functions of several variables and to establish it on a firm basis, without reference to intuition, we proceed in exactly the same way as in the case of functions of one variable. It is sufficient to discuss these matters in the case of two variables only, since the methods are essentially the same for functions of more than two independent variables.

1. The Principle of the Point of Accumulation.

We again base our discussion on Bolzano and Weierstrass's principle of the point of accumulation. A pair of numbers (x, y) will be called a point P in space of two dimensions, and may be represented in the usual way by means of a point with the rectangular co-ordinates x and y in an xy-plane. We now consider a bounded infinite set of such points $P(x, y)$; that is, the set is to contain an infinite number of points, and all the points are to lie in a bounded part of the plane, so that $|x| < C$ and $|y| < C$, where C is a constant. The principle of the point of accumulation can then be stated as follows: *every bounded infinite set of points has at least one point of accumulation.* That is, there exists a point Q with co-ordinates (ξ, η) such that an infinite number of points of the given set lie in every neighbourhood of the point Q, say in every region

$$|x - \xi| < \delta, \quad |y - \eta| < \delta$$

where δ is any positive number. Or, in other words, *out of the infinite set of points we can choose a sequence* $P_1, P_2, P_3, \ldots$ *in such a way that these points approach a limit point* Q.

This principle of the point of accumulation is just as intuitively clear for several dimensions as it is for one dimension. It can be proved analytically by the method used in the corresponding proof in Vol. I (p. 58), merely by substituting rectangular regions for the intervals used there. An easier proof can be constructed, however, by using the principle of the point of accumulation for one dimension. To do this we notice that by hypothesis every point $P(x, y)$ of the set has an abscissa x for which the inequality $|x| < C$ holds. Either there is an $x = x_0$ which is the abscissa of an infinite number of points P (which therefore lie vertically above one another) or else each x belongs only to a finite number of points P. In the first case, we fix upon x_0 and consider the infinite number of values of y such that (x_0, y) belongs to our set. These values of y have a point of accumulation η_0, by the principle of the point of accumulation for one dimension. Hence we can find a sequence of values of y, say $y_1, y_2, \ldots$, such that $y_n \to \eta_0$, from which it follows that the points (x_0, y_n) of the set tend to the limit point (x_0, η_0), which is thus a point of accumulation of the set. In the second case, there must be an infinite number of distinct values of x which are the abscissæ of points of the set, and we can choose a sequence $x_1, x_2, \ldots$ of these abscissæ tending to a unique limit ξ. For each x_n let $P_n(x_n, y_n)$ be a point of the set with abscissa x_n. The numbers y_n are an infinite bounded set of numbers; hence we can choose a sub-sequence $y_{n_1}, y_{n_2}, \ldots$ tending to a limit η. The corresponding sub-sequence of abscissæ $x_{n_1}, x_{n_2}, \ldots$ still tends to the limit ξ; hence the points $P_{n_1}, P_{n_2}, \ldots$ tend to the limit point (ξ, η). In either case, therefore, we can find a sequence of points of the set tending to a limit point, and the theorem is proved.

A first and important consequence of the principle of the point of accumulation is Cauchy's *convergence test*, which can be expressed as follows:

A sequence of points $P_1, P_2, P_3, \ldots$ *with the co-ordinates* (x_1, y_1), (x_2, y_2), (x_3, y_3), ... *tends to a limit point if, and only if, for every* $\epsilon > 0$ *there is a suffix* $N = N(\epsilon)$ *such that the distance between the points* P_n *and* P_m, $\sqrt{(x_n - x_m)^2 + (y_n - y_m)^2}$, *is less than* ϵ *whenever both* n *and* m *are greater than* N.

2. Some Concepts of the Theory of Sets of Points.

The general concept of a limit point is fundamental in many of the more refined investigations of the foundations of analysis based on the theory of sets of points. Although these matters are not essential for most of the purposes of this book, we shall mention some of them here for the sake of completeness.

A bounded set of points, consisting of an infinite number of points, is said to be *closed* if it contains all its limit points; that is, limit points of sequences of points of the set are again points of the set. For example, all the points lying on a closed curve or surface form a closed set. For functions defined in closed sets we can state the two following fundamental theorems:

A function which is continuous in a bounded closed set of points assumes a greatest and a least value in that set.

A function which is continuous in a bounded closed set is uniformly continuous in that set.

The proofs of these theorems are so like the corresponding proofs for functions of one variable that we shall omit them.

The least upper bound of the distance between the points P_1 and P_2 for all pairs of points P_1, P_2, where both points belong to a set, is called the *diameter* of that set. If the set is closed, this upper bound will actually be assumed for a pair of points of the set. The student will be able to prove this easily, remembering that the distance between two points is a continuous function of the co-ordinates of the points.

By using the theorem that a continuous function on a bounded closed set does assume its least value, we can readily establish the following fact: if a point P does not belong to a closed set M, a positive *least distance from* P *to* M *exists*; that is, a point Q of M exists such that no point of M has a smaller distance from P than Q has. This enables us to show that the closed regions defined in section 1 (p. 41) are actually closed sets according to the definition here. For let C be a closed curve, and let R be the closed region consisting of all points interior to C or on C; we have to show that all the limit points of R belong to R. We assume the contrary, i.e. that there is a point P not belonging to R which is a limit point of R. Then, in particular, P does not lie on C; hence by the theorem above it has a positive least distance from C (C being a closed set). We can therefore describe a circle about P as centre, so small that no point of C lies in the circle; we have only to make the radius of the circle less than the least distance from P to C. The point P is outside C, since otherwise it would belong to R; and since every point in the small circle can be joined to P by a line-segment which does not cross the curve C, every point of the circle lies outside C, and so no point of the circle belongs to R. But we assumed that P is a limit point of R, which requires that the circle should contain an infinite number of points of R. Hence the assumption that there is a limit point of R which does not itself belong to R leads to a contradiction, and our assertion is proved. The extension to closed regions R bounded by several closed curves is obvious.

A useful property of closed sets is contained in the *theorem on shrinking sequences of closed sets*:

If the sets $M_1, M_2, M_3, \ldots$ *are all closed, and each set is contained in the preceding one, then there is a point* (ξ, η) *which belongs to all the sets.*

In each of the sets M_n let us choose a point P_n. The sequence P_n must either contain an infinite number of repetitions of some one point, or else an infinite number of distinct points. If one point P is repeated an infinite number of times, then it belongs to all the sets; for if M_n is any one of the sets, P belongs to a set M_{n_1}, where $n_1 > n$, and M_{n_1} is contained in M_n. If there are an infinite number of distinct points P_n, then by the principle of the point of accumulation they possess a point of accumulation (ξ, η). This point belongs to each M_n. For whenever $m > n$ the point P_m belongs to M_n, since it is a point of M_m which is contained in M_n. Hence (ξ, η) is a limit point of points P_m of M_n, and since M_n is closed, (ξ, η) is a point of M_n. Thus in either case there exists a point common to all the sets M_n, and the theorem is proved.*

A set is said to be *open* if for every point of the set we can find a circle about the point as centre which belongs completely to the set. An open set is *connected* if every pair of points A and B of the set can be joined by a broken (polygonal) line which lies entirely in the set.

The word "domain" is often used with the restricted meaning of a connected open set. As examples we have the interior of a closed curve, or the interior of a circle with the points of a radius removed. The points of accumulation of a domain which do not themselves belong to the domain are called the *boundary points*. *The boundary B of a domain D is a closed set.* Here we shall sketch the proof of this statement.

* The assumption that the sets M_n are closed is essential, as the following example shows. Let M_n be the set $0 < x < \frac{1}{n}$. Each set is contained in the preceding, but no point belongs to all the sets. For if $x = 0$ the point belongs to no set, while if $x > 0$ it belongs to no set M_n for which $\frac{1}{n} < x$.

A point P which is a limit point of B does not belong to D, for every point of D lies in a circle composed only of points of D and hence devoid of points of B. It is also a limit point of D, for arbitrarily close to P we can find a point Q of B, and arbitrarily close to Q we can find points of D. Hence P belongs to B.

If to a domain D we add its boundary points B, we obtain a closed set. For every limit point of the combined set is either a limit point of B and belongs to B, or is a limit point of D and belongs either to D or to B. Such sets are called *closed regions*, and are particularly useful for our purposes.

Finally, we define a *neighbourhood* of a point P as any open set containing P. If we denote the co-ordinates of P by (ξ, η), the two simplest examples of neighbourhoods of P are the circular neighbourhood, consisting of all points (x, y) such that

$$(x - \xi)^2 + (y - \eta)^2 < \delta^2,$$

and the square neighbourhood, consisting of all points (x, y) such that

$$|x - \xi| < \delta \quad \text{and} \quad |y - \eta| < \delta.$$

3. The Heine-Borel Covering Theorem.

A further consequence of the principle of the point of accumulation, which is useful in many proofs and refined investigations, is the *Heine-Borel covering theorem*, which runs as follows:

If corresponding to every point of a bounded closed set M *a neighbourhood of the point, say a square or a circle, is assigned, it is possible to choose a finite number of these neighbourhoods in such a way that they completely cover* M. The last statement of course means that every point of M belongs to at least one of the finite number of selected neighbourhoods.

By an indirect method the proof can be derived almost immediately from the theorem on shrinking closed sets. We suppose that the theorem is false. The set M, being bounded, lies in a square Q. This square we subdivide into four equal squares. For at least one of these four squares, the part of M lying in or on the boundary of that square cannot be covered by a finite number of the neighbourhoods; for if each of the four parts of M could be covered in this way, M itself would be covered. This part of M we call M_1, and we see at once that M_1 is closed.

We now subdivide the square containing M_1 into four equal squares. By the same argument, the part M_2 of M_1 lying in or on the boundary of one of these squares cannot be covered by a finite number of the neighbourhoods. Continuing the process, we obtain a sequence of closed sets $M_1, M_2, M_3, \ldots$, each enclosed in the preceding; each of these is contained in a square whose side tends to zero, and none of them can be covered by a finite number of the neighbourhoods. By the theorem on shrinking sequences of closed sets we know that there is a point (ξ, η) which belongs to all these sets, and hence a fortiori belongs to M. To the point (ξ, η) there accordingly corresponds one of the neighbourhoods, containing a small square about (ξ, η). But since each M_n contains (ξ, η) and is itself contained in a square whose side tends to 0 as $1/n$ does, each M_n after a certain n is completely contained in the small square about (ξ, η), and is therefore covered by one neighbourhood of the set. The assumption that the theorem is false has therefore led to a contradiction, and the theorem is proved.

EXAMPLES

1. A convex region R may be defined as a bounded and closed region with the property that if A, B are any two points belonging to R, all points of the segment AB belong to R. Prove the following statements:

(*a*)* If A is a point not belonging to R, there is a straight line passing through A which has no point in common with R.

(*b*)* Through every point P on the boundary of R there is a straight line l (a so-called "line of support") such that all points of R lie on one and the same side of l or on l itself.

(*c*) If a point A lies on the same side of every line of support as the points of R, then A is also a point of R.

(*d*) The centre of mass of R is a point of R.

(*e*) A closed curve forms the boundary of a convex region, provided that it has not more than two points in common with any straight line.

(*f*)* A closed curve forms the boundary of a convex region, provided that its curvature is everywhere positive. (It is assumed that if the whole curve is traversed the tangent makes one complete revolution.)

2. (*a*) If S is an arbitrary closed and bounded set, there is one "least convex envelope" E of S, i.e. a set which

(1) contains all points of S,

(2) is contained in all convex sets containing S.

(3) is convex.

(*b*) E may also be described in the following way:

A point P is in E if, and only if, for every straight line which leaves all points of S on one and the same side, P is also on this side.

(*c*) The centre of mass of S is a point of E.

2. The Concept of Limit for Functions of Several Variables

We shall find it useful to refine our conceptions of the various limiting processes connected with several variables and to consider them from a single point of view. Here we again restrict ourselves to the typical case of two variables.

1. Double Sequences and their Limits.

In the case of one variable we began with the study of sequences of numbers a_n, where the suffix n could be any integer. Here *double sequences* have a corresponding importance. These are sets of numbers a_{nm} with two suffixes, where the suffixes m and n run through the sequence of all the integers independently of one another, so that we have e.g. the numbers

$$a_{11}, a_{12}, a_{21}, a_{13}, a_{22}, a_{31}, a_{14}, a_{23}, \ldots .$$

Examples of such sequences are the sets of numbers

$$a_{nm} = \frac{1}{n+m}, \quad a_{nm} = \frac{1}{n^2+m^2}, \quad a_{nm} = \frac{n}{n+m}.$$

We now make the following statement:

The double sequence a_{nm} *converges as* $n \to \infty$ *and* $m \to \infty$ *to a limit, or more precisely a "double limit", l if the absolute difference* $| a_{nm} - l |$ *is less than an arbitrarily small pre-assigned positive number* ϵ *whenever* n *and* m *are both sufficiently large, that is, whenever they are both larger than a certain number* N *depending only on* ϵ. We then write

$$\lim_{\substack{n\to\infty \\ m\to\infty}} a_{nm} = l.$$

Thus, for example,

$$\lim_{\substack{n\to\infty \\ m\to\infty}} \frac{1}{n+m} = 0$$

and $$\lim_{\substack{n\to\infty\\ m\to\infty}} \frac{m+n^2}{mn^2} = \lim_{\substack{n\to\infty\\ m\to\infty}} \left(\frac{1}{n^2}+\frac{1}{m}\right) = 0.$$

Following Cauchy, we can determine, without referring to the limit, whether the sequence converges or not, by using the following criterion:

The sequence a_{nm} *converges if, and only if, for every* $\epsilon > 0$ *a number* $N = N(\epsilon)$ *exists such that* $| a_{nm} - a_{n'm'} | < \epsilon$ *whenever the four suffixes* n, m, n′, m′ *are all greater than* N.

Many problems in analysis involving several variables depend on the resolution of these double limiting processes into two successive ordinary limiting processes. In other words, instead of allowing n and m to increase simultaneously beyond all bounds, we first attempt to keep one of the suffixes, say m, fixed, and let n alone tend to ∞. The limit thus found (if it exists) will in general depend on m; let us say that it has the value l_m. We now let m tend to ∞. The question now arises whether, and if so when, the limit of l_m is identical with the original double limit, and also the question whether we obtain the same result, no matter which variable we first allow to increase; that is, whether we could have first formed the limit $\lim_{m\to\infty} a_{nm} = \lambda_n$ and then the limit $\lim_{n\to\infty} \lambda_n$ and still have obtained the same result

We shall begin by gaining a general idea of the position from a few examples. In the case of the double sequence $a_{nm} = \dfrac{1}{n+m}$, when m is fixed we obviously obtain the result $\lim_{n\to\infty} a_{nm} = l_m = 0$, and therefore $\lim_{m\to\infty} l_m = 0$; the same result is obtained if we perform the passages to the limit in the reverse order. For the sequence

$$a_{nm} = \frac{n}{n+m} = \frac{1}{1+\dfrac{m}{n}},$$

however, we obtain

$$\lim_{n\to\infty} a_{nm} = l_m = 1$$

and consequently

$$\lim_{m\to\infty} l_m = 1;$$

while on performing the passages to the limit in the reverse order we first obtain

$$\lim_{m\to\infty} a_{nm} = \lambda_n = 0$$

and then

$$\lim_{n\to\infty} \lambda_n = 0.$$

In this case, then, the result of the successive limiting processes is not independent of their order:

$$\lim_{m\to\infty} (\lim_{n\to\infty} a_{nm}) \neq \lim_{n\to\infty} (\lim_{m\to\infty} a_{nm}).$$

In addition, if we let n and m increase beyond all bounds simultaneously, we find that the double limit fails to exist.*

Another example is given by the sequence

$$a_{nm} = \frac{\sin n}{m}.$$

Here the double limit $\lim_{\substack{n\to\infty \\ m\to\infty}} a_{nm}$ exists and has the value 0, since the numerator of the fraction can never exceed 1 in absolute value, while the denominator increases beyond all bounds. We obtain the same limit if we first let m tend to ∞; we find that $\lim_{m\to\infty} a_{nm} = \lambda_n = 0$, so that $\lim_{n\to\infty} \lambda_n = 0$. If, however, we wish to perform the passages to the limit in the reverse order, keeping m fixed and letting n increase beyond all bounds, we encounter the difficulty that $\lim_{n\to\infty} \sin n$ does not exist. Hence the resolution of the double limiting process into two ordinary limiting processes cannot be carried out in both ways.

The position can be summarized by means of two theorems. The first of these is as follows:

If the double limit $\lim_{\substack{n\to\infty \\ m\to\infty}} a_{nm} = l$ *exists, and the simple limit* $\lim_{n\to\infty} a_{nm} = l_m$ *exists for every value of* m, *then the limit* $\lim_{m\to\infty} l_m$ *also exists, and* $\lim_{m\to\infty} l_m = l$. *Again, if the double limit exists and has the value* l, *and the limit* $\lim_{m\to\infty} a_{nm} = \lambda_n$ *exists for every value of* n, *then* $\lim_{n\to\infty} \lambda_n$ *also exists and has the value* l. In symbols:

$$l = \lim_{\substack{n\to\infty \\ m\to\infty}} a_{nm} = \lim_{m\to\infty} (\lim_{n\to\infty} a_{nm}) = \lim_{n\to\infty} (\lim_{m\to\infty} a_{nm});$$

* For if such a limit existed it would necessarily have the value 0, since we can make a_{nm} arbitrarily close to 0 by choosing n large enough and choosing $m = n^2$. On the other hand, $a_{nm} = \frac{1}{2}$ whenever $n = m$, no matter how large n is. These two facts contradict the assumption that the double limit exists. But even when $\lim_{m\to\infty} (\lim_{n\to\infty} a_{nm}) = \lim_{n\to\infty} (\lim_{m\to\infty} a_{nm})$ the double limit $\lim_{\substack{n\to\infty \\ m\to\infty}} a_{nm}$ may fail to exist, as is shown by the example $a_{nm} = \dfrac{1}{(n - m) + \frac{1}{2}}$.

the double limit can be resolved into simple limiting processes and this resolution is independent of the order of the simple limiting processes.

The proof follows almost at once from the definition of the double limit. In virtue of the existence of $\lim\limits_{\substack{n\to\infty\\m\to\infty}} a_{nm} = l$, for every positive ϵ there is an $N = N(\epsilon)$ such that the relation $|a_{nm} - l| < \epsilon$ holds whenever n and m are both larger than N. If we now keep m fixed and let n increase beyond all bounds, we find that $|\lim\limits_{n\to\infty} a_{nm} - l| = |l_m - l| \leqq \epsilon$. This inequality holds for any positive ϵ provided only that m is larger than $N(\epsilon)$; in other words, it is equivalent to the statement $\lim\limits_{m\to\infty} (\lim\limits_{n\to\infty} a_{nm}) = l$. The other part of the theorem can be proved in a similar way.

The second theorem is in some respects a converse of the first. It gives a sufficient condition for the equivalence of a repeated limiting process and a double limit. This theorem is based on the concept of uniform convergence, which we define as follows:

The sequence a_{nm} *converges as* $n \to \infty$ *to the limit* l_m uniformly *in* m, *provided that the limit* $\lim\limits_{n\to\infty} a_{nm} = l_m$ *exists for every* m *and in addition for every positive* ϵ *it is possible to find an* $N = N(\epsilon)$, *depending on* ϵ *but not on* m, *such that* $|l_m - a_{nm}| < \epsilon$ *whenever* $n > N$.

For example, the sequence $a_{nm} = \dfrac{n}{m(n+m)} = \dfrac{1}{m} - \dfrac{1}{n+m}$ converges uniformly to the limit $l_m = \dfrac{1}{m}$, as we see immediately from the estimate

$$\left| a_{nm} - \frac{1}{m} \right| = \frac{1}{n+m} < \frac{1}{n};$$

we have only to put $N \geqq \dfrac{1}{\varepsilon}$. On the other hand, the condition for uniform convergence does not hold in the case of the sequence $a_{nm} = \dfrac{m}{m+n}$. For fixed values of m the equation $\lim\limits_{n\to\infty} a_{nm} = l_m = 0$ is always true; but the convergence is not uniform. For if any particular value, say 1/100, is assigned to ε, then no matter how large a value of n we choose there are always values of m for which $|a_{nm} - l_m| = a_{nm}$ exceeds ε. We have only to take $m = 2n$ to obtain $a_{nm} = \frac{2}{3}$, which is a value differing from the limit 0 by more than 1/100.

We now have the following theorem:

If the limit $\lim_{n\to\infty} a_{nm} = l_m$ *exists uniformly with respect to* m, *and if further the limit* $\lim_{m\to\infty} l_m = l$ *exists, then the double limit* $\lim_{\substack{n\to\infty\\ m\to\infty}} a_{nm}$ *exists and has the value* l:

$$\lim_{m\to\infty}(\lim_{n\to\infty} a_{nm}) = \lim_{\substack{n\to\infty\\ m\to\infty}} a_{nm}.$$

We can then reverse the order of the passages to the limit, provided that $\lim_{m\to\infty} a_{nm} = \lambda_n$ *exists.*

By making use of the inequality

$$|a_{nm} - l| \leqq |a_{nm} - l_m| + |l_m - l|$$

the proof can be carried out just as for the previous theorem, and we accordingly leave it to the reader.

2. Double Limits in the Case of Continuous Variables.

In many cases limiting processes occur in which certain suffixes, e.g. n, are integers and increase beyond all bounds, while at the same time one or more continuous variables $x, y, \ldots$, tend to limiting values $\xi, \eta, \ldots$. Other processes involve continuous variables only and not suffixes. Our previous discussions apply to such cases without essential modification. We point out in the first instance that the concept of the limit of a sequence of functions $f_n(x)$ or $f_n(x, y)$ as $n \to \infty$ can be classified as one of these limiting processes. We have already seen (Vol. I, Chap. VIII, p. 393—the definition and proofs can be applied unaltered to functions of several variables) that if the convergence of the sequence $f_n(x)$ is uniform the limit function $f(x)$ is continuous, provided that the functions $f_n(x)$ are continuous. This continuity gives the equations

$$f(\xi) = \lim_{x\to\xi} f(x) = \lim_{x\to\xi}(\lim_{n\to\infty} f_n(x)) = \lim_{n\to\infty} f_n(\xi) = \lim_{n\to\infty}(\lim_{x\to\xi} f_n(x)),$$

which express the reversibility of the order of the passages to the limit $n \to \infty$ and $x \to \xi$.

Further examples of the part played by the question of the reversibility of the order of passages to the limit have already occurred, e.g. in the theorem on the order of partial differentiation, and we shall meet with other examples later. Here we mention only the case of the function

$$f(x, y) = \frac{x^2 - y^2}{x^2 + y^2}.$$

For fixed non-zero values of y we obtain the limit $\lim_{x \to 0} f(x, y) = -1$, while for fixed non-zero values of x we have $\lim_{y \to 0} f(x, y) = +1$. Thus

$$\lim_{y \to 0} (\lim_{x \to 0} f(x, y)) \neq \lim_{x \to 0} (\lim_{y \to 0} f(x, y)),$$

and the order of the passages to the limit is not immaterial. This is of course connected with the discontinuity of the function at the origin.

In conclusion we remark that *for continuous variables the resolution of a double limit into successive ordinary limiting processes and the reversibility of the order of the passages to the limit are controlled by theorems which correspond exactly to those established on p.* 103 *for double sequences.*

3. Dini's Theorem on the Uniform Convergence of Monotonic Sequences of Functions.

In many refined analytical investigations it is useful to be able to apply a certain general theorem on uniform convergence, which we shall state and prove here. We already know (Vol. I, p. 387 *et seq.*) that a sequence of functions may converge to a continuous limit function, even though the convergence is not uniform. In an important special case, however, we can conclude from the continuity of the limit that the convergence is uniform. This is the case in which the sequence of functions is monotonic, that is, when for all fixed values of x the value of the function $f_n(x)$ either increases steadily or decreases steadily as n increases. Without loss of generality we may assume that the values *increase*, or do not decrease, monotonically; we can then state the following theorem:

If in the closed region R *the sequence of continuous functions* $f_n(x, y)$ *converges to the continuous limit function* $f(x, y)$, *and if at each point* (x, y) *of the region the inequality*

$$f_{n+1}(x, y) \geqq f_n(x, y)$$

holds, then the convergence is uniform in R.

The proof is indirect, and is a typical example of the use of the principle of the point of accumulation. If the convergence is not uniform, a positive number a will exist such that for arbitrarily large values of n—say for all the values of n belonging to the infinite set $n_1, n_2, \ldots$—the value of the function at a point P_n in the region, $f_n(P_n)$, differs from $f(P_n)$ by more than a. If we let n run through the sequence of values $n_1, n_2, \ldots$, the points $P_{n_1}, P_{n_2}, \ldots$ will have at least one point of accumulation Q; and since R is closed, Q will belong to R. Now for every point P in R and every whole number μ we have

$$f(P) = f_\mu(P) + R_\mu(P),$$

where $f_\mu(P)$ and the "remainder" $R_\mu(P)$ are continuous functions of the point P. In addition,

$$R_\mu(P) \geqq R_n(P),$$

whenever $n > \mu$, as we assumed that the sequence increases monotonically. In particular, for $n > \mu$ the inequality

$$R_\mu(P_n) \geqq R_n(P_n) \geqq a$$

will hold. If we consider the sub-sequence $P_{n_1}, P_{n_2}, P_{n_3}, \ldots$ of the sequence which tends to the limit point Q, on account of the continuity of R_μ for fixed values of μ we also have $R_\mu(Q) \geqq a$. Since in this limiting process the suffix n increases beyond all bounds, we may take the index μ as large as we please, for the above inequality holds whenever $n > \mu$, and in the sequence of points P_n tending to Q there are an infinite number of values of the suffix n, hence an infinite number of values of n greater than μ. But the relation $R_\mu(Q) \geqq a$ for all values of μ contradicts the fact that $R_\mu(Q)$ tends to 0 as μ increases. Thus the assumption that the convergence is non-uniform leads to contradiction, and the theorem is proved.

Examples

1. State whether the following limits exist:

(a) $$\lim_{\substack{n\to\infty \\ m\to\infty}} \frac{(\log n)^2 - (\log m)^2}{(\log n)^2 + (\log m)^2},$$

(b) $$\lim_{\substack{n\to\infty \\ m\to\infty}} \frac{\tan n + \tan m}{1 - \tan n \tan m},$$

(c) $$\lim_{\substack{n\to\infty \\ m\to\infty}} \frac{1}{m^2} \sum_{\nu=1}^{n} \cos \frac{\nu}{m}.$$

2. Prove that a function $f(x, y)$ is continuous, if

(*a*) when y is fixed f is a continuous function in x;

(*b*) when x is fixed f is uniformly continuous in y, in the sense that for every ε there is a δ, independent of x *and* y, such that

$$|f(x, y_1) - f(x, y)| \leqq \varepsilon$$

when

$$|y_1 - y| \leqq \delta.$$

3. Prove that $f(x, y)$ is continuous at $x = 0$, $y = 0$, if the function $\Phi(t, \varphi) = f(t \cos\varphi, t \sin\varphi)$ is

(*a*) a continuous function of t when φ is fixed;

(*b*) uniformly continuous in φ when t is fixed, so that for every ε there is a δ, independent of t and φ, such that

$$|\Phi(t, \varphi_1) - \Phi(t, \varphi)| \leqq \varepsilon$$

when

$$|\varphi_1 - \varphi| \leqq \delta.$$

4. Prove that the complementary set of a closed set S (i.e. the set of all points not in S) is an open set.

3. Homogeneous Functions

We finally touch on one other special point, the theory of *homogeneous functions.* The simplest homogeneous functions occurring in analysis and its applications are the homogeneous polynomials in several variables. We say that a function of the form $ax + by$ is a homogeneous function of the first degree in x and y, that a function of the form $ax^2 + bxy + cy^2$ is a homogeneous function of the second degree, and in general that *a polynomial in* x *and* y *(or in a greater number of variables) is a homogeneous function of degree* h *if in each term the sum of the indices of the independent variables is equal to* h, that is, if the terms (apart from constant coefficients) are of the form x^h, $x^{h-1}y$, $x^{h-2}y^2$, . . . , y^h. These homogeneous polynomials have the property that the equation

$$f(tx, ty) = t^h f(x, y)$$

holds for every value of t. We now say in general that *a function* f(x, y, . . .) *is homogeneous of degree* h *if it satisfies the equation*

$$f(tx, ty, \ldots) = t^h f(x, y, \ldots).$$

Examples of homogeneous functions which are *not* polynomials are

$$\tan\left(\frac{y}{x}\right), \quad (h = 0),$$

$$x^2 \sin\frac{x}{y} + y\sqrt{x^2 + y^2} \log\frac{x + y}{x}, \quad (h = 2).$$

Another example is the cosine of the angle between two vectors with the respective components x, y, z and u, v, w:

$$\frac{xu + yv + zw}{\sqrt{x^2 + y^2 + z^2}\ \sqrt{u^2 + v^2 + w^2}}, \quad (h = 0).$$

The length of the vector with components x, y, z,

$$\sqrt{x^2 + y^2 + z^2},$$

is an example of a function which is *positively* homogeneous and of the first degree; that is, the equation defining homogeneous functions does not hold for this function unless t is positive or zero.

Homogeneous functions which are also differentiable satisfy the characteristic *Euler's relation*

$$xf_x + yf_y + zf_z + \ldots = hf(x, y, z, \ldots).$$

To prove this we differentiate both sides of the equation $f(tx, ty, \ldots) = t^h f(x, y, \ldots)$ with respect to t; this is permissible, since the equation is an identity in t. Applying the chain rule to the function on the left, we obtain

$$xf_x(tx, ty, \ldots) + yf_y(tx, ty, \ldots) + \ldots = ht^{h-1}f(x, y, \ldots).$$

If we substitute $t = 1$ in this, the statement follows.

Conversely, it is easy to show that not only is the validity of Euler's relation merely a consequence of the homogeneity of the function $f(x, y, \ldots)$, but also the homogeneity of the function is a consequence of Euler's relation, so that *Euler's relation is a necessary and sufficient condition for the homogeneity of the function.* The fact that a function is homogeneous of degree h can also be expressed by saying that the value of the function divided by x^h depends only on the ratios y/x, z/x, ... It is therefore sufficient to show that it follows from the Euler relation that if new variables $\xi = x$, $\eta = \dfrac{y}{x}$, $\zeta = \dfrac{z}{x}$, ... are introduced, the function

$$\frac{1}{x^h} f(x, y, z, \ldots) = \frac{1}{\xi^h} f(\xi, \eta\xi, \zeta\xi, \ldots) = g(\xi, \eta, \zeta, \ldots)$$

no longer depends on the variable ξ, i.e. that the equation $g_\xi = 0$ is an identity. In order to prove this we use the chain rule:

$$g_\xi = (f_x + \eta f_y + \dots) \frac{1}{\xi^h} - \frac{h}{\xi^{h+1}} f$$

$$= (xf_x + yf_y + \dots) \frac{1}{x^{h+1}} - \frac{h}{x^{h+1}} f;$$

The expression on the right vanishes in virtue of Euler's relation, and our statement is proved.

This last statement can also be proved in a more elegant but less direct way. We wish to show that from Euler's relation it follows that the function

$$g(t) = t^h f(x, y, \dots) - f(tx, ty, \dots)$$

has the value 0 for all values of t. It is obvious that $g(1) = 0$. Again,

$$g'(t) = ht^{h-1} f(x, y, \dots) - xf_x(tx, ty, \dots) - yf_y(tx, ty, \dots) - \dots .$$

On applying Euler's relation to the arguments tx, ty, ... we find that

$$xf_x(tx, ty, \dots) + yf_y(tx, ty, \dots) + \dots = \frac{h}{t} f(tx, ty, \dots),$$

and thus $g(t)$ satisfies the differential equation

$$g'(t) = g(t) \frac{h}{t}.$$

If we write $g(t) = \gamma(t) t^h$ we obtain $g'(t) = \frac{h}{t} g(t) + t^h \gamma'(t)$, so that $\gamma(t)$ satisfies the differential equation

$$t^h \gamma'(t) = 0,$$

which has the unique solution $\gamma = \text{const.} = c$. Since for $t = 1$ it is obvious that $\gamma(t) = 0$, the constant c is 0, and so $g(t) = 0$ for all values of t, as was to be proved.

EXAMPLES

1. Prove that if $f(x, y, z, \dots)$ is a homogeneous function of degree h, any k-th derivative of f is a homogeneous function of degree $h - k$.

2. Prove that for a homogeneous function f of the first degree

$$x^2 f_{xx} + y^2 f_{yy} + z^2 f_{zz} + \dots + 2xy f_{xy} + \dots = 0.$$

CHAPTER III

Developments and Applications of the Differential Calculus

1. Implicit Functions

1. General Remarks.

In analytical geometry it frequently happens that the equation of a curve is given, not in the form $y = f(x)$, but in the form $F(x, y) = 0$. Accordingly, a straight line may be represented by the equation $ax + by + c = 0$, or an ellipse by the equation $x^2/a^2 + y^2/b^2 = 1$. To obtain the equation of the curve in the form $y = f(x)$ we must "solve" the equation $F(x, y) = 0$ for y.

Again, in Vol. I we considered the problem of finding the inverse function of a function $y = f(x)$, in other words, the problem of solving the equation $F(x, y) = y - f(x) = 0$ for the variable x. These examples suggest the importance of studying the notion of solving an equation $F(x, y) = 0$ for x or for y. We shall now proceed to this investigation, and in section 3 (p. 153) we shall extend the results to functions of several variables.

In the simplest cases, such as the equations mentioned above, the solution can readily be found in terms of elementary functions. In other cases the solution can be approximated to as closely as we desire. For many purposes, however, it is preferable not to work with the solved form of the equation or with these approximations, but instead to draw conclusions about the solution by studying the function $F(x, y)$ itself, in which neither of the variables x, y is given preference over the other.

The idea that every function $F(x, y)$ yields a function $y = f(x)$ or $x = \phi(y)$ given implicitly by means of the equation $F(x, y) = 0$ is erroneous. On the contrary, it is easy to give examples of functions $F(x, y)$ which, when equated to zero, permit of no solution in terms of functions of one variable. Thus, for example, the equation $x^2 + y^2 = 0$ is satisfied by the single pair of values $x = 0$, $y = 0$ only, while the equation $x^2 + y^2 + 1 = 0$ is satisfied by no (real) values at all. It is therefore necessary to investigate the matter more closely in order to find out whether an equation $F(x, y) = 0$ defines a function $y = f(x)$, and what are the properties of this function.

2. Geometrical Interpretation.*

In order to clarify the situation we think of the function $u = F(x, y)$ as represented by a surface in three-dimensional space. The solutions of the equation $F(x, y) = 0$ are the same as the simultaneous solutions of the two equations $u = F(x, y)$ and $u = 0$. Geometrically, our problem is to find whether curves $y = f(x)$ or $x = \phi(y)$ exist in which the surface $u = F(x, y)$ intersects the xy-plane. (How *far* such a curve of intersection may extend does not concern us here.)

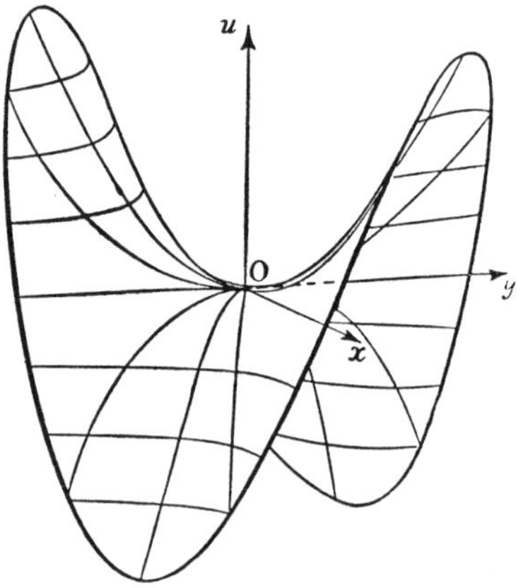

Fig. 1.—The surface $u = xy$

A first possibility is that the surface and the plane may have no point in common. For example, the paraboloid $u = F(x, y) = x^2 + y^2 + 1$ lies entirely above the xy-plane. In such a case there is obviously no curve of intersection. We therefore need only consider cases in which there is a point (x_0, y_0) at which $F(x_0, y_0) = 0$; the values x_0, y_0 are called an " initial solution ".

If an initial solution exists, two possibilities remain. Either the tangent plane at the point (x_0, y_0) is horizontal or it is not. If it is, we can readily show by means of examples that the solution $y = f(x)$ or $x = \phi(y)$ may fail to exist. For example, the paraboloid $u = x^2 + y^2$ has the initial solution $x = 0$, $y = 0$, but has no other point in the xy-plane. Again, the surface $u = xy$ has the initial solution $x = 0$, $y = 0$, and in fact

* Cf. also Vol. I, Chap. X, section 5 (pp. 481-5).

intersects the xy-plane along the lines $x=0$ and $y=0$ (cf. figs. 1, 2). But in no neighbourhood of the origin can we represent the *whole* intersection by a function $y=f(x)$ or by a function $x=\phi(y)$. On the other hand, it is quite possible for the equation $F(x, y)=0$ to have a solution, even when the tangent plane at the initial solution is horizontal, as, for example, in the case $(y-x)^4=0$. In the (exceptional) case of a horizontal tangent plane, therefore, no definite general statement can be made.

The remaining possibility is that at the initial solution the tangent plane is not horizontal. Then intuition tells us, roughly speaking, that the surface $u=F(x, y)$ cannot bend fast enough

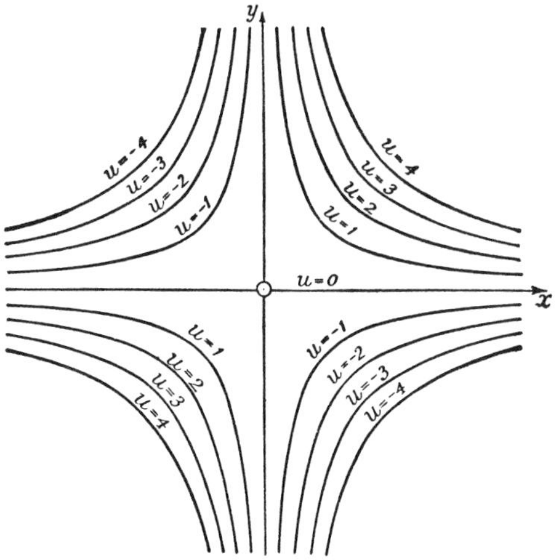

Fig. 2.—Contour lines of $u = xy$

to avoid cutting the xy-plane near (x_0, y_0) in a single well-defined curve of intersection, and that a portion of the curve near the initial solution can be represented by the equation $y=f(x)$ or $x=\phi(y)$. The statement that the tangent plane is not horizontal is the same as the statement that $F_x(x_0, y_0)$ and $F_y(x_0, y_0)$ are not both zero. This is the case which we shall discuss analytically in the next sub-section.

3. The Theorem of Implicit Functions.

The general theorem which states sufficient conditions for the existence of implicit functions and at the same time gives a rule for differentiating them is as follows:

If $F(x, y)$ *has continuous derivatives* F_x *and* F_y, *and if at the point* (x_0, y_0) *within its region of definition the equation* $F(x_0, y_0) = 0$ *is satisfied, while* $F_y(x_0, y_0)$ *is not zero, then we can mark off about the point* (x_0, y_0) *a rectangle* $x_1 \leqq x \leqq x_2$, $y_1 \leqq y \leqq y_2$ *such that for every* x *in the interval* $x_1 \leqq x \leqq x_2$ *the equation* $F(x, y) = 0$ *determines exactly one value* $y = f(x)$ *lying in the interval* $y_1 \leqq y \leqq y_2$. *This function satisfies the equation* $y_0 = f(x_0)$, *and for every* x *in the interval the equation*

$$F(x, f(x)) = 0$$

is satisfied. The function $f(x)$ *is continuous and differentiable, and its derivative and differential are given by the equations*

$$y' = f'(x) = -\frac{F_x}{F_y} \quad \text{and} \quad dy = df(x) = -\frac{F_x}{F_y}\,dx$$

respectively.

We shall assume for the present that the first part of the theorem, relating to the existence and continuity of the implicitly-defined function, is already proved, and shall confine ourselves to proving the differentiability of the function and the differentiation formulæ; the proof of the existence and continuity of the solution we shall postpone to sub-section 6 (p. 119).

If we could differentiate the terms of the equation $F(x, f(x)) = 0$ by the chain rule, the above equation would follow at once.* Since, however, the differentiability of $f(x)$ must first be proved, we must consider the matter in somewhat greater detail.

As the derivatives F_x and F_y have been assumed continuous, the function $F(x, y)$ is differentiable. We can therefore write

$$F(x+h, y+k) = F(x, y) + hF_x(x, y) + kF_y(x, y) + \epsilon_1 h + \epsilon_2 k,$$

where ϵ_1 and ϵ_2 are two quantities which tend to zero as h and k do or as $\rho = +\sqrt{(h^2 + k^2)}$ does. We now confine our attention to pairs of values (x, y) and $(x + h, y + k)$ for which both x and $x + h$ lie in the interval $x_1 \leqq x \leqq x_2$ and for which $y = f(x)$ and $y + k = f(x + h)$. For such pairs of values we have $F(x, y) = 0$ and $F(x + h, y + k) = 0$, so that the preceding equation reduces to

$$0 = hF_x + kF_y + \epsilon_1 h + \epsilon_2 k.$$

We assume that $f(x)$ has been proved continuous. Hence as h

* Cf. Vol. I, p. 483.

tends to 0, so does k, and with them ϵ_1 and ϵ_2 also tend to 0. If we divide by hF_y (which by hypothesis is not zero), the last equation gives

$$\left(1+\frac{\epsilon_2}{F_y}\right)\frac{k}{h}+\frac{F_x}{F_y}+\frac{\epsilon_1}{F_y}=0,$$

and on performing the passage to the limit $h \to 0$ we have

$$\lim_{h\to 0}\frac{k}{h}+\frac{F_x}{F_y}=0.$$

But

$$\frac{k}{h}=\frac{f(x+h)-f(x)}{h};$$

this proves the differentiability of $f(x)$ and gives the required rule for differentiation,

$$y'=\lim_{h\to 0}\frac{f(x+h)-f(x)}{h}=\lim_{h\to 0}\frac{k}{h}=-\frac{F_x}{F_y}.$$

We can also write this rule in the form

$$F_x+F_y y'=0$$

or

$$dF=F_x dx+F_y dy=0.$$

This last equation states that in virtue of the equation $F(x, y) = 0$ the differentials dx and dy cannot be chosen independently of one another.

An implicit function can usually be differentiated more easily by using this rule than by first writing down the explicit form of the function. The rule can be used whenever the explicit representation of the function is theoretically possible according to the theorem of implicit functions, even in cases where the practical solution in terms of the ordinary functions (rational functions, trigonometric functions, &c.) is extremely complicated or impossible.

Suppose that the second order partial derivatives of $F(x, y)$ exist and are continuous. In the equation $y' = -\frac{F_x}{F_y}$, whose right-hand side is a compound function of x, we can differentiate according to the chain rule and then substitute for y' its value $-\frac{F_x}{F_y}$. This gives

$$y'' = -\frac{F_{xx}F_y^2 - 2F_{xy}F_xF_y + F_{yy}F_x^2}{F_y^3}$$

as the formula for the second derivative of $y = f(x)$.

In the same way we can obtain the higher derivatives of $f(x)$ by repeated differentiation.

4. Examples.

1. For the function $y = f(x)$ obtained from the equation of the circle

$$F(x, y) = x^2 + y^2 - 1 = 0$$

we obtain the derivative

$$y' = -\frac{F_x}{F_y} = -\frac{x}{y}.$$

This can easily be verified directly. If we solve for y, the equation of the circle gives either the function $y = \sqrt{(1 - x^2)}$ or the function $y = -\sqrt{(1 - x^2)}$, representing the upper and lower semicircles respectively. In the first case differentiation gives

$$y' = -\frac{x}{\sqrt{(1 - x^2)}},$$

and in the second case

$$y' = \frac{x}{\sqrt{(1 - x^2)}}.$$

Thus in both cases $y' = -\frac{x}{y}$.

2. In the case of the *lemniscate* (Vol. I, p. 72)

$$F(x, y) = (x^2 + y^2)^2 - 2a^2(x^2 - y^2) = 0$$

it is not easy to solve for y. For $x = 0$, $y = 0$ we obtain $F = 0$, $F_x = 0$, $F_y = 0$. Here our theorem fails, as might be expected from the fact that two different branches of the lemniscate pass through the origin. For all points of the curve for which $y \neq 0$, however, our rule applies, and the derivative of the function $y = f(x)$ is given by

$$y' = -\frac{F_x}{F_y} = -\frac{4x(x^2 + y^2) - 4a^2x}{4y(x^2 + y^2) + 4a^2y}.$$

We can obtain important information about the curve from this equation, without bringing in the explicit expression for y. For example, maxima or minima may occur where $y' = 0$, that is, for $x = 0$ or for $x^2 + y^2 = a^2$. From the equation of the lemniscate, $y = 0$ when $x = 0$; but at the origin there is no extreme value (cf. fig. 26, Vol. I, p. 72). The two equations therefore give the four points $\left(\pm\frac{a}{2}\sqrt{3}, \pm\frac{a}{2}\right)$ as the maxima and minima.

3. In the case of the *folium of Descartes*

$$F(x, y) = x^3 + y^3 - 3axy = 0$$

(cf. fig. 3), the explicit solution would be exceedingly inconvenient. At the

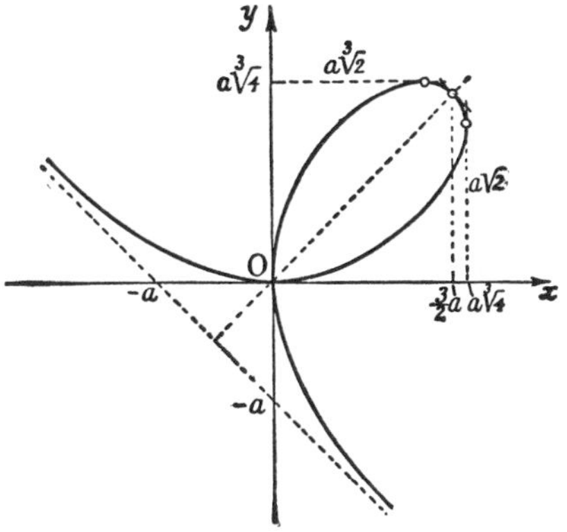

Fig. 3.—Folium of Descartes

origin, where the curve intersects itself, our rule again fails, since at that point $F = F_x = F_y = 0$. For all points at which $y^2 \neq ax$ we have

$$y' = -\frac{F_x}{F_y} = -\frac{x^2 - ay}{y^2 - ax}.$$

Accordingly, there is a zero of the derivative when $x^2 - ay = 0$, or, if we use the equation of the curve, when

$$x = a\sqrt[3]{2}, \quad y = a\sqrt[3]{4}.$$

5. The Theorem of Implicit Functions for more than Two Independent Variables.

The general theorem of implicit functions can be extended to the case of several independent variables as follows:

Let $F(x, y, \ldots, z, u)$ *be a continuous function of the independent variables* $x, y, \ldots, z, u$, *and let it possess continuous partial derivatives* $F_x, F_y, \ldots, F_z, F_u$. *For the system of values* $x_0, y_0, \ldots, z_0, u_0$ *corresponding to an interior point of the region of definition of* F, *let* $F(x_0, y_0, \ldots, z_0, u_0) = 0$ *and*

$$F_u(x_0, y_0, \ldots, z_0, u_0) \neq 0.$$

Then we can mark off an interval $u_1 \leqq u \leqq u_2$ *about* u_0 *and a region* R *containing* $(x_0, y_0, \ldots, z_0)$ *in its interior such that for every* (x, y, . . . , z) *in* R *the equation* F(x, y, . . . , z, u) = 0 *is satisfied by exactly one value of* u *in the interval* $u_1 \leqq u \leqq u_2$. *For this value of* u, *which we denote by* u = f(x, y, . . . , z), *the equation*

$$F(x, y, \ldots, z, f(x, y, \ldots, z)) = 0$$

holds identically in R; *in addition,*

$$u_0 = f(x_0, y_0, \ldots, z_0).$$

The function f *is a continuous function of the independent variables* x, y, . . . , z, *and possesses continuous partial derivatives given by the equations*

$$F_x + F_u f_x = 0,$$
$$F_y + F_u f_y = 0,$$
$$\cdot \quad \cdot \quad \cdot \quad \cdot \quad \cdot \quad \cdot \quad \cdot$$
$$F_z + F_u f_z = 0.$$

For the proof of the existence and continuity of $f(x, y, \ldots, z)$ we refer the reader to the next sub-section (p. 121). The formulæ of differentiation follow from those for the case of one independent variable, since we can e.g. let $y, \ldots, z$ remain constant and thus find the formula for f_x.

If we wish, we can combine our differentiation formulæ in the single equation

$$F_x dx + F_y dy + \ldots + F_z dz + F_u du = 0.$$

In words:

If in a function F(x, y, . . . , z, u) *the variables are not independent of one another, but are subject to the condition* F = 0, *then the linear parts of the increments of these variables are likewise not independent of one another, but are connected by the condition* dF = 0, *that is, by the linear equation*

$$F_x dx + F_y dy + \ldots + F_z dz + F_u du = 0.$$

If we here replace du by the expression $u_x dx + u_y dy + \ldots + u_z dz$ and then equate the coefficient of each of the mutually independent differentials $dx, dy, \ldots, dz$ to zero, we again obtain the above differentiation formulæ.

Incidentally, the concept of implicit functions enables us to give a general definition of the concept of an *algebraic function.*

We say that $u = f(x, y, \ldots)$ is an algebraic function of the independent variables $x, y, \ldots$ if u can be defined implicitly by an equation $F(x, y, \ldots, u) = 0$, where F is a polynomial in the arguments $x, y, \ldots, u$; briefly, if u "satisfies an algebraic equation". All functions which do not satisfy an algebraic equation are called *transcendental*.

As an example of our differentiation formulæ we consider the equation of the sphere,

$$x^2 + y^2 + u^2 - 1 = 0.$$

For the partial derivatives we obtain

$$u_x = -\frac{x}{u}, \quad u_y = -\frac{y}{u},$$

and by further differentiation

$$u_{xx} = -\frac{1}{u} + \frac{x}{u^2} u_x = -\frac{x^2 + u^2}{u^3},$$

$$u_{xy} = \frac{x}{u^2} u_y = -\frac{xy}{u^3},$$

$$u_{yy} = -\frac{1}{u} + \frac{y}{u^2} u_y = -\frac{y^2 + u^2}{u^3}.$$

6. Proof of the Existence and Continuity of the Implicit Functions.

Although in many special cases the existence and continuity of implicit functions follows from the fact that the equation $F(x, y) = 0$ can actually be solved in terms of the usual functions by means of some special device, yet it is still necessary to give a general analytical proof of the existence theorem stated above.

As a first step we mark out a rectangle $x_1 \leqq x \leqq x_2$, $y_1 \leqq y \leqq y_2$ in which the equation $F(x, y) = 0$ determines a unique function $y = f(x)$. We shall make no attempt to find the *largest* rectangle of this type; we only wish to show that such a rectangle *exists*.

Since $F_y(x, y)$ is continuous and $F_y(x_0, y_0) \neq 0$, we can find a rectangle R, with the point $P(x_0, y_0)$ as centre, so small that in the whole of R the function F_y remains different from zero and thus is always of the same sign. Without loss of generality we can assume that this sign is positive, so that F_y is positive everywhere in R; otherwise, we should merely have to replace the function F by $-F$, which leaves the equation $F(x, y) = 0$ unaltered. Since $F_y > 0$ on every line-segment $x = \text{const.}$ parallel to the y-axis and lying in R, the function $F(x, y)$, considered as a function of y alone, is monotonic increasing. But $F(x_0, y_0) = 0$; hence if A is a point of R with co-ordinates x_0 and y_1 $(y_1 < y_0)$ on the vertical line through P (cf. fig. 4), the value of the function at A, $F(x_0, y_1)$, is negative, while at the point B with co-ordinates x_0 and y_2 $(y_2 > y_0)$ the value of the function, $F(x_0, y_2)$, is positive. Owing to the continuity of $F(x, y)$, it follows that $F(x, y)$ has negative values along a certain horizontal line-segment $y = y_1$ through A and lying in R, and has positive values along a line-segment $y = y_2$ through B and lying in R. We can therefore mark off an interval $x_1 \leqq x \leqq x_2$ about x_0 so small that for values of x in that interval the function $F(x, y)$ remains negative along the horizontal through A and positive along the horizontal through B. In other words, for $x_1 \leqq x \leqq x_2$ the inequalities $F(x, y_1) < 0$ and $F(x, y_2) > 0$ hold.

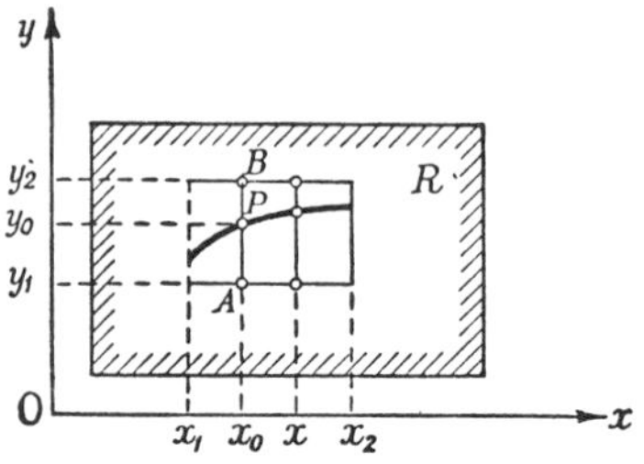

Fig. 4

We now suppose that x is fixed at any value in the interval $x_1 \leqq x \leqq x_2$, and let y increase from y_1 to y_2. The point (x, y) then remains in the rectangle

$$x_1 \leqq x \leqq x_2, \quad y_1 \leqq y \leqq y_2,$$

which we assume to be completely within R. Since $F_y(x, y) > 0$, the value of the function $F(x, y)$ increases monotonically and continuously from a negative to a positive value, and can never have the same value for two points with the same abscissa. Hence for each value of x in the interval $x_1 \leqq x \leqq x_2$ there is a *uniquely determined* * value of y for which the equation $F(x, y) = 0$ is satisfied. This value of y is thus a function of x; we have accordingly proved the existence and the uniqueness of the solution of the equation $F(x, y) = 0$. At the same time the part played by the condition $F_y \neq 0$ has been clearly brought out.

* If the restriction $y_1 \leqq y \leqq y_2$ is omitted, this will not necessarily remain true. For example, let F be $x^2 + y^2 - 1$ and let $x_0 = 0$, $y_0 = 1$. Then for $-\frac{1}{2} \leqq x \leqq \frac{1}{2}$ there is just one solution, $y = f(x)$, in the interval $0 \leqq y \leqq 2$; but if y is unrestricted, there are two solutions, $y = \sqrt{(1 - x^2)}$ and $y = -\sqrt{(1 - x^2)}$.

If this condition were not fulfilled, the values of the function at A and at B might not have opposite signs, so that $F(x, y)$ need not pass through zero on vertical line-segments. Or, if the signs at A and at B were different, the derivative F_y could change sign, so that for a fixed value of x the function $F(x, y)$ would not increase monotonically with y and might assume the value zero more than once, thus destroying the uniqueness of the solution.

This proof merely tells us that the function $y = f(x)$ exists. It is a typical case of a pure " existence theorem ", in which the practical possibility of calculating the solution does not come under consideration at all.*

The *continuity* of the function $f(x)$ follows almost at once from the above considerations. Let $R(x_1' \leqq x \leqq x_2',\ y_1' \leqq y \leqq y_2')$ be a rectangle lying entirely within the rectangle $x_1 \leqq x \leqq x_2$, $y_1 \leqq y \leqq y_2$ found above. For this smaller rectangle we can carry out exactly the same process as before in order to obtain a solution $y = f(x)$ of the equation $F(x, y) = 0$. In the larger rectangle, however, this solution was uniquely determined; hence the newly-found function $f(x)$ is the same as the old one. If we now wish e.g. to prove the continuity of the function $f(x)$ at the point $x = x_0$, we must show that for any small positive number ϵ $|f(x) - f(x_0)| < \epsilon$, provided only that x lies sufficiently near the point x_0. For this purpose we put

$$y_1' = y_0 + \epsilon \quad \text{and} \quad y_2' = y_0 - \epsilon,$$

and for these values y_1' and y_2' we determine the corresponding x-interval $x_1' \leqq x \leqq x_2'$. Then by the above construction, for each x in this interval the corresponding $f(x)$ lies between the bounds y_1' and y_2', and therefore differs from y_0 by less than ϵ. This expresses the continuity of $f(x)$ at the point x_0. Since we can apply the above argument to any point x in the interval $x_1 \leqq x \leqq x_2$, we have proved that the function is continuous at each point of this interval.

The proof of the general theorem for $F(x, y, \ldots, z, u)$, a function with a greater number of independent variables, follows exactly the same lines as the proof just completed, and offers no further difficulties.

* The sacrifice of the statement of such practical methods in a general proof is sometimes an essential step towards the simplification of proofs.

EXAMPLES

1. Prove that the following equations have unique solutions for y near the points indicated:

(a) $x^2 + xy + y^2 = 7$ (2, 1).

(b) $x \cos xy = 0$ (1, $\pi/2$).

(c) $xy + \log xy = 1$ (1, 1).

(d) $x^5 + y^5 + xy = 3$ (1, 1).

2. Find the first derivatives of the solutions in Ex. 1.

3. Find the second derivatives of the solutions in Ex. 1.

4. Find the maximum and minimum values of the function $y = f(x)$ defined by the equation $x^2 + xy + y^2 = 27$.

5. Show that the equation $x + y + z = \sin xyz$ can be solved for z near (0, 0, 0). Find the partial derivatives of the solution.

2. Curves and Surfaces in Implicit Form

1. Plane Curves in Implicit Form.

We have previously expressed plane curves in the form $y = f(x)$, which is unsymmetrical, giving the preference to one of the co-ordinates. The tangent and the normal to the curve are found to be given by the equations

$$(\eta - y) - (\xi - x)f'(x) = 0$$

and

$$(\eta - y)f'(x) + (\xi - x) = 0$$

respectively, where ξ and η are the current co-ordinates of the tangent and the normal, and x and y are the co-ordinates of the point of the curve. We have also found an expression for the curvature, and criteria for points of inflection (Vol. I, Chap. V). We shall now obtain the corresponding formulæ for curves which are represented implicitly by equations of the type $F(x, y) = 0$. We do this under the assumption that at the point in question F_x and F_y are not both zero, so that $F_x^2 + F_y^2 \neq 0$.

If we suppose that $F_y \neq 0$, say, we can substitute for y' in the equation of the tangent at the point (x, y) of the curve its value $-F_x/F_y$, and at once obtain the equation of the tangent in the form

$$(\xi - x)F_x + (\eta - y)F_y = 0.$$

Similarly, for the normal we have

$$(\xi - x)F_y - (\eta - y)F_x = 0.$$

Without going out of our way to use the explicit form of the equation of the curve, we can also obtain the equation of the tangent directly in the following way. If a and b are any two constants, the equation

$$a(\xi - x) + b(\eta - y) = 0$$

with current co-ordinates ξ and η represents a straight line passing through the point $P(x, y)$. If now P is any point of the curve, i.e. if $F(x, y) = 0$, we wish to find the line through P with the property that if P_1 is a point of the curve with co-ordinates $x_1 = x + h$ and $y_1 = y + k$, the distance from the line to P_1 tends to zero to a higher order than $\rho = \sqrt{(h^2 + k^2)}$. In virtue of the differentiability of the function F we can write

$$F(x + h, y + k) = F(x, y) + hF_x + kF_y + \epsilon\rho,$$

where ρ tends to 0 as ϵ does. Since the two points P and P_1 both lie on the curve, this equation reduces to $hF_x + kF_y = -\epsilon\rho$. As we have assumed that $F_x^2 + F_y^2 \neq 0$, we can write this last in the form

$$h\frac{F_x}{\sqrt{(F_x^2 + F_y^2)}} + k\frac{F_y}{\sqrt{(F_x^2 + F_y^2)}} = \epsilon_1\rho,$$

where $\epsilon_1 = -\dfrac{\epsilon}{\sqrt{(F_x^2 + F_y^2)}}$ also tends to zero as ρ does. If we write $a = \dfrac{F_x}{\sqrt{(F_x^2 + F_y^2)}}$ and $b = \dfrac{F_y}{\sqrt{(F_x^2 + F_y^2)}}$, the left-hand side of this equation may be regarded as the expression obtained when we substitute the co-ordinates of the point $(x_1 = x + h, y_1 = y + k)$ for ξ and η in the canonical form of the equation of the line, $a(\xi - x) + b(\eta - y) = 0$. This is the distance of the point P_1 from the line. Thus the distance of P_1 from the line is numerically equal to $|\epsilon_1\rho|$, which vanishes as ρ does to a higher order than ρ. The equation

$$\frac{F_x}{\sqrt{(F_x^2 + F_y^2)}}(\xi - x) + \frac{F_y}{\sqrt{(F_x^2 + F_y^2)}}(\eta - y) = 0$$

or

$$F_x(\xi - x) + F_y(\eta - y) = 0$$

is the same as the equation of the tangent found in the preceding paragraph. We can therefore regard the tangent at P as that line * whose distance from neighbouring points P_1 of the curve vanishes to a higher order than the distance PP_1.

The *direction cosines of the normal* to the curve are given by the two equations

$$\cos \alpha = \frac{F_x}{\sqrt{(F_x^2 + F_y^2)}}, \quad \sin \alpha = \frac{F_y}{\sqrt{(F_x^2 + F_y^2)}},$$

which represent the components of a unit vector in the direction of the normal; that is, of a vector with length 1 in the direction of the normal at the point $P(x, y)$ of the curve.

The *direction cosines of the tangent* at the point $P(x, y)$ are given by

$$\cos \beta = \frac{F_y}{\sqrt{(F_x^2 + F_y^2)}}; \quad \sin \beta = -\frac{F_x}{\sqrt{(F_x^2 + F_y^2)}}.$$

More generally, if instead of the curve $F(x, y) = 0$ we consider the curve

$$F(x, y) = c,$$

where c is any constant, everything in the above discussion remains unchanged. We have only to replace the function $F(x, y)$ by $F(x, y) - c$, which has the same derivatives as the original function. Thus for these curves the equation of the tangent and the normal have exactly the same forms as above.

The class of all the curves which we obtain when we allow c to range through all the values in an interval is called a *family of curves*. The plane vector with components F_x and F_y, which is the *gradient* of the function $F(x, y)$, is at each point of the plane *perpendicular to the curve of the family passing through that point*, as we have already seen on p. 90. This again yields the equation of the tangent. For the vector with components $(\xi - x)$ and $(\eta - y)$ in the direction of the tangent must be perpendicular to the gradient, so that the scalar product

$$(\xi - x)F_x + (\eta - y)F_y$$

must vanish.

* The reader will find it easy to prove for himself that two such lines cannot exist, so that our condition determines the tangent uniquely.

While we have taken the positive sign for the square root occurring in the above formulæ, we could equally well have taken the negative root. This arbitrariness corresponds to the fact that we can call the direction towards either side of the curve the positive direction at will. We shall continue to choose the positive square root and thereby fix a definite direction of the normal. It is, however, to be observed that if we replace the function $F(x, y)$ by $-F(x, y)$ this direction is reversed, although the geometrical nature of the curve is unaffected. (As regards the sign of the normal, cf. Chap. V, section 2 (pp. 363-4)).

We have already seen (Vol. I, p. 159) that for a curve explicitly represented in the form $y = f(x)$ the condition $f''(x) = 0$ is a *necessary* condition for the occurrence of a *point of inflection*. If we replace this expression by its equivalent,

$$f''(x) = -\frac{F_{xx}F_y^2 - 2F_{xy}F_xF_y + F_{yy}F_x^2}{F_y^3},$$

we obtain the equation

$$F_{xx}F_y^2 - 2F_{xy}F_xF_y + F_{yy}F_x^2 = 0$$

as a necessary condition for the occurrence of a point of inflection. In this condition there is no longer any preference given to either of the two variables x, y. It has a completely symmetrical character and no longer depends on the assumption that $F_y \neq 0$.

If we substitute for y' and y'' in the formula for the *curvature* found previously (Vol. I, p. 281)

$$k = \frac{y''}{\sqrt{(1 + y'^2)^3}},$$

we obtain the formula

$$k = \frac{F_{xx}F_y^2 - 2F_{xy}F_xF_y + F_{yy}F_x^2}{(F_x^2 + F_y^2)^{3/2}},$$

which is likewise perfectly symmetrical.* For the co-ordinates (ξ, η) of the centre of curvature we obtain the expressions

$$\xi = x + \rho \frac{F_x}{\sqrt{(F_x^2 + F_y^2)}},$$

* For the sign of the curvature cf. Vol. I, p. 282.

$$\eta = y + \rho \frac{F_y}{\sqrt{(F_x^2 + F_y^2)}},$$

where

$$\rho = \frac{1}{k}.$$

If the two curves $F(x, y) = 0$ and $G(x, y) = 0$ intersect one another at the point with co-ordinates x, y, the *angle between the curves* is defined as the angle ω formed by their tangents (or normals) at the point of intersection. If we recall the expressions given above for the direction cosines of the normals and the formula for the scalar product (Chap. I, section 1, p. 8), we obtain the expression

$$\cos\omega = \frac{F_x G_x + F_y G_y}{\sqrt{(F_x^2 + F_y^2)}\sqrt{(G_x^2 + G_y^2)}}$$

for the cosine of this angle. Since we have taken the positive square roots here, the cosine is uniquely determined; this corresponds to the fact that we have thereby chosen definite directions for the normals and have thus determined the angle between them uniquely.

By putting $\omega = \pi/2$ in the last formula we obtain the condition for *orthogonality*, i.e. that the curves intersect at right angles,

$$F_x G_x + F_y G_y = 0.$$

If the curves are to *touch* one another, the ratio of the differentials, $dy : dx$, must be the same for the two curves. That is, the condition

$$dy : dx = -F_x : F_y = -G_x : G_y$$

must be fulfilled. This may also be written in the form

$$F_x G_y - F_y G_x = 0.$$

As an example we consider the parabolas

$$y^2 - 2p\left(x + \frac{p}{2}\right) = 0$$

(cf. fig. 9, p. 137), all of which have the origin as focus ("confocal" parabolas). If $p_1 > 0$ and $p_2 < 0$, the two parabolas

$$F = y^2 - 2p_1\left(x + \frac{p_1}{2}\right) = 0 \quad \text{and} \quad G = y^2 - 2p_2\left(x + \frac{p_2}{2}\right) = 0$$

intersect one another, and at the intersection they are at right angles to one another, for

$$F_x G_x + F_y G_y = 4p_1 p_2 + 4y^2 = 4\frac{p_2 F - p_1 G}{p_2 - p_1} = 0,$$

since

$$F = G = 0, \quad p_2 - p_1 \neq 0.$$

As a second example we consider the ellipse

$$\frac{x^2}{a^2} + \frac{y^2}{b^2} = 1.$$

The equation of the tangent at the point (x, y) is

$$(\xi - x)\frac{x}{a^2} + (\eta - y)\frac{y}{b^2} = 0$$

or

$$\xi \frac{x}{a^2} + \eta \frac{y}{b^2} - 1 = 0,$$

as we know from analytical geometry.

We find that the curvature is

$$k = \frac{a^4 b^4}{(a^4 y^2 + b^4 x^2)^{3/2}}.$$

If $a > b$, this has its greatest value a/b^2 at the vertices $y = 0$, $x = \pm a$. Its least value b/a^2 occurs at the other vertices $x = 0$, $y = \pm b$.

2. Singular Points of Curves.

We now add a few remarks on the *singular points of a curve*. Here we shall content ourselves with giving a number of typical examples; for a more thorough investigation we refer the reader to the appendix to this chapter (p. 209).

In the formulæ obtained above the expression $F_x^2 + F_y^2$ frequently occurs in the denominator. Accordingly we may expect something unusual to happen when this quantity vanishes, i.e. when $F_x = 0$ and $F_y = 0$ at a point of the curve. This is especially brought out by the fact that at such a point the expression $y' = -F_x/F_y$ for the slope of the tangent to the curve loses its meaning.

We say that a point of a curve is a *regular point* if in the neighbourhood of this point either the co-ordinate y can be represented as a continuously differentiable function of x, or else x can be represented as a continuously differentiable function of y. In either case the curve has a tangent, and in the neighbourhood of the point in question the curve differs but little from that tangent. All other points of a curve are called *singular points* (or *singularities*).

From the theory of implicit functions we know that a point of the curve $F(x, y) = 0$ is regular if at that point $F_y \neq 0$, since we can then solve the equation so as to obtain a unique differentiable solution $y = f(x)$. Similarly, the point is regular if $F_x \neq 0$. The singular points of the curve are accordingly to be sought for among those points of the curve at which the equations

$$F_x = 0, \quad F_y = 0$$

are satisfied in addition to the equation of the curve.

An important type of singularity is a *multiple point*, that is, a point through which two or more branches of the curve pass. For example, the origin is a multiple point of the lemniscate

$$(x^2 + y^2)^2 - 2a^2(x^2 - y^2) = 0.$$

In the neighbourhood of such a point it is impossible to express the equation of the curve uniquely in the form $y = f(x)$ or $x = \phi(y)$.

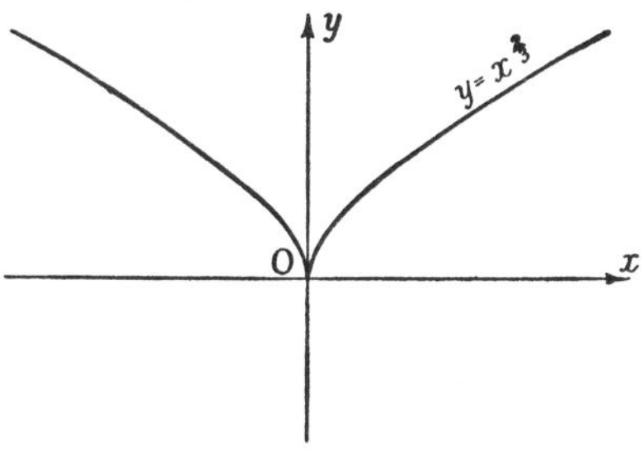

Fig. 5.—The surface $y^3 - x^2 = 0$

The truth of the relations $F_x = 0$ and $F_y = 0$ is a necessary, but by no means a sufficient, condition for a multiple point; on the contrary, quite a different type of singularity may occur, such as a *cusp*.

As an example we consider the curve

$$y^3 - x^2 = 0$$

(cf. fig. 5), which has a cusp at the origin. At that point both the first partial derivatives of F vanish.

Moreover, cases may occur in which F_x and F_y both vanish, and yet there is no striking peculiarity of the curve at the point, the curve being regular there.

This is exemplified by the curve

$$y^3 - x^4 = 0$$

or, in explicit form,

$$y = x^{4/3}.$$

From the equations $(-x)^{4/3} = x^{4/3}$, $y' = \frac{4}{3}x^{1/3}$ we see at once that the curve is symmetrical with respect to the y-axis and touches the x-axis at the origin, like a parabola. Yet the origin is a somewhat special point on the curve, since the second derivative is infinite there. The curvature is therefore infinite, while the direction of the tangent exhibits no peculiarity. Another example is the curve $(y - x)^2 = 0$, which is a straight line and therefore regular throughout, even though $F_x = 0$ and $F_y = 0$ for every point of the line.

As a result of this discussion we see that in the investigation and discussion of singular points of a curve it is not enough to verify that the two equations $F_x = 0$ and $F_y = 0$ are satisfied; on the contrary, each case must be studied specially (cf. Appendix, section 2, p. 209).

3. Implicit Representation of Surfaces.

Hitherto we have usually represented a function $z = f(x, y)$ (here we write z instead of the symbol u employed above) by means of a surface in xyz-space. If, however, we are originally given not the function, but a surface in space, the preference which this form of expression gives to the co-ordinate z may prove inconvenient, just as in the case of the expression of plane curves in the form $y = f(x)$. It is more natural and more general to represent surfaces in space by equations of the form $F(x, y, z) = 0$ or $F(x, y, z) = \text{const.}$, e.g. to represent the sphere by the equation $x^2 + y^2 + z^2 - r^2 = 0$, and not by $z = \pm\sqrt{(r^2 - x^2 - y^2)}$. The form $z - f(x, y) = 0$ can then be treated as a special case.

In order to establish the equation of the tangent plane to the surface $F(x, y, z) = 0$ at the point (x, y, z), we first make the assumption * that at that point $F_x^2 + F_y^2 + F_z^2 \neq 0$; i.e. that at least one of the partial derivatives, say F_z, is not zero. Then from the equation of the surface we can determine $z = f(x, y)$ explicitly as a function of x and y. If in the equation of the tangent plane

$$\zeta - z = (\xi - x)z_x + (\eta - y)z_y$$

we substitute for the derivatives z_x and z_y their values

* The vanishing of this expression indicates the possibility that certain singularities may occur; this, however, we shall not discuss.

$z_x = -F_x/F_z$ and $z_y = -F_y/F_z$, we obtain the equation of the tangent plane in the form

$$(\xi - x)F_x + (\eta - y)F_y + (\zeta - z)F_z = 0,$$

where ξ, η, ζ are current co-ordinates.

As in the case of the tangent to a plane curve, we can derive this equation directly from the implicit representation of the surface, by setting ourselves the problem of finding a plane through the point (x, y, z) of the surface with the property that the distance from the plane to the point $(x + h, y + k, z + l)$ of the surface vanishes as $\rho = \sqrt{(h^2 + k^2 + l^2)}$ does, to a higher order than ρ.

Elementary theorems of analytical geometry (cf. Chap. I, section 1, p. 9) show that the *direction cosines of the normal to the surface*, that is, of the normal to the tangent plane, are given by the expressions

$$\cos\alpha = \frac{F_x}{\sqrt{(F_x^2 + F_y^2 + F_z^2)}}, \quad \cos\beta = \frac{F_y}{\sqrt{(F_x^2 + F_y^2 + F_z^2)}},$$

$$\cos\gamma = \frac{F_z}{\sqrt{(F_x^2 + F_y^2 + F_z^2)}}.$$

In taking the positive square root in the denominator we have assigned a definite sense of direction to the normal (cf. p. 125).

If two surfaces $F(x, y, z) = 0$ and $G(x, y, z) = 0$ intersect one another at a point, the *angle ω between the surfaces* is defined as the angle between their tangent planes, or, what is the same thing, the angle between their normals. This is given by

$$\cos\omega = \frac{F_xG_x + F_yG_y + F_zG_z}{\sqrt{(F_x^2 + F_y^2 + F_z^2)}\sqrt{(G_x^2 + G_y^2 + G_z^2)}}.$$

In particular, the condition for perpendicularity (orthogonality) is

$$F_xG_x + F_yG_y + F_zG_z = 0.$$

Instead of the single surface $F(x, y, z) = 0$ we may consider the whole *family of surfaces* $F(x, y, z) = c$, where c is a constant different for each surface of the family. Here we assume that through each point of space, or at least through every point of a certain region of space, there passes one and only one surface of the family; or, as we say, that the family *covers* the region *simply*. The individual surfaces are then called the *level surfaces* of the function $F(x, y, z)$. In Chap. II, section 7 (p. 88) we considered the gradient of this function, that is, the vector with the components F_x, F_y, F_z. We see that these components have the same ratios as the direction cosines of the normal; hence we conclude that the *gradient* at the point with the co-ordinates (x, y, z) *is perpendicular to the level surface passing through that point.* (If we accept this fact as already proved in Chap. II, section 7 (p. 90), we at once have a new and simple method for deriving the equation of the tangent plane, just like that given above (p. 124) for the equation of the tangent line.)

As an example we consider the *sphere*

$$x^2 + y^2 + z^2 - r^2 = 0.$$

At the point (x, y, z) the tangent plane is

$$(\xi - x)2x + (\eta - y)2y + (\zeta - z)2z = 0$$

$$\xi x + \eta y + \zeta z - r^2 = 0.$$

The direction cosines of the normal are proportional to x, y, z; that is, the normal coincides with the radius vector drawn from the origin to the point (x, y, z).

For the most general *ellipsoid* with the co-ordinate axes as principal axes,

$$\frac{x^2}{a^2} + \frac{y^2}{b^2} + \frac{z^2}{c^2} = 1,$$

the equation of the tangent plane is

$$\xi \frac{x}{a^2} + \eta \frac{y}{b^2} + \zeta \frac{z}{c^2} - 1 = 0.$$

Examples

1. Find the tangent plane

(*a*) of the surface

$$x^3 + 2xy^2 - 7z^3 + 3y + 1 = 0$$

at the point (1, 1, 1);

(*b*) of the surface

$$(x^2 + y^2)^2 + x^2 - y^2 + 7xy + 3x + z^4 - z = 14$$

at the point (1, 1, 1);

(c) of the surface

$$\sin^2 x + \cos(y + z) = \tfrac{3}{4}$$

at the point $\left(\frac{\pi}{6}, \frac{\pi}{3}, 0\right)$.

2. Calculate the curvature of the curve

$$\sin x + \cos y = 1$$

at the origin.

3*. Find the curvature at the origin of each of the two branches of the curve

$$y(ax + by) = cx^3 + ex^2y + fxy^2 + gy^3.$$

4. Find the curvature of a curve which is given in polar co-ordinates by the equation $f(r, \theta) = 0$.

5. Prove that the three surfaces of the family of surfaces

$$\frac{xy}{z} = u, \quad \sqrt{(x^2 + z^2)} + \sqrt{(y^2 + z^2)} = v, \quad \sqrt{(x^2 + z^2)} - \sqrt{(y^2 + z^2)} = w$$

which pass through a single point are orthogonal to one another.

6. The points A and B move uniformly with the same velocity, A starting from the origin and moving along the z-axis, B starting from the point $(a, 0, 0)$ and moving parallel to the y-axis. Find the surface enveloped by the straight lines joining them.

7. Prove that the intersections of the curve

$$(x + y - a)^3 + 27axy = 0$$

with the line $x + y = a$ are inflections of the curve.

8. Discuss the singular points of the following curves:

$$\begin{aligned}
&(a)\ F(x, y) = ax^3 + by^3 - cxy = 0;\\
&(b)\ F(x, y) = (y^2 - 2x^2)^2 - x^5 = 0;\\
&(c)\ F(x, y) = (1 + e^{1/x})y - x = 0;\\
&(d)\ F(x, y) = y^2(2a - x) - x^3 = 0;\\
&(e)\ F(x, y) = (y - 2x)^2 - x^5 = 0.
\end{aligned}$$

9. Let (x, y) be a double point of the curve $F(x, y) = 0$. Calculate the angle φ between the two tangents at (x, y), assuming that not all the second derivatives of F vanish at (x, y).

Find the angle between the tangents at the double point (a) of the lemniscate, (b) of the folium of Descartes (cf. p. 116).

10. Determine a and b so that the conics

$$\begin{gathered}
4x^2 + 4xy + y^2 - 10x - 10y + 11 = 0\\
(y + bx - 1 - b)^2 - a(by - x + 1 - b) = 0
\end{gathered}$$

cut one another orthogonally at the point (1, 1) and have the same curvature at this point.

11. If $F(x, y, z) = 1$ is the equation of a surface, F being a homogeneous function of degree h, then the tangent plane at the point (x, y, z) is given by

$$\xi F_x + \eta F_y + \zeta F_z = h.$$

12. Let K' and K'' be two circles having two points A and B in common. If a circle K is orthogonal to K' and K'', then it is also orthogonal to every circle passing through A and B.

13. Let z be defined as a function of x and y by the equation

$$x^3 + y^3 + z^3 - 3xyz = 0.$$

Express z_x and z_y as functions of x, y, z.

3. Systems of Functions, Transformations, and Mappings

1. General Remarks.

The results we have obtained for implicit functions now enable us to consider systems of functions, that is, to discuss several functions simultaneously. In this section we shall consider the particularly important case of systems where the number of functions is the same as the number of independent variables. We begin by investigating the meaning of such systems in the case of two independent variables. If the two functions

$$\xi = \phi(x, y) \quad \text{and} \quad \eta = \psi(x, y)$$

are both differentiable in a region R of the xy-plane, we can interpret this system of functions in two different ways. The first interpretation (the second will be given in sub-section 2, p. 138) is by means of a *mapping* or *transformation*. To the point P with co-ordinates (x, y) in the xy-plane there corresponds the image point Π with the co-ordinates (ξ, η) in the $\xi\eta$-plane.

An example of such a mapping is the *affine* mapping or transformation

$$\begin{aligned} \xi &= ax + by \\ \eta &= cx + dy \end{aligned}$$

of Chapter I (p. 28), where a, b, c, d are constants.

Frequently (x, y) and (ξ, η) are interpreted as points of one and the same plane. In this case we speak of a *mapping of the* xy-*plane on itself*, or a *transformation of the* xy-*plane into itself*.*

* It is also possible to interpret a single function $\xi = f(x)$ of a single variable as a mapping, if we think of a point with co-ordinate x on an x-axis as being brought by means of the function into correspondence with a point ξ

[*Continued overleaf*.

The fundamental problem connected with a mapping is that of its inversion; that is, the question whether and how x and y can in virtue of the equations $\xi = \phi(x, y)$ and $\eta = \psi(x, y)$ be regarded as functions of ξ and η, and how these inverse functions are to be differentiated.

If when the point (x, y) ranges over the region R its image point (ξ, η) ranges over a region B of the $\xi\eta$-plane, we call B the *image region* of R. If *two different points of* R *always correspond to two different points of* B, then for each point of B we can always find a single point of R of which it is the image. Thus to each point of B we can assign the point of R of which it is the image. (This point of R is sometimes called the " model ", as opposed to the " image ".) That is, we can invert the mapping uniquely, or determine x and y uniquely as functions

$$x = g(\xi, \eta), \quad y = h(\xi, \eta)$$

of ξ and η, which are defined in B. We then say that the original mapping can be *uniquely inverted*, or has a *unique inverse*, or is a *one-to-one* * *mapping*, and we call $x = g(\xi, \eta)$, $y = h(\xi, \eta)$ the transformation *inverse* to the original transformation or mapping.

If in this mapping the point P with co-ordinates (x, y) describes a curve in the region R, its image point will likewise describe a curve in the region B, which is called the *image curve* of the first. For example, the curve $x = c$, which is parallel to the y-axis, corresponds to a curve in the $\xi\eta$-plane which is given in parametric form by the equations

$$\xi = \phi(c, y), \quad \eta = \psi(c, y),$$

where y is the parameter. Again, to the curve $y = k$ there corresponds the curve

$$\xi = \phi(x, k), \quad \eta = \psi(x, k).$$

If to c and k we assign sequences of neighbouring values c_1, c_2, c_3, . . . and k_1, k_2, k_3, . . . , then the rectangular " co-ordinate

on a ξ-axis. By this point-to-point correspondence the whole or a part of the x-axis is mapped on the whole or a part of the ξ-axis. A uniform " scale " of equidistant x-values on the x-axis will in general be expanded or contracted into a non-uniform scale of ξ-values on the ξ-axis. The ξ-scale may be regarded as a representation of the function $\xi = f(x)$. Such a point of view is frequently found useful in applications (e.g. in nomography).

* Often written (1, 1).

net" consisting of the lines $x =$ const. and $y =$ const. (e.g. the network of lines on ordinary graph paper) usually gives rise to a corresponding *curvilinear net* of curves in the $\xi\eta$-plane

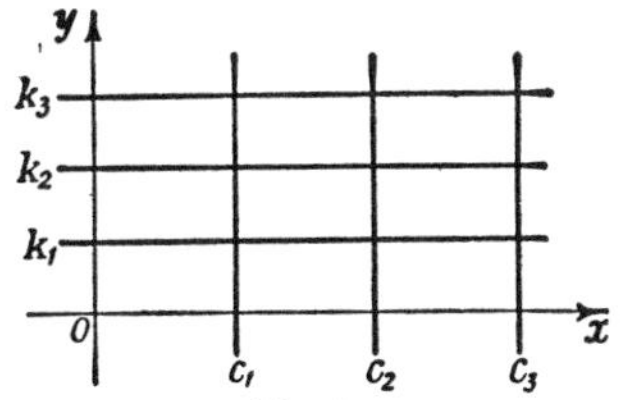

Fig. 6

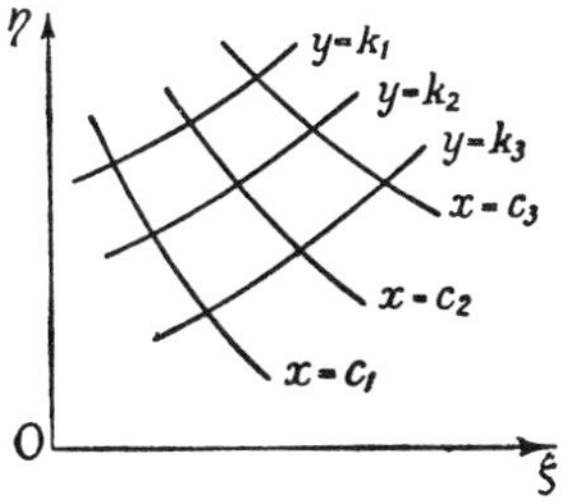

Fig. 7

Nets of curves $x =$ const. and $y =$ const. in the xy-plane and the $\xi\eta$-plane

(figs. 6, 7). The two families of curves composing this net of curves can be written in implicit form. If we represent the inverse mapping by the equations

$$x = g(\xi, \eta), \quad y = h(\xi, \eta),$$

the equations of the curves are simply

$$g(\xi, \eta) = c \quad \text{and} \quad h(\xi, \eta) = k$$

respectively.

In the same way, the two families of lines $\xi = \gamma$ and $\eta = \kappa$ in the $\xi\eta$-plane correspond to the two families of curves

$$\phi(x, y) = \gamma, \quad \psi(x, y) = \kappa$$

in the xy-plane.

As an example we consider *inversion*, or the mapping by *reciprocal radii* or *reflection in the unit circle.* This transformation is given by the equations

$$\xi = \frac{x}{x^2 + y^2}, \quad \eta = \frac{y}{x^2 + y^2}.$$

To the point P with co-ordinates (x, y) there corresponds the point Π with co-ordinates (ξ, η) lying on the same line OP and satisfying the equation $\xi^2 + \eta^2 = \dfrac{1}{x^2 + y^2}$ or $O\Pi = \dfrac{1}{OP}$, so that the radius vector to P is the reciprocal of the radius vector to Π. Points inside the unit circle are mapped on points outside the circle and vice versa.

From the relation $\xi^2 + \eta^2 = \dfrac{1}{x^2 + y^2}$ we find that the *inverse transformation is*

$$x = \frac{\xi}{\xi^2 + \eta^2}, \quad y = \frac{\eta}{\xi^2 + \eta^2},$$

which is again inversion.

For the region R we may take the whole xy-plane with the exception of the origin, and for the region B we may take the whole $\xi\eta$-plane with the exception of the origin. The lines $\xi = c$ and $\eta = k$ in the $\xi\eta$-plane correspond to the circles $x^2 + y^2 - \frac{1}{c}x = 0$ and $x^2 + y^2 - \frac{1}{k}y = 0$ in the xy-plane respectively; at the origin these circles touch the y-axis and the x-axis respectively. In the same way, the rectilinear co-ordinate net in the xy-plane corresponds to the two families of circles touching the ξ-axis and the η-axis respectively at the origin.

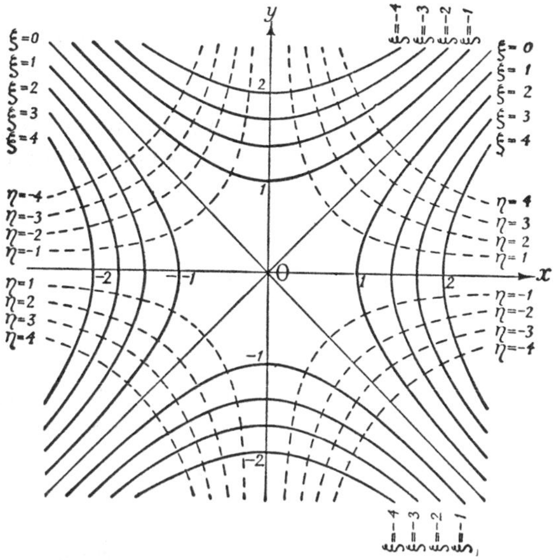

Fig. 8.—Orthogonal families of rectangular hyperbolas

As a further example we consider the mapping

$$\xi = x^2 - y^2, \quad \eta = 2xy.$$

The curves $\xi = \text{const.}$ give rise in the xy-plane to the rectangular hyperbolas $x^2 - y^2 = \text{const.}$, whose asymptotes are the lines $x = y$ and $x = -y$; the lines $\eta = \text{const.}$ also correspond to a family of rectangular hyperbolas, having the co-ordinate axes as asymptotes. The hyperbolas of each family cut those of the other family at right angles (cf. fig. 8). The lines parallel to the axes in the xy-plane correspond to two families of parabolas in the $\xi\eta$-plane, the parabolas $\eta^2 = 4c^2(c^2 - \xi)$ corresponding to the lines $x = c$ and the parabolas $\eta^2 = 4c^2(c^2 + \xi)$ corresponding to the lines $y = c$. All these parabolas have the origin as focus and the ξ-axis as axis (a family of confocal and coaxial parabolas; cf. fig. 9). For systems of confocal ellipses and hyperbolas cf. Ex. 5, p. 158.

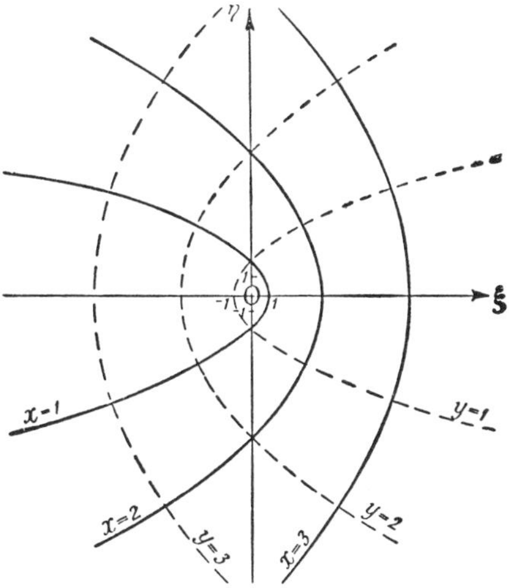

Fig. 9.—Orthogonal families of confocal parabolas

One-to-one transformations have an important interpretation and application in the representation of deformations or motions of continuously-distributed substances, such as fluids. If we think of such a substance as spread out at a given time over a region R and then deformed by a motion, the substance originally spread over R will in general cover a region B different from R. Each particle of the substance can be distinguished at the beginning of the motion by its co-ordinates (x, y) in R, and at the end of the motion by its co-ordinates (ξ, η) in B. The one-to-one character of the transformation obtained by bringing (x, y) into correspondence with (ξ, η) is simply the mathematical expression of the physically obvious fact that the separate particles must remain recognizable after the motion, i.e. that separate particles remain separate.

2. Introduction of New Curvilinear Co-ordinates.

Closely connected with the first interpretation (as a mapping) which we can give to a system of equations $\xi = \phi(x, y)$, $\eta = \psi(x, y)$ is the second interpretation, as a *transformation of co-ordinates* in the plane. If the functions ϕ and ψ happen not to be linear, this is no longer an "affine" transformation, but a *transformation to general curvilinear co-ordinates.*

We again assume that when (x, y) ranges over a region R of the xy-plane the corresponding point (ξ, η) ranges over a region B of the $\xi\eta$-plane, and also that for each point of B the corresponding (x, y) in R can be uniquely determined; in other words, that the transformation is one-to-one. The inverse transformation we again denote by $x = g(\xi, \eta)$, $y = h(\xi, \eta)$.

By the *co-ordinates of a point* P in a region R we can mean any number-pair which serves to specify the position of the point P in R uniquely. Rectangular co-ordinates are the simplest case of co-ordinates which extend over the whole plane. Another typical case is the system of polar co-ordinates in the xy-plane, introduced by the equations

$$\xi = r = \sqrt{(x^2 + y^2)},$$
$$\eta = \theta = \text{arc}\tan(y/x) \qquad (0 \leqq \theta < 2\pi).$$

When we are given a system of functions $\xi = \phi(x, y)$, $\eta = \psi(x, y)$ as above, we can in general assign to each point P (x, y) the corresponding values (ξ, η) as new co-ordinates. For each pair of values (ξ, η) belonging to the region B uniquely determines the pair (x, y), and thus uniquely determines the position of the point P in R; this entitles us to call ξ, η the co-ordinates of the point P. The "co-ordinate lines" $\xi =$ const. and $\eta =$ const. are then represented in the xy-plane by two families of curves, which are defined implicitly by the equations $\phi(x, y) =$ const. and $\psi(x, y) =$ const. respectively. These co-ordinate curves cover the region R with a co-ordinate net (usually curved), for which reason the co-ordinates (ξ, η) are also called *curvilinear co-ordinates* in R.

We shall once again point out how closely these two interpretations of our system of equations are interrelated. The curves in the $\xi\eta$-plane which in the mapping correspond to straight lines parallel to the axes in the xy-plane can be directly regarded as the co-ordinate curves for the curvilinear co-ordinates $x = g(\xi, \eta)$, $y = h(\xi, \eta)$ in the $\xi\eta$-plane; conversely, the co-ordinate curves of the curvilinear co-ordinate system $\xi = \phi(x, y)$, $\eta = \psi(x, y)$ in the xy-plane in the mapping are the images of the straight lines parallel to the axes in the $\xi\eta$-plane. Even in the interpretation of (ξ, η) as curvilinear co-ordinates in the xy-plane we must consider a $\xi\eta$-plane and a region B of that plane in which the point with the co-ordinates (ξ, η) can vary, if we wish to keep the situation clear. The difference is mainly in the point of view.* If we are chiefly interested in the region R of the xy-plane, we regard ξ, η simply as a new means of locating points in the region R, the region B of the $\xi\eta$-plane being then merely subsidiary; while if we are equally interested in the two regions R and B in the xy-plane and the $\xi\eta$-plane respectively, it is preferable to regard the system of equations as specifying a correspondence between the two regions, that is, a mapping of one on the other. It is, however, always desirable to keep the two interpretations, mapping and transformation of co-ordinates, both in mind at the same time.

If, for example, we introduce polar co-ordinates (r, θ) and interpret r and θ as rectangular co-ordinates in an $r\theta$-plane, the circles $r =$ const. and the lines $\theta =$ const. are mapped on straight lines parallel to the axes in the $r\theta$-plane. If the region R of the xy-plane is the circle $x^2 + y^2 \leqq 1$, the point (r, θ) of the $r\theta$-plane will range over a rectangle $0 \leqq r \leqq 1$, $0 \leqq \theta < 2\pi$, where corresponding points of the sides $\theta = 0$ and $\theta = 2\pi$ are associated with one and the same point of R and the whole side $r = 0$ is the image of the origin $x = 0$, $y = 0$.

Another example of a curvilinear co-ordinate system is the system of parabolic co-ordinates. We arrive at these by considering the family of confocal parabolas in the xy-plane (cf. also p. 126 and fig. 9)

$$y^2 = 2p\left(x + \frac{p}{2}\right),$$

all of which have the origin as focus and the x-axis as axis. Through each point of the plane there pass two parabolas of the family, one corresponding to a positive parameter value $p = \xi$ and the other to a negative parameter value $p = \eta$. We obtain these two values by solving for p the quadratic equation which results when in the equation $y^2 = 2p(x + p/2)$ we substitute the values of x and y corresponding to the point; this gives

$$\xi = -x + \sqrt{(x^2 + y^2)}, \quad \eta = -x - \sqrt{(x^2 + y^2)}.$$

These two quantities may be introduced as curvilinear co-ordinates in the xy-plane, the confocal parabolas then becoming the co-ordinate curves. These are indicated in fig. 9, if we imagine the symbols (x, y) and (ξ, η) interchanged.

* There is, however, a real difference, in that the equations always define a *mapping*, no matter how many points (x, y) correspond to one point (ξ, η), while they define a *transformation of co-ordinates* only when the correspondence is one-to-one.

In introducing parabolic co-ordinates (ξ, η) we must bear in mind that the *one* pair of values (ξ, η) corresponds to the *two* points (x, y) and $(x, -y)$ which are the two intersections of the corresponding parabolas. Hence in order to obtain a one-to-one correspondence between the pair (x, y) and the pair (ξ, η) we must restrict ourselves to the half-plane $y \geqq 0$, say. Then every region R in this half-plane is in a one-to-one correspondence with a region B of the $\xi\eta$-plane, and the rectangular co-ordinates (ξ, η) of each point in this region B are exactly the same as the parabolic co-ordinates of the corresponding point in the region R.

3. Extension to More than Two Independent Variables.

In the case of three or more independent variables the state of affairs is analogous. Thus a system of three continuously-differentiable functions

$$\xi = \phi(x, y, z), \quad \eta = \psi(x, y, z), \quad \zeta = \chi(x, y, z),$$

defined in a region R of xyz-space, may be regarded as the mapping of the region R on a region B of $\xi\eta\zeta$-space. If we assume that this mapping of R on B is one-to-one, so that for each image point (ξ, η, ζ) of B the co-ordinates (x, y, z) of the corresponding point (" model " point) in R can be uniquely calculated by means of functions

$$x = g(\xi, \eta, \zeta), \quad y = h(\xi, \eta, \zeta), \quad z = l(\xi, \eta, \zeta),$$

then (ξ, η, ζ) may also be regarded as general co-ordinates of the point P in the region R. The surfaces $\xi = \text{const.}$, $\eta = \text{const.}$, $\zeta = \text{const.}$, or, in other symbols,

$$\phi(x, y, z) = \text{const.}, \quad \psi(x, y, z) = \text{const.}, \quad \chi(x, y, z) = \text{const.}$$

then form a system of three families of surfaces which cover the region R and may be called curvilinear co-ordinate surfaces.

Just as in the case of two independent variables, we can interpret one-to-one transformations in three dimensions as deformations of a substance spread continuously throughout a region of space.

A very important case of transformation of co-ordinates is given by *polar co-ordinates in space.* These specify the position of a point P in space by three numbers: (1) the distance $r = \sqrt{(x^2 + y^2 + z^2)}$ from the origin, (2) the geographical longitude ϕ, that is, the angle between the xz-plane and the plane determined by P and the z-axis, and (3) the polar distance θ, that is, the angle between the radius vector OP and the positive z-axis. As we see from fig. 10, the three polar co-ordinates r, ϕ, θ are related to the rectangular co-ordinates by the equations of transformation

$$x = r\cos\phi\sin\theta,$$
$$y = r\sin\phi\sin\theta,$$
$$z = r\cos\theta,$$

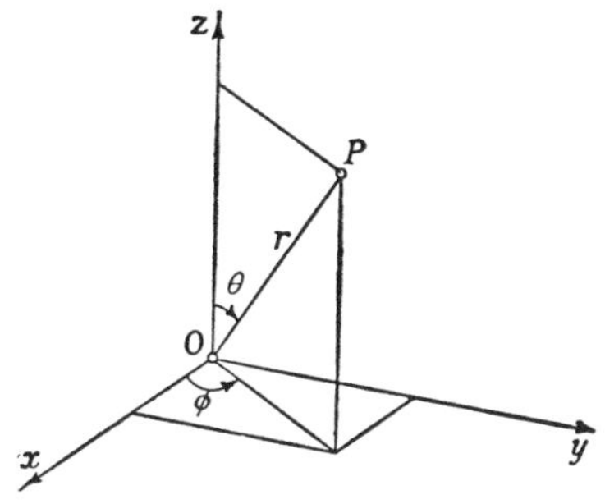

Fig. 10.—Three-dimensional polar co-ordinates

from which we obtain the inverse relations

$$r = \sqrt{(x^2+y^2+z^2)},$$
$$\phi = \text{arc}\cos\frac{x}{\sqrt{(x^2+y^2)}} = \text{arc}\sin\frac{y}{\sqrt{(x^2+y^2)}},$$
$$\theta = \text{arc}\cos\frac{z}{\sqrt{(x^2+y^2+z^2)}} = \text{arc}\sin\frac{\sqrt{(x^2+y^2)}}{\sqrt{(x^2+y^2+z^2)}}.$$

For polar co-ordinates in the plane the origin is an exceptional point, at which the one-to-one correspondence fails, since the angle is indeterminate there. In the same way, for polar co-ordinates in space the whole of the z-axis is an exception, since the longitude ϕ is indeterminate there. At the origin itself the polar distance θ is also indeterminate.

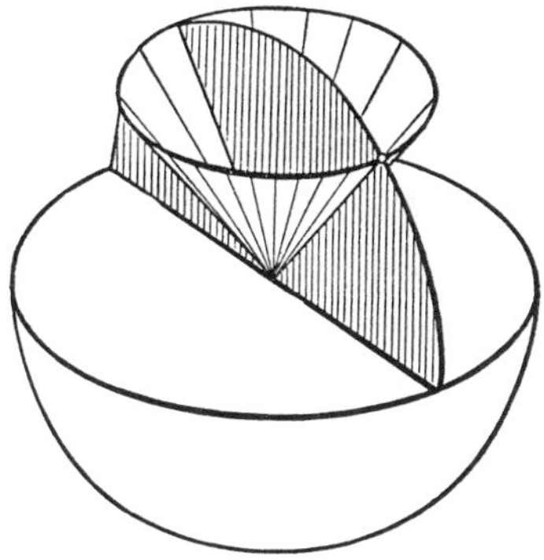
Fig. 11.—Co-ordinate surfaces for three-dimensional polar co-ordinates

The co-ordinate surfaces for three-dimensional polar co-ordinates are as follows: (1) for constant values of r, the concentric spheres about the origin; (2) for constant values of ϕ, the family of half-planes through the z-axis; (3) for constant values of θ, the circular cones with the z-axis as axis and the origin as vertex (fig. 11).

Another co-ordinate system which is often used is the system of *cylindrical co-ordinates*. These are obtained by introducing polar co-ordinates ρ, ϕ in the xy-plane and retaining z as the third co-ordinate. Then the formulæ of transformation from rectangular co-ordinates to cylindrical co-ordinates are

$$\begin{aligned} x &= \rho\cos\phi, \\ y &= \rho\sin\phi, \\ z &= z \end{aligned}$$

and the inverse transformation is

$$\begin{aligned} \rho &= \sqrt{(x^2+y^2)}, \\ \phi &= \operatorname{arc\,cos}\frac{x}{\sqrt{(x^2+y^2)}} = \operatorname{arc\,sin}\frac{y}{\sqrt{(x^2+y^2)}}. \\ z &= z \end{aligned}$$

The co-ordinate surfaces $\rho =$ const. are the vertical circular cylinders which intersect the xy-plane in concentric circles with the origin as centre; the surfaces $\phi =$ const. are the half-planes through the z-axis, and the surfaces $z =$ const. are the planes parallel to the xy-plane.

4. Differentiation Formulæ for the Inverse Functions.

In many cases of practical importance it is possible to solve the given system of equations directly, as in the above examples, and thus to recognize that the inverse functions are continuous and possess continuous derivatives. For the time being, therefore, let us assume the existence and differentiability of the inverse functions. Then without actually solving the equations explicitly we can calculate the derivatives of the inverse functions in the following way. We substitute the inverse functions $x = g(\xi, \eta)$, $y = h(\xi, \eta)$ in the given equations $\xi = \phi(x, y)$, $\eta = \psi(x, y)$. On the right we obtain the compound functions $\phi(g(\xi, \eta), h(\xi, \eta))$ and $\psi(g(\xi, \eta), h(\xi, \eta))$ of ξ and η; but these must be equal to ξ and η respectively. We now differentiate each of the equations

$$\begin{aligned} \xi &= \phi(g(\xi, \eta),\ h(\xi, \eta)), \\ \eta &= \psi(g(\xi, \eta),\ h(\xi, \eta)) \end{aligned}$$

with respect to ξ and to η, regarding ξ and η as independent variables.* If on the right we apply the chain rule for the differentiation of compound functions, we obtain the system of equations

$$1 = \phi_x g_\xi + \phi_y h_\xi, \quad 0 = \phi_x g_\eta + \phi_y h_\eta,$$
$$0 = \psi_x g_\xi + \psi_y h_\xi, \quad 1 = \psi_x g_\eta + \psi_y h_\eta.$$

Solving these equations, we obtain

$$g_\xi = \frac{\psi_y}{D}, \quad g_\eta = -\frac{\phi_y}{D}, \quad h_\xi = -\frac{\psi_x}{D}, \quad h_\eta = \frac{\phi_x}{D},$$

or

$$x_\xi = \frac{\eta_y}{D}, \quad x_\eta = -\frac{\xi_y}{D}, \quad y_\xi = -\frac{\eta_x}{D}, \quad y_\eta = \frac{\xi_x}{D},$$

i.e. the partial derivatives of the inverse functions $x = g(\xi, \eta)$ and $y = h(\xi, \eta)$ with respect to ξ and η, expressed in terms of the derivatives of the original functions $\phi(x, y)$ and $\psi(x, y)$ with respect to x and y. For brevity we have here written

$$D = \xi_x \eta_y - \xi_y \eta_x = \begin{vmatrix} \frac{\partial \xi}{\partial x} & \frac{\partial \xi}{\partial y} \\ \frac{\partial \eta}{\partial x} & \frac{\partial \eta}{\partial y} \end{vmatrix}.$$

This expression D, which we assume is not zero at the point in question, is called the *Jacobian* or *functional determinant* of the functions $\xi = \phi(x, y)$ and $\eta = \psi(x, y)$ with respect to the variables x and y.

In the above, as occasionally elsewhere, we have used the shorter notation $\xi(x, y)$ instead of the more detailed notation $\xi = \phi(x, y)$, which distinguishes between the quantity ξ and its functional expression $\phi(x, y)$. We shall often use similar abbreviations in the future when there is no risk of confusion.

For polar co-ordinates in the plane expressed in terms of rectangular co-ordinates,

$$\xi = r = \sqrt{(x^2 + y^2)} \quad \text{and} \quad \eta = \theta = \arctan\frac{y}{x},$$

* These equations hold for all values of ξ and η under consideration; as we say, they hold *identically*, in contrast to equations between variables which are satisfied only for *some* of the values of these variables. Such identical equations or *identities*, when differentiated with respect to any of the variables occurring in them, again yield identities, as follows immediately from the definition.

for example, the partial derivatives are

$$r_x = \frac{x}{\sqrt{(x^2+y^2)}} = \frac{x}{r}, \qquad r_y = \frac{y}{\sqrt{(x^2+y^2)}} = \frac{y}{r},$$

$$\theta_x = -\frac{y}{x^2+y^2} = -\frac{y}{r^2}, \quad \theta_y = \frac{x}{x^2+y^2} = \frac{x}{r^2}.$$

Hence the Jacobian has the value

$$D = \frac{x}{r}\frac{x}{r^2} - \frac{y}{r}\left(-\frac{y}{r^2}\right) = \frac{1}{r},$$

and the partial derivatives of the inverse functions (rectangular co-ordinates expressed in terms of polar co-ordinates) are

$$x_r = \frac{x}{r}, \quad x_\theta = -y, \qquad y_r = \frac{y}{r}, \quad y_\theta = x,$$

as we could have found more easily by direct differentiation of the inverse formulæ $x = r\cos\theta$, $y = r\sin\theta$.

The Jacobian occurs so frequently that a special symbol is often used for it:

$$D = \frac{\partial(\xi, \eta)}{\partial(x, y)}.$$

The appropriateness of this abbreviation will soon be obvious. From the formulæ

$$x_\xi = \frac{\eta_y}{D}, \qquad x_\eta = -\frac{\xi_y}{D},$$

$$y_\xi = -\frac{\eta_x}{D}, \qquad y_\eta = \frac{\xi_x}{D}$$

for the derivatives of the inverse functions we find that the Jacobian of the functions $x = x(\xi, \eta)$ and $y = y(\xi, \eta)$ with respect to ξ and η is given by the expression

$$\frac{\partial(x, y)}{\partial(\xi, \eta)} = x_\xi y_\eta - x_\eta y_\xi = \frac{\xi_x\eta_y - \xi_y\eta_x}{D^2} = \frac{1}{D} = 1 \div \frac{\partial(\xi, \eta)}{\partial(x, y)}.$$

That is, *the Jacobian of the inverse system of functions is the reciprocal of the Jacobian of the original system.*

In the same way we can also express the second derivatives of the inverse functions in terms of the first and second derivatives of the given functions. We have only to differentiate the linear equations given above with respect to ξ and to η by means of the chain rule. (We assume, of course, that the given function possesses continuous derivatives of the second order.) We then obtain linear equations from which the required derivatives can readily be calculated.

For example, to calculate the derivatives

$$\frac{\partial^2 x}{\partial \xi^2} = g_{\xi\xi} \quad \text{and} \quad \frac{\partial^2 y}{\partial \xi^2} = h_{\xi\xi}$$

we differentiate the two equations

$$1 = \xi_x x_\xi + \xi_y y_\xi$$
$$0 = \eta_x x_\xi + \eta_y y_\xi$$

once again with respect to ξ and by the chain rule obtain

$$0 = \xi_{xx} x_\xi^2 + 2\xi_{xy} x_\xi y_\xi + \xi_{yy} y_\xi^2 + \xi_x x_{\xi\xi} + \xi_y y_{\xi\xi},$$
$$0 = \eta_{xx} x_\xi^2 + 2\eta_{xy} x_\xi y_\xi + \eta_{yy} y_\xi^2 + \eta_x x_{\xi\xi} + \eta_y y_{\xi\xi}.$$

If we solve this system of linear equations, regarding the quantities $x_{\xi\xi}$ and $y_{\xi\xi}$ as unknowns (the determinant of the system is again D, and therefore, by hypothesis, not zero) and then replace x_ξ and y_ξ by the values already known for them, a brief calculation gives

$$x_{\xi\xi} = -\frac{1}{D^3} \begin{vmatrix} \xi_{xx}\eta_y^2 - 2\xi_{xy}\eta_x\eta_y + \xi_{yy}\eta_x^2 & \xi_y \\ \eta_{xx}\eta_y^2 - 2\eta_{xy}\eta_x\eta_y + \eta_{yy}\eta_x^2 & \eta_y \end{vmatrix}$$

and

$$y_{\xi\xi} = \frac{1}{D^3} \begin{vmatrix} \xi_{xx}\eta_y^2 - 2\xi_{xy}\eta_x\eta_y + \xi_{yy}\eta_x^2 & \xi_x \\ \eta_{xx}\eta_y^2 - 2\eta_{xy}\eta_x\eta_y + \eta_{yy}\eta_x^2 & \eta_x \end{vmatrix}.$$

The third and higher derivatives can be obtained in the same way, by repeated differentiation of the linear system of equations; at each stage we obtain a system of linear equations with the (non-vanishing) determinant D.

5. Resolution and Combination of Mappings and Transformations.

In Chapter I we saw that every affine transformation can be analysed into simple or, as we say, *primitive* transformations, the first of which deforms the plane in one direction only and the second deforms the already deformed plane again in another direction. In each of these transformations there is really only *one* new variable introduced.

We can now do exactly the same thing for transformations in general.

We begin with some remarks on the combination of transformations. If the transformation

$$\xi = \phi(x, y), \quad \eta = \psi(x, y)$$

gives a one-to-one mapping of the point (x, y), which ranges over a region R, on the point (ξ, η) of the region B in the $\xi\eta$-plane, and if the equations

$$u = \Phi(\xi, \eta), \quad v = \Psi(\xi, \eta)$$

give a one-to-one mapping of the region B on a region R' in the uv-plane, then a one-to-one mapping of R on R' simultaneously occurs. This mapping we naturally call the *resultant mapping* or *resultant transformation*, and say that it is obtained by combining the two given mappings. The resultant transformation is given by the equations

$$u = \Phi(\phi(x, y), \psi(x, y)), \quad v = \Psi(\phi(x, y), \psi(x, y));$$

from the definition it follows at once that this mapping is one-to-one.

By the rules for differentiating compound functions we obtain

$$\frac{\partial u}{\partial x} = \Phi_\xi \phi_x + \Phi_\eta \psi_x.$$

$$\frac{\partial u}{\partial y} = \Phi_\xi \phi_y + \Phi_\eta \psi_y,$$

$$\frac{\partial v}{\partial x} = \Psi_\xi \phi_x + \Psi_\eta \psi_x,$$

$$\frac{\partial v}{\partial y} = \Psi_\xi \phi_y + \Psi_\eta \psi_y.$$

On comparing this with the law for the multiplication of determinants (cf. p. 36) we find * that the Jacobian of u and v with respect to x and y is

$$\frac{\partial u}{\partial x}\frac{\partial v}{\partial y} - \frac{\partial u}{\partial y}\frac{\partial v}{\partial x} = (\Phi_\xi \Psi_\eta - \Phi_\eta \Psi_\xi)(\phi_x \psi_y - \phi_y \psi_x).$$

* The same result can of course be obtained by straightforward multiplication.

In words:

The Jacobian of the resultant transformation is equal to the product of the Jacobians of the individual transformations.

In symbols:

$$\frac{\partial(u, v)}{\partial(x, y)} = \frac{\partial(u, v)}{\partial(\xi, \eta)} \frac{\partial(\xi, \eta)}{\partial(x, y)}.$$

This equation brings out the appropriateness of our symbol for the Jacobian. *When transformations are combined, the Jacobians behave in the same way as the derivatives behave when functions of one variable are combined.* The Jacobian of the resultant transformation differs from zero, provided the same is true for the individual (or component) transformations.

If, in particular, the second transformation

$$u = \Phi(\xi, \eta), \quad v = \Psi(\xi, \eta)$$

is the inverse of the first,

$$\xi = \phi(x, y), \quad \eta = \psi(x, y),$$

and if both transformations are differentiable, the resultant transformation will simply be the identical transformation, that is, $u = x$, $v = y$. The Jacobian of this last transformation is obviously 1, so that we again obtain the relation of p. 144,

$$\frac{\partial(\xi, \eta)}{\partial(x, y)} \frac{\partial(x, y)}{\partial(\xi, \eta)} = 1.$$

From this, incidentally, it follows that neither of the two Jacobians can vanish.

Before we take up the question of the resolution of an arbitrary transformation into primitive transformations, we shall consider the following primitive transformation:

$$\xi = \phi(x, y), \quad \eta = y.$$

We assume that the Jacobian $D = \phi_x$ of this transformation differs from zero throughout the region R, i.e. we assume that $\phi_x > 0$, say, in the region. The transformation deforms the region R into a region B; and we may imagine that the effect of the transformation is to move each point in the direction of the x-axis, since the ordinate is unchanged. After deformation the point (x, y) has a new abscissa which depends on both x and y. The condition $\phi_x > 0$ means that when y is fixed ξ varies monotonically with x. This ensures the one-to-one correspondence of the points on a line $y =$ const. before and

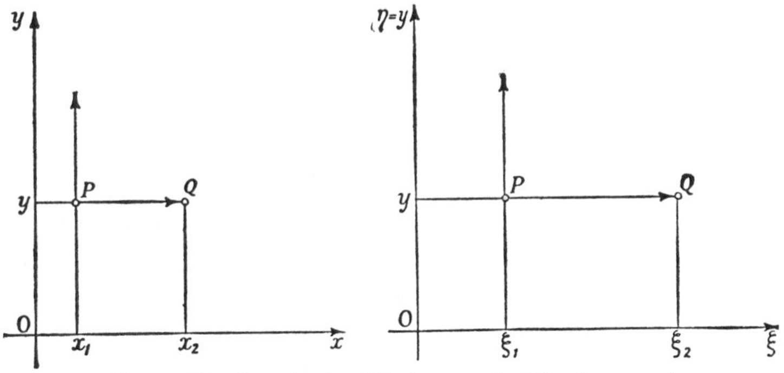

Fig. 12.—Transformation in which the sense of rotation is preserved

after the transformation; in fact, two points $P(x_1, y)$ and $Q(x_2, y)$ with the same ordinate y and $x_2 > x_1$ are transformed into two points P' and Q' which again have the same ordinate and whose abscissæ satisfy the inequality $\xi_2 > \xi_1$ (cf. fig. 12). This fact also

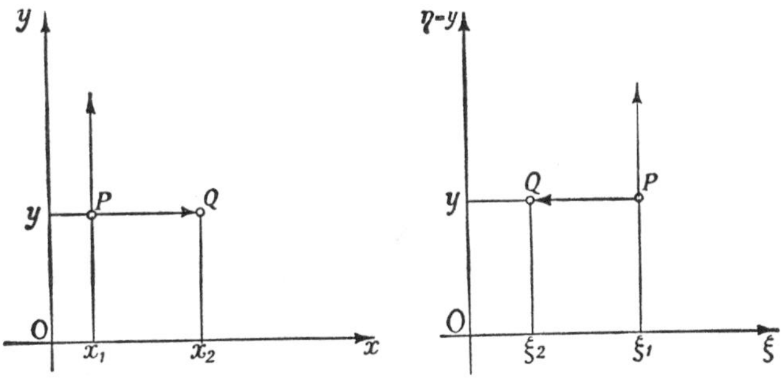

Fig. 13.—Transformation in which the sense of rotation is reversed

shows that after the transformation the sense of rotation is the same as that in the xy-plane.

If ϕ_x were negative, the two points P and Q would correspond to points with the same ordinate and with abscissæ ξ_1 and ξ_2, but this time we should have $\xi_1 > \xi_2$ (cf. fig. 13). The sense of rotation would therefore be reversed, as we have already seen in Chapter I (p. 35) for the simple case of affine transformations.

If the primitive transformation

$$\xi = \phi(x, y), \quad \eta = y$$

is continuously differentiable, and its Jacobian ϕ_x *differs from zero at a point* $P(x_0, y_0)$, *then in a neighbourhood of* P *the transformation has a unique inverse, and this inverse is also a primitive transformation of the same type.* In virtue of the hypothesis $\phi_x \neq 0$ we can apply the theorem on implicit functions given in section 1, No. 3 (p. 114), and thus find that in a neighbourhood of (x_0, y_0) the equation $\xi = \phi(x, y)$ determines the quantity x uniquely as a continuously differentiable function $x = g(\xi, y)$ of ξ and y.* The two formulæ

$$x = g(\xi, \eta), \quad y = \eta$$

therefore give us the inverse transformation, whose determinant is $g_\xi = 1/\phi_x \neq 0$.

If we now think of the region B in the $\xi\eta$-plane as itself mapped on a region R in the uv-plane by means of a primitive transformation

$$u = \xi, \quad v = \Psi(\xi, \eta),$$

where we assume that Ψ_η is positive, the state of affairs is just as above, except that the deformation takes place in the direction of the other co-ordinate. This transformation likewise preserves the sense of rotation (or reverses it if the relation $\Psi_\eta < 0$ holds instead of $\Psi_\eta > 0$).

By combining the two primitive transformations we obtain the transformation

$$\begin{aligned} u &= \phi(x, y), \\ v &= \Psi(\phi(x, y), y) = \psi(x, y), \end{aligned}$$

and from the theorem on Jacobians we see that

$$\frac{\partial(\phi, \psi)}{\partial(x, y)} = \phi_x \Psi_\eta.$$

* Here we use the fact that a function with two continuous derivatives is differentiable.

We now assert that an arbitrary one-to-one continuously differentiable transformation

$$u = \phi(x, y), \quad v = \psi(x, y)$$

of the region R in the xy-plane on a region R' in the uv-plane can be resolved in the neighbourhood of any point interior to R into continuously differentiable primitive transformations, provided that throughout the whole region R the Jacobian

$$\frac{\partial(u, v)}{\partial(x, y)} = \phi_x\psi_y - \phi_y\psi_x$$

differs from zero.

From the non-vanishing of the Jacobian it follows that at no point can we have both $\phi_x = 0$ and $\phi_y = 0$. We consider a point with co-ordinates (x_0, y_0) and assume that at that point $\phi_x \neq 0$. Then by the main theorem of section 1, No. 5 (p. 117) we can mark off intervals $x_1 \leqq x \leqq x_2$, $y_1 \leqq y \leqq y_2$, $u_1 \leqq u \leqq u_2$ about x_0, y_0, and $u_0 = u(x_0, y_0)$ respectively, in such a way that within these bounds the equation $u = \phi(x, y)$ can be solved uniquely for x and defines $x = g(u, y)$ as a continuously differentiable function of u and y. If we substitute this expression in $v = \psi(x, y)$, we obtain $v = \psi(g(u, y), y) = \Psi(u, y)$. Hence in any neighbourhood of the point (x_0, y_0) we may regard the given transformation as composed of the two primitive transformations

$$\xi = \phi(x, y), \quad \eta = y$$

and

$$u = \xi, \quad v = \Psi(\xi, \eta).$$

Similarly, in a neighbourhood of a point (x_0, y_0) at which $\phi_y \neq 0$ we can resolve the given transformation into two primitive transformations of the form

$$\xi = x, \quad \eta = \phi(x, y)$$
$$u = \eta, \quad v = \Psi_1(\xi, \eta) \quad (= \psi\{x, y(u, x)\}).$$

This pair of transformations is not exactly identical in form with the pairs considered above, each of which leaves one of the co-ordinate directions unaltered. It can easily be brought into that form, however, by interchanging the letters u and v (this interchange is itself the resultant of three very simple primitive transformations (cf. the footnote on p. 31)). For the purposes of the present chapter, however, it is more convenient not to carry out this resolution; instead, we write the last set of equations in the form

$$\xi = x,\ \eta = \phi(x, y),$$
$$\bar{u} = -\Psi_1(\xi, \eta),\ \bar{v} = \eta,$$
$$u = \bar{v},\ v = -\bar{u}.$$

These last represent two primitive transformations, each affecting one co-ordinate direction only, and also a rotation of the axes in the uv-plane through an angle of 90°. The rotation is so easy to deal with that it need not be split up into primitive transformations.

It is not to be expected that we can resolve a transformation into primitive transformations in one and the same way throughout the whole region. Since, however, one of the two types of resolution can be carried out for every interior point of R, every closed region interior to R can be subdivided into a finite number of sub-regions* in such a way that in each sub-region one of the resolutions is possible.

From the possibility of this resolution into primitive transformations we can draw an interesting conclusion. We have seen that in the case of a primitive transformation the sense of rotation is reversed or preserved according as the Jacobian is negative or positive. From this it follows that *in the case of general transformations the sense of rotation is reversed or preserved according as the sign of the Jacobian is negative or positive.* For if the sign of the Jacobian is positive, when the resolution into primitive transformations is carried out the Jacobians of the primitive transformations will either be both positive or both negative. (The rotation of the u- and v-axes through 90°, required in some cases, has $+1$ for its Jacobian and leaves the sense of rotation unchanged, and accordingly does not affect the discussion at all.) In the first case it is obvious that the sense of rotation is preserved; in the second case this follows from the fact that two reversals of the sense bring us back to the original sense. If the Jacobian is negative, however, one, and only one, of the primitive transformations will have a negative Jacobian and will therefore reverse the sense, while the other will not affect it.

* This follows from the covering theorem (cf. p. 99).

6. General Theorem on the Inversion of Transformations and Systems of Implicit Functions.

The possibility of inverting a transformation depends on the following general theorem:

If in the neighbourhood of a point (x_0, y_0) *the functions* $\phi(x, y)$ *and* $\psi(x, y)$ *are continuously differentiable,* and* $u_0 = \phi(x_0, y_0)$, $v_0 = \psi(x_0, y_0)$, *and if in addition the Jacobian* $D = \phi_x\psi_y - \phi_y\psi_x$ *is not zero at* (x_0, y_0), *then in a neighbourhood of the point* (x_0, y_0) *the system of equations* $u = \phi(x, y)$, $v = \psi(x, y)$ *has a unique inverse*; that is, there is a uniquely determined pair of functions $x = g(u, v)$, $y = h(u, v)$ such that $x_0 = g(u_0, v_0)$ and $y_0 = h(u_0, v_0)$ and also the equations

$$u = \phi(g(u, v),\ h(u, v)) \quad \text{and} \quad v = \psi(g(u, v),\ h(u, v))$$

hold in some neighbourhood of the point (u_0, v_0).

In the neighbourhood of (u_0, v_0) *the so-called inverse functions* $x = g(u, v)$, $y = h(u, v)$ *possess continuous derivatives which are given by the expressions*

$$\frac{\partial x}{\partial u} = \frac{1}{D}\frac{\partial v}{\partial y}, \qquad \frac{\partial x}{\partial v} = -\frac{1}{D}\frac{\partial u}{\partial y},$$

$$\frac{\partial y}{\partial u} = -\frac{1}{D}\frac{\partial v}{\partial x}, \qquad \frac{\partial y}{\partial v} = \frac{1}{D}\frac{\partial u}{\partial x}.$$

The proof follows from the discussions in No. 5 (p. 149). For in a sufficiently small neighbourhood of the point (x_0, y_0) we can resolve the transformation $u = \phi(x, y)$, $v = \psi(x, y)$ into continuously differentiable primitive transformations, possibly with a rotation of the u- and v-axes through 90° in addition. Each of these has a unique inverse, which is itself a continuously differentiable transformation. The combination of these inverse transformations at once gives us the transformation which is the inverse of the given one. This, being a combination of continuously differentiable transformations, is itself continuously differentiable. It then follows from No. 4 (p. 143) that the differentiation formulæ hold as stated.

This inversion theorem is a special case of a more general theorem which may be regarded as an extension of the theorem of implicit functions to systems of functions. The theorem of

* I.e. are continuous and possess continuous derivatives.

implicit functions (section 1, p. 117) applies to the solution of one equation for one of the variables. The general theorem is as follows:

If $\phi(x, y, u, v, \ldots, w)$ *and* $\psi(x, y, u, v, \ldots, w)$ *are continuously differentiable functions of* x, y, u, v, . . . , w, *and the equations*

$$\phi(x, y, u, v, \ldots, w) = 0 \quad \text{and} \quad \psi(x, y, u, v, \ldots, w) = 0$$

are satisfied by a certain set of values $x_0, y_0, u_0, v_0, \ldots, w_0$, *and if in addition the Jacobian of* ϕ *and* ψ *with respect to* x *and* y *differs from zero at that point (that is,* $D = \phi_x\psi_y - \phi_y\psi_x \neq 0$), *then in the neighbourhood of that point the equations* $\phi = 0$ *and* $\psi = 0$ *can be solved in one, and only one, way for* x *and* y, *and this solution gives* x *and* y *as continuously differentiable functions of* u, v, . . . , w.

The proof of this theorem is similar to that of the inversion theorem above. From the assumption that $D \neq 0$ we can conclude without loss of generality that at the point in question $\phi_x \neq 0$. Then by the main theorem of section 1 (p. 117), if we restrict $x, y, u, v, \ldots, w$ to sufficiently small intervals about $x_0, y_0, u_0, v_0, \ldots, w_0$ respectively, the equation $\phi(x, y, u, v, \ldots, w)$ can be solved in exactly one way for x as a function of the other variables, and this solution $x = g(y, u, v, \ldots, w)$ is a continuously differentiable function of its arguments, and has the partial derivative $g_y = -\phi_y/\phi_x$. If we substitute this function $x = g(y, u, v, \ldots, w)$ in $\psi(x, y, u, v, \ldots, w)$, we obtain a function $\psi(x, y, u, v, \ldots, w) = \Psi(y, u, v, \ldots, w)$, and

$$\Psi_y = -\psi_x\frac{\phi_y}{\phi_x} + \psi_y = \frac{D}{\phi_x}.$$

Hence in virtue of the assumption that $D \neq 0$ we see that the derivative Ψ_y is not zero. Thus if we restrict $y, u, v, \ldots, w$ to intervals about $y_0, u_0, v_0, \ldots, w_0$ (which we take to be smaller than the intervals to which they were previously restricted), we can solve the equation $\Psi = 0$ in exactly one way for y as a function of $u, v, \ldots, w$, and this solution is continuously differentiable. Substituting this expression for y in the equation $x = g(y, u, v, \ldots, w)$ now gives x as a function of $u, v, \ldots, w$, and this solution is continuously differentiable and unique, subject to the restriction of $x, y, u, v, \ldots, w$ to sufficiently small intervals about $x_0, y_0, u_0, v_0, \ldots, w_0$ respectively.

7. Non-independent Functions.

It is worth mentioning that if the Jacobian D vanishes at a point (x_0, y_0), no general statement can be made about the possibility of solving the equations in the neighbourhood of that point. Even if the inverse functions do happen to exist, however, they cannot be differentiable, for then the product $\dfrac{\partial(\xi, \eta)}{\partial(x, y)} \dfrac{\partial(x, y)}{\partial(\xi, \eta)}$ would vanish, while by p. 147 it must be equal to 1.

For example, the equations

$$u = x^3, \quad v = y$$

can be solved uniquely, the solutions being

$$x = \sqrt[3]{u}, \quad y = v,$$

although the Jacobian vanishes at the origin; but the function $\sqrt[3]{u}$ is not differentiable at the origin.

On the other hand, the equations

$$u = x^2 - y^2, \quad v = 2xy$$

cannot be solved uniquely in the neighbourhood of the origin, since the two points (x, y) and $(-x, -y)$ of the xy-plane both correspond to the same point of the uv-plane.

If, however, the Jacobian vanishes *identically*, that is, not merely at the single point (x, y), but at every point in a whole neighbourhood of the point (x, y), then the transformation is of the type called *degenerate*. In this case we say that the functions $u = \phi(x, y)$ and $v = \psi(x, y)$ are *dependent*. We first consider the special, almost trivial, case in which the equations $\phi_x = 0$ and $\phi_y = 0$ hold everywhere, so that the function $\phi(x, y)$ is a constant.

We then see that while the point (x, y) ranges over a whole region its image (u, v) always remains on the line $u = \text{const.}$ That is, our region is mapped only on a line, instead of on a region, so that there is no possibility here of speaking of a one-to-one mapping of two two-dimensional regions on one another. A similar situation arises in the general case in which at least one of the derivatives ϕ_x or ϕ_y does not vanish, but the Jacobian D is still zero. We suppose that at a point (x_0, y_0) of the region under consideration we have $\phi_x \neq 0$. It is then possible to resolve our transformation into two primitive transformations $\xi = \phi(x, y)$, $\eta = y$ and $u = \xi$, $v = \psi(\xi, \eta)$ just as in No. 5 (p. 150), for there we made use only of the assumption $\phi_x \neq 0$. In virtue of the equation $D = \phi_x \psi_\eta = 0$, however, ψ_η must be identically zero in the region where $\phi_x \neq 0$; that is, the quantity $\psi = v$ does not depend on η at all, and v is a function of $\xi = u$ alone. Our result is therefore as follows:

If the Jacobian of the transformation vanishes identically, a region of the xy*-plane is mapped by the transformation on a curve in the* uv*-plane instead of on a region,* since in a certain interval of values of u only one value of v corresponds to each value of u. *Thus if the Jacobian vanishes identically the functions are not independent, i.e. a relation*

$$F(\phi, \psi) = 0$$

exists which is satisfied for all systems of values (x, y) *in the above-mentioned region.* For if $F(u, v) = 0$ is the equation of the curve in the uv-plane on which the region of the xy-plane is mapped, then for all points of this region the equation

$$F(\phi(x, y), \psi(x, y)) = 0$$

is satisfied, i.e. this equation is an identity in x and y.

The exceptional case discussed separately at the beginning is obviously included in this general statement. The curve in question is then just the curve $u =$ const., which is a parallel to the v-axis.

An example of a degenerate transformation is

$$\xi = x + y, \quad \eta = (x + y)^2.$$

According to this transformation all the points of the xy-plane are mapped on the points of the parabola $\eta = \xi^2$ in the $\xi\eta$-plane. An inversion of the transformation is out of the question, for all the points of the line $x + y$ = const. are mapped on a single point (ξ, η). As we can easily verify, the value of the Jacobian is zero. The relation between the functions ξ and η, in accordance with the general theorem, is given by the equation $F(\xi, \eta) = \xi^2 - \eta = 0$.

8. Concluding Remarks.

The generalization of the theory for three or more independent variables offers no particular difficulties. The chief difference is that instead of the two-rowed determinant D we have determinants with three or more rows. In the case of transformations with three independent variables,

$$\xi = \phi(x, y, z), \quad \eta = \psi(x, y, z), \quad \zeta = \chi(x, y, z),$$
$$x = g(\xi, \eta, \zeta), \quad y = h(\xi, \eta, \zeta), \quad z = l(\xi, \eta, \zeta),$$

the Jacobian is given by the equation

$$D = \frac{\partial(\xi, \eta, \zeta)}{\partial(x, y, z)} = \begin{vmatrix} \phi_x & \psi_x & \chi_x \\ \phi_y & \psi_y & \chi_y \\ \phi_z & \psi_z & \chi_z \end{vmatrix}.$$

In the same way, for transformations

$$\begin{matrix} \xi_i = \phi_i(x_1, x_2, \ldots, x_n) \\ x_i = g_i(\xi_1, \xi_2, \ldots, \xi_n) \end{matrix} \qquad i = 1, 2, \ldots, n$$

with n independent variables the Jacobian is

$$\frac{\partial(\xi_1, \xi_2, \ldots, \xi_n)}{\partial(x_1, x_2, \ldots, x_n)} = \begin{vmatrix} \frac{\partial\phi_1}{\partial x_1}, & \frac{\partial\phi_2}{\partial x_1}, & \ldots, & \frac{\partial\phi_n}{\partial x_1} \\ \frac{\partial\phi_1}{\partial x_2}, & \frac{\partial\phi_2}{\partial x_2}, & \ldots, & \frac{\partial\phi_n}{\partial x_2} \\ \cdot & \cdot & \cdot & \cdot \\ \frac{\partial\phi_1}{\partial x_n}, & \frac{\partial\phi_2}{\partial x_n}, & \ldots, & \frac{\partial\phi_n}{\partial x_n} \end{vmatrix}.$$

For more than two independent variables it is still true that when transformations are combined the Jacobians are multiplied together. In symbols,

$$\frac{\partial(\xi_1, \xi_2, \ldots, \xi_n)}{\partial(\eta_1, \eta_2, \ldots, \eta_n)} \cdot \frac{\partial(\eta_1, \eta_2, \ldots, \eta_n)}{\partial(x_1, x_2, \ldots, x_n)} = \frac{\partial(\xi_1, \xi_2, \ldots, \xi_n)}{\partial(x_1, x_2, \ldots, x_n)}.$$

In particular, the Jacobian of the inverse transformation is the reciprocal of the Jacobian of the original transformation.

The theorems on the resolution and combination of transformations, on the inversion of a transformation, and on the dependence of transformations remain valid for three and more independent variables. The proofs are similar to those for the case $n = 2$; to avoid unnecessary repetition we shall omit them here.

In the preceding section we have seen that the behaviour of

a general transformation in many ways resembles that of an affine transformation, and that the Jacobian plays the same part as the determinant does in the case of affine transformations. The following remark makes this even clearer. Since the functions $\xi = \phi(x, y)$ and $\eta = \psi(x, y)$ are differentiable in the neighbourhood of (x_0, y_0), we can express them in the form

$$\xi - \xi_0 = (x - x_0)\,\phi_x(x_0, y_0) + (y - y_0)\,\phi_y(x_0, y_0) + \epsilon\sqrt{(x - x_0)^2 + (y - y_0)^2},$$

$$\eta - \eta_0 = (x - x_0)\,\psi_x(x_0, y_0) + (y - y_0)\,\psi_y(x_0, y_0) + \delta\sqrt{(x - x_0)^2 + (y - y_0)^2},$$

where ϵ and δ tend to zero with $\sqrt{\{(x - x_0)^2 + (y - y_0)^2\}}$. This shows that for sufficiently small values of $|x - x_0|$ and $|y - y_0|$ the transformation may be regarded, to a first approximation, as affine, since it can be represented approximately by the affine transformation

$$\xi = \xi_0 + (x - x_0)\phi_x(x_0, y_0) + (y - y_0)\phi_y(x_0, y_0),$$
$$\eta = \eta_0 + (x - x_0)\psi_x(x_0, y_0) + (y - y_0)\psi_y(x_0, y_0),$$

whose determinant is the Jacobian of the original transformation.

Examples

1. If $f(x)$ is a continuously differentiable function, then the transformation

$$u = f(x), \; v = -y + xf(x)$$

has a single inverse in every region of the xy-plane in which $f'(x) \neq 0$. The inverse transformation has the form

$$x = g(u), \; y = -v + ug(u).$$

2. A transformation is said to be "conformal" (see p. 166) if the angle between any two curves is preserved.

(*a*) Prove that the inversion

$$\xi = \frac{x}{x^2 + y^2}, \; \eta = \frac{y}{x^2 + y^2}$$

is a conformal transformation.

(*b*) Prove that the inverse of any circle is another circle or a straight line.

(*c*) Find the Jacobian of the inversion.

3. Prove that in a curvilinear triangle which is formed by three circles passing through one point O, the sum of the angles is π.

4. A transformation of the plane

$$u = \varphi(x, y),\ v = \psi(x, y)$$

is conformal if the functions φ and ψ satisfy the identities

$$\varphi_x = \psi_y,\ \varphi_y = -\psi_x.$$

5. The equation

$$\frac{x^2}{a-t} + \frac{y^2}{b-t} = 1 \qquad (a > b)$$

determines two values of t, depending on x and y:

$$t_1 = \lambda(x, y),$$
$$t_2 = \mu(x, y).$$

(*a*) Prove that the curves $t_1 =$ const. and $t_2 =$ const. are ellipses and hyperbolas all having the same foci (confocal conics).

(*b*) Prove that the curves $t_1 =$ const. and $t_2 =$ const. are orthogonal.

(*c*) t_1 and t_2 may be used as curvilinear co-ordinates (so-called "focal" co-ordinates). Express x and y in terms of these co-ordinates.

(*d*) Express the Jacobian $\frac{\partial(t_1, t_2)}{\partial(x, y)}$ in terms of x and y.

(*e*) Find the condition that two curves, which are represented parametrically in the system of focal co-ordinates by the equations

$$t_1 = f_1(\lambda),\ t_2 = f_2(\lambda) \quad \text{and} \quad t_1 = g_1(\mu),\ t_2 = g_2(\mu),$$

are orthogonal to one another.

6. (*a*) Prove that the equation in t

$$\frac{x^2}{a-t} + \frac{y^2}{b-t} + \frac{z^2}{c-t} = 1 \qquad (a > b > c)$$

has three distinct real roots t_1, t_2, t_3, which lie respectively in the intervals

$$-\infty < t < c,\ c < t < b,\ b < t < a,$$

provided that the point (x, y, z) does not lie on a co-ordinate plane.

(*b*) Prove that the three surfaces $t_1 =$ const., $t_2 =$ const., $t_3 =$ const. passing through an arbitrary point are orthogonal to one another.

(*c*) Express x, y, z in terms of the " focal co-ordinates " t_1, t_2, t_3.

7. Prove that the transformation of the xy-plane given by the equations

$$\xi = \tfrac{1}{2}\left(x + \frac{x}{x^2 + y^2}\right),\quad \eta = \tfrac{1}{2}\left(y - \frac{y}{x^2 + y^2}\right)$$

(*a*) is conformal;

(*b*) transforms straight lines through the origin and circles with the origin as centre in the xy-plane into confocal conics $t =$ const. given by

$$\frac{\xi^2}{t + \frac{1}{2}} + \frac{\eta^2}{t - \frac{1}{2}} = 1.$$

8. Inversion in three dimensions is defined by the formulæ

$$\xi = \frac{x}{x^2 + y^2 + z^2}, \quad \eta = \frac{y}{x^2 + y^2 + z^2}, \quad \zeta = \frac{z}{x^2 + y^2 + z^2}.$$

Prove that

(a) the angle between any two surfaces is unchanged;

(b) spheres are transformed either into spheres or into planes.

9. Prove that if all the normals of a surface $z = u(x, y)$ meet the z-axis, then the surface is a surface of revolution.

4. Applications

1. Applications to the Theory of Surfaces.

In the study of surfaces, as in that of curves, *parametric representation* is frequently to be preferred to other types of representation. Here we need two parameters instead of one; we denote them by u and v. A parametric representation may be expressed in the form

$$x = \phi(u, v), \quad y = \psi(u, v), \quad z = \chi(u, v),$$

where ϕ, ψ, and χ are given functions of the parameters u and v and the point (u, v) ranges over a given region R in the uv-plane. The corresponding point with the three rectangular co-ordinates (x, y, z) then ranges over a configuration in xyz-space. In general this configuration is a surface, which can be represented in the form $z = f(x, y)$, say. For we can seek to solve two of our three equations for u and v in terms of the two corresponding rectangular co-ordinates. If we substitute the expressions thus found for u and v in the third equation, we obtain an unsymmetrical representation of the surface, $z = f(x, y)$, say.* Hence in order to ensure that the equations really do represent a surface, we have only to assume that the three Jacobians

$$\begin{vmatrix} \phi_u & \phi_v \\ \psi_u & \psi_v \end{vmatrix}, \quad \begin{vmatrix} \psi_u & \psi_v \\ \chi_u & \chi_v \end{vmatrix}, \quad \begin{vmatrix} \chi_u & \chi_v \\ \phi_u & \phi_v \end{vmatrix}$$

do not all vanish at once; in a single formula, that

$$(\phi_u\psi_v - \phi_v\psi_u)^2 + (\psi_u\chi_v - \psi_v\chi_u)^2 + (\chi_u\phi_v - \chi_v\phi_u)^2 > 0.$$

Then in some neighbourhood of each point in space represented

* This is actually a special case of the parametric form, as we see by putting $x = u$ and $y = v$.

by our three equations it is certainly possible to express one of the three co-ordinates uniquely in terms of the other two.

A simple example of parametric representation is the representation of the spherical surface $x^2 + y^2 + z^2 = r^2$ of radius r by the equations

$$x = r \cos u \sin v, \quad y = r \sin u \sin v, \quad z = r \cos v$$
$$(0 \leqq u < 2\pi, \quad 0 \leqq v \leqq \pi),$$

where $v = \theta$ is the polar distance and $u = \varphi$ is the geographical longitude of the point on the sphere (cf. p. 141).

This example exhibits one of the advantages of parametric representation. The three co-ordinates are given explicitly as functions of u and v, and these functions are single-valued. If v runs from $\pi/2$ to π we obtain the lower hemisphere, i.e. $z = -\sqrt{(r^2 - x^2 - y^2)}$, while values of v from 0 to $\pi/2$ give the upper hemisphere. Thus with the parametric representation it is not necessary, as it is with the representation $z = \pm\sqrt{(r^2 - x^2 - y^2)}$, to consider two "single-valued branches" of the function in order to obtain the whole sphere.

We obtain another parametric representation of the sphere by means of *stereographic projection*. In order to project the sphere $x^2 + y^2 + z^2 - r^2 = 0$

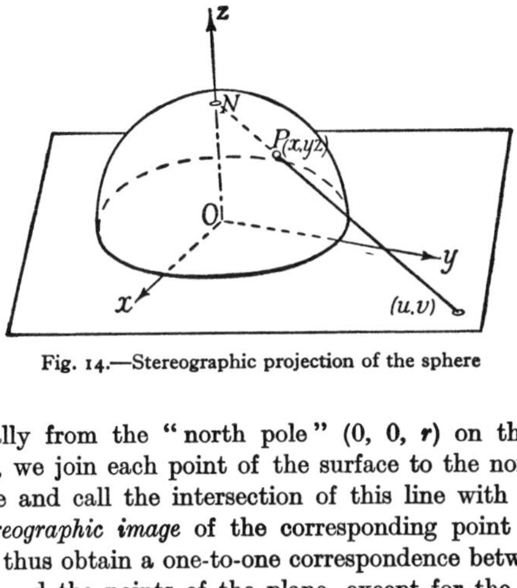

Fig. 14.—Stereographic projection of the sphere

stereographically from the "north pole" $(0, 0, r)$ on the "equatorial plane" $z = 0$, we join each point of the surface to the north pole N by a straight line and call the intersection of this line with the equatorial plane the *stereographic image* of the corresponding point of the sphere (fig. 14). We thus obtain a one-to-one correspondence between the points of the sphere and the points of the plane, except for the north pole N. Using elementary geometry, we readily find that this correspondence is expressed by the formulæ

$$x = \frac{2r^2u}{u^2 + v^2 + r^2}, \quad y = \frac{2r^2v}{u^2 + v^2 + r^2}, \quad z = \frac{(u^2 + v^2 - r^2)r}{u^2 + v^2 + r^2},$$

where (u, v) are the rectangular co-ordinates of the image-point in the plane. These equations may be regarded as a parametric representation of the sphere, the parameters u and v being rectangular co-ordinates in the uv-plane.

As a further example we give parametric representations of the surfaces

$$\frac{x^2}{a^2}+\frac{y^2}{b^2}-\frac{z^2}{c^2}=1 \quad \text{and} \quad \frac{x^2}{a^2}-\frac{y^2}{b^2}-\frac{z^2}{c^2}=1,$$

which are called the *hyperboloid of one sheet* and the *hyperboloid of two*

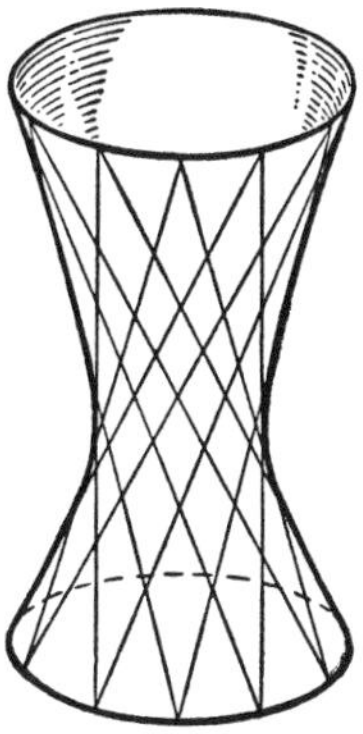

Fig. 15.—Hyperboloid of one sheet

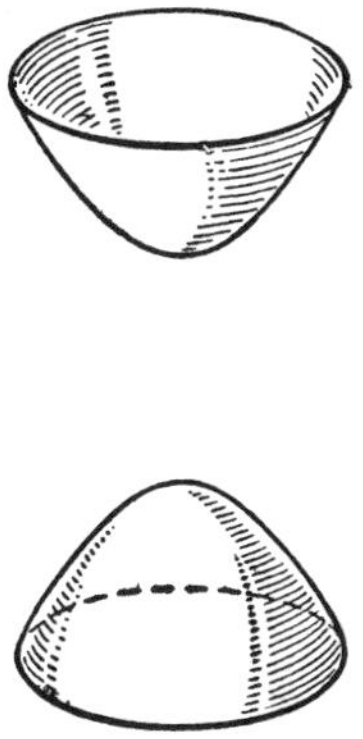

Fig. 16.—Hyperboloid of two sheets

sheets respectively (cf. figs. 15 and 16). The hyperboloid of one sheet is represented by

$$\begin{aligned} x &= a\cos u\,\frac{e^v+e^{-v}}{2} = a\cos u\cosh v, \\ y &= b\sin u\,\frac{e^v+e^{-v}}{2} = b\sin u\cosh v, \qquad \begin{aligned} &0 \leqq u < 2\pi \\ -&\infty < v < +\infty \end{aligned} \\ z &= c\,\frac{e^v-e^{-v}}{2} = c\sinh v; \end{aligned}$$

the hyperboloid of two sheets by

$$\begin{aligned} x &= a\,\frac{e^v+e^{-v}}{2} = a\cosh v, \\ y &= b\cos u\,\frac{e^v-e^{-v}}{2} = b\cos u\sinh v, \qquad \begin{aligned} &0 \leqq u < 2\pi \\ -&\infty < v < +\infty \end{aligned} \\ z &= c\sin u\,\frac{e^v-e^{-v}}{2} = c\sin u\sinh v. \end{aligned}$$

In general, we may regard the *parametric representation* of surface as the *mapping of the region* R *of the* uv-*plane on the* *corresponding surface*, where, as always, the word mapping is understood to mean a point-to-point correspondence. To each point of the region R of the uv-plane there corresponds one point of the surface, and in general the converse is also true.*

In the same way, a curve $u = u(t)$, $v = v(t)$ in the uv-plane corresponds in virtue of the equations $x = \phi(u(t), v(t)) = x(t), \ldots$ to a curve on the surface (cf. p. 85). In particular, in the representation of the sphere by means of polar co-ordinates the meridians are represented by the equation $u =$ const. and the parallels of latitude by $v =$ const. This net of curves thus corresponds to the system of parallels to the axes in the uv-plane.

The representation of a curve on a given surface is one of the most important methods for thorough investigation of the properties of the surface. Here we shall give only the expression for s, the length of arc of such a curve. As we mentioned in Chap. II, section 7 (p. 86), we have

$$\left(\frac{ds}{dt}\right)^2 = \left(\frac{dx}{dt}\right)^2 + \left(\frac{dy}{dt}\right)^2 + \left(\frac{dz}{dt}\right)^2,$$

so that in virtue of the equations

$$\frac{dx}{dt} = x_u \frac{du}{dt} + x_v \frac{dv}{dt}, \text{ \&c.},$$

we obtain

$$\left(\frac{ds}{dt}\right)^2 = E\left(\frac{du}{dt}\right)^2 + 2F\frac{du}{dt}\frac{dv}{dt} + G\left(\frac{dv}{dt}\right)^2,$$

where for the sake of compactness we have introduced the *Gaussian fundamental quantities of the surface,*

$$E = \left(\frac{\partial x}{\partial u}\right)^2 + \left(\frac{\partial y}{\partial u}\right)^2 + \left(\frac{\partial z}{\partial u}\right)^2,$$

$$F = \frac{\partial x}{\partial u}\frac{\partial x}{\partial v} + \frac{\partial y}{\partial u}\frac{\partial y}{\partial v} + \frac{\partial z}{\partial u}\frac{\partial z}{\partial v},$$

$$G = \left(\frac{\partial x}{\partial v}\right)^2 + \left(\frac{\partial y}{\partial v}\right)^2 + \left(\frac{\partial z}{\partial v}\right)^2.$$

* This, of course, is not always the case. For example, in the representation of the sphere by polar co-ordinates (p. 160) the poles of the sphere correspond to the whole line-segments $v = 0$ and $v = \pi$ respectively.

These are independent of the particular choice of the curve on the surface, and depend only on the surface itself and its parametric representation. The above expressions for the derivative of the length of arc with respect to the parameter are usually expressed symbolically by omitting the reference to the parameter t and saying that the "line element" ds on the surface is given by the "quadratic differential form"

$$ds^2 = E\,du^2 + 2F\,du\,dv + G\,dv^2.$$

For the direction cosines of the normal to a surface given in the form $\Phi(x, y, z) = 0$ we have already obtained (p. 130) the expressions

$$\cos\alpha = \frac{\Phi_x}{\sqrt{(\Phi_x^2 + \Phi_y^2 + \Phi_z^2)}}, \quad \cos\beta = \frac{\Phi_y}{\sqrt{(\Phi_x^2 + \Phi_y^2 + \Phi_z^2)}},$$

$$\cos\gamma = \frac{\Phi_z}{\sqrt{(\Phi_x^2 + \Phi_y^2 + \Phi_z^2)}}.$$

To obtain these direction cosines in the case of parametric representation, we suppose that the surface given by the equations $x = \phi(u, v)$, $y = \psi(u, v)$, $z = \chi(u, v)$ is written in the form $\Phi(x, y, z) = 0$. The equation

$$\Phi(\phi(u, v), \psi(u, v), \chi(u, v)) = 0$$

is then an identity in u and v, and by differentiation we obtain

$$\Phi_x\phi_u + \Phi_y\psi_u + \Phi_z\chi_u = 0,$$
$$\Phi_x\phi_v + \Phi_y\psi_v + \Phi_z\chi_v = 0.$$

From these it follows at once that (cf. Chap. I, section 3, p. 26)

$$\Phi_x = \rho(\psi_u\chi_v - \chi_u\psi_v); \quad \Phi_y = \rho(\chi_u\phi_v - \phi_u\chi_v);$$
$$\Phi_z = \rho(\phi_u\psi_v - \psi_u\phi_v),$$

where ρ is a suitably chosen multiplier. From the definition of E, F, G we find by direct expansion that

$$(\psi_u\chi_v - \chi_u\psi_v)^2 + (\chi_u\phi_v - \phi_u\chi_v)^2 + (\phi_u\psi_v - \psi_u\phi_v)^2 = EG - F^2,$$

and combining this with the preceding equation, we have

$$\Phi_x^2 + \Phi_y^2 + \Phi_z^2 = \rho^2(EG - F^2).$$

Thus we finally obtain the formulæ for the direction cosines of the normal to the surface in the form

$$\cos\alpha = \frac{\psi_u\chi_v - \chi_u\psi_v}{\sqrt{(EG - F^2)}}, \quad \cos\beta = \frac{\chi_u\phi_v - \phi_u\chi_v}{\sqrt{(EG - F^2)}},$$

$$\cos\gamma = \frac{\phi_u\psi_v - \psi_u\phi_v}{\sqrt{(EG - F^2)}}.$$

The equations $u = g(t)$, $v = h(t)$, as we have seen, represent a curve on the surface. The direction cosines of the tangent to this curve are given according to the chain rule by the expressions

$$\cos\alpha = \frac{dx}{ds} = \frac{dx}{dt}\frac{dt}{ds} = \frac{x_u u' + x_v v'}{\sqrt{(Eu'^2 + 2Fu'v' + Gv'^2)}},$$

$$\cos\beta = \frac{y_u u' + y_v v'}{\sqrt{(Eu'^2 + 2Fu'v' + Gv'^2)}}, \quad \cos\gamma = \frac{z_u u' + z_v v'}{\sqrt{(Eu'^2 + 2Fu'v' + Gv'^2)}}.$$

Here for brevity we have put $\dfrac{dg(t)}{dt} = u'$, $\dfrac{dh(t)}{dt} = v'$. If we now consider a second curve on the surface, given by the equations $u = g_1(t)$, $v = h_1(t)$, whose tangent has the direction cosines $\cos\alpha_1$, $\cos\beta_1$, $\cos\gamma_1$, and if we use the abbreviations

$$\frac{dg_1(t)}{dt} = \dot{u}, \quad \frac{dh_1(t)}{dt} = \dot{v},$$

then the cosine of the *angle between the two curves* is given by the cosine of the angle between their tangents, that is, by

$$\cos\omega = \cos\alpha\cos\alpha_1 + \cos\beta\cos\beta_1 + \cos\gamma\cos\gamma_1$$
$$= \frac{E\dot{u}u' + F(\dot{u}v' + u'\dot{v}) + G\dot{v}v'}{\sqrt{(E\dot{u}^2 + 2F\dot{u}\dot{v} + G\dot{v}^2)}\sqrt{(Eu'^2 + 2Fu'v' + Gv'^2)}},$$

where all the quantities on the right are to be given the values which they have at the point of intersection of the two curves.

In particular, we may consider those curves on the surface which are given by equations $u =$ const. or $v =$ const. If in our parametric representation we substitute a definite fixed value for u, we obtain a three-dimensional or twisted curve lying on the surface and having v as parameter; and a corresponding statement holds good if we substitute a fixed value for v and allow u to vary. These curves $u =$ const. and $v =$ const. are the *parametric curves* on the surface. The net of parametric curves corresponds to the net of parallels to the axes in the uv-plane (fig. 17).

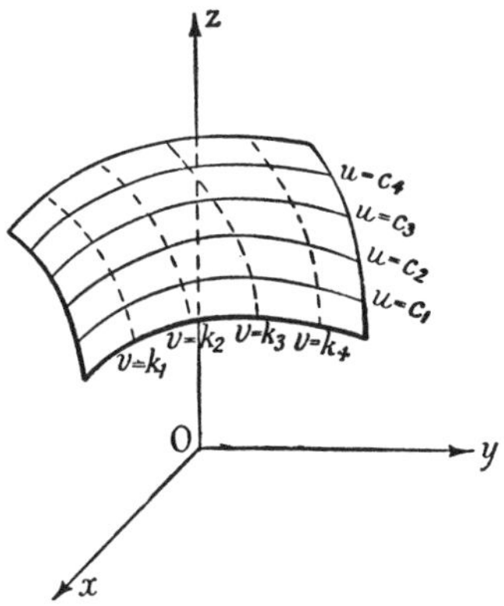

Fig. 17.—Parametric curves $u =$ const., $v =$ const.

The mapping of one plane region on another may be regarded as a special case of parametric representation. For if the third of our functions $\chi(u, v)$ vanishes for all values of u and v under consideration, then as the point (u, v) ranges over its given region the point (x, y, z) will range over a region in the xy-plane. Hence our equations merely represent the mapping of a region of the uv-plane on a region of the xy-plane; or if we prefer to think in terms of transformations of co-ordinates, the equations define a system of curvilinear co-ordinates in the uv-region, and the inverse functions (if they exist) define a curvilinear uv-system of co-ordinates in the plane xy-region. In terms of the curvilinear co-ordinates (u, v) the line element in the xy-plane is simply

$$ds^2 = E\,du^2 + 2F\,du\,dv + G\,dv^2,$$

where

$$E = \left(\frac{\partial x}{\partial u}\right)^2 + \left(\frac{\partial x}{\partial v}\right)^2,$$

$$F = \frac{\partial x}{\partial u}\frac{\partial x}{\partial v} + \frac{\partial y}{\partial u}\frac{\partial y}{\partial v},$$

$$G = \left(\frac{\partial x}{\partial v}\right)^2 + \left(\frac{\partial y}{\partial v}\right)^2.$$

As a further example of the representation of a surface in parametric form we consider the *anchor ring* or *torus*. This is obtained by rotating a circle about a line which lies in the plane of the circle and does not intersect it (cf. fig. 18). If we take this axis of rotation as the z-axis and choose the y-axis in such a way that it passes through the centre of the circle, whose y-co-ordinate we denote by a, and if the radius of the circle is $r < |a|$, we obtain in the first instance

$$x = 0, \quad y - a = r\cos\theta, \quad z = r\sin\theta \qquad (0 \leq \theta < 2\pi)$$

as a parametric representation of the circle in the yz-plane. Now letting the circle rotate about the z-axis, we find that for each point of the circle $x^2 + y^2$ remains constant, that is, $x^2 + y^2 = (a + r\cos\theta)^2$. Thus if the angle of rotation about the z-axis is denoted by φ we have

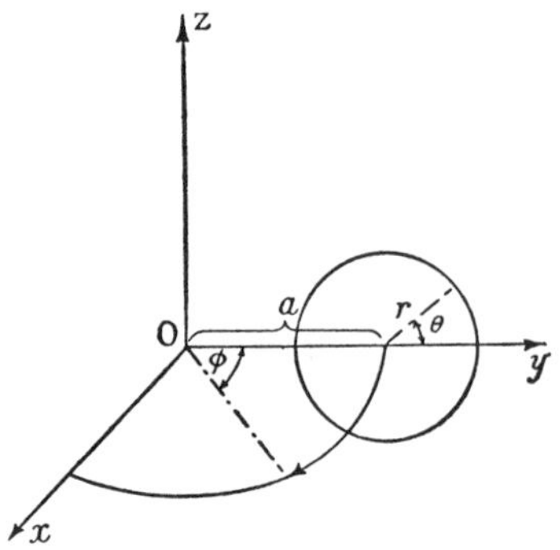

Fig. 18.—Generation of an anchor ring by the rotation of a circle

$$\begin{aligned} x &= (a + r\cos\theta)\sin\varphi, \\ y &= (a + r\cos\theta)\cos\varphi, \quad & 0 \leqq \varphi < 2\pi \\ z &= \qquad r\sin\theta & 0 \leqq \theta < 2\pi \end{aligned}$$

as a parametric representation of the anchor ring in terms of the parameters θ and φ. In this representation the anchor ring appears as the image of a square of side 2π in the $\theta\varphi$-plane, where any pair of boundary points lying on the same line $\theta =$ const. or $\varphi =$ const. corresponds to only one point on the surface, and the four corners of the square all correspond to the same point.

For the line element on the anchor ring we have

$$ds^2 = r^2 d\theta^2 + (a + r\cos\theta)^2 d\varphi^2.$$

2. Conformal Representation in General.

A transformation

$$\xi = \phi(x, y), \ \eta = \psi(x, y)$$

is called a *conformal transformation* if any two curves are transformed by it into two others which make the same angle with each other as the original ones do.

Theorem.—A necessary and sufficient condition that our (continuously differentiable) transformation should be conformal is that the *Cauchy-Riemann equations*

$$\phi_x - \psi_y = 0, \ \phi_y + \psi_x = 0$$

or

$$\phi_x + \psi_y = 0, \ \phi_y - \psi_x = 0$$

hold. In the first case the direction of the angles is preserved, in the second case the direction is reversed.*

Proof.—We assume that the transformation is conformal.

* This last statement follows directly from the statements on p. 151 concerning the sign of the Jacobian $\phi_x\psi_y - \phi_y\psi_x$.

Then the two orthogonal curves $\xi = \text{const.}$, $\eta = \text{const.}$ in the $\xi\eta$-plane must correspond to orthogonal curves $\phi(x, y) = \text{const.}$ and $\psi(x, y) = \text{const.}$ in the xy-plane.

Hence from the formula for the angle between two curves (p. 126) it follows immediately that

$$\phi_x\psi_x + \phi_y\psi_y = 0.$$

In the same way, the curves corresponding to $\xi + \eta = \text{const.}$ and $\xi - \eta = \text{const.}$ must be orthogonal. This gives

$$(\phi_x + \psi_x)(\phi_x - \psi_x) + (\phi_y + \psi_y)(\phi_y - \psi_y) = 0,$$

and therefore

$$\phi_x^2 + \phi_y^2 = \psi_x^2 + \psi_y^2.$$

The first of our equations can be written in the form

$$\phi_x = \lambda\psi_y, \; \phi_y = -\lambda\psi_x,$$

where λ denotes a constant of proportionality. Introducing this in the second equation, we immediately get $\lambda^2 = 1$, so that one or other of our two systems of Cauchy-Riemann equations holds.

That the equations are a *sufficient* condition is confirmed by the following remark:

If two curves in the xy-plane are given by equations $F(x, y) = 0$, $G(x, y) = 0$ and if according to our transformation $F(x, y) = \Phi(\xi, \eta)$, $G(x, y) = \Gamma(\xi, \eta)$, then by using the Cauchy-Riemann equations we readily obtain

$$F_x^2 + F_y^2 = (\Phi_\xi^2 + \Phi_\eta^2)(\phi_x^2 + \phi_y^2),$$
$$G_x^2 + G_y^2 = (\Gamma_\xi^2 + \Gamma_\eta^2)(\phi_x^2 + \phi_y^2),$$
$$F_xG_x + F_yG_y = (\Phi_\xi\Gamma_\xi + \Phi_\eta\Gamma_\eta)(\phi_x^2 + \phi_y^2);$$

therefore

$$\frac{F_xG_x + F_yG_y}{\sqrt{(F_x^2 + F_y^2)}\sqrt{(G_x^2 + G_y^2)}} = \frac{\Phi_\xi\Gamma_\xi + \Phi_\eta\Gamma_\eta}{\sqrt{(\Phi_\xi^2 + \Phi_\eta^2)}\sqrt{(\Gamma_\xi^2 + \Gamma_\eta^2)}}.$$

That is, the curves $F = 0$, $G = 0$ and their images $\Phi = 0$, $\Gamma = 0$ make the same angle with each other.

Examples

1. (*a*) Prove that the stereographic projection of the unit sphere on the plane is conformal.

(*b*) Prove that circles on the sphere are transformed either into circles or into straight lines in the plane.

(c) Prove that in stereographic projection reflection of the spherical surface in the equatorial plane corresponds to an inversion in the uv-plane.

(d) Find the expression for the line element on the sphere in terms of the parameters u, v.

2. Calculate the line element

(a) on the sphere

$$x = \cos u \sin v,\ y = \sin u \sin v,\ z = \cos v;$$

(b) on the hyperboloid

$$x = \cos u \cosh v,\ y = \sin u \cosh v,\ z = \sinh v;$$

(c) on a surface of revolution given by

$$r = \sqrt{(x^2 + y^2)} = f(z),$$

using the cylindrical co-ordinates z and $\theta = \arctan \frac{y}{x}$ as co-ordinates on the surface;

(d)* on the quadric $t_3 =$ const. of the family of confocal quadrics given by

$$\frac{x^2}{a-t} + \frac{y^2}{b-t} + \frac{z^2}{c-t} = 1,$$

using t_1 and t_2 as co-ordinates on the quadric (cf. Ex. 6, p. 158).

3. Prove that if a new system of curvilinear co-ordinates r, s is introduced on a surface with parameters u, v by means of the equations

$$u = u(r, s),\ v = v(r, s),$$

then

$$E'G' - F'^2 = (EG - F^2)\left\{\frac{\partial(u, v)}{\partial(r, s)}\right\}^2,$$

where E', F', G' denote the fundamental quantities taken with respect to r, s and E, F, G those taken with respect to u, v.

4. Let t be a tangent to a surface S at the point P, and consider the sections of S made by all planes containing t. Prove that the centres of curvature of the different sections lie on a circle.

5. If t is a tangent to the surface S at the point P, we call the curvature of the normal plane section through t (i.e. the section through t and the normal) at that point the "curvature (k) of S in the direction t". For every tangent at P we take the vector with the direction of t, initial point P, and length $\frac{1}{\sqrt{k}}$. Prove that the final points of these vectors lie on a conic.

6*. A curve is given as the intersection of the two surfaces

$$x^2 + y^2 + z^2 = 1$$
$$ax^2 + by^2 + cz^2 = 0.$$

Find the equations of

(a) the tangent,

(b) the osculating plane, at any point of the curve.

5. Families of Curves, Families of Surfaces, and their Envelopes

1. General Remarks.

On various occasions we have already considered curves or surfaces not as individual configurations, but as members of a *family* of curves or surfaces, such as $f(x, y) = c$, where to each value of c there corresponds a different curve of the family.

For example, the lines parallel to the y-axis in the xy-plane, that is, the lines $x = c$, form a family of curves. The same is true for the family of concentric circles $x^2 + y^2 = c^2$ about the origin; to each value of c there corresponds a circle of the family, namely the circle with radius c. Similarly, the rectangular hyperbolas $xy = c$ form a family of curves, sketched in fig. 2, p. 113. The particular value $c = 0$ corresponds to the degenerate hyperbola consisting of the two co-ordinate axes. Another example of a family of curves is the set of all the normals to a given curve. If the curve is given in terms of the parameter t by the equations $\xi = \varphi(t)$, $\eta = \psi(t)$, we obtain the equation of the family of normals in the form

$$(x - \varphi(t))\,\varphi'(t) + (y - \psi(t))\,\psi'(t) = 0,$$

where t is used instead of c to denote the parameter of the family.

The general concept of a family of curves can be expressed analytically in the following way. Let

$$f(x, y, c)$$

be a continuously differentiable function of the two independent variables x and y and of the *parameter* c, this parameter varying in a given interval. (Thus the parameter is really a third independent variable, which is lettered differently simply because it plays a different part.) Then if the equation

$$f(x, y, c) = 0$$

for each value of the parameter c represents a curve, the aggregate of the curves obtained as c describes its interval is called a *family of curves* depending on the parameter c.

The curves of such a family may also be represented in parametric form by means of a *parameter t of the curve*, in the form

$$x = \phi(t, c), \quad y = \psi(t, c),$$

where c is again the parameter of the family. If we assign c a fixed value, these equations represent a curve with the parameter t.

For example, the equations

$$x = c \cos t, \quad y = c \sin t$$

represent the family of concentric circles mentioned above; again, the equations

$$x = ct, \quad y = \frac{1}{t}$$

represent the family of rectangular hyperbolas mentioned above, except for the degenerate hyperbola consisting of the co-ordinate axes.

Occasionally we are led to consider families of curves which depend not on one parameter but on several parameters. For example, the aggregate of all circles $(x - a)^2 + (y - b)^2 = c^2$ in the plane is a family of curves depending on the three parameters a, b, c. If nothing is said to the contrary, we shall always understand a family of curves to be a " one-parameter " family, depending on a single parameter. The other cases we shall distinguish by speaking of two-parameter, three-parameter, or multi-parameter families of curves.

Similaı statements of course hold for families of surfaces in space. If we are given a continuously differentiable function $f(x, y, z, c)$, and if for each value of the parameter c in a certain definite interval the equation

$$f(x, y, z, c) = 0$$

represents a surface in the space with rectangular co-ordinates x, y, z, then the aggregate of the surfaces obtained by letting c describe its interval is called a *family of surfaces*, or, more precisely, a one-parameter family of surfaces with the parameter c. For example, the spheres $x^2 + y^2 + z^2 = c^2$ about the origin form such a family. As with curves, we can also consider families of surfaces depending on several parameters.

Thus the planes defined by the equation

$$ax + by + \sqrt{1 - a^2 - b^2}\, z + 1 = 0$$

form a two-parameter family, depending on the parameters a and b, if the parameters a and b range over the region $a^2 + b^2 \leq 1$. This family of surfaces consists of the class of all planes which are at unit distance from the origin.*

* Sometimes a one-parametric family of surfaces is referred to as ∞^1 surfaces, a two-parametric family as ∞^2 surfaces, and so on.

2. Envelopes of One-Parameter Families of Curves.

If a family of straight lines is identical with the aggregate of the tangents to a plane curve E—as e.g. the family of normals of a curve C is identical with the family of tangents to the evolute E of C (cf. Vol. I, p. 308)—we shall say that the curve E is the *envelope* of the family of lines. In the same way we shall say that the family of circles with radius 1 and centre on the x-axis, that is, the family of circles with the equation $(x - c)^2 + y^2 - 1 = 0$, has the pair of lines $y = 1$ and $y = -1$, which touch each of

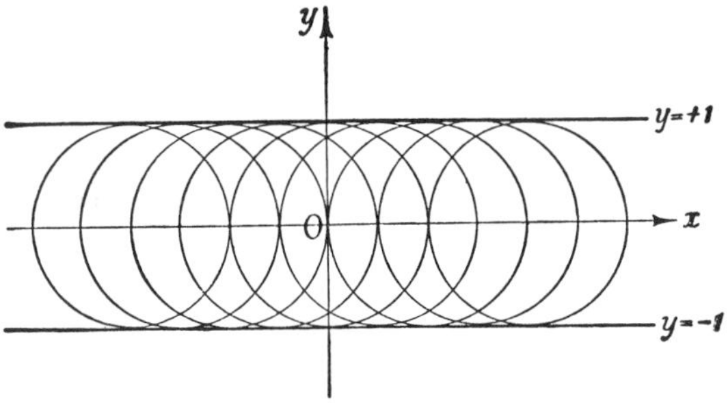

Fig. 19.—Family of circles with envelope

the circles, as its envelope (fig. 19). In these cases we can obtain the point of contact of the envelope and the curve of the family by finding the intersection of two curves of the family with parameter values c and $c + h$ and then letting h tend to zero. We may express this briefly by saying that the envelope is the locus of the intersections of neighbouring curves.

With other families of curves it may again happen that a curve E exists which at each of its points touches some one of the curves of the family, the particular curve depending of course on the point of E in question. We then call E the *envelope* of the family of curves. The question now arises of finding the envelope E of a given family of curves $f(x, y, c) = 0$. We first make a few plausible remarks, in which we assume that an envelope E does exist and that it can be obtained, as in the above cases, as the locus of the intersections of neighbouring curves.* We then obtain the point of contact of the curve

* Since this last assumption will be shown by examples to be too restrictive, we shall shortly replace these plausibilities by a more complete discussion.

$f(x, y, c) = 0$ with the curve E in the following way. In addition to this curve we consider a neighbouring curve $f(x, y, c + h) = 0$, find the intersection of these two curves, and then let h tend to zero. The point of intersection must then approach the point of contact sought. At the point of intersection the equation

$$\frac{f(x, y, c + h) - f(x, y, c)}{h} = 0$$

is true as well as the equations $f(x, y, c + h) = 0$ and $f(x, y, c) = 0$ In the first equation we perform the passage to the limit $h \to 0$. Since we have assumed the existence of the partial derivative f_c, this gives the two equations

$$f(x, y, c) = 0, \quad f_c(x, y, c) = 0$$

for the point of contact of the curve $f(x, y, c) = 0$ with the envelope. If we can determine x and y as functions of c by means of these equations, we obtain the parametric representation of a curve with the parameter c, and this curve is the envelope. By elimination of the parameter c it can also be represented in the form $g(x, y) = 0$. This equation is called the " discriminant " of the family, and the curve given by the equation $g(x, y) = 0$ is called the " discriminant curve ".

We are thus led to the following rule: *in order to obtain the envelope of a family of curves* f(x, y, c) = 0, *we consider the two equations* f(x, y, c) = 0 *and* $\mathrm{f_c}$(x, y, c) = 0 *simultaneously and attempt to express* x *and* y *as functions of* c *by means of them or to eliminate the quantity* c *between them.*

We shall now replace the above heuristic considerations by a more complete and more general discussion, based on the definition of the envelope as the curve of contact. At the same time we shall learn under what conditions our rule actually does give the envelope, and what other possibilities present themselves.

We assume to begin with that E is an envelope which can be represented in terms of the parameter c by two continuously differentiable functions

$$x = x(c), \quad y = y(c),$$

where $\left(\frac{dx}{dc}\right)^2 + \left(\frac{dy}{dc}\right)^2 \neq 0$, and which at the point with parameter c touches the curve of the family with the same value of the parameter c. In the first place, the equation $f(x, y, c) = 0$ is satisfied at the point of contact. If in this equation we substitute the expressions $x(c)$ and $y(c)$ for x and y, it remains valid for all values of c in the interval. On differentiating with respect to c we at once obtain

$$f_x \frac{dx}{dc} + f_y \frac{dy}{dc} + f_c = 0.$$

Now the condition of tangency is

$$f_x \frac{dx}{dc} + f_y \frac{dy}{dc} = 0;$$

for the quantities dx/dc and dy/dc are proportional to the direction cosines of the tangent to E and the quantities f_x and f_y are proportional to the direction cosines of the normal to the curve $f(x, y, c) = 0$ of the family, and these directions must be at right angles to one another. It follows that the envelope satisfies the equation $f_c = 0$, and we thus see that the rule given above is a *necessary condition* for the envelope.

In order to find out how far this condition is also *sufficient*, we assume that a curve E represented by two continuously differentiable functions $x = x(c)$ and $y = y(c)$ satisfies the two equations $f(x, y, c) = 0$ and $f_c(x, y, c) = 0$. In the first equation we again substitute $x(c)$ and $y(c)$ for x and y; this equation then becomes an identity in c. If we differentiate with respect to c and remember that $f_c = 0$, we at once obtain the relation

$$f_x \frac{dx}{dc} + f_y \frac{dy}{dc} = 0,$$

which therefore holds for all points of E. If the two expressions $f_x^2 + f_y^2$ and $(dx/dc)^2 + (dy/dc)^2$ both differ from zero at a point of E, so that at that point both the curve E and the curve of the family have well-defined tangents, this equation states that the envelope and the curve of the family touch one another. With these additional assumptions our rule is a *sufficient* condition for the envelope as well as a necessary one. If, however, f_x and f_y both vanish, the curve of the family may have a singular point (cf. section 2, p. 128), and we can draw no conclusions about the contact of the curves.

Thus after we have found the discriminant curve it is still

necessary to make a further investigation in each case, in order to discover whether it is really an envelope or to what extent it fails to be one.

In conclusion we state the condition for the discriminant curve of a family of curves given in parametric form

$$x = \phi(t, c), \quad y = \psi(t, c),$$

with the curve parameter t. This is

$$\phi_t\psi_c - \phi_c\psi_t = 0.$$

We can readily obtain it e.g. if we pass from the parametric representation of the family to the original expression by elimination of t.

3. Examples.

1. $(x - c)^2 + y^2 = 1$. As we have seen on p. 171, this equation represents the family of circles of unit radius whose centres lie on the x-axis (fig. 19). Geometrically we see at once that the envelope must consist of the two lines $y = 1$ and $y = -1$. We can verify this by means of our rule; for the two equations $(x - c)^2 + y^2 = 1$ and $-2(x - c) = 0$ immediately give us the envelope in the form $y^2 = 1$.

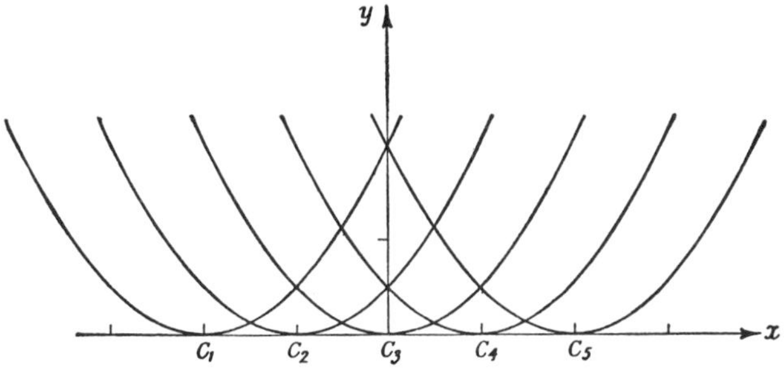

Fig. 20.—Family of parabolas with envelope

2. The family of circles of unit radius passing through the origin, whose centres, therefore, must lie on the circle of unit radius about the origin, is given by the equation

$$(x - \cos c)^2 + (y - \sin c)^2 = 1$$

or

$$x^2 + y^2 - 2x \cos c - 2y \sin c = 0.$$

The derivative with respect to c equated to zero gives $x \sin c - y \cos c = 0$. These two equations are satisfied by the values $x = 0$ and $y = 0$.

If, however, $x^2 + y^2 \neq 0$, it readily follows from our equations that $\sin c = y/2$, $\cos c = x/2$, so that on eliminating c we obtain $x^2 + y^2 = 4$. Thus for the envelope our rule gives us the circle of radius 2 about the origin, as is anticipated by geometrical intuition; but it also gives us the isolated point $x = 0$, $y = 0$.

3. The family of parabolas $(x - c)^2 - 2y = 0$ (cf. fig. 20) also has an envelope, which both by intuition and by our rule is found to be the x-axis.

4. We next consider the family of circles $(x - 2c)^2 + y^2 - c^2 = 0$

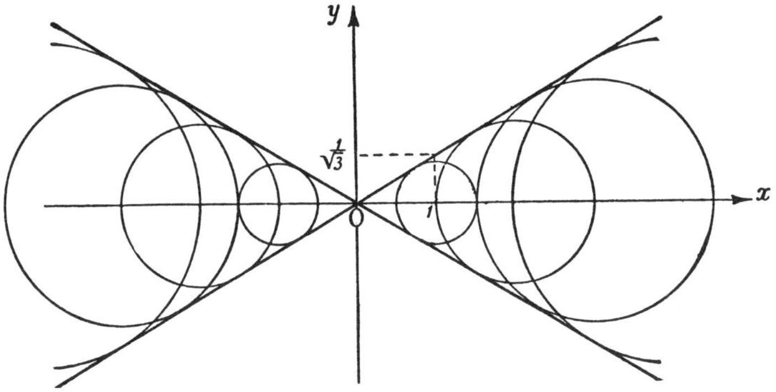

Fig. 21.—The family $(x - 2c)^2 + y^2 - c^2 = 0$

(cf. fig. 21). Differentiation with respect to c gives $2x - 3c = 0$, and by substitution we find that the equation of the envelope is

$$y^2 = \frac{x^2}{3};$$

that is, the envelope consists of the two lines $y = \frac{1}{\sqrt{3}}x$ and $y = -\frac{1}{\sqrt{3}}x$. The origin is an exception, in that contact does not occur there.

5. Another example is the family of straight lines on which unit length is intercepted by the x- and y-axes. If $\alpha = c$ is the angle indicated in fig. 22, these lines are given by the equation

$$\frac{x}{\cos\alpha} + \frac{y}{\sin\alpha} = 1.$$

The condition for the envelope is

$$\frac{\sin\alpha}{\cos^2\alpha}x - \frac{\cos\alpha}{\sin^2\alpha}y = 0,$$

which, in conjunction with the equation of the lines, gives the envelope in parametric form,

$$x = \cos^3\alpha, \quad y = \sin^3\alpha.$$

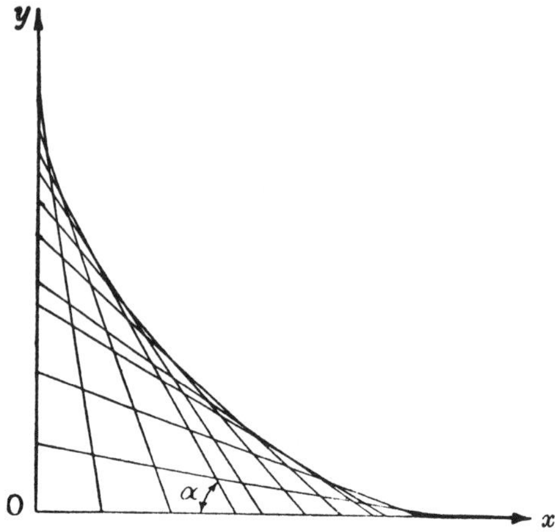

Fig. 22.—Arc of the astroid as envelope of straight lines

From these we obtain the further equation

$$x^{2/3} + y^{2/3} = 1.$$

This curve is called the *astroid* (cf. Vol. I, Chap. V, Ex. 6, p. 267). It consists (figs. 23, 24) of four symmetrical branches meeting in four cusps.

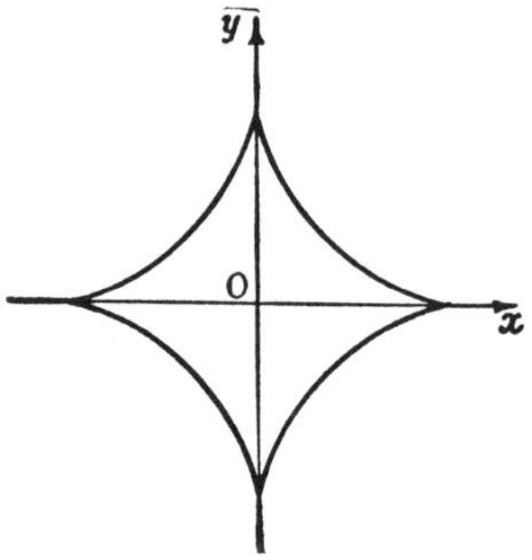

Fig. 23.—Astroid

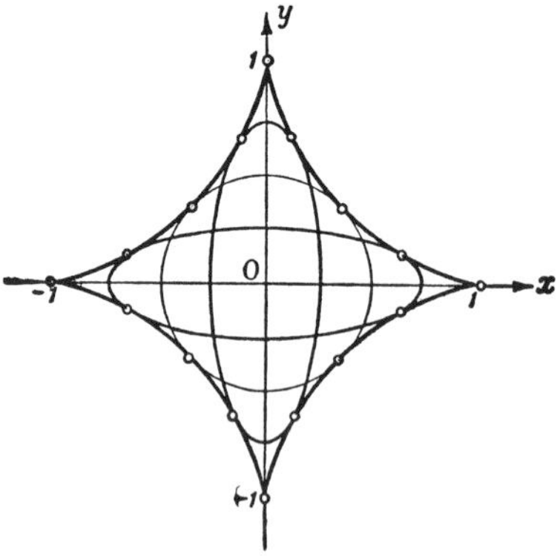

Fig. 24.—Astroid as envelope of ellipses

6. The astroid $x^{2/3} + y^{2/3} = 1$ also appears as the envelope of the family of ellipses

$$\frac{x^2}{c^2} + \frac{y^2}{(1 - c)^2} = 1$$

whose semi-axes c and $(1 - c)$ have the constant sum 1 (fig. 24).

7. The family of curves $(x - c)^2 - y^3 = 0$ shows that in certain circumstances our process may fail to give an envelope. Here the rule gives

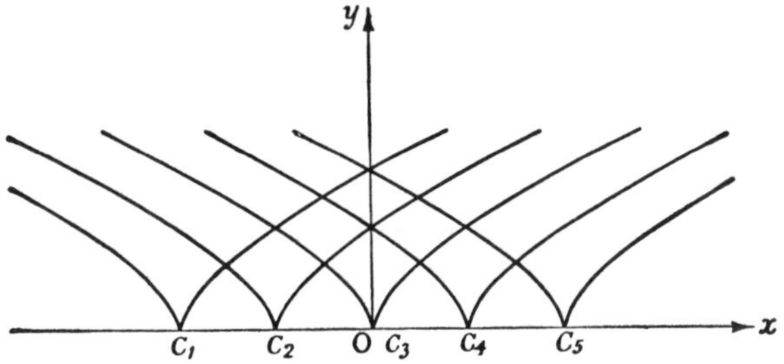

Fig. 25.—The family $(x - c)^2 - y^3 = 0$

the x-axis. But, as fig. 25 shows, this is not an envelope; it is the locus of the cusps of the curves of the family.

8. In the case of the family

$$(x - c)^3 - y^2 = 0$$

we again find that the discriminant curve is the x-axis (cf. fig. 26). This

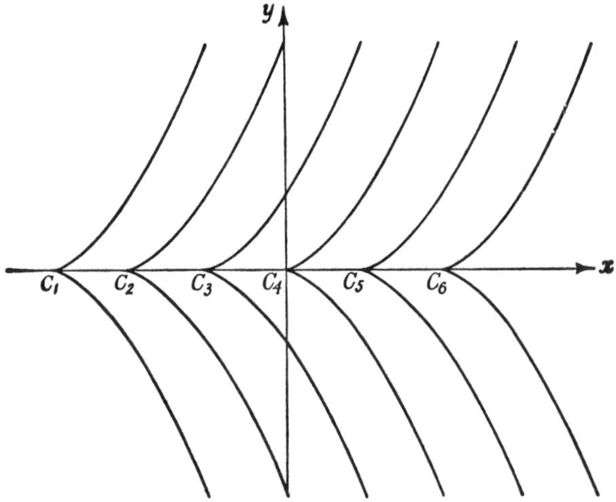

Fig. 26.—The family $(x - c)^3 - y^2 = 0$

is again the cusp-locus; but it touches each of the curves, and in this sense must be regarded as the envelope.

9. Another example, in which the discriminant curve consists of the envelope plus the locus of the double points, is given by the family of *strophoids*

$$[x^2 + (y - c)^2] (x - 2) + x = 0$$

(cf. fig. 27). All the curves of the family are similar to each other and arise from one another by translation parallel to the y-axis. By differentiation we obtain $f_c = -2(y-c)(x-2) = 0$, so that we must have either $x = 2$ or $y = c$. The line $x = 2$ does not enter into the matter, however, for no finite value of y corresponds to $x = 2$. We therefore have $y = c$, so that the discriminant curve is $x^2(x-2) + x = 0$. This curve consists of the two straight lines $x = 0$ and $x = 1$. As we see from fig. 27, only $x = 0$ is the envelope; the line $x = 1$ passes through the double points of the curves.

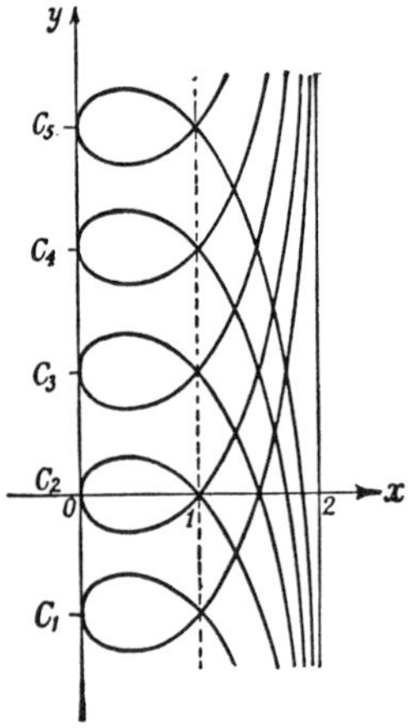

Fig. 27.—Family of strophoids

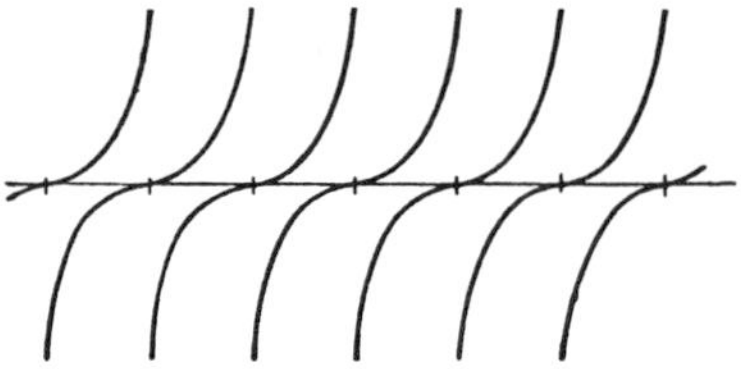

Fig. 28.—Family of cubical parabolas

10. The envelope need not be the locus of the points of intersection of neighbouring curves; this is shown by the family of identical parallel cubical parabolas $y - (x-c)^3 = 0$. No two of these curves intersect each other. The rule gives the equation $f_c = 3(x-c)^2 = 0$, so that the x-axis $y = 0$ is the discriminant curve. Since all the curves of the family are touched by it, it is also the envelope (fig. 28).

11. The notion of the envelope enables us to give a new definition for the evolute of a curve C (cf. Vol. I, pp. 283, 307 *et seq.*). Let C be given by $x = \varphi(t)$, $y = \psi(t)$. We then define the evolute E of C as the envelope of the normals of C. As the normals of C are given by

$$\{x - \varphi(t)\}\varphi'(t) + \{y - \psi(t)\}\psi'(t) = 0,$$

the envelope is found by differentiating this equation with respect to t:

$$0 = \{x - \varphi(t)\}\varphi''(t) + \{y - \psi(t)\}\psi''(t) - \varphi'^2(t) - \psi'^2(t).$$

From this equation and the preceding one we obtain the parametric representation of the envelope,

$$x = \varphi(t) - \psi'(t)\frac{\varphi'^2 + \psi'^2}{\psi''\varphi' - \varphi''\psi'} = \varphi - \frac{\psi'\rho}{\sqrt{(\varphi'^2 + \psi'^2)}},$$

$$y = \psi(t) + \varphi'(t)\frac{\varphi'^2 + \psi'^2}{\psi''\varphi' - \varphi''\psi'} = \psi + \frac{\varphi'\rho}{\sqrt{(\varphi'^2 + \psi'^2)}},$$

where

$$\rho = \frac{(\varphi'^2 + \psi'^2)^{\frac{3}{2}}}{\psi''\varphi' - \varphi''\psi'}$$

denotes the radius of curvature (cf. Vol. I, p. 281). These equations are identical with those given in Vol I, p. 283 for the evolute.

12. Let a curve C be given by $x = \varphi(t)$, $y = \psi(t)$. We form the envelope E of the circles having their centres on C and passing through the origin O. Since the circles are given by

$$x^2 + y^2 - 2x\varphi(t) - 2y\psi(t) = 0,$$

the equation of E is

$$x\varphi'(t) + y\psi'(t) = 0.$$

Hence if P is the point $(\varphi(t), \psi(t))$ and $Q(x, y)$ the corresponding point of E, then OQ is perpendicular to the tangent to C at P. Since by definition $PQ = PO$, PO and PQ make equal angles with the tangent to C at P.

If we imagine O to be a luminous point and C a reflecting curve, then QP is the reflected ray corresponding to OP. The envelope of the reflected rays is called the *caustic* of C with respect to O. *The caustic is the evolute of E.* For the reflected ray PQ is normal to E, since a circle with centre P touches E at Q, and the envelope of the normals of E is its evolute, as we saw in the preceding example.

For example, let C be a circle passing through O. Then E is the path described by the point O' of a circle C' congruent to C which rolls on C and starts with O and O' coincident. For during the motion O and O' always occupy symmetrical positions with respect to the common tangent of the two circles. Thus E will be a special epicycloid, in fact, a cardioid (cf. Vol. I, p. 267, Ex. 2 and 3). As the evolute of an epicycloid is a similar epicycloid (cf. Vol. I, p. 311, Ex. 1), the caustic of C with respect to O is in this case a cardioid.

4. Envelopes of Families of Surfaces.

The remarks made about the envelopes of families of curves apply with but little alteration to families of surfaces also. If in the first instance we consider a one-parameter family of surfaces $f(x, y, z, c) = 0$ in a definite interval of parameter values c, we shall say that a surface E is the envelope of the family if it touches each surface of the family along a whole curve, and if further these curves of contact form a one-parameter family of curves on E which completely cover E.

An example is given by the family of all spheres of unit radius with centres on the z-axis. We see intuitively that the envelope is the cylinder $x^2 + y^2 - 1 = 0$ with unit radius and axis along the z-axis; the family of curves of contact is simply the family of circles parallel to the xy-plane, with unit radius and centre on the z-axis.*

* The envelopes of spheres of constant radius whose centres lie along curves are called *tube-surfaces*.

As in sub-section 2 (p. 172), if we assume that the envelope does exist we can find it by the following heuristic method. We first consider the surfaces $f(x, y, z, c) = 0$ and $f(x, y, z, c+h) = 0$ corresponding to two different parameter values c and $c + h$. These two equations determine the curve of intersection of the two surfaces (we expressly assume that such a curve of intersection exists). In addition to the two equations above, this curve also satisfies the third equation

$$\frac{f(x, y, z, c+h) - f(x, y, z, c)}{h} = 0.$$

If we let h tend to zero, the curve of intersection will approach a definite limiting position, and this limit curve is determined by the two equations

$$f(x, y, z, c) = 0, \quad f_c(x, y, z, c) = 0.$$

This curve is often referred to in a non-rigorous but intuitive way as the intersection of " neighbouring " surfaces of the family. It is still a function of the parameter c, so that all the curves of intersection for the different values of c form a one-parameter family of curves in space. If we eliminate the quantity c from the two equations above we obtain an equation, which is called the " discriminant ". As in sub-section 2 (p. 172), we can show that the envelope must satisfy this discriminant equation.

Just as in the case of plane curves, we may readily convince ourselves that a plane touching the discriminant surface also touches the corresponding surface of the family, provided that $f_x^2 + f_y^2 + f_z^2 \neq 0$. Hence the discriminant surface again gives the envelopes of the family and the loci of the singularities of the surfaces of the family.

As a first example we consider the family of spheres

$$x^2 + y^2 + (z - c)^2 - 1 = 0$$

mentioned above. To find the envelope we have the additional equation

$$-2(z - c) = 0.$$

For fixed values of c these two equations obviously represent the circle of unit radius parallel to the xy-plane at the height $z = c$. If we eliminate the parameter c between the two equations, we obtain the equation of the envelope in the form $x^2 + y^2 - 1 = 0$, which is the equation of the right circular cylinder with unit radius and the z-axis as axis.

While for families of curves the formation of the envelope has a meaning only for one-parameter families, in the case of families of surfaces it is also possible to find envelopes of two-parameter families $f(x, y, z, c_1, c_2) = 0$. If, for example, we consider the family of all spheres with unit radius and centre on the xy-plane, represented by the equation

$$(x - c_1)^2 + (y - c_2)^2 + z^2 - 1 = 0,$$

intuition at once tells us that the two planes $z = 1$ and $z = -1$ touch a surface of the family at every point. In general we shall say that a surface E is the envelope of a two-parameter family of surfaces if at every point P of E the surface E touches a surface of the family in such a way that as P ranges over E the parameter values c_1, c_2 corresponding to the surface touching E at P range over a region of the c_1c_2-plane, and in addition different points (c_1, c_2) correspond to different points P of E. A surface of the family then touches the envelope in a *point*, and not, as before, along a whole curve.

With assumptions similar to those made in the case of plane curves, we find that the point of contact of a surface of the family with the envelope, if it exists, must satisfy the equations

$$f(x, y, z, c_1, c_2) = 0, \quad f_{c_1}(x, y, z, c_1, c_2) = 0, \quad f_{c_2}(x, y, z, c_1, c_2) = 0.$$

From these three equations we can in general find the point of contact of each separate surface by assigning the corresponding values to the parameters. If, conversely, we eliminate the parameters c_1 and c_2, we obtain an equation which the envelope must satisfy.

For example, the family of spheres with unit radius and centre on the xy-plane is given by the equation

$$f(x, y, z, c_1, c_2) = (x - c_1)^2 + (y - c_2)^2 + z^2 - 1 = 0$$

with the two parameters c_1 and c_2. The rule for forming the envelope gives the two equations

$$f_{c_1} = -2(x - c_1) = 0 \quad \text{and} \quad f_{c_2} = -2(y - c_2) = 0.$$

Thus for the discriminant equation we have $z^2 - 1 = 0$, and in fact the two planes $z = 1$ and $z = -1$ *are* envelopes, as we have already seen intuitively.

EXAMPLES

1. Let $z = u(x, y)$ be the equation of a tube-surface, i.e. the envelope of a family of spheres of unit radius with their centres on some curve $y = f(x)$ in the xy-plane. Prove that

$$u^2(u_x^2 + u_y^2 + 1) = 1.$$

2. (*a*) Find the envelope of the two-parameter family of planes for which

$$OP + OQ + OR = \text{const.} = 1,$$

where P, Q, R denote the points of intersection of the planes with the co-ordinate axes and O the origin.

(*b*) Find the envelope of the planes for which

$$OP^2 + OQ^2 + OR^2 = 1.$$

3. Let C be an arbitrary curve in the plane, and consider the circles of radius p whose centres lie on C. Prove that the envelope of these circles is formed by the two curves parallel to C at the distance p (cf. the definition of parallel curves, Vol. I, p. 291).

4*. A family of straight lines in space may be given as the intersection of two planes depending on a parameter t:

$$a(t)x + b(t)y + c(t)z = 1$$
$$d(t)x + e(t)y + f(t)z = 1.$$

Prove that if these straight lines are tangents to some curve, i.e. possess an envelope, then

$$\begin{vmatrix} a-d & b-e & c-f \\ a' & b' & c' \\ d' & e' & f' \end{vmatrix} = 0.$$

5*. A family of planes is given by

$$x \cos t + y \sin t + z = t,$$

where t is a parameter.

(*a*) Find the equation of the envelope of the planes in cylindrical co-ordinates (r, z, θ).

(*b*) Prove that the envelope consists of the tangents to a certain curve.

6. If a body is always thrown from the same initial position with the same initial velocity but at different angles, its trajectories form a family of parabolas (it is assumed that the motion always takes place in the same vertical plane). Prove that the envelope of these parabolas is another parabola.

7*. Find the envelope of the family of spheres which touch the three spheres

$$S_1\colon (x-\tfrac{3}{2})^2+y^2+z^2=\tfrac{9}{4},$$
$$S_2\colon x^2+(y-\tfrac{3}{2})^2+z^2=\tfrac{9}{4},$$
$$S_3\colon x^2+y^2+(z-\tfrac{3}{2})^2=\tfrac{9}{4}.$$

8. If a plane curve C is given by $x=f(t)$, $y=g(t)$, its "polar reciprocal" C' is defined as the envelope of the family of straight lines

$$\xi f(t)+\eta g(t)=1,$$

where (ξ, η) are current co-ordinates.

(*a*) Prove that C is the polar reciprocal of C' also.

(*b*) Find the polar reciprocal of the circle

$$(x-a)^2+(y-b)^2=1.$$

(*c*) Find the polar reciprocal of the ellipse

$$\frac{x^2}{a^2}+\frac{y^2}{b^2}=1.$$

6. Maxima and Minima

1. Necessary Conditions.

The theory of maxima and minima for functions of several variables, like that for functions of a single variable, forms one of the most important applications of differentiation.

We shall begin by considering a function $u=f(x, y)$ of two independent variables x, y, which we shall represent by a surface in xyu-space. We say that this surface has a maximum with the co-ordinates (x_0, y_0) if all the other values of u in a neighbourhood of that point (all round the point) are less than $u(x_0, y_0)$. Geometrically, such a maximum corresponds to a "hill-top" on the surface. In the same way, we shall call the point (x_0, y_0) a minimum if all other values of the function in a certain neighbourhood of $P_0(x_0, y_0)$ are greater than $u_0=u(x_0, y_0)$. Just as with functions of one variable, these concepts always refer only to a sufficiently small neighbourhood of the point in question. Considered as a whole, the surface may very well have points which are higher than the hill-tops. Analytically, we formulate our definition as follows, so that it applies to functions of more than two independent variables:

A function $u=f(x, y, \ldots)$ *has a maximum (or a minimum) at the point* $(x_0, y_0, \ldots)$ *if at every point in a neighbourhood of* $(x_0, y_0, \ldots)$ *the function assumes a smaller value (or a larger value) than at the point itself.*

If in the neighbourhood of $(x_0, y_0, \ldots)$ the function assumes values which are not greater than the value of the function at the point (but may be equal to it), we say that the function has an *improper maximum* at the point. We define an *improper minimum* in a similar way.

We again emphasize that this definition refers to a suitably chosen neighbourhood of the point, extending in all directions about the point. Thus in a closed region the value of a maximum may very well lie below the greatest value assumed by the function in the region.* If the greatest value is reached at a point P_0 of the boundary, it need not be a maximum in the sense defined above, as we have already seen for functions of one variable. For if the function is defined in the closed region only, we cannot find a complete neighbourhood of P_0 in which the function is defined; and if, on the other hand, the closed region is contained in a larger region in which the function *is* defined, then in this larger region the function may not have a maximum at P_0, as the following example shows. The function $u = -x - y$ is defined over the whole xy-plane, but we consider it only in the square $0 \leqq x \leqq 1$, $0 \leqq y \leqq 1$. In this closed region it reaches its greatest value 0 at the origin. This greatest value, however, is not a maximum. For if we consider a neighbourhood all round the origin, we find that the function assumes values greater than zero. If, however, we know that the greatest or least value of a function is assumed at a single point *interior* to the region, that point must necessarily be a maximum or a minimum in the sense defined above.

We shall first give *necessary* conditions for the occurrence of an extreme value. (As in the case of functions of one variable, we use the terms † *extreme value, extreme point* when we do not wish to distinguish between maxima and minima.) That is, we find conditions which must be satisfied at a point $(x_0, y_0, \ldots)$ if there is to be an extreme value at that point. *The equations*

$$\begin{aligned} f_x(x_0, y_0, z_0, \ldots) &= 0, \\ f_y(x_0, y_0, z_0, \ldots) &= 0, \\ f_z(x_0, y_0, z_0, \ldots) &= 0, \\ \cdots\cdots\cdots \end{aligned}$$

are necessary conditions for the occurrence of a maximum or minimum of a differentiable function $u = f(x, y, z, \ldots)$ *at the point* P_0 *with co-ordinates* $(x_0, y_0, z_0, \ldots)$.

* We already know (cf. p. 97) that a continuous function always assumes a greatest and a least value in a closed region.

† On the other hand, as will be seen later (p. 186), the terms *stationary value, stationary point* include points which are neither maxima nor minima.

In fact, these conditions follow at once from the known conditions for functions of one independent variable. If we consider the variables $y, z, \ldots$ as fixed at the values $y_0, z_0, \ldots$ and regard the function in the neighbourhood of P_0 as a function of the single variable x, this function of x must have an extreme value at the point $x = x_0$, and by our previous results we must have $f_x(x_0, y_0, z_0, \ldots) = 0$.

Geometrically, the vanishing of the partial derivatives in the case of functions of two independent variables means that at the point (x_0, y_0) the tangent plane to the surface $u = f(x, y)$ is parallel to the xy-plane.

For many purposes it is more convenient to combine the conditions in one equation. This equation is

$$df(x_0, y_0, z_0, \ldots) = f_x(x_0, y_0, z_0, \ldots)dx + f_y(x_0, y_0, z_0, \ldots)dy \\ + f_z(x_0, y_0, z_0, \ldots)dz + \ldots = 0.$$

In words: *at an extreme point the differential (linear part of the increment) of the function must vanish,* no matter what values we assign to the differentials $dx, dy, dz, \ldots$ of the independent variables $x, y, z, \ldots$. Conversely, if the above equation is satisfied for arbitrary values of $dx, dy, \ldots$ it follows that at the given point $f_x = f_y = \ldots = 0$. We have only to take all but one of the (mutually independent) variables equal to zero.

In the equations

$$\begin{aligned} f_x(x_0, y_0, z_0, \ldots) &= 0, \\ f_y(x_0, y_0, z_0, \ldots) &= 0, \\ f_z(x_0, y_0, z_0, \ldots) &= 0, \\ \cdots\cdots \end{aligned}$$

there are as many unknowns $x_0, y_0, z_0, \ldots$ as there are equations. As a rule, therefore, we can calculate the position of the extreme points by means of them. But a point obtained in this way need not by any means be an extreme point.

We consider e.g. the function $u = xy$. Our two equations at once give $x = 0$, $y = 0$. In the neighbourhood of the point $x = 0$, $y = 0$, however, the function assumes both positive and negative values, according to the quadrant. The function therefore has not an extreme value there. The geometrical representation of the surface $u = xy$, which is a hyperbolic paraboloid, shows that the origin is a *saddle point* (cf. fig. 1, p. 112).

It is useful to have a simple expression for a point at which the above equations are satisfied, irrespective of whether the function has an extreme point or not. We accordingly say that if there is a point $(x_0, y_0, z_0, \ldots)$ at which $f_x = 0, f_y = 0, f_z = 0, \ldots$, or at which

$$df = f_x dx + f_y dy + f_z dz + \ldots = 0,$$

the function has a *stationary value* at that (stationary) point (cf. footnote, p. 184).

Every point interior to a closed region at which a differentiable function assumes its greatest or its least value is a stationary point.

To decide whether and when our system of equations really gives an extreme value, we must make further investigations. In many cases, however, the state of affairs is clear from the outset, in particular, if we know that the greatest or least value of the function must be assumed at an interior point P of the region and find that our equations determine only a *single* stationary system $x = x_0$, $y = y_0$, ... This system of values must then determine the point P, which is necessarily a stationary point. If such considerations do not apply, however, we must investigate the matter more closely; this we postpone to the appendix to this chapter (p. 204). Meanwhile we shall illustrate the foregoing results by means of some examples.

2. Examples.

1. For the function $u = x^2 + y^2$ the partial derivatives vanish only at the origin, so that this point alone can be an extreme point. The function actually has a minimum, for at all points (x, y) different from $(0, 0)$ the function $u = x^2 + y^2$ must be positive, being a sum of squares.

2. The function

$$u = \sqrt{(1 - x^2 - y^2)}, \quad (x^2 + y^2 < 1)$$

has the partial derivatives

$$u_x = -\frac{x}{\sqrt{(1 - x^2 - y^2)}}, \quad u_y = -\frac{y}{\sqrt{(1 - x^2 - y^2)}},$$

and these vanish only at the origin. Here we have a maximum, for at all other points (x, y) in the neighbourhood of the origin the quantity $1 - x^2 - y^2$ under the square root is less than it is at the origin.

3. We wish to construct the triangle for which the product of the sines of the three angles is greatest; that is, we wish to find the maximum of the function

$$f(x, y) = \sin x \sin y \sin (x + y)$$

in the region $0 \leqq x \leqq \pi$, $0 \leqq y \leqq \pi$, $0 \leqq x + y \leqq \pi$. Since f is positive in the interior of this region, its greatest value is positive. On the boundary of the region, where the equality sign holds in at least one of the inequalities defining the region, we have $f(x, y) = 0$, so that the greatest value must lie in the interior.

If we equate the derivatives to zero, we obtain the two equations

$$\cos x \sin y \sin(x + y) + \sin x \sin y \cos(x + y) = 0,$$
$$\sin x \cos y \sin(x + y) + \sin x \sin y \cos(x + y) = 0.$$

Since $0 < x < \pi$, $0 < y < \pi$, $0 < x + y < \pi$, these give $\tan x = \tan y$, or $x = y$. If we substitute this value in the first equation, we obtain the relation $\sin 3x = 0$; hence $x = \frac{\pi}{3}$, $y = \frac{\pi}{3}$ is the only stationary point, and the required triangle is equilateral.

4. Three points P_1, P_2, P_3, with co-ordinates (x_1, y_1), (x_2, y_2), and (x_3, y_3) respectively, are the vertices of an acute-angled triangle. We wish to find a fourth point P with co-ordinates (x, y) such that the sum of its distances from P_1, P_2, and P_3 is the least possible. This sum of distances is a continuous function of x and y, and at some point P inside a large circle enclosing the triangle it has a least value. This point P cannot lie at a vertex of the triangle, for then the foot of the perpendicular from one of the other two vertices on to the opposite side would give a smaller sum of distances. Again, P cannot lie on the circumference of the circle, if this is sufficiently far away from the triangle. With the distances

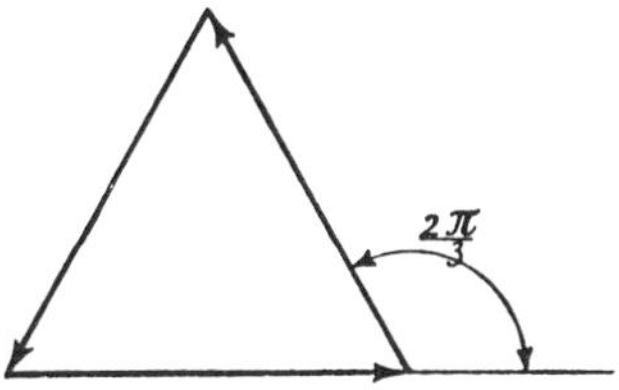

Fig. 29.—Three vectors with equal magnitudes and sum zero

$$r_i = \sqrt{(x - x_i)^2 + (y - y_i)^2}$$

we now form the function

$$f(x, y) = r_1 + r_2 + r_3,$$

which is differentiable everywhere except at P_1, P_2, and P_3. We know that at the point P the partial derivatives with respect to x and y must vanish. Thus by differentiating f we obtain the conditions

$$\frac{x - x_1}{r_1} + \frac{x - x_2}{r_2} + \frac{x - x_3}{r_3} = 0,$$

$$\frac{y - y_1}{r_1} + \frac{y - y_2}{r_2} + \frac{y - y_3}{r_3} = 0$$

for P. According to these equations the three plane vectors $\boldsymbol{u}_1$, $\boldsymbol{u}_2$, $\boldsymbol{u}_3$, with components

$$\frac{x-x_1}{r_1}, \frac{y-y_1}{r_1}, \quad \frac{x-x_2}{r_2}, \frac{y-y_2}{r_2}, \quad \frac{x-x_3}{r_3}, \frac{y-y_3}{r_3}$$

respectively, have the vector sum 0. Also, these vectors are each of unit length. When combined geometrically, then, they form an equilateral triangle; that is, each vector is brought into the direction of the next by a rotation through $\frac{2}{3}\pi$ (fig. 29). Since these three vectors have the same directions as the three vectors from P_1, P_2, P_3 to P, it follows that each of the three sides of the triangle must subtend the same angle $\frac{2}{3}\pi$ at the point P.

3. Maxima and Minima with Subsidiary Conditions.

The problem of determining the maxima and minima of functions of several variables frequently presents itself in a form differing from that treated above. If e.g. we wish to find the point of a given surface $\phi(x, y, z) = 0$ which is at the least distance from the origin, then we have to determine the minimum of the function

$$f(x, y, z) = \sqrt{(x^2 + y^2 + z^2)},$$

where the quantities x, y, z, however, are no longer three independent variables, but are connected by the equation of the surface $\phi(x, y, z) = 0$ as a subsidiary condition. Such "maxima and minima with subsidiary conditions" do not, indeed, represent a fundamentally new problem. Thus in our example we need only solve for one of the variables, say z, in terms of the other two, and then substitute this expression in the formula for the distance $\sqrt{(x^2 + y^2 + z^2)}$, to reduce the problem to that of determining the stationary values of a function of the two variables x, y.

It is, however, more convenient, and also more elegant, to express the conditions for a stationary value in a symmetrical form, in which no preference is given to any one of the variables.

As a very simple case, which is nevertheless typical, we consider the following problem: *to find the stationary values of a function* f(x, y) *when the two variables* x, y *are not mutually independent, but are connected by a subsidiary condition*

$$\phi(x, y) = 0.$$

In order to give geometrical plausibility to the analytical treatment, we assume first that the subsidiary condition is represented, as in fig. 30, by a curve in the xy-plane without singularities and that in addition the family of curves $f(x, y) = c =$ const. covers a portion of the plane, as in the figure. The problem is then as follows: among the curves of the family which intersect the curve $\phi = 0$, to find that one for which the constant c is the

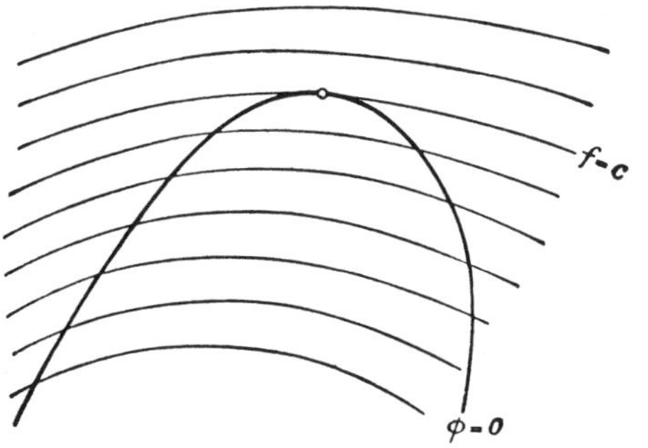

Fig. 30.—Extreme value of f with subsidiary condition $\phi = 0$

greatest possible or the least possible. As we describe the curve $\phi = 0$ we cross the curves $f(x, y) = c$, and in general c changes monotonically; at the point where the sense in which we run through the c-scale is reversed we may expect an extreme value. From fig. 30 we see that this occurs for the curve of the family which touches the curve $\phi = 0$. The co-ordinates of the point of contact will be the required values $x = \xi$, $y = \eta$ corresponding to the extreme value of $f(x, y)$. If the two curves $f =$ const. and $\phi = 0$ touch, they have the same tangent. Thus at the point $x = \xi$, $y = \eta$ the proportional relation

$$f_x : f_y = \phi_x : \phi_y$$

holds; or, if we introduce the constant of proportionality λ, the two equations

$$f_x + \lambda\phi_x = 0$$
$$f_y + \lambda\phi_y = 0$$

are satisfied. These, with the equation

$$\phi(x, y) = 0,$$

serve to determine the co-ordinates (ξ, η) of the point of contact and also the constant of proportionality λ.

This argument may fail, e.g. when the curve $\phi = 0$ has a singular point, say a cusp as in fig. 31, at the point (ξ, η) at which it meets a curve $f = c$ with the greatest or least possible c. In this case, however, we have both

$$\phi_x(\xi, \eta) = 0 \quad \text{and} \quad \phi_y(\xi, \eta) = 0.$$

In any case we are led intuitively to the following rule, which we shall prove in the next sub-section:

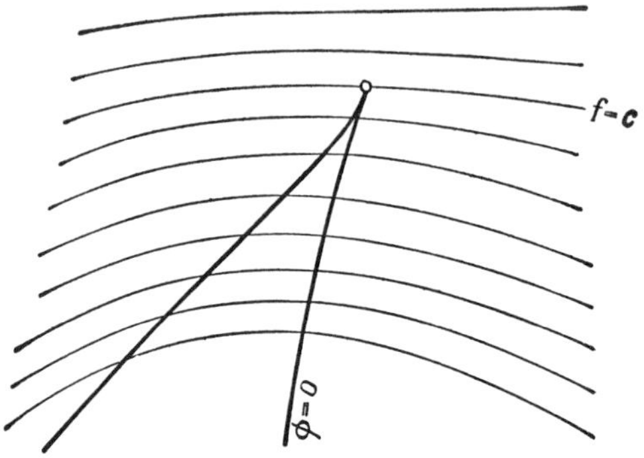

Fig. 31.—Extreme value at a singular point of $\phi = 0$

In order that an extreme value of the function $f(x, y)$ *may occur at the point* $x = \xi, y = \eta$, *with the subsidiary condition* $\phi(x, y) = 0$, *the point* (ξ, η) *being such that the two equations*

$$\phi_x(\xi, \eta) = 0 \quad \text{and} \quad \phi_y(\xi, \eta) = 0$$

are not *both satisfied, it is necessary that there should be a constant of proportionality such that the two equations*

$$f_x(\xi, \eta) + \lambda\phi_x(\xi, \eta) = 0 \quad \text{and} \quad f_y(\xi, \eta) + \lambda\phi_y(\xi, \eta) = 0$$

are satisfied, together with the equation

$$\phi(\xi, \eta) = 0.$$

This rule is known as *Lagrange's method of undetermined multipliers*, and the factor λ is known as *Lagrange's multiplier*.

We observe that for the determination of the quantities ξ, η, and λ this rule gives as many equations as there are unknowns. We have therefore replaced the problem of finding the positions of the extreme values (ξ, η) by a problem in which there is an additional unknown λ, but in which we have the advantage of complete symmetry. Lagrange's rule is usually expressed as follows:

To find the extreme values of the function f(x, y) *subject to the subsidiary condition* $\phi(x, y) = 0$, *we add to* f(x, y) *the product of* $\phi(x, y)$ *and an unknown factor* λ *independent of* x *and* y, *and write down the known necessary conditions,*

$$f_x + \lambda\phi_x = 0, \quad f_y + \lambda\phi_y = 0,$$

for an extreme value of $F = f + \lambda\phi$. In conjunction with the subsidiary condition $\phi = 0$ these serve to determine the co-ordinates of the extreme value and the constant of proportionality λ.

Before proceeding to prove the rule of undetermined multipliers rigorously we shall illustrate its use by means of a simple example. We wish to find the extreme values of the function

$$u = xy$$

on the circle with unit radius and centre the origin, that is, with the subsidiary condition

$$x^2 + y^2 - 1 = 0.$$

According to our rule, by differentiating $xy + \lambda(x^2 + y^2 - 1)$ with respect to x and to y we find that at the stationary points the two equations

$$y + 2\lambda x = 0$$
$$x + 2\lambda y = 0$$

have to be satisfied. In addition we have the subsidiary condition

$$x^2 + y^2 - 1 = 0.$$

On solving we obtain the four points

$$\xi = \tfrac{1}{2}\sqrt{2}, \qquad \eta = \tfrac{1}{2}\sqrt{2},$$
$$\xi = -\tfrac{1}{2}\sqrt{2}, \qquad \eta = -\tfrac{1}{2}\sqrt{2},$$
$$\xi = \tfrac{1}{2}\sqrt{2}, \qquad \eta = -\tfrac{1}{2}\sqrt{2},$$
$$\xi = -\tfrac{1}{2}\sqrt{2}, \qquad \eta = \tfrac{1}{2}\sqrt{2}.$$

The first two of these give a maximum value $u = \tfrac{1}{2}$, the second two a minimum value $u = -\tfrac{1}{2}$, of the function $u = xy$. That the first two do really give the greatest value and the second two the least value of the function u can be seen as follows: on the circumference the function must assume a greatest and a least value (cf. p. 97), and since the circumference has no boundary point, these points of greatest and least value must be stationary points for the function.

4. Proof of the Method of Undetermined Multipliers in the Simplest Case.

As we should expect, we arrive at an analytical proof of the method of undetermined multipliers by reducing it to the known case of "free" extreme values. We assume that at the extreme point the two partial derivatives $\phi_x(\xi, \eta)$ and $\phi_y(\xi, \eta)$ do not both vanish; to be specific, we assume that $\phi_y(\xi, \eta) \neq 0$. Then by section 1, No. 3 (p. 114), in a neighbourhood of this point the equation $\phi(x, y) = 0$ determines y uniquely as a continuously differentiable function of x, $y = g(x)$. If we substitute this expression in $f(x, y)$, the function

$$f(x, g(x))$$

must have a free extreme value at the point $x = \xi$. For this the equation

$$f'(x) = f_x + f_y g'(x) = 0$$

must hold at $x = \xi$. In addition, the implicitly defined function $y = g(x)$ satisfies the relation $\phi_x + \phi_y g'(x) = 0$ identically. If we multiply this equation by $\lambda = -f_y/\phi_y$ and add it to $f_x + f_y g'(x) = 0$, then we obtain

$$f_x + \lambda\phi_x = 0,$$

and by the definition of λ the equation

$$f_y + \lambda\phi_y = 0$$

holds. This establishes the method of undetermined multipliers.

This proof brings out the importance of the assumption that the derivatives ϕ_x and ϕ_y do not both vanish at the point (ξ, η). If both these derivatives vanish the rule breaks down, as is shown analytically by the following example. We wish to make the function

$$f(x, y) = x^2 + y^2$$

a minimum, subject to the condition

$$\phi(x, y) = (x - 1)^3 - y^2 = 0.$$

By fig. 32, the shortest distance from the origin to the curve $(x - 1)^3 - y^2 = 0$ is obviously given by the line joining the origin to the cusp S of the curve (we can easily prove that the circle with unit radius and centre the origin has no other point in common with the curve). The co-ordinates of S, that is,

$x = 1$ and $y = 0$, satisfy the equations $\phi(x, y) = 0$ and $f_y + \lambda\phi_y = 0$, no matter what value is assigned to λ, but

$$f_x + \lambda\phi_x = 2x + 3\lambda(x - 1)^2 = 2 \neq 0.$$

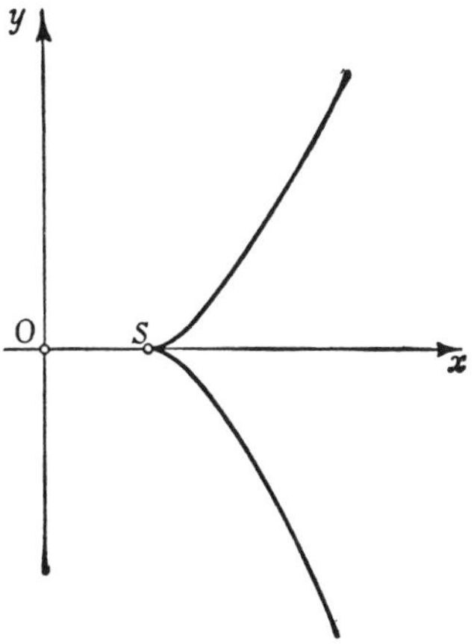

Fig. 32.—The surface $(x - 1)^3 - y^2 = 0$

We can state the proof of the method of undetermined multipliers in a slightly different way, which is particularly convenient for generalization. We have seen that the vanishing of the differential of a function at a given point is a necessary condition for the occurrence of an extreme value of the function at that point. For the present problem we can also make the following statement:

In order that the function f(x, y) *may have an extreme value at the point* (ξ, η), *subject to the subsidiary condition* $\phi(x, y) = 0$, *it is necessary that the differential* df *shall vanish at that point, it being assumed that the differentials* dx *and* dy *are not independent of one another, but are chosen in accordance with the equation*

$$d\phi = \phi_x dx + \phi_y dy = 0$$

deduced from $\phi = 0$. Thus at the point (ξ, η) the differentials dx and dy must satisfy the equation

$$df = f_x(\xi, \eta)\,dx + f_y(\xi, \eta)\,dy = 0$$

whenever they satisfy the equation $d\phi = 0$. If we multiply the first of these equations by a number λ, undetermined in the first instance, and add it to the second, we obtain

$$(f_x + \lambda\phi_x)dx + (f_y + \lambda\phi_y)dy = 0.$$

If we determine λ so that

$$f_y + \lambda\phi_y = 0,$$

as is possible in virtue of the assumption that $\phi_y \neq 0$, it necessarily follows that $(f_x + \lambda\phi_x)\,dx = 0$, and since the differential dx can be chosen arbitrarily, e.g. equal to 1, we have

$$f_x + \lambda\phi_x = 0.$$

5. Generalization of the Method of Undetermined Multipliers.

We can extend the method of undetermined multipliers to a greater number of variables and also to a greater number of subsidiary conditions. We shall consider a special case which includes every essential feature. We seek the extreme values of the function

$$u = f(x, y, z, t),$$

when the four variables x, y, z, t satisfy the two subsidiary conditions

$$\phi(x, y, z, t) = 0, \quad \psi(x, y, z, t) = 0.$$

We assume that at the point (ξ, η, ζ, τ) the function takes a value which is an extreme value when compared with the values at all neighbouring points satisfying the subsidiary conditions. We assume further that in the neighbourhood of the point $P(\xi, \eta, \zeta, \tau)$ two of the variables, say z and t, can be represented as functions of the other two, x and y, by means of the equations

$$\phi(x, y, z, t) = 0 \quad \text{and} \quad \psi(x, y, z, t) = 0.$$

In fact, to ensure that such solutions $z = g(x, y)$ and $t = h(x, y)$ can be found, we assume that at the point P the Jacobian

$$\frac{\partial(\phi, \psi)}{\partial(z, t)} = \phi_z\psi_t - \phi_t\psi_z$$

is not zero (cf. p. 153). If we now substitute the functions

$$z = g(x, y) \quad \text{and} \quad t = h(x, y)$$

in the function $u = f(x, y, z, t)$, then $f(x, y, z, t)$ becomes a function of the two independent variables x and y, and this function must have a free extreme value at the point $x = \xi$, $y = \eta$; that is, its two partial derivatives must vanish at that point. The two equations

$$f_x + f_z\frac{\partial z}{\partial x} + f_t\frac{\partial t}{\partial x} = 0,$$

$$f_y + f_z\frac{\partial z}{\partial y} + f_t\frac{\partial t}{\partial y} = 0$$

must therefore hold. In order to calculate from the subsidiary conditions the four derivatives $\frac{\partial z}{\partial x}, \frac{\partial z}{\partial y}, \frac{\partial t}{\partial x}, \frac{\partial t}{\partial y}$ occurring here, we could write down the two pairs of equations

$$\phi_x + \phi_z \frac{\partial z}{\partial x} + \phi_t \frac{\partial t}{\partial x} = 0,$$

$$\psi_x + \psi_z \frac{\partial z}{\partial x} + \psi_t \frac{\partial t}{\partial x} = 0$$

and

$$\phi_y + \phi_z \frac{\partial z}{\partial y} + \phi_t \frac{\partial t}{\partial y} = 0,$$

$$\psi_y + \psi_z \frac{\partial z}{\partial y} + \psi_t \frac{\partial t}{\partial y} = 0$$

and solve them for the unknowns $\frac{\partial z}{\partial x}, \ldots, \frac{\partial t}{\partial y}$, which is possible because the Jacobian $\frac{\partial(\phi, \psi)}{\partial(z, t)}$ does not vanish. The problem would then be solved.

Instead, we prefer to retain formal symmetry and clarity by proceeding as follows. We determine two numbers λ and μ in such a way that the two equations

$$f_z + \lambda\phi_z + \mu\psi_z = 0,$$
$$f_t + \lambda\phi_t + \mu\psi_t = 0$$

are satisfied at the point where the extreme value occurs. The determination of these "multipliers" λ and μ is possible, since we have assumed that the Jacobian $\frac{\partial(\phi, \psi)}{\partial(z, t)}$ is not zero. If we multiply the equations

$$\phi_x + \phi_z \frac{\partial z}{\partial x} + \phi_t \frac{\partial t}{\partial x} = 0 \quad \text{and} \quad \psi_x + \psi_z \frac{\partial z}{\partial x} + \psi_t \frac{\partial t}{\partial x} = 0$$

by λ and μ respectively and add them to the equation

$$f_x + f_z \frac{\partial z}{\partial x} + f_t \frac{\partial t}{\partial x} = 0,$$

we have

$$f_x + \lambda\phi_x + \mu\psi_x + (f_z + \lambda\phi_z + \mu\psi_z)\frac{\partial z}{\partial x} + (f_t + \lambda\phi_t + \mu\psi_t)\frac{\partial t}{\partial x} = 0$$

Hence by the definition of λ and μ

$$f_x + \lambda\phi_x + \mu\psi_x = 0.$$

Similarly, if we multiply the equations

$$\phi_y + \phi_z \frac{\partial z}{\partial y} + \phi_t \frac{\partial t}{\partial y} = 0$$

and

$$\psi_y + \psi_z \frac{\partial z}{\partial y} + \psi_t \frac{\partial t}{\partial y} = 0$$

by λ and μ respectively and add them to the equation

$$f_y + f_z \frac{\partial z}{\partial y} + f_t \frac{\partial t}{\partial y} = 0,$$

we obtain the further equation

$$f_y + \lambda\phi_y + \mu\psi_y = 0.$$

We thus arrive at the following result:

If the point (ξ, η, ζ, τ) *is an extreme point of* f(x, y, z, t) *subject to the subsidiary conditions*

$$\phi(x, y, z, t) = 0,$$
$$\psi(x, y, z, t) = 0,$$

and if at that point $\dfrac{\partial(\phi, \psi)}{\partial(z, t)}$ *is not zero, then two numbers* λ *and* μ *exist such that at the point* (ξ, η, ζ, τ) *the equations*

$$f_x + \lambda\phi_x + \mu\psi_x = 0,$$
$$f_y + \lambda\phi_y + \mu\psi_y = 0,$$
$$f_z + \lambda\phi_z + \mu\psi_z = 0,$$
$$f_t + \lambda\phi_t + \mu\psi_t = 0,$$

and also the subsidiary conditions, are satisfied.

These last conditions are perfectly symmetrical. Every trace of emphasis on the two variables x and y has disappeared from them, and we should equally well have obtained them if, instead of assuming that $\dfrac{\partial(\phi, \psi)}{\partial(z, t)} \neq 0$, we had merely assumed that any one of the Jacobians $\dfrac{\partial(\phi, \psi)}{\partial(x, y)}, \dfrac{\partial(\phi, \psi)}{\partial(x, z)}, \ldots, \dfrac{\partial(\phi, \psi)}{\partial(z, t)}$ did not vanish, so that in the neighbourhood of the point in question a certain pair of the quantities x, y, z, t (although possibly not z and t) could be expressed in terms of the other pair. For this symmetry of our equations we have of course paid the price; in addition to the unknowns ξ, η, ζ, τ we now have λ and μ also. Thus instead of four unknowns we now have six, determined by the six equations above.

Here too we could have carried out the proof somewhat more elegantly by using the differential notation. In this notation, the necessary condition for the occurrence of an extreme value at the point P is the equation

$$df = 0,$$

where the differentials dz and dt are to be expressed in terms of dx and dy. These differentials are connected by the relations

$$d\phi = \phi_x dx + \phi_y dy + \phi_z dz + \phi_t dt = 0,$$
$$d\psi = \psi_x dx + \psi_y dy + \psi_z dz + \psi_t dt = 0,$$

obtained by differentiating the subsidiary conditions. If we assume that the two-rowed determinants occurring here do not all vanish at the point (ξ, η, ζ, τ), e.g. if we assume that the expression $\dfrac{\partial(\phi, \psi)}{\partial(z, t)}$ is not zero, then we can determine two numbers λ and μ which satisfy the two equations

$$f_z + \lambda\phi_z + \mu\psi_z = 0,$$
$$f_t + \lambda\phi_t + \mu\psi_t = 0.$$

If we multiply the equation $d\phi = 0$ by λ and the equation $d\psi = 0$ by μ and add them to the equation $df = 0$, then by the last two equations we obtain

$$d(f + \lambda\phi + \mu\psi) = (f_x + \lambda\phi_x + \mu\psi_x)dx + (f_y + \lambda\phi_y + \mu\psi_y)dy.$$

Since here dx and dy are independent differentials (that is, arbitrary numbers), it follows that the numbers λ and μ also satisfy the equations

$$f_x + \lambda\phi_x + \mu\psi_x = 0,$$
$$f_y + \lambda\phi_y + \mu\psi_y = 0,$$

and we are once again led to the method of undetermined multipliers.

In exactly the same way we can state and prove the method of undetermined multipliers for an arbitrary number of variables and an arbitrary number of subsidiary conditions. The general rule is as follows:

If in a function

$$u = f(x_1, x_2, \ldots, x_n)$$

the n *variables* $x_1, x_2, \ldots, x_n$ *are not all independent, but are connected by the* m *subsidiary conditions* $(m < n)$

$$\begin{aligned} \phi_1(x_1, x_2, \ldots, x_n) &= 0, \\ \phi_2(x_1, x_2, \ldots, x_n) &= 0, \\ \cdots\cdots & \\ \cdots\cdots & \\ \phi_m(x_1, x_2, \ldots, x_n) &= 0, \end{aligned}$$

then we introduce m *multipliers* $\lambda_1, \lambda_2, \ldots, \lambda_m$ *and equate the derivatives of the function*

$$F = f + \lambda_1\phi_1 + \lambda_2\phi_2 + \ldots + \lambda_m\phi_m$$

with respect to $x_1, x_2, \ldots, x_n$, *when* $\lambda_1, \lambda_2, \ldots, \lambda_n$ *are constant, to zero. The equations*

$$\frac{\partial F}{\partial x_1} = 0, \ \ldots, \ \frac{\partial F}{\partial x_n} = 0$$

thus obtained, together with the m *subsidiary conditions*

$$\phi_1 = 0, \ \ldots, \ \phi_m = 0,$$

represent a system of $m + n$ *equations for the* $m + n$ *unknown quantities* $x_1, x_2, \ldots, x_n, \lambda_1, \ldots, \lambda_m$. *These equations must be satisfied at every extreme value of* f, *unless at that extreme value every one of the Jacobians of the* m *functions* $\phi_1, \phi_2, \ldots, \phi_m$ *with respect to* m *of the variables* $x_1, \ldots, x_n$ *has the value zero.*

In connexion with the method of undetermined multipliers we have still to make the following important remark. The rule gives us an elegant formal method for determining the points where extreme values occur, but it merely gives us a *necessary* condition. The further question arises whether and when the points which we find by means of the multiplier method do actually give us a maximum or a minimum of the function.

Into this question we shall not enter; its discussion would lead us much too far afield. As in the case of free extreme values, when we apply the method of undetermined multipliers we usually know beforehand that an extreme value does *exist*. If, then, the method determines the point P uniquely and the exceptional case (all the Jacobians zero) does not occur anywhere in the region under discussion, we can be sure that we have really found the point where the extreme value occurs.

6. Examples.

1. As a first example we attempt to find the maximum of the function $f(x, y, z) = x^2y^2z^2$ subject to the subsidiary condition $x^2 + y^2 + z^2 = c^2$. On the spherical surface $x^2 + y^2 + z^2 = c^2$ the function must assume a greatest value, and since the spherical surface has no boundary points this greatest value must be a maximum in the sense defined above. According to the rule we form the expression

$$F = x^2y^2z^2 + \lambda(x^2 + y^2 + z^2 - c^2),$$

and by differentiation obtain

$$\begin{aligned} 2xy^2z^2 + 2\lambda x &= 0, \\ 2x^2yz^2 + 2\lambda y &= 0, \\ 2x^2y^2z + 2\lambda z &= 0. \end{aligned}$$

The solutions with $x = 0$, $y = 0$, or $z = 0$ can be excluded, for at these points the function f takes on its least value, zero. The other solutions of the equation are $x^2 = y^2 = z^2$, $\lambda = -x^4$. Using the subsidiary condition, we obtain the values

$$x = \pm\frac{c}{\sqrt{3}}, \quad y = \pm\frac{c}{\sqrt{3}}, \quad z = \pm\frac{c}{\sqrt{3}}$$

for the required co-ordinates.

At all these points the function assumes the same value $c^6/27$, which is accordingly the required maximum value. Hence any triad of numbers satisfies the relation

$$\sqrt[3]{x^2y^2z^2} \leq \frac{c^2}{3} = \frac{x^2 + y^2 + z^2}{3};$$

that is, the geometric mean of three positive numbers x^2, y^2, z^2 is never greater than their arithmetic mean.

In fact, it is true that for any arbitrary number of positive numbers the geometric mean never exceeds the arithmetic mean. The proof is similar to that just given.*

2. As a second example we shall seek to find the triangle (with sides

* For another proof, see Vol. I, Ex. 19, p. 167.

x, y, z) with given perimeter $2s$, and the greatest possible area. By a well-known formula the square of the area is given by

$$f(x, y, z) = s(s - x)(s - y)(s - z).$$

We have therefore to find the maximum of this function subject to the subsidiary condition

$$\varphi = x + y + z - 2s = 0,$$

where x, y, z are restricted by the inequalities

$$x \geqq 0, \quad y \geqq 0, \quad z \geqq 0, \quad x + y \geqq z, \quad x + z \geqq y, \quad y + z \geqq x.$$

On the boundary of this closed region, i.e. whenever one of these inequalities becomes an equation, we always have $f = 0$. Consequently the greatest value of f occurs in the interior and is a maximum. We form the function

$$F(x, y, z) = s(s - x)(s - y)(s - z) + \lambda(x + y + z - 2s),$$

and by differentiation obtain the three conditions

$$-s(s - y)(s - z) + \lambda = 0, \quad -s(s - x)(s - z) + \lambda = 0,$$
$$-s(s - x)(s - y) + \lambda = 0.$$

By solving each of these for λ and equating the three resulting expressions we obtain $x = y = z = 2s/3$; that is, the solution is an equilateral triangle.

3. We shall now prove the following theorem: the inequality

$$uv \leqq \frac{1}{\alpha} u^{\alpha} + \frac{1}{\beta} v^{\beta}$$

holds for every $u \geqq 0$, $v \geqq 0$ and every $\alpha > 0$, $\beta > 0$ for which $\frac{1}{\alpha} + \frac{1}{\beta} = 1.$

The inequality is certainly valid if either u or v vanishes. We may therefore restrict ourselves to values of u and v such that $uv \neq 0$. If the inequality holds for a pair of numbers u, v, it also holds for all numbers $ut^{1/\alpha}$, $vt^{1/\beta}$, where t is an arbitrary positive number. We need therefore consider only values of u, v for which $uv = 1$. Hence we have to show that the inequality

$$\frac{1}{\alpha} u^{\alpha} + \frac{1}{\beta} v^{\beta} \geqq 1$$

holds for all positive numbers u, v such that $uv = 1$.

To do this we solve the problem of finding the minimum of $\frac{1}{\alpha} u^{\alpha} + \frac{1}{\beta} v^{\beta}$ subject to the subsidiary condition $uv = 1$. This minimum obviously exists and occurs at a point (u, v) where $u \neq 0$, $v \neq 0$. A multiplier $-\lambda$ for which the equations

$$u^{\alpha-1} - \lambda v = 0, \quad v^{\beta-1} - \lambda u = 0$$

hold therefore exists. On multiplication by u and v respectively these at once yield $u^\alpha = \lambda$, $v^\beta = \lambda$. Taken with $uv = 1$, these imply that $u = v = 1$. The minimum value of the function $\frac{1}{\alpha} u^\alpha + \frac{1}{\beta} v^\beta$ is therefore $\frac{1}{\alpha} + \frac{1}{\beta} = 1$. That is, the statement that

$$\frac{1}{\alpha} u^\alpha + \frac{1}{\beta} v^\beta \geqq 1$$

when $uv = 1$ is proved.

If in the inequality $uv \leqq \frac{1}{\alpha} u^\alpha + \frac{1}{\beta} v^\beta$ just proved we replace u and v by

$$u = \frac{u_i}{(\sum_{i=1}^{n} u_i^\alpha)^{1/\alpha}} \quad \text{and} \quad v = \frac{v_i}{(\sum_{i=1}^{n} v_i^\beta)^{1/\beta}},$$

respectively, where $u_1, u_2, \ldots, u_n, v_1, v_2, \ldots, v_n$ are arbitrary non-negative numbers and at least one u and at least one v is not zero, and if we then sum the inequalities thus obtained for $i = 1, \ldots, n$, we obtain *Hölder's inequality*

$$\sum_{i=1}^{n} u_i v_i \leqq (\sum_{i=1}^{n} u_i^\alpha)^{1/\alpha} (\sum_{i=1}^{n} v_i^\beta)^{1/\beta}.$$

This holds for any $2n$ numbers u_i, v_i where $u_i \geqq 0$, $v_i \geqq 0$ $(i = 1, 2, \ldots, n)$, not all the u's and not all the v's are zero, and the indices α, β are such that $\alpha > 0$, $\beta > 0$, $\frac{1}{\alpha} + \frac{1}{\beta} = 1$.

4. Finally, we seek to find the point on the closed surface

$$\varphi(x, y, z) = 0$$

which is at the least distance from the fixed point (ξ, η, ζ). If the distance is a minimum its square is also a minimum; we accordingly consider the function

$$F(x, y, z) = (x - \xi)^2 + (y - \eta)^2 + (z - \zeta)^2 + \lambda\varphi(x, y, z).$$

Differentiation gives the conditions

$$2(x - \xi) + \lambda\varphi_x = 0, \quad 2(y - \eta) + \lambda\varphi_y = 0, \quad 2(z - \zeta) + \lambda\varphi_z = 0,$$

or, in another form,

$$\frac{x - \xi}{\varphi_x} = \frac{y - \eta}{\varphi_y} = \frac{z - \zeta}{\varphi_z}.$$

These equations state that the fixed point (ξ, η, ζ) lies on the normal to the surface at the point of extreme distance (x, y, z). Therefore in order to travel along the shortest path from a point to a (differentiable) surface, we must travel in a direction normal to the surface. Of course further discussion is required to decide whether we have found a maximum or a minimum or neither. (Consider, e.g., a point within a spherical surface. The points of extreme distance lie at the ends of the diameter through the point; the distance to one of these points is a minimum, to the other a maximum.)

EXAMPLES

1. Find the greatest and least distances of a point on the ellipse

$$\frac{x^2}{4} + \frac{y^2}{1} = 1$$

from the straight line $x + y - 4 = 0$.

2. The sum of the lengths of the twelve edges of a rectangular block is a; the sum of the areas of the six faces is $a^2/25$. Calculate the lengths of the edges when the excess of the volume of the block over that of a cube whose edge is equal to the least edge of the block is greatest.

3. Determine the maxima and minima of the function

$$(ax^2 + by^2)e^{-x^2-y^2} \qquad (0 < a < b).$$

4. Show that the maximum value of the expression

$$\frac{ax^2 + 2bxy + cy^2}{ex^2 + 2fxy + gy^2} \qquad (eg - f^2 > 0)$$

is equal to the greater of the roots of the equation in λ

$$(ac - b^2) - \lambda(ag - 2bf + ec) + \lambda^2(eg - f^2) = 0.$$

5. Calculate the maximum values of the following expressions:

$$(a)\ \frac{x^2 + 6xy + 3y^2}{x^2 - xy + y^2}, \qquad (b)\ \frac{x^4 + 2x^3y}{x^4 + y^4}.$$

6. Determine the stationary points of the function

$$f(x, y) = y^2\left(\sin x - \frac{x}{2}\right)$$

and state their nature.

7*. Find the values of a and b for the ellipse

$$\frac{x^2}{a^2} + \frac{y^2}{b^2} = 1$$

of least area containing the circle

$$(x - 1)^2 + y^2 = 1$$

in its interior.

8. Find the quadrilateral with given edges a, b, c, d which includes the greatest area.

9. Which point of the sphere $x^2 + y^2 + z^2 = 1$ is at the greatest distance from the point (1, 2, 3)?

10. Let $P_1P_2P_3P_4$ be a convex quadrilateral. Find the point O for which the sum of the distances from P_1, P_2, P_3, P_4 is a minimum.

11. Find the point (x, y, z) of the ellipsoid

$$\frac{x^2}{a^2} + \frac{y^2}{b^2} + \frac{z^2}{c^2} = 1$$

for which

$$(a) \quad A + B + C,$$
$$(b) \quad \sqrt{(A^2 + B^2 + C^2)}$$

is a minimum, where A, B, C denote the intercepts which the tangent plane at (x, y, z) $(x > 0,\ y > 0,\ z > 0)$ makes on the co-ordinate axes.

12. Find the rectangular parallelepiped of greatest volume inscribed in the ellipsoid

$$\frac{x^2}{a^2} + \frac{y^2}{b^2} + \frac{z^2}{c^2} = 1.$$

13. Find the rectangle of greatest perimeter inscribed in the ellipse

$$\frac{x^2}{a^2} + \frac{y^2}{b^2} = 1.$$

14. Find the point of the ellipse

$$5x^2 - 6xy + 5y^2 = 4$$

for which the tangent is at the greatest distance from the origin.

15*. Prove that the length l of the greatest axis of the ellipsoid

$$ax^2 + by^2 + cz^2 + 2dxy + 2exz + 2fyz = 1$$

is given by the greatest real root of the equation

$$\begin{vmatrix} a - \frac{1}{l^2} & d & e \\ d & b - \frac{1}{l^2} & f \\ e & f & c - \frac{1}{l^2} \end{vmatrix} = 0.$$

Appendix to Chapter III

1. Sufficient Conditions for Extreme Values

In the theory of maxima and minima in the preceding chapter we have contented ourselves with finding *necessary* conditions for the occurrence of an extreme value. In many cases occurring in actual practice the nature of the "stationary" point thus found can be determined from the special nature of the problem, and we can thus decide whether it is a maximum or a minimum. Yet it is important to have general *sufficient* conditions for the occurrence of an extreme value. Such criteria will be developed here for the typical case of two independent variables.

If we consider a point (x_0, y_0) at which the function is stationary, that is, a point at which both first partial derivatives of the function vanish, the occurrence of an extreme value is connected with the question whether the expression

$$f(x_0 + h, y_0 + k) - f(x_0, y_0)$$

has or has not the same sign for all sufficiently small values of h and k. If we expand this expression by Taylor's theorem (Chap. II, p. 80), with the remainder of the third order, in virtue of the equations $f_x(x_0, y_0) = 0$ and $f_y(x_0, y_0) = 0$ we at once obtain

$$f(x_0 + h, y_0 + k) - f(x_0, y_0) = \tfrac{1}{2}(h^2 f_{xx} + 2hk f_{xy} + k^2 f_{yy}) + \epsilon\rho^2,$$

where $\rho^2 = h^2 + k^2$ and ϵ tends to zero with ρ.

From this we see that in a sufficiently small neighbourhood of the point (x_0, y_0) the behaviour of the functional difference $f(x_0 + h, y_0 + k) - f(x_0, y_0)$ is essentially determined by the expression

$$Q(h, k) = ah^2 + 2bhk + ck^2,$$

where for brevity we have put

$$a = f_{xx}(x_0, y_0), \quad b = f_{xy}(x_0, y_0), \quad c = f_{yy}(x_0, y_0).$$

In order to study the problem of extreme values we must investigate this homogeneous quadratic expression in h and k, or, as we say, the *quadratic form* Q. We assume that the coefficients a, b, c do not all vanish. In the exceptional case where they do all vanish, which we shall not consider, we must begin with a Taylor series extending to terms of higher order.

With regard to the quadratic form Q there are three different possible cases:

1. The form is *definite*. That is, when h and k assume all values, Q assumes values of one sign only, and vanishes only for $h = 0$, $k = 0$. We say that the form is *positively definite* or *negatively definite* according as this sign is positive or negative. For example, the expression $h^2 + k^2$, which we obtain when $a = c = 1$, $b = 0$, is positively definite, while the expression $-h^2 + 2hk - 2k^2 = -(h - k)^2 - k^2$ is negatively definite.

2. The form is *indefinite*. That is, it can assume values of different sign, e.g. the form $Q = 2hk$, which has the value 2 for $h = 1$, $k = 1$ and the value -2 for $h = -1$, $k = 1$.

3. Finally, there is still a third possibility, namely that in which the form vanishes for values of h, k other than $h = 0$, $k = 0$, but otherwise assumes values of one sign only, e.g. the form $(h + k)^2$, which vanishes for all sets of values h, k such that $h = -k$. Such forms are called *semi-definite*.

The quadratic form $Q = ah^2 + 2bhk + ck^2$ is definite if, and only if, the condition

$$ac - b^2 > 0$$

is satisfied; it is then positively definite if $a > 0$ (so that $c > 0$ also), otherwise it is negatively definite.

In order that the form may be indefinite it is necessary and sufficient that

$$ac - b^2 < 0,$$

while the semi-definite case is characterized by the equation *

$$ac - b^2 = 0.$$

* These conditions are easily obtained as follows. Either $a = c = 0$, in which case we must have $b \neq 0$, and the form is, as already remarked, indefinite; the criterion therefore holds for this case: or else we must have, say, $a \neq 0$; we can then write

$$ah^2 + 2bhk + ck^2 = a\left[\left(h + \frac{b}{a}k\right)^2 + \frac{ca - b^2}{a^2}k^2\right].$$

This form is obviously definite if $ca - b^2 > 0$, and it then has the same sign as a. It is semi-definite if $ca - b^2 = 0$, for then it vanishes for all values of h, k that satisfy the equation $h/k = -b/a$, but for all other values it has the same sign. It is indefinite if $ca - b^2 < 0$, for it then assumes values of different sign when k vanishes and when $h + \frac{b}{a}k$ vanishes.

We shall now prove the following statements. If the quadratic form $Q(h, k)$ is positively definite, the stationary value assumed for $h = 0$, $k = 0$ is a *minimum*. If the form is negatively definite, the stationary value is a *maximum*. If the form is indefinite, we have neither a maximum nor a minimum; the point is a *saddle point*. Thus, definite character of the form Q is a sufficient condition for an extreme value, while indefinite character of Q excludes the possibility of an extreme value. We shall not consider the semi-definite case, which leads to involved discussions.

In order to prove the first statement we have only to use the fact that if Q is a positively definite form there is a positive number m, independent of h and k, such that*

$$Q \geqq 2m(h^2 + k^2) = 2m\rho^2.$$

Therefore

$$f(x_0 + h, y_0 + k) - f(x_0, y_0) = \tfrac{1}{2}Q(h, k) + \epsilon\rho^2 \geqq (m + \epsilon)\rho^2.$$

If we now choose ρ so small that the number ϵ is less in absolute value than $\frac{1}{2}m$, we obviously have

$$f(x_0 + h, y_0 + k) - f(x_0, y_0) \geqq \frac{m}{2}\rho^2.$$

Thus for this neighbourhood of the point (x_0, y_0) the value of the function is everywhere greater than $f(x_0, y_0)$, except of course at (x_0, y_0) itself. In the same way, when the form is negatively definite the point is a maximum.

Finally, if the form is indefinite, there is a pair of values (h_1, k_1) for which Q is negative and another pair (h_2, k_2) for which Q is positive. We can therefore find a positive number m such that

$$Q(h_1, k_1) < -2m\rho_1^2,$$
$$Q(h_2, k_2) > 2m\rho_2^2.$$

If we now put $h = th_1$, $k = tk_1$, $\rho^2 = h^2 + k^2$ $(t \neq 0)$, that is, if

* To see this we consider the quotient $\dfrac{Q(h, k)}{h^2 + k^2}$ as a function of the two quantities $u = \dfrac{h}{\sqrt{(h^2 + k^2)}}$ and $v = \dfrac{k}{\sqrt{(h^2 + k^2)}}$. Then $u^2 + v^2 = 1$, and the form becomes a continuous function of u and v, which must have a least value $2m$ on the circle $u^2 + v^2 = 1$. This value m obviously satisfies our conditions; it is not zero, for on the circle u and v never vanish simultaneously.

we consider a point $(x_0 + h, y_0 + k)$ on the line joining (x_0, y_0) to $(x_0 + h_1, y_0 + k_1)$, then from $Q(h, k) = t^2 Q(h_1, k_1)$ and $\rho^2 = t^2 \rho_1^2$ we have

$$Q(h, k) < -2m\rho^2.$$

Thus by choice of a sufficiently small t (and corresponding ρ) we can make the expression $f(x_0 + h, y_0 + k) - f(x_0, y_0)$ negative. We need only choose t so small that for $h = th_1$, $k = tk_1$ the absolute value of the quantity ϵ is less than $\frac{1}{2}m$. For such a set of values we have $f(x_0 + h, y_0 + k) - f(x_0, y_0) < -m\rho^2/2$, so that the value $f(x_0 + h, y_0 + k)$ is less than the stationary value $f(x_0, y_0)$. In the same way, on carrying out the corresponding process for the system $h = th_2$, $k = tk_2$, we find that in an arbitrarily small neighbourhood of (x_0, y_0) there are points at which the value of the function is greater than $f(x_0, y_0)$. Thus we have neither a maximum nor a minimum, but instead what we may call a saddle value.

If $a = b = c = 0$ at the stationary point, so that the quadratic form vanishes identically, and also in the semi-definite case, this discussion fails to apply. To obtain sufficient conditions for these cases would lead to involved calculations.

Thus we have the following rule for distinguishing maxima and minima:

If at a point (x_0, y_0) *the equations*

$$f_x(x_0, y_0) = 0, \quad f_y(x_0, y_0) = 0$$

hold, and also the inequality

$$f_{xx} f_{yy} - f_{xy}^2 > 0,$$

then at that point the function has an extreme value. This is a maximum if $f_{xx} < 0$ *(and consequently* $f_{yy} < 0$*), and a minimum if* $f_{xx} > 0$.

If, on the other hand,

$$f_{xx} f_{yy} - f_{xy}^2 < 0,$$

the stationary value is neither a maximum nor a minimum. The case

$$f_{xx} f_{yy} - f_{xy}^2 = 0$$

remains undecided.

These conditions permit of a simple geometrical interpretation. The necessary conditions $f_x = f_y = 0$ state that the tangent plane to the surface $z = f(x, y)$ is horizontal. If we really have an extreme value, then in the neighbourhood of the point in question the tangent plane does not intersect the surface. In the case of a saddle point, on the contrary, the plane cuts the surface in a curve which has several branches at the point. This matter will be clearer after the discussion of singular points in the next section.

As an example we seek to find the extreme values of the function

$$f(x, y) = x^2 + xy + y^2 + ax + by.$$

If we equate the first derivatives to zero, we obtain the equations

$$2x + y + a = 0, \quad x + 2y + b = 0,$$

which have the solution $x = \frac{1}{3}(b - 2a)$, $y = \frac{1}{3}(a - 2b)$. The expression

$$f_{xx}f_{yy} - f_{xy}{}^2 = 3$$

is positive, as is $f_{xx} = 2$. The function therefore has a minimum at the point in question.

The function

$$f(x, y) = (y - x^2)^2 + x^5$$

has a stationary point at the origin. There the expression $f_{xx}f_{yy} - f_{xy}{}^2$ vanishes, and our criterion fails. We readily see, however, that the function has not an extreme value there, for in the neighbourhood of the origin the function assumes both positive and negative values.

On the other hand, the function

$$f(x, y) = (x - y)^4 + (y - 1)^4$$

has a minimum at the point $x = 1$, $y = 1$, though the expression $f_{xx}f_{yy} - f_{xy}{}^2$ vanishes there. For

$$f(1 + h, 1 + k) - f(1, 1) = (h - k)^4 + k^4,$$

and this quantity is positive when $\rho \neq 0$.

EXAMPLE

If $\phi(a) = k \neq 0$, $\phi'(a) \neq 0$, and x, y, z satisfy the relation

$$\phi(x)\phi(y)\phi(z) = k^3,$$

prove that the function

$$f(x) + f(y) + f(z)$$

has a maximum when $x = y = z = a$, provided that

$$f'(a)\left(\frac{\phi''(a)}{\phi'(a)} - \frac{\phi'(a)}{\phi(a)}\right) > f''(a).$$

2. Singular Points of Plane Curves

In Chap. III, section 2 (p. 128) we saw that a curve $f(x, y) = 0$ in general has a singular point at a point $x = x_0$, $y = y_0$ such that the three equations

$$f(x_0, y_0) = 0, \quad f_x(x_0, y_0) = 0, \quad f_y(x_0, y_0) = 0$$

hold. In order to study these singular points systematically, we assume that in the neighbourhood of the point in question the function $f(x, y)$ has continuous derivatives up to the second order, and that at that point the second derivatives do not all vanish. By expanding in a Taylor series up to terms of the second order we obtain the equation of the curve in the form

$$2f(x, y) = (x - x_0)^2 f_{xx}(x_0, y_0) + 2(x - x_0)(y - y_0) f_{xy}(x_0, y_0) + (y - y_0)^2 f_{yy}(x_0, y_0) + \epsilon\rho^2 = 0,$$

where we have put $\rho^2 = (x - x_0)^2 + (y - y_0)^2$ and ϵ tends to zero with ρ.

Using a parameter t, we can write the equation of the general straight line through the point (x_0, y_0) in the form

$$x - x_0 = at, \quad y - y_0 = bt,$$

where a and b are two arbitrary constants, which we may suppose to be so chosen that $a^2 + b^2 = 1$. To determine the point of intersection of this line with the curve $f(x, y) = 0$ we substitute these expressions in the above expansion for $f(x, y)$; for the point of intersection we thus obtain the equation

$$a^2t^2 f_{xx} + 2abt^2 f_{xy} + b^2t^2 f_{yy} + \epsilon t^2 = 0.$$

A first solution is $t = 0$, i.e. the point (x_0, y_0) itself, as is obvious. It is, however, worthy of notice that the left-hand side of the equation is divisible by t^2, so that t is a "double root" of the equation. For this reason the singular points are also sometimes called "double points" of the curve.

If we remove the factor t^2, we are left with the equation

$$a^2 f_{xx} + 2ab f_{xy} + b^2 f_{yy} + \epsilon = 0.$$

We now inquire whether it is possible for the line to intersect the curve in another point which tends to (x_0, y_0) as the line tends to some particular limiting position. Such a limiting position of a secant we of course call a tangent. To discuss this, we observe that as a point tends to (x_0, y_0) the quantity t tends to zero, and therefore ϵ also tends to zero. If the equation above is still to be satisfied, the expression $a^2 f_{xx} + 2ab f_{xy} + b^2 f_{yy}$ must also tend to zero; that is, for the limiting position of the line we must have

$$a^2 f_{xx} + 2ab f_{xy} + b^2 f_{yy} = 0.$$

This equation gives us a quadratic condition determining the ratio a/b which fixes the line.

If the discriminant of the equation is negative, that is, if

$$f_{xx} f_{yy} - f_{xy}^2 < 0,$$

we obtain *two distinct real tangents*. The curve has a *double point* or *node*, like that exhibited by the lemniscate $(x^2 + y^2)^2 - (x^2 - y^2) = 0$ at the origin or the strophoid $(x^2 + y^2)(x - 2a) + a^2 x = 0$ at the point $x_0 = a$, $y_0 = 0$.

If the discriminant vanishes, that is, if

$$f_{xx} f_{yy} - f_{xy}^2 = 0,$$

we obtain two coincident tangents; it is then possible e.g. that two branches of the curve touch one another, or that the curve has a cusp.

Finally, if

$$f_{xx} f_{yy} - f_{xy}^2 > 0,$$

there is no (real) tangent at all. This occurs e.g. in the case of the so-called *isolated points* or *conjugate points* of an algebraic curve. These are points at which the equation of the curve is satisfied, but in whose neighbourhood no other point of the curve lies.

The curve $(x^2 - a^2)^2 + (y^2 - b^2)^2 = a^4 + b^4$ exemplifies this. The values $x = 0$, $y = 0$ satisfy the equation, but for all other values in the region $|x| < a\sqrt{2}$, $|y| < b\sqrt{2}$ the left-hand side is less than the right.

We have omitted the case in which all the derivatives of the second order vanish. This case leads to involved investigations, and we shall not consider it. Through such a point several branches of the curve may pass, or singularities of other types may occur.

Finally, we shall briefly mention the connexion between these matters and the theory of maxima and minima. Owing to the vanishing of the first derivatives, the equation of the tangent plane to the surface $z = f(x, y)$ at a stationary point (x_0, y_0) is simply

$$z - f(x_0, y_0) = 0.$$

The equation

$$f(x, y) - f(x_0, y_0) = 0$$

therefore gives us the projection on the xy-plane of the curve of intersection of the tangent plane with the surface, and we see that the point (x_0, y_0) is a singular point of this curve. If this is an isolated point, in a certain neighbourhood the tangent plane has no other point in common with the surface, and the function $f(x, y)$ has a maximum or a minimum at the point (x_0, y_0) (cf. p. 208). If, however, the singular point is a multiple point, the tangent plane cuts the surface in a curve with two branches, and the point corresponds to a saddle value. These remarks lead us precisely to the sufficient conditions which we have already found in section 1 (p. 207).

3. Singular Points of Surfaces

In a similar way we can discuss a singular point of a surface $f(x, y, z) = 0$, i.e. a point for which

$$f = 0,\ f_x = f_y = f_z = 0.$$

Without loss of generality we may take the point as the origin O. If we write

$$f_{xx} = \alpha,\ f_{yy} = \beta,\ f_{zz} = \gamma,\ f_{xy} = \lambda,\ f_{yz} = \mu,\ f_{xz} = \nu$$

for the values at this point, we obtain the equation

$$\alpha x^2 + \beta y^2 + \gamma z^2 + 2\lambda xy + 2\mu yz + 2\nu xz = 0$$

for a point (x, y, z) which lies on a tangent to the surface at O.

This equation represents a quadratic cone touching the

surface at the singular point—instead of the tangent plane at an ordinary point of the surface—if we assume that not all of the quantities $\alpha, \beta, \ldots, \nu$ vanish and that the above equation has real solutions other than $x = y = z = 0$.

4. Connection between Euler's and Lagrange's Representations of the Motion of a Fluid

Let (a, b, c) be the co-ordinates of a particle at the time $t = 0$ in a moving continuum (liquid or gas). Then the motion can be represented by three functions

$$x = x(a, b, c, t),$$
$$y = y(a, b, c, t),$$
$$z = z(a, b, c, t),$$

or in terms of a position vector $\mathbf{x} = \mathbf{x}(a, b, c, t)$. Velocity and acceleration are given by the derivatives with respect to the time t. Thus the velocity vector is $\dot{\mathbf{x}}$ with components $\dot{x}, \dot{y}, \dot{z}$, and the acceleration vector is $\ddot{\mathbf{x}}$ with components $\ddot{x}, \ddot{y}, \ddot{z}$, all of which appear as functions of the initial position (a, b, c) and the parameter t. For each value of t we have a transformation of the co-ordinates (a, b, c) belonging to the different points of the moving continuum into the co-ordinates (x, y, z) at the time t. This is the so-called Lagrange representation of the motion. Another representation introduced by Euler is based upon the knowledge of three functions

$$u(x, y, z, t), \quad v(x, y, z, t), \quad w(x, y, z, t)$$

representing the components $\dot{x}, \dot{y}, \dot{z}$ of the velocity $\dot{\mathbf{x}}$ of the motion at the point (x, y, z) at the time t.

In order to pass from the first representation to the second we have to use the first representation to calculate a, b, c as functions of x, y, z, and t, and to substitute these expressions in the expressions for $\dot{x}(a, b, c, t)$, $\dot{y}(a, b, c, t)$, $\dot{z}(a, b, c, t)$:

$$u(x, y, z, t) = \dot{x}\{a(x, y, z, t), b(x, y, z, t), c(x, y, z, t), t\}, \text{ \&c.}$$

We then get the components of the acceleration from

$$\dot{x}(a, b, c, t) = u\{x(a, b, c, t), y(a, b, c, t), z(a, b, c, t), t\}, \text{ \&c.}$$

as follows:

$$\ddot{x} = u_x\dot{x} + u_y\dot{y} + u_z\dot{z} + u_t, \text{ \&c.},$$

or

$$\ddot{x} = u_x u + u_y v + u_z w + u_t,$$
$$\ddot{y} = v_x u + v_y v + v_z w + v_t,$$
$$\ddot{z} = w_x u + w_y v + w_z w + w_t.$$

In the mechanics of a continuum the following equation connecting Euler's and Lagrange's representations is fundamental:

$$\operatorname{div} \dot{x} = u_x + v_y + w_z = \frac{\dot{D}}{D},$$

where

$$D(x, y, z, t) = \frac{\partial(x, y, z)}{\partial(a, b, c)}$$

is the Jacobian characterizing the motion.

The reader may complete the proof of this and the corresponding theorem in two dimensions by using the various rules for the differentiation of implicit functions.

5. Tangential Representation of a Closed Curve

A family of straight lines with parameter α may be given by

$$x \cos \alpha + y \sin \alpha - p(\alpha) = 0, \quad . \quad . \quad . \quad . \quad (1)$$

where $p(\alpha)$ denotes a function which is twice continuously differentiable and periodic of period 2π (a so-called *tangential function*). The envelope C of these lines is a closed curve satisfying (1) and the further equation

$$-x \sin \alpha + y \cos \alpha - p'(\alpha) = 0.$$

Hence

$$\left.\begin{aligned} x &= p \cos \alpha - p' \sin \alpha \\ y &= p \sin \alpha + p' \cos \alpha \end{aligned}\right\} \quad . \quad . \quad . \quad . \quad . \quad (2)$$

is the parametric representation of C (α being the parameter). Formula (1) gives the equation of the tangents of C and is referred to as the *tangential equation* of C.

Since

$$x' = -(p + p'') \sin \alpha, \quad y' = (p + p'') \cos \alpha,$$

we at once have the following expressions for the length L and area A of C:

$$L = \int_0^{2\pi} (p + p'')\,d\alpha = \int_0^{2\pi} p\,d\alpha,$$

$$A = \tfrac{1}{2}\int_0^{2\pi} (xy' - yx')\,d\alpha = \tfrac{1}{2}\int_0^{2\pi} (p + p'')p\,d\alpha = \tfrac{1}{2}\int_0^{2\pi} (p^2 - p'^2)\,d\alpha,$$

since $p'(\alpha)$ is also a function of period 2π.*

From this we deduce the isoperimetric inequality

$$L^2 \geqq 4\pi A,$$

where the equality sign holds for the circle only. This may also be expressed by the statement: among all closed curves of given length the circle has the greatest area.

For the proof we make use of the Fourier expansion of $p(\alpha)$ (Vol. I, Chap. IX, p. 447),

$$p(\alpha) = \frac{a_0}{2} + \sum_{\nu=1}^{\infty} (a_\nu \cos\nu\alpha + b_\nu \sin\nu\alpha);$$

then

$$p'(\alpha) = \sum_{\nu=1}^{\infty} \nu(b_\nu \cos\nu\alpha - a_\nu \sin\nu\alpha),$$

so that (using the orthogonality relations of Vol. I, p. 438) we have

$$L = \pi a_0,$$

$$A = \frac{\pi}{2}\left(\frac{a_0^2}{2} - \sum_{\nu=2}^{\infty} (\nu^2 - 1)(a_\nu^2 + b_\nu^2)\right).$$

Thus

$$A \leqq \frac{\pi a_0^2}{4} = \frac{L^2}{4\pi};$$

in particular, $A = \dfrac{L^2}{4\pi}$ only if $a_\nu = b_\nu = 0$ for $\nu \geqq 2$, i.e. $p(\alpha) = \dfrac{a_0}{2} + a_1 \cos\alpha + b_1 \sin\alpha$; the latter equation defines a circle, as is easily proved from (2).

* Since $p(\alpha) + c$ is obviously the tangential function of the parallel curve at a distance c from C, the formulæ for the area and the length of a parallel curve (cf. Vol. I, p. 291, Ex. 22, and p. 553) are easily derived from these expressions.

CHAPTER IV

Multiple Integrals

The idea of differentiation and the operations with derivatives in the case of functions of several variables are obtained almost immediately by reduction to their analogues for functions of one variable. As regards integration and its relation to differentiation, on the other hand, the case of several variables is more involved, since the concept of integral can be generalized for functions of several variables in a variety of ways. In this chapter we shall study multiple integrals such as we have already met in Vol. I, Chap. X (p. 486). In addition to these, however, we have also to consider the so-called line integrals in the plane, and surface integrals, as well as line integrals, in three dimensions (Chap. V, p. 343). In the end, however, it is found that all questions of integration can be reduced to the original concept of the integral in the case of one independent variable.

1. Ordinary Integrals as Functions of a Parameter

Before we study the new situations which arise with functions of more than one variable, we shall discuss some concepts which are directly related to matters already familiar to us.

1. Examples and Definitions.

If $f(x, y)$ is a continuous function of x and y in the rectangular region $\alpha \leqq x \leqq \beta$, $a \leqq y \leqq b$, we may in the first instance think of the quantity x as fixed, and we can then integrate the function $f(x, y)$, which is now a function of y alone, over the interval $a \leqq y \leqq b$. We thus arrive at the expression

$$\int_a^b f(x, y)\,dy,$$

which still depends on the choice of the quantity x. In a sense, therefore, we are considering not an integral but the family of integrals $\int_a^b f(x, y)\,dy$ which we obtain for different values of x. This quantity, which is kept fixed during the integration and to which we can assign any value in its interval, we call a *parameter*. Our ordinary *integral* therefore appears as a *function of the parameter x*.

Integrals which are functions of a parameter frequently occur in analysis and its applications.

Thus, as the substitution $xy = u$ readily shows,

$$\int_0^1 \frac{x\,dy}{\sqrt{(1 - x^2y^2)}} = \text{arc sin}\,x.$$

Again, in integrating the general power function we may regard the index as a parameter and write accordingly

$$\int_0^1 y^x\,dy = \frac{1}{x+1},$$

where we assume that $x > -1$.

If we represent the region of definition of the function $f(x, y)$ geometrically, and make the parallel to the y-axis corresponding to the fixed value of x intersect the rectangle as in fig. 1, then we obtain the function of y which is to be integrated by considering the values of the function $f(x, y)$ as a function of y along the line of intersection AB. We may also speak of integrating the function $f(x, y)$ *along the segment AB*.

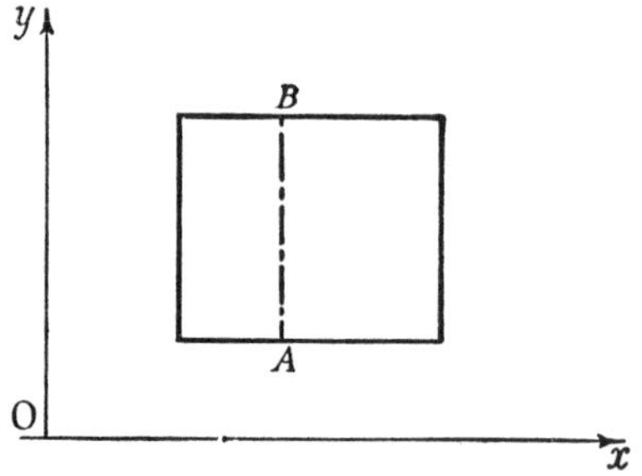

Fig. 1

This geometrical point of view suggests a generalization. If the region of definition R in which the function $f(x, y)$ is considered is not a rectangle, but instead has the shape shown in fig. 2 (that is, if any parallel to the y-axis cuts the boundary in at most two points), then for a fixed value of x we can again integrate the values of the function $f(x, y)$ along the line AB in which the parallel to the y-axis intersects the region of definition R. The initial and final points of the interval of integration

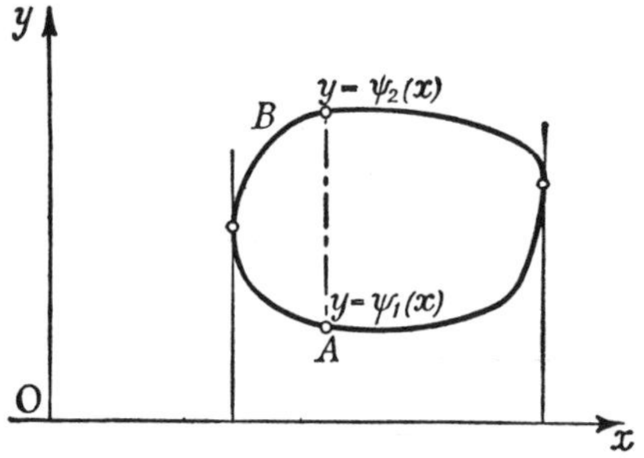

Fig. 2

will themselves vary as x varies. In other words, we have to consider an integral of the type

$$\int_{\psi_1(x)}^{\psi_2(x)} f(x, y)\, dy = F(x),$$

that is, an integral with the variable of integration y and the parameter x, in which the parameter occurs both in the integrand and in the limits of integration.

If, for example, the region of definition is a circle with unit radius and centre the origin, we shall have to consider integrals of the type

$$\int_{-\sqrt{(1-x^2)}}^{+\sqrt{(1-x^2)}} f(x, y)\, dy.$$

2. Continuity and Differentiability of an Integral with respect to the Parameter.

The integral

$$F(x) = \int_a^b f(x, y)\, dy$$

is a continuous function of the parameter x, *if* f(x, y) *is continuous in the region in question.*

For

$$\left| F(x+h) - F(x) \right| = \left| \int_a^b (f(x+h, y) - f(x, y))\, dy \right|$$
$$\leq \int_a^b \left| f(x+h, y) - f(x, y) \right| dy.$$

In virtue of the (uniform) continuity of $f(x, y)$, for sufficiently small values of h the integrand on the right, considered as a function of y, may be made uniformly as small as we please, and the statement follows immediately. In particular, therefore, we can integrate the function $F(x)$ *with respect to the parameter* x between the limits α and β, obtaining

$$\int_\alpha^\beta F(x)\,dx = \int_\alpha^\beta \left(\int_a^b f(x, y)\,dy\right) dx.$$

The integral on the right we also write in the form

$$\int_\alpha^\beta \int_a^b f(x, y)\,dy\,dx;$$

we call it a *repeated integral* or *multiple integral* (in this case a *double integral*).

We now investigate the possibility of differentiating $F(x)$. In the first place, we consider the case where the limits are fixed and assume that the function $f(x, y)$ has a continuous partial derivative f_x throughout the closed rectangle R. It is natural to try to form the x-derivative of the integral in the following way: instead of first integrating and then differentiating we reverse the order of these two processes, that is, we first differentiate f with respect to x and then integrate with respect to y. As a matter of fact, the following theorem is true:

If in the closed rectangle $\alpha \leqq x \leqq \beta$, $a \leqq y \leqq b$ *the function* $f(x, y)$ *has a continuous derivative with respect to* x, *we may differentiate the integral with respect to the parameter under the integral sign,* that is, if* $\alpha \leqq x \leqq \beta$,

$$\frac{d}{dx} F(x) = \frac{d}{dx}\int_a^b f(x, y)\,dy = \int_a^b f_x(x, y)\,dy.$$

* From this we obtain a simple proof of the fact, which we have already proved (Chap. II, p. 56), that in the formation of the mixed derivative g_{xy} of a function $g(x, y)$ the order of differentiation can be changed, provided that g_{xy} is continuous and g_x exists. For if we put $f(x, y) = g_y(x, y)$, we have

$$g(x, y) = g(x, a) + \int_a^y f(x, \eta)\,d\eta.$$

Since $f(x, y)$ has a continuous derivative with respect to x in the rectangle $\alpha \leqq x \leqq \beta$, $a \leqq y \leqq b$, it follows that

$$g_x = g_x(x, a) + \int_a^y f_x(x, \eta)\,d\eta,$$

Proof. If both x and $x + h$ belong to the interval $\alpha \leqq x \leqq \beta$, we can write

$$F(x+h) - F(x) = \int_a^b f(x+h, y)\, dy - \int_a^b f(x, y)\, dy$$
$$= \int_a^b \{f(x+h, y) - f(x, y)\}\, dy.$$

Since we have assumed that $f(x, y)$ is differentiable, the mean value theorem of the differential calculus in its usual form gives *

$$f(x+h, y) - f(x, y) = hf_x(x + \theta h, y), \quad 0 < \theta < 1.$$

Moreover, since the derivative f_x is assumed to be continuous in the closed region and therefore uniformly continuous, the absolute value of the difference

$$f_x(x + \theta h, y) - f_x(x, y)$$

is less than a positive quantity ϵ which is independent of x and y and tends to zero with h. Thus

$$\left| \frac{F(x+h) - F(x)}{h} - \int_a^b f_x(x, y)\, dy \right|$$
$$= \left| \int_a^b f_x(x + \theta h, y)\, dy - \int_a^b f_x(x, y)\, dy \right| \leqq \int_a^b \epsilon\, dy = \epsilon(b - a).$$

If we now let h tend to zero, ϵ also tends to zero, and the relation

$$\lim_{h \to 0} \frac{F(x+h) - F(x)}{h} = \int_a^b f_x(x, y)\, dy = F'(x)$$

at once follows; our statement is thus proved.

In a similar way we can establish the continuity of the integral and the rule for differentiating the integral with respect to a

and therefore

$$g_{yx} = f_x(x, y).$$

In the same way, $g_{xy} = f_x(x, y)$, and therefore $g_{xy} = g_{yx}$.

* Here the quantity θ depends on y, and may even vary discontinuously with y. This does not matter, for by the equation $f_x(x + \theta h, y) = h^{-1}(f(x + h, y) - f(x, y))$ we see at once that $f_x(x + \theta h, y)$ is a continuous function of x and y, and is therefore integrable.

parameter when the parameter occurs in the *limits*. If, for example, we wish to differentiate

$$F(x) = \int_{\psi_1(x)}^{\psi_2(x)} f(x, y)\,dy,$$

we start with the expression

$$F(x) = \int_u^v f(x, y)\,dy = \Phi(u, v, x),$$

where $u = \psi_1(x)$, $v = \psi_2(x)$. Here we assume that $\psi_1(x)$ and $\psi_2(x)$ have continuous derivatives with respect to x throughout the interval and that $f(x, y)$ is continuously differentiable (cf. p. 62) in a region wholly enclosing the region R. By the chain rule we now obtain

$$F'(x) = \frac{\partial \Phi}{\partial x} + \frac{\partial \Phi}{\partial u}\frac{du}{dx} + \frac{\partial \Phi}{\partial v}\frac{dv}{dx}.$$

If we apply the fundamental theorem of the integral calculus (Vol. I, p. 111), this gives the formula

$$F'(x) = \int_{\psi_1(x)}^{\psi_2(x)} f_x(x, y)\,dy - \psi_1'(x) f(x, \psi_1(x)) + \psi_2'(x) f(x, \psi_2(x)).$$

Thus if for $F(x)$ we take the function

$$F(x) = \int_0^x \sin(xy)\,dy,$$

we obtain

$$\frac{dF(x)}{dx} = \int_0^x y \cos(xy)\,dy + \sin(x^2).$$

If we take

$$F(x) = \int_0^1 \frac{x\,dy}{\sqrt{(1 - x^2y^2)}} = \arcsin x,$$

we obtain the relation

$$F'(x) = \int_0^1 \frac{dy}{\sqrt{(1 - x^2y^2)^3}} = \frac{1}{\sqrt{(1 - x^2)}},$$

as the reader can verify directly.

Other examples are given by the integrals

$$F_n(x) = \int_0^x \frac{(x - y)^n}{n!} f(y)\,dy,$$

$$F_0(x) = \int_0^x f(y)\,dy,$$

where n is any positive integer and $f(y)$ is a continuous function of y only in the interval under consideration. Since the expression arising from differentiation with respect to the upper limit x vanishes, the rule gives us

$$F_n'(x) = F_{n-1}(x).$$

Since $F_0'(x) = f(x)$, this at once gives

$$F_n^{(n+1)}(x) = f(x).$$

Therefore $F_n(x)$ is the function whose $(n+1)$-th derivative is equal to $f(x)$ and which, together with its first n derivatives, vanishes when $x = 0$; it arises from $F_{n-1}(x)$ by integration from 0 to x. Hence $F_n(x)$ is the function which is obtained from $f(x)$ by integrating $n+1$ times between the limits 0 and x. This repeated integration can therefore be replaced by a single integration of the function $\dfrac{(x-y)^n}{n!}f(y)$ with respect to y.

The rules for differentiating an integral with respect to a parameter often remain valid even when differentiation under the integral sign gives a function which is not continuous everywhere. In such cases, instead of applying general criteria, it is more convenient to verify whether such a differentiation is permissible in each special case.

As an example we consider the elliptic integral (cf. Vol. I, p. 243)

$$F(k) = \int_{-1}^{+1} \frac{dx}{\sqrt{(1-x^2)(1-k^2x^2)}}; \quad (k^2 < 1).$$

The function

$$f(k, x) = \frac{1}{\sqrt{(1-x^2)(1-k^2x^2)}}$$

is discontinuous at $x = +1$ and at $x = -1$, but the integral (as an improper integral) has a meaning. Formal differentiation with respect to the parameter k gives

$$F'(k) = \int_{-1}^{+1} \frac{kx^2\,dx}{\sqrt{(1-x^2)(1-k^2x^2)^3}}.$$

To investigate whether this equation is correct, we repeat the argument by which we obtained our differentiation formula. This gives

$$\frac{F(k+h)-F(k)}{h} = \int_{-1}^{+1} f_k(k+\theta h, x)\,dx = \int_{-1}^{+1} \frac{(k+\theta h)x^2\,dx}{\sqrt{(1-x^2)(1-(k+\theta h)^2x^2)^3}}.$$

The difference between this expression and the integral obtained by formal differentiation is

$$\Delta = \int_{-1}^{+1} \frac{x^2}{\sqrt{1-x^2}} \left(\frac{k+\theta h}{\sqrt{(1-(k+\theta h)^2x^2)^3}} - \frac{k}{\sqrt{(1-k^2x^2)^3}} \right) dx.$$

We must show that this integral tends to zero with h. For this purpose we mark off about k an interval $k_0 \leqq k \leqq k_1$ not containing the values ± 1, and we choose h so small that $k + \theta h$ lies in this interval. The function

$$\frac{k}{\sqrt{(1-k^2x^2)^3}}$$

is continuous in the closed region $-1 \leqq x \leqq 1$, $k_0 \leqq k \leqq k_1$, and is therefore uniformly continuous. The difference

$$\left| \frac{k+\theta h}{\sqrt{(1-(k+\theta h)^2x^2)^3}} - \frac{k}{\sqrt{(1-k^2x^2)^3}} \right|$$

consequently remains below a bound ε which is independent of x and k and which tends to zero with h. Hence the integral Δ also remains less in absolute value than

$$\int_{-1}^{+1} \frac{x^2\,dx}{\sqrt{1-x^2}}\,\varepsilon = M\varepsilon,$$

where M is a constant independent of ε. That is, the integral Δ tends to zero as h does, which is what we wished to show.

Differentiation under the integral sign is therefore permissible in this case. Similar considerations lead to the required result in other cases.

Improper integrals with an infinite range of integration are discussed in the Appendix to this chapter, § 4, p. 307.

EXAMPLES

1. Evaluate

$$F(y) = \int_0^1 x^{y-1}(y \log x + 1)\,dx.$$

2. Let $f(x, y)$ be twice continuously differentiable, and let $u(x, y, z)$ be defined as follows:

$$u(x, y, z) = \int_0^{2\pi} f(x + z\cos\varphi,\, y + z\sin\varphi)\,d\varphi.$$

Prove that

$$z(u_{xx} + u_{yy} - u_{zz}) - u_z = 0.$$

3 *. If $f(x)$ is twice continuously differentiable and

$$u(x, t) = \frac{1}{t^{p-2}}\int_{-t}^{+t} f(x+y)\,(t^2-y^2)^{\frac{p-3}{2}}\,dy \qquad (p > 1),$$

prove that

$$u_{xx} = \frac{p-1}{t}\,u_t + u_{tt}.$$

4. The Bessel function $J_0(x)$ may be defined by

$$J_0(x) = \frac{1}{\pi}\int_{-1}^{+1} \frac{\cos xt}{\sqrt{(1-t^2)}}\,dt.$$

Prove that

$$J_0'' + \frac{1}{x}J_0' + J_0 = 0.$$

5. For any non-negative integral index n the Bessel function $J_n(x)$ may be defined by

$$J_n(x) = \frac{x^n}{1 \cdot 3 \cdot 5 \ldots (2n-1)\pi}\int_{-1}^{+1} \cos xt\,(1-t^2)^{n-\frac{1}{2}}\,dt.$$

Prove that

(a) $$J_n'' + \frac{1}{x}J_n' + \left(1 - \frac{n^2}{x^2}\right)J_n = 0 \qquad (n \geqq 0),$$

(b) $$J_{n+1} = J_{n-1} - 2J_n' \qquad (n \geqq 1)$$

and $$J_1 = -J_0'.$$

2. The Integral of a Continuous Function over a Region of the Plane or of Space

1. The Double Integral (Domain Integral) as a Volume.

The first and most important generalization of the ordinary integral, like the ordinary integral itself, is suggested by geometrical intuition. Let R be a closed region of the xy-plane, bounded—as we assume all along—by one or more arcs of curves with continuously turning tangents, and let $z = f(x, y)$ be a function which is continuous in R. We assume in the first instance that f is non-negative, and represent it by a surface in xyz-space vertically above the region R. We now wish to find (or, more precisely, to *define*, since we have not yet done so) the volume V below the surface. This has been done in detail for rectangular regions in Vol. I, Chap. X (p. 486), and, moreover, the case is so similar to that of the ordinary integral that we feel justified in mentioning it somewhat briefly here. The student will see at once that a natural way of arriving at this volume is to subdivide R into N sub-regions $R_1, R_2, \ldots, R_N$, each having boundaries that are sectionally smooth (p. 41), and to find the greatest value M_i and the least value m_i of f in each region R_i. The areas of the regions R_i we denote by ΔR_i. On each region R_i as base we construct a cylinder of altitude M_i. This set of cylinders completely encloses the volume under the surface. Again, with each region R_i as base we construct a cylinder of altitude m_i, and hence with volume $m_i \Delta R_i$; these cylinders lie completely within the volume under the surface. Then

$$\sum_1^N m_i \Delta R_i \leqq V \leqq \sum_1^N M_i \Delta R_i.$$

These sums $\Sigma m_i \Delta R_i$ and $\Sigma M_i \Delta R_i$ we call the *lower* and *upper* sums respectively.

If we now make our subdivision finer and finer, so that the number N increases beyond all bounds, while the greatest diameter of the regions R_i (that is, the greatest distance between two points of R_i) at the same time tends to zero, we see intuitively (and shall later prove rigorously) that the upper and lower sums must approach one another more and more closely, so that *the volume* V *can be regarded as the common limit of the upper and lower sums as* N *tends to* ∞.

We can obviously obtain the same limiting value if instead of m_i or M_i we take any number between m_i and M_i, e.g. $f(x_i, y_i)$, the value of the function at a point (x_i, y_i) in the region R_i.

2. The General Analytical Concept of the Integral.

These concepts suggested by geometry must now be studied analytically and made more precise without direct reference to intuition. We accordingly proceed as follows. We consider a closed region R with area ΔR, and a function $f(x, y)$ which is defined and continuous everywhere in R, including the boundary. As before, we subdivide the region by sectionally smooth arcs * into N sub-regions $R_1, R_2, \ldots, R_N$ with areas $\Delta R_1, \ldots, \Delta R_N$. In R_i we choose an arbitrary point (ξ_i, η_i) where the function has the value $f_i = f(\xi_i, \eta_i)$ and we form the sum

$$V_N = \sum_1^N f_i \Delta R_i.$$

The fundamental theorem is then as follows:

If the number N *increases beyond all bounds and at the same*

* I.e. arcs which are given in a suitable co-ordinate system by an equation $y = \phi(x)$, where ϕ is a continuous function whose derivative is continuous except for a finite number of jump discontinuities (cf. p. 41).

time the greatest of the diameters of the sub-regions tends to zero, then V_N *tends to a limit* V. *This limit is independent of the particular nature of the subdivision of the regions* R *and of the choice of the point* (ξ_i, η_i) *in* R_i. *The limit* V *we call the (double) integral of the function* f(x, y) *over the region* R: *in symbols,*

$$\iint_R f(x, y)\, dS.$$

Corollary. We obtain the same limit if we take the sum only over those sub-regions R_i which lie entirely in the interior of R, that is, which have no points in common with the boundary of R.

This existence theorem for the integral * of a continuous function must be proved in a purely analytical way. The proof, which is very similar to the corresponding proof for one variable, is given in the appendix to this chapter (p. 293).

We shall now illustrate this concept of an integral by considering some special subdivisions. The simplest case is that in which R is a rectangle $a \leqq x \leqq b$, $c \leqq y \leqq d$ and the sub-regions R_i are also rectangles, formed by subdividing the x-interval into n equal parts and the y-interval into m equal parts, of lengths

$$h = \frac{b-a}{n} \quad \text{and} \quad k = \frac{d-c}{m}.$$

The points of subdivision we call $x_0 = a, x_1, x_2, \ldots, x_n = b$ and

* We can refine this theorem further in a way which is useful for many purposes. In the subdivision into N sub-regions it is not necessary to choose a value which is actually assumed by the function $f(x, y)$ at a definite point (ξ_i, η_i) of the corresponding sub-region; it is sufficient to choose values which differ from the values of the function $f(\xi_i, \eta_i)$ by quantities which tend uniformly to zero as the subdivision is made finer. In other words, instead of the values of the function $f(\xi_i, \eta_i)$ we can consider the quantities

$$f_i = f(\xi_i, \eta_i) + \epsilon_{i, N}$$

where $|\epsilon_{i, N}| < \epsilon_N$, $\lim_{N\to\infty} \epsilon_N = 0$. (The number $\epsilon_{i, N}$ is therefore the difference between the value of the function at a point of the i-th sub-region of the subdivision into N sub-regions and the quantity f_i with which we form the sum.) This theorem is almost trivial; for, since the numbers $\epsilon_{i, N}$ tend uniformly to zero, the absolute value of the difference between the two sums

$$\sum_1^N f_i \Delta R_i \quad \text{and} \quad \sum_1^N (f_i + \epsilon_{i, N}) \Delta R_i$$

is less than $\epsilon_N \Sigma \Delta R_i$, and can be made as small as we please if we take the number N sufficiently large. E.g. if we have $f(x, y) = P(x, y)\, Q(x, y)$ we may take $f_i = P_i Q_i$, where P_i and Q_i are the maxima of P and Q in R, which are in general not assumed at the same point.

$y_0 = c$, y_1, y_2, . . . , $y_m = d$ respectively, and through these points we draw parallels to the y-axis and the x-axis respectively. We then have $N = nm$. All the sub-regions are rectangles with area $\Delta R_i = hk = \Delta x \Delta y$, if we put $h = \Delta x$, $k = \Delta y$. For the point (ξ_i, η_i) we can take any point in the corresponding rectangle, and we then form the sum

$$\sum_i f(\xi_i, \eta_i)\Delta x \Delta y$$

for all the rectangles of the subdivision.

If we now let n and m simultaneously increase beyond all bounds, the sum will tend to the integral of the function f over the rectangle R.

These rectangles can also be characterized by two suffixes μ and ν, corresponding to the co-ordinates $x = a + \nu h$ and $y = c + \mu k$ of the lower left-hand corner of the rectangle in question. Here ν assumes integral values from 0 to $(n - 1)$ and μ from 0 to $(m - 1)$. With this identification of the rectangles by the suffixes ν and μ we may appropriately write the sum as a double sum *

$$\sum_{\nu=0}^{n-1} \sum_{\mu=0}^{m-1} f(\xi_\nu, \eta_\mu)\Delta x \Delta y.$$

Even when R is not a rectangle, it is often convenient to subdivide the region into rectangular sub-regions R_i. To do this we superpose on the plane the rectangular net formed by the lines

$$x = \nu h \quad (\nu = 0, \pm 1, \pm 2, \ldots)$$
$$y = \mu k \quad (\mu = 0, \pm 1, \pm 2, \ldots),$$

where h and k are numbers chosen arbitrarily. We now consider all those rectangles of the division which lie entirely within R. These rectangles we call R_i. Of course they do not completely fill the region; on the contrary, in addition to these rectangles R also contains certain regions R_l adjacent to the boundary which are bounded partly by lines of the net and partly by portions of the boundary of R. But by the corollary on p. 225 we can calculate the integral of the function f over the region R by summing over the interior rectangles only and then passing to the limit.

* If we are to write the sum in this way, we must suppose that the points (ξ_i, η_i) are chosen so as to lie in vertical or horizontal straight lines.

Another type of subdivision which is frequently applied is the subdivision by a polar co-ordinate net (fig. 3). Let the origin O of the polar co-ordinate system lie in the interior of our region. We subdivide the entire angle 2π into n parts of magnitude

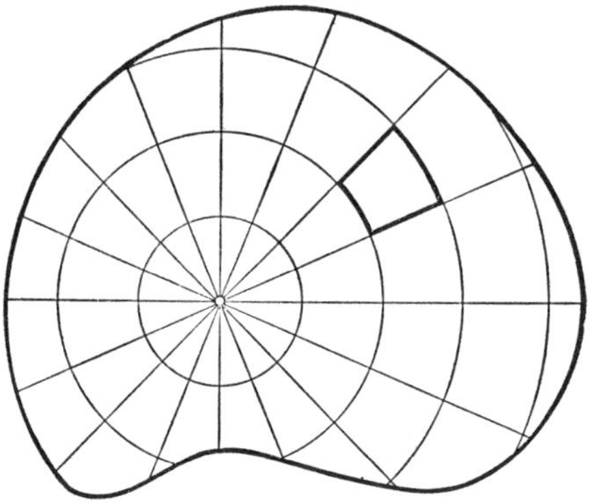

Fig. 3.—Subdivision by polar co-ordinate nets

$\Delta\theta = 2\pi/n = h$, and we also choose a second quantity $k = \Delta r$. We now draw the lines $\theta = \nu h (\nu = 0, 1, 2, \ldots, n-1)$ through the origin and also the concentric circles $r_\mu = \mu k (\mu = 1, 2, \ldots)$. Those which lie entirely in the interior of R we denote by R_i and their areas by ΔR_i. We can then regard the integral of the function $f(x, y)$ over the region R as the limit of the sum

$$\Sigma f(\xi_i, \eta_i)\Delta R_i$$

where (ξ_i, η_i) is a point chosen arbitrarily in R_i. The sum is taken over all the sub-regions R_i in the interior of R, and the passage to the limit consists in letting h and k tend simultaneously to zero.

By elementary geometry the area ΔR_i is given by the equation

$$\Delta R_i = \tfrac{1}{2}({r^2}_{\mu+1} - {r_\mu}^2)h = \tfrac{1}{2}(2\mu + 1)k^2 h,$$

if we assume that R_i lies in the ring bounded by the circles with radii μk and $(\mu + 1)k$.

3. Examples.

The simplest example is the function $f(x, y) = 1$. Here the limit of the sum is obviously independent of the mode of subdivision and is always equal to the area of the region R. Consequently, the integral of the function

$f(x, y) = 1$ over the region is also equal to this area. This might have been expected, for the integral is the volume of the cylinder of unit altitude with the region R as base.

As a further example we consider the integral of the function $f(x, y) = x$ over the square $0 \leqq x \leqq 1$, $0 \leqq y \leqq 1$. The intuitive interpretation of the integral as a volume shows that the value of our integral must be $\frac{1}{2}$. We can verify this by means of the analytical definition of the integral. We subdivide the rectangle into squares of side $h = 1/n$, and for the point (ξ_i, η_i) we choose the lower left-hand corner of the small square. Then each one of the squares in the vertical column whose left-hand side has the abscissa νh contributes the amount νh^3 to the sum. This expression occurs n times. Thus the contribution of the whole column of squares amounts to $n\nu h^3 = \nu h^2$. If we now form the sum from $\nu = 0$ to $\nu = n - 1$, we obtain

$$\sum_{\nu=0}^{n-1} \nu h^2 = \frac{n(n-1)}{2} h^2 = \frac{1}{2} - \frac{h}{2}.$$

The limit of this expression as $h \to 0$ is $\frac{1}{2}$, as we stated.

In a similar way we can integrate the product xy, or more generally any function $f(x, y)$ which can be represented as a product of a function of x and a function of y in the form $f(x, y) = \varphi(x)\psi(y)$, provided that the region of integration is a rectangle with sides parallel to the axes, say

$$a \leqq x \leqq b,$$
$$c \leqq y \leqq d.$$

We use the same division of the rectangle as on p. 225, and for the value of the function in each sub-rectangle we take the value of the function at the lower left-hand corner. The integral is then the limit of the sum

$$hk \sum_{\nu=0}^{n-1} \sum_{\mu=0}^{m-1} \varphi(\nu h)\psi(\mu k),$$

which may also be written as the product of two sums in the form

$$\left\{\sum_{\nu=0}^{n-1} h\,\varphi(\nu h)\right\} \left\{\sum_{\mu=0}^{m-1} k\,\psi(\mu k)\right\}.$$

But in accordance with the definition of the ordinary integral, as $h \to 0$ and $k \to 0$ each of these factors tends to the integral of the corresponding function over the interval from a to b or from c to d respectively. We thus obtain the general rule: *if a function* $f(x, y)$ *can be represented as a product of two functions* $\varphi(x)$ *and* $\psi(y)$, *its double integral over a rectangle* $a \leqq x \leqq b$, $c \leqq y \leqq d$ *can be resolved into the product of two integrals:*

$$\iint_R f(x, y)\,dx\,dy = \int_a^b \varphi(x)\,dx \cdot \int_c^d \psi(y)\,dy.$$

In virtue of this rule and the summation rule (cf. p. 231) we can, for example, integrate any polynomial over a rectangle with sides parallel to the axes.

As a last example we consider a case in which it is convenient to use a subdivision by the polar co-ordinate net instead of a subdivision into rectangles. Let the region R be the circle with unit radius and centre the origin, given by $x^2 + y^2 \leq 1$, and let

$$f(x, y) = \sqrt{(1 - x^2 - y^2)};$$

in other words, we wish to find the volume of a hemisphere of unit radius.

We construct the polar co-ordinate net as before. From the sub-region lying between the circles with radii $r_\mu = \mu k$ and $r_{\mu+1} = (\mu + 1)k$ and between the lines $\theta = \nu h$ and $\theta = (\nu + 1)h$ $\left(h = \frac{2\pi}{n}\right)$ we obtain the contribution

$$\frac{1}{2}\sqrt{1 - \left(\frac{r_{\mu+1} + r_\mu}{2}\right)^2}\,(r^2_{\mu+1} - r_\mu^2)h = \sqrt{1 - \rho_\mu^2}\,\rho_\mu kh,$$

where for the value of the function in the sub-region R_i we have taken the value which the function assumes on an intermediate circle with the radius $\rho_\mu = \frac{r_{\mu+1} + r_\mu}{2}$. All sub-regions which lie in the same ring give the same contribution, and since there are $n = 2\pi/h$ such regions the contribution of the whole ring is

$$2\pi\sqrt{1 - \rho_\mu^2}\,\rho_\mu k.$$

The integral is therefore the limit of the sum

$$\sum_{\mu=0}^{m-1} 2\pi\sqrt{1 - \rho_\mu^2}\,\rho_\mu k,$$

and, as we already know, this sum tends to the single integral

$$2\pi\int_0^1 r\sqrt{1 - r^2}\,dr = -\frac{2\pi}{3}\sqrt{(1 - r^2)^3}\,\Big|_0^1 = \frac{2\pi}{3};$$

we therefore obtain

$$\iint_R \sqrt{1 - x^2 - y^2}\,dS = \frac{2\pi}{3},$$

in agreement with the known formula for the volume of a sphere.

4. Notation. Extensions. Fundamental Rules.

The rectangular subdivision of the region R is associated with the symbol for the double integral which has been in use since Leibnitz's time. Starting with the symbol

$$\sum_{\nu=0}^{n-1}\sum_{\mu=0}^{m-1} f(\xi_\nu, \eta_\mu)\Delta x\Delta y$$

for the sum over the rectangles, we indicate the passage to the limit, from the sum to the integral, by replacing the double summation sign by a double integral sign and writing the *symbol* $dx\,dy$ instead of the *product* of the quantities Δx and Δy. Accordingly, the double integral is frequently written in the form

$$\iint_R f(x, y)\,dx\,dy$$

instead of in the form

$$\iint_R f(x, y)\,dS$$

in which the area of ΔR is replaced by the symbol dS. We again emphasize that the symbol $dx\,dy$ does not mean a product, but merely refers symbolically to the passage to the limit of the above sums of nm terms as $n \to \infty$ and $m \to \infty$.

It is clear that in double integrals, just as in ordinary integrals of a single variable, the notation for the " variables of integration " is immaterial, so that we could equally well have written

$$\iint_R f(u, v)\,du\,dv \quad \text{or} \quad \iint_R f(\xi, \eta)\,d\xi\,d\eta.$$

In introducing the concept of integral we saw that for a positive function $f(x, y)$ the integral represents the volume under the surface $z = f(x, y)$. In the analytical definition of integral, however, it is quite unnecessary that the function $f(x, y)$ should be positive everywhere; it may be negative, or it may change sign, in which last case the surface intersects the region R. Thus in the general case the integral gives the volume in question with a definite sign, the sign being positive for surfaces or portions of surfaces which lie above the xy-plane. If the whole of the surface corresponding to the region R consists of several such portions, the integral represents the sum of the corresponding volumes taken with their proper signs. In particular, a double integral may vanish although the function under the integral sign does not vanish everywhere.

For double integrals, as for single integrals, the following fundamental rules hold, the proofs being a simple repetition of those in Vol. I (p. 81). If c is a constant, then

$$\iint_R cf(x, y)\,dS = c\iint_R f(x, y)\,dS.$$

Also,

$$\iint_R (f(x, y) + \phi(x, y))\, dS = \iint_R f(x, y)\, dS + \iint_R \phi(x, y)\, dS,$$

that is: *the integral of the sum of two functions is equal to the sum of their two integrals.* Finally, if the region R consists of two subregions R' and R'' that have at most portions of the boundary in common, then

$$\iint_R f(x, y)\, dS = \iint_{R'} f(x, y)\, dS + \iint_{R''} f(x, y)\, dS,$$

that is: *when regions are joined together the corresponding integrals are added.*

5. Integral Estimates and the Mean Value Theorem.

As in the case of one independent variable, there are some very useful estimation theorems for the double integral. Since the proofs are practically the same as those of Vol. I, Chap. II, section 7 (p. 126), we shall here be content with a statement of the facts.

If $f(x, y) \geqq 0$ in R, then

$$\iint_R f(x, y)\, dS \geqq 0;$$

similarly, if $f(x, y) \leqq 0$,

$$\iint_R f(x, y)\, dS \leqq 0.$$

This leads to the following result:

If the inequality

$$f(x, y) \geqq \phi(x, y)$$

holds everywhere in R, *then*

$$\iint_R f(x, y)\, dS \geqq \iint_R \phi(x, y)\, dS.$$

A direct application of this theorem gives the relations

$$\iint_R f(x, y)\, dS \leqq \iint_R \left| f(x, y) \right| dS$$

and

$$\iint_R f(x, y)\, dS \geqq -\iint_R \left| f(x, y) \right| dS.$$

We can also combine these two inequalities in a single formula:

$$\left|\iint_R f(x, y)\,dS\right| \leqq \iint_R \left|f(x, y)\right| dS.$$

If m is the lower bound and M the upper bound of the values of the function $f(x, y)$ in R, then

$$m\Delta R \leqq \iint f(x, y)\,dS \leqq M\Delta R,$$

where ΔR is the area of the region R. The integral can then be expressed in the form

$$\iint_R f(x, y)\,dS = \mu\Delta R,$$

where μ is a number intermediate between m and M, the exact value of which cannot in general be specified more exactly.*

This form of the estimation formula we again call the *mean value theorem of the integral calculus.*

Here again the following generalization holds: if $p(x, y)$ is an arbitrary positive continuous function in R, then

$$\iint_R p(x, y) f(x, y)\,dS = \mu \iint_R p(x, y)\,dS,$$

where μ denotes a number between the greatest and least values of f, which cannot be further specified.

These integral estimates show as before that *the integral varies continuously with the function.* More precisely, if $f(x, y)$ and $\phi(x, y)$ are two functions which satisfy the inequality

$$|f(x, y) - \phi(x, y)| < \epsilon,$$

where ϵ is a fixed positive number in the whole region R with area ΔR, then the integrals $\iint_R f(x, y)\,dS$ and $\iint_R \phi(x, y)\,dS$ differ by less than $\epsilon\Delta R$, that is, by less than a number which tends to zero with ϵ.

In the same way we see that *the integral of a function varies continuously with the region.* For suppose that two regions R' and R'' are obtained from one another by the addition or removal of portions whose total area is less than ϵ, and suppose that

* Just as in the case of continuous functions of one variable, we can state that the value μ is certainly assumed at *some* point of the region R by the *continuous* function $f(x, y)$.

$f(x, y)$ is a function which is continuous in both regions and such that $|f(x, y)| < M$, where M is a fixed number. Then the two integrals $\iint_{R'} f(x, y)\,dS$ and $\iint_{R''} f(x, y)\,dS$ differ by less than $M\epsilon$, that is, by less than a number which tends to zero with ϵ. The proof of this fact follows at once from the last theorem of the preceding sub-section.

We can therefore calculate the integral over a region R as accurately as we please by taking it over a sub-region of R whose total area differs from the area of R by a sufficiently small amount. For example, in the region R we can construct a polygon whose total area differs by as little as we please from the area of R. In particular, we may suppose this polygon to be bounded by lines parallel to the x- and y-axes alternately, that is, to be pieced together out of rectangles with sides parallel to the axes.

6. Integrals over Regions in Three and More Dimensions.

Every statement we have made for integrals over regions of the xy-plane can be extended without further complication or the introduction of new ideas to regions in three or more dimensions. If e.g. we consider the case of the integral over a three-dimensional region R, we have only to subdivide this region R by means of a finite number of surfaces with continuously varying tangent planes into sub-regions which completely fill R and which we denote by $R_1, R_2, \ldots, R_N$. If $f(x, y, z)$ is a function which is continuous in the closed region R, and if (ξ_i, η_i, ζ_i) denotes an arbitrary point in the region R_i, we again form the sum

$$\sum_{i=1}^{N} f(\xi_i, \eta_i, \zeta_i)\Delta R_i,$$

in which ΔR_i denotes the volume of the region R_i. The sum is taken over all the regions R_i, or, if it is more convenient, only over those sub-regions which do not adjoin the boundary of R. If we now let the number of sub-regions increase beyond all bounds in such a way that the diameter of the largest of them tends to zero, we again find a limit independent of the particular mode of subdivision and of the choice of the intermediate points. This limit we call the *integral of* f(x, y, z) *over the region* R, and we denote it symbolically by

$$\iiint_R f(x, y, z)\,dV.$$

If, in particular, we effect a subdivision of the region into rectangular regions with sides Δx, Δy, Δz, the volumes of the inner regions R_i will all have the same value $\Delta x \Delta y \Delta z$. As on p. 230, we indicate the possibility of this type of subdivision and the passage to the limit by introducing the symbolic notation

$$\iiint_R f(x, y, z)\, dx\, dy\, dz$$

in addition to the one above. All the facts which we have mentioned for double integrals remain valid for triple integrals apart from the necessary changes in notation.

For regions of more than three dimensions the multiple integral can be defined in exactly the same way, once we have suitably defined the concept of volume for such regions. If in the first instance we restrict ourselves to rectangular regions and subdivide these into similarly oriented rectangular subregions, and if we further define the volume of a rectangle

$$a_1 \leqq x_1 \leqq a_1 + h_1, \quad a_2 \leqq x_2 \leqq a_2 + h_2, \ldots, \quad a_n \leqq x_n \leqq a_n + h_n,$$

as the product $h_1 h_2 \ldots h_n$, the definition of integral involves nothing new. We denote an integral over the n-dimensional region R by

$$\int\int \cdots\cdots \int_R f(x_1, x_2, \ldots, x_n)\, dx_1\, dx_2 \ldots dx_n.$$

For more general regions and more general subdivisions we must rely on the abstract definition of volume which we shall give in section 1 of the appendix (p. 287).

In what follows, apart from section 3 of the appendix, we can confine ourselves to integrals in at most three dimensions.

7. Space Differentiation. Mass and Density.

In the case of single integrals and functions of one variable, we obtain the integrand from the integral by a process of differentiation, taking the integral over an interval of length h, dividing by the length h, and then letting h tend to zero. For functions of one variable this fact represents the fundamental connexion between the differential calculus and the integral calculus, and we interpreted it intuitively in terms of the concepts of total mass and density. For the multiple integrals of functions of several variables the same connexion exists; but here it is not so fundamental in character.

We consider the multiple integral (domain integral)

$$\iint_B f(x, y)\, dS \quad \text{or} \quad \iiint_B f(x, y, z)\, dV$$

of a continuous function of two or more variables over a region B which contains a fixed point P with co-ordinates (x_0, y_0)—or (x_0, y_0, z_0), as the case may be—and which has the content * ΔB. If we then divide the value of this integral by the content ΔB, it follows from the considerations of sub-section 5 (p. 232) that the quotient will be an intermediate value of the integrand, that is, a number between the greatest and the least values of the integrand in the region. If we now let the diameter of the region B about the point P tend to zero, so that the content ΔB also tends to zero, this intermediate value of the function $f(x, y)$—or $f(x, y, z)$—must tend to the value of the function at the point P. Thus the passage to the limit yields the relations

$$\lim_{\Delta B \to 0} \frac{1}{\Delta B} \iint_B f(x, y)\, dS = f(x_0, y_0)$$

and

$$\lim_{\Delta B \to 0} \frac{1}{\Delta B} \iiint_B f(x, y, z)\, dV = f(x_0, y_0, z_0).$$

This limiting process, which corresponds to the differentiation described above for integrals with one independent variable, we call the *space differentiation* of the integral. We see, then, that the *space differentiation of a multiple integral gives the integrand.*

This connexion enables us to interpret the relation of integrand to integral in the case of several independent variables, as before, by means of the physical concepts of density and total mass. We think of a mass of any substance whatever as distributed over a two- or three-dimensional region R in such a way that an arbitrarily small mass is contained in each sufficiently small sub-region. In order to define the specific mass or density at a point P, we first consider a neighbourhood B of the point P with content ΔB, and divide the total mass in this neighbourhood by the content. The quotient we shall call the *mean density* or *average density* in this sub-region. If we now let the diameter of B tend to zero, from the average density in the region B we obtain in the limit the *density at the point* P,

* The word *content* is used as a general word to include the idea of length in one dimension, area in two dimensions, volume in three dimensions, and so on.

provided always that such a limit exists independently of the choice of the sequence of regions. If we denote this density by $\mu(x, y)$—or by $\mu(x, y, z)$—and assume that it is continuous, we see at once that the process described above is simply the space differentiation of the integral

$$\iint_R \mu(x, y)\, dS,$$

or

$$\iiint_R \mu(x, y, z)\, dV,$$

taken over the whole region R. This integral taken over the whole region therefore gives us the *total mass* of the substance of density μ in the region * R.

From the physical point of view such a representation of the mass of a substance is naturally an idealization. That this idealization is reasonable, i.e. that it approximates to the actual situation with sufficient accuracy, is one of the assumptions of *physics*.

These ideas, moreover, retain their mathematical significance even when μ is not positive everywhere. Such negative densities and masses may also have a physical interpretation, e.g. in the study of the distribution of electric charge.

3. Reduction of the Multiple Integral to Repeated Single Integrals

The fact that every multiple integral can be reduced to single integrals is of fundamental importance in the evaluation of multiple integrals. It enables us to apply all the methods which we have previously developed for finding indefinite integrals to the evaluation of multiple integrals.

1. Integrals over a Rectangle.

In the first place we take the region R as a rectangle $a \leqq x \leqq b$, $\alpha \leqq y \leqq \beta$ in the xy-plane, and we consider a continuous function $f(x, y)$ in R. In Vol. I, Chap. X (pp. 490–1) we used a process of cutting the volume under the surface $z = f(x, y)$ into slices in order to make the following statement appear plausible:

* What we have shown here is that the distribution given by the multiple integral has the same space-derivative as the mass-distribution originally given. It remains to be proved that this implies that the two distributions are actually identical; in other words, that the statement "space differentiation gives the density μ" can be satisfied by only one distribution of mass. The proof, which is not difficult, is passed over here. (It closely resembles the proof of the Heine-Borel covering theorem.)

To find the double integral of f(x, y) *over the region* R, *we first regard* y *as constant and integrate* f(x, y) *with respect to* x *between the limits* a *and* b. *This integral*

$$\phi(y) = \int_a^b f(x, y)\,dx$$

is a function of the parameter y, *and we have then to integrate it between the limits* α *and* β. *In symbols,*

$$\iint_R f(x, y)\,dS = \int_\alpha^\beta \phi(y)\,dy, \quad \phi(y) = \int_a^b f(x, y)\,dx,$$

or, more briefly,

$$\iint_R f(x, y)\,dS = \int_\alpha^\beta dy \int_a^b f(x, y)\,dx.$$

In order to prove this statement analytically, we return to the definition of the multiple integral on p. 226. Taking

$$h = \frac{b-a}{n} \quad \text{and} \quad k = \frac{\alpha - \beta}{m},$$

we have

$$\iint_R f(x, y)\,dS = \lim_{\substack{m\to\infty \\ n\to\infty}} \sum_{\nu=1}^{n} \sum_{\mu=1}^{m} f(a + \mu h, \alpha + \nu k)\,hk,$$

where the limit is to be understood to mean that the sum on the right-hand side differs from the value of the integral by less than an arbitrarily small pre-assigned positive quantity ϵ, provided only that the numbers m and n are *both* larger than a bound * N depending only on ϵ. By introducing the expression

$$\Phi_\nu = \sum_{\mu=1}^{m} f(a + \mu h, \alpha + \nu k)h$$

we can write this sum in the form

$$\sum_{\nu=1}^{n} \Phi_\nu k.$$

If we now choose an arbitrary fixed value for ϵ, e.g. $\frac{1}{100}$ or $\frac{1}{10{,}000}$,

* The root idea of the following proof is simply that of resolving the double limit as m and n increase simultaneously into the two successive single limiting processes, first $m \to \infty$ when n is fixed and then $n \to \infty$ (cf. Chap. II, Appendix, section 2 (p. 103)).

and for n choose any definite fixed number greater than N, we know that

$$\left| \iint_R f(x, y)\, dS - k \sum_{\nu=1}^{n} \Phi_\nu \right| < \epsilon$$

no matter how large the number m is, provided only that it is greater than N. If we keep n fixed during the limiting process, the above expression will never exceed ϵ. In accordance with the definition of the ordinary integral, however, in this limiting process the expression Φ_ν tends to the integral

$$\int_a^b f(x, \alpha + \nu k)\, dx = \phi(\alpha + \nu k),$$

and we therefore obtain

$$\left| \iint_R f(x, y)\, dS - k \sum_{\nu=1}^{n} \phi(\alpha + \nu k) \right| \leqq \epsilon.$$

For arbitrarily small values of ϵ this inequality holds for all values of n which are greater than a fixed number N depending only on ϵ. If we now let n tend to ∞ (i.e. let k tend to zero), then by the definition of the single integral and the continuity of $\int_a^b f(x, y)\, dx = \phi(y)$ we obtain

$$\lim_{n \to \infty} k \sum_{\nu=1}^{n} \phi(\alpha + \nu k) = \int_\alpha^\beta \phi(y)\, dy,$$

whence

$$\left| \iint_R f(x, y)\, dS - \int_\alpha^\beta \phi(y)\, dy \right| \leqq \epsilon.$$

Since ϵ can be chosen as small as we please and the left-hand side is a fixed number, this inequality can only hold if the left-hand side vanishes, i.e. if

$$\iint_R f(x, y)\, dS = \int_\alpha^\beta dy \int_a^b f(x, y)\, dx.$$

This gives the required transformation.

This result accordingly *reduces double integration to the performance of two successive single integrations. The double integral can be represented as a repeated single integral.*

Since the parts played by x and y are interchangeable, no further proof is required to show that the equation

$$\iint_B f(x, y)\,dS = \int_a^b dx \int_\alpha^\beta f(x, y)\,dy$$

is also true.

2. Results. Change of Order of Integration. Differentiation under the Integral Sign.

From the last two formulæ of the preceding sub-section we obtain the relation

$$\int_\alpha^\beta dy \int_a^b f(x, y)\,dx = \int_a^b dx \int_\alpha^\beta f(x, y)\,dy,$$

or, in words:

In the repeated integration of a continuous function with constant limits of integration the order of integration can be reversed.

This theorem can also be stated as follows:

If the function $f(x, y)$ *is continuous in the closed rectangle, then in this rectangle we can perform the integration of the integral* $\int_a^b f(x, y)\,dx$ *with respect to the parameter* y *by integrating with respect to* y *under the integral sign, that is, by integrating first with respect to* y *and then with respect to* x.

This theorem corresponds exactly to the rule for the differentiation of an integral with respect to a parameter (cf. section 1, p. 219).

We obtain a further result if we regard one of the above limits, say b, as a variable parameter. We can then differentiate the double integral with respect to this parameter; by the fundamental theorem of the differential and integral calculus we obtain the result

$$\frac{\partial}{\partial b}\iint_B f(x, y)\,dx\,dy = \int_\alpha^\beta f(b, y)\,dy.$$

Similarly, if we regard β as a variable parameter we obtain

$$\frac{\partial}{\partial \beta}\iint_B f(x, y)\,dx\,dy = \int_a^b f(x, \beta)\,dx.$$

Finally, from the two equations we obtain

$$\frac{\partial^2}{\partial b\,\partial \beta}\iint_B f(x, y)\,dx\,dy = f(b, \beta)$$

by repeated differentiation.

In other words:

*Differentiation of the integral with respect to one of the upper limits leads to an ordinary integral over the corresponding side of the rectangle; mixed differentiation with respect to the two upper limits gives the integrand at the corresponding corner of the rectangle.**

The theorem on the change of order in integration has many applications. In particular, it is frequently used in the explicit calculation of simple definite integrals for which no indefinite integral can be found.

As an example—for further examples see the appendix, section 3, pp. 313–6—we consider the integral

$$I = \int_0^\infty \frac{e^{-ax} - e^{-bx}}{x}\, dx,$$

which converges for $a > 0$, $b > 0$. We can express I as a repeated integral in the form

$$I = \int_0^\infty dx \int_a^b e^{-xy}\, dy.$$

In this improper repeated integral we cannot at once apply our theorem on change of order. If, however, we write

$$I = \lim_{T \to \infty} \int_0^T dx \int_a^b e^{-xy}\, dy,$$

by changing the order of integration we obtain

$$I = \lim_{T \to \infty} \int_a^b \frac{1 - e^{-Ty}}{y}\, dy = \log \frac{b}{a} - \lim_{T \to \infty} \int_a^b \frac{e^{-Ty}}{y}\, dy.$$

Since in virtue of the relation

$$\int_c^b \frac{e^{-Ty}}{y}\, dy = \int_{Ta}^{Tb} \frac{e^{-y}}{y}\, dy$$

the second integral tends to zero as T increases, we have

$$I = \int_0^\infty \frac{e^{-ax} - e^{-bx}}{x}\, dx = \log \frac{b}{a}.$$

In a similar way we can prove the following general theorem: if $f(t)$ is sectionally smooth for $t \geqq 0$, and if the integral $\int_1^\infty \frac{f(t)}{t}\, dt$ exists, then

$$I = \int_0^\infty \frac{f(ax) - f(bx)}{x}\, dx = f(0) \log \frac{b}{a} \quad (a > 0,\ b > 0).$$

* The reader's attention may be drawn to the connexion between this formula and the theorem on change of order of differentiation (cf. p. 55); he should investigate for himself to what extent the two facts are equivalent.

For here we can again express the single integral as a repeated integral

$$I = \int_0^\infty dx \int_b^a f'(xy)\,dy$$

and change the order of integration.

3. Extension of the Result to More General Regions.

By a simple extension of the results already obtained we can prove that our result holds for regions more general than rectangles. We begin by considering a *convex region R*, that is, a region whose boundary curve is not cut by any straight line in more than two points unless the whole straight line between

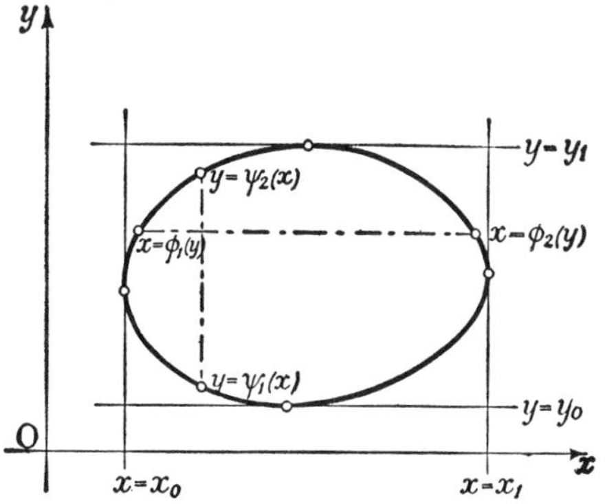

Fig. 4.—General convex region of integration

these two points is a part of the boundary (fig. 4). We suppose that the region lies between the "lines of support" (cf. ex. 1 (*b*), p. 100) $x = x_0$, $x = x_1$ and $y = y_0$, $y = y_1$ respectively. Since for points of R the x-co-ordinate lies in the interval $x_0 \leqq x \leqq x_1$ and the y-co-ordinate in the interval $y_0 \leqq y \leqq y_1$, we consider the integrals

$$\int_{\phi_1(y)}^{\phi_2(y)} f(x, y)\,dx$$

and

$$\int_{\psi_1(x)}^{\psi_2(x)} f(x, y)\,dy,$$

which are taken along the segments in which the lines $y =$ const. and $x =$ const. respectively intersect the region. Here $\phi_2(y)$ and

$\phi_1(y)$ denote the abscissæ of the points in which the boundary of the region is intersected by the line $y = \text{const.}$, and $\psi_2(x)$ and $\psi_1(x)$ the ordinates of the points in which the boundary is intersected by the lines $x = \text{const.}$ The integral $\int_{\phi_1(y)}^{\phi_2(y)} f(x, y)\,dx$ is therefore a function of the parameter y, where the parameter appears both under the integral sign and in the upper and lower limits, and a similar statement holds for the integral $\int_{\psi_1(x)}^{\psi_2(x)} f(x, y)\,dy$ as a function of x. The resolution into repeated integrals is then given by the equations

$$\iint_R f(x, y)\,dS = \int_{y_0}^{y_1} dy \int_{\phi_1(y)}^{\phi_2(y)} f(x, y)\,dx$$

$$= \int_{x_0}^{x_1} dx \int_{\psi_1(x)}^{\psi_2(x)} f(x, y)\,dy.$$

To prove this we first choose a sequence of points on the arc $y = \psi_2(x)$, the distance between successive points being less than

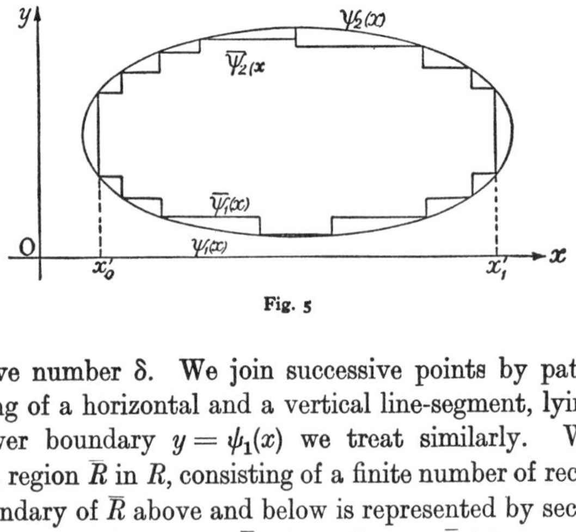

Fig. 5

a positive number δ. We join successive points by paths each consisting of a horizontal and a vertical line-segment, lying in R. The lower boundary $y = \psi_1(x)$ we treat similarly. We thus obtain a region $\bar{R}$ in R, consisting of a finite number of rectangles. The boundary of $\bar{R}$ above and below is represented by sectionally continuous functions $y = \bar{\psi}_2(x)$ and $y = \bar{\psi}_1(x)$ respectively (cf. fig. 5). By the known theorem for rectangles we have

$$\iint_{\bar{R}} f(x, y)\,dS = \int_{x_0'}^{x_1'} dx \int_{\bar{\psi}_1(x)}^{\bar{\psi}_2(x)} f(x, y)\,dy.$$

Since $\psi_1(x)$ and $\psi_2(x)$ are uniformly continuous, as $\delta \to 0$ the functions $\bar{\psi}_1(x)$ and $\bar{\psi}_2(x)$ tend uniformly to $\psi_1(x)$ and $\psi_2(x)$ respectively, and so

$$\lim_{\delta \to 0} \int_{\bar{\psi}_1(x)}^{\bar{\psi}_2(x)} f(x, y)\, dy = \int_{\psi_1(x)}^{\psi_2(x)} f(x, y)\, dy$$

uniformly in x. It follows that

$$\lim_{\delta \to 0} \int_{x_0'}^{x_1'} dx \int_{\bar{\psi}_1(x)}^{\bar{\psi}_2(x)} f(x, y)\, dx = \int_{x_0}^{x_1} dx \int_{\psi_1(x)}^{\psi_2(x)} f(x, y)\, dx.$$

On the other hand, as $\delta \to 0$ the region $\bar{R}$ tends to R. Hence

$$\lim_{\delta \to 0} \iint_{\bar{R}} f(x, y)\, dS = \iint_{R} f(x, y)\, dS.$$

Combining the three equations, we have

$$\iint_{R} f(x, y)\, dS = \int_{x_0}^{x_1} dx \int_{\psi_1(x)}^{\psi_2(x)} f(x, y)\, dy.$$

The other statement can be established in a similar way.

A similar argument is available if we abandon the hypothesis

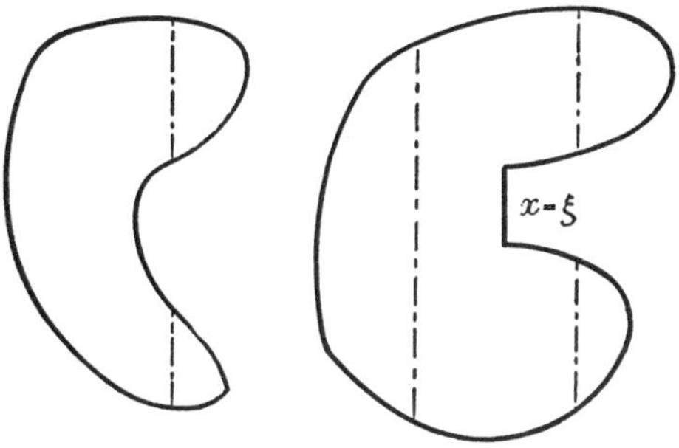

Fig. 6.—Non-convex regions of integration

of convexity and consider regions of the form indicated in fig. 6. We assume merely that the boundary curve of the region is intersected by every parallel to the x-axis and by every parallel to the y-axis in a bounded number of points or intervals. By $\int f(x, y)\, dy$ we then mean the sum of the integrals of the function $f(x, y)$ for a fixed x, taken over all the intervals which the line $x =$ const. has in common with the closed region. For non-convex regions the number of these intervals may exceed unity. It may change suddenly at a point $x = \xi$ (as in fig. 6, right) in such a way that the expression $\int f(x, y)\,dy$ has a jump discontinuity at this point. Without essential changes in the proof, however, the resolution of the double integral

$$\iint_R f(x, y)\,dS = \int dx \int f(x, y)\,dy$$

remains valid, the integration with respect to x being taken along the whole interval $x_0 \leqq x \leqq x_1$ over which the region R lies. Naturally the corresponding resolution

$$\iint_R f(x, y)\,dS = \int dy \int f(x, y)\,dx$$

also holds.

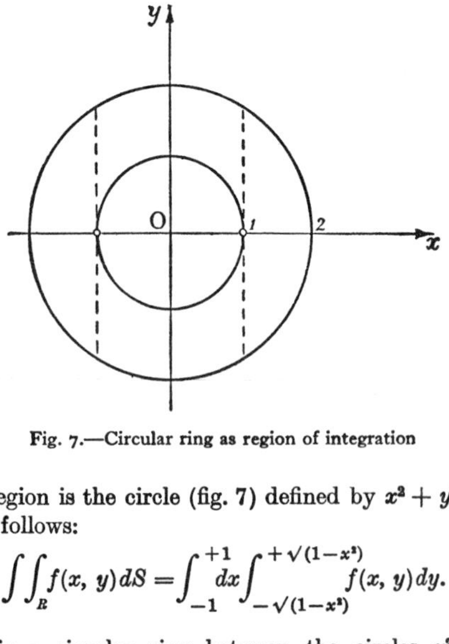

Fig. 7.—Circular ring as region of integration

If e.g. the region is the circle (fig. 7) defined by $x^2 + y^2 \leqq 1$, then the resolution is as follows:

$$\iint_R f(x, y)\,dS = \int_{-1}^{+1} dx \int_{-\sqrt{(1-x^2)}}^{+\sqrt{(1-x^2)}} f(x, y)\,dy.$$

If the region is a circular ring between the circles $x^2 + y^2 = 1$ and $x^2 + y^2 = 4$ (fig. 7), then

$$\iint_R f(x, y)\,dx\,dy = \int_{-2}^{-1} dx \int_{-\sqrt{(4-x^2)}}^{+\sqrt{(4-x^2)}} f(x, y)\,dy + \int_{1}^{2} dx \int_{-\sqrt{(4-x^2)}}^{+\sqrt{(4-x^2)}} f(x, y)\,dy$$

$$+ \int_{-1}^{+1} dx \int_{-\sqrt{(4-x^2)}}^{-\sqrt{(1-x^2)}} f(x, y)\,dy + \int_{-1}^{+1} dx \int_{\sqrt{(1-x^2)}}^{+\sqrt{(4-x^2)}} f(x, y)\,dy.$$

As a final example we take as the region R a triangle (fig. 8) bounded by the lines $x = y$, $y = 0$, and $x = a(a > 0)$. If we integrate first with respect to x,

$$\iint_R f(x, y)\,dS = \int_0^a dy \int_y^a f(x, y)\,dx,$$

and if we integrate first with respect to y,

$$\iint_R f(x, y)\,dS = \int_0^a dx \int_0^x f(x, y)\,dy.$$

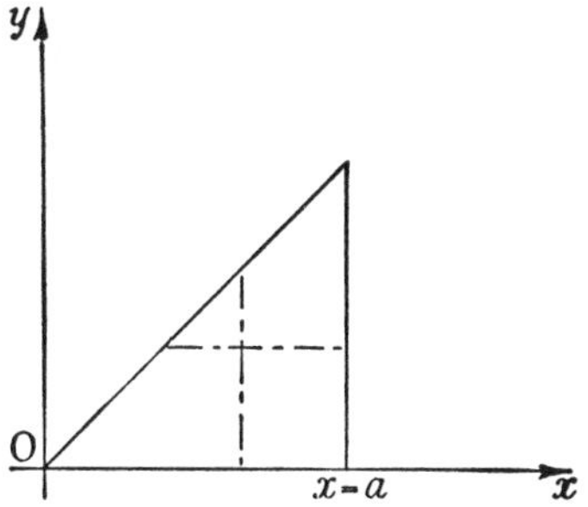

Fig. 8.—Triangle as region of integration

Comparing the two results, we have

$$\int_0^a dx \int_0^x f(x, y)\,dy = \int_0^a dy \int_y^a f(x, y)\,dx.$$

In particular, if $f(x, y)$ depends on y only, our formula gives

$$\int_0^a dx \int_0^x f(y)\,dy = \int_0^a f(y)\,(a - y)\,dy.$$

From this we see that if the indefinite integral $\int_0^x f(y)\,dy$ of a function $f(x)$ is integrated again, the result can be expressed by a single integral (cf. p. 221).

4. Extension of the Results to Regions in Several Dimensions.

The corresponding theorems in more than two dimensions are so closely analogous to those already given that it will be sufficient to state them without proof. If we first consider the rectangular region $x_0 \leqq x \leqq x_1$, $y_0 \leqq y \leqq y_1$, $z_0 \leqq z \leqq z_1$, and a function $f(x, y, z)$ which is continuous in this region, we can reduce the triple integral

$$V = \iiint_R f(x, y, z)\,dV$$

in several ways to single integrals or double integrals. Thus

$$\iiint_R f(x, y, z)\,dV = \int_{z_0}^{z_1} dz \iint_B f(x, y, z)\,dx\,dy.$$

Here

$$\iint_B f(x, y, z)\,dx\,dy$$

is the double integral of the function taken over the rectangle $x_0 \leqq x \leqq x_1$, $y_0 \leqq y \leqq y_1$, z being kept constant as a parameter during this integration, so that the double integral is a function of the parameter z. Either of the remaining co-ordinates x and y can be singled out in the same way.

Moreover, the triple integral V can also be represented as a repeated integral in the form of a succession of three single integrations. In this representation we first consider the expression

$$\int_{z_0}^{z_1} f(x, y, z)\,dz,$$

x and y being fixed, and then consider

$$\int_{y_0}^{y_1} dy \int_{z_0}^{z_1} f(x, y, z)\,dz,$$

x being fixed. We finally obtain

$$V = \int_{x_0}^{x_1} dx \int_{y_0}^{y_1} dy \int_{z_0}^{z_1} f(x, y, z)\,dz.$$

In this repeated integral we could equally well have integrated first with respect to x and then with respect to y and finally with respect to z, or we could have made any other change in the order of integration; this follows at once from the fact that the repeated integral is always equal to the triple integral. We therefore have the following theorem:

A repeated integral of a continuous function throughout a closed rectangular region is independent of the order of integration.

The way in which the resolution is to be performed for non-rectangular regions in three dimensions scarcely requires special mention. We content ourselves with writing down the resolution for a spherical region $x^2 + y^2 + z^2 \leqq 1$:

$$\iiint_R f(x, y, z)\,dx\,dy\,dz = \int_{-1}^{+1} dx \int_{-\sqrt{(1-x^2)}}^{+\sqrt{(1-x^2)}} dy \int_{-\sqrt{(1-x^2-y^2)}}^{+\sqrt{(1-x^2-y^2)}} f(x, y, z)\,dz.$$

Examples

Evaluate the integrals in Ex. 1–8:

1. $\iint x^2y^2\,dx\,dy$ over the circle $x^2 + y^2 \leqq 1$.

2. $\iint \dfrac{x^3 + y^3 - 3xy(x^2 + y^2)}{(x^2 + y^2)^{\frac{3}{2}}}\,dx\,dy$ over the circle $x^2 + y^2 \leqq 1$.

3. $\iiint (x^2 + y^2 + z^2)\,xyz\,dx\,dy\,dz$ throughout the sphere $x^2 + y^2 + z^2 \leqq r^2$.

4. $\iiint z\,dx\,dy\,dz$ throughout the region defined by the inequalities $x^2 + y^2 \leqq z^2$, $x^2 + y^2 + z^2 \leqq 1$.

5. $\iiint (x + y + z)\,x^2y^2z^2\,dx\,dy\,dz$ throughout the region $x + y + z \leqq 1$, $x \geqq 0$, $y \geqq 0$, $z \geqq 0$.

6. $\iiint \dfrac{dx\,dy\,dz}{x^2 + y^2 + (z - 2)^2}$ throughout the sphere $x^2 + y^2 + z^2 \leqq 1$.

7. $\iiint \dfrac{dx\,dy\,dz}{x^2 + y^2 + (z - \frac{1}{2})^2}$ throughout the sphere $x^2 + y^2 + z^2 \leqq 1$.

8. $\iint \dfrac{dx\,dy}{\sqrt{x^2 + y^2}}$ over the square $|x| \leqq 1$, $|y| \leqq 1$.

9. Prove that if $f(x)$ is a continuous function

$$\lim_{h \to +0} \int_{-1}^{+1} \frac{h}{h^2 + x^2} f(x)\,dx = \pi f(0).$$

4. Transformation of Multiple Integrals

In the case of single integrals the introduction of a new variable of integration is one of the chief methods for transforming and simplifying given integrals. The introduction of new variables is likewise of great importance in the case of several variables. In the case of multiple integrals, in spite of their reduction to single integrals, explicit evaluation is generally more difficult than in the case of one independent variable, and the integration in terms of elementary functions is less often possible. Yet in many cases we can evaluate such integrals by introducing new variables in place of the original variables under the integral sign. Quite apart from the question of the explicit evaluation of double integrals, the change of variables is of fundamental importance, since the transformation theory gives us a more complete mastery of the concept of integral.

The most important special case is the transformation to polar co-ordinates, which has already been carried out in Vol. I, Chap. X (p. 494). Here we shall at once proceed to general transformations. We first consider the case of a double integral

$$\iint_R f(x, y)\,dS = \iint f(x, y)\,dx\,dy,$$

taken over a region R of the xy-plane. Let the equations

$$x = \phi(u, v)$$
$$y = \psi(u, v)$$

give a one-to-one mapping of the region R on the closed region R' of the uv-plane. We assume that in the region R' the functions ϕ and ψ have continuous partial derivatives of the first order and that their Jacobian

$$D = \begin{vmatrix} \phi_u & \phi_v \\ \psi_u & \psi_v \end{vmatrix} = \phi_u\psi_v - \psi_u\phi_v$$

never vanishes in the closed region R'; to be specific, we assume that it is everywhere *positive*. We then know that with these assumptions the system of functions $x = \phi(u, v)$, $y = \psi(u, v)$ possesses a unique inverse $u = g(x, y)$, $v = h(x, y)$ (p. 152). Moreover, the two families of curves $u =$ const. and $v =$ const. form a net over the region R.

Heuristic considerations readily suggest how the integral $\iint_R f(x, y)\,dx\,dy$ can be expressed as an integral with respect to u and v. We naturally think of calculating the double integral $\iint_R f(x, y)\,dS$ by abandoning the rectangular subdivision of the region R and instead using a subdivision into sub-regions R_i by means of curves of the net $u =$ const. or $v =$ const. We therefore consider the values $u = \nu h$ and $v = \mu k$, where $h = \Delta u$ and $k = \Delta v$ are given numbers and ν and μ take all integer values such that the lines $u = \nu h$ and $v = \mu k$ intersect R' (so that their images are curves in R). These curves define a number of meshes, and for the sub-regions R_i we choose those meshes which lie in the interior of R (figs. 9, 10). We now have to find the area of such a mesh.

If the mesh, instead of being bounded by curves, were an

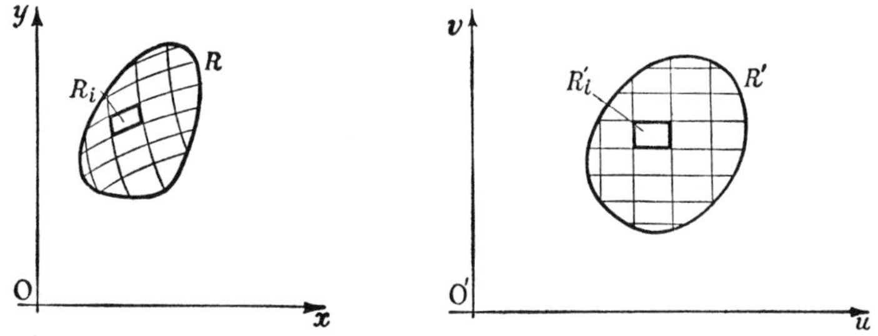

Figs. 9, 10.—Decomposition of regions in a transformation

ordinary parallelogram, half the parallelogram being formed by the triangle with the vertices corresponding to the values (u_ν, v_μ), $(u_\nu + h, v_\mu)$, and $(u_\nu, v_\mu + k)$, then by a formula of elementary analytical geometry (cf. Chap. I, p. 14) the area of the parallelogram would be given by the determinant

$$\begin{vmatrix} \phi(u_\nu + h, v_\mu) - \phi(u_\nu, v_\mu) & \phi(u_\nu, v_\mu + k) - \phi(u_\nu, v_\mu) \\ \psi(u_\nu + h, v_\mu) - \psi(u_\nu, v_\mu) & \psi(u_\nu, v_\mu + k) - \psi(u_\nu, v_\mu) \end{vmatrix},$$

which is approximately equal to

$$\begin{vmatrix} \phi_u(u_\nu, v_\mu) & \phi_v(u_\nu, v_\mu) \\ \psi_u(u_\nu, v_\mu) & \psi_v(u_\nu, v_\mu) \end{vmatrix} hk = hkD.$$

On multiplying this expression by the value of the function f in the corresponding mesh, summing over all the regions R_i lying entirely within R, and then performing the passage to the limit $h \to 0$ and $k \to 0$, we obtain the expression

$$\iint_{R'} f(\phi(u, v), \psi(u, v))\, D\, du\, dv$$

for the integral transformed to the new variables.

This discussion is incomplete, however, since we have not shown that it is permissible to replace the curvilinear meshes by parallelograms or to replace the area of such a parallelogram by the expression $(\phi_u\psi_v - \psi_u\phi_v)hk$; that is, we have not shown that the error thus caused vanishes in the limit as $h \to 0$ and $k \to 0$. Instead of completing this method of proof by making these estimates, we prefer to develop the proof of the transformation formula in a somewhat different way, which can subsequently be extended directly to regions of higher dimensions.

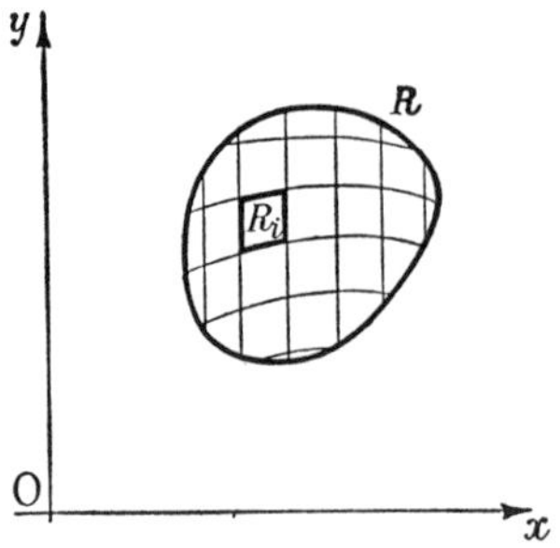

Fig. 11

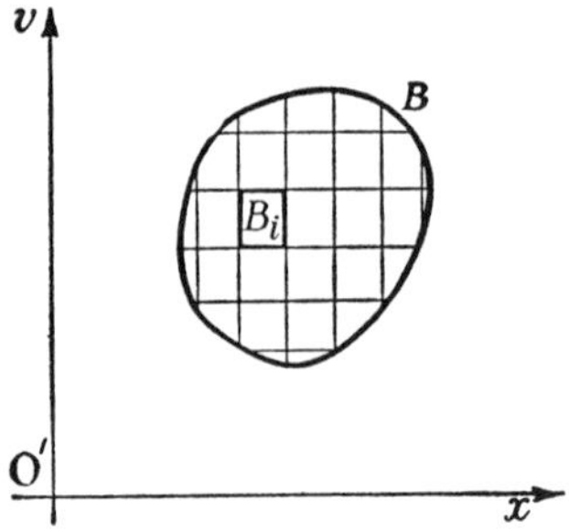

Fig. 12

For this purpose we use the results of Chap. III, section 3 (p. 150) and perform the transformation from the variables x, y to the new variables u, v in two steps instead of in one. We replace the variables x, y by new variables x, v by means of the equations

$$x = x$$
$$y = \Phi(v, x).$$

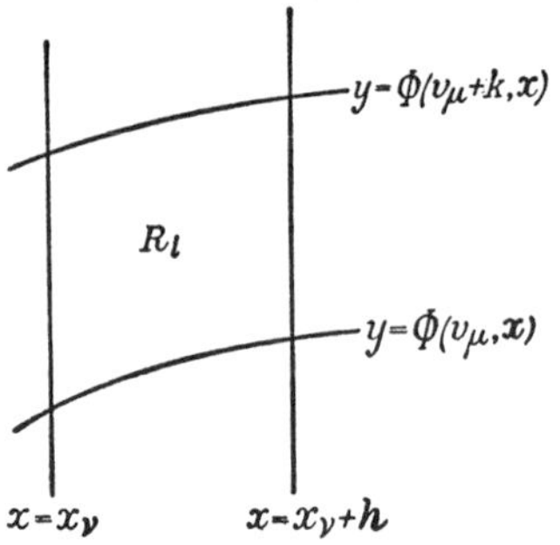

Fig. 13

Here we assume that the expression Φ_v vanishes nowhere in the region R, i.e. that Φ_v is everywhere greater than zero, say, and that the whole region R can be mapped in a one-to-one way on the region B of the xv-plane. We then map this region B in a one-to-one way on the region R' of the uv-plane by means of a second transformation

$$x = \Psi(u, v)$$
$$v = v,$$

where we further assume that the expression Ψ_u is positive throughout the region B. We now effect the transformation of the integral $\iint_R f(x, y)\, dx\, dy$ in two steps. We start with a subdivision of the region B into rectangular sub-regions of sides $\Delta x = h$ and $\Delta v = k$ bounded by the lines $x = \text{const.} = x_\nu$ and $v = \text{const.} = v_\mu$ in the xv-plane. This subdivision of B corresponds to a subdivision of the region R into sub-regions R_i, each sub-region being bounded by two parallel lines $x = x_\nu$ and $x = x_\nu + h$ and by two arcs of curves $y = \Phi(v_\mu, x)$ and $y = \Phi(v_\mu + k, x)$ (figs. 11, 12). By the elementary interpretation of the single integral, the area of the sub-region (fig. 13) is

$$\Delta R_i = \int_{x_\nu}^{x_\nu + h} [\Phi(v_\mu + k, x) - \Phi(v_\mu, x)]\, dx,$$

which by the mean value theorem of the integral calculus can be written in the form

$$\Delta R_i = h[\Phi(v_\mu + k, \bar{x}_\nu) - \Phi(v_\mu, \bar{x}_\nu)],$$

where $\bar{x}_\nu$ is a number between x_ν and $x_\nu + h$. By the mean value theorem of the differential calculus this finally becomes

$$\Delta R_i = hk\Phi_v(\bar{v}_\mu, \bar{x}_\nu),$$

in which $\bar{v}_\mu$ denotes a value between v_μ and $v_\mu + k$, so that $(\bar{v}_\mu, \bar{x}_\mu)$ are the co-ordinates of a point of the sub-region in B under consideration. The integral over R is therefore the limit of the sum

$$\Sigma f_i \Delta R_i = \Sigma hk f(\bar{x}_\nu, \Phi(\bar{v}_\mu, \bar{x}_\nu))\Phi_v(\bar{v}_\mu, \bar{x}_\nu)$$

as $h \to 0$, $k \to 0$. We see at once that the expression on the right tends to the integral

$$\iint_B f(x, y)\Phi_v\, dx\, dv \qquad (y = \Phi(v, x))$$

taken over the region B. Therefore

$$\iint_R f(x, y)\, dx\, dy = \iint_B f(x, y)\Phi_v\, dx\, dv.$$

To the integral on the right we now apply exactly the same argument as that just employed for $\iint_R f(x, y)\, dx\, dy$, transforming the region B into the region R' by means of the equations $x = \Psi(u, v)$, $v = v$.

The integral over B then becomes an integral over R' with an integrand of the form $f\Phi_v\Psi_u$, and we finally obtain

$$\iint_{R'} f(x, y)\Phi_v\Psi_u\,du\,dv.$$

Here the quantities x and y are to be expressed in terms of the independent variables u and v by means of the two transformations above. We have therefore proved the transformation formula

$$\iint_B f(x, y)\,dx\,dy = \iint_{R'} f(x, y)\Phi_v\Psi_u\,du\,dv.$$

By introducing the direct transformation $x = \phi(u, v)$, $y = \psi(u, v)$ the formula can at once be put in the form stated previously. For $\dfrac{\partial(x, y)}{\partial(x, v)} = \Phi_v$ and $\dfrac{\partial(x, v)}{\partial(u, v)} = \Psi_u$, and so by Chap. III, section 3 (p. 147) we have

$$D = \frac{\partial(x, y)}{\partial(u, v)} = \Phi_v\Psi_u.$$

We have therefore established the transformation formula for all cases in which the transformation $x = \phi(u, v)$, $y = \psi(u, v)$ can be resolved into a succession of two primitive transformations of the forms * $x = x$, $y = \Phi(v, x)$ and $v = v$, $x = \Psi(u, v)$.

In Chap. III, section 3 (p. 151), however, we saw that we can subdivide a closed region R into a finite number of regions in each of which such a resolution is possible, except perhaps that it may also be necessary to replace u by v and v by $-u$; this substitution is merely a rotation of the axes, and we see that it does not affect the value of the integral; in fact, even the simple heuristic argument at the beginning of this sub-section is perfectly rigorous for this case. We thus arrive at the following general result:†

* We have assumed above that the two derivatives Φ_v and Ψ_u are positive, but we easily see that this is not a serious restriction. For the inequality $\dfrac{\partial(x, y)}{\partial(u, v)} > 0$ shows that these two derivatives must have the same sign. If they were both negative, we should merely have to replace x by $-x$ and y by $-y$, which leaves the integral unchanged. The two primitive transformations then have positive Jacobians.

† The above proof in the first instance holds only for every closed region R_1 lying entirely within R. Since, however, R_1 can be chosen so as to occupy all of R except a portion of arbitrarily small area, the transformation formula continues to hold for R itself.

If the transformation $x = \phi(u, v)$, $y = \psi(u, v)$ *represents a continuous one-to-one mapping of the closed region* R *of the* xy-*plane on a region* R′ *of the* uv-*plane, and if the functions* ϕ *and* ψ *have continuous first derivatives and their Jacobian* $\frac{\partial(x, y)}{\partial(u, v)} = \phi_u\psi_v - \psi_u\phi_v$ *is everywhere positive, then*

$$\iint_R f(x, y)\,dx\,dy = \iint_{R'} f(\phi(u, v), \psi(u, v))\frac{\partial(x, y)}{\partial(u, v)}\,du\,dv.$$

For completeness we add that the transformation formula remains valid if the determinant $\frac{\partial(x, y)}{\partial(u, v)}$ vanishes without, however, changing its sign, at a finite number of isolated points of the region. For then we have only to cut these points out of R by enclosing them in small circles of radius ρ. The proof is valid for the residual region. If we then let ρ tend to zero,* the transformation formula continues to hold for the region R in virtue of the continuity of all the functions involved.

We make use of this fact whenever we introduce polar coordinates with the origin in the interior of the region; for the Jacobian, being equal to r, vanishes at the origin.

In Chap. V, section 4 (p. 377) we shall return to transformations with negative Jacobians, and we shall see that the argument remains essentially the same. Nevertheless, we would point out here that provided the Jacobian D does not vanish the hypothesis $D > 0$ in a sense involves no loss of generality, for we can always change the sign of D by interchanging u and v. A different method of proving the transformation formula will be given in Chap. V, § 3, p. 373.

Regions of more than Two Dimensions.

We can of course proceed in the same way with regions of more than two dimensions, e.g. with regions in three-dimensional space, and obtain the following general result:

If a closed region R *of* xyz . . . *-space is mapped on a region* R′ *of* uvw . . . *-space by a one-to-one transformation whose Jacobian*

$$\frac{\partial(x, y, z, \ldots)}{\partial(u, v, w, \ldots)}$$

* For another application of this method, see section 5, p. 262.

is everywhere positive, then the transformation formula

$$\iint \cdots \int_R f(x, y, z, \ldots)\, dx\, dy\, dz \ldots$$

$$= \iint \cdots \int_{R'} f(x, y, z, \ldots) \frac{\partial(x, y, z, \ldots)}{\partial(u, v, w, \ldots)}\, du\, dv\, dw \ldots$$

holds. In n dimensions the Jacobian is an n-rowed determinant of similar construction to the Jacobian in two dimensions.

As a special application, we can obtain the *transformation formula for polar co-ordinates* in another way. In the case of polar co-ordinates in the plane we must write r and θ instead of u and v, and we at once obtain the expression $\frac{\partial(x, y)}{\partial(r, \theta)} = r$ (cf. p. 144). In the case of *polar co-ordinates in space*, defined by the equations

$$x = r \cos\varphi \sin\theta$$
$$y = r \sin\varphi \sin\theta$$
$$z = r \cos\theta,$$

in which φ ranges from 0 to 2π, θ from 0 to π, and r from 0 to $+\infty$, we must identify u, v, w with r, θ, φ; as the expression for the Jacobian we obtain

$$\frac{\partial(x, y, z)}{\partial(r, \theta, \varphi)} = \begin{vmatrix} \cos\varphi \sin\theta & r\cos\varphi\cos\theta & -r\sin\varphi\sin\theta \\ \sin\varphi\sin\theta & r\sin\varphi\cos\theta & r\cos\varphi\sin\theta \\ \cos\theta & -r\sin\theta & 0 \end{vmatrix} = r^2 \sin\theta.$$

(This value $r^2 \sin\theta$ is obtained by expanding in terms of the elements of the third column.) The transformation to polar co-ordinates in space is therefore given by the formula

$$\iiint_R f(x, y, z)\, dx\, dy\, dz = \iiint_{R'} f(x, y, z)\, r^2 \sin\theta\, dr\, d\theta\, d\varphi.$$

As in the corresponding case in the plane, we can also arrive at the transformation formula without using the general theory. We have only to start with a subdivision of space given by the spheres $r = \text{const.}$, the cones $\theta = \text{const.}$, and the planes $\varphi = \text{const.}$ The details of this elementary method are similar to those of Vol. I, Chap. X, section 2 (p. 494) and can be left to the reader.

In the case of polar co-ordinates in space our assumptions are not satisfied when $r = 0$ or $\theta = 0$, since the Jacobian then vanishes. The validity of the transformation formula, however, is not thereby destroyed. We can easily convince ourselves of this, as we did in the case of the plane.

Examples

1*. Evaluate the integral $\iint e^{\frac{y-x}{y+x}}\,dx\,dy$ taken over the triangle with vertices (0, 0), (0, 1), (1, 0).

2. Evaluate the integral

$$\iint \frac{dx\,dy}{(1+x^2+y^2)^2}$$

taken

(a) over one loop of the lemniscate $(x^2+y^2)^2-(x^2-y^2)=0$,

(b) over the triangle with vertices (0, 0), (2, 0), (1, $\sqrt{3}$).

3. Evaluate the integral

$$\iiint xyz\,dx\,dy\,dz$$

taken throughout the ellipsoid $\dfrac{x^2}{a^2}+\dfrac{y^2}{b^2}+\dfrac{z^2}{c^2}\leqq 1$.

4. Prove that

$$\iint_R e^{-(x^2+y^2)}\,dx\,dy = ae^{-a^2}\int_0^\infty \frac{e^{-u^2}}{a^2+u^2}\,du$$

(where R denotes the half-plane $x \geqq a > 0$), by applying the transformation

$$x^2+y^2=u^2+a^2, \quad y=vx.$$

5. Prove that

$$\left|\iint (u_x^2+u_y^2)\,dx\,dy\right|$$

is invariant on inversion.

6. Evaluate the integral of Ex. 4, p. 247, by using three-dimensional polar co-ordinates.

7. Evaluate the integral

$$I=\iiint \cos(x\xi+y\eta+z\zeta)\,d\xi\,d\eta\,d\zeta$$

taken throughout the sphere $\xi^2+\eta^2+\zeta^2\leqq 1$.

8. Prove that

$$\iint \cos(x\xi+y\eta)\,d\xi\,d\eta = 2J_1(r)/r, \qquad (r=\sqrt{(x^2+y^2)})$$

where the integral is to be extended over the circle $\xi^2+\eta^2\leqq 1$ and J_1 denotes the Bessel function defined in Ex. 5, p. 223.

5. Improper Integrals

In the case of functions of one variable we found it necessary to extend the concept of integral to functions other than those which are continuous in a closed region of integration. In fact, we did consider the integrals of functions with jump discontinuities and of functions with infinite values, and also integrals over infinite intervals of integration. The corresponding extensions of the concept of integral for functions of several variables must now be discussed.

1. Functions with Jump Discontinuities.

For functions which have jump discontinuities in the region of integration R the extension of the concept of integral is immediate. We assume that the region of integration can be subdivided by a finite number of smooth arcs of curves * into a finite number of sub-regions $R_1, R_2, \ldots, R_n$ in such a way that the integrand f is continuous in the interior of each sub-region, and as the boundary of such a sub-region is approached from the interior the values of the function tend to definite continuous boundary values; but the limiting values obtained as we approach a point on a curve separating two sub-regions may differ according as we approach the point from one sub-region or the other. The integral of the function f over the region R we shall then define as the sum of the integrals of the function f over the sub-regions R_ν. The integrals of f over the regions R_ν are at once given by our original definition if for each sub-region we suppose that the function is extended by including the boundary values, so that it becomes a continuous function in the closed region R_ν.

As an example we consider a function $f(x, y)$ which is defined in the square $0 \leqq x \leqq 1$, $0 \leqq y \leqq 1$ by

$$f(x, y) = 1 \quad \text{for} \quad y < x,$$
$$f(x, y) = 2 \quad \text{for} \quad y \geqq x.$$

For this function the line $y = x$ is a line of discontinuity, and by the process described we find that the improper integral $\iint f(x, y)\,dx\,dy$ taken over the square has the value $\frac{3}{2}$.

* By a smooth arc of a curve we mean an arc with a continuously turning tangent.

2. Functions with Isolated Infinite Discontinuities.

If the integrand becomes infinite at a single point P of the region of integration, we define the integral of the function f over the region R by a process analogous to that for one independent variable. We mark off a neighbourhood U_ν about the point of discontinuity P, so that the closed residue $R_\nu = R - U_\nu$ no longer contains the point P. There are many possible sequences of neighbourhoods U_ν whose diameters tend to zero as ν increases, e.g. the sequence of circles or spheres about the point P with radius $c = 1/\nu$. If the sequence of the integrals over the residual region R_ν tends to a limit I, i.e. if

$$\lim_{\nu\to\infty} \iint_{R_\nu} f(x, y)\, dS = I,$$

and if this limit is independent of the particular choice of the sequence R_ν, then its value is called the integral or, more accurately, the *improper integral* of the function f over the region R, and we write

$$I = \iint_R f(x, y)\, dS.$$

Such an integral taken over the region R is sometimes called a *convergent integral* (or is said to *converge*). If no limiting value I exists, the integral is said to be *divergent* (or to *diverge*). The definition of course remains valid if P is an isolated point of indeterminacy, such as the origin for the function $\sin\left(\frac{1}{x^2 + y^2}\right)$.

If in the neighbourhood of P the absolute value of the function remains below a fixed bound, the integral is always convergent.

The *general conditions* for the *convergence* of an integral can therefore be stated as follows. To every positive ϵ there corresponds a bound $\delta = \delta(\epsilon)$ for which the following condition is satisfied: if U and U' are any two (open) sub-regions of R which contain the point of discontinuity P and whose diameter is smaller than δ, then the integrals of the function f over the closed residual regions $G - U$ and $G - U'$ differ in absolute value by less than ϵ. We shall illustrate these ideas by means of a few examples.

The function

$$f(x, y) = \log\sqrt{x^2 + y^2}$$

becomes infinite at the origin of the xy-plane. Therefore in order to calculate the integral over a region R containing the origin, e.g. over the circle $x^2 + y^2 \leq 1$, we must cut out the origin by surrounding it with a region U_δ whose diameter is less than δ, and we must then investigate the convergence of the integral taken over the residual region $R_\delta = R - U_\delta$ as $\delta \to 0$. The neighbourhood U_δ certainly lies within a circle of radius δ about the origin. In accordance with section 4 (p. 254) we transform the integral to polar co-ordinates and obtain

$$\iint_{R_\delta} \log\sqrt{x^2+y^2}\,dx\,dy = \iint_{R_\delta'} r\log r\,dr\,d\theta,$$

where the integral on the right is taken over the region R_δ' of the $r\theta$-plane corresponding to the region R_δ. In our case this is a region which contains the rectangle $\delta \leq r \leq 1$, $0 \leq \theta \leq 2\pi$ but does not include the straight line $r = 0$. The function $r\log r$ is continuous for $r = 0$, however, if we assign the value 0 to it at that point; for $\lim\limits_{r\to 0} r\log r = 0$. We can therefore let δ tend to 0 and regard the transformed integral

$$\iint_{R'} r\log r\,dr\,d\theta = \lim_{\delta\to 0}\iint_{R_\delta'} r\log r\,dr\,d\theta$$

as an ordinary integral in the sense of section 2 (p. 224). The convergence of the integral is therefore established.

At the same time this example shows that, as in the case of one independent variable, a properly chosen transformation of co-ordinates sometimes changes an improper integral into a proper integral. This fact clearly shows how inadmissible a restriction we should lay upon ourselves if we refused to consider improper integrals.

As a further example we consider the integral

$$\iint_R \frac{dx\,dy}{(\sqrt{x^2+y^2})^\alpha},$$

taken over the same region. If we first think of the integral as taken over the region R_δ obtained from R by cutting out a circle with radius δ, and then transforming to polar coordinates, we obtain

$$\iint_{R_\delta} \frac{1}{r^{\alpha-1}}\,dr\,d\theta,$$

or, as a repeated integral,

$$\int_0^{2\pi} d\theta \int_\delta^R \frac{dr}{r^{\alpha-1}} = 2\pi\int_\delta^R \frac{dr}{r^{\alpha-1}}.$$

From Vol. I (p. 246) we know that the integral $\int_0^R \frac{dr}{r^{a-1}}$ is convergent if, and only if, $a < 2$. We therefore conclude that the double integral $\iint_R \frac{dx\,dy}{(\sqrt{x^2+y^2})^a}$ is likewise convergent if, and only if, $a < 2$. As in the preceding example, the convergence is independent of the particular choice of the sub-regions U_δ.

This remark can readily be used to obtain a *sufficient* (by no means a *necessary*) criterion for the convergence of improper double integrals, which is applicable in many special cases.

If in the closed region R *the function* f(x, y) *is continuous everywhere except at one point* P, *which we take as the origin* x = 0, y = 0, *and* f *becomes infinite at* P, *and if there is a fixed bound* M *and a positive number* a < 2 *such that*

$$|f(x, y)| \leq \frac{M}{(\sqrt{x^2+y^2})^a}$$

everywhere in R, *then the integral*

$$\iint_R f(x, y)\,dx\,dy$$

converges.

The proof is obtained from the above by considering the relation

$$\left|\iint_B f(x, y)\,dx\,dy\right| \leq \iint_B \left|f(x, y)\right| dx\,dy \leq M \iint_B \frac{dx\,dy}{(\sqrt{x^2+y^2})^a},$$

where B is a region not containing P and lying within a small circular neighbourhood of P.

We can deal with the triple integral

$$\iiint_R \frac{dx\,dy\,dz}{(\sqrt{x^2+y^2+z^2})^a}$$

in a similar way. If R contains the origin, we introduce polar co-ordinates and obtain

$$\iiint_{R'} \frac{1}{r^{a-2}} \sin\theta\,dr\,d\theta\,d\phi.$$

A discussion similar to the preceding shows us that convergence occurs when $a < 3$. As a general criterion we have the following:

The integral of a function f(x, y, z) *which becomes infinite at the origin, but is continuous at every other point of a region* R *containing the origin, is convergent, if there is a fixed bound* M *and a positive number* $\alpha < 3$ *such that the inequality*

$$|f(x, y, z)| \leqq \frac{M}{(\sqrt{x^2 + y^2 + z^2})^\alpha}$$

holds everywhere in the region.

From these criteria we conclude more generally that integrals of the form

$$\iint_R \frac{g(x, y)\, dx\, dy}{(\sqrt{(x-a)^2 + (y-b)^2})^\alpha}, \quad (\alpha < 2)$$

over a two-dimensional region and integrals of the form

$$\iiint_R \frac{g(x, y, z)\, dx\, dy\, dz}{(\sqrt{(x-a)^2 + (y-b)^2 + (z-c)^2})^\alpha}, \quad (\alpha < 3)$$

over a three-dimensional region converge, where (a, b), or (a, b, c), is a fixed point in the interior of the region R and g is a continuous function in the closed region R. We have only to transfer this point to the origin by translation of the co-ordinate system and then to apply our criterion.

3. Functions with Lines of Infinite Discontinuity.

If a function $f(x, y)$ becomes infinite not only at a single point but along whole curves C in the region R, we can proceed to define the integral of a function f over the region R in an exactly analogous way. We cut the curve of discontinuity C out of the region R by enclosing it in a region U_ϵ of area less than ϵ. If then as ϵ tends to 0 the integral of the function f over the region $R - U_\epsilon$ tends to a limit I independent of the particular choice of the region U_ϵ, we say that *the integral of* f *over the region* R *is convergent* (or *converges*) and we take this limiting value as the value of the integral.

The simplest example is the case in which the curve C consists of a portion of a straight line, say a segment of the y-axis. If the relation

$$|f(x, y)| < \frac{M}{x^\alpha},$$

where M is a fixed bound and a is less than 1, is valid everywhere in the region R, then the integral over the region R converges. The proof is similar to the proofs of the preceding sub-section. For example, we may cut the y-axis out of the region by means of straight lines parallel to it.

4. Infinite Regions of Integration.

If the region R extends to infinity, we approximate to it by a sequence of sub-regions $R_1, R_2, \ldots, R_\nu, \ldots$, which are all bounded and have the property that every arbitrary bounded sub-region of R is contained in every R_n for which n is greater than a certain m. (If, for example, R is the whole plane, for R_ν we can choose the circular region of radius ν about the origin.) If the limit

$$\lim_{\nu\to\infty}\iint_{R_\nu} f(x, y)\,dS$$

exists and is independent of the particular choice of the sequence of sub-regions R_ν, we call it the integral of the function f over the region R.

To illustrate this statement by an example, we consider the integral

$$\iint_R e^{-x^2-y^2}\,dx\,dy,$$

where the region of integration is the whole xy-plane. In order to establish the convergence of this integral we first choose the sub-regions R_ν as the circles K_ν with radius ν,

$$x^2 + y^2 \leqq \nu^2;$$

these obviously satisfy the above requirements. We have therefore to investigate the limit of the integral

$$\iint_{K_\nu} e^{-x^2-y^2}\,dx\,dy$$

as $\nu \to \infty$. But we have already evaluated this integral (Vol. I. p. 496) and have found it to be equal to $\pi(1 - e^{-\nu^2})$. Now

$$\lim_{\nu\to\infty} \pi(1 - e^{-\nu^2}) = \pi.$$

If we also show that not only the sequence of circles, but also every other sequence of sub-regions R with the properties mentioned, leads to the same value π, then according to our definition the number π will be the value of the improper integral.

Let any sequence of such regions $R_1, R_2, \ldots$ be given. By hypothesis, each circle K_m is contained in the interior of R_ν provided ν is sufficiently large; on the other hand, every R_ν is bounded and is therefore contained in a circle K_M of sufficiently large radius M. Since the integrand $e^{-x^2-y^2}$ is positive everywhere, it follows that

$$\iint_{K_m} e^{-x^2-y^2} dx\, dy \leq \iint_{R_\nu} e^{-x^2-y^2} dx\, dy \leq \iint_{K_M} e^{-x^2-y^2} dx\, dy.$$

As m and M increase, the integrals over K_m and K_M have the same limit π, so that the integral over R_ν must have the same limit; this proves that the integral must converge to the limit π.

We obtain a particularly interesting result if for the regions R_ν we choose the squares $|x| \leq \nu$, $|y| \leq \nu$. The integral $\iint_{R_\nu} e^{-x^2-y^2} dx\, dy$ can then be reduced to two simple integrations (cf. section 2, p. 228):

$$\iint_{R_\nu} e^{-x^2-y^2} dx\, dy = \int_{-\nu}^{\nu} e^{-x^2} dx \int_{-\nu}^{\nu} e^{-y^2} dy = \left(\int_{-\nu}^{\nu} e^{-x^2} dx\right)^2 = \left(2\int_0^{\nu} e^{-x^2} dx\right)^2.$$

If we now let ν tend to ∞, we must again obtain the same limit π. Hence

$$\left(2\int_0^{\infty} e^{-x^2} dx\right)^2 = \pi$$

or

$$\int_0^{\infty} e^{-x^2} dx = \tfrac{1}{2}\sqrt{\pi},$$

in agreement with Vol. I, p. 496.

5. Summary and Extensions.

It is useful to consider the concepts of this section again from a single unifying point of view. Our extension of the concept of integral to cases in which the definitions in section 2 (p. 224) are not immediately applicable consists in regarding the value of the integral as the limiting value of a sequence of integrals over regions R_ν, which approximate to the original region of integration R as ν increases. For this purpose we regard the region R as *open* instead of closed; we assign all the points of discontinuity of the function f to the boundary and consider the boundary as not belonging to R. We then say that *the region is approximated to by a sequence of regions* $R_1, R_2, \ldots, R_n, \ldots$ *if all the closed regions* R_n *lie in* R *and every arbitrarily chosen closed sub-region in the interior of* R *is also a sub-region of the region* R_n, *provided only that* n *is sufficiently large.* If in particular the sub-regions R_n are so chosen that each one contains the preceding one in its interior, we say that they converge *monotonically* to the region R.

For the sub-regions R_n we can at once apply the original definition of the integral given in section 2, p. 224. We now say that *the integral of* f *over the region* R *converges if the integral over* R_n *has a limiting value independent of the particular choice of the sequence of regions* R_n. It is useful to state specifically the following general facts which have been illustrated by the previous examples.

(1) If the function f is *nowhere negative* in the region R, it is sufficient to show that for a *single monotonic sequence* R_ν the sequence of values of the integral converges, in order to ensure convergence to the same limit for an *arbitrary* sequence R_ν'.

Proof. R_ν, being a closed region in the interior of R, is contained in all regions R_n' from a certain $n(\nu)$ onward. Conversely, every region R_n' is contained in a certain R_m, for the same reason. Since the function is nowhere negative, it follows that

$$\iint_{R_\nu} f(x, y)\, dx\, dy \leqq \iint_{R'_n} f(x, y)\, dx\, dy \leqq \iint_{R_m} f(x, y)\, dx\, dy.$$

As ν increases the two outer bounds tend to the same limit; the sequence of integrals $\iint_{R'_n} f(x, y)\, dx\, dy$ must therefore converge to that limit, and our statement is proved.

In particular, if for R_ν we choose a monotonic sequence of regions tending to R, it follows that the function f, which is nowhere negative, has a convergent integral over the region R, provided only that the sequence of integrals over R_ν remains below a bound M. For these integrals then form a sequence of numbers which is monotonic non-decreasing and bounded, and therefore convergent.

The case in which f is nowhere positive in R can at once be reduced to the preceding if we replace f by $-f$.

(2) If f changes sign in the region R, we can apply the previous theorem to $|f|$. If the integral of this absolute value converges, it is certain that the integral of the function f itself converges This is most easily proved by the following device. We put

$$f = f_1 - f_2,$$

where $f_1 = f$ if $f \geqq 0$, otherwise $f_1 = 0$, and $f_2 = -f$ if $f \leqq 0$, otherwise $f_2 = 0$. The two functions f_1, f_2 are nowhere negative, are continuous where f is continuous, and in absolute value never exceed f itself. Hence, if the integral of $|f|$ remains bounded for a monotonic sequence R_ν, the integrals of f_1 and f_2 converge, and with them the integral of their difference, $f_1 - f_2$.

6. Geometrical Applications

1. Elementary Calculation of Volumes.

The concept of volume forms the starting-point of our definition of integral. It is immediately obvious, therefore, how we can use multiple integrals in order to calculate volumes.

For example, in order to calculate the volume of the *ellipsoid of revolution*

$$\frac{x^2 + y^2}{a^2} + \frac{z^2}{b^2} = 1$$

we write the equation in the form

$$z = \pm \frac{b}{a}\sqrt{(a^2 - x^2 - y^2)}.$$

The volume of the half of the ellipsoid above the xy-plane is therefore given by the double integral

$$\frac{V}{2} = \frac{b}{a}\iint \sqrt{(a^2 - x^2 - y^2)}\,dx\,dy$$

taken over the circle $x^2 + y^2 \leq a^2$. If we transform to polar co-ordinates, the double integral becomes

$$\iint r\sqrt{(a^2 - r^2)}\,dr\,d\theta,$$

or on resolution into single integrals

$$\frac{V}{2} = \frac{b}{a}\int_0^{2\pi} d\theta \int_0^a r\sqrt{(a^2 - r^2)}\,dr = 2\pi\frac{b}{a}\int_0^a r\sqrt{(a^2 - r^2)}\,dr,$$

which gives the required value,

$$V = \frac{4}{3}\pi a^2 b.$$

To calculate the volume of the *general ellipsoid*

$$\frac{x^2}{a^2} + \frac{y^2}{b^2} + \frac{z^2}{c^2} = 1$$

we make the transformation

$$x = a\rho\cos\theta, \quad y = b\rho\sin\theta,$$

$$\frac{\partial(x, y)}{\partial(\rho, \theta)} = ab\rho$$

and for half the volume obtain

$$\frac{V}{2} = c\iint_R \sqrt{\left(1 - \frac{x^2}{a^2} - \frac{y^2}{b^2}\right)}\,dx\,dy = abc\iint_{R'} \rho\sqrt{(1 - \rho^2)}\,d\rho\,d\theta.$$

Here the region R' is the rectangle $0 \leqq \rho \leqq 1$, $0 \leqq \theta \leqq 2\pi$. Thus

$$\frac{V}{2} = abc\int_0^{2\pi} d\theta \int_0^1 \rho\sqrt{(1 - \rho^2)}\,d\rho = \frac{2}{3}\pi abc$$

or

$$V = \frac{4}{3}\pi abc.$$

Finally, we shall calculate the volume of the pyramid enclosed by the three co-ordinate planes and the plane $ax + by + cz - 1 = 0$, where we assume that a, b, and c are positive. For the volume we obtain

$$V = \frac{1}{c}\iint_R (1 - ax - by)\,dx\,dy,$$

where the region of integration is the triangle $0 \leqq x \leqq \frac{1}{a}$, $0 \leqq y \leqq \frac{1}{b}(1 - ax)$ in the xy-plane. Therefore

$$V = \frac{1}{c}\int_0^{1/a} dx \int_0^{(1-ax)/b} (1 - ax - by)\,dy.$$

Integration with respect to y gives

$$(1 - ax)y - \frac{b}{2}y^2 \Big|_0^{(1-ax)/b} = \frac{(1 - ax)^2}{2b},$$

and if we integrate again by means of the substitution $1 - ax = t$, we obtain

$$V = \frac{1}{2bc}\int_0^{1/a} (1 - ax)^2\,dx = -\frac{1}{6abc}(1 - ax)^3 \Big|_0^{1/a} = \frac{1}{6abc}.$$

We could of course have obtained the result from the theorem of elementary geometry that the volume of a pyramid is one-third of the product of base and altitude.

In order to calculate the volume of a more complicated solid we can subdivide the solid into pieces whose volumes can be expressed directly by double integrals. Later, however (in particular in the next chapter), we shall obtain expressions for the volume bounded by a closed surface which do not involve this subdivision.

2. General Remarks on the Calculation of Volumes. Solids of Revolution. Volumes in Polar Co-ordinates.

Just as we can express the area of a plane region R by the double integral

$$\iint_R dS = \iint_R dx\, dy,$$

we may also express the volume of a three-dimensional region R by the integral

$$V = \iiint_R dx\, dy\, dz$$

over the region R. In fact this point of view exactly corresponds to our definition of integral (cf. Appendix, p. 291) and expresses the geometrical fact that we can find the volume of a region by cutting the space into identical parallelepipeds, finding the total volume of the parallelepipeds contained entirely in R, and then letting the diameter of the parallelepipeds tend to zero. The resolution of this integral for V into an integral $\int dz \iint dx\, dy$ expresses *Cavalieri's principle*, known to us from elementary geometry, according to which the volume of a solid is determined if we know the area of every plane cross-section which is perpendicular to a definite line, say the z-axis. The general expression given above for the volume of a three-dimensional region at once enables us to find various formulæ for calculating volumes. For this purpose we have only to introduce new independent variables into the integral instead of x, y, z.

The most important examples are given by polar co-ordinates and by cylindrical co-ordinates; the latter will be defined below. We shall calculate e.g. the *volume of a solid of revolution* obtained by rotating a curve $x = \phi(z)$ about the z-axis. We assume that the rotating curve does not intersect the z-axis and that the solid of revolution is bounded above and below by planes $z =$ const. The solid is therefore defined by inequalities of the form $a \leqq z \leqq b$ and $0 \leqq \sqrt{(x^2 + y^2)} \leqq \phi(z)$. Its volume is given by the integral above. If we now introduce the cylindrical co-ordinates z, $\rho = \sqrt{(x^2 + y^2)}$, $\theta = \text{arc cos}\dfrac{x}{\rho} = \text{arc sin}\dfrac{y}{\rho}$ instead of x, y, z, we at once obtain the expression

$$V = \iiint_R dx\,dy\,dz = \int_a^b dz \int_0^{2\pi} d\theta \int_0^{\phi(z)} \rho\, d\rho$$

for the volume. If we perform the single integrations, we at once obtain

$$V = \pi \int_a^b \phi(z)^2 dz$$

(cf. Vol. I, Chap. V, section 2, p. 285).

We can also obtain this expression intuitively. If we cut the solid of revolution into small slices $z_\nu \leqq z \leqq z_{\nu+1}$ by planes perpendicular to the z-axis, and if by m_ν we denote the minimum and by M_ν the maximum of the distance $\phi(z)$ from the axis in this slice, then the volume of the slice lies between the volumes of two cylinders with altitude $\Delta z = z_{\nu+1} - z_\nu$ and radii m_ν and M_ν respectively. Hence

$$\Sigma m_\nu^2 \pi \Delta z \leqq V \leqq \Sigma M_\nu^2 \pi \Delta z.$$

By the definition of the ordinary integral, therefore,

$$V = \pi \int_a^b \phi(z)^2 dz.$$

If the region R contains the origin O of a polar co-ordinate system (r, θ, ϕ) and if the surface is given in polar co-ordinates by an equation

$$r = f(\theta, \phi)$$

where the function $f(\theta, \phi)$ is single-valued, it is frequently advantageous to use these polar co-ordinates instead of (x, y, z) in calculating the volume. If we substitute the value of the Jacobian $\dfrac{\partial(x, y, z)}{\partial(r, \theta, \phi)} = r^2 \sin\theta$ (as calculated on p. 254) in the transformation formula, we at once obtain the expression

$$V = \iiint_R r^2 \sin\theta\, dr\, d\theta\, d\phi = \int_0^{2\pi} d\phi \int_0^{\pi} \sin\theta\, d\theta \int_0^{f(\theta, \phi)} r^2 dr$$

for the volume. Integration with respect to r gives

$$V = \frac{1}{3} \int_0^{2\pi} d\phi \int_0^{\pi} f^3(\theta, \phi) \sin\theta\, d\theta.$$

In the special case of the sphere, in which $f(\theta, \varphi) = R$ is constant, we at once obtain the value $\frac{4}{3}\pi R^3$ for the volume of the sphere.

3. Area of a Curved Surface.

We have already expressed the length of arc of a curve by an ordinary integral (Vol. I, p. 279). We now wish to find an analogous expression for the area of a curved surface by means of a double integral. We regard the length of a curve as the limiting value of the length of an inscribed polygon when the lengths of the individual sides tend to zero. For the measurement of areas a direct analogy with this measurement of length would be as follows: in the curved surface we inscribe a polyhedron formed of plane triangles, determine the area of the polyhedron, make the inscribed net of triangles finer by letting the length of the longest side tend to zero, and seek to find the limiting value of the area of the polyhedron. This limiting value would then be called the area of the curved surface. It turns out, however, that such a definition of area would have no precise meaning, for in general this process does not yield a definite limiting value. This phenomenon may be explained in the following way: a polygon inscribed in a smooth curve always has the property, expressed by the mean value theorem of the differential calculus, that the direction of the individual side of the polygon approaches the direction of the curve as closely as we please if the subdivision is fine enough. With curved surfaces the situation is quite different. The sides of a polyhedron inscribed in a curved surface may be inclined to the tangent plane to the surface at a neighbouring point as steeply as we please, even if the polyhedral faces have arbitrarily small diameters. The area of such a polyhedron, therefore, cannot by any means be regarded as an approximation to the area of the curved surface. In the appendix we shall consider an example of this state of affairs in detail (pp. 341–2).

In the definition of the length of a smooth curve, however, instead of using an inscribed polygon we can equally well use a circumscribed polygon, that is, a polygon of which every side touches the curve. This definition of the length of a curve as the limit of the length of a circumscribed polygon can easily be extended to curved surfaces. The extension is even easier if we start from the following remark: we can obtain the length of a curve $y = f(x)$ which has a continuous derivative $f'(x)$ and lies between the abscissæ a and b by subdividing the interval between a and b at the points $x_0, x_1, \ldots, x_n$ into n parts of equal or different lengths, choosing an arbitrary point ξ_ν in the ν-th sub-interval, constructing the tangent to the curve at this point, and measuring the length l_ν of the portion of this tangent lying in the strip $x_\nu \leqq x \leqq x_{\nu+1}$. The sum $\sum_{\nu=1}^{n} l_\nu$ then tends to the length of the curve, i.e. to the integral $\int_a^b \sqrt{\{1 + f'(x)^2\}}\, dx$, if we let n increase beyond all bounds and at the same time let the length of the longest sub-interval tend to zero. This statement follows from the fact that $l_\nu = (x_{\nu+1} - x_\nu)\sqrt{\{1 + f'(\xi_\nu)^2\}}$.

We can now define the area of a curved surface in a similar way. We begin by considering a surface which lies above the region R of the xy-plane and is represented by a function $z = f(x, y)$ with continuous derivatives. We subdivide R into n sub-regions $R_1, R_2, \ldots, R_n$ with the areas $\Delta R_1, \ldots, \Delta R_n$, and in these sub-regions we choose points $(\xi_1, \eta_1), \ldots, (\xi_n, \eta_n)$. At the point of the surface with the co-ordinates ξ_ν, η_ν and $\zeta_\nu = f(\xi_\nu, \eta_\nu)$ we construct the tangent plane and find the area of the portion of this plane lying above the region R_ν. If α_ν is the angle which the tangent plane

$$z - \zeta_\nu = f_x(\xi_\nu, \eta_\nu)(x - \xi_\nu) + f_y(\xi_\nu, \eta_\nu)(y - \eta_\nu)$$

makes with the xy-plane, and if $\Delta\tau_\nu$ is the area of the portion τ_ν of the tangent plane above R_ν, then the region R_ν is the projection of τ_ν on the xy-plane, so that

$$\Delta R_\nu = \Delta\tau_\nu \cos\alpha_\nu.$$

Again (cf. Chap. III, section 2, p. 130),

$$\cos\alpha_\nu = \frac{1}{\sqrt{1 + f_x^2(\xi_\nu, \eta_\nu) + f_y^2(\xi_\nu, \eta_\nu)}},$$

and therefore

$$\Delta\tau_\nu = \sqrt{1 + f_x^2(\xi_\nu, \eta_\nu) + f_y^2(\xi_\nu, \eta_\nu)}\,.\,\Delta R_\nu.$$

If we now form the sum of all these areas

$$\sum_{\nu=1}^{n} \Delta\tau_\nu$$

and let n increase beyond all bounds, at the same time letting the diameter (and consequently the area) of the largest subdivision tend to zero, then according to our definition of integral this sum will have the limit

$$A = \iint_R \sqrt{1 + f_x^2 + f_y^2}\, dS.$$

This integral, which is independent of the mode of subdivision of the region R, we shall *define* as the *area of the given surface.* If the surface happens to be a plane surface, this definition agrees with the preceding; for example, if $z = f(x, y) = 0$, we have

$$A = \iint_R dS.$$

It is occasionally convenient to call the symbol

$$d\sigma = \sqrt{1 + f_x^2 + f_y^2}\, dS = \sqrt{1 + f_x^2 + f_y^2}\, dx\, dy$$

the *element of area* of the surface $z = f(x, y)$. The area integral can then be written symbolically in the form

$$\iint_R d\sigma.$$

We arrive at another form of the expression for the area if we think of the surface as given by an equation $\phi(x, y, z) = 0$ instead of $z = f(x, y)$. If we assume that on the surface $\phi_z \neq 0$, say $\phi_z > 0$, then the equations

$$\frac{\partial z}{\partial x} = -\frac{\phi_x}{\phi_z}, \quad \frac{\partial z}{\partial y} = -\frac{\phi_y}{\phi_z}$$

at once give the expression

$$\iint_R \sqrt{\phi_x^2 + \phi_y^2 + \phi_z^2}\, \frac{1}{\phi_z}\, dx\, dy$$

for the area, the region R again being the projection of the surface on the xy-plane.

As an example of the application of the area formula we consider the area of a spherical surface. The equation $z = \sqrt{(R^2 - x^2 - y^2)}$ represents a hemisphere of radius R. We have

$$\frac{\partial z}{\partial x} = -\frac{x}{\sqrt{(R^2 - x^2 - y^2)}}, \quad \frac{\partial z}{\partial y} = -\frac{y}{\sqrt{(R^2 - x^2 - y^2)}}.$$

The area of the hemisphere is therefore given by the integral

$$\tfrac{1}{2}A = R \iint_{R'} \frac{dx\, dy}{\sqrt{(R^2 - x^2 - y^2)}},$$

where the region of integration R' is the circle of radius R lying in the xy-plane and having the origin as its centre. By introducing polar co-ordinates and resolving the integral into single integrals we further obtain

$$\tfrac{1}{2}A = R\int_0^{2\pi} d\theta \int_0^R \frac{r\,dr}{\sqrt{(R^2 - r^2)}} = 2\pi R \int_0^R \frac{r\,dr}{\sqrt{(R^2 - r^2)}}.$$

The ordinary integral on the right can easily be evaluated by means of the substitution $R^2 - r^2 = u$; we have

$$\tfrac{1}{2}A = -2\pi R\sqrt{R^2 - r^2}\,\Big|_0^R = 2\pi R^2,$$

in agreement with the fact, known from elementary geometry, that the area of the surface of a sphere is $4\pi R^2$.

In the definition of area we have hitherto singled out the co-ordinate z. If, however, the surface had been given by an equation of the form $x = x(y, z)$ or $y = y(x, z)$, we could equally well have represented the area by integrals of the form

$$\iint \sqrt{(1 + x_y^2 + x_z^2)}\,dy\,dz \quad \text{or} \quad \iint \sqrt{(1 + y_x^2 + y_z^2)}\,dz\,dx,$$

or, if the surface were given implicitly, we should have

$$\iint \sqrt{(\phi_x^2 + \phi_y^2 + \phi_z^2)}\,\frac{1}{\phi_y}\,dz\,dx$$

or

$$\iint \sqrt{(\phi_x^2 + \phi_y^2 + \phi_z^2)}\,\frac{1}{\phi_x}\,dy\,dz.$$

That all these expressions do actually define the same area is self-evident. The equality of the different expressions can, however, be verified directly. For example, we apply the transformation

$$x = x(y, z),$$
$$y = y$$

to the integral

$$\iint \frac{\sqrt{(\phi_x^2 + \phi_y^2 + \phi_z^2)}}{\phi_z}\,dx\,dy.$$

Here $x = x(y, z)$ is found by solving the equation $\phi(x, y, z) = 0$ for x. The Jacobian is $\dfrac{\partial(x, y)}{\partial(y, z)} = \dfrac{\phi_z}{\phi_x}$, and therefore

$$\iint_R \frac{\sqrt{(\phi_x^2+\phi_y^2+\phi_z^2)}}{\phi_z}\,dx\,dy = \iint_{R'} \frac{\sqrt{(\phi_x^2+\phi_y^2+\phi_z^2)}}{\phi_x}\,dy\,dz.$$

The integral on the right is to be taken over the projection R' of the surface on the yz-plane.

If in expressing the area of a surface we wish to get rid of any special assumption about the position of the surface relative to the co-ordinate system, we must represent the surface in the parametric form

$$x = \phi(u, v), \quad y = \psi(u, v), \quad z = \chi(u, v).$$

A definite region R' of the uv-plane then corresponds to the surface. In order to introduce the parameters u and v in the above formulæ we first consider a portion of the surface and assume that for this portion the Jacobian $\dfrac{\partial(x, y)}{\partial(u, v)} = D$ is everywhere positive. According to Chap. III, section 3, p. 153, for this portion we can then solve for u and v as functions of x and y, obtaining

$$u_x = \frac{\psi_v}{D}, \qquad v_x = -\frac{\psi_u}{D},$$

$$u_y = -\frac{\phi_v}{D}, \qquad v_y = \frac{\phi_u}{D}.$$

for the partial derivatives.

In virtue of the equations

$$\frac{\partial z}{\partial x} = \frac{\partial z}{\partial u}u_x + \frac{\partial z}{\partial v}v_x \quad \text{and} \quad \frac{\partial z}{\partial y} = \frac{\partial z}{\partial u}u_y + \frac{\partial z}{\partial v}v_y$$

we obtain the expression

$$\sqrt{\left\{1+\left(\frac{\partial z}{\partial x}\right)^2+\left(\frac{\partial z}{\partial y}\right)^2\right\}}$$

$$= \frac{1}{D}\sqrt{\{(\phi_u\psi_v-\psi_u\phi_v)^2+(\psi_u\chi_v-\chi_u\psi_v)^2+(\chi_u\phi_v-\phi_u\chi_v)^2\}}.$$

If we now introduce u and v as new independent variables and apply the rules for the transformation of double integrals (p. 253), we find that the area of the portion of the surface corresponding to R' is

$$A = \iint_{R'} \sqrt{\{(\phi_u\psi_v - \psi_u\phi_v)^2 + (\psi_u\chi_v - \chi_u\psi_v)^2 + (\chi_u\phi_v - \phi_u\chi_v)^2\}}\, du\, dv.$$

In this expression there is no longer any distinction between the co-ordinates x, y, and z. Since we arrive at the same integral expression for the area no matter which one of the special non-parametric representations we start with, it follows that all these expressions are equal and represent the area.

So far we have only considered a portion of the surface on which one particular Jacobian does not vanish. We reach the same result, however, no matter which of the three Jacobians does not vanish. If then we suppose that at each point of the surface *one* of the Jacobians is not zero, we can subdivide the whole surface into portions like the above, and thus find that the preceding integral still gives the area of the whole surface:

$$A = \iint_{R'} \sqrt{\{(\phi_u\psi_v - \psi_u\phi_v)^2 + (\psi_u\chi_v - \chi_u\psi_v)^2 + (\chi_u\phi_v - \phi_u\chi_v)^2\}}\, du\, dv.$$

The expression for the area of a surface in parametric representation can be put in another noteworthy form if we make use of the coefficients of the line element (cf. Chap. III, section 4, p. 163)

$$ds^2 = E\,du^2 + 2F\,du\,dv + G\,dv^2,$$

that is, of the expressions

$$E = \phi_u^2 + \psi_u^2 + \chi_u^2,$$
$$F = \phi_u\phi_v + \psi_u\psi_v + \chi_u\chi_v,$$
$$G = \phi_v^2 + \psi_v^2 + \chi_v^2.$$

A simple calculation shows that

$$EG - F^2 = (\phi_u\psi_v - \psi_u\phi_v)^2 + (\psi_u\chi_v - \chi_u\psi_v)^2 + (\chi_u\phi_v - \phi_u\chi_v)^2.$$

Thus for the area we obtain the expression

$$\iint \sqrt{(EG - F^2)}\, du\, dv,$$

and for the element of area

$$d\sigma = \sqrt{(EG - F^2)}\, du\, dv.$$

As an example we again consider the area of a sphere with radius R, which we now represent parametrically by the equations

$$x = R \cos u \sin v,$$
$$y = R \sin u \sin v,$$
$$z = R \cos v,$$

where u and v range over the region $0 \leqq u \leqq 2\pi$ and $0 \leqq v \leqq \pi$. A simple calculation once more gives us the expression

$$R^2 \int_0^{2\pi} du \int_0^{\pi} \sin v \, dv = 4\pi R^2$$

for the area.

In particular, we can apply our result to the surface of revolution formed by rotating the curve $z = \phi(x)$ about the z-axis. If we refer the surface to polar co-ordinates (u, v) in the xy-plane as parameters, we obtain

$$x = u \cos v, \quad y = u \sin v, \quad z = \phi(\sqrt{x^2 + y^2}) = \phi(u).$$

Then

$$E = 1 + \phi'^2(u), \quad F = 0, \quad G = u^2,$$

and the area is given in the form

$$\int_0^{2\pi} dv \int_{u_0}^{u_1} u\sqrt{1 + \phi'^2(u)}\, du = 2\pi \int_{u_0}^{u_1} u\sqrt{1 + \phi'^2(u)}\, du.$$

If instead of u we introduce the length of arc s of the meridian curve $z = \phi(u)$ as parameter, we obtain the *area of the surface of revolution* in the form

$$2\pi \int_{s_0}^{s_1} u\, ds,$$

where u is the distance from the axis of the point on the rotating curve corresponding to s (Guldin's rule; cf. Vol. I, p. 285).

As an example we calculate the surface area of the *torus* or *anchor ring* (cf. Chap. III, section 4, p. 165) obtained by rotating the circle $(y - a)^2 + z^2 = r^2$ about the z-axis. If we introduce the length of arc s of the circle as a parameter we have $u = a + r \cos \frac{s}{r}$, and the area is therefore

$$2\pi \int_0^{2\pi r} u\, ds = 2\pi \int_0^{2\pi r} \left(a + r \cos \frac{s}{r}\right) ds = 2\pi a \, . \, 2\pi r.$$

The area of an anchor ring is therefore equal to the product of the circumference of the generating circle and the length of the path described by the centre of the circle.

Examples

1. Calculate the volume of the solid defined by

$$\frac{\{\sqrt{(x^2+y^2)}-1\}^2}{a^2}+\frac{z^2}{b^2}\leqq 1 \qquad (b<1).$$

2. Find the volume cut off from the paraboloid

$$\frac{x^2}{a^2}+\frac{y^2}{b^2}=z$$

by the plane $z=h$.

3. Find the volume cut off from the ellipsoid

$$\frac{x^2}{a^2}+\frac{y^2}{b^2}+\frac{z^2}{c^2}=1$$

by the plane

$$lx+my+nz=p.$$

4. (*a*) Show that if any closed curve $\theta=f(\varphi)$ is drawn on the surface

$$r^2=a^2\cos 2\theta$$

(r, θ, φ being polar co-ordinates in space), the area of the surface so enclosed is equal to the area enclosed by the projection of the curve on the sphere $r=a$, the origin of co-ordinates being the vertex of projection.

(*b*) Express the area by a simple integral.

(*c*) Find the area of the whole surface.

5. Find the area of the surface of the spheroid formed by rotating an ellipse about its major axis, and show that if the fourth and higher powers of the eccentricity e may be neglected, this area is equal to that of the sphere whose volume is equal to that of the spheroid.

6. Find the volume and surface area of the solid generated by rotating the triangle ABC about the side AB.

7*. A tube-surface is generated by the spheres of unit radius whose centres form the closed plane curve L. Prove that the area A of the surface is 2π times the length of L.

8*. (*a*) Calculate the volume of the region defined by

$$\begin{aligned} x^2+y^2+z^2 &\leqq r^2 \\ x^2+y^2-rx &\geqq 0 \\ x^2+y^2+rx &\geqq 0. \end{aligned}$$

(*b*) Calculate the area of the spherical part of the boundary of this region, i.e. the area of the surface

$$\begin{aligned} x^2+y^2+z^2 &= r^2 \\ x^2+y^2-rx &\geqq 0 \\ x^2+y^2+rx &\geqq 0. \end{aligned}$$

9. Calculate the area of that part of the screw surface

$$y - x \tan \frac{z}{h} = 0$$

for which

$$r^2 \leqq x^2 + y^2 \leqq R^2, \qquad |z| \leqq \frac{\pi}{2} h.$$

10. Calculate the area of the surface

$$(x^2 + y^2 + z^2)^2 = x^2 - y^2.$$

7. Physical Applications

In section 2, No. 7 (p. 235) we have already seen how the concept of mass is connected with that of a multiple integral. Here we shall study some of the other concepts of mechanics. We begin with a more detailed study of moment and of moment of inertia than was possible in Vol. I, Chap. X (p. 496).

1. Moments and Centre of Mass.

The moment with respect to the xy*-plane of a particle with mass* m *is defined as the product* mz *of the mass and the* z*-co-ordinate.* Similarly, the moment with respect to the yz-plane is mx and that with respect to the zx-plane is my. The *moments of several particles combine additively*; that is, the three moments of a system of particles with masses $m_1, m_2, \ldots, m_n$ and co-ordinates (x_1, y_1, z_1), $\ldots, (x_n, y_n, z_n)$ are given by the expressions

$$T_x = \sum_{\nu=1}^{n} m_\nu x_\nu, \quad T_y = \sum_{\nu=1}^{n} m_\nu y_\nu, \quad T_z = \sum_{\nu=1}^{n} m_\nu z_\nu.$$

If instead of a finite number of particles we are dealing with a mass distributed continuously with density $\mu = \mu(x, y, z)$ through a region in space or over a surface or curve, we define the moment of the mass-distribution by a limiting process, as in Vol. I, Chap. X, section 6 (p. 497), and thus express the moments by integrals. For example, with a distribution in space we subdivide the region R into n sub-regions, imagine the total mass of each sub-region concentrated at any one of its points, and then form the moment of the system of these n particles. We see at once that as $n \to \infty$ and at the same time the greatest diameter of the sub-regions tends to zero the sums tend to the limits

$$T_x = \iiint_R \mu x\,dx\,dy\,dz, \quad T_y = \iiint_R \mu y\,dx\,dy\,dz,$$
$$T_z = \iiint_R \mu z\,dx\,dy\,dz,$$

which we call the *moments of the volume-distribution.*

Similarly, if the mass is distributed over a surface S given by the equations $x = \phi(u, v)$, $y = \psi(u, v)$, $z = \chi(u, v)$ with surface density $\mu(u, v)$, we define the *moments of the surface distribution* by the expressions

$$T_x = \iint_S \mu x\,d\sigma = \iint_R \mu x\sqrt{EG - F^2}\,du\,dv,$$

$$T_y = \iint_S \mu y\,d\sigma = \iint_R \mu y\sqrt{EG - F^2}\,du\,dv,$$

$$T_z = \iint_S \mu z\,d\sigma = \iint_R \mu z\sqrt{EG - F^2}\,du\,dv.$$

Finally, the *moments of a curve* x(s), y(s), z(s) *in space* with mass density $\mu(s)$ are defined by the expressions

$$T_x = \int_{s_0}^{s_1} \mu x\,ds, \quad T_y = \int_{s_0}^{s_1} \mu y\,ds, \quad T_z = \int_{s_0}^{s_1} \mu z\,ds,$$

where s denotes the length of arc.

The *centroid* (*centre of mass*) of a mass of total amount M distributed through a region R is defined as the point with co-ordinates

$$\xi = \frac{T_x}{M}, \quad \eta = \frac{T_y}{M}, \quad \zeta = \frac{T_z}{M}.$$

For a distribution in space the co-ordinates of the centre of mass are therefore given by the expressions

$$\xi = \frac{1}{M}\iiint_R \mu x\,dx\,dy\,dz, \text{ \&c., where } M = \iiint_R \mu\,dx\,dy\,dz.$$

As an example we first consider the uniform hemispherical region H with mass density 1:

$$x^2 + y^2 + z^2 \leqq 1,$$
$$z \geqq 0.$$

The first two moments

$$T_x = \iiint_H x\,dx\,dy\,dz,$$

$$T_y = \iiint_H y\,dx\,dy\,dz$$

are zero, since the integration with respect to x or with respect to y gives the value zero. For the third,

$$T_z = \iiint_H z\,dx\,dy\,dz,$$

we introduce cylindrical co-ordinates (r, z, θ) by means of the equations

$$z = z,$$
$$x = r\cos\theta,$$
$$y = r\sin\theta$$

and obtain

$$T_z = \int_0^1 z\,dz \int_0^{\sqrt{(1-z^2)}} r\,dr \int_0^{2\pi} d\theta = 2\pi \int_0^1 \frac{1-z^2}{2} z\,dz = \pi\left(\frac{z^2}{2} - \frac{z^4}{4}\right)\Big|_0^1 = \frac{\pi}{4}.$$

Since the total mass is $2\pi/3$, the co-ordinates of the centre of mass are $x = 0$, $y = 0$, $z = \frac{3}{8}$.

We shall next calculate the centre of mass of a hemispherical surface of unit radius over which a mass of unit density is uniformly distributed. For the parametric representation

$$x = \cos u \sin v, \quad y = \sin u \sin v, \quad z = \cos v$$

we calculate the surface element from the formula on p. 273 and find that

$$\sqrt{EG - F^2}\,du\,dv = \sin v\,du\,dv$$

We accordingly obtain

$$T_x = \int_0^{\pi/2} \sin^2 v\,dv \int_0^{2\pi} \cos u\,du = 0,$$

$$T_y = \int_0^{\pi/2} \sin^2 v\,dv \int_0^{2\pi} \sin u\,du = 0,$$

$$T_z = \int_0^{\pi/2} \sin v \cos v\,dv \int_0^{2\pi} du = 2\pi \frac{\sin^2 v}{2}\Big|_0^{\pi/2} = \pi$$

for the three moments. Since the total mass is obviously 2π, we see that the centre of mass lies at the point with co-ordinates $x = 0$, $y = 0$, $z = \frac{1}{2}$.

2. Moment of Inertia.

The generalization of the concept of moment of inertia is equally obvious. *The moment of inertia of a particle with respect to the* x*-axis is the product of its mass and* $\rho^2 = y^2 + z^2$, that is, the square of the distance of the point from the x-axis. In the same way, we define the moment of inertia about the x-axis of a mass distributed with density $\mu(x, y, z)$ through a region R by the expression

$$\iiint_R \mu(y^2 + z^2)\,dx\,dy\,dz.$$

The moments of inertia about the other axes are represented by similar expressions. Occasionally the moment of inertia with respect to a *point*, say the *origin*, is defined by the expression

$$\iiint_R \mu(x^2 + y^2 + z^2)\,dx\,dy\,dz,$$

and the moment of inertia with respect to a *plane*, say the yz-plane, by

$$\iiint_R \mu x^2\,dx\,dy\,dz.$$

Similarly, the moment of inertia, with respect to the x-axis, of a surface distribution is given by

$$\iint_S \mu(y^2 + z^2)\,d\sigma,$$

where $\mu(u, v)$ is a continuous function of two parameters u and v.

The moment of inertia of a mass distributed with density $\mu(x, y, z)$ through a region R, with respect to an axis parallel to the x-axis and passing through the point (ξ, η, ζ), is given by the expression

$$\iiint_R \mu[(y - \eta)^2 + (z - \zeta)^2]\,dx\,dy\,dz.$$

If in particular we let (ξ, η, ζ) be the centre of mass (cf. p. 277) and recall the relations for the co-ordinates of the centre of mass (given on p. 277), we at once obtain the equation

$$\iiint_R \mu(y^2 + z^2)\,dx\,dy\,dz = \iiint_R \mu[(y - \eta)^2 + (z - \zeta)^2]\,dx\,dy\,dz$$
$$+ (\eta^2 + \zeta^2)\iiint_R \mu\,dx\,dy\,dz.$$

Since any arbitrary axis of rotation of a body can be chosen as the x-axis, the meaning of this equation can be expressed as follows:

The moment of inertia of a rigid body with respect to an arbitrary axis of rotation is equal to the moment of inertia of the body about a parallel axis through its centre of mass plus the product of the total mass and the square of the distance between the centre of mass and the axis of rotation (Steiner's theorem).

The physical meaning of the moment of inertia for regions in several dimensions is exactly the same as that already stated in Vol. I, Chap. V, section 2 (p. 286):

The kinetic energy of a body rotating uniformly about an axis is equal to half the product of the square of the angular velocity and the moment of inertia.

The following examples may serve to illustrate the concept and the actual calculation of the moment of inertia in simple cases.

For the sphere V with centre at the origin, unit radius and unit density, we see by symmetry that the moment of inertia with respect to any axis through the origin is

$$I = \iiint_V (x^2 + y^2)\,dx\,dy\,dz = \iiint_V (x^2 + z^2)\,dx\,dy\,dz$$
$$= \iiint_V (y^2 + z^2)\,dx\,dy\,dz.$$

If we add the three integrals, we obtain

$$3I = \iiint_V 2(x^2 + y^2 + z^2)\,dx\,dy\,dz,$$

or, if we introduce polar co-ordinates,

$$I = \frac{2}{3}\int_0^1 r^4\,dr \int_0^\pi \sin v\,dv \int_0^{2\pi} du = \frac{2}{3}\cdot\frac{1}{5}\cdot 2\cdot 2\pi = \frac{8\pi}{15}.$$

For a beam with edges a, b, c, parallel to the x-axis, the y-axis, and the z-axis respectively, with unit density and centre of mass at the origin we find that the moment of inertia with respect to the xy-plane is

$$\int_{-a/2}^{a/2} dx \int_{-b/2}^{b/2} dy \int_{-c/2}^{c/2} z^2\,dz = ab\,\frac{c^3}{12}.$$

3. The Compound Pendulum.

The above ideas find an application in the mathematical treatment of the compound pendulum, that is, of a rigid body which oscillates about a fixed axis under the influence of gravity.

We consider a plane through G, the centre of mass of the rigid body, perpendicular to the axis of rotation; let this plane cut the axis in the point O (fig. 14). Then the motion of the body is obviously given if we state the angle $\varphi = \varphi(t)$ which OG makes at time t with the downward vertical line through O. In order to determine this function $\varphi(t)$ and also the period of oscillation of the pendulum, we require to assume a knowledge of certain physical facts (cf. Chap. VI, section 1, p. 412). We make use of the law of conservation of energy, which states that during the motion of the body the sum of its kinetic and potential energies remains constant. Here V, the potential energy of the body, is the product Mgh, where M is the total mass, g the gravitational acceleration, and h the height of the centre of mass above an arbitrary horizontal line, e.g. above the horizontal line through the lowest position reached by the centre of mass during the motion. If we denote OG, the distance of the centre of mass from the axis, by s, then $V = Mgs(1 - \cos\varphi)$. By p. 280 the kinetic energy is given by $T = \frac{1}{2}I\dot{\varphi}^2$, where I is the moment of inertia of the body with respect to the axis of rotation and we have written $\dot{\varphi}$ for $d\varphi/dt$. The law of conservation of energy therefore gives the equation

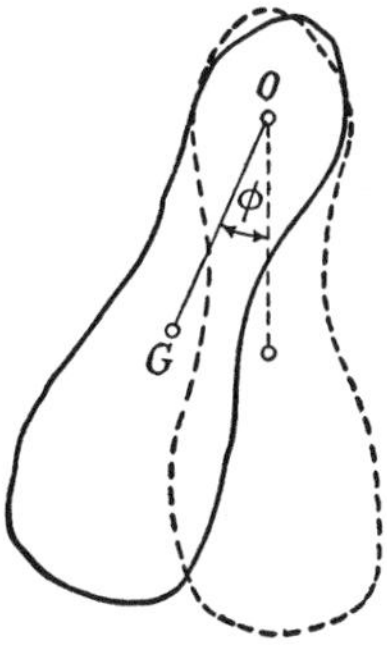

Fig. 14.—The compound pendulum

$$\frac{1}{2}I\dot{\varphi}^2 - Mgs\cos\varphi = \text{const.}$$

If we introduce the constant $l = I/Ms$, this is exactly the same as the equation previously found (Vol. I, Chap. V, p. 302) for the simple pendulum; l is accordingly known as the *length of the equivalent simple pendulum.*

We can now directly apply the formulæ previously obtained (*loc. cit.*). The period of oscillation is given by the formula

$$T = 2\sqrt{\frac{l}{2g}}\int_{-\phi_0}^{\phi_0}\frac{d\varphi}{\sqrt{\cos\varphi - \cos\varphi_0}},$$

where φ_0 corresponds to the greatest displacement of the centre of mass; for small angles this is approximately

$$T = 2\pi\sqrt{\frac{l}{g}} = 2\pi\sqrt{\frac{I}{Mgs}}.$$

The formula for the simple pendulum is of course included in this as a special case. For if the whole mass M is concentrated at the centre of mass, then $I = Ms^2$, so that $l = s$.

Investigating further, we recall that I, the moment of inertia about the axis of rotation, is connected with I_0, the moment of inertia about a parallel axis through the centre of mass, by the relation (cf. p. 279)

$$I = I_0 + Ms^2.$$

Hence

$$l = s + \frac{I_0}{Ms},$$

or, if we introduce the constant $a = I_0/M$,

$$l = s + \frac{a}{s}.$$

We see at once that in a compound pendulum l always exceeds s, so that the period of a compound pendulum is always greater than that of the simple pendulum obtained by concentrating the mass M at the centre of mass. Moreover, we note that the period is the same for all parallel axes at the same distance s from the centre of mass. For the length of the equivalent simple pendulum depends only on the two quantities s and $a = I_0/M$, and therefore remains the same provided neither the direction of the axis of rotation nor its distance from the centre of mass is altered.

If in the formula $l = s + a/s$ we replace the quantity s by a/s, that is, if the axis is moved from the distance s to the distance a/s from the centre of mass, then l remains unchanged. This means that a compound pendulum has the same period of oscillation for all parallel axes which have the distance s or a/s from the centre of mass.

The formula $T = 2\pi\sqrt{\left(\frac{s + a/s}{g}\right)}$ shows at once that the period T increases beyond all bounds as s tends to zero or to infinity. It must therefore have a minimum for some value s_0. By differentiating we obtain

$$s_0 = \sqrt{a} = \sqrt{\frac{I_0}{M}}.$$

A pendulum whose axis is at a distance $s_0 = \sqrt{I_0/M}$ from the centre of mass will be relatively insensitive to small displacements of the axis. For in this case dT/ds vanishes, so that first-order changes in s produce only second-order changes in T. This fact has been applied by Prof. Schuler of Göttingen in the construction of very accurate clocks.

4. Potential of Attracting Masses.

We have seen in Chap. II, section 7 (p. 90) that according to Newton's law of gravitation the force which a fixed particle Q with co-ordinates (ξ, η, ζ) and mass m exerts on a second particle P with co-ordinates (x, y, z) and unit mass is given, apart from the gravitational constant γ, by

$$m \operatorname{grad} \frac{1}{r},$$

where $r = \sqrt{(x - \xi)^2 + (y - \eta)^2 + (z - \zeta)^2}$ is the distance between the points P and Q. The direction of the force is along the line joining the two particles, and its magnitude is inversely proportional to the square of the distance. If we now consider the force exerted on P by a number of points $Q_1, Q_2, \ldots, Q_n$ with respective masses $m_1, m_2, \ldots, m_n$, we can express the total force as the gradient of the quantity

$$\frac{m_1}{r_1} + \frac{m_2}{r_2} + \ldots + \frac{m_n}{r_n},$$

where r_ν denotes the distance of the point Q_ν from the point P. If a force can be expressed as a gradient of a function, it is customary to call this function the *potential of the force*; we accordingly define the gravitational potential of the system of particles $Q_1, Q_2, \ldots, Q_n$ at the point P as the expression

$$\sum_{\nu=1}^{n} \frac{m_\nu}{\sqrt{(x-\xi_\nu)^2+(y-\eta_\nu)^2+(z-\zeta_\nu)^2}}.$$

We now suppose that instead of being concentrated at a finite number of points the gravitating masses are distributed with continuous density μ over a portion R of space or a surface S or a curve C. Then the potential of this mass-distribution at a point with co-ordinates (x, y, z) outside the system of masses is defined as

$$\iiint_R \frac{\mu(\xi, \eta, \zeta)}{r}\, d\xi\, d\eta\, d\zeta,$$

or

$$\iint_S \frac{\mu}{r}\, d\sigma,$$

or

$$\int_{s_0}^{s_1} \frac{\mu}{r}\, ds.$$

In the first case the integration is taken throughout the region R with rectangular co-ordinates (ξ, η, ζ), in the second case over the surface S with the element of surface $d\sigma$, and in the third case along the curve with length of arc s. In all three formulæ r denotes the distance of the point P from the point (ξ, η, ζ) of the region of integration and μ the mass density at the point (ξ, η, ζ).

Thus e.g. the potential at a point P with co-ordinates (x, y, z), due to a sphere K of constant density equal to unity, with unit radius and centre the origin, is given by the integral

$$\iiint_K \frac{d\xi\, d\eta\, d\zeta}{\sqrt{(x-\xi)^2+(y-\eta)^2+(z-\zeta)^2}}$$
$$= \int_{-1}^{+1} d\xi \int_{-\sqrt{(1-\xi^2)}}^{+\sqrt{(1-\xi^2)}} d\eta \int_{-\sqrt{(1-\xi^2-\eta^2)}}^{+\sqrt{(1-\xi^2-\eta^2)}} \frac{1}{r}\, d\zeta.$$

In all these expressions the co-ordinates (x, y, z) of the point P appear, *not* as variables of integration, but as parameters, and the potentials are functions of these parameters.

To obtain the components of the force from the potential we have to differentiate the integral with respect to the parameters. The rules for differentiation with respect to a parameter extend directly to multiple integrals, and by section 1 (p. 218) the differentiation can be performed under the integral sign, provided that the point P does not belong to the region of integration, that is, provided that we are certain that there is no point of the closed region of integration for which the distance r has the value zero. Thus, for example, we find that the *components of the gravitational force* on unit mass due to a mass distributed with unit density through a region R in space are given by the expressions

$$F_1 = -\iiint_R \frac{x-\xi}{r^3}\, d\xi\, d\eta\, d\zeta, \quad F_2 = -\iiint_R \frac{y-\eta}{r^3}\, d\xi\, d\eta\, d\zeta,$$

$$F_3 = -\iiint_R \frac{z-\zeta}{r^3}\, d\xi\, d\eta\, d\zeta.$$

Finally, we point out that the expressions for the potential and its first derivatives continue to have a meaning if the point P lies in the interior of the region of integration. The integrals are then improper integrals, and, as is easily shown, their convergence follows from the criteria of section 5 (p. 257).

As an example we shall calculate the potential at an internal point and at an external point, due to a spherical surface S with radius a and unit surface density. If we take the centre of the sphere as origin and make the x-axis pass through the point P (inside or outside the sphere), the point P will have the co-ordinates $(x, 0, 0)$, and the potential will be

$$U = \iint \frac{d\sigma}{\sqrt{(x-\xi)^2 + \eta^2 + \zeta^2}}.$$

If we introduce polar co-ordinates on the sphere by means of the equations

$$\xi = a\cos\theta,$$
$$\eta = a\sin\theta\cos\varphi,$$
$$\zeta = a\sin\theta\sin\varphi,$$

then

$$U = \int_0^\pi \frac{a^2\sin\theta}{\sqrt{(x-a\cos\theta)^2 + a^2\sin^2\theta}}\, d\theta \int_0^{2\pi} d\varphi$$

$$= 2\pi \int_0^\pi \frac{a^2\sin\theta}{\sqrt{x^2+a^2-2ax\cos\theta}}\, d\theta.$$

If we put $x^2 + a^2 - 2ax\cos\theta = r^2$, so that $ax\sin\theta\, d\theta = r\,dr$, then (provided that $x \neq 0$) the integral becomes

$$U = \frac{2\pi a}{x}\int_{|x-a|}^{|x+a|} \frac{r\,dr}{r} = \frac{2\pi a}{x}(|x+a| - |x-a|).$$

For $|x| > a$ we therefore have

$$U = \frac{4\pi a^2}{|x|},$$

and for $|x| < a$

$$U = 4\pi a.$$

Hence the potential at an external point is the same as if the whole mass $4\pi a^2$ were concentrated at the centre of the sphere. On the other hand, throughout the interior the potential is constant. At the surface of the sphere the potential is continuous; the expression for U is still defined (as an improper integral) and has the value $4\pi a$. The component of force F_x in the x-direction, however, has a jump of amount -4π at the surface of the sphere, for if $|x| > a$, we have

$$F_x = -\frac{4\pi a^2}{x^2},$$

while $F_x = 0$ if $|x| < a$.

The potential of a solid sphere of unit density is found from the above by multiplying by da and then integrating with respect to a. This gives the value

$$\frac{4\pi a^3}{3|x|}$$

for the potential at an external point, which is again the same as if the total mass $\frac{4}{3}\pi a^3$ were concentrated at the centre.

Examples

1. Find the position of the centre of mass of the curved surface of a right cone.

2. Find the co-ordinates of the centre of mass of the portion of the paraboloid

$$z^2 + y^2 = px$$

cut off by the plane $x = x_0$.

3*. A tube-surface is generated by a family of spheres of unit radius with their centres in the xy-plane. Let S be a portion of the surface lying above the xy-plane and Π the area of the projection of S on the xy-plane. Prove that the z-co-ordinate of the centre of mass of S is equal to Π/S.

4. Calculate the moment of inertia of the solid enclosed between the two cylinders

$$x^2 + y^2 = R \quad \text{and} \quad x^2 + y^2 = R' \qquad (R > R')$$

and the two planes $z = h$ and $z = -h$, with respect to (a) the z-axis, (b) the x-axis.

5. If A, B, C denote the moments of inertia of an arbitrary solid of positive density with respect to the x-, y-, z-axes, then the " triangle inequalities "

$$A + B > C, \quad A + C > B, \quad B + C > A$$

are satisfied.

6. Find the moment of inertia of the ellipsoid

$$\frac{x^2}{a^2} + \frac{y^2}{b^2} + \frac{z^2}{c^2} = 1$$

with respect to

(a) the z-axis,

(b) an arbitrary axis through the origin, given by

$$x : y : z = \alpha : \beta : \gamma \qquad (\alpha^2 + \beta^2 + \gamma^2 = 1).$$

7*. Find the envelopes of the planes with respect to which the ellipsoid

$$\frac{x^2}{a^2} + \frac{y^2}{b^2} + \frac{z^2}{c^2} = 1$$

has the same moment of inertia h.

8. Let O be an arbitrary point and S an arbitrary body. On every ray from O we take the point at the distance $1/\sqrt{I}$ from O, where I denotes the moment of inertia of S with respect to the straight line coinciding with the ray. Prove that the points so constructed form an ellipsoid (the so-called *momental ellipsoid*).

9. Find the momental ellipsoid of the ellipsoid

$$\frac{x^2}{a^2} + \frac{y^2}{b^2} + \frac{z^2}{c^2} = 1$$

at the point (ξ, η, ζ).

10. Find the co-ordinates of the centre of mass of the surface of the sphere $x^2 + y^2 + z^2 = 1$, the density being given by

$$\mu = \frac{1}{\sqrt{(x-1)^2 + y^2 + z^2}}.$$

11. Find the x-co-ordinate of the centre of mass of the octant of the ellipsoid

$$\frac{x^2}{a^2} + \frac{y^2}{b^2} + \frac{z^2}{c^2} = 1, \qquad x \geqq 0,\ y \geqq 0,\ z \geqq 0.$$

12. A system of masses S consists of two parts S_1 and S_2; I_1, I_2, I are the respective moments of inertia of S_1, S_2, S about three parallel axes passing through the respective centres of mass. Prove that

$$I = I_1 + I_2 + \frac{m_1 m_2}{m_1 + m_2} d^2,$$

where m_1 and m_2 are the masses of S_1 and S_2 and d the distance between the axes passing through their centres of mass.

13. Calculate the potential of the ellipsoid of revolution

$$\frac{x^2 + y^2}{a^2} + \frac{z^2}{b^2} = 1$$

at its centre $(b > a)$.

14. Calculate the potential of a solid of revolution

$$r = \sqrt{(x^2 + y^2)} \leqq f(z), \qquad a \leqq z \leqq b,$$

at the origin.

Appendix to Chapter IV

1. The Existence of the Multiple Integral

1. The Content of Plane Regions and Regions of Higher Dimensions.

In order to obtain the analytical proof of the existence of the multiple integral of a continuous function, we must begin with a study of the idea of *content*.

In Vol. I, Chap. V (p. 269) we saw how the content of a plane region can in general be expressed by an integral. Without making use of that fact, and without considering the existence of the area as guaranteed by intuition, we shall now proceed to give a general definition of the idea of " content " and investigate under what conditions this concept has a meaning.

We begin with a rectangle with sides parallel to the x- and y-axes, and define the area of such a rectangle as the product of the base and the altitude. If the given rectangle is subdivided into smaller rectangles by a number of parallels to the sides, it is clear from this definition that the area of the rectangle is equal to the sum of the areas of all the sub-rectangles. The area of a region which is composed of a finite number of rectangles * can now be defined as the sum of the areas of these rectangles.

The area thus defined is independent of the way in which the region is subdivided (or resolved) into rectangles. For if we are given two different resolutions, we can find a third resolution which is a finer subdivision of the two original ones. We do this by prolonging throughout the region all the lines which occur in either of the resolutions. These lines subdivide the two subdivisions into still smaller rectangles. The sum of the areas of these small rectangles is equal to the sum of the areas of the rectangles both of the first resolution and of the second resolution.

Now in order to define the area of an arbitrary bounded region B we form an inner approximation and an outer approximation to the region, that is, we find two regions B_i and B_e, each consisting of rectangles, the region B_i being entirely within B and

* Throughout this section the word rectangle will always be understood to mean a rectangle with sides parallel to the axes.

the exterior region B_e containing B. For this purpose we first enclose the region B in a large square. Then we divide this square into small rectangles by drawing parallels to the axes. Those rectangles having points in common with B together form a region B_e which encloses B; those rectangles which lie wholly within B form a region B_i which is contained in B.

We now wish to define the area $C(B)$ of B in such a way that for every choice of B_i and B_e the area of B lies between that of B_i and that of B_e:

$$C(B_i) \leqq C(B) \leqq C(B_e).$$

If we make the subdivisions finer, so that the diameters of the rectangles tend to zero, then the areas $C(B_i)$ form a monotonic increasing sequence and the areas $C(B_e)$ form a monotonic decreasing sequence. For to the regions B_i rectangles can only be added, and from B_e rectangles can only be removed. Therefore $C(B_i)$ has a limit and so has $C(B_e)$. *If these two limits are equal, we call this common limit the area of the region* B.

Under what conditions are the two limits, $C(B_i)$ and $C(B_e)$, equal? Of course the answer is, when the difference $C(B_e) - C(B_i)$ tends to zero as the fineness of the subdivisions increases. The region $B_e - B_i$ consists of those rectangles which have points in common with the boundary of B. Therefore if the area of this region $B_e - B_i$ tends to zero, it follows that the boundary of B can be enclosed in a region composed of rectangles and having as small an area as we please, namely in $B_e - B_i$. Conversely, if the boundary of B can be enclosed in the interior of a region S consisting of rectangles with a total area as small as we please, and if the subdivision is sufficiently fine, the rectangles $B_e - B_i$ will all lie in S; the area of $B_e - B_i$ will then be less than that of S, so that it tends to zero.

The result is as follows: *the limits of* C(B$_i$) *and* C(B$_e$) *are equal if, and only if, the boundary of* B *can be enclosed in a region consisting of rectangles of total area as small as we please. In this case our definition actually does assign a content* * *to* B.

* From the geometrical point of view it is somewhat unsatisfactory that in defining the content we have singled out a particular co-ordinate system. As a matter of fact, however, there is no difficulty in showing that the content is independent of the co-ordinate system, not only for two dimensions but also for n dimensions. We shall, however, omit this discussion here. For, on the one hand, it is not necessary for our particular purpose, which is the

In the next sub-section (p. 291) we shall prove the intuitively plausible fact that *every sectionally smooth continuous curve* (that is, every continuous curve which has a continuously turning tangent except at a finite number of points) *can be enclosed in a region formed from rectangles, whose area is as small as we please.* The condition is therefore satisfied whenever the region B consists of a finite number of parts, each bounded by a finite number of sectionally smooth curves. Such regions have a unique area; others do not arise in practical applications.

We shall show on p. 292 that if a region B is subdivided by sectionally smooth curves the sum of the contents of the sub-regions is equal to the content of the whole region B. Here we shall merely show that the present definition of area agrees with the integral formulæ obtained previously.

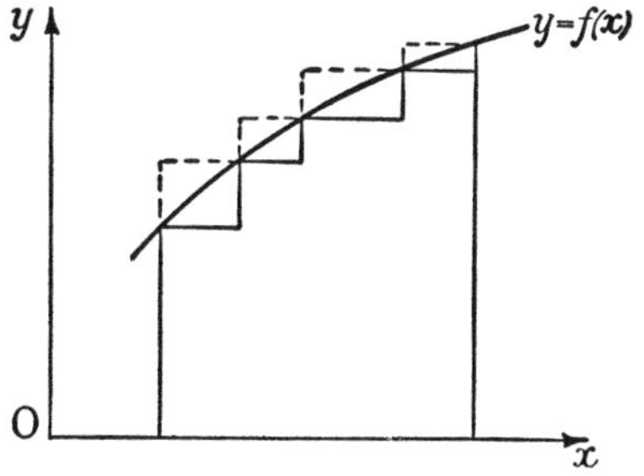

Fig. 15.—Approximation to a region by sets of rectangles

We begin by considering a region B bounded by the x-axis, the lines $x = a$, $x = b$, and a curve $y = f(x)$. For the regions B_e and B_i, respectively contained in and containing B, we can take the regions composed of rectangles shown in fig. 15 (the one by dotted lines and the other by continuous lines). According to the definition of a simple integral in Vol. I, Chap. II, section 1 (p. 78), the areas B_i and B_e are respectively an upper sum $\overline{F}_n$ and a lower sum $\underline{F}_n$ for the integral $\int_a^b y\,dx$. In addition to our formula

$$C(B_i) \leqq C(B) \leqq C(B_e)$$

proof of the existence of the double integral; and, on the other hand, the fact that the content is independent of the co-ordinate system follows immediately when we represent the content by a multiple integral and recall that the transformation formula shows that the value of this integral is unchanged when new rectangular co-ordinates are introduced.

we accordingly have the further inequality

$$C(B_i) \leqq \int_a^b f(x)\,dx \leqq C(B_e),$$

by the definition of integral. Since $\lim C(B_i) = \lim C(B_e)$, it follows that $C(B) = \int_a^b f(x)\,dx$, in agreement with what we have said already.

In the case of an arbitrary region B, subdivision of the region by lines parallel to the axes shows that our definition of content agrees with the expression for the area:

$$\iint_B dx\,dy.$$

The present definition of the area can immediately be extended to three-dimensional regions, and in fact to regions in n dimensions. The content of a parallelepiped with sides parallel to the axes is defined as the product of the lengths of the three sides. We then extend the definition to regions composed of a finite number of such parallelepipeds. For an arbitrary region B we then find regions B_i composed of parallelepipeds and lying in B and similar regions B_e containing B. The definition of the content of the region B as the common limit of the content of B_e and that of B_i again has a meaning, provided that the boundary of the region B can be enclosed in a set of parallelepipeds of arbitrarily small total content. In the next sub-section (p. 292) we shall show that this can always be done for regions bounded by surfaces having sectionally continuous tangent planes. As before, we shall henceforth restrict ourselves to such regions. The word region is always to mean a bounded closed region whose boundary consists of a finite number of surfaces with sectionally continuous derivatives.

The volume of a cylinder with its axis in the direction of the z-axis and its base in the xy-plane is the product of the area of the base and the altitude. This is at once clear when the base is composed of rectangles with sides parallel to the axes. In the general case the cylinder can be enclosed between two cylinders whose bases are regions composed of rectangles and whose volumes differ from that of the given cylinder by arbitrarily small amounts. The theorem therefore holds for cylinders with any base. From this it follows as before that the double integral

$$\iint_B f(x, y)\,dx\,dy$$

gives the volume of a portion of space bounded above by the surface $z = f(x, y)$, below by the plane region B, and at the sides by the vertical lines by which the edge of the surface is projected into the boundary of B. Further, we see that the definition of volume for a general region in space R agrees with the integral expression

$$\iiint_B dx\,dy\,dz.$$

2. A Theorem on Smooth Arcs.

In discussing areas we used the theorem that a continuous curve with a continuously turning tangent at all but a finite number of points can always be enclosed in a region composed of rectangles with sides parallel to the axes and having an arbitrarily small total content. It is obviously sufficient to prove the theorem for the individual arcs with continuous tangents. Let such an arc be given by the equations

$$\begin{matrix} x = \phi(s) \\ y = \psi(s) \end{matrix} \quad a \leqq s \leqq b,$$

where the parameter s is the length of arc and $\phi(s)$ and $\psi(s)$ are continuously differentiable functions. Then

$$|\phi'(s)| \leqq 1,$$
$$|\psi'(s)| \leqq 1.$$

By the mean value theorem of the differential calculus, for any two values s and s_1 of s in the interval $a \leqq s \leqq b$ we have

$$|x - x_1| = |\phi(s) - \phi(s_1)| \leqq |s - s_1|,$$
$$|y - y_1| = |\psi(s) - \psi(s_1)| \leqq |s - s_1|.$$

If, therefore, we subdivide the curve into n arcs of length $\epsilon = (b - a)/n$ and denote the initial point of the ν-th arc by (x_ν, y_ν) and an arbitrary point of that arc by (x, y), we have

$$|x - x_\nu| \leqq \epsilon \quad \text{or} \quad x_\nu - \epsilon \leqq x \leqq x_\nu + \epsilon,$$
$$|y - y_\nu| \leqq \epsilon \quad \text{or} \quad y_\nu - \epsilon \leqq y \leqq y_\nu + \epsilon.$$

The points of the ν-th arc therefore all lie in a square with side

2ϵ and area $4\epsilon^2$. The whole curve is included in n such squares, whose total area is at most

$$4\epsilon^2 n = 4\epsilon(b - a).$$

This quantity can be made as small as we please by taking ϵ sufficiently small.

There is no difficulty in proving the corresponding theorem for surfaces in space defined by the equations

$$x = \phi(u, v)$$
$$y = \psi(u, v)$$
$$z = \chi(u, v),$$

where the functions ϕ, ψ, χ have sectionally continuous derivatives. It is found that every such surface can be enclosed in a region of arbitrarily small volume, consisting of a number of parallelepipeds.

A consequence of this theorem is that if a plane region R bounded by a sectionally smooth curve is subdivided into two subregions R', R'' which are separated by sectionally smooth arcs, the area of R is equal to the sum of the areas of R' and R''. For we can subdivide the plane by straight lines parallel to the coordinate axes and so close together that all the rectangles which have points in common with the boundary of R or with the arcs separating R' and R'' have an arbitrarily small total area. As before, we define R_e as the region consisting of all rectangles having points in common with R, and R_i as the region consisting of all rectangles entirely within R; the regions R_e', R_i', R_e'', R_i'' are similarly defined. The regions R_e' and R_e'' together cover R_e, some rectangles being counted twice; hence $C(R_e') + C(R_e'') \geqq C(R_e) \geqq C(R)$. Again, R_i' and R_i'' are contained in R_i, and are completely separate; hence $C(R) \geqq C(R_i) \geqq C(R_i') + C(R_i'')$. Since $C(R_e')$ and $C(R_e'')$ can be made to approximate as closely as we desire to $C(R')$ and $C(R'')$ by making the subdivision fine enough, the first of these inequalities gives $C(R') + C(R'') \geqq C(R)$; the second similarly gives $C(R') + C(R'') \leqq C(R)$. Taken together, these inequalities prove our statement.

It is clear that this addition theorem still holds when the region R is subdivided into any finite number of regions $R^{(1)}$, $R^{(2)}$, . . . , $R^{(n)}$. The extension to more than two dimensions follows the same lines and offers no difficulty at all.

3. The Existence of the Multiple Integral of a Continuous Function.

Let the function $f(x, y)$ be continuous in the interior and on the boundary of a region R. We wish to show that as the diameters of the sub-regions R_ν tend to zero the upper and lower sums $\Sigma m_\nu \Delta R_\nu$, $\Sigma M_\nu \Delta R_\nu$ (defined in Chap. IV, section 2, p. 224) tend to a common limit which is independent of the mode of subdivision. The proof is essentially the same as the corresponding proof in Vol. I, Chap. II, Appendix (p. 131), and can therefore be given quite briefly here.

We first suppose that the subdivision of R into sub-regions R_ν is effected by polygonal paths. We choose the maximum diameter δ of the sub-regions R_ν so small that for every two points whose distance apart is less than δ the values of the function differ by less than ϵ. Then in each of these regions we have

$$M_\nu - m_\nu < \epsilon.$$

Thus for the difference between the upper sum and the lower sum we have

$$\Sigma M_\nu \Delta R_\nu - \Sigma m_\nu \Delta R_\nu < \Sigma \epsilon \Delta R_\nu = \epsilon C(R).$$

Every subdivision obtained by subdividing the given subdivision further obviously has a lower sum which is between the upper and lower sums of the original subdivision.

The proof is complete once we show that for every two subdivisions of R into sub-regions with diameters less than δ the corresponding upper and lower sums of the two subdivisions differ from one another by as little as we please, provided only that δ is chosen sufficiently small.

If we are given a second subdivision into sub-regions R_ν' which have diameters less than δ, then in this subdivision also the upper and lower sums will differ by less than $\epsilon C(R)$:

$$\Sigma M_\nu' \Delta R_\nu' - \Sigma m_\nu' \Delta R_\nu' < \epsilon C(R).$$

The two subdivisions together define a new subdivision which is a further subdivision of each of the two and which is obtained by collecting the common points of each pair of regions R_ν and R_μ' (if such points exist) into a region $R_{\nu\mu}''$. By the previous remark, the lower sum of this third subdivision is not smaller than the lower sum of the two original subdivisions, and differs from each of them by less than $\epsilon C(R)$. Therefore the lower sums $\Sigma m_\nu \Delta R_\nu$ and $\Sigma m_\nu' \Delta R_\nu'$ differ by less than $2\epsilon C(R)$. If we now let ϵ tend to zero, it follows from Cauchy's test that the lower sums have a limit independent of the mode of subdivision. Since we have already seen that the upper sums differ from the lower sums by as little as we please, the upper sums have the same limit. This proves the existence of the double integral $\iint_R f(x, y)\, dS$ for polygonal subdivisions of R.

We made this assumption in order to be sure that a common subdivision into a finite number of regions $R_{\nu\mu}''$ really exists. If, for example, the boundaries of the sub-regions are curves, and a portion of a boundary curve in one subdivision consists of the line $x = 0$ and a portion of a boundary in the other consists of the curve $x^2 \sin \frac{1}{x} = y$, then the common subdivision will have an infinite number of cells in the neighbourhood of $x = 0$. We can, however, easily get rid of this assumption of polygonal subdivision. For by p. 291 we can replace every curvilinear subdivision by a polygonal subdivision such that the total difference of the areas, and hence the difference of the corresponding lower sums, is arbitrarily small. This obviously reduces the case of sub-regions of arbitrary boundary to the special case already discussed.

The proof is clearly independent of the number of dimensions.

The corollaries on the existence of the double integral stated in Chap. IV, section 2 (p. 225) follow immediately from the approximation formula developed there and require no further proof here.

2. General Formula for the Area (or Volume) of a Region bounded by Segments of Straight Lines or Plane Areas (Guldin's Formula). The Polar Planimeter.

The transformations on pp. 299–300 enable us to give a simple proof of the following theorems:

If a straight-line segment S of constant or variable length l is in motion in a plane, and if t represents time, then the area swept out by the moving segment is

$$A = \int_{t_0}^{t_1} l(t) \frac{dn}{dt} dt,$$

where t_0 and t_1 correspond to the initial and final positions of the segment S, and dn/dt is the component of the velocity of the mean centre of S in the direction perpendicular to S.

Again, the volume V swept out by a moving plane area P of area A is

$$V = \int_{t_0}^{t_1} A(t) \frac{dn}{dt} dt,$$

where dn/dt is the component velocity of the mean centre of the area A perpendicular to the plane of P.

Both in these formulæ and in the proofs, we assume to begin with that the moving segment S or plane area A passes once and

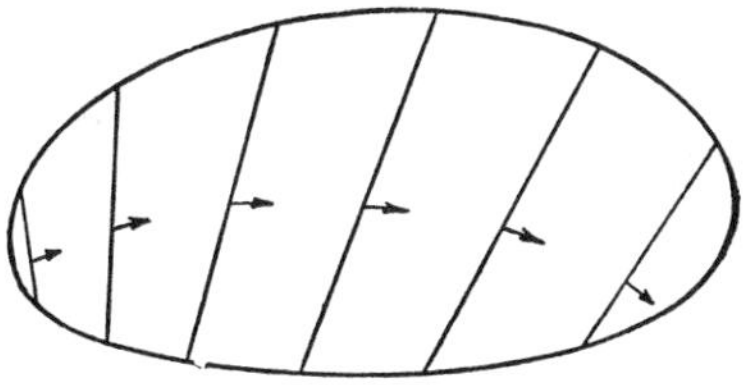

Fig. 16

once only through each point of the region swept out (see fig. 16).

We first give the proof in the case of a segment moving in a plane. The generating segment must be represented by an equation of the form

$$\alpha(t)x + \beta(t)y + \gamma(t) = 0, \qquad \alpha^2 + \beta^2 = 1, \qquad (a)$$

or else in the form obtained by solving this equation for the variable t:

$$t = \phi(x, y).$$

We first carry out the transformation of $A = \iint dx\, dy$ by means of the formula on p. 299 for the special case

$$f(x, y) = 1.$$

Denoting by ds the line element taken along the segment S, we obtain the expression

$$A = \int_{t_0}^{t_1} dt \int_s \frac{ds}{|\operatorname{grad} \phi|}$$

for the area. It is easy to see, by substituting $t = \phi(x, y)$ in formula (a) and differentiating with respect to x and y, that

$$\left|\frac{1}{\operatorname{grad}\phi}\right| = \pm(\alpha' x + \beta' y + \gamma').$$

Hence the area is given by

$$\pm A = \int_{t_0}^{t_1} dt \int_S (\alpha' x + \beta' y + \gamma')\, ds.$$

Here α', β', γ' denote the derivatives of α, β, γ with respect to t. The integration with respect to s is to be taken along the segment S.

The single integral with respect to s is equal to

$$l(t)\,(\alpha' X + \beta' Y + \gamma'),$$

where (X, Y) are the co-ordinates of the mean centre of S. But X and Y satisfy the equation $\alpha X + \beta Y + \gamma = 0$. On differentiating this equation with respect to t, we obtain

$$\alpha' X + \beta' Y + \gamma' + \alpha X' + \beta Y' = 0.$$

Thus

$$-(\alpha' X + \beta' Y + \gamma') = \alpha X' + \beta Y'.$$

Here α, β are the components of the unit vector perpendicular to the segment S, and X', Y' are the velocity components of the mean centre at the time t. The expression $\alpha' X + \beta' Y + \gamma'$ is thus equal to the velocity of the mean centre perpendicular to S. This proves our formula.

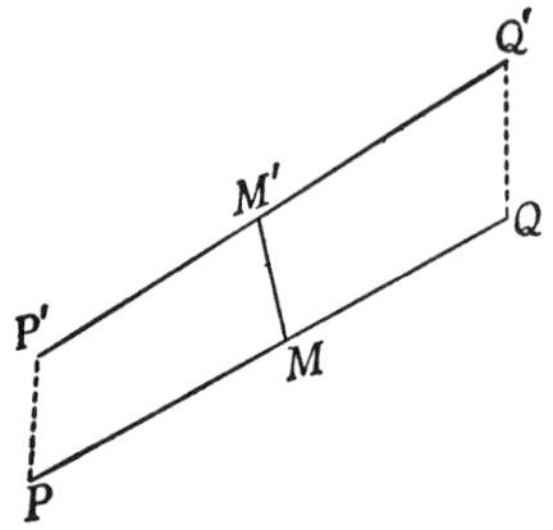

Fig. 17

This result can be shown to be intuitively plausible by the following argument. We consider two neighbouring positions of the segment S, PQ and $P'Q'$, say (fig. 17). These two segments determine an area which is given approximately by the product of the length PQ of S and the distance $M'M$ of the mean centre of one segment from that of the other. The error in this approximation is of higher order than that of the increment of time δt corresponding to the displacement. It would be an instructive example for the reader to try to fill in the details of this geometrical argument and provide a strict proof.

The corresponding theorem in three dimensions can be proved in the same way by the use of the transformation formulæ for volume integrals given on p. 300. There is no need to go through the proof here.

In the special case of a plane region which is rotated about an axis while retaining its original size and shape, we have the problem already considered in Vol. I (Chap. V, p. 285), where Guldin's rule for the volume of a solid of revolution was given.

Our formulæ associate a definite sign with the area of the region swept out. In the two-dimensional case the sign depends on which of the two directions normal to S is regarded as positive. (The same is true in three dimensions.) The area obtained is positive if the segment S, as it passes through any point, moves in the direction of the positive normal; otherwise it is negative.

These observations allow us to extend our results to cases in which the segment or plane area does not always move in the same sense, or covers part of the plane (or space) more than once. The integrals given above will then express the algebraic sum of the areas (or volumes) of the parts of the region described, each taken with the appropriate sign. We leave it to the reader to work out how this may be taken account of in practice.

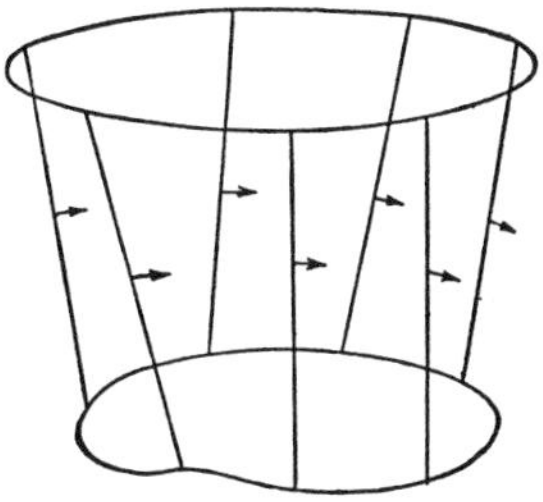

Fig. 18

As an example, let a segment of constant length move so as to have its end-points always on two fixed curves C and C' in a plane, as in fig. 18. From the arrows showing the positive direction of the normal we can determine the sign with which each area appears in the integral, and we find that the integral gives the difference between the areas enclosed by C and C'. If C' contains zero area, as when it degenerates into a single segment of a curve, multiply-described, the integral gives the area enclosed by C.

This principle is used in the construction of the well-known polar planimeter (Amsler's planimeter). This is a mechanical apparatus for measuring plane areas. It consists of a rigid rod at the centre of which is a measuring-wheel which can roll on the drawing-paper. The plane of the wheel is perpendicular to the rod. When the instrument is to be used to measure the area enclosed by a curve C drawn on the paper, one end of the rod is moved round the curve, while the other is connected to a fixed point O, the pole, by means of a rigid member jointed to it. This end of the rod therefore describes (multiply) an arc of a circle: that is, a closed curve containing zero area. It follows that the normal motion of the mean centre of the rod gives the area enclosed by C, apart from the constant multiplier l. But this normal component is proportional to the angle through which the measuring-wheel turns, provided that the circumference of the wheel moves on the paper as the rod moves, in which case the position of the wheel is only affected by the motion normal to the rod.

In the instrument as usually constructed the wheel is not exactly at the centre of the rod, but this only alters the factor of proportionality in the result, and the factor can be determined directly by a calibration of the instrument.

EXAMPLE

Let S be a tube-surface (cf. p. 182) generated by a family of unit spheres whose centres lie on a closed curve C in the xy-plane. Prove that the volume enclosed by S is π times the length of C.

3. Volumes and Areas in Space of any Number of Dimensions

1. Resolution of Multiple Integrals.

If the region R of the xy-plane is covered by a family of curves $\phi(x, y) = \text{const.}$ in such a way that through each point of R there passes one, and only one, curve of the family, we can take the quantity $\phi(x, y) = \xi$ as a new independent variable, that is, we can take the curves represented by $\phi(x, y) = \text{const.}$ as a family of parametric curves.

For the second independent variable we can take the quantity $\eta = y$, provided that we restrict ourselves to a region R in which the curves $\phi(x, y) = \text{const.}$ and $y = \text{const.}$ determine the points uniquely.

If we introduce these new variables, a double integral $\iint_R f(x, y)\,dx\,dy$ is transformed as follows:

$$\iint f(x, y)\,dx\,dy = \iint \frac{f(x, y)}{\phi_x}\,d\xi\,d\eta.$$

If we keep ξ constant and integrate the right-hand side with respect to η, the integral with respect to η can be written in the form

$$\int \frac{f(x, y)}{\sqrt{(\phi_x^2 + \phi_y^2)}} \frac{\sqrt{(\phi_x^2 + \phi_y^2)}}{\phi_x}\,d\eta.$$

Since

$$\frac{ds}{d\eta} = \frac{\sqrt{(\phi_x^2 + \phi_y^2)}}{\phi_x}$$

this integral may be regarded as an integral along the curve $\phi(x, y) = \xi$, the length of arc s being the variable of integration. Thus we obtain the resolution

$$\iint f(x, y)\,dx\,dy = \int d\xi \int \frac{f(x, y)}{\sqrt{(\phi_x^2 + \phi_y^2)}}\,ds$$

for our double integral. The intuitive meaning of this resolution is very easily recognized if we suppose that corresponding to the curves $\phi(x, y) = \text{const.}$ there is a family of orthogonal curves which intersect each separate curve $\phi = \text{const.}$ at right angles, in the direction of the vector grad ϕ. If the orthogonal curves are represented by the functions $x(\sigma)$ and $y(\sigma)$, where σ is the length of arc on them, then

$$\frac{dx}{d\sigma} = \frac{\phi_x}{\sqrt{(\phi_x^2 + \phi_y^2)}}, \quad \frac{dy}{d\sigma} = \frac{\phi_y}{\sqrt{(\phi_x^2 + \phi_y^2)}}.$$

Since

$$\frac{d\xi}{d\sigma} = \phi_x \frac{dx}{d\sigma} + \phi_y \frac{dy}{d\sigma},$$

we obtain

$$\frac{d\xi}{d\sigma} = \sqrt{(\phi_x^2 + \phi_y^2)} = \sqrt{(\text{grad } \phi)^2}.$$

We now consider the quadrilateral mesh bounded by two curves $\phi(x, y) = \xi$, $\phi(x, y) = \xi + \Delta\xi$, and two orthogonal curves which cut off a portion of length Δs from $\phi(x, y) = \xi$. The area of this mesh is given approximately by the product $\Delta s \Delta\sigma$, and this in turn is approximately equal to

$$\frac{\Delta s \Delta \xi}{\sqrt{(\phi_x^2 + \phi_y^2)}}.$$

The transformation of the double integral,

$$\iint f(x, y)\, dx\, dy = \iint f(x, y) \frac{ds\, d\xi}{\sqrt{(\phi_x^2 + \phi_y^2)}},$$

simply means this: *instead of calculating the double integral by subdividing the region into small squares, we may use a subdivision determined by the curves* $\phi(\mathrm{x}, \mathrm{y}) = $ *const. and their orthogonal curves.*

A similar resolution can be effected in three-dimensional space. If the region R is covered by a family of surfaces $\phi(x, y, z) = \text{const.}$ in such a way that through every point there passes one, and only one, surface, then we can take the quantity $\xi = \phi(x, y, z)$ as a variable of integration. In this way we resolve a triple integral

$$\iiint_R f(x, y, z)\, dx\, dy\, dz$$
$$= \int d\xi \iint \frac{f(x, y, z)}{\sqrt{(\phi_x^2 + \phi_y^2 + \phi_z^2)}} \frac{\sqrt{(\phi_x^2 + \phi_y^2 + \phi_z^2)}}{\phi_x}\, dy\, dz$$

into an integration

$$\iint \frac{f(x, y, z)}{\sqrt{(\phi_x^2 + \phi_y^2 + \phi_z^2)}}\, d\sigma$$

over the surface $\phi = \xi$ and a subsequent integration with respect to ξ:

$$\iiint f(x, y, z)\, dx\, dy\, dz = \int d\xi \iint \frac{f(x, y, z)}{\sqrt{(\phi_x^2 + \phi_y^2 + \phi_z^2)}}\, d\sigma.$$

2. Areas of Surfaces and Integration over Surfaces in more than Three Dimensions.

In n-dimensional space, that is, in the region of sets of values with n co-ordinates, an $(n - 1)$-dimensional surface is defined by an equation

$$\phi(x_1, x_2, \ldots, x_n) = \text{const.}$$

We suppose that a portion of this surface corresponds to a certain region B of the variables $x_1, x_2, \ldots, x_{n-1}$, where x_n is to be calculated from the equation $\phi(x_1, x_2, \ldots, x_n) = \text{const.}$

We now define the area of this portion of surface as the absolute value of the integral

$$A = \int\int \cdots \int_B \frac{\sqrt{(\phi_{x_1}^2 + \phi_{x_2}^2 + \ldots + \phi_{x_n}^2)}}{\phi_{x_n}} dx_1 dx_2 \ldots dx_{n-1}.$$

In the first instance this definition is only a formal generalization of the formulæ for the area obtained by intuition in the case of three dimensions. Nevertheless, it has a certain justification in the fact that the quantity A is independent of the choice of the co-ordinate x_n. This may be proved in the same way as for the three-dimensional case (cf. Chap. IV, section 6, p. 271).

The integral of a function $f(x_1, x_2, \ldots, x_n)$ over this $(n-1)$-dimensional surface we define as

$$\int\int \cdots \int f(x_1, x_2, \ldots, x_n)\, d\sigma$$

$$= \int\int \cdots \int_B f(x_1, x_2, \ldots, x_n) \frac{\sqrt{(\phi_{x_1}^2 + \ldots + \phi_{x_n}^2)}}{\phi_{x_n}} dx_1 dx_2 \ldots dx_{n-1},$$

where, as before, we suppose that x_n is expressed in terms of $x_1, \ldots, x_{n-1}$ by means of the equation $\phi(x_1, \ldots, x_n) = \text{const.}$ We again find that the expression is independent of the choice of the variable x_n.

As in the case of two or three dimensions, a multiple integral over an n-dimensional region R,

$$\int\int \cdots \int_R f(x_1, \ldots, x_n)\, dx_1 \ldots dx_n,$$

can be resolved as on p. 300. We assume that the region R is covered by a family of surfaces

$$\phi(x_1, x_2, \ldots, x_n) = \text{const.}$$

in such a way that through each point $(x_1, \ldots, x_n)$ of R there passes one, and only one, surface. If instead of $x_1, \ldots, x_{n-1}, x_n$, we introduce

$$x_1, \ldots, x_{n-1},\ \xi = \phi(x_1, \ldots, x_n),$$

as independent variables, the multiple integral becomes

$$\int d\xi \int \cdots \int \frac{f(x_1, \ldots, x_n)}{\sqrt{(\phi_{x_1}{}^2 + \ldots + \phi_{x_n}{}^2)}} \frac{\sqrt{(\phi_{x_1}{}^2 + \ldots + \phi_{x_n}{}^2)}}{\phi_{x_n}} dx_1 \ldots dx_{n-1}$$
$$= \int d\xi \int \cdots \int \frac{f(x_1, \ldots, x_n)}{\sqrt{(\phi_{x_1}{}^2 + \ldots + \phi_{x_n}{}^2)}} d\sigma.$$

3. Area and Volume of the n-Dimensional Unit Sphere.

As an example we shall calculate the area and volume of the sphere in n-dimensional space, that is, the area of the $(n - 1)$-dimensional surface determined by the equation

$$x_1{}^2 + \ldots + x_n{}^2 = R^2$$

and the volume interior to the $(n - 1)$-dimensional surface, which is the volume given by the inequality

$$x_1{}^2 + \ldots + x_n{}^2 \leqq R^2.$$

Let a continuous function $f(r)$ of $r = \sqrt{(x_1{}^2 + \ldots + x_n{}^2)}$ be given inside the sphere. We shall first find the multiple integral $\int \cdots \int \int f(r)\, dx_1 \ldots dx_n$ over the sphere $x_1{}^2 + \ldots + x_n{}^2 \leqq R^2$. We introduce the new variable

$$r^2 = \phi(x_1, \ldots, x_n) = x_1{}^2 + \ldots + x_n{}^2,$$

and in virtue of the relations

$$\sqrt{(\phi_{x_1}{}^2 + \ldots + \phi_{x_n}{}^2)} = 2r,$$
$$d(r^2) = 2r\, dr$$

we obtain the resolution

$$\int \int \cdots \int f(r)\, dx_1 \ldots dx_n = \int_0^R f(r)\, dr \int \cdots \int d\sigma = \int_0^R f(r) \Omega_n(r)\, dr,$$

where $\Omega_n(r)$ is the area of the sphere $x_1{}^2 + \ldots + x_n{}^2 = r^2$.

According to our general definition, the area of a hemisphere of radius r is given by the integral

$$\frac{1}{2} \Omega_n(r) = r \int \cdots \int \frac{dx_1 \ldots dx_{n-1}}{x_n},$$

where the integration is extended throughout the interior of the $(n - 1)$-dimensional sphere

$$x_1{}^2 + \ldots + x_{n-1}{}^2 \leqq r^2.$$

If instead of the variables x_ν we introduce the quantities

$$\xi_\nu = \frac{x_\nu}{r}; \quad \sum_1^n \xi_\nu^2 = 1,$$

we obtain

$$\Omega_n(r) = 2r^{n-1}\int \ldots \int \frac{d\xi_1 \ldots d\xi_{n-1}}{\xi_n} = r^{n-1}\omega_n,$$

where we denote the area of the unit sphere $\xi_1^2 + \ldots + \xi_n^2 = 1$ by

$$\omega_n = 2\int\int \ldots \int \frac{d\xi_1 \ldots d\xi_{n-1}}{\xi_n}.$$

Then it follows that

$$\int\int \ldots \int f(r)\, dx_1 \ldots dx_n = \omega_n \int_0^R f(r) r^{n-1}\, dr.$$

We can now calculate ω_n conveniently from this formula; we extend the integration on the left throughout the whole $x_1 x_2 \ldots x_n$-space (i.e. we let R increase beyond all bounds) and for $f(r)$ we choose a function for which both the n-tuple integral on the left and the single integral on the right can be explicitly evaluated. Such a function is

$$f(r) = e^{-(x_1^2 + x_2^2 + \ldots + x_n^2)} = e^{-r^2}.$$

With this function the equation takes the form

$$\left(\int_{-\infty}^{\infty} e^{-x^2}\, dx\right)^n = \omega_n \int_0^{\infty} e^{-r^2} r^{n-1}\, ds.$$

Since

$$\int_{-\infty}^{\infty} e^{-x^2}\, dx = \sqrt{\pi}$$

(p. 262) and

$$\int_0^{\infty} e^{-r^2} r^{n-1}\, dr = \frac{1}{2}\Gamma\left(\frac{n}{2}\right)$$

(p. 324), we obtain

$$\omega_n = \frac{2(\sqrt{\pi})^n}{\Gamma(n/2)}.$$

Here $\Gamma\left(\frac{n}{2}\right)$ means the elementary expression $\left(\frac{n-2}{2}\right)!$ if n

is even and $\frac{(n-2)(n-4)\ldots 1}{2^{(n-1)/2}}\sqrt{\pi}$ if n is odd. For the general definition of the gamma function, see Vol. I, p. 250, and pp. 323–5 of the present volume. In order to find the volume of the n-dimensional unit sphere we now put $f(r)=1$ and obtain

$$v_n = \int\cdots\int\int dx_1\,dx_2\ldots dx_n = \omega_n\int_0^1 r^{n-1}\,dr = \frac{\omega_n}{n};$$

hence

$$v_n = \frac{(\sqrt{\pi})^n}{\Gamma((n+2)/2)}.$$

4. Generalizations. Parametric Representations.

In n-dimensional space we can consider an r-dimensional manifold for any $r \leqq n$ and seek to define its content. For this purpose a parametric representation is advantageous. Let the r-dimensional manifold be given by the equations

$$\begin{aligned} x_1 &= \phi_1(u_1, \ldots, u_r) \\ &\cdots\cdots\cdots \\ x_n &= \phi_n(u_1, \ldots, u_r), \end{aligned}$$

where the functions ϕ_ν possess continuous derivatives in a region B of the variables $(u_1, \ldots, u_r)$. As the variables $u_1, \ldots, u_r$ range over this region, the point $(x_1, \ldots, x_n)$ describes an r-dimensional surface.

From the rectangular array

$$\begin{bmatrix} \frac{\partial x_1}{\partial u_1} & \frac{\partial x_2}{\partial u_1} & \cdots & \frac{\partial x_n}{\partial u_1} \\ \frac{\partial x_1}{\partial u_2} & \frac{\partial x_2}{\partial u_2} & \cdots & \frac{\partial x_n}{\partial u_2} \\ \cdot & \cdot & \cdots & \cdot \\ \frac{\partial x_1}{\partial u_r} & \frac{\partial x_2}{\partial u_r} & \cdots & \frac{\partial x_n}{\partial u_r} \end{bmatrix}$$

we now form all possible r-rowed determinants

$$D_\nu \left(\nu = 1, \ldots, \binom{n}{r}\right),$$

the first of which, for example, is the determinant

$$\begin{vmatrix} \frac{\partial x_1}{\partial u_1} & \frac{\partial x_2}{\partial u_1} & \cdots & \frac{\partial x_r}{\partial u_1} \\ \frac{\partial x_1}{\partial u_2} & \frac{\partial x_2}{\partial u_2} & \cdots & \frac{\partial x_r}{\partial u_2} \\ \cdot & \cdot & \cdot & \cdot \\ \frac{\partial x_1}{\partial u_r} & \frac{\partial x_2}{\partial u_r} & \cdots & \frac{\partial x_r}{\partial u_r} \end{vmatrix}.$$

The content of the r-dimensional surface is then given by the integral

$$\int \ldots \int \sqrt{D_1^2 + D_2^2 + \ldots + D_{\binom{n}{r}}^2}\, du_1 \ldots du_r.$$

By means of the theorem on the transformation of multiple integrals (Chap. IV, section 4, p. 254), and simple calculations with determinants which we shall omit here, we can prove that this expression for the content remains unchanged if we replace the parameters $u_1, \ldots, u_r$ by other parameters. We likewise see that in the case $r = 1$ this reduces to the usual formula for the length of arc, and in the case $r = 2$ in a space of three dimensions it becomes the formula for the area.

We shall give a proof for the case $r = n - 1$, where n is arbitrary; i.e. we shall prove the following theorem: If $\phi(x_1, \ldots, x_n) = 0$ is an arbitrary $(n - 1)$-dimensional portion of surface in n-dimensional space, and if this portion can also be represented parametrically by the equations

$$x_i = \psi_i(u_1, \ldots, u_{n-1}), \qquad (i = 1, \ldots, n),$$

then its area is given by

$$A = \int \ldots \int \sqrt{D_1^2 + \ldots + D_n^2}\, du_1 \ldots du_{n-1},$$

where D_i is a Jacobian of $(n - 1)$ rows:

$$D_i = \frac{\partial(x_1, \ldots, x_{i-1}, x_{i+1}, \ldots, x_n)}{\partial(u_1, \ldots, u_{n-1})} = 1 \Big/ \frac{\partial(u_1, \ldots, u_{n-1})}{\partial(x_1, \ldots, x_{i-1}, x_{i+1}, \ldots, x_n)}.$$

Here, as always, we assume the existence and continuity of all the derivatives involved.

Without loss of generality we may assume that $\phi_{x_n} \neq 0$. Then, since by p. 301 A is given by

$$A = \int \cdots \int \frac{|\operatorname{grad}\phi|}{\phi_{x_n}} dx_1 \ldots dx_{n-1},$$

we have only to show that

$$\frac{1}{\phi_{x_n}} |\operatorname{grad}\phi|\, dx_1 \ldots dx_{n-1} = \sqrt{\sum_i D_i^2}\, du_1 \ldots du_{n-1},$$

or

$$|\operatorname{grad}\phi|^2 = \phi_{x_n}^2 (\sum_i D_i^2) \left(\frac{\partial(u_1, \ldots, u_{n-1})}{\partial(x_1, \ldots, x_{n-1})}\right)^2 = \frac{\phi_{x_n}^2}{D_n^2} \sum_i D_i^2.$$

Now from the properties of Jacobians we have

$$\frac{D_\nu}{D_n} = \frac{\partial(x_1, \ldots, x_{\nu-1}, x_{\nu+1}, \ldots, x_n)}{\partial(u_1, \ldots, u_{n-1})} \Big/ \frac{\partial(x_1, \ldots, x_{n-1})}{\partial(u_1, \ldots, u_{n-1})}$$

$$= \frac{\partial(x_1, \ldots, x_{\nu-1}, x_{\nu+1}, \ldots, x_n)}{\partial(x_1, \ldots, x_{n-1})}.$$

This last Jacobian corresponds to the introduction of $(x_1, \ldots, x_{\nu-1}, x_{\nu+1}, \ldots, x_n)$ instead of $(x_1, \ldots, x_{n-1})$ as independent variables. But as the partial derivatives $\frac{\partial x_n}{\partial x_i}$ are obtained from the equations

$$\phi_{x_n} \frac{\partial x_n}{\partial x_i} + \phi_{x_n} = 0, \quad (i = 1, \ldots, n-1),$$

we have $\frac{D_\nu}{D_n} = \pm \frac{\phi_{x_\nu}}{\phi_{x_n}}$. Hence

$$\frac{D_\nu^2}{D_n^2} = \frac{\phi_{x_\nu}^2}{\phi_{x_n}^2},$$

which proves the formula for A.

It may be mentioned here that the expression $\sum_i D_i^2$ may be represented as a determinant of $(n-1)$ rows,

$$\sum_{i=1}^{n} D_i^2 = |x_{u_i} x_{u_k}| = \begin{vmatrix} x_{u_1}^2 & x_{u_1} x_{u_2} & \ldots & x_{u_1} x_{u_{n-1}} \\ \ldots & \ldots & \ldots & \ldots \\ x_{u_{n-1}} x_{u_1} & \ldots & \ldots & x_{u_{n-1}}^2 \end{vmatrix} = G,$$

so that

$$A = \int \ldots \int \sqrt{G}\, du_1 \ldots du_{n-1}.$$

Here the elements of the determinant are the inner products of the vectors $x_{u_i} = \left(\frac{\partial x_1}{\partial u_i}, \ldots, \frac{\partial x_n}{\partial u_i}\right)$ and $x_{u_k} = \left(\frac{\partial x_1}{\partial u_k}, \ldots, \frac{\partial x_n}{\partial u_k}\right)$, i.e. the expressions

$$\sum_l \frac{\partial \phi_l}{\partial u_i} \frac{\partial \phi_l}{\partial u_k}.$$

Examples

1. Calculate the volume of the n-dimensional ellipsoid

$$\frac{x_1^2}{a_1^2} + \ldots + \frac{x_n^2}{a_n^2} = 1.$$

2. Express the integral I of a function of x_1, depending on x_1 alone, over the unit sphere $x_1^2 + \ldots + x_n^2 = 1$ in n-dimensional space, as a single integral.

4. Improper Integrals as Functions of a Parameter

1. Uniform Convergence. Continuous Dependence on the Parameter.

Improper integrals frequently appear as functions of a parameter; thus e.g. the integral of the general power

$$\int_0^1 y^x \, dy = \frac{1}{x+1}$$

in the interval $-1 < x < 0$ is an improper integral.

We have seen that an integral over a finite interval is continuous when regarded as a function of a parameter, provided that the integrand is continuous. In the case of an infinite interval, however, the situation is not so simple. Let us consider e.g. the integral

$$F(x) = \int_0^\infty \frac{\sin xy}{y} dy.$$

According as $x > 0$ or $x < 0$, this is transformed by the substitution $xy = z$ into

$$\int_0^\infty \frac{\sin z}{z} dz \quad \text{or} \quad \int_0^{-\infty} \frac{\sin z}{z} dz = -\int_0^\infty \frac{\sin z}{z} dz.$$

The integral $\int_0^\infty \frac{\sin z}{z}\,dz$ converges, as we have seen in Vol. I (pp. 252, 418), and in fact it has the value $\pi/2$ (Vol. I, p. 450, and p. 315 below). Thus in spite of the fact that the function $(\sin xy)/y$, regarded as a function of x and y, is continuous everywhere and its integral converges for every value of x, the function $F(x)$ is discontinuous; it is equal to $\pi/2$ for positive values of x, to $-\pi/2$ for negative values of x, and to zero for $x = 0$.

In itself this fact is not at all surprising, for it is analogous to the situation which we have already met with in the case of infinite series (Vol. I, Chap. VIII, p. 394), and we must remember that the process of integration is a generalized summation. In the case of an infinite series of continuous functions we required, if we were to be sure that the series represented a continuous function, that the convergence should be *uniform.* Here, in the case of convergent integrals depending on a parameter, we shall again have to introduce the concept of uniform convergence.

We say that *the convergent integral*

$$F(x) = \int_0^\infty f(x, y)\,dy$$

converges uniformly (in x) *in the interval* $\alpha \leqq x \leqq \beta$, *provided that the "remainder" of the integral can be made arbitrarily small, simultaneously for all values of* x *in the interval under consideration*; more precisely: provided that for a given positive number ϵ there is a positive number $A = A(\epsilon)$, which does *not* depend on x and is such that whenever $B \geqq A$

$$\left| \int_B^\infty f(x, y)\,dy \right| < \epsilon.$$

As a useful test we mention the fact that *the integral* $\int_0^\infty f(x, y)\,dy$ *converges uniformly (and absolutely) if from a point* $y = y_0$ *onward the relation*

$$|f(x, y)| < \frac{M}{y^a}$$

holds, where M *is a positive constant and* $a > 1$. For in this case

$$\left|\int_B^\infty f(x, y)\,dy\right| < M\int_B^\infty \frac{dy}{y^a} = M\frac{1}{(a-1)B^{a-1}} \leq M\frac{1}{(a-1)A^{a-1}};$$

the right-hand side can be made as small as we please by choosing A sufficiently large, and it is independent of x. This is a straightforward analogue of the test for the uniform convergence of series given in Vol. I, p. 392.

We readily see that *a uniformly convergent integral of a continuous function is itself a continuous function.* For if we choose a number A such that

$$\left|\int_A^\infty f(x, y)\,dy\right| < \epsilon$$

for all values of x in the interval under consideration, we have

$$\left|F(x+h) - F(x)\right| < \left|\int_0^A \{f(x+h, y) - f(x, y)\}dy\right| + 2\epsilon.$$

In virtue of the continuity of the function $f(x, y)$ we can choose h so small that the finite integral on the right is less than ϵ, which proves the continuity of the integral.

A similar result holds when the region of integration is finite, but the integrand has a point of infinite discontinuity. Suppose e.g. that the function $f(x, y)$ tends to infinity as $y \to a$. We then say that *the convergent integral*

$$F(x) = \int_a^b f(x, y)\,dy$$

converges uniformly in $\alpha \leqq x \leqq \beta$ *if for every positive number* ϵ *we can find a number* k *such that*

$$\left|\int_a^{a+h} f(x, y)\,dy\right| < \epsilon,$$

provided $h \leqq k$, *where* k *is independent of* x. *Uniform convergence in this sense occurs if in the neighbourhood of the point* $y = a$ *the relation*

$$|f(x, y)| < \frac{M}{(y-a)^\nu}$$

holds, where as before M is a positive constant and $\nu < 1$. Just as above, we show that in the case of uniform convergence $F(x)$ is a continuous function.

If the convergence is uniform, the improper integrals $F(x)$ are continuous in a certain interval, say in $\alpha \leqq x \leqq \beta$. We can then integrate them over this interval and thus form the corresponding improper repeated integral

$$\int_\alpha^\beta dx \int_0^\infty f(x, y)\, dy$$

or

$$\int_\alpha^\beta dx \int_a^b f(x, y)\, dy.$$

Instead of the finite interval $\alpha \leqq x \leqq \beta$ we can of course also consider an infinite interval of integration.

2. **Integration and Differentiation of Improper Integrals with respect to a Parameter.**

It is not true in general that improper integrals may be differentiated or integrated under the sign of integration with respect to a parameter. In other words, these operations are not interchangeable in order with the original integration (cf. the example on p. 316).

In order to determine whether the order of integration in improper repeated integrals is reversible, we can often use the following test, or else make a special investigation on the lines of the following proof.

If the improper integral

$$F(x) = \int_0^\infty f(x, y)\, dy$$

converges uniformly in the interval $\alpha \leqq x \leqq \beta$, *then*

$$\int_\alpha^\beta dx \int_0^\infty f(x, y)\, dy = \int_0^\infty dy \int_\alpha^\beta f(x, y)\, dx.$$

To prove this we put

$$\int_0^\infty f(x, y)\, dy = \int_0^A f(x, y)\, dy + R_A(x).$$

Then by hypothesis $| R_A(x) | < \epsilon(A)$, where $\epsilon(A)$ is a number depending only on A and not on x and tending to zero as $A \to \infty$. In virtue of the elementary theorem for ordinary integrals we have

$$\int_{\alpha}^{\beta} dx \int_{0}^{\infty} f(x, y)\, dy = \int_{\alpha}^{\beta} dx \int_{0}^{A} f(x, y)\, dy + \int_{\alpha}^{\beta} R_A(x)\, dx$$
$$= \int_{0}^{A} dy \int_{\alpha}^{\beta} f(x, y)\, dx + \int_{\alpha}^{\beta} R_A(x)\, dx,$$

whence by the mean value theorem of the integral calculus

$$\left| \int_{\alpha}^{\beta} dx \int_{0}^{\infty} f(x, y)\, dy - \int_{0}^{A} dy \int_{\alpha}^{\beta} f(x, y)\, dx \right| \leqq \epsilon(A) \left| \beta - \alpha \right|.$$

If we now let A tend to infinity, we obtain the formula stated above.

If the integration with respect to a parameter also takes place over an infinite interval of integration, the change of order is not always possible, even though the convergence is uniform. It can, however, be performed if the corresponding improper *double* integral exists (cf. Chap. IV, section 5, p. 262 *et seq.*). Thus e.g.

$$\int_{0}^{\infty} dx \int_{0}^{\infty} f(x, y)\, dy = \int_{0}^{\infty} dy \int_{0}^{\infty} f(x, y)\, dx,$$

if the double integral $\int\int f(x, y)\, dx\, dy$ over the whole first quadrant exists.

The proof of this follows from the fact that the improper double integral is independent of the mode of approximation to the region of integration. In one case we perform this approximation by means of infinite strips parallel to the x-axis, in the other by strips parallel to the y-axis.

A similar result also holds if the interval of integration is finite, but the integrand is discontinuous along a finite number of straight lines $y =$ const. or on a finite number of more general curves in the region of integration. The corresponding theorem is as follows:

If when x *lies in the interval* $\alpha \leqq \mathrm{x} \leqq \beta$ *the function* $\mathrm{f(x, y)}$ *is discontinuous only along a finite number of straight lines* $\mathrm{y} = \mathrm{a}_1$, $\mathrm{y} = \mathrm{a}_2, \ldots, \mathrm{y} = \mathrm{a}_r$, *and if the integral*

$$\int_{a}^{b} f(x, y)\, dy$$

converges uniformly in x, *then in this interval it represents a continuous function of* x, *and*

$$\int_a^\beta dx \int_a^b f(x, y)\, dy = \int_a^b dy \int_a^\beta f(x, y)\, dx.$$

That is, under these hypotheses the order of integration can be changed. The proof of the theorem is analogous to that given above.

It is equally easy to extend the rules for differentiation with respect to a parameter. The following theorem holds:

If the function f(x, y) *has a sectionally continuous derivative with respect to* x *in the interval* $\alpha \leqq x \leqq \beta$ *and the two integrals*

$$F(x) = \int_0^\infty f(x, y)\, dy \quad \text{and} \quad \int_0^\infty f_x(x, y)\, dy$$

converge uniformly, then

$$F'(x) = \int_0^\infty f_x(x, y)\, dy.$$

That is, under these hypotheses the order of the processes of integration and of differentiation with respect to a parameter can be interchanged. For if we put

$$G(x) = \int_0^\infty f_x(x, y)\, dy,$$

then, using the theorem of interchangeability just proved, we have

$$\int_a^\xi G(x)\, dx = \int_a^\xi dx \int_0^\infty f_x(x, y)\, dy = \int_0^\infty dy \int_a^\xi f_x(x, y)\, dx.$$

The integrand on the right has the value

$$\int_a^\xi f_x(x, y)\, dx = f(\xi, y) - f(\alpha, y);$$

therefore

$$\int_a^\xi G(x)\, dx = F(\xi) - F(\alpha);$$

hence if we differentiate and then replace ξ by x we obtain

$$\frac{dF(x)}{dx} = G(x) = \int_0^\infty f_x(x, y)\, dy,$$

as was to be proved.

We can similarly extend the rule for differentiation when one of the limits depends on the parameter x. For we can write

$$\int_{\phi(x)}^{\infty} f(x, y)\,dy = \int_{\phi(x)}^{a} f(x, y)\,dy + \int_{a}^{\infty} f(x, y)\,dy,$$

where a is any fixed value in the interval of integration, and we can then apply rules previously proved to each of the two terms on the right.

As above, our rules of differentiation also hold for improper integrals with finite intervals of integration.

3. Examples.

1. As an example we consider the integral

$$\int_0^{\infty} e^{-xy}\,dy = \frac{1}{x}. \qquad (x > 0)$$

If $x \geqq 1$ this integral converges uniformly, since for positive values of A we have

$$\int_A^{\infty} e^{-xy}\,dy \leqq \int_A^{\infty} e^{-y}\,dy = e^{-A},$$

where the right-hand side no longer depends on x and can be made as small as we please if we choose A sufficiently large. The same is true of the integrals of the partial derivatives of the function with respect to x. By repeated differentiation we thus obtain

$$\int_0^{\infty} ye^{-xy}\,dy = \frac{1}{x^2}, \quad \int_0^{\infty} y^2e^{-xy}\,dy = \frac{2}{x^3}, \ldots, \quad \int_0^{\infty} y^ne^{-xy}\,dy = \frac{n!}{x^{n+1}}.$$

If, in particular, we put $x = 1$, we have

$$\Gamma(n+1) = \int_0^{\infty} y^n e^{-y}\,dy = n!.$$

This formula has already been established in a different way in Vol. I, Chap. IV (p. 251).

2. Further, let us consider the integral

$$\int_0^{\infty} \frac{dy}{x^2 + y^2} = \frac{\pi}{2}\frac{1}{x}.$$

Again it is easy for us to convince ourselves that if $x \geq a$, where a is any positive number, all the assumptions required for differentiation under the integral sign are satisfied. By repeated differentiation we therefore obtain the sequence of formulæ

$$\int_0^{\infty} \frac{dy}{(x^2+y^2)^2} = \frac{\pi}{2}\cdot\frac{1}{2}\cdot\frac{1}{x^3}, \quad \int_0^{\infty} \frac{dy}{(x^2+y^2)^3} = \frac{\pi}{2}\cdot\frac{1.3}{2.4}\cdot\frac{1}{x^5}, \ldots$$

$$\int_0^{\infty} \frac{dy}{(x^2+y^2)^n} = \frac{\pi}{2}\cdot\frac{1\cdot 3.\ \ldots (2n-3)}{2.4.\ \ldots (2n-2)}\cdot\frac{1}{x^{2n-1}}.$$

From these formulæ we can derive another proof of Wallis's product for π (cf. Vol. I, Chap. IV, section 4, p. 224). For if we put $x = \sqrt{n}$, we have

$$\int_0^\infty \frac{dy}{(1 + y^2/n)^n} = \frac{\pi}{2} \cdot \frac{1 . 3 \ldots (2n - 3)}{2 . 4 \ldots (2n - 2)} \sqrt{n}.$$

As n increases the left-hand side tends to the integral $\int_0^\infty e^{-y^2} dy = \frac{1}{2}\sqrt{\pi}$. For the difference

$$\int_0^\infty e^{-y^2} dy - \int_0^\infty \frac{dy}{(1 + y^2/n)^n}$$

satisfies the inequality

$$\left| \int_0^\infty e^{-y^2} dy - \int_0^\infty \frac{dy}{(1 + y^2/n)^n} \right| \leq \int_0^T \left| e^{-y^2} - \frac{1}{(1 + y^2/n)^n} \right| dy$$
$$+ \int_T^\infty e^{-y^2} dy + \int_T^\infty \frac{dy}{(1 + y^2/n)^n},$$

or, since $(1 + y^2/n)^n > y^2$,

$$\left| \int_0^\infty e^{-y^2} dy - \int_0^\infty \frac{dy}{(1 + y^2/n)^n} \right| \leq \int_0^T \left| e^{-y^2} - \frac{1}{(1 + y^2/n)^n} \right| dy + \int_T^\infty e^{-y^2} dy + \frac{1}{T}.$$

But if we choose T so large that $\int_T^\infty e^{-y^2} dy + \frac{1}{T} < \frac{\varepsilon}{2}$, and then choose n so large that

$$\int_0^T \left| e^{-y^2} - \frac{1}{(1 + y^2/n)^n} \right| dy < \frac{\varepsilon}{2},$$

as is possible in virtue of the uniform convergence of the process

$$\lim_{n \to \infty} (1 + y^2/n)^{-n} = e^{-y^2},$$

it follows at once that

$$\left| \int_0^\infty \left(e^{-y^2} - \frac{1}{(1 + y^2/n)^n} \right) dy \right| < \varepsilon.$$

This establishes the relation

$$\lim_{n \to \infty} \frac{1 . 3 \ldots (2n - 3)}{2 . 4 \ldots (2n - 2)} \sqrt{n} = \frac{1}{\sqrt{\pi}},$$

which is equivalent to that obtained in Vol. I, p. 224.

3. With a view to calculating the integral $\int_0^\infty \frac{\sin y}{y} dy$, we shall discuss the function $F(x) = \int_0^\infty e^{-xy} \frac{\sin y}{y} dy$. This integral converges uniformly if $x \geq 0$, while the integral

$$\int_0^\infty e^{-xy} \sin y \, dy$$

converges uniformly if $x \geqq \delta > 0$, where δ is an arbitrarily small positive number. Both these statements will be proved below. Therefore $F(x)$ is continuous if $x \geqq 0$, and if $x \geqq \delta$ we have

$$F'(x) = -\int_0^\infty e^{-xy} \sin y\, dy.$$

We can easily evaluate this last integral by integrating by parts twice; we obtain

$$F'(x) = -\frac{1}{1+x^2}.$$

From this we can find the value of $F(x)$ by integration; this value is

$$F(x) = \operatorname{arc\,cot} x + C,$$

where C is a constant. In virtue of the relation

$$\left| \int_0^\infty e^{-xy} \frac{\sin y}{y}\, dy \right| \leqq \int_0^\infty e^{-xy}\, dy = \frac{e^{-xy}}{x}\Big|_\infty^0 = \frac{1}{x},$$

which holds if $x \geqq \delta$, we see that $\lim_{x\to\infty} F(x) = 0$. Since $\lim_{x\to\infty} \operatorname{arc\,cot} x = 0$, C must also be 0, and we obtain

$$F(x) = \operatorname{arc\,cot} x.$$

On account of the continuity of $F(x)$ for $x \geqq 0$,

$$\lim_{x\to 0} F(x) = F(0) = \int_0^\infty \frac{\sin y}{y}\, dy,$$

which, since $\lim_{x\to 0} \operatorname{arc\,cot} x = \frac{\pi}{2}$, gives the required formula

$$\int_0^\infty \frac{\sin y}{y}\, dy = \frac{\pi}{2}$$

(cf. Vol. I, p. 450, footnote).

We now return to the proof that

$$\int_0^\infty e^{-xy} \frac{\sin y}{y}\, dy$$

converges uniformly if $x \geqq 0$. If A is an arbitrary number and $k\pi$ is the least multiple of π which exceeds A, we can write the "remainder" of the integral in the form

$$\int_A^\infty e^{-xy} \frac{\sin y}{y}\, dy = \int_A^{k\pi} e^{-xy} \frac{\sin y}{y}\, dy + \sum_{\nu=k}^{\infty} \int_{\nu\pi}^{(\nu+1)\pi} e^{-xy} \frac{\sin y}{y}\, dy.$$

The terms of the series on the right have alternating signs and their absolute values tend monotonically to zero. By Leibnitz's test (Vol. I, p. 370), therefore, the series converges, and the absolute value of its sum is less than that of its first term. Hence we have the inequality

$$\left| \int_A^\infty e^{-xy} \frac{\sin y}{y}\, dy \right| < \int_A^{(k+1)\pi} e^{-xy} \frac{|\sin y|}{y}\, dy < \int_A^{(k+1)\pi} \frac{1}{A}\, dy < \frac{2\pi}{A},$$

in which the right-hand side is independent of x and can be made as small as we please. This establishes the uniformity of the convergence. The uniform convergence of

$$\int_0^\infty e^{-xy} \sin y\, dy$$

for $x \geqq \delta > 0$ follows at once from the relation

$$\int_A^\infty \left| e^{-xy} \sin y \right| dy \leqq \int_A^\infty e^{-xy}\, dy = \frac{e^{-Ax}}{x} \leqq \frac{e^{-A\delta}}{\delta}.$$

On p. 310 we learned that uniform convergence of the integrals is a sufficient condition for interchangeability of the order of integration. Mere *convergence* is not sufficient, as the following example shows:

If we put $f(x, y) = (2 - xy)xy e^{-xy}$, then since

$$f(x, y) = \frac{\partial}{\partial y}(xy^2 e^{-xy}),$$

the integral $\int_0^\infty f(x, y)\, dy$ exists for every x in the interval $0 \leqq x \leqq 1$, and in fact for every such value of x it has the value 0. Therefore

$$\int_0^1 dx \int_0^\infty f(x, y)\, dy = 0.$$

On the other hand, since

$$f(x, y) = \frac{\partial}{\partial x}(x^2 y e^{-xy}),$$

for every $y \geqq 0$ we have

$$\int_0^1 f(x, y)\, dx = y e^{-y},$$

and therefore

$$\int_0^\infty dy \int_0^1 f(x, y)\, dx = \int_0^\infty y e^{-y}\, dy = \int_0^\infty e^{-y}\, dy = 1.$$

Hence

$$\int_0^1 dx \int_0^\infty f(x, y)\, dy \neq \int_0^\infty dy \int_0^1 f(x, y)\, dx.$$

4. Evaluation of Fresnel's Integrals.

The integrals

$$F_1 = \int_{-\infty}^{+\infty} \sin(\tau^2)\, d\tau, \quad F_2 = \int_{-\infty}^{+\infty} \cos(\tau^2)\, d\tau,$$

which are of importance in optics, are known as Fresnel's integrals. In order to evaluate them, we apply the substitution $\tau^2 = t$, obtaining

$$F_1 = \int_0^{\infty} \frac{\sin t}{\sqrt{t}}\, dt, \quad F_2 = \int_0^{\infty} \frac{\cos t}{\sqrt{t}}\, dt.$$

Here we put

$$\frac{1}{\sqrt{t}} = \frac{2}{\sqrt{\pi}} \int_0^{\infty} e^{-x^2 t}\, dx$$

(this follows from the substitution $x = \tau/\sqrt{t}$) and change the order of integration, as is permissible by our rules. Then

$$F_1 = \frac{2}{\sqrt{\pi}} \int_0^{\infty} dx \int_0^{\infty} e^{-x^2 t} \sin t\, dt, \quad F_2 = \frac{2}{\sqrt{\pi}} \int_0^{\infty} dx \int_0^{\infty} e^{-x^2 t} \cos t\, dt.$$

The inner integrals are easily evaluated by integration by parts, and F_1 and F_2 reduce to the elementary rational integrals

$$F_1 = \frac{2}{\sqrt{\pi}} \int_0^{\infty} \frac{1}{1 + x^4}\, dx, \quad F_2 = \frac{2}{\sqrt{\pi}} \int_0^{\infty} \frac{x^2}{1 + x^4}\, dx.$$

The integrals may be evaluated by the methods given in Vol. I (cf. Vol. I, p. 234); the second integral can be reduced to the first by means of the substitution $x' = \dfrac{1}{x}$, and both have the value $\dfrac{\pi}{2\sqrt{2}}$. That is,

$$F_1 = F_2 = \sqrt{\frac{\pi}{2}}.$$

EXAMPLES

1. Evaluate $\int_0^{\infty} x^n e^{-x^2}\, dx$.

2. How must a, b, c be chosen in order that

$$\int_{-\infty}^{\infty} \int_{-\infty}^{\infty} e^{-(ax^2 + 2bxy + cy^2)}\, dx\, dy = 1?$$

3. Evaluate

$$(a)\quad \int_{-\infty}^{+\infty}\int_{-\infty}^{+\infty} e^{-(ax^2+2bxy+cy^2)}(Ax^2 + 2Bxy + Cy^2)\,dx\,dy.$$

$$(b)\quad \int_{-\infty}^{+\infty}\int_{-\infty}^{+\infty} e^{-(ax^2+2bxy+cy^2)}(ax^2 + 2bxy + cy^2)\,dx\,dy.$$

$$(a > 0,\ ac - b^2 > 0).$$

4. Evaluate the following integrals:

$$(a)\quad K(a) = \int_0^{\infty} e^{-ax^2}\cos x\,dx.$$

$$(b)\quad \int_0^{\infty} \frac{e^{-bx} - e^{-ax}}{x}\cos x\,dx.$$

$$(c)\quad I(a) = \int_0^{\infty} e^{-x^2 - a^2/x^2}\,dx.$$

$$(d)\quad \int_0^{\infty} \frac{\sin(ax)J_0(bx)}{x}\,dx$$ (where J_0 denotes the Bessel function defined in Ex. 4, p. 223).

5*. Prove that $\int_0^{n\pi} \frac{\sin^2 ax}{x}\,dx$ is of the order of $\log n$ when n is large, and that

$$\int_0^{\infty} \frac{\sin^2 ax - \sin^2 bx}{x}\,dx = \tfrac{1}{2}\log\frac{a}{b}.$$

6. Replace the statement "the integral $\int_0^{\infty} f(x, y)\,dy$ is not uniformly convergent" by an equivalent statement not involving any form of the words "uniformly convergent". (Cf. Vol. I, p. 45, Ex. 1.)

5. The Fourier Integral

1. Introduction.

The theory given in section 2, p. 310 *et seq.* is illustrated by the important example known as Fourier's integral theorem. It will be remembered that Fourier series give a representation of a sectionally smooth but otherwise arbitrary periodic function in terms of trigonometric functions. Fourier's integral gives a corresponding trigonometrical representation of a function $f(x)$ which is defined in the whole interval $-\infty < x < +\infty$ and is not subject to any condition of periodicity.

We shall make the following assumptions about the function $f(x)$:

(1) $f(x)$ is sectionally smooth; that is, the function $f(x)$ and its first derivative are continuous in any finite interval, except possibly for a finite number of jump discontinuities.

(2) The integral

$$\int_{-\infty}^{\infty} |f(x)|\, dx = C$$

is convergent.

(3) At a discontinuity x of the function it is assumed that $f(x)$ is the arithmetic mean of the limits on the right and left. Thus

$$f(x) = \tfrac{1}{2}(f(x+0) + f(x-0)).$$

Fourier's integral theorem may then be stated as follows:

$$f(x) = \frac{1}{\pi}\int_0^{\infty} d\tau \int_{-\infty}^{\infty} f(t) \cos \tau(t-x)\, dt,$$

or, in complex notation, we have the equivalent formula

$$f(x) = \frac{1}{2\pi}\int_{-\infty}^{\infty} d\tau \int_{-\infty}^{\infty} f(t) e^{-i\tau(t-x)}\, dt.$$

We may also state the theorem in the following form: if

$$g(\tau) = \frac{1}{\sqrt{2\pi}} \int_{-\infty}^{\infty} f(t) e^{-it\tau}\, dt,$$

then

$$f(x) = \frac{1}{\sqrt{2\pi}} \int_{-\infty}^{\infty} g(\tau) e^{i\tau x}\, d\tau.$$

The two formulæ last written are reciprocal equations for $f(x)$ and $g(x)$, each equation being the solution of the other. If the variable $\rho = \tau/2\pi$ is introduced and finally replaced by τ again, we can express Fourier's integral theorem by means of the two reciprocal formulæ

$$h(\tau) = \int_{-\infty}^{\infty} f(t) e^{-2\pi i t\tau}\, dt, \quad f(x) = \int_{-\infty}^{\infty} h(t) e^{2\pi i t x}\, dt,$$

where

$$h(\tau) = \sqrt{2\pi}\, g(2\pi\tau).$$

We shall give some examples to illustrate this theorem and then proceed to the proof. We first observe that if $f(x)$ is an even function—i.e. if $f(x) = f(-x)$—then a short calculation shows that the theorem may be stated in the simplified form

$$f(x) = \frac{2}{\pi}\int_0^\infty \cos(\tau x)\,d\tau \int_0^\infty f(t)\cos(\tau t)\,dt.$$

If, on the other hand, $f(x)$ is an odd function—i.e. if $f(x) = -f(-x)$—we obtain in the same way

$$f(x) = \frac{2}{\pi}\int_0^\infty \sin(\tau x)\,d\tau \int_0^\infty f(t)\sin(\tau t)\,dt.$$

EXAMPLES

1. Let $f(x) = 1$ when $x^2 < 1$, $f(x) = 0$ when $x^2 > 1$. Then

$$\begin{aligned} f(x) &= \frac{2}{\pi}\int_0^\infty \cos(\tau x)\,d\tau \int_0^1 \cos(t\tau)\,dt \\ &= \frac{2}{\pi}\int_0^\infty \frac{\sin\tau\cos(\tau x)}{\tau}\,d\tau = \begin{cases} 0, & x^2 > 1, \\ \frac{1}{2}, & x^2 = 1, \\ 1, & x^2 < 1. \end{cases} \end{aligned}$$

The integral on the right has played a part in mathematical literature under the name of Dirichlet's discontinuous factor.

2. Let $f(x) = e^{-kx}$ $(k > 0)$ when $x > 0$ and $f(x) = f(-x)$. It is easy to show that

$$f(x) = \frac{2}{\pi}\int_0^\infty \cos(\tau x)\,d\tau \int_0^\infty e^{-kt}\cos(t\tau)\,dt = \frac{2}{\pi}\int_0^\infty \frac{k\cos(\tau x)}{k^2 + \tau^2}\,d\tau.$$

But if we put $f(-x) = -f(x)$, we obtain

$$f(x) = \frac{2}{\pi}\int_0^\infty \sin(\tau x)\,d\tau \int_0^\infty e^{-kt}\sin(t\tau)\,dt = \frac{2}{\pi}\int_0^\infty \frac{\tau\sin(\tau x)}{k^2 + \tau^2}\,d\tau.$$

Hence we obtain the two integral formulæ

$$\int_0^\infty \frac{\cos(\tau x)}{k^2 + \tau^2}\,d\tau = \frac{\pi}{2}\,\frac{e^{-kx}}{k}, \quad \int_0^\infty \frac{\tau\sin(\tau x)}{k^2 + \tau^2}\,d\tau = \frac{\pi}{2}\,e^{-kx}, \quad k > 0.$$

3. The function $f(x) = e^{-x^2/2}$ gives an interesting illustration of the reciprocal formulæ. Since

$$\sqrt{\frac{2}{\pi}}\int_0^\infty e^{-t^2/2}\cos\tau t\,dt = e^{-\tau^2/2}$$

(see p. 318, Ex. 4*a*), the two reciprocal formulæ for $g(\tau)$ and $f(x)$ coincide.

2. **Proof of Fourier's Integral Theorem.**

The essential steps in the proof of Fourier's integral formula are a transformation and a simple limit operation applied to Dirichlet's limit formula

$$\pi f(x) = \lim_{\lambda \to \infty} \int_{-a}^{a} f(x+t) \frac{\sin(\lambda t)}{t}\, dt,$$

which holds for arbitrary positive values of a. We shall first prove this formula, although the substance of the proof has been given in Vol. I, Chap. IX, § 5 (p. 450). We rely on the elementary limit formula (cf. Vol. I, Chap. IX, p. 448)

$$\lim_{\lambda \to \infty} \int_{\alpha}^{\beta} \sin(\lambda t)\, s(t)\, dt = 0,$$

which holds when $s(t)$ is continuous or sectionally continuous in the interval $\alpha \leqq t \leqq \beta$ but is otherwise arbitrary.

Let us first consider the interval from 0 to a. In this interval

$$s(t) = \frac{f(x+t) - f(x+0)}{t}$$

is a sectionally continuous function which, by the assumptions about $f(x)$, must have the limit $f'(x+0)$ as t tends to zero. Thus

$$\int_0^a f(x+t) \frac{\sin(\lambda t)}{t}\, dt = \int_0^a f(x+0) \frac{\sin(\lambda t)}{t}\, dt + \int_0^a s(t) \sin(\lambda t)\, dt,$$

and the elementary formula given above shows that the last integral on the right tends to zero as λ tends to infinity.

The first integral on the right has the limit

$$\lim_{\lambda \to \infty} f(x+0) \int_0^{a\lambda} \frac{\sin\sigma}{\sigma}\, d\sigma = f(x+0) \int_0^{\infty} \frac{\sin\sigma}{\sigma}\, d\sigma = \frac{\pi}{2} f(x+0)$$

(cf. p. 315). If we now apply the corresponding argument to the integral from $-a$ to 0, we obtain Dirichlet's formula.

The next step in the proof of Fourier's theorem is the substitution of the expression

$$\frac{\sin(\lambda t)}{t} = \int_0^{\lambda} \cos(t\tau)\, d\tau$$

in Dirichlet's formula. We also introduce the notation

$$\int_{-a}^{a} f(x+t)\,\frac{\sin(\lambda t)}{t}\,dt = \int_{-a}^{a} f(x+t)\,dt \int_{0}^{\lambda} \cos(t\tau)\,d\tau$$

$$= \int_{0}^{\lambda} d\tau \int_{-a}^{a} f(x+t)\cos(t\tau)\,dt = F(\lambda, a).$$

Dirichlet's formula then states that

$$\pi f(x) = \lim_{\lambda \to \infty} F(\lambda, a).$$

Since this limit is independent of a, we may write

$$\pi f(x) = \lim_{a \to \infty} \lim_{\lambda \to \infty} F(\lambda, a).$$

If it were permissible to interchange the order of the limit operations in this formula, that is, if we might take the limit as a tends to infinity under the sign of integration, we should at once have

$$\pi f(x) = \lim_{\lambda \to \infty} \int_{0}^{\lambda} d\tau \int_{-\infty}^{\infty} f(x+t)\cos(t\tau)\,dt = \int_{0}^{\infty} d\tau \int_{-\infty}^{\infty} f(x+t)\cos(t\tau)\,dt.$$

This immediately gives Fourier's integral formula if we write $x + t = t'$ and then replace t' by t. Thus the proof will be complete if we establish the change of order of limit operations

$$\lim_{a \to \infty} \lim_{\lambda \to \infty} F(\lambda, a) = \lim_{\lambda \to \infty} \lim_{a \to \infty} F(\lambda, a).$$

Our previous work (p. 310; cf. also p. 104) shows that it is sufficient to prove that the limit

$$\lim_{a \to \infty} F(\lambda, a) = \int_{-\infty}^{\infty} f(x+t)\,\frac{\sin(\lambda t)}{t}\,dt$$

exists uniformly with respect to λ.

To prove this, we must show that if ϵ is given in advance we can find A independent of λ, such that $|F(\lambda, a) - F(\lambda, b)| < \epsilon$ whenever a and b both exceed A. But

$$|F(\lambda, a) - F(\lambda, b)| \leq \int_{a}^{b} |f(x+t)|\,\frac{|\sin(\lambda t)|}{t}\,dt$$

$$+ \int_{-b}^{-a} |f(x+t)|\,\frac{|\sin(\lambda t)|}{|t|}\,dt \leq \frac{2}{A}\int_{-\infty}^{\infty} |f(t)|\,dt.$$

It follows at once that

$$| F(\lambda, a) - F(\lambda, b) | < \frac{2C}{A},$$

so that it is only necessary to take $A = 2C/\epsilon$. This gives the proof of uniform convergence, and completes the proof of Fourier's integral theorem.

6. The Eulerian Integrals (Gamma Function) *

One of the most important examples of a function defined by an improper integral involving a parameter is the gamma function $\Gamma(x)$. Here we shall give a fairly detailed discussion of this function.

1. Definition and Functional Equation.

The function $\Gamma(x)$ is defined for every $x > 0$ by the improper integral

$$\Gamma(x) = \int_0^\infty e^{-t} t^{x-1} \, dt.$$

In Vol. I, Chap. IV, pp. 250–1, we studied this integral for integral arguments $x = n$. The method used there shows at once that the integral converges for any $x > 0$, the convergence being uniform in every closed interval of the positive x-axis which does not include the point $x = 0$. *The function* $\Gamma(x)$ *is therefore continuous for* $x > 0$.

By simple substitutions we can transform the integral for $\Gamma(x)$ into other forms which are often used. Here we only mention the substitution $t = u^2$, which transforms the gamma function into the form

$$\Gamma(x) = 2\int_0^\infty e^{-u^2} u^{2x-1} \, du.$$

Thus the frequently-occurring integral

$$\int_0^\infty e^{-u^2} u^a \, du \qquad (a > -1)$$

* A discussion closely related to the present one is given by E. Artin, *Einführung in die Theorie der* Γ*-Funktion* (Leipzig, 1931).

can be expressed in terms of the gamma function as

$$\int_0^\infty e^{-u^2} u^a \, du = \frac{1}{2}\Gamma\left(\frac{1+a}{2}\right)$$

(cf. section 3, p. 303).

Integration by parts shows, as in Vol. I, p. 251, that the relation

$$\Gamma(x+1) = x\Gamma(x)$$

holds for any $x > 0$. This equation is called the *functional equation of the gamma function.*

Of course $\Gamma(x)$ is not uniquely defined by the property of being a solution of this functional equation. In fact, we obtain another solution merely by multiplying $\Gamma(x)$ by an arbitrary periodic function $p(x)$ with period unity. On the other hand, the functions

$$u(x) = \Gamma(x)p(x), \qquad p(x+1) = p(x)$$

represent the aggregate of all solutions of the equation; for if $u(x)$ is any solution, the quotient

$$f(x) = \frac{u(x)}{\Gamma(x)},$$

which can always be formed since $\Gamma(x) \neq 0$, satisfies the equation

$$f(x+1) = f(x).$$

Instead of the function $\Gamma(x)$, it is frequently more convenient to consider the function $u(x) = \log \Gamma(x)$; since $\Gamma(x) > 0$ for $x > 0$, this is always defined. The function satisfies the functional equation (difference equation)

$$u(x+1) - u(x) = \log x.$$

We obtain other solutions of this equation by adding to $\log \Gamma(x)$ an arbitrary periodic function with period unity. In order to specify the function $\log \Gamma(x)$ uniquely, therefore, we must supplement the functional equation by other conditions. One very simple condition of this type is given by the following theorem, due to H. Bohr:

Every convex solution of the difference equation

$$u(x+1) - u(x) = \log x$$

is identical with the function $\log\Gamma(x)$ *in the interval* $0 < x < \infty$, *except perhaps for an additive constant.*

2. **Convex Functions: Proof of Bohr's Theorem.**

We say that a function $f(x)$ is *convex* in a region $a \leqq x \leqq b$ if for every two points x_1 and x_2 of the region and every two positive numbers α, β, where $\alpha + \beta = 1$, the expression

$$\alpha f(x_1) + \beta f(x_2) - f(\alpha x_1 + \beta x_2)$$

never changes sign; or, intuitively speaking, if the chord joining two points of the curve $y = f(x)$ either never lies beneath or

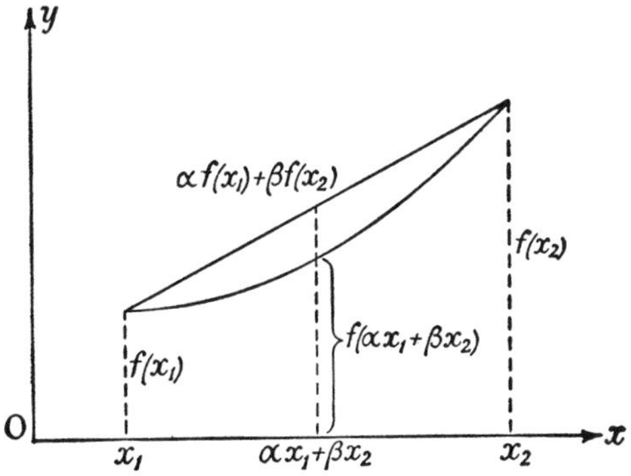

Fig. 19.—A function which is convex downwards

never lies above the arc of the curve itself between x_1 and x_2 (cf. fig. 19). (Cf. also Chap. I, section 1, p. 8 and Chap. II, p. 100.)

Before proving this theorem we shall establish certain properties of convex functions. We restrict ourselves to functions which are "convex downwards", for which

$$\alpha f(x_1) + \beta f(x_2) - f(\alpha x_1 + \beta x_2) \geqq 0$$

holds; functions which are "convex upwards" can always be changed into functions which are "convex downwards" by multiplying by -1.

If a convex function $f(x)$ is twice continuously differentiable, the expression

$$\alpha f(x_1) + \beta f(x_2) - f(\alpha x_1 + \beta x_2)$$

can be represented by the double integral

$$\beta(x_2 - x_1)^2 \int_0^1 dt \int_{\beta t}^{t} f''\{x_1 + (x_2 - x_1)\tau\}\, d\tau,$$

as is easily verified. Thus the inequality of the definition is certainly satisfied, provided that

$$f''(x) \geqq 0.$$

On the other hand, the passage to the limit $x_2 \to x_1$ shows that this condition is also necessary, and it is therefore a characteristic property of convex functions which are twice continuously differentiable.

A fact which is noteworthy and useful in applications is that we need not assume the continuity of a convex function $f(x)$; on the contrary, this property follows from the definition of convexity. We can in fact replace the above inequality by an apparently weaker one, which, however, is equivalent to it, expressed in the following theorem:

If for every x *and* h *for which the arguments* x ± h *still lie in the region of definition the bounded function* f(x) *satisfies the inequality*

$$f(x + h) + f(x - h) - 2f(x) \geqq 0,$$

that is, if the mid-point of every chord of the curve y = f(x) *lies above or on the curve, then* f(x) *is convex.*

We first show that *every bounded function* f(x) *which satisfies the inequality*

$$f(x + h) + f(x - h) - 2f(x) \geqq 0$$

is continuous.

To prove this we write the condition in the form

$$f(x) - f(x - h) \leqq f(x + h) - f(x),$$

from which we derive the inequalities

$$\begin{aligned} f(x - \nu h) - f(x - (\nu + 1)h) &\leqq f(x + h) - f(x) \\ &\leqq f(x + (\nu + 1)h) - f(x + \nu h), \end{aligned}$$

valid for every integer $\nu \geqq 0$. If we add these for values of ν from $\nu = 0$ to $\nu = n - 1$, we obtain the estimate

$$\frac{f(x) - f(x - nh)}{n} \leqq f(x + h) - f(x) \leqq \frac{f(x + nh) - f(x)}{n},$$

and hence if we assume that $|f(x)| \leqq C$,

$$|f(x+h)-f(x)| \leqq \frac{2C}{n}.$$

Here n can be any positive integer such that the argument $x \pm nh$ lies in the interval of definition. If we let h tend to zero, the largest possible number n increases without limit, that is, the expression $f(x+h)-f(x)$ tends to zero. This proves the continuity of $f(x)$.

From the continuity of $f(x)$ we can now easily prove its convexity, that is, we can establish the inequality

$$\alpha f(x_1) + \beta f(x_2) - f(\alpha x_1 + \beta x_2) \geqq 0.$$

From the inequality

$$f(x) - f(x - nh) \leqq n\{f(x+h) - f(x)\},$$

by means of the substitution

$$\xi = x - nh,$$

we obtain the relation

$$\frac{f(\xi + nh) - f(\xi)}{n} \leqq \frac{f(\xi + (n+1)h) - f(\xi)}{n+1},$$

and hence in general

$$\frac{f(\xi + mh) - f(\xi)}{m} \leqq \frac{f(\xi + nh) - f(\xi)}{n}, \qquad 0 < m \leqq n.$$

If we put $\xi + nh = \xi_1$, a few transformations give

$$\left(1 - \frac{m}{n}\right) f(\xi) + \frac{m}{n} f(\xi_1) \geqq f\left(\left(1 - \frac{m}{n}\right)\xi + \frac{m}{n}\xi_1\right),$$

which is exactly the inequality we require for rational values of α and β. We then deduce its validity for any values of α and β from the continuity * of $f(x)$.

* From the inequalities

$$\frac{f(x+c) - f(x)}{c} \leqq \frac{f(x+a) - f(x)}{a} \leqq \frac{f(x+b) - f(x)}{b},$$

whose validity for any numbers $c \leqq a \leqq b$ differing from zero is a direct consequence of the definition of convexity, we see that the difference quotient $\frac{f(x+a) - f(x)}{a}$ is bounded and monotonic if a tends to zero through positive values or through negative values, and it therefore possesses a limit. Thus a convex function has a derivative on the right and on the left at every point.

Finally, we apply the following inequality, which is obvious from the geometrical interpretation of a convex function:

$$f(x+h)+f(x-h)-\{f(x+\delta)+f(x-\delta)\} \geqq 0.$$

Here δ and h are two positive numbers, $\delta \leqq h$.

This is proved by adding the two relations

$$\frac{1}{2}\left(1+\frac{\delta}{h}\right)f(x-h)+\frac{1}{2}\left(1-\frac{\delta}{h}\right)f(x+h)-f(x-\delta) \geqq 0,$$

$$\frac{1}{2}\left(1-\frac{\delta}{h}\right)f(x-h)+\frac{1}{2}\left(1+\frac{\delta}{h}\right)f(x+h)-f(x+\delta) \geqq 0.$$

We now return to the *theorem of Bohr* stated above (p. 324–5). We see at once that $\log\Gamma(x)$ is convex. For if we write $\Gamma(x)$ in the form

$$\Gamma(x)=\int_0^\infty e^{-t/2}t^{(x-1+h)/2}\,.\,e^{-t/2}t^{(x-1-h)/2}\,dt,$$

where h has any positive value and x any value greater than h, and apply Schwarz's inequality (cf. Vol. I, Chap. IX, p. 451), we have

$$\{\Gamma(x)\}^2 \leqq \Gamma(x+h)\Gamma(x-h),$$

and therefore *

$$\log\Gamma(x+h)+\log\Gamma(x-h)-2\log\Gamma(x) \geqq 0.$$

Again, if $f(x)$ and $g(x)$ are two continuous convex solutions of the functional equation

$$u(x+1)-u(x)=\log x,$$

* This fact is a special case of a general theorem. If the functions $f_\nu(x)$, $\nu = 1, 2, \ldots, n$ satisfy the conditions

$$f_\nu(x) \geqq 0 \quad \text{and} \quad \{f_\nu(x)\}^2 \leqq f_\nu(x-h)f_\nu(x+h),$$

so that the functions $\log f_\nu(x)$ are convex, then the sum $\sum_1^n f_\nu(x)$ also satisfies these conditions.

For if we write $\Sigma f_\nu(x)$ in the form

$$\sum_1^n f_\nu(x)=\sum_1^n \frac{f_\nu(x)}{\sqrt{f_\nu(x-h)}\sqrt{f_\nu(x+h)}}\sqrt{f_\nu(x-h)}\sqrt{f_\nu(x+h)},$$

and use the relation

$$\frac{f_\nu(x)}{\sqrt{f_\nu(x-h)}\sqrt{f_\nu(x+h)}} \leqq 1,$$

the difference

$$\phi(x) = f(x) - g(x)$$

is a continuous periodic function of period unity. Moreover, since

$$f(x+1) - f(x) = \log x$$

and

$$f(x) - f(x-1) = \log(x-1),$$

$f(x)$ satisfies the relation

$$f(x+1) + f(x-1) - 2f(x) = \log \frac{x}{x-1}.$$

Since $f(x)$ is convex, the inequality

$$f(x+h) + f(x-h) - 2f(x) \leqq \log \frac{x}{x-1},$$

holds for every h in the range $0 < h \leqq 1$ (cf. p. 328).

We likewise obtain

$$g(x+h) + g(x-h) - 2g(x) \leqq \log \frac{x}{x-1},$$

and therefore

$$|\,\phi(x+h) + \phi(x-h) - 2\phi(x)\,| \leqq 2 \log \frac{x}{x-1}.$$

If we now let x increase beyond all bounds, the expression $\log \dfrac{x}{x-1}$ tends to zero, and so does the function

$$\phi(x+h) + \phi(x-h) - 2\phi(x).$$

we obtain the inequality

$$\{\sum_1^n f_\nu(x)\}^2 \leqq (\sum_1^n \sqrt{f_\nu(x-h)}\,\sqrt{f_\nu(x+h)}\,)^2;$$

if we now apply Schwarz's inequality to the right-hand side, we have

$$\{\sum_1^n f_\nu(x)\}^2 \leqq \sum_1^n f_\nu(x-h) \sum_1^n f_\nu(x+h).$$

An analogous theorem holds for integrals of the form

$$\int_a^b f(x, t)\,dt,$$

if for all values of the parameter t the functions $f(x, t)$ satisfy the conditions

$$f(x, t) \geqq 0 \quad \text{and} \quad \{f(x, t)\}^2 \leqq f(x-h, t)\,f(x+h, t).$$

The gamma function is of this type.

Since this function is periodic, we obtain the equation

$$\phi(x+h)+\phi(x-h)-2\phi(x)=0,$$

valid for every $x>0$.

A continuous periodic function $\phi(x)$ which satisfies such a condition for every positive value of h and every value of x greater than h must be a constant.* This, however, proves that any continuous convex solution of the equation $u(x+1)-u(x)=\log x$ can differ from $\log\Gamma(x)$ only by an additive constant.

3. The Infinite Product for the Gamma Function.

In this sub-section we shall give the *infinite products* for the gamma function found by Gauss and Weierstrass.

We first show that the relation

$$\Gamma(x)=\lim_{n\to\infty}G_n(x)$$

holds, where

$$G_n(x)=\frac{1\,.\,2\ldots(n-1)}{x(x+1)\ldots(x+n-1)}\,n^x.$$

This statement is plausible, since for integers $x=\nu$ we have

$$G_n(\nu)=(\nu-1)!\,\frac{n}{n}\,\frac{n}{n+1}\cdots\frac{n}{n+\nu-1},$$

and as n increases this obviously tends to the value $(\nu-1)!$.

We must show in general that the sequence $G_n(x)$ converges for every $x\neq 0,-1,-2,\ldots$, and then that the limit function $G(x)$ coincides with the gamma function for positive values of x. To prove this last statement we notice that if $x>0$ the function $\log G(x)$ satisfies the functional equation

$$u(x+1)-u(x)=\log x.$$

By Bohr's theorem we have only to show that $\log G(x)$ is convex.

* If, say, $\phi(1)=\phi(2)=a$, then by the equation $\phi(\frac{3}{2})=\frac{1}{2}(\phi(1)+\phi(2))$ we have $\phi(\frac{3}{2})=a$, and likewise $\phi(x_\nu)=a$ at all points x_ν of the interval $1\leq x\leq 2$ which are obtained by repeated bisection of the sub-intervals. Since these points x_ν are everywhere dense, from the continuity of $\phi(x)$ it follows that $\phi(x)=a$ throughout the interval $1\leq x\leq 2$, and by the periodicity of $\phi(x)$ this holds for every $x>0$.

In order to prove the convergence of the sequence $G_n(x)$ for $x \neq 0, -1, -2, \ldots$ we introduce the expression

$$n = (1+1)\left(1+\frac{1}{2}\right)\left(1+\frac{1}{3}\right)\cdots\left(1+\frac{1}{n-1}\right)$$

for the number n, and accordingly write

$$G_n(x) = \frac{1}{x}\prod_{1}^{n-1}\frac{(1+1/\nu)^x}{1+x/\nu}.$$

By a test proved in Vol. I, p. 421, the product

$$\prod_{1}^{n-1}\frac{(1+1/\nu)^x}{1+x/\nu}$$

converges absolutely and uniformly provided that the series

$$\sum_{1}^{\infty}\left|\frac{(1+1/\nu)^x}{1+x/\nu}-1\right|$$

converges uniformly. If we use the Taylor expansion in powers of $1/\nu$ up to the terms of second order, the general term of this series can be written in the form

$$\frac{(1+1/\nu)^x-(1+x/\nu)}{1+x/\nu} = \frac{1}{2\nu^2}\,\frac{x(x-1)(1+\theta/\nu)^{x-2}}{1+x/\nu},$$

where θ is a number between 0 and 1. From this it follows that

$$\left|\frac{(1+1/\nu)^x-(1+x/\nu)}{1+x/\nu}\right| \leq \frac{C}{\nu^2},$$

where C is a constant independent of ν. In every closed region which contains none of the points $x = -1, -2, \ldots$ we can replace the estimate C by a number which is also independent of x. In every such region the series converges uniformly, and therefore the product does so too.

The limit function

$$G(x) = \lim_{n\to\infty}\frac{1.2\ldots(n-1)}{x(x+1)\ldots(x+n-1)}\,n^x$$

is continuous for every $x \neq 0, -1, -2, \ldots$ and, as we see at once, satisfies the functional equation

$$G(x+1) = xG(x).$$

In order to show that if $x > 0$ the function $G(x)$ is identical with the function $\Gamma(x)$, we consider the function $\log G(x)$ for $x > 1$. It is the limit function of the sequence

$$\log G_n(x) = \log(n-1)! + x\log n - \sum_{\nu=0}^{n-1} \log(x+\nu).$$

For any positive value of h and any value of x greater than h the functions $\log G_n(x)$ satisfy the condition for convexity,

$$\begin{aligned}\log G_n(x+h) &+ \log G_n(x-h) - 2\log G_n(x) \\ &= \sum_{\nu=0}^{n-1} (2\log(x+\nu) - \log(x+h+\nu) - \log(x-h+\nu)) \geqq 0,\end{aligned}$$

which consequently applies to the function $\log G(x)$ also. Since in addition

$$\log G(1) = 0 = \log \Gamma(1),$$

by the general theorem $G(x)$ must be identical with $\Gamma(x)$. We have therefore obtained Gauss's *infinite product* for $\Gamma(x)$:

$$\begin{aligned}\Gamma(x) &= \lim_{n\to\infty} \frac{1 \cdot 2 \ldots (n-1)}{x(x+1)\ldots(x+n-1)} n^x \\ &= \frac{1}{x} \prod_{\nu=1}^{\infty} \frac{(1+1/\nu)^x}{1+x/\nu}.\end{aligned}$$

The theoretical importance of this expression arises from the fact that we can regard it as defining the gamma function, not only for all positive values of x, but also for all negative non-integral values of x.

This product can easily be put in a somewhat different form. If in the expression

$$n^x = e^{x \log n}$$

we substitute for $\log n$ the value

$$\log n = 1 + \frac{1}{2} + \ldots + \frac{1}{n} - \gamma + \epsilon_n,$$

where γ is Euler's constant (cf. Vol. I, p. 381) and ϵ_n tends to zero as $n \to \infty$, we obtain an expression for $\frac{1}{\Gamma(x)}$,

$$\frac{1}{\Gamma(x)} = x \lim (1+x)\left(1+\frac{x}{2}\right)\cdots\left(1+\frac{x}{n-1}\right)e^{-x-\frac{x}{2}-\cdots-\frac{x}{n}+\gamma x-\epsilon_n x}$$

$$= xe^{\gamma x} \lim e^{-\epsilon_n x-\frac{x}{n}} \prod_1^{n-1}\left(1+\frac{x}{\nu}\right)e^{-\frac{x}{\nu}}.$$

Since the factor $e^{-\epsilon_n x-\frac{x}{n}}$ tends to 1 as n increases, the product $\prod_1^{\infty}\left(1+\frac{x}{\nu}\right)e^{-\frac{x}{\nu}}$ also converges and gives Weierstrass's infinite product for $\frac{1}{\Gamma(x)}$,

$$\frac{1}{\Gamma(x)} = xe^{\gamma x}\prod_1^{\infty}\left(1+\frac{x}{\nu}\right)e^{-x/\nu},$$

from which we see at once that $\frac{1}{\Gamma(x)}$ has zeros of the first order at the points $x = 0, -1, -2, \ldots$.

4. The Function $\log\Gamma(x)$ and its Derivatives.

If we form the logarithm of Weierstrass's infinite product

$$\frac{1}{\Gamma(x)} = xe^{\gamma x}\prod_{\nu=1}^{\infty}\left(1+\frac{x}{\nu}\right)e^{-x/\nu},$$

we obtain an expression for the function $\log\Gamma(x)$:

$$\log\Gamma(x) = -\log x - \gamma x - \sum_{\nu=1}^{\infty}\left(\log\left(1+\frac{x}{\nu}\right) - \frac{x}{\nu}\right).$$

By the relation

$$\log\left(1+\frac{x}{\nu}\right) - \frac{x}{\nu} = -\frac{1}{\nu}\int_0^x \frac{t\,dt}{\nu+t},$$

whence

$$\left|\log\left(1+\frac{x}{\nu}\right) - \frac{x}{\nu}\right| \leq \frac{1}{\nu^2}\int_0^x t\,dt = \frac{x^2}{2\nu^2},$$

the right-hand side of the equation for $\log\Gamma(x)$ is dominated by the series $\frac{x^2}{2}\sum_1^{\infty}\frac{1}{\nu^2}$, and therefore converges absolutely and uniformly in every closed interval of the positive x-axis.

The derivatives of the function $\log\Gamma(x)$ are of particular interest, since they provide an explicit representation of the values of the series $\sum_0^\infty \left(\frac{1}{x+\nu}\right)^m$.

If we differentiate the expression for $\log\Gamma(x)$ term by term with respect to x, we again obtain a series which, since

$$\frac{1}{x+\nu} - \frac{1}{\nu} = -\frac{x}{\nu(x+\nu)},$$

converges absolutely and uniformly in every closed interval of the positive x-axis. Hence, by known theorems on the differentiation of infinite series,

$$\frac{\Gamma'(x)}{\Gamma(x)} = \frac{d}{dx}\log\Gamma(x) = -\frac{1}{x} - \gamma - \sum_1^\infty\left(\frac{1}{x+\nu} - \frac{1}{\nu}\right).$$

If we again differentiate term by term, we similarly obtain

$$\frac{d^2}{dx^2}\log\Gamma(x) = \sum_{\nu=0}^\infty \frac{1}{(x+\nu)^2}$$

and finally, forming the higher derivatives,

$$\sum_0^\infty \frac{1}{(x+\nu)^m} = \frac{(-1)^m}{(m-1)!}\frac{d^m}{dx^m}\log\Gamma(x) \qquad (m \geqq 2).$$

5. **The Extension Theorem.**

The values of the gamma function for negative values of x can easily be obtained from the values for positive values of x by means of the so-called extension theorem. If we form the product $\Gamma(x)\Gamma(-x)$, which is

$$\lim_{n\to\infty}\frac{1.2\ldots(n-1)}{x(x+1)\ldots(x+n-1)}n^x \lim_{n\to\infty}\frac{1.2\ldots(n-1)}{-x(1-x)(2-x)\ldots(n-1-x)}n^{-x}$$

and combine the two limiting processes into one, we obtain

$$\Gamma(x)\Gamma(-x) = -\frac{1}{x^2}\lim_{n\to\infty}\frac{1}{\{1-(x/1)^2\}\{1-(x/2)^2\}\ldots\{1-(x/(n-1))^2\}}.$$

But by the infinite product for the sine,

$$\frac{\sin\pi x}{\pi x} = \prod_{\nu=1}^\infty\left(1-\left(\frac{x}{\nu}\right)^2\right),$$

deduced in Vol. I, p. 445, we have

$$\Gamma(x)\Gamma(-x) = -\frac{\pi}{x\sin\pi x}.$$

Hence

$$\Gamma(-x) = -\frac{\pi}{x\sin\pi x}\frac{1}{\Gamma(x)}.$$

We can put this relation in a somewhat different form by calculating the product $\Gamma(x)\Gamma(1-x)$. Since

$$\Gamma(1-x) = -x\Gamma(-x),$$

$\Gamma(x)\Gamma(1-x) = -x\Gamma(x)\Gamma(-x)$, and we obtain the *extension theorem*

$$\Gamma(x)\Gamma(1-x) = \frac{\pi}{\sin\pi x}.$$

Thus if we put $x = \frac{1}{2}$, we have $\Gamma(\frac{1}{2}) = \sqrt{\pi}$. Since $\Gamma(\frac{1}{2}) = 2\int_0^\infty e^{-u^2}du$, here is a new proof for the fact that the integral $\int_0^\infty e^{-u^2}du$ has the value $\frac{1}{2}\sqrt{\pi}$. In addition, we can calculate the gamma function for the arguments $x = n + \frac{1}{2}$, where n is any positive integer:

$$\Gamma\left(n+\frac{1}{2}\right) = \left(n-\frac{1}{2}\right)\left(n-\frac{3}{2}\right)\cdots\frac{3}{2}\frac{1}{2}\Gamma\left(\frac{1}{2}\right)$$

$$= \frac{(2n-1)(2n-3)\ldots3.1}{2^n}\sqrt{\pi}.$$

6. The Beta Function.

Another important function defined by an improper integral involving a parameter is Euler's beta function. The beta function is defined by

$$B(x, y) = \int_0^1 t^{x-1}(1-t)^{y-1}dt.$$

If either x or y is less than unity, the integral is improper. By the criterion of section 4, p. 307, however, it converges uniformly in x and y, provided we restrict ourselves to intervals $x \geqq \epsilon$, $y \geqq \eta$, where ϵ and η are arbitrary positive numbers. It therefore represents a continuous function for all positive values of x and y.

We obtain a somewhat different expression by using the substitution $t = \tau + \frac{1}{2}$:

$$B(x, y) = \int_{-\frac{1}{2}}^{\frac{1}{2}} \left(\frac{1}{2} + \tau\right)^{x-1} \left(\frac{1}{2} - \tau\right)^{y-1} d\tau,$$

or, in general, if we now put $\tau = t/2s$, where $s > 0$,

$$(2s)^{x+y-1} B(x, y) = \int_{-s}^{s} (s + t)^{x-1} (s - t)^{y-1} dt.$$

If, finally, we put $t = \sin^2\phi$ in the original formula, we obtain

$$B(x, y) = 2\int_0^{\pi/2} \sin^{2x-1}\phi \cos^{2y-1}\phi \, d\phi.$$

We shall now show how the beta function can be expressed in terms of the gamma function, by using a few transformations which may seem strange at first sight.

If we multiply both sides of the equation

$$(2s)^{x+y-1} B(x, y) = \int_{-s}^{s} (s + t)^{x-1} (s - t)^{y-1} dt$$

by e^{-2s} and integrate with respect to s from 0 to A, we have

$$B(x, y) \int_0^A e^{-2s} (2s)^{x+y-1} ds = \int_0^A e^{-2s} ds \int_{-s}^{s} (s + t)^{x-1} (s - t)^{y-1} dt.$$

The double integral on the right may be regarded as an integral of the function $e^{-2s}(s + t)^{x-1}(s - t)^{y-1}$, the region of integration being the isosceles triangle bounded by the lines

$$s \pm t = 0 \quad \text{and} \quad s = A.$$

If we apply the transformation

$$\sigma = s + t,$$
$$\tau = s - t,$$

this integral becomes

$$\frac{1}{2} \iint_R e^{-\sigma-\tau} \sigma^{x-1} \tau^{y-1} \, d\sigma \, d\tau.$$

As the region of integration we now have the triangle in the $\sigma\tau$-plane bounded by the lines $\sigma = 0$, $\tau = 0$, and $\sigma + \tau = 2A$.

If we now let A increase beyond all bounds, the left-hand side tends to the function

$$\frac{1}{2}\,\mathrm{B}(x, y)\,\Gamma(x+y).$$

The right-hand side must therefore converge also, and its limit is the double integral over the whole first quadrant of the $\sigma\tau$-plane, the quadrant being approximated to by means of isosceles triangles. Since the integrand is positive in this region and the integral converges for a monotonic sequence of regions, by Chap. IV (p. 263) this limit is independent of the mode of approximation to the quadrant.

In particular, we can use squares of side A, and accordingly write

$$\mathrm{B}(x, y)\,\Gamma(x+y) = \lim_{A\to\infty} \int_0^A \int_0^A e^{-\sigma-\tau}\sigma^{x-1}\tau^{y-1}\,d\sigma\,d\tau$$

$$= \int_0^\infty e^{-\sigma}\sigma^{x-1}\,d\sigma \int_0^\infty e^{-\tau}\tau^{y-1}\,d\tau.$$

We therefore obtain the important relation *

$$\mathrm{B}(x, y) = \frac{\Gamma(x)\,\Gamma(y)}{\Gamma(x+y)}.$$

From this relation we see that the beta function is related to the binomial coefficients $\binom{n+m}{n} = \frac{(n+m)!}{n!\;m!}$ in roughly the same

* This equation can also be obtained from Bohr's theorem. We first show that $\mathrm{B}(x, y)$ satisfies the functional equation

$$\mathrm{B}(x+1, y) = \frac{x}{x+y}\,\mathrm{B}(x, y),$$

so that the function

$$u(x, y) = \Gamma(x+y)\,\mathrm{B}(x, y),$$

considered as a function of x, satisfies the functional equation of the gamma function,

$$u(x+1) = xu(x).$$

Since by the theorem in the footnote on p. 328 it follows that $\log u(x, y)$ is a convex function of x, we have

$$\Gamma(x+y)\,\mathrm{B}(x, y) = \Gamma(x)\,.\,a(y),$$

and finally, if we put $x = 1$, $a(y) = \Gamma(y)$.

way as the gamma function is related to the numbers $n!$. For integers $x = n$, $y = m$, in fact, the function

$$\frac{1}{(x+y+1)\,\mathrm{B}(x+1,\, y+1)}$$

has the value $\binom{n+m}{n}$.

Finally, we mention that the definite integrals

$$\int_0^{\pi/2} \sin^a t\, dt \quad \text{and} \quad \int_0^{\pi/2} \cos^a t\, dt,$$

which are identical with the functions

$$\frac{1}{2}\mathrm{B}\left(\frac{a+1}{2}, \frac{1}{2}\right) = \frac{1}{2}\mathrm{B}\left(\frac{1}{2}, \frac{a+1}{2}\right),$$

can be simply expressed in terms of the gamma function:

$$\int_0^{\pi/2} \sin^a t\, dt = \int_0^{\pi/2} \cos^a t\, dt = \frac{\sqrt{\pi}}{a}\,\frac{\Gamma((1+a)/2)}{\Gamma(a/2)}.$$

Examples

1. Prove that the volume of the positive octant bounded by the planes $x = 0$, $y = 0$, $z = h$, and the surface

$$\frac{x^m}{a^m} + \frac{y^m}{b^m} = \frac{z}{c} \qquad (m > 0)$$

is

$$abh\left(\frac{h}{c}\right)^{\frac{2}{m}} \frac{\Gamma\left(1+\frac{1}{m}\right)^2}{\Gamma\left(2+\frac{2}{m}\right)}.$$

2. Prove that

$$\iiint f\left(\frac{x^2}{a^2} + \frac{y^2}{b^2} + \frac{z^2}{c^2}\right) x^{p-1} y^{q-1} z^{r-1}\, dx\, dy\, dz$$

taken throughout the positive octant of the ellipsoid $\frac{x^2}{a^2} + \frac{y^2}{b^2} + \frac{z^2}{c^2} \leqq 1$ is equal to

$$\frac{a^p b^q c^r}{8}\,\frac{\Gamma\left(\frac{p}{2}\right)\Gamma\left(\frac{q}{2}\right)\Gamma\left(\frac{r}{2}\right)}{\Gamma\left(\frac{p+q+r}{2}\right)} \int_0^1 f(\xi)\,\xi^{\frac{p+q+r}{2}-1}\, d\xi.$$

(Hint: Introduce new variables ξ, η, ζ by writing

$$\frac{x^2}{a^2}+\frac{y^2}{b^2}+\frac{z^2}{c^2}=\xi \qquad x=a\sqrt{\xi(1-\eta)}$$

$$\frac{y^2}{b^2}+\frac{z^2}{c^2}=\xi\eta \qquad \text{or} \qquad y=b\sqrt{\xi\eta(1-\zeta)}$$

$$\frac{z^2}{c^2}=\xi\eta\zeta \qquad z=c\sqrt{\xi\eta\zeta},$$

and perform the integrations with respect to η and ζ.)

3. Find the x-co-ordinate of the centre of mass of the solid

$$\left(\frac{x}{a}\right)^{\frac{1}{n}}+\left(\frac{y}{b}\right)^{\frac{1}{n}}+\left(\frac{z}{c}\right)^{\frac{1}{n}}\leqq 1, \quad x\geqq 0, y\geqq 0, z\geqq 0.$$

4. Find the moment of inertia of the area enclosed by the astroid

$$x^{\frac{2}{3}}+y^{\frac{2}{3}}=R^{\frac{2}{3}}$$

with respect to the x-axis.

5. Prove that

$$2^{2x}\frac{\Gamma(x)\Gamma(x+\frac{1}{2})}{\Gamma(2x)}=2\sqrt{\pi}.$$

7. Differentiation and Integration to Fractional Order. Abel's Integral Equation

Using our knowledge of the gamma function, we shall now carry out a simple process of generalization of the concepts of differentiation and integration. We have already seen (p. 221) that the formula

$$F(x)=\int_0^x\frac{(x-t)^{n-1}}{(n-1)!}f(t)\,dt=\frac{1}{\Gamma(n)}\int_0^x(x-t)^{n-1}f(t)\,dt$$

gives the n-times-repeated integral of the function $f(x)$ between the limits 0 and x. If D symbolically denotes the operator in differentiation and if D^{-1} denotes the operator $\int_0^x \cdot\cdot\, dx$, which is the inverse of differentiation, we may write

$$F(x)=D^{-n}f(x).$$

The mathematical statement conveyed by this formula is that the function $F(x)$ and its first $(n-1)$ derivatives vanish at $x=0$ and the n-th derivative of $F(x)$ is $f(x)$. But it is now very natural to construct a definition for the operator $D^{-\lambda}$ even when the positive number λ is not necessarily an integer. *The integral of order* λ *of the function* f(x) *between the limits* 0 *and* x *is defined by the expression*

$$D^{-\lambda}f(x) = \frac{1}{\Gamma(\lambda)}\int_0^x (x-t)^{\lambda-1}f(t)\,dt.$$

This definition may now be used to generalize nth-order differentiation, symbolized by the operator D^n or $\frac{d^n}{dx^n}$, to μth-order differentiation, where μ is an arbitrary non-negative number. Let m be the least integer greater than μ, so that $\mu = m - \rho$, where $0 < \rho < 1$. Then our definition is

$$D^{\mu}f(x) = D^m D^{-\rho}f(x) = \frac{d^m}{dx^m}\frac{1}{\Gamma(\rho)}\int_0^x (x-t)^{\rho-1}f(t)\,dt.$$

A reversal of the order of the two processes would give the definition

$$D^{\mu}f(x) = D^{-\rho}D^m f(x) = \frac{1}{\Gamma(\rho)}\int_0^x (x-t)^{\rho-1}f^{(m)}(t)\,dt.$$

It may be left as an exercise for the reader to employ the formulæ for the gamma function to prove that

$$D^{\alpha}D^{\beta}f(x) = D^{\beta}D^{\alpha}f(x),$$

where α and β are arbitrary real numbers. He should show that these relations and the generalized process of differentiation have a meaning whenever the function $f(x)$ is differentiable in the ordinary way to a sufficiently high order. In general $D^{\mu}f(x)$ exists if $f(x)$ has continuous derivatives up to and including the mth order.

In connexion with these ideas we may mention Abel's *integral equation*, which has important applications. Since $\Gamma(\frac{1}{2}) = \sqrt{\pi}$, the integral of a function $f(x)$ to the order $\frac{1}{2}$ is given by the formula

$$D^{-\frac{1}{2}}f(x) = \frac{1}{\sqrt{\pi}}\int_0^x \frac{f(t)}{\sqrt{x-t}}\,dt = \psi(x).$$

If we assume that the function $\psi(x)$ on the right-hand side is given and that it is required to find $f(x)$, then the above formula is Abel's integral equation. If the function $\psi(x)$ is continuously differentiable and vanishes at $x = 0$, the solution of the equation is given by the formula

$$f(x) = D^{\frac{1}{2}}\psi(x),$$

or

$$f(x) = \frac{1}{\sqrt{\pi}} \frac{d}{dx} \int_0^x \frac{\psi(t)}{\sqrt{x-t}}\, dt.$$

8. Note on the Definition of the Area of a Curved Surface

In section 6 of Chap. IV (p. 269) we defined the area of a curved surface in a way somewhat dissimilar to that in which we defined the length of arc in Vol. I, Chap. V (p. 277). In the definition of length we started with inscribed polygons, while in the definition of area we used tangent planes instead of inscribed polyhedra.

In order to see why we cannot use inscribed polyhedra, we may consider a cylindrical surface in xyz-space with the equation $x^2 + y^2 = 1$, lying between the planes $z = 0$ and $z = 1$. The area of this cylindrical surface is 2π. In it we now inscribe a polyhedral surface, all of whose faces are identical triangles, as follows. We first subdivide the circumference of the unit circle into n equal parts, and on the cylinder we consider the m equidistant horizontal circles $z = 0$, $z = h$, $z = 2h$, . . . , $z = (m - 1)h$, where $h = 1/m$. We perform the subdivision of each of these circles into n equal parts in such a way that the points of division of each circle lie above the centres of the arcs of the preceding circle. We now consider a polyhedron inscribed in the cylinder whose edges consist of the chords of the circles and of the lines joining neighbouring points of division of neighbouring circles. The faces of this polyhedron are congruent isosceles triangles, and if n and m are chosen sufficiently large this polyhedron will lie as close as we please to the cylindrical surface. If we now keep n fixed, we can choose m so large that each of the triangles is as nearly parallel as we please to the xy-plane and therefore makes an arbitrarily steep angle with the surface of the cylinder Then we can no longer expect that the sum of the areas of the triangles will be an approximation to the area of the cylinder.

In fact, for the bases of the individual triangles we have the value $2\sin\pi/n$, and for the altitude, by Pythagoras' theorem, we have

$$\sqrt{\frac{1}{m^2}+\left(1-\cos\frac{\pi}{n}\right)^2}=\sqrt{\frac{1}{m^2}+4\sin^4\frac{\pi}{2n}}.$$

Since the number of triangles is obviously $2mn$, the surface area of the polyhedron is

$$F_{n,m}=2mn\sin\frac{\pi}{n}\sqrt{\frac{1}{m^2}+4\sin^4\frac{\pi}{2n}}=2n\sin\frac{\pi}{n}\sqrt{1+4m^2\sin^4\frac{\pi}{2n}}.$$

The limit of this expression is not independent of the way in which m and n tend to infinity. If, for example, we keep n fixed and let $m\to\infty$, the expression increases beyond all bounds. If, however, we make m and n tend to ∞ together, putting $m=n$, the expression tends to 2π. If we put $m=n^2$, we obtain the limit $2\pi\sqrt{1+\pi^4/4}$, and so on. From the above expression $F_{n,m}$ for the area of the polyhedron we see that the lower limit (lower point of accumulation; cf. Vol. I, p. 62) of the set of numbers $F_{n,m}$ is 2π; this follows at once from $F_{n,m}\geqq 2n\sin\pi/n$ and $\lim\limits_{n\to\infty}2n\sin\pi/n=2\pi$.

In conclusion we mention—without proof—a theoretically interesting fact of which the example just given is a particular instance. If we have any arbitrary sequence of polyhedra tending to a given surface, we have seen that the areas of the polyhedra need not tend to the area of the surface. But the limit of the areas of the polyhedra (if it exists), or, more generally, any point of accumulation of the values of these areas, is always greater than, or at least equal to, the area of the curved surface. If for every sequence of such polyhedral surfaces we find the lower limit of the area, these numbers form a definite set of numbers associated with the curved surface. *The area of the surface can be defined as the lower limit* (lower point of accumulation) *of this set of numbers.**

* This remarkable property of the area is called *semi-continuity,* or more precisely *lower semi-continuity.*

CHAPTER V

Integration over Regions in Several Dimensions

The multiple integrals discussed in the previous chapter are not the only possible extension of the idea of integral to the case of more than one independent variable. On the contrary, there are other generalizations, corresponding to the fact that regions of several dimensions may enclose other manifolds of fewer dimensions and we can consider integrals over such manifolds. In the case of two independent variables, in addition to integrals over two-dimensional regions we can consider integrals along curves, which are one-dimensional manifolds. In the case of three independent variables, besides integrals throughout three-dimensional regions and integrals along curves, we have to consider integrals over curved surfaces, which are two-dimensional manifolds enclosed in three-dimensional space. These concepts of integrals along curves (curvilinear integrals), integrals over surfaces, and so on, with many straightforward applications, will be introduced and their mutual relations will be investigated in the present chapter.

1. Line Integrals

We associate the definition of the single integral with the intuitive idea of *area* (Vol. I, Chap. II, p. 77) and arrive at the multiple integral by straightforward generalization to the case of a greater number of dimensions. On the other hand, the physical idea of *work* also leads us to the single integral (Vol. I, Chap. V, p. 304). If we seek to give a mathematical definition of work for an arbitrary field of force in space of more than one dimension, we obtain the curvilinear or line integral as a new generalization of the original concept of the integral of a function of a single variable.

1. Definition of the Line Integral. Notation.

We begin with the purely mathematical definition of the integral along a curve (*line integral, curvilinear integral*), in three-dimensional xyz-space. Let a sectionally smooth * curve C in this space be given parametrically by the equations

$$x = x(t), \quad y = y(t), \quad z = z(t),$$

where, as usual, $x(t)$, $y(t)$, $z(t)$ are continuous functions with sectionally continuous first derivatives. We consider an arc of this curve joining the points P_0 and P with co-ordinates (x_0, y_0, z_0) and $(\bar{x}, \bar{y}, z)$ respectively and corresponding, say, to the values of the parameter t in the interval $\alpha \leqq t \leqq \beta$. If a continuous function $f(x, y, z)$ is defined in any region containing this arc, then along the arc this function will be a function $f(x(t), y(t), z(t))$ of the parameter t alone. In order to define, in analogy with the ordinary integral, a line integral of the function along the curve C, we divide up the arc into small pieces by means of the points $P_0, P_1, P_2, \ldots, P_n$, $(P_n = P)$ and denote the difference of the abscissæ of P_i and P_{i+1} by Δx_i. We now form the sum

$$\sum_{i=0}^{n-1} f(x(t_i), y(t_i), z(t_i))\Delta x_i,$$

where t_i can be given any value in that interval of the parameter which corresponds to the arc between P_i and P_{i+1}. If we let the number of points of subdivision increase beyond all bounds and assume that the length of the longest of the arcs $P_i P_{i+1}$ tends to zero, then we may expect that the above sum will tend to a definite limit. This limit we denote by

$$\int_C f(x, y, z)\, dx$$

and call it a *line integral of the function* f(x, y, z) *along the curve* C. That this limit does exist and is actually independent of the

* Here, as before (cf. p. 41), we say that a curve is sectionally smooth (Ger. *stückweise glatt*) if it consists of a finite number of arcs, each one of which has a continuously turning tangent at each of its points, including the end-points.

choice of the points of division can be proved directly, just as we proved the existence of the ordinary integral. It can be proved even more simply, however, by writing the sum in the form

$$\sum_{i=0}^{n-1} f(x(t_i), y(t_i), z(t_i)) \frac{\Delta x_i}{\Delta t_i} \Delta t_i,$$

where Δt_i denotes the increment of the parameter t as we pass from one point of subdivision to the next. By the definition of the ordinary integral, in the passage to the limit the right-hand side tends to

$$\int_\alpha^\beta f(x(t), y(t), z(t)) \frac{dx}{dt} dt,$$

and for the line integral we obtain the expression

$$\int_C f(x, y, z) dx = \int_\alpha^\beta f(x, y, z) \frac{dx}{dt} dt,$$

which expresses the line integral as an ordinary integral with respect to the parameter t.

The ordinary integral is a special case of the line integral, which arises if we take an interval of the x-axis as the path of integration.

We can now define the line integrals

$$\int_C f(x, y, z) dy = \int_\alpha^\beta f(x, y, z) \frac{dy}{dt} dt$$

and

$$\int_C f(x, y, z) dz = \int_\alpha^\beta f(x, y, z) \frac{dz}{dt} dt$$

just as above. Using the right-hand side of the formulæ, we can verify the fact that *the line integral depends only on the curve itself* and not on the way in which it is expressed, i.e. not on the choice of parameter. For if we use the continuously differentiable function $\phi(t)$ to introduce a new parameter $\tau = \phi(t)$ and if in the interval in question $d\phi(t)/dt > 0$, then we have a one-to-one transformation of the parameter interval into a parameter interval $\alpha_1 \leqq \tau \leqq \beta_1$, and

$$\int_\alpha^\beta f(x, y, z) \frac{dx}{dt} dt = \int_{\alpha_1}^{\beta_1} f(x, y, z) \frac{dx}{d\tau} d\tau.$$

In applications line integrals usually occur in the following combination. Let $a(x, y, z)$, $b(x, y, z)$, $c(x, y, z)$ be three functions which are continuous in a region containing C. We consider the sum of the three line integrals

$$\int_C a(x, y, z)\,dx + \int_C b(x, y, z)\,dy + \int_C c(x, y, z)\,dz,$$

which can also be written in the form

$$\int_C \{a\,dx + b\,dy + c\,dz\} = \int_\alpha^\beta (a\dot{x} + b\dot{y} + c\dot{z})\,dt,$$

where, as before, $dx/dt = \dot{x}$, and so on. We suppose that the functions a, b, c are respectively the x-, y-, and z-components of a vector $\boldsymbol{A}$ and that $\boldsymbol{x}$ is the position vector of the point (x, y, z) of the curve. Then the quantities $\dot{x}$, $\dot{y}$, $\dot{z}$ are the components of the vector $\dot{\boldsymbol{x}} = d\boldsymbol{x}/dt$, and we can write the integrand as the scalar product $\boldsymbol{A}\dot{\boldsymbol{x}}$. For the line integral we thus have the expression

$$\int_\alpha^\beta \boldsymbol{A}\dot{\boldsymbol{x}}\,dt = \int_C \boldsymbol{A}\,d\boldsymbol{x},$$

where the meaning of the notation is obvious.

Just as we have considered line integrals in three-dimensional space, so we can of course consider similar integrals in the plane:

$$\int_C f(x, y)\,dx, \quad \int_C f(x, y)\,dy, \quad \int_C \{a\,dx + b\,dy\}.$$

Moreover, these ideas can be extended to line integrals of functions of n variables. In this general case we can most simply define a line integral

$$\int_C f(x_1, x_2, \ldots, x_n)\,dx_i$$

by supposing that in n-dimensional space the n quantities $x_1, x_2, \ldots, x_n$ are all given as functions of a parameter t in the interval $\alpha \leqq t \leqq \beta$. The values $x_1(t), x_2(t), \ldots, x_n(t)$ in this interval then correspond to a curve C in n-dimensional space. *We then define the line integral*

$$\int_C f(x_1, x_2, \ldots, x_n)\,dx_i$$

by the expression

$$\int_\alpha^\beta f(x_1(t), x_2(t), \ldots, x_n(t)) \frac{dx_i}{dt} dt.$$

If we consider n functions $a_1, a_2, \ldots, a_n$ of the n variables $x_1, x_2, \ldots, x_n$, then we can again form the general line integral

$$\int_C \{a_1 dx_1 + a_2 dx_2 + \ldots + a_n dx_n\}$$

and express it in vector notation in the form

$$\int_\alpha^\beta \boldsymbol{A}\dot{\boldsymbol{x}}\, dt = \int_C \boldsymbol{A}\, d\boldsymbol{x},$$

where, as above, by $\boldsymbol{A}$ we mean the " vector " with components $(a_1, a_2, \ldots, a_n)$ and by $\boldsymbol{x}$ the position vector of the point $(x_1, x_2, \ldots, x_n)$.

The formulæ for the area of a region bounded by a closed curve C (Vol. I, Chap. V, section 2, p. 273) provide an instance where a line integral occurs naturally. If the closed sectionally smooth curve C in the xy-plane is given by the equations $x = x(t)$, $y = y(t)$, the area A of the region bounded by the curve is given by

$$A = -\int_\alpha^\beta y\dot{x}\, dt = \int_\alpha^\beta x\dot{y}\, dt = -\frac{1}{2}\int_\alpha^\beta \{y\dot{x} - x\dot{y}\}\, dt.$$

In our new terminology these are simply the line integrals

$$A = -\int_C y\, dx = \int_C x\, dy = -\frac{1}{2}\int_C \{y\, dx - x\, dy\},$$

taken round C in the direction in which the value of the parameter increases.

2. Fundamental Rules.

From the expression for line integrals in terms of ordinary integrals we may draw several immediate conclusions.

The value of the line integral depends on the sense in which the curve C *is described, and in fact is multiplied by* -1 *if the sense of description is reversed*, i.e. if the curve is described from P to P_0 instead of from P_0 to P. The proof of this is self-evident. This sign property makes it always convenient to think of the curve C as having a definite direction; we then call it an *oriented* curve (cf. Vol. I, Chap. V, section 2, p. 268). We shall occasionally use the symbol $-C$ to denote the curve obtained by describing C in the reverse direction.

If the curve C is formed by joining together two curves C_1 and C_2 described in succession (which we may indicate by writing $C = C_1 + C_2$), then the relation

$$\int_C = \int_{C_1} + \int_{C_2}$$

holds for the corresponding line integrals, the meaning of the notation being obvious.

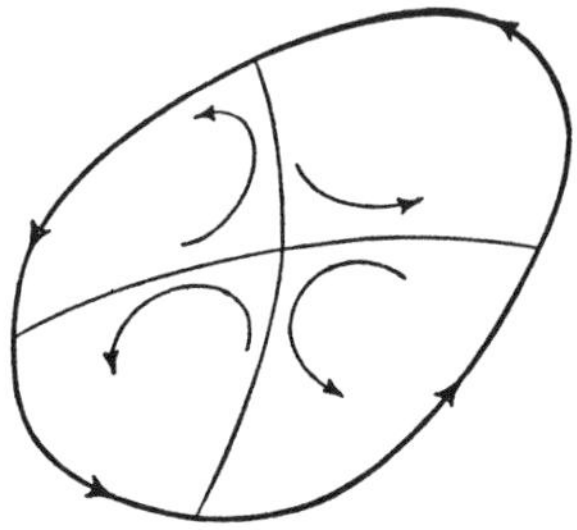

Fig. 1

The following rule is particularly important. If we restrict ourselves to the case of two variables x, y and consider a line integral

$$\int_C \{a\,dx + b\,dy\}$$

along a *closed* curve C (like that in fig. 1) within which the vector field a, b is everywhere defined and continuous, then *the formula*

$$\int_C \{a\,dx + b\,dy\}$$
$$= \int_{C_1} \{a\,dx + b\,dy\} + \int_{C_2} \{a\,dx + b\,dy\} + \dots + \int_{C_n} \{a\,dx + b\,dy\}$$

holds for every resolution of the closed region R *bounded by the oriented curve* C *into similarly bounded sub-regions* $R_1, R_2, \dots, R_n$ *with boundary curves* $C_1, C_2, \dots, C_n$. Here we assume that all the regions are described in the same sense. To prove the statement, we notice that in the addition of the integrals on the right the parts which are taken over a portion of the boundary C add together as is required to form the integral round C, while every boundary curve lying within R is the common boundary of two sub-regions and is consequently described twice, once in each direction, so that the integrals along these arcs cancel one another.

Exactly the same result applies to the resolution of a line integral along a curve C in three (or more) dimensions, provided that the curve forms the boundary of a portion of a surface and this portion is subdivided by the curves $C_1, C_2, \dots, C_n$.

A somewhat different application of this principle occurs in the following theorem. Let two oriented closed curves C and C' (cf. fig. 2) be subdivided by the points $A_1, \ldots, A_n$ and $A_1', \ldots, A_n'$

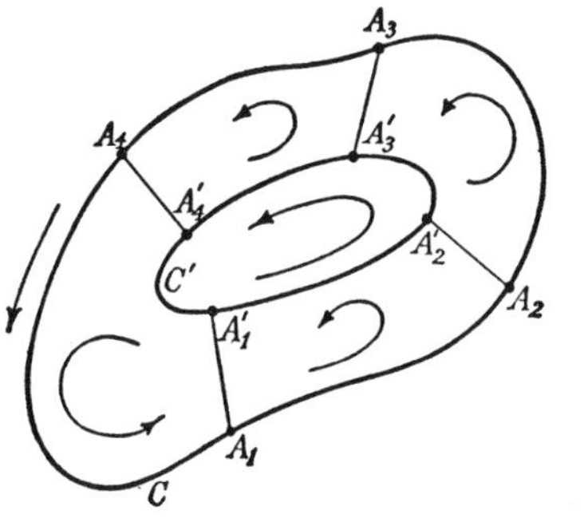

Fig. 2.—$\sum_{i=1}^{n} \int_{C_i} = \int_C - \int_{C'}$

respectively, in the order of the sense of orientation, and let each pair of corresponding points A_i and A_i' be joined by a curved line. If by C_i we denote the closed oriented curve $A_i A_{i+1} A_{i+1}' A_i'$, then

$$\sum_{i=1}^{n} \int_{C_i} (a\,dx + b\,dy) = \int_C (a\,dx + b\,dy) - \int_{C'} (a\,dx + b\,dy).$$

The proof of this theorem is immediately suggested by the figure. In order that it may hold, it is not necessary to assume that the two curves C and C' never intersect themselves or one another.

Finally, we mention an *integral estimate for line integrals*:

$$\int_C \{a\,dx + b\,dy + c\,dz\} \leqq ML,$$

where M is an upper bound of $\sqrt{(a^2 + b^2 + c^2)}$ on C and L is the length of C. The proof follows at once from the inequality

$$\left| a\frac{dx}{dt} + b\frac{dy}{dt} + c\frac{dz}{dt} \right| \leqq \sqrt{a^2+b^2+c^2}\sqrt{\left(\frac{dx}{dt}\right)^2 + \left(\frac{dy}{dt}\right)^2 + \left(\frac{dz}{dt}\right)^2},$$

which is obtained by applying Schwarz's inequality (Vol. I, p. 12).

3. Interpretation of Line Integrals in Mechanics.

As we have already mentioned, the line integral is closely related to the idea of *work*. If a particle moves along a curve under the influence of a field of force—which in general may vary from point to point—and if the field of force is given by the vector $\boldsymbol{A}$ with components a, b, c, the line integral represents the work done by the field on the particle. For, if the force is constant and the motion takes place in a straight line, the work is defined as the scalar product of the force vector and the " displacement " vector. In order to generalize this definition convincingly, we replace the path C by the polygon with vertices $P_0, P_1, P_2, \ldots, P_n = P$, and instead of the actual force we take a " substitute force " which is constant along each of these segments P_iP_{i+1}, being equal to the actual value of the force at the initial point P_i. The work performed by this substitute force along the segment from P_i to P_{i+1} is

$$a(x_i, y_i, z_i)\Delta x_i + b(x_i, y_i, z_i)\Delta y_i + c(x_i, y_i, z_i)\Delta z_i,$$

since the displacement vector from P_i to P_{i+1} has the components $\Delta x_i, \Delta y_i, \Delta z_i$. If we sum over the whole polygon, we obtain an expression which tends to the line integral as we pass to the limit $n \to \infty$. Thus the line integral is actually the expression for the work done during the motion.

Other physical interpretations of the line integral will be given later (cf. section 3, pp. 370–1).

4. Integration of Total Differentials.

A particularly important case is that in which the vector $\boldsymbol{A}$ with components (a, b, c) is the *gradient of a potential*,* i.e. there exists a function $F(x, y, z)$ of the co-ordinates such that

$$\boldsymbol{A} = \operatorname{grad} F$$

or

$$a = F_x, \quad b = F_y, \quad c = F_z.$$

Although in general the value of a line integral in a vector field depends not only on the end-points but also on the entire course of the curve C, the following theorem is valid here:

The line integral over a gradient field is equal to the difference between the values of the potential function at the end-points and does not depend on the course of C *between the end-points.* That is, we obtain the same value for all curves which join the two end-points and remain entirely within the region in which the potential function F is defined.

* If $A = \operatorname{grad} F$, then the function F is often called the *potential* of the vector field.

In this case the line integral takes the form

$$\int_C \{a\,dx + b\,dy + c\,dz\} = \int_\alpha^\beta \{F_x\dot{x} + F_y\dot{y} + F_z\dot{z}\}\,dt,$$

and the expression in brackets on the right is simply the derivative dF/dt of the function F with respect to the parameter t. We can therefore perform the integration explicitly, and obtain on the right the difference of the values of F at the end point and the initial point of the path of integration. In this case, therefore, we at once have the formula

$$\int_C \{a\,dx + b\,dy + c\,dz\} = F(x(\beta),\ y(\beta),\ z(\beta)) - F(x(\alpha),\ y(\alpha),\ z(\alpha)).$$

This applies e.g. to the field of force due to a gravitating particle, which we have already (Chap. II, section 7, p. 91) recognized as the gradient field of the potential $1/r$. The work done by this gravitational force when another particle moves from its initial position to its final position is therefore independent of the path.

The expression $a\,dx + b\,dy + c\,dz$ is formally identical with what we have (p. 66) called the *total differential* of the function $F(x, y, z)$,

$$a\,dx + b\,dy + c\,dz = dF.$$

We may therefore write our formula in the form

$$\int_C dF = F(x(\beta), y(\beta), z(\beta)) - F(x(\alpha), y(\alpha), z(\alpha))$$

and speak of *integrating the total differential* $a\,dx + b\,dy + c\,dz$.

The following fact is of fundamental importance. The statement "*the integral is independent of the path*" is equivalent to the statement "*the integral round a closed curve has the value zero*". For if we subdivide a closed curve by means of two points P_0 and P into two arcs C and C_1, the equality of the two line integrals taken along C and C_1 from P_0 to P means exactly the same thing as the vanishing of the sum of the integral taken along C in the direction from P_0 to P and the integral taken along C_1 in the direction from P to P_0; and this sum is the integral taken round the closed curve.

5. The Main Theorem on Line Integrals.

As we have already emphasized, it is only under very special conditions that a line integral is independent of the path, or, what is equivalent, that the line integral round a closed curve is zero. For example, if a closed curve C forms the boundary of a region of positive area, then by p. 347 the line integral $\int x\,dy$ or $\int (x\,dy - y\,dx)$ is not zero. The chief problem of the theory of line integrals is to show that the *sufficient* condition for independence of the path, given on p. 350, is also *necessary*, and then to express this necessary and sufficient condition in a convenient and useful form.

We shall first investigate this question of independence of the path in the case of plane curves. We may add in advance that the results in the case of three or more variables are exactly analogous.

We now make the following assumptions. Let the functions $a(x, y)$ and $b(x, y)$ (which we shall again interpret as components of a plane vector field $\boldsymbol{A}$), together with their partial derivatives a_y and b_x, be continuous in a region R of the plane. The following theorem then holds:

The line integral

$$\int_C \{a\,dx + b\,dy\}$$

taken along the curve C *in* R *is independent of the particular choice of the path* C *and is determined solely by the initial and final points of the curve* C, *if, and only if,* a dx + b dy *is the total differential of a function* U(x, y), *that is, if, and only if, a function* U(x, y) *exists in* R *such that the relations*

$$U_x = a, \qquad U_y = b$$

or

$$\boldsymbol{A} = \operatorname{grad} U$$

hold everywhere in R.

We have already proved on p. 351 that this condition is *sufficient*, i.e. that from this it does actually follow that the integral is independent of the path.

It is easy to see that the condition is *necessary*. If the integral is independent of the path, then for a fixed initial point P_0 of C it is a (one-valued) function $U(\xi, \eta)$ of the co-ordinates (ξ, η) of the end-point P. $U(\xi, \eta)$ is differentiable with respect to ξ and η, and in fact for every interior point of R we have

$$U_\xi(\xi, \eta) = \lim_{h \to 0} \frac{1}{h}\{U(\xi + h, \eta) - U(\xi, \eta)\}$$

$$= \lim_{h \to 0} \frac{1}{h}\left[\int_{C+C_h}\{a\,dx + b\,dy\} - \int_C\{a\,dx + b\,dy\}\right]$$

$$= \lim_{h \to 0} \frac{1}{h}\int_{C_h}\{a\,dx + b\,dy\}.$$

Here C is *any* sectionally smooth curve whatever joining P_0 to the point P in R, and C_h is a sectionally smooth curve in R joining P to the point P_1 with co-ordinates $(\xi + h, \eta)$. Since for sufficiently small values of h the line-segment PP_1 belongs to R, this segment can be taken as the path of integration C_h. Then the parametric representation $x = t$, $y = \eta$, $\xi \leqq t \leqq \xi + h$ of this curve C_h gives

$$U_\xi(\xi, \eta) = \lim_{h \to 0} \frac{1}{h}\int_\xi^{\xi+h} a(t, \eta)\,dt = a(\xi, \eta).$$

Similarly, we find that

$$U_\eta(\xi, \eta) = \lim_{h \to 0} \frac{1}{h}\int_\eta^{\eta+h} b(\xi, t)\,dt = b(\xi, \eta).$$

Hence it is actually true that $U_x(x, y) = a$, $U_y(x, y) = b$, as was stated. This result, which has so far been proved only for interior points of R, holds on the boundary also, in virtue of the continuity of all our functions.

The above theorem, however, is of no great value, since as yet we have no general way of finding whether the vector field $\boldsymbol{A}$ is a gradient field or not. Instead of the gradient character of the vector field, we therefore attempt to state some other condition referring only to the functions a and b themselves. This is given in the following main theorem:

If R *is a simply-connected (open) region, a necessary and at the same time a sufficient condition that the integral* $\int_C (a\,dx + b\,dy)$ *shall be independent of the path* C *joining two given points in* R *is that the "condition of integrability"*

$$a_y = b_x$$

is satisfied for all points of R. *For a fixed initial point of* C *the integral* $\int_C (a\,dx + b\,dy)$ *then represents a function* $U(\xi, \eta)$ *of the co-ordinates* (ξ, η) *of the end-point, and the vector field* $\boldsymbol{A}$ *is the gradient field of this function* U, which may therefore be called the *potential* of the field.

That the condition is necessary follows from the theorem which we first stated and proved. For by this theorem, if the integral is independent of the path, a function $U(x, y)$ exists in R for which $U_x = a$ and $U_y = b$. Since the derivatives

$$U_{yx} = a_y(x, y) \quad \text{and} \quad U_{xy} = b_x(x, y)$$

are continuous, by Chap. II, section 3 (p. 55) the equation $U_{xy} = U_{yx}$ holds, and therefore

$$a_y(x, y) = b_x(x, y),$$

as stated.

In order to show that the condition $a_y = b_x$ is also sufficient, and consequently equivalent to the condition that $\boldsymbol{A}$ is a gradient, we must now use the assumption $a_y = b_x$ to construct a function $U(x, y)$ in R such that $U_x = a(x, y)$ and $U_y = b(x, y)$. We first consider the simple case in which R is a rectangle with sides parallel to the axes, given by the inequalities $\alpha < x < \beta$, $\gamma < y < \delta$. The fixed point P_0 of the region with co-ordinates (ξ_0, η_0) is joined to the point P with co-ordinates (ξ, η) by means of two line-segments P_0P', $P'P$ parallel to the axes, meeting at the point P' with co-ordinates (ξ_0, η). The line P_0P' is parametrically represented by $x = \xi_0$, $y = t$, where $\eta_0 \leqq t \leqq \eta$, and $P'P$ by $x = t$, $y = \eta$, where $\xi_0 \leqq t \leqq \xi$ (cf. fig. 3). Hence the integral $\int(a\,dx + b\,dy)$ from P_0 to P taken along this pair of lines is given by

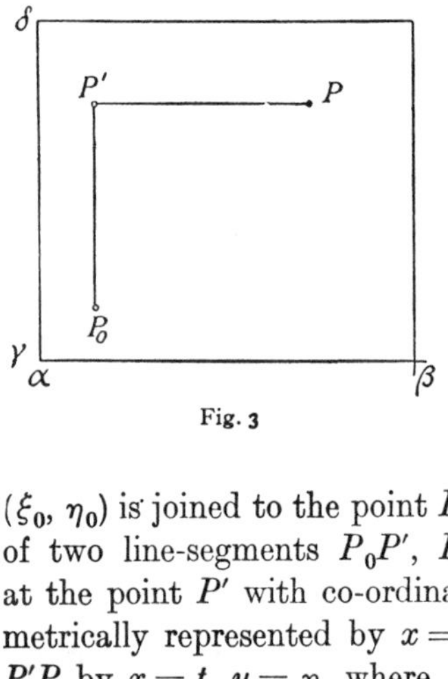

Fig. 3

$$\int_C \{a\,dx + b\,dy\} = \int_{\eta_0}^{\eta} b(\xi_0, t)\,dt + \int_{\xi_0}^{\xi} a(t, \eta)\,dt.$$

The function

$$U(\xi, \eta) = \int_{\eta_0}^{\eta} b(\xi_0, t)\,dt + \int_{\xi_0}^{\xi} a(t, \eta)\,dt$$

defined in this way is the function required. For by differentiation we at once have

$$U_\xi(\xi, \eta) = a(\xi, \eta)$$

and

$$U_\eta(\xi, \eta) = b(\xi_0, \eta) + \frac{\partial}{\partial \eta}\int_{\xi_0}^{\xi} a(t, \eta)\,dt.$$

Since $a_\eta(t, \eta)$ is continuous, we may differentiate under the integral sign on the right:

$$U_\eta(\xi, \eta) = b(\xi_0, \eta) + \int_{\xi_0}^{\xi} a_\eta(t, \eta)\,dt.$$

As $a_y(x, y) = b_x(x, y)$, we have

$$\begin{aligned} U_\eta(\xi, \eta) &= b(\xi_0, \eta) + \int_{\xi_0}^{\xi} b_t(t, \eta)\,dt \\ &= b(\xi_0, \eta) + b(\xi, \eta) - b(\xi_0, \eta) = b(\xi, \eta). \end{aligned}$$

Thus the statement about the derivatives of $U(\xi, \eta)$ is proved, and from this it follows at once that the line integral is independent of the path. In general, therefore,

$$U(\xi, \eta) = \int_C (a\,dx + b\,dy),$$

where C is an arbitrary sectionally smooth curve joining P_0 to P and lying in the rectangle. The theorem is accordingly proved for the case of a rectangular region R.

To generalize the result for any simply-connected region R we have merely to extend the construction of the function U to such a general region. We say that a two-dimensional open region is *simply-connected* if every closed polygon within it can, by a continuous deformation within the region, be made to shrink up to a point. This pictorial idea of shrinking to a point can be made precise in the following way. Let the vertices of the polygon Π be $P_0, P_1, \ldots, P_n$ with co-ordinates (x_0, y_0), $(x_1, y_1), \ldots, (x_n, y_n)$ respectively. We now think of these vertices as moving continuously with the time, starting at $P_0, P_1, \ldots, P_n$ respectively when $t = 0$ and all coming together at time $t = 1$

at one and the same point (ξ, η) in R. That is, we suppose that there are points $P_0(t), P_1(t), \ldots, P_n(t)$, whose co-ordinates $(x_0(t), y_0(t)), (x_1(t), y_1(t)), \ldots, (x_n(t), y_n(t))$ are continuous functions of t for $0 \leqq t \leqq 1$, and also that

$$P_0(0) = (x_0(0), y_0(0)) = P_0, \ldots, P_n(0) = (x_n(0), y_n(0)) = P_n$$

and

$$P_0(1) = (x_0(1), y_0(1)) = (\xi, \eta), \ldots, P_n(1) = (x_n(1), y_n(1)) = (\xi, \eta).$$

Of course any closed polygon can be made to shrink to a point if we do not restrict its position in any way. The essential feature of our definition of a simply-connected region is that *every* closed polygon *in the region* can be shrunk to a point, *the polygon* $\Pi(t)$ *with vertices* $P_0(t), P_1(t) \ldots, P_n(t), P_0(t)$ *remaining in the region during the whole process of shrinking*, i.e. for all values of t in the interval $0 \leqq t \leqq 1$.

It is intuitively clear that this definition agrees with that on p. 41. For if our region R is multiply-connected in the sense of p. 41, there is a " hole " in it, and a closed polygon in R enclosing this " hole " cannot be shrunk to a point without crossing the " hole ", i.e. without leaving R. Conversely, if there are no " holes " in R, any closed polygon can be shrunk to a point. We shall not prove this analytically, however, as the proof is lengthy and, moreover, we require only the definition given here.

We shall see that in the generalization of our main theorem the limitation to a simply-connected region R is essential.

This generalization for any simply-connected region follows the same lines as the proof for rectangles, in that we again construct a function $U(x, y)$ in the region R for which $U_x = a$ and $U_y = b$. Starting from an arbitrary point P_0 in R, we define $U(x, y)$ by the statement

$$U(x, y) = \int_{P_0}^{P} (a\,dx + b\,dy),$$

where the path of integration is any polygonal path in R joining the point P_0 to the point $P(x, y)$. If we can show that the value $U(x, y)$ thus defined is independent of the particular polygonal path which we have chosen, then we have actually constructed a function which satisfies the conditions $U_x = a$, $U_y = b$.

We therefore have merely to prove that the integral is independent of the path, or instead, that the integral $\int(a\,dx + b\,dy)$ round a closed polygon Π containing the point P_0 vanishes. For this purpose we make Π shrink to a point in R; that is, in R we form the polygon $\Pi(t)$ with vertices $P_0(t)$, $P_1(t)$, . . . , $P_n(t)$ which coincides with Π at $t = 0$ and reduces to a single point at $t = 1$. Since the " line integral " for a single point—a curve of zero length—clearly has the value zero, our problem is merely that of showing that the line integral along $\Pi(t)$ remains constant as t varies from 0 to 1; we shall then know that the integral along $\Pi(t)$ is 0 for all values of t, and, in particular, that the integral along Π is 0 for $t = 0$.

Now consider any value t' of t. Since the polygon $\Pi(t')$ lies within R, we can choose a sequence of points (not necessarily vertices) $A_0' = P_0(t')$, A_1', A_2', . . . , $A_m' = A_0'$ on $\Pi(t')$ so close together that each pair A_i', A_{i+1}' lies within a rectangle R_i interior to R. If t is any parametric value close enough to t', the polygon $\Pi(t)$ lies so close to $\Pi(t')$ that on $\Pi(t)$ we can choose points $A_0, A_1, \ldots, A_m = A_0$ for which the segments $A_i'A_i$ and $A_{i+1}'A_{i+1}$ and the whole arc A_iA_{i+1} all lie in the rectangle R_i. Then by what we have already proved for rectangles, the integral round the closed polygonal path $A_i'A_{i+1}'A_{i+1}A_iA_i'$ is zero. Thus if we denote that polygonal path by C_i, we have (cf. p. 349)

$$\int_{\Pi(t)}(a\,dx + b\,dy) - \int_{\Pi(t')}(a\,dx + b\,dy) = \sum_{i=0}^{m-1}\int_{C_i}(a\,dx + b\,dy) = 0.$$

For all values of t close enough to t', therefore, the integral round $\Pi(t)$ is equal to the integral round $\Pi(t')$. Thus if we think of the integral round $\Pi(t)$ as a function $\phi(t)$ of the parameter t, it follows that $\phi(t)$ is a constant; that is, the integral round $\Pi(t)$ has the same value for every value of t, which is what was required to complete the proof of the theorem.

Finally, we emphasize that for three or more dimensions an exactly analogous theorem holds and is proved in an exactly analogous way. We content ourselves by stating the theorem for three variables:

If in an open region R *within which any closed polygon can be made to shrink continuously to a point we are given a continuous vector field* $\boldsymbol{A}$ *with components* a(x, y, z), b(x, y, z), c(x, y, z) *and continuous partial derivatives* a_y, a_z, b_z, b_x, c_x, c_y, *then a necessary and sufficient condition that the line integral*

$$\int_{C}\{a\,dx + b\,dy + c\,dz\}$$

may be independent of the path C *in* R *is that the conditions*

$$a_y = b_x,\ b_z = c_y,\ c_x = a_z,$$

or, in vector notation, the condition

$$\operatorname{curl} \boldsymbol{A} = 0,$$

shall be satisfied.

For a fixed initial point P_0 the line integral is a function $U(x, y, z)$ of the co-ordinates of the end-point, and in fact

$$\int_{P_0}^{P}\{a\,dx + b\,dy + c\,dz\} = U(x, y, z) - U(x_0, y_0, z_0),$$

or, in vector notation,

$$\int_{P_0}^{P} \boldsymbol{A}\,d\boldsymbol{x} = U(P) - U(P_0),$$

where the convenient abbreviation $U(P)$ denotes the value of the function U at a point P.

6. The Significance of Simple Connectivity.

Throughout the above discussion it is essential that the region under consideration should be simply-connected. If the connectivity of the region were not simple, we should not be certain that the function U could everywhere be determined uniquely by integration along polygonal paths.

We give the following example to show that in multiply-connected regions the conditions of integrability are not sufficient to ensure that the integral is independent of the path.

The functions

$$a(x, y) = -\frac{y}{x^2 + y^2},\quad b(x, y) = \frac{x}{x^2 + y^2}$$

are defined and continuous for all values of x, y except $x = 0$, $y = 0$. Their derivatives

$$a_y(x, y) = -\frac{1}{x^2 + y^2} + \frac{2y^2}{(x^2 + y^2)^2},\quad b_x(x, y) = \frac{1}{x^2 + y^2} - \frac{2x^2}{(x^2 + y^2)^2}$$

are also continuous, except at the origin, and satisfy the condition

$$a_y(x, y) = b_x(x, y) = \frac{y^2 - x^2}{(x^2 + y^2)^2}.$$

If we now take the integral

$$\int_C \{a\,dx + b\,dy\}$$

round the circle C with centre at the origin given by $x = \cos t$, $y = \sin t$, then C cannot be enclosed in a simply-connected region R in which the assumptions are satisfied; for the region R we must take a ring-shaped region that does not contain the point (0, 0). Then

$$\int_C \{a\,dx + b\,dy\} = \int_0^{2\pi} \{-\sin t(-\sin t) + \cos t \,.\, \cos t\}\,dt = \int_0^{2\pi} dt = 2\pi,$$

and the integral round the closed curve is therefore not zero.*

EXAMPLES

1. Evaluate the integral

$$\int_C (e^x \sin y\,dx + e^x \cos y\,dy),$$

where C is a curve joining the points (0, 0) and (ξ, η).

2.* Evaluate the integral

$$\int_C \left(\frac{e^x}{x^2 + y^2} (x \cos y + y \sin y)\,dy + \frac{e^x}{x^2 + y^2} (x \sin y - y \cos y)\,dx \right)$$

along a closed curve enclosing the origin, which does not intersect itself.

2. Connexion between Line Integrals and Double Integrals in the Plane. (The Integral Theorems of Gauss, Stokes, and Green.)

1. Statement and Proof of Gauss's Theorem.

For functions of a single independent variable one of the fundamental formulæ stating the relation between differentiation and integration is

$$\int_{x_0}^{x_1} f'(x)\,dx = f(x_1) - f(x_0).$$

* We may remark in passing that the value of the integral $\int(a\,dx + b\,dy)$ for any curve which does not intersect itself and which encloses the origin is the same, namely, 2π. This follows immediately from the general theorem on subdivisions (cf. p. 349) if we subdivide the ring-shaped region between two such curves C and C' into a number of simply-connected regions by cross-curves C_i and apply the theorem to each of these.

An analogous formula—Gauss's theorem—holds in two dimensions. Here again a differentiation is cancelled by an integration, in the sense that double integrals of the form

$$\iint_R f_x \, dx \, dy \quad \text{or} \quad \iint_R g_y \, dx \, dy$$

are transformed into integrals that are only taken round the boundary curve C of R. We here regard the boundary C as an oriented curve and indicate the sense of description by means of a sign. Gauss's theorem is then as follows:

If the functions f(x, y) *and* g(x, y) *are continuous and have continuous derivatives in a region* R *bounded by a sectionally smooth curve* C, *then the formula*

$$\iint_R [f_x(x, y) + g_y(x, y)] \, dx \, dy = \int_{+0} \{f(x, y) \, dy - g(x, y) \, dx\}$$

holds, where the integral on the right is a line integral round the closed boundary C of the region, taken in the positive sense of description, i.e. in such a way that the interior of the region R remains on the left as the boundary is described.

In the proof we first restrict ourselves to the case in which the boundary C is cut by every line parallel to one of the axes in two points at most; in addition, we assume that $g(x, y)$ is zero everywhere in R. Then by the results of the previous chapter, section 3 (p. 243), we can express the integral

$$\iint_R f_x(x, y) \, dx \, dy$$

as a repeated integral in the form

$$\iint_R f_x(x, y) \, dx \, dy = \int dy \int f_x(x, y) \, dx,$$

where y ranges over the interval to which points of R correspond and the integral $\int f_x(x, y) \, dx$ is to be taken along the segments common to the lines $y =$ const. and the region R. If $x_0(y)$ (fig. 4) denotes the point of entry and $x_1(y)$ the point of emergence of the parallel at the distance y from the x-axis, where $x_1 \geqq x_0$, then

$$\int_{x_0(y)}^{x_1(y)} f_x(x, y) \, dx = f(x_1(y), y) - f(x_0(y), y).$$

If, further, we denote the least and greatest values of y to which points of R correspond by η_0 and η_1, then by integrating this equation with respect to y from η_0 to η_1 we obviously obtain

$$\iint_R f_x(x, y)\,dx\,dy = \int_{\eta_0}^{\eta_1} f(x_1(y), y)\,dy + \int_{\eta_1}^{\eta_0} f(x_0(y), y)\,dy.$$

For the special case $g(x, y) = 0$, however, this equation is

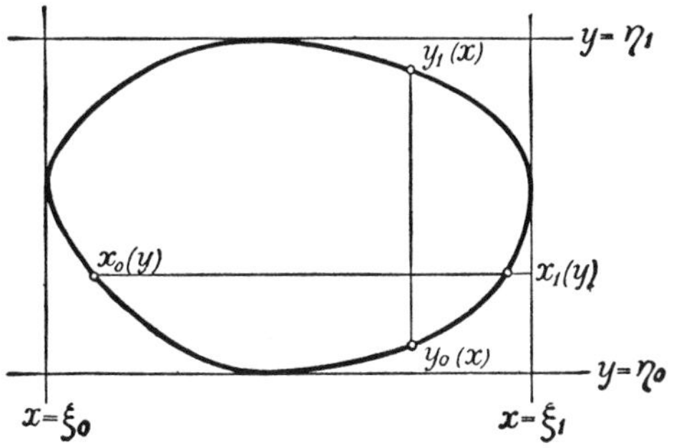

Fig. 4

equivalent to the theorem of Gauss stated above, as follows immediately from the definition of the line integral

$$\int_{+C} f(x, y)\,dy.$$

It is to be noted that the case in which the boundary of R contains portions parallel to the x-axis is included in the above. These portions contribute nothing to the boundary integral, for along every such portion the line integral $\int f(x, y)\,dy$ vanishes, since y is constant there.

If we make use of our assumption that no parallel to the y-axis cuts the boundary of R in more than two points, the same considerations lead us to the formula

$$\iint_R g_y(x, y)\,dx\,dy = \int_{\xi_0}^{\xi_1} \{g(x, y_1(x)) - g(x, y_0(x))\}\,dx$$

or *

$$\iint_R g_y(x, y)\,dx\,dy = -\int_{+C} g(x, y)\,dx.$$

* The occurrence of the negative sign on the right-hand side should not cause surprise; the x-axis and y-axis in the plane are not exactly equivalent, as the x-axis is transformed into the y-axis by a *positive* rotation of $\pi/2$, while the y-axis is transformed into the x-axis by a *negative* rotation of $\pi/2$.

Addition of the two formulæ finally gives Gauss's theorem in the general form

$$\iint_R [f_x(x, y) + g_y(x, y)]\,dx\,dy = \int_{+C} \{f(x, y)\,dy - g(x, y)\,dx\}$$

stated above.

We can now extend our formula to more general regions, which do not possess the property of being cut by every parallel to the axes in two points at most. We start from the fact that by piecing together a finite number of regions with that property we can construct regions which in general do not possess such a property (cf. fig. 5). For each separate region Gauss's theorem holds; and, on addition, the parts of the line integrals along the internal connecting lines cancel one another in the usual way (p. 349), since each of these is traversed twice, once in each direction, and we are left with Gauss's theorem for the entire region. Conversely, this proves Gauss's theorem for all regions R which can be divided into a finite number of sub-regions in such a way that the boundary of each of these sub-regions is intersected by parallels to the co-ordinate axes in not more than two points. We mention without proof that Gauss's theorem does actually hold for *any* region with sectionally smooth boundaries.* The proof can be obtained by a passage to the limit.

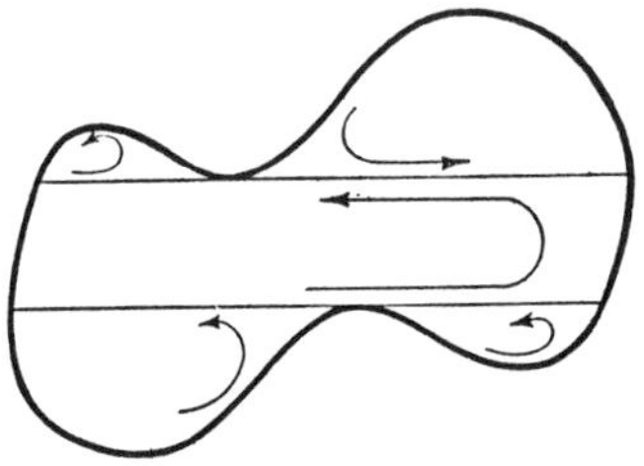

Fig. 5.—Non-convex region formed from convex regions

In conclusion we remark that the condition that the region can be divided into a finite number of sub-regions, each of which is cut by every line parallel to an axis in two points at most, can be replaced by the following condition: the boundary of the region can be subdivided into a finite number of portions, each of which has a unique projection on the two co-ordinate axes; here, however, we allow the projection on one of the two axes to consist of a single point, i.e. we allow the boundary to contain portions parallel to the axes.

* For such regions our assumption is not necessarily satisfied. For example, the boundary may partly consist of the curve $y = x^2 \sin 1/x$, which is cut by the x-axis in an infinite number of points.

As a special application of Gauss's theorem we deduce our previous formulæ for the area of the region R. We put $f(x, y) = x$ and $g(x, y) = 0$, and at once obtain

$$A = \iint_R dx\,dy = \int_{+C} x\,dy.$$

for the area A. In exactly the same way, if $f(x, y) = 0$ and $g(x, y) = y$, we obtain

$$A = -\int_{+C} y\,dx,$$

in agreement with previous results (Vol. I, p. 273). For the sign, see section 4, 1, below (pp. 374 *et seq.*).

2. Vector Form of Gauss's Theorem. Stokes's Theorem.

Gauss's theorem can be stated in a particularly simple way if we make use of the notation of vector analysis. For this purpose we consider the two functions $f(x, y)$ and $g(x, y)$ as the components of a plane vector field $\boldsymbol{A}$. The integrand is then given, by the equation

$$f_x(x, y) + g_y(x, y) = \operatorname{div} \boldsymbol{A},$$

as the divergence of the vector $\boldsymbol{A}$ (cf. p. 91). In order to obtain a vector expression for the line integral on the right-hand side of Gauss's theorem, we introduce the length of arc s of the boundary curve C; the positive sense of description is to be taken as the direction in which s increases. The right-hand side then becomes

$$\int_C \{f(x, y)\dot{y} - g(x, y)\dot{x}\}\,ds,$$

where we put $dx/ds = \dot{x}$ and $dy/ds = \dot{y}$.

We now recall that the plane vector $\boldsymbol{t}$ with x-component $\dot{x}$ and y-component $\dot{y}$ has the absolute value unity and the direction of the tangent, and points in the direction in which s increases, while the vector $\boldsymbol{n}$ with x-component $\dot{y}(s)$ and y-component $-\dot{x}(s)$ has the absolute value unity and is perpendicular to the tangent, and, moreover, has the same position relative to the vector $\boldsymbol{t}$ as the positive x-axis has relative to the positive y-axis.* Hence if the direction in which the length of arc increases

* We see this from considerations of continuity; we may suppose that the tangent to the curve is made to coincide with the y-axis in such a way that the $\boldsymbol{t}$-direction is the same as the direction in which y increases. Then $\dot{x} = 0$, $\dot{y} = 1$; and from this it follows that the normal vector $\boldsymbol{n}$ must point in the direction of the positive x-axis.

is that in which the boundary of the region is positively described, $\boldsymbol{n}$ is the unit vector in the direction of the outward-drawn normal (fig. 6). It is useful to notice that we can also write the components of the normal vector $\boldsymbol{n}$ in the form

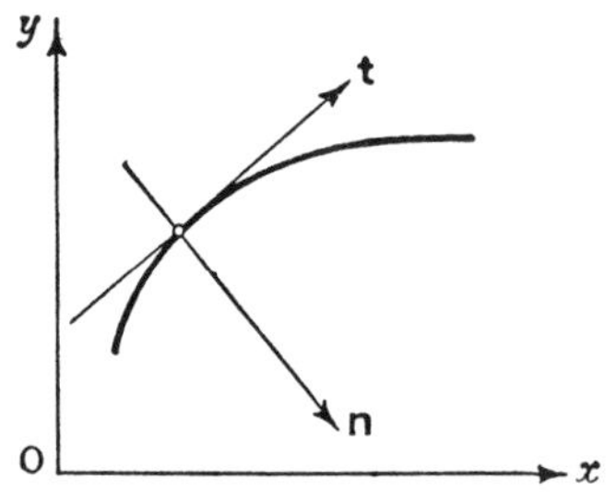

Fig. 6.—Tangent and normal directions

$$\dot{y}(s) = \frac{\partial x}{\partial n}, \quad -\dot{x}(s) = \frac{\partial y}{\partial n},$$

where $\partial/\partial n$ denotes differentiation in the direction of the outward-drawn normal; * Gauss's theorem can therefore also be written in the form

$$\iint_R (f_x + g_y)\,dx\,dy = \int_{+C} \left(f\frac{\partial x}{\partial n} + g\frac{\partial y}{\partial n} \right) ds.$$

We now see that the integrand is simply the scalar product $\boldsymbol{An}$ or the normal component of the vector $\boldsymbol{A}$. Consequently we obtain Gauss's theorem in the important form

$$\iint_R \operatorname{div} \boldsymbol{A}\,dx\,dy = \int_C \boldsymbol{An}\,ds = \int_C A_n\,ds.$$

In words: *the integral of the divergence of a plane vector field over a closed region* R *is equal to the line integral, along the boundary, of the component of the vector field in the direction of the outward-drawn normal.*

In order to arrive at an entirely different vector interpretation of Gauss's theorem in the plane, we first replace $g(x, y)$ by $-g(x, y)$. Gauss's theorem then gives

$$\iint_R [f_x(x, y) - g_y(x, y)]\,dx\,dy = \int_{+C} [g(x, y)\dot{x} + f(x, y)\dot{y}]\,ds.$$

If the two functions $f(x, y)$ and $g(x, y)$ are again taken as components of a vector field $\boldsymbol{A}$, g this time being the x-component and f the y-component, and if we again interpret $\dot{x}(s)$ and $\dot{y}(s)$ as the components of the tangential unit vector $\boldsymbol{t}$, we see that the integrand on the right can be written in the form $\boldsymbol{At} = A_t$, where $\boldsymbol{At}$ is the scalar product of the vectors $\boldsymbol{A}$ and $\boldsymbol{t}$, i.e. the

* For " differentiation in a given direction " see Chap. II, section 4 (p. 62).

tangential component of the vector $\boldsymbol{A}$. The integrand on the left we have already met with (p. 92) in forming the curl. In order to apply the concept of curl here we imagine the vector field $\boldsymbol{A}$ extended in any way in space, e.g. by taking the z-component everywhere equal to zero. The integrand on the left is then just the component of the vector curl $\boldsymbol{A}$ in the z-direction, so that the above equation for the plane can be written in the following form:

$$\iint_R (\text{curl } \boldsymbol{A})_z \, dx \, dy = \int_C A_t \, ds.$$

If by the curl of a vector field in the xy-plane we mean the z-component of the vector curl $\boldsymbol{A}$, where $\boldsymbol{A}$ is any vector field obtained by extension as above, we can formulate Gauss's theorem as follows:

The integral of the curl of a plane vector field over a closed region is equal to the integral of the tangential component taken round the boundary. This statement is commonly referred to as Stokes's theorem in the plane.*

If we now make use of the vector character of the curl of a vector field in space and observe that the above result involves the components of the vector field in the xy-plane only, we can free Stokes's theorem for plane regions from the restriction that these plane regions lie in the xy-plane. We thus arrive at the following more general statement of Stokes's theorem:

$$\iint_T (\text{curl } \boldsymbol{A})_n \, dS = \int_C A_t \, ds,$$

where $\boldsymbol{T}$ is any plane region in space, bounded by the curve C, and $(\text{curl } \boldsymbol{A})_n$ is the component of the vector curl $\boldsymbol{A}$ in the direction of the normal to the plane containing T.

* We remark in passing that Gauss's theorem or Stokes's theorem can be used to give a new simple proof for the main theorem on line integrals (section 1, p. 353), in particular, for the fact that the condition $f_x = g_y$ is sufficient to ensure that the line integral is independent of the path. We have seen that this independence of the path is equivalent to the vanishing of the integral round every closed path. If such a path is the boundary of a region R of the type considered, Stokes's theorem transforms the line integral

$$\int_{+C} \{g(x, y)\,dx + f(x, y)\,dy\}$$

into the integral of the expression $f_x - g_y$ over the region; and if this expression vanishes, the vanishing of the line integral immediately follows.

3. Green's Theorem. Integral of the Jacobian.

Certain other integral transformations, usually known as Green's theorems, are closely related to Gauss's theorem. They have many applications in the theory of differential equations. In order to obtain these theorems we consider two functions $u(x, y)$ and $v(x, y)$, which we assume to have continuous derivatives of the first and second order in the region R. In virtue of the equations

$$\frac{\partial}{\partial x}(uv_x) = u_x v_x + uv_{xx}, \quad \frac{\partial}{\partial y}(uv_y) = u_y v_y + uv_{yy},$$

Gauss's theorem gives the formula

$$\iint_R (u_x v_x + uv_{xx} + u_y v_y + uv_{yy})\,dx\,dy = \int_{+C} \{uv_x\,dy - uv_y\,dx\}$$

or

$$\iint_R (u_x v_x + u_y v_y)\,dx\,dy = -\iint_R u\,\Delta v\,dx\,dy + \int_{+C} \{-uv_y\,dx + uv_x\,dy\},$$

where, as in Chap. II (p. 93), we use the symbol

$$\Delta v = v_{xx} + v_{yy}.$$

This last integral formula is called Green's (first) theorem. It has been proved above, subject to the assumption that the functions u_x, v_x, u_y, v_y, v_{xx}, v_{yy} are continuous in the closed region. If in addition we assume the continuity of the functions u_{xx} and u_{yy}, we can in a similar way obtain the formula

$$\iint_R (u_x v_x + u_y v_y)\,dx\,dy = -\iint_R v\,\Delta u\,dx\,dy + \int_{+C} \{-vu_y\,dx + vu_x\,dy\},$$

and from these two formulæ we obtain by subtraction the relation known as Green's (second) theorem:

$$\iint_R (u\,\Delta v - v\,\Delta u)\,dx\,dy = \int_{+C} \{(vu_y - uv_y)\,dx - (vu_x - uv_x)\,dy\}.$$

We can write the line integral in Green's theorem somewhat differently if we recall that the derivative of a function $f(x, y)$ in the direction of the outward-drawn normal to the curve is given by the equation

$$\frac{\partial}{\partial n} f(x, y) = f_x \dot{y} - f_y \dot{x},$$

provided that the direction in which s increases is that corresponding to positive description of the boundary. Thus, if in general we use the symbol $\partial/\partial n$ to denote differentiation with respect to the outward-drawn normal to the curve, Green's theorems can be written in the form

$$\iint_R (u_x v_x + u_y v_y)\,dx\,dy = -\iint_R v\Delta u\,dx\,dy + \int_{+C} v\frac{\partial u}{\partial n}\,ds$$

and

$$\iint_R (u\Delta v - v\Delta u)\,dx\,dy = \int_{+C}\left(u\frac{\partial v}{\partial n} - v\frac{\partial u}{\partial n}\right)ds.$$

We can also express the first form of Green's theorem in yet another way, by means of the vector notation:

$$\iint_R (\operatorname{grad} u \operatorname{grad} v)\,dx\,dy = -\iint_R v \operatorname{div} \operatorname{grad} u\,dx\,dy + \int_C v\frac{\partial u}{\partial n}\,ds.$$

Here the quantity under the integral sign on the left is the scalar product of the two gradients, grad u and grad v, and the symbol Δu is replaced by the equivalent symbol div grad u.

We obtain another remarkable relation between integrals if we transform the double integrals of the products $u_x v_y$ and $u_y v_x$ respectively into line integrals by means of Gauss's theorem and then subtract:

$$\iint_R (u_x v_y - u_y v_x)\,dx\,dy = \int_{+C}\{uv_x\,dx + uv_y\,dy\}.$$

This formula gives us a new insight into the nature of the Jacobian. As the integrand on the left we have the Jacobian $\dfrac{\partial(u, v)}{\partial(x, y)}$. We assume that the Jacobian is positive throughout the region R and that the region R of the xy-plane is mapped on a region R' of the uv-plane by means of the equations

$$u = u(x, y), \quad v = v(x, y),$$

the sense of description of the boundary being preserved since $\dfrac{\partial(u, v)}{\partial(x, y)} > 0$. The area of the region R, as we already know, is given by the line integral

$$\int_{+C} u\,dv = \int_{+C} u(v_x\,dx + v_y\,dy)$$

taken round this boundary in the positive sense. Thus the integral of the Jacobian

$$\iint_R \frac{\partial(u, v)}{\partial(x, y)}\, dx\, dy$$

gives the area of the image region, and

$$\iint_{R'} du\, dv = \iint_R \frac{\partial(u, v)}{\partial(x, y)}\, dx\, dy.$$

Thus we have once again obtained the transformation formula of Chap. IV (p. 253) for the special case in which the integrand on the left is unity. If we divide the integral

$$\iint_R \frac{\partial(u, v)}{\partial(x, y)}\, dx\, dy$$

by the area of the region R and then let the diameter of R tend to zero, in other words, if we carry out a space-differentiation of this integral, in the limit we obtain the integrand, that is, the Jacobian $\frac{\partial(u, v)}{\partial(x, y)}$. *The Jacobian is therefore the limit of the quotient of the area of the image region and the area of the original region as the diameter tends to zero, or, as we may say, it is the local ratio of areal distortion.**

4. The Transformation of Δu to Polar Co-ordinates.

A process like that of the last sub-section enables us to transform the expression $\Delta u = u_{xx} + u_{yy}$ to new co-ordinates, e.g. to polar co-ordinates (r, θ). For this purpose we use the formula

$$\iint_R \Delta u\, dx\, dy = \int_{+C} \frac{\partial u}{\partial n}\, ds,$$

which arises from Green's theorem if we put $v = 1$. If we divide

* Since by the mean value theorem of the integral calculus the ratio of the area of a region to the area of its image is given by an intermediate value of the Jacobian, the definition of the double integral now leads us almost at once to the general transformation formula

$$\iint_R f(u, v)\, du\, dv = \iint_{R'} f \frac{\partial(u, v)}{\partial(x, y)}\, dx\, dy;$$

the reader may work out the details for himself. For another complete proof of the transformation formula cf. section 3, No. 3 (p. 373).

both sides of this equation by the area of the region R and let the diameter of R tend to zero—that is, if we carry out a space-differentiation—in the limit we again obtain the expression for Δu.

In order to transform Δu to other co-ordinates, we therefore have only to apply the corresponding transformation to the simple line integral $\int_C \frac{\partial u}{\partial n}\, ds$, divide by the area, and perform a passage to the limit. The advantage over the direct calculation is that we need not carry out the somewhat complicated calculation of the second derivatives of u, since only the first derivatives occur in the line integral.

As an important example we shall work out the transformation of Δu to polar co-ordinates (r, θ). For the region R we choose a small mesh of the polar co-ordinate net, say that between the circles r and $r + h$ and the lines θ and $\theta + k$, whose area, as we know, has the value $kh\rho$, where $\rho = r + \frac{1}{2}h$.

By our general discussion we then have

$$\Delta u = \lim_{\substack{h \to 0 \\ k \to 0}} \frac{1}{\rho k h} \int_{+C} \frac{\partial u}{\partial n}\, ds,$$

or, if we calculate the line integral for our special boundary,

$$\Delta u = \lim_{\substack{h \to 0 \\ k \to 0}} \frac{1}{\rho} \left\{ \frac{1}{k} \int_\theta^{\theta+k} \frac{(r+h)u_r(r+h, \theta) - r u_r(r, \theta)}{h}\, d\theta \right.$$
$$\left. + \frac{1}{h} \int_r^{r+h} \frac{u_\theta(r, \theta+k) - u_\theta(r, \theta)}{kr}\, dr \right\}.$$

If we use the mean value theorems, we can also write this equation in the form

$$\Delta u = \lim_{\substack{h \to 0 \\ k \to 0}} \frac{1}{\rho} \left\{ r_1 u_{rr}(r_1, \theta_1) + u_r(r_1, \theta_1) + \frac{1}{r} u_{\theta\theta}(r_2, \theta_2) \right\},$$

where r_1, r_2 and θ_1, θ_2 denote values of the variables r, θ which lie between r and $r + h$ and between θ and $\theta + k$. For the limit as $h \to 0$, $k \to 0$ we at once obtain

$$\Delta u = \frac{1}{r}(r u_r)_r + \frac{1}{r^2} u_{\theta\theta},$$

which is the required transformation formula.

3. INTERPRETATION AND APPLICATIONS OF THE INTEGRAL THEOREMS FOR THE PLANE

1. Interpretation of Gauss's Theorem. Divergence and Intensity of Flow.

We shall now interpret the integral theorems given in the previous section in terms of the steady flow of an incompressible fluid in two dimensions. Such a flow (which of course is only an idealization of actual physical conditions) occurs when a fluid distributed over a plane with constant surface density unity moves in such a way that the state of motion, that is, the velocity vector at each point, is independent of the time (which is what we mean by the term "steady"). Such a flow is therefore determined by the field of its velocity vector $\boldsymbol{v}$. We shall call the components of this velocity vector v_1 and v_2. If we consider any curve C to which we arbitrarily assign a positive direction of the normal—we denote the unit vector in the direction of the normal by $\boldsymbol{n}$—then the total amount of the fluid which passes across the curve in the positive direction of the normal in unit time is given by the integral

$$\int_C \boldsymbol{v}\boldsymbol{n}\, ds,$$

if we denote * the length of arc on C by s. If the curve is closed and encloses a region R, and if $\boldsymbol{n}$ is the outward-drawn normal, then Gauss's theorem

$$\int_C \boldsymbol{v}\boldsymbol{n}\, ds = \iint_R \operatorname{div} \boldsymbol{v}\, dx\, dy$$

states that the total amount of fluid leaving the region R in unit time is equal to the integral over the region of the divergence of the velocity field. This statement at once leads us to the intuitive interpretation of the concept of divergence. The line integral on the left will not in general vanish. If it has a positive value, the total amount of fluid in the region is decreasing; if it has a negative value, the amount of fluid is increasing. If the whole phenomenon is steady, i.e. independent of the time, so that there can be no increase or decrease in the amount of the fluid in the region, the substance is necessarily being created or destroyed in the region itself. We say that the region encloses *sources* or *sinks*; the steady character of the flow is then expressed by the fact that the sources or sinks regulate the entry or exit of the fluid in the interior in such a way that the amount of fluid remains constant within each region. The total amount of fluid leaving the region may be called the *total flow* out of the region. This is positive or negative according as the sources or the sinks predominate. If we divide the total flow by the area of the region, we obtain the average

* In order to see that the integral actually has this meaning, we first think of the curve as replaced by a polygon with sides of length $\Delta s_1, \Delta s_2, \dots, \Delta s_n$, assume that on each side of the polygon the velocity vector is constant, and then perform the usual passage to the limit from polygon to curve.

or mean intensity of flow. If we now let the diameter of the region tend to zero, that is, if we carry out a space-differentiation, we obtain in the limit the *intensity of flow* at the point in question. Gauss's theorem tells us that div $\mathbf{v}$, the *divergence of the velocity field, is equal to the intensity of flow.* Gauss's theorem accordingly leads to an intuitive interpretation of the hitherto purely formal concept of divergence.

This interpretation of the divergence can also be roughly expressed in the following way: we think of the flow as divided into a flow in the direction of the x-axis with velocity v_1 and a flow in the direction of the y-axis with velocity v_2, and consider a rectangle with corners $P_1(\xi, \eta)$, $P_2(\xi + h, \eta)$, $P_3(\xi, \eta + k)$, $P_4(\xi + h, \eta + k)$. If the velocity v_1 were constant along each of the two sides P_1P_3 and P_2P_4 and had the respective values $v_1(\xi, \eta)$ and $v_1(\xi + h, \eta)$ there, the total amount of fluid leaving the rectangle in the x-direction in unit time would be given by the difference $kv_1(\xi + h, \eta) - kv_1(\xi, \eta)$. If we divide by hk, the area of the rectangle, we obtain

$$\frac{v_1(\xi + h, \eta) - v_1(\xi, \eta)}{h}.$$

The average net flow out of the region in the direction of the y-axis is obtained in the same way. The expression

$$\frac{v_1(\xi + h, \eta) - v_1(\xi, \eta)}{h} + \frac{v_2(\xi, \eta + k) - v_2(\xi, \eta)}{k}$$

therefore gives an approximation to the average net flow out of the region, and the passage to the limit $h \to 0$, $k \to 0$ again leads to the meaning of the divergence given above.

Special interest attaches to the case of a *source-free flow*, that is, a flow in which fluid is neither created nor destroyed in the region under consideration. This type of flow is characterized by the condition

$$\operatorname{div} \mathbf{v} = 0,$$

which by Gauss's theorem is equivalent to the condition

$$\int_C v_n \, ds = 0,$$

where the integral is taken round any closed curve.

2. Interpretation of Stokes's Theorem.

Stokes's theorem can also be interpreted in a simple way in terms of the flow of an incompressible fluid in two dimensions. Let the velocity of flow be given by the vector $\mathbf{v}$ with components v_1, v_2. The integral $\int_{+C} v_t \, ds$ taken round a closed curve C we shall call the *circulation* of the fluid along this curve. By Stokes's theorem this can at once be expressed in the form

$$\int_C v_t \, ds = \iint_R \operatorname{curl} \mathbf{v} \, dx \, dy,$$

and this equation further shows us that the expression curl $\boldsymbol{v}$ is to be regarded as the specific circulation or *circulation-density* at a given point. Stokes's theorem then states that the circulation along the curve C is equal to the integral of the circulation-density over the region enclosed by the curve.

Here again special interest attaches to cases of flow for which the circulation along *every* closed curve is zero, so that by Stokes's theorem the circulation-density vanishes everywhere. Such flows are said to be *irrotational*, and are characterized by the equation

$$\operatorname{curl} \boldsymbol{v} = 0.$$

If a steady flow is both source-free and irrotational, it satisfies the two systems of equations

$$\operatorname{curl} \boldsymbol{v} = \frac{\partial v_1}{\partial y} - \frac{\partial v_2}{\partial x} = 0,$$

$$\operatorname{div} \boldsymbol{v} = \frac{\partial v_1}{\partial x} + \frac{\partial v_2}{\partial y} = 0.$$

These two equations, by the way, are of special interest in that they occur in other branches of mathematics, in particular, in the theory of functions of a complex variable*, thus forming the connexion between the latter subject and hydrodynamics.

We shall mention yet another interpretation of Stokes's theorem. If we think of $\boldsymbol{v}$ as representing a field of *force* instead of a velocity field, the line integral

$$\int_C v_t\, ds = \int_{+C} \{v_1\, dx + v_2\, dy\},$$

taken round any curve, closed or not, gives the work done by the field of force on a particle describing the curve C. If C is a closed curve which forms the boundary of a region R, then Stokes's theorem states that the work done in describing the boundary of R is equal to the integral over R of the curl of the field of force. If the work done in describing a closed path is always to have the value zero, the equation

$$\operatorname{curl} \boldsymbol{v} = \frac{\partial v_1}{\partial y} - \frac{\partial v_2}{\partial x} = 0$$

must be true everywhere. Conversely, if this equation is true everywhere, it follows from Stokes's theorem that the integral

$$\int_{+C} v_t\, ds = \int_{+C} (v_1\, dx + v_2\, dy)$$

vanishes everywhere (cf. p. 365, footnote).

* Cf. Chapter VIII, pp. 532, 550.

This result shows, in accordance with section 1, p. 358, that *the work done is independent of the path if, and only if,*

$$\text{curl } \boldsymbol{v} = 0$$

throughout the region.

3. Transformation of Double Integrals.

As an application of Gauss's theorem we give another method for deriving the transformation formula for double integrals (cf. Chap. IV, section 4, p. 253, and p. 368, footnote). Let us suppose that R is a closed region of the xy-plane bounded by the curve C and that the transformation $x = x(u, v)$, $y = y(u, v)$ gives a one-to-one mapping of R on the region R' of the uv-plane bounded by the curve C', the sense of description of the boundary being preserved. Let the two regions satisfy the conditions for the applicability of Gauss's theorem. In order to transform the integral

$$I = \iint_R f(x, y)\,dx\,dy$$

into an integral over the region R' we first transform it into a line integral round the boundary C. This line integral, being a simple integral, can at once be transformed into a line integral round C', the boundary of R', and the latter, by Gauss's theorem, can be transformed into a double integral over R'. In order to carry out this process we consider *any* function $A(x, y)$, obtained from f by indefinite integration, for which

$$A_x = f.$$

Then by Gauss's theorem

$$I = \iint_R A_x\,dx\,dy = \int_{+C} A\,dy.$$

If in the line integral on the right we now introduce the variables u, v instead of x, y, i.e. if we transform it by means of the functions $x(u, v)$ and $y(u, v)$ into an integral along the boundary C' of R', we at once obtain

$$I = \int_{+C'} A(y_u\,du + y_v\,dv).$$

To the boundary integral on the right we apply Gauss's theorem in the reverse direction, transforming it into a double integral over R':

$$\int_{+C'} (Ay_u)\,du + (Ay_v)\,dv = \iint_{R'} [(Ay_v)_u - (Ay_u)_v]\,du\,dv.$$

From the equations

$$(Ay_u)_v = A_v y_u + A y_{vu}$$

and

$$(Ay_v)_u = A_u y_v + A y_{uv},$$

as well as

$$A_u = A_x x_u + A_y y_u, \quad A_v = A_x x_v + A_y y_v, \quad A_x = f,$$

we find after a short calculation that

$$(Ay_v)_u - (Ay_u)_v = (x_u y_v - x_v y_u) f,$$

so that finally

$$I = \iint_R f\,dx\,dy = \iint_{R'} (x_u y_v - x_v y_u) f\,du\,dv,$$

as was to be proved.

4. Surface Integrals

The theory of integration for three independent variables includes not only triple integrals and line integrals but the third concept of the *surface integral*. In order to explain the latter we begin with some considerations of a general nature, which at the same time will serve to refine our previous ideas, in particular those relating to double integrals.

1. Oriented Regions and Integration Over Them.

We start from the ordinary integral $\int_a^b f(x)\,dx$ of a function $f(x)$ of the independent variable x. The region of integration is the interval between $x = a$ and $x = b$. We are necessarily led (Vol. I, p. 81) to the convention

$$\int_a^b f(x)\,dx = -\int_b^a f(x)\,dx,$$

which we can also express in the following way: the region of integration, that is, the interval R under consideration, is given a definite direction, or, as we say, a definite orientation. If we reverse the orientation, that is, if we describe the interval in the opposite direction, the value of the integral is multiplied by -1. This convention may also be expressed by the equation

$$\int_{+C} f(x)\,dx = -\int_{-C} f(x)\,dx,$$

where the region of integration is denoted by $+C$ when it is described in the direction $a \to b$ and by $-C$ when it is described in the direction $b \to a$.

In the case of line integrals in the plane and in space we have likewise seen that it is necessary to assign a definite sense of description to the curve along which we are integrating, and that if this orientation is reversed the integral is multiplied by -1. It is now evident that a full treatment of the case of integration over regions of several dimensions demands the adoption of analogous conventions, and that our previous definitions should be extended accordingly.

In Vol. I, p. 268, we gave a definite sign to the area of a region R, the sign being positive or negative according as the sense of description of the boundary is positive or negative. A plane region to which we attach a definite sign in this way we call an oriented region (fig. 7); in accordance with what we have just said, we shall call it positively oriented if the sense of description of the boundary is positive, otherwise negatively oriented. Now we have represented the area of a region R by the double integral $\iint_R dx\,dy$. If this area is to be taken as positive, we shall attach to the region a positive sense of description of the boundary, and we accordingly represent the absolute value of the area symbolically by the expression

$$\iint_{+R} dx\,dy = |A|.$$

If we think of the region as negatively oriented, so that its area is negative, we express the actual value of the area by the symbol $\iint_{-R} dx\,dy$, and accordingly have the definition

$$\iint_{-R} dx\,dy = -|A|.$$

Again, the area is expressed * as a line integral by the formula

$$|A| = -\int_{+C} y\,dx = \int_{+C} x\,dy.$$

If nothing special is said to the contrary, we shall always take R as a positively oriented region.

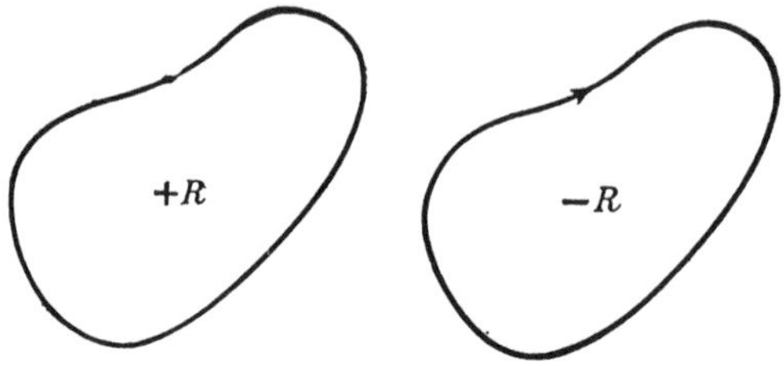

Fig. 7.—Oriented regions

In the same way, *we now state the general definition for any double integral whatever*:

$$\iint_R f(x, y)\,dx\,dy = \iint_{+R} f(x, y)\,dx\,dy;$$

$$\iint_R f(x, y)\,dx\,dy = -\iint_{-R} f(x, y)\,dx\,dy.$$

This definition corresponds exactly to the convention already adopted in the case of ordinary integrals and line integrals. The equations do not represent any newly-proved fact: they are simply *definitions* and are justified solely on grounds of convenience.

An example will illustrate the usefulness of this convention. We saw (p. 253) that in the one-to-one mapping of the region R of the xy-plane on a region R' of the uv-plane the area of the region R is given in the new co-ordinates by the integral

$$\iint_R dx\,dy = \iint_{R'} \frac{\partial(x, y)}{\partial(u, v)}\,du\,dv,$$

provided that the Jacobian is positive everywhere in R. We know that

* It is useful to verify by an example that the integral $-\int_{+R} y\,dx$ is really a positive number. If, for example, R is a square $0 \leqq x \leqq 1$, $0 \leqq y \leqq 1$, on both the vertical sides we have $dx = 0$. The side $y = 0$ likewise contributes nothing to the line integral; and on the third side we have $dx < 0$ and $y = 1$.

if the Jacobian is positive, the orientation (i.e. the sense of description of the boundary) of R and R' is the same, while if the Jacobian is negative, the regions have opposite orientations. The above formula therefore would not hold if the Jacobian were negative, if we considered the double integral without regard to the orientation. But *it remains true for the case of a negative Jacobian* if by R we mean a (positively or negatively) *oriented* region and by R' the *oriented* region which arises from R as a result of the transformation. For if the orientation is reversed, the effect of the negative sign of the Jacobian is cancelled by the above convention.

In the same way, we can now regard the general transformation equation

$$\iint_R f(x, y)\, dx\, dy = \iint_{R'} f(x, y) \frac{\partial(x, y)}{\partial(u, v)}\, du\, dv$$

as valid, whether the Jacobian $\dfrac{\partial(x, y)}{\partial(u, v)}$ is positive everywhere or * negative everywhere in the region R, it being assumed that the integrals are taken as integrals over *oriented* regions and that in the mapping the oriented region R becomes the oriented region R'. *Thus only by introducing orientation and the sign principle do we arrive at transformation formulæ for double integrals which are valid without exception.*

The orientation of a region can also be defined geometrically without reference to the boundary in the following way. We first consider any point of the region whatever, and to this point assign a sense of rotation, which we can represent e.g. as the sense of description of a small circle with this point as centre. We now say that the region R is oriented if such a sense of rotation is assigned to every point of R and if on continuous passage from one point to another the sense of rotation is preserved.

By means of this remark we can now assign an orientation to a surface lying in xyz-space. On the surface we can first assign a sense of rotation to a point by surrounding it by a small curve lying on the surface and assigning a definite sense of description to this curve. If we now move the point continuously over the surface to any other position and along with the point move the oriented curve with its orientation, we assign a sense of rotation to every point of the surface in this way (exceptional cases will

* The formula does not hold, however, if the Jacobian changes sign in the region; in this case the assumption that the mapping is one-to-one cannot be satisfied.

be discussed later). We call the surface with this sense of rotation an *oriented surface* (fig. 8).

We can get a better grasp of this orientation of a surface in space as follows. A portion of a surface in space will have two different sides, which we can best distinguish as the positive side and the negative side. (Which of the two sides we call positive and which negative is of no intrinsic importance.) For example, as the positive side of the xy-plane we can take the side indicated by the positive z-axis. We now mark the positive side of a surface S by constructing at each point of the surface a vector pointing out into space on the positive side; e.g. the normal to the surface, if a unique normal exists at the point. If we think of ourselves as standing on the surface with our heads on the positive side, we say that *the surface is positively oriented*

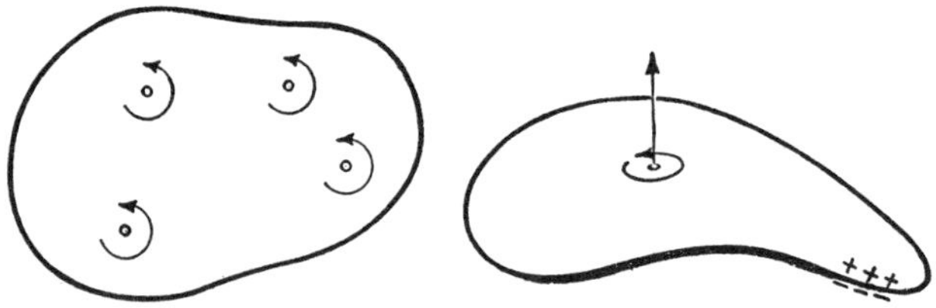

Fig. 8.—Orientation of a curved surface

if the orientation of the surface and the line from feet to head together form a right-handed screw (cf. Chap. I, p. 2), or, in other words, if the surface together with its orientation can be continuously deformed in such a way that it becomes the positively oriented xy-plane and at the same time the direction of the positive normal becomes the direction of the positive z-axis. Otherwise, we say that the surface is negatively oriented. We thus see that there is a natural way of determining the sign of the orientation of a surface, provided that the two sides of the surface are given signs to begin with. Any difficulty which the beginner may find in these matters lies simply in the fact that here we are discussing not proofs but *definitions*, which are justified solely by their convenience in simplifying subsequent discussion.

We must not omit to mention that curved surfaces exist to which *no* orientation can possibly be assigned, since on them it is not possible to distinguish two separate sides. The simplest surface of this type was discovered by Möbius and is called the

Möbius band; it is shown in fig. 9. We can easily make such a surface from a long strip of paper by fastening the ends of the strip together after rotating one end through an angle of 180° from its original position. The Möbius band has the property

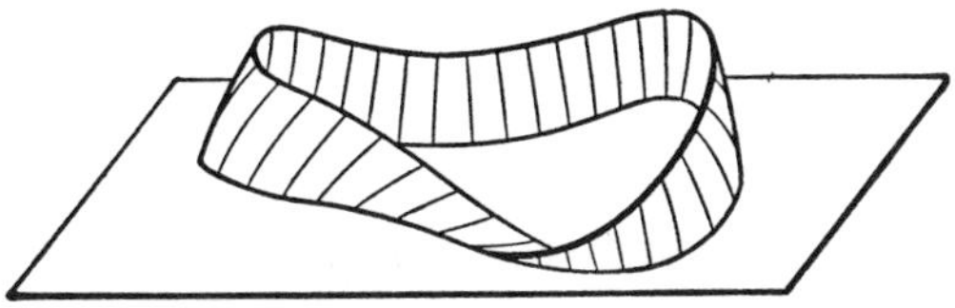

Fig. 9.—The Möbius band

that if we start from a definite point, say on the centre line of the band, and move along the centre line, after a complete circuit we come back to the same point, but on the *opposite* side of the surface.* If during this motion we carry with us a small oriented curve, without altering its orientation, we shall find that we return to the starting-point with the orientation reversed. We see that with such a surface we can pass from one side to the other without crossing the boundary, and hence that it is impossible to assign to the surface an orientation in the sense described above. Such *non-orientable surfaces* are definitely excluded from the subsequent discussion.

* We can obtain a parametric representation for the Möbius band as follows. Consider first the circle $x = 2\cos u$, $y = 2\sin u$. At the point of the circle corresponding to the value u of the parameter we construct the unit vector $\boldsymbol{j}$, which starts from the point of the circle, lies in the same plane as the z-axis and the radius to the point, and makes the angle $\frac{1}{2}u$ with the positive z-axis. At the same point we also construct the vector $-\boldsymbol{j}$. Thus we have a line segment composed of the two vectors, with length 2 and its mid-point on the circle. As u goes from 0 to 2π this line segment travels with u, turning through an angle π, so that finally $\boldsymbol{j}$ comes to the original position of $-\boldsymbol{j}$. It is therefore clear that the line segment describes a Möbius band. For each value of u the point on the line at the distance v from the circumference in the direction of $\boldsymbol{j}$ (where $-1 \leqq v \leqq +1$) has the co-ordinates

$$x = 2\cos u + v\sin\frac{u}{2}\cos u,$$

$$y = 2\sin u + v\sin\frac{u}{2}\sin u,$$

$$z = v\cos\frac{u}{2},$$

where

$$0 \leqq u \leqq 2\pi$$
$$-1 \leqq v \leqq 1.$$

These equations therefore represent the Möbius band parametrically.

We can also express the orientation of a surface by thinking of the surface as represented parametrically by two parameters u and v. Then a definite region R of the uv-plane will be mapped on the surface S. If in the region R we choose any orientation, the mapping transfers this orientation to the surface S, thus defining an orientation of the surface.

Just as we can assign an orientation to a region in the plane or to a surface, we can also assign an orientation to a three-dimensional region. For this purpose the following convention is advantageous. We consider a region of space R bounded by a closed surface S. We take the side of the surface towards the *interior* of the region as the positive side. If we give the surface an orientation which with the direction from negative to positive across the surface determines a right-handed screw, we say that the *region of space* R *is positively oriented* (cf. fig. 10); if, on the other hand, we give the surface an orientation which with the negative to positive direction determines a left-handed screw, we say that the region is negatively oriented. For example, the cube $0 \leqq x \leqq 1$, $0 \leqq y \leqq 1$, $0 \leqq z \leqq 1$ is positively oriented if we give its base in the xy-plane a positive orientation.

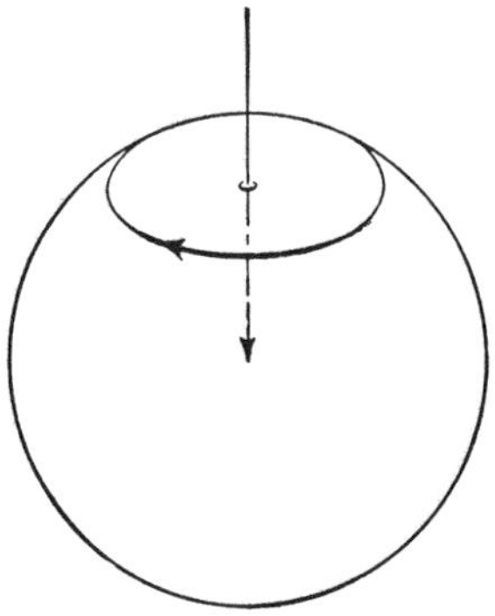

Fig. 10.—Positively oriented sphere

For regions in space, just as for regions in the plane, it is convenient to assign a positive or a negative sign to the *volume* according as the region is positively or negatively oriented (cf. p. 376). We shall again agree that an integral taken over an oriented region has its sign changed if the orientation of the region is reversed:

$$\iiint_{+R} f(x, y, z)\,dx\,dy\,dz = -\iiint_{-R} f(x, y, z)\,dx\,dy\,dz.$$

The same argument as we have already developed for two dimensions shows again that the transformation formula

$$\iiint_{R} f(x, y, z)\,dx\,dy\,dz = \iiint_{R'} f(x, y, z)\,\frac{\partial(x, y, z)}{\partial(u, v, w)}\,du\,dv\,dw$$

only acquires full validity when these conventions are adopted, since it now continues to hold when the Jacobian $\frac{\partial(x, y, z)}{\partial(u, v, w)}$ is negative everywhere in the region. For, as we explained for two dimensions in Chap. III (p. 151), a mapping of R on R' with a negative Jacobian reverses the orientation.

2. Definition of the Integral over a Surface in Space.

Having made these preliminary remarks, we can now give a general definition of the concept of surface integral. We consider a region of xyz-space in which the three continuous functions $a(x, y, z)$, $b(x, y, z)$, $c(x, y, z)$ are defined as components of a vector field $\boldsymbol{A} = \boldsymbol{A}(x, y, z)$. We first consider a surface S which has a one-to-one projection on a closed region R of the xy-plane and is defined by an equation $z = z(x, y)$; we assume that this surface is given an orientation which is transferred by projection on the xy-plane to the region R. We use the letter $\boldsymbol{n}'$ to denote the unit vector in that direction normal to the surface S which in conjunction with the orientation of the surface forms a right-handed screw. We now divide the surface S into n portions * $S_1, S_2, \ldots, S_n$ with areas $\Delta S_1, \Delta S_2, \ldots, \Delta S_n$. The projections of these portions on the xy-plane form a number of sub-regions R_ν of the region R, with areas $\Delta R_1, \Delta R_2, \ldots, \Delta R_n$, and these regions cover the region R exactly once. We take the areas ΔS_ν as positive, and accordingly have to assign a positive or negative sign to the area ΔR_ν according as the projection gives a positive or a negative orientation to the corresponding regions R or R_ν in the xy-plane. The areas $\Delta S_1, \Delta S_2, \ldots, \Delta S_n$ and $\Delta R_1, \Delta R_2, \ldots, \Delta R_n$ are connected by an equation of the form

$$\Delta R_\nu = q_\nu \Delta S_\nu,$$

where q_ν denotes a quantity which tends to the cosine of the angle $\gamma(x, y, z)$ between the positive normal direction $\boldsymbol{n}'$ and the positive z-axis as the diameter of the portion S_ν approaches zero. Now let (x_ν, y_ν, z_ν) be a point in the ν-th sub-region of the surface; i.e. $z_\nu = z(x_\nu, y_\nu)$. Then if the greatest diameter of the portions S_ν (and with it the diameters of the sub-regions R_ν) tends to zero, the sum

* In this connexion see Chap. IV, section 6 (p. 269).

$$\sum_{\nu=1}^{n} c(x_\nu, y_\nu, z_\nu)\Delta R_\nu = \sum_{\nu=1}^{n} c(x_\nu, y_\nu, z_\nu) q_\nu \Delta S_\nu$$

tends to a quantity which we denote by the symbol

$$\iint_S c(x, y, z)\, dx\, dy$$

or

$$\iint_S c(x, y, z) \cos\gamma\, dS.$$

We call this expression the *surface integral taken over the oriented surface* S. This limit does actually exist, since we may regard the integral as an ordinary integral over the two-dimensional oriented region R, namely as the integral

$$\iint_R c\, dx\, dy,$$

where the integrand is the function $c\{x, y, z(x, y)\}$.

For the generalization to which we now proceed and for applications it is essential that in this integration the region R should be regarded as *oriented.*

If the surface S also has a one-to-one projection on the yz-plane or the xz-plane, that is, if it can be represented by a single-valued function $x = x(y, z)$ or $y = y(z, x)$, we can in the same way define the integrals

$$\iint_S a(x, y, z)\, dy\, dz = \iint_{R'} a\{x(y,z), y, z\}\, dy\, dz = \iint_S a(x, y, z) \cos a\, dS$$

and

$$\iint_S b(x, y, z)\, dz\, dx = \iint_{R''} b\{x, y(z, x), z\}\, dz\, dx = \iint_S b(x, y, z) \cos\beta\, dS,$$

where R' and R'' are the oriented projections of the oriented surface S on the corresponding co-ordinate planes and a and β are the angles between the positive normal to the surface and the positive x- or y-axis respectively.

Adding these expressions, we obtain the general definition of the surface integral taken over the surface S,

$$\iint_S \{a(x, y, z)\, dy\, dz + b(x, y, z)\, dz\, dx + c(x, y, z)\, dx\, dy\}$$
$$= \iint_S \{a(x, y, z) \cos a + b(x, y, z) \cos\beta + c(x, y, z) \cos\gamma\} dS.$$

If by $\partial/\partial n'$ we denote differentiation in the direction of the unit vector $\boldsymbol{n}'$ in the positive normal direction,* we can also write

$$\cos\alpha = \frac{\partial x}{\partial n'}, \quad \cos\beta = \frac{\partial y}{\partial n'}, \quad \cos\gamma = \frac{\partial z}{\partial n'},$$

and consequently we can express the surface integral in the form

$$\iint_S \left\{ a\frac{\partial x}{\partial n'} + b\frac{\partial y}{\partial n'} + c\frac{\partial z}{\partial n'} \right\} dS.$$

If a, b, c are the components of a vector $\boldsymbol{A}$, the quantity in brackets under the integral sign is the component of the vector $\boldsymbol{A}$ in the direction of the positive normal to the surface, which we can also write in the form $\boldsymbol{A}\boldsymbol{n}'$ or $A_{n'}$.

Incidentally, if we think of the surface as given parametrically by the equations $x = x(u, v)$, $y = y(u, v)$, $z = z(u, v)$, where the oriented surface S corresponds to the oriented region B in the uv-plane, we can write the surface integral in the form

$$\iint_B \left\{ a(x, y, z)\frac{\partial(y, z)}{\partial(u, v)} + b(x, y, z)\frac{\partial(z, x)}{\partial(u, v)} + c(x, y, z)\frac{\partial(x, y)}{\partial(u, v)} \right\} du\,dv$$

and thus once again express it as an ordinary integral, namely, as a double integral over B.

It is now easy to get rid of the special assumptions about the position of the surface S relative to the co-ordinate planes. We assume that the oriented surface S can be divided by a finite number of smooth arcs of curves into a finite number of portions S_1, S_2, ... in such a way that each portion satisfies the assumptions made above. The exceptional case in which a portion of the surface S or the whole surface S is normal to a co-ordinate plane, so that its projection on that plane is only a curve instead of a two-dimensional region, can be dealt with by disregarding this projection in the formation of the integral, since a double integral vanishes when the region of integration shrinks down to a curve. We can now form the surface integral for each of the portions S according to the above definition, and we can define the integral over the oriented surface S as the sum of the integrals thus defined.

If, for example, the surface S is a closed surface, a sphere, say, we recognize that the projections of the various portions S_ν

* The letter $\boldsymbol{n}'$ is used here for the *positive normal* because $\boldsymbol{n}$ has been used for the *outward-drawn* normal in two dimensions.

lie partly above one another and have opposite orientations. If the parametric representation $x = x(u, v)$, $y = y(u, v)$, $z = z(u, v)$ gives a one-to-one mapping of a bounded surface S on an oriented region B of the uv-plane, the parametric expression given above for the surface integral is always valid; if we make use of this parametric expression in defining the surface integral, there is no need to subdivide the surface S.

3. Physical Interpretation of Surface Integrals.

The concept of surface integral can also be interpreted intuitively in terms of the steady flow of an incompressible fluid (this time in three dimensions), whose density we take as unity. Let the vector $\boldsymbol{A}$ be the velocity vector of this flow; then at each point of a surface S the product $\boldsymbol{A}\boldsymbol{n}'$ gives the component of the velocity of flow in the direction of the positive normal to the surface; the expression

$$\boldsymbol{A}\boldsymbol{n}'\Delta S_\nu = \Delta S_\nu\{a(x_\nu, y_\nu, z_\nu)\cos\alpha_\nu + b(x_\nu, y_\nu, z_\nu)\cos\beta_\nu + c(x_\nu, y_\nu, z_\nu)\cos\gamma_\nu\}$$

is therefore approximately equal to the amount of fluid which flows in unit time across the element of surface S from the negative side of the surface to the positive side (this quantity may of course be negative). The surface integral

$$\iint_S \{a\,dy\,dz + b\,dz\,dx + c\,dx\,dy\} = \iint_S A_{n'}\,dS$$

therefore represents the total amount of fluid flowing across the surface S from the negative side to the positive in unit time. We notice here that an important part is played in the mathematical description of the motion of fluid by the distinction between the positive and negative sides of a surface, i.e. by the introduction of orientation.

In other physical applications the vector $\boldsymbol{A}$ denotes the force, due to a field, acting at a point (x, y, z). The direction of the vector $\boldsymbol{A}$ then gives the direction of the *lines of force* and its absolute value gives the *magnitude* of the force. In this interpretation the integral

$$\iint_S \{a\,dy\,dz + b\,dz\,dx + c\,dx\,dy\}$$

is called the total flux of force across the surface from the negative side to the positive.

5. Gauss's Theorem and Green's Theorem in Space

1. Gauss's Theorem and its Physical Interpretation.

By means of the concept of a surface integral we can extend Gauss's theorem, which we proved in section 2 (p. 360) for two dimensions, to three dimensions. The essential point in the statement of Gauss's theorem in two dimensions is that an integral taken over a plane region is reduced to a line integral taken round the boundary of the region. We now consider a closed three-dimensional region R in xyz-space and assume—as always—that its boundary surface S can be divided into a finite number of portions with continuously turning tangent planes. In addition, we assume to begin with that each line parallel to a co-ordinate axis which has internal points in common with R cuts the boundary of R in exactly two points; this last assumption will be removed later.

Let the three functions $a(x, y, z)$, $b(x, y, z)$, $c(x, y, z)$, together with their first partial derivatives, be continuous in the region R and on its boundary; we take them to be the components of a vector field $\boldsymbol{A} = \boldsymbol{A}(x, y, z)$. We now consider the integral

$$\iiint_R \frac{\partial c(x, y, z)}{\partial z}\, dx\, dy\, dz$$

taken over the region R. We suppose that the region R is projected on the xy-plane; we thus obtain a region B in that plane. If we erect the normal to the xy-plane at a point (x, y) of B and if we denote the z-co-ordinates of its point of entrance and point of exit by $z = z_0(x, y)$, $z = z_1(x, y)$ respectively, we can transform the volume integral over R by means of the formula

$$\iiint_R f\, dx\, dy\, dz = \iint_B dx\, dy \int_{z_0}^{z_1} f\, dz.$$

Since $f = \partial c/\partial z$, the integration with respect to z can be carried out, giving

$$\int_{z_0}^{z_1} \frac{\partial c}{\partial z}\, dz = c(x, y, z_1) - c(x, y, z_0) = c_1 - c_0,$$

so that

$$\iiint_R \frac{\partial c(x, y, z)}{\partial z}\, dx\, dy\, dz = \iint_B c_1\, dx\, dy - \iint_B c_0\, dx\, dy.$$

If we think of the surface S as positively oriented with respect to the region R, then the portion of the surface S consisting of the points of entry $z = z_0(x, y)$ has a positive orientation when projected on B, while the portion $z = z_1(x, y)$ consisting of the points of exit has a negative orientation. Hence the last two integrals combine to form one integral

$$-\iint_S c(x, y, z)\, dx\, dy$$

taken over the whole surface S. We thus obtain the formula

$$\iiint_R \frac{\partial c(x, y, z)}{\partial z} dx\,dy\,dz = -\iint_S c(x, y, z)\,dx\,dy.$$

This formula obviously remains valid if S contains cylindrical portions perpendicular to the xy-plane; for these contribute nothing to the surface integral, as the regions obtained by projecting them orthogonally on to the xy-plane merely consist of curved lines.

If we obtain the corresponding formulæ for the components a and b and add the three formulæ, we obtain the general formula

$$\iiint_R \left\{\frac{\partial a(x, y, z)}{\partial x} + \frac{\partial b(x, y, z)}{\partial y} + \frac{\partial c(x, y, z)}{\partial z}\right\} dx\,dy\,dz$$

$$= -\iint_S \{a(x, y, z)\,dy\,dz + b(x, y, z)\,dz\,dx + c(x, y, z)\,dx\,dy\},$$

which is known as Gauss's theorem. Using the notation of p. 382, we can also write this in the form

$$\iiint_R (a_x + b_y + c_z)\,dx\,dy\,dz = -\iint_S (a \cos\alpha + b \cos\beta + c \cos\gamma)\,dS.$$

Here the surface is to be positively oriented with respect to R; α, β, γ are accordingly the angles which the inward-drawn normal n' makes with the positive co-ordinate axes.

This formula can easily be extended to more general regions. We have only to require that the region R is capable of being subdivided by a finite number of portions of surfaces with continuously turning tangent planes into sub-regions R_ν, each of which has the properties assumed above, in particular, is such that every line, parallel to an axis, having points in common with the interior of R_ν cuts the boundary of R_ν in only two points. Gauss's theorem holds for each region R_ν. On adding, we obtain on the left a triple integral over the whole region R; on the right, some of the surface integrals combine to form the surface integral over S, while the others (namely, those taken over the surfaces by which R is subdivided) cancel one another, as we have already seen in the case of the plane (pp. 348, 362). Finally, we remark that, as before (p. 362), it is sufficient to require that the boundary of R consists of a finite number of portions of surfaces, each of which has a unique projection on all three co-ordinate planes, except that cylindrical portions whose projections are curves are again permissible.

As a special case of Gauss's theorem we obtain the formula for *the volume of a region* R *bounded by an oriented closed surface.* If, for example, we put $a = 0$, $b = 0$, $c = z$, we immediately obtain the expression

$$V = \iiint_R dx\,dy\,dz = -\iint_S z\,dx\,dy$$

for the volume.

In the same way, we also obtain the expressions

$$V = -\iint_S x\,dy\,dz = -\iint_S y\,dz\,dx$$

for the volume.*

As in the case of the corresponding formula in the plane, it is usual to express Gauss's theorem in another form. In the first place, if a, b, c are the components of a vector field $\boldsymbol{A}$, we can write the expression

$$\frac{\partial a}{\partial x} + \frac{\partial b}{\partial y} + \frac{\partial c}{\partial z}$$

in the abbreviated form introduced in Chap. II, section 7 (p. 91),

$$\operatorname{div} \boldsymbol{A} = \frac{\partial a}{\partial x} + \frac{\partial b}{\partial y} + \frac{\partial c}{\partial z}.$$

In the second place, the discussion on p. 383 enables us to express the surface integral as the integral of the normal component $A_{n'}$ of the vector $\boldsymbol{A}$ in the direction of the inward-drawn normal $\boldsymbol{n}'$. Thus we obtain the *vector form of Gauss's theorem*,

$$\iiint_R \operatorname{div} \boldsymbol{A}\,dx\,dy\,dz = -\iint_S \boldsymbol{A}\boldsymbol{n}'dS = -\iint_S A_{n'}\,dS.$$

* It is noteworthy that cyclical interchange of x, y, z in these expressions brings about no change of sign, whereas in the case of the corresponding formulæ for the area of a two-dimensional region the formula

$$A = \int_{+C} x\,dy = -\int_{+C} y\,dx$$

shows that interchanging x and y causes a change of sign in the integral expression. This is due to the fact that in two dimensions an interchange of the positive x-direction with the positive y-direction reverses the sense of rotation of the plane, while in three dimensions a cyclical interchange of the positive co-ordinate directions, that is, replacement of x by y, of y by z, and of z by x, does not change a right-handed system of axes into a left-handed system.

In Gauss's theorem for space, as in the case of the plane, it is convenient to introduce the outward-drawn normal instead of the positive normal $\boldsymbol{n}'$. We denote this normal unit vector by $\boldsymbol{n}$, so that

$$\boldsymbol{n} = -\boldsymbol{n}',$$

and on introducing $\boldsymbol{n}$ instead of $\boldsymbol{n}'$ in our formulæ we have to make corresponding changes of sign. We can now express Gauss's theorem in the following form:

$$\iiint_R \operatorname{div} \boldsymbol{A}\, dx\, dy\, dz = \iint_S A_n\, dS = \iint_S \boldsymbol{A}\boldsymbol{n}\, dS,$$

or, if we denote the cosines of the angles which the outward-drawn normal $\boldsymbol{n}$ makes with the positive co-ordinate axes by $\frac{\partial x}{\partial n}, \frac{\partial y}{\partial n}, \frac{\partial z}{\partial n}$, we can write

$$\iiint_R (a_x + b_y + c_z)\, dx\, dy\, dz = \iint_S \left(a \frac{\partial x}{\partial n} + b \frac{\partial y}{\partial n} + c \frac{\partial z}{\partial n}\right) dS.$$

As in the case of the plane, we here obtain an intuitive interpretation of Gauss's theorem by taking the vector $\boldsymbol{A}$ as the velocity field of a steady flow of an incompressible fluid of unit density. The total mass of fluid which in unit time flows across a small surface ΔS from the interior of R to the exterior is given approximately by the expression $A_n \Delta S$, where A_n is the component of the velocity vector $\boldsymbol{A}$ in the direction of the outward normal $\boldsymbol{n}$ at a point of the surface element. Accordingly, the total amount of fluid which flows across a surface S from the inside to the outside in unit time is given by the integral $\iint_S A_n\, dS$ taken over the surface. In this interpretation, therefore, the right-hand side of Gauss's theorem represents the total amount of fluid leaving the region R in unit time. This amount of fluid is transformed into the integral of the divergence throughout the interior of the region R. From this we obtain the intuitive interpretation of the expression $\operatorname{div}\boldsymbol{A}$. Since we have taken the flow as incompressible and steady, that is, independent of the time, the total amount of fluid flowing outwards must be continuously supplied; that is, in the interior of the region there must be sources producing a (positive or negative) quantity of fluid. The surface integral on the right represents the total flow out of the region R; if we divide by the volume of the region, we obtain the average flow out of R. If we think of the region R as shrinking to a point, so that its diameter tends to zero, in other words, if we carry out a space-differentiation of the integral $\iiint \operatorname{div}\boldsymbol{A}\, dx\, dy\, dz$, we obtain the source-intensity at the point under consideration. On the other hand, this space-differentiation gives the integrand div $\boldsymbol{A}$ at that point, and we thus see that *the divergence of the vector* $\boldsymbol{A}$ *is the source-intensity of the steady incompressible flow represented by* $\boldsymbol{A}$.

Particular interest attaches to cases of flow which are *source-free*, so that fluid is neither created nor annihilated at any point of the region. A flow of this type is characterized by the fact that the equation

$$\operatorname{div}\boldsymbol{A} = \frac{\partial a}{\partial x} + \frac{\partial b}{\partial y} + \frac{\partial c}{\partial z} = 0$$

is satisfied everywhere. It then follows that *for every closed surface* S *the integral over* S *of the normal component* $\iint_S A_n\, dS$ *has the value zero.* We consider two surfaces S_1 and S_2, both bounded by the same oriented curve C in space, which together enclose a simply-connected region of space R, and we apply Gauss's theorem to the region R. For the positive normal direction on the surface S_1, however, we shall take the normal pointing towards the *inside* of the region R (as in fig. 11) instead of that towards the outside, so that the sense of description of C in conjunction with the positive normal for either surface forms a right-handed screw. In Gauss's theorem, then, we must insert different signs for the surfaces S_1 and S_2. We thus obtain

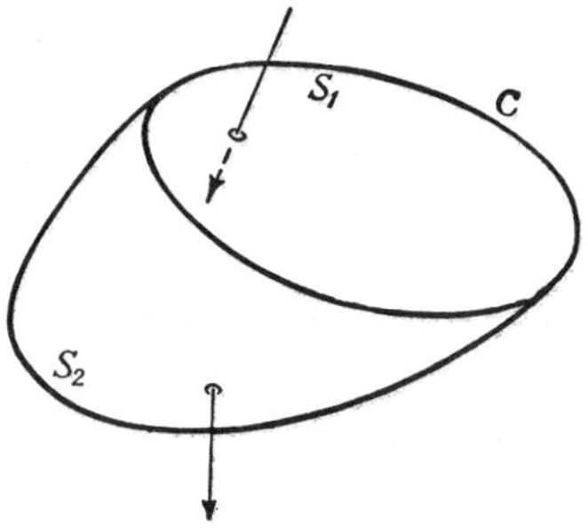

Fig. 11

$$\iiint_R \operatorname{div}\boldsymbol{A}\, dx\, dy\, dz = \iint_{S_1} A_n\, dS - \iint_{S_2} A_n\, dS.$$

Since, by hypothesis, the left-hand side is zero, we have

$$\iint_{S_1} A_n\, dS = \iint_{S_2} A_n\, dS.$$

In words: if a flow is source-free, the same amount of fluid flows in unit time across any two surfaces with the same boundary curve. This amount of fluid, therefore, no longer depends on the choice of the surface S with the closed boundary curve C. It can therefore only depend on the choice of C, and the problem arises how the amount of fluid can be expressed in terms of the curve C. This question is answered in the next section (p. 396) by means of Stokes's theorem.

2. Green's Theorem.

Just as in the case of two independent variables (p. 366), Gauss's theorem leads to some important consequences, which are known as Green's theorem.

We arrive at these formulæ by applying Gauss's theorem (in vector form) to a vector field $\boldsymbol{A}$ which is given in the special form

$$\boldsymbol{A} = u \operatorname{grad} v$$

and therefore has the components uv_x, uv_y, uv_z. Then in R we have

$$\operatorname{div} \boldsymbol{A} = \frac{\partial}{\partial x}(uv_x) + \frac{\partial}{\partial y}(uv_y) + \frac{\partial}{\partial z}(uv_z),$$

and on the boundary

$$A_n = u \frac{\partial v}{\partial n}.$$

Then if we use the familiar symbol

$$\Delta v = v_{xx} + v_{yy} + v_{zz},$$

Gauss's theorem immediately gives us Green's theorem:

$$\iiint_R (u_x v_x + u_y v_y + u_z v_z)\, dx\, dy\, dz$$
$$= -\iiint_R u \Delta v\, dx\, dy\, dz + \iint_S u \frac{\partial v}{\partial n}\, dS.$$

If we apply the same argument to the vector field $\boldsymbol{A} = v \operatorname{grad} u$, we obtain the formula

$$\iiint_R (u_x v_x + u_y v_y + u_z v_z)\, dx\, dy\, dz$$
$$= -\iiint_R v \Delta u\, dx\, dy\, dz + \iint_S v \frac{\partial u}{\partial n}\, dS.$$

If we subtract this last formula from the first one, we obtain the second form of Green's theorem,

$$\iiint_R (u\Delta v - v\Delta u)\, dx\, dy\, dz = \iint_S \left(u \frac{\partial v}{\partial n} - v \frac{\partial u}{\partial n}\right) dS.$$

3. Application of Gauss's Theorem and Green's Theorem in Space.

1. *Transformation of* Δu *to Polar Co-ordinates.*

If in the second form of Green's theorem we substitute the special function $v = 1$, we obtain

$$\iiint \Delta u\, dx\, dy\, dz = \iint \frac{\partial u}{\partial n}\, dS.$$

Just as in the plane, we can use this formula to transform Δu to polar co-ordinates (r, φ, θ) by choosing for the region R a cell of the polar co-ordinate net in space between the co-ordinate surfaces r and $r + h$, φ and $\varphi + k$, θ and $\theta + l$. We obtain

$$\Delta u = \frac{1}{r^2 \sin\theta}\left\{\frac{\partial}{\partial r}(r^2 u_r \sin\theta) + \frac{\partial}{\partial \varphi}\left(\frac{u_\phi}{\sin\theta}\right) + \frac{\partial}{\partial \theta}(u_\theta \sin\theta)\right\}.$$

The calculations, which are analogous to those for the plane case (cf. p. 369), are left to the reader.

2. *Space Forces and Surface Forces.*

The forces acting in a continuum may be regarded either as space forces or as surface forces. The connexion between these two points of view is given by Gauss's theorem.

We content ourselves by considering a special case, namely, the force in a fluid of constant density, say $\rho = 1$, in which there is a pressure $p(x, y, z)$ which in general depends on the point (x, y, z). This means that on every surface element through the point (x, y, z) the fluid exerts a force which is perpendicular to the surface element and has the surface density $p(x, y, z)$. If we consider a region R bounded by the surface S and lying in the fluid, the volume R will be subject to a force whose total x-component is given by the surface integral

$$X = -\iint_S p \frac{\partial x}{\partial n}\, dS,$$

where $\partial x/\partial n$ is the cosine of the angle between the x-axis and the outward-drawn normal to the surface. In the same way, the y- and z-components of the total force are given by

$$Y = -\iint_S p \frac{\partial y}{\partial n}\, dS,$$

$$Z = -\iint_S p \frac{\partial z}{\partial n}\, dS.$$

Gauss's theorem now gives

$$X = -\iiint_R p_x \, dx\, dy\, dz,$$

$$Y = -\iiint_R p_y \, dx\, dy\, dz,$$

$$Z = -\iiint_R p_z \, dx\, dy\, dz,$$

and we thus obtain

$$F = -\iiint_R \operatorname{grad} p \, dx\, dy\, dz$$

for F, the total force exerted on R.

We can express this result as follows. The forces in a fluid due to a pressure $p(x, y, z)$ may on the one hand be regarded as surface forces (pressures) which act with density $p(x, y, z)$ perpendicular to each surface element through the point (x, y, z), and on the other hand as volume forces, that is, as forces which act on every element of volume with volume density $-\operatorname{grad} p$.

EXAMPLE

1*. Let the equations

$$x_i = x_i(p_1, p_2, p_3) \qquad (i = 1, 2, 3)$$

define an arbitrary "orthogonal" co-ordinate system p_1, p_2, p_3; that is, if we put $a_{ik} = \frac{\partial x_i}{\partial p_k}$, then the equations

$$\begin{aligned} a_{11}a_{21} + a_{12}a_{22} + a_{13}a_{23} &= 0 \\ a_{11}a_{31} + a_{12}a_{32} + a_{13}a_{33} &= 0 \\ a_{21}a_{31} + a_{22}a_{32} + a_{23}a_{33} &= 0 \end{aligned}$$

are to hold.

(a) Prove that

$$\frac{\partial(x_1, x_2, x_3)}{\partial(p_1, p_2, p_3)} = \sqrt{e_1e_2e_3},$$

where

$$e_i = a_{1i}^2 + a_{2i}^2 + a_{3i}^2.$$

(b) Prove that

$$\frac{\partial p_i}{\partial x_k} = \frac{1}{e_i}\frac{\partial x_k}{\partial p_i} = \frac{1}{e_i}a_{ki}.$$

(c) Express $\Delta u = u_{x_1x_1} + u_{x_2x_2} + u_{x_3x_3}$ in terms of p_1, p_2, p_3, using Gauss's theorem.

(d) Express Δu in the focal co-ordinates t_1, t_2, t_3 defined in Ex. 6, p. 158.

6. STOKES'S THEOREM IN SPACE

1. Statement and Proof of the Theorem.

In this section we shall give a discussion of Stokes's theorem for any curved surface. We have already (p. 365) met with Stokes's theorem in two dimensions.

Let C be a closed sectionally smooth oriented curve in space, and let S be a surface, bounded by C, whose positive normal is continuous or sectionally continuous and in conjunction with the sense of description of the boundary curve forms a right-handed screw. Further, let $\boldsymbol{B}$ be a vector field defined in a neighbourhood of S, with components $\phi(x, y, z)$, $\psi(x, y, z)$, $\chi(x, y, z)$.

Stokes's theorem then states that

$$\iint (\operatorname{curl} \boldsymbol{B})_n \, dS = \int_C B_t \, ds,$$

where the arc s of the curve C increases in the direction in which C is described, and B_t is the tangential component of $\boldsymbol{B}$ along C. Written in full, Stokes's formula is

$$\iint_S \left\{ \left(\frac{\partial \chi}{\partial y} - \frac{\partial \psi}{\partial z} \right) dy\, dz + \left(\frac{\partial \phi}{\partial z} - \frac{\partial \chi}{\partial x} \right) dz\, dx + \left(\frac{\partial \psi}{\partial x} - \frac{\partial \phi}{\partial y} \right) dx\, dy \right\}$$

$$= \int_C (\phi\, dx + \psi\, dy + \chi\, dz).$$

This transforms a surface integral taken over the oriented surface S into a line integral taken round the correspondingly oriented boundary of the surface.

The truth of Stokes's theorem can immediately be made plausible by the following train of thought. The theorem has already been proved for a plane surface (p. 365). Then if S is a polyhedral surface composed of plane polygonal surfaces, so that the boundary curve C is a polygon, we can apply Stokes's theorem to each of the plane portions and add the corresponding formulæ. In this process the line integrals along all the interior edges of the polyhedra cancel, and we at once obtain Stokes's theorem for the polyhedral surface. In order to obtain the general statement of Stokes's theorem we have only to perform a passage to the limit, leading from polyhedra to arbitrary surfaces S and to arbitrary sectionally smooth boundary curves C.

The rigorous performance of this passage to the limit, however, would be troublesome; having made these heuristic remarks, therefore, we shall carry out the proof by means of a simple calculation.

If for brevity we put

$$\boldsymbol{A} = \operatorname{curl} \boldsymbol{B},$$

the components of $\boldsymbol{A}$ are given by

$$a(x, y, z) = \frac{\partial \chi}{\partial y} - \frac{\partial \psi}{\partial z}, \quad b(x, y, z) = \frac{\partial \phi}{\partial z} - \frac{\partial \chi}{\partial x}, \quad c(x, y, z) = \frac{\partial \psi}{\partial x} - \frac{\partial \phi}{\partial y},$$

and (cf. p. 93)

$$\operatorname{div} \boldsymbol{A} = \operatorname{div} \operatorname{curl} \boldsymbol{B} = 0.$$

We take the oriented surface S bounded by the oriented curve C and consider the problem of changing the integral

$$\iint_S A_n\, dS = \iint_S (a\, dy\, dz + b\, dz\, dx + c\, dx\, dy)$$

taken over S into an expression depending only on the boundary curve C. To do this, we imagine the surface represented in the usual way by two parameters u, v, so that the surface corresponds to a closed region D in the uv-plane. By the general rule, the transformation of the surface integral to the region D gives the expression

$$\iint_S \{a\, dy\, dz + b\, dz\, dx + c\, dx\, dy\}$$

$$= \iint_D \left\{ \left(\frac{\partial \chi}{\partial y} - \frac{\partial \psi}{\partial z}\right)\left(\frac{\partial y}{\partial u}\frac{\partial z}{\partial v} - \frac{\partial z}{\partial u}\frac{\partial y}{\partial v}\right) + \left(\frac{\partial \phi}{\partial z} - \frac{\partial \chi}{\partial x}\right)\left(\frac{\partial z}{\partial u}\frac{\partial x}{\partial v} - \frac{\partial x}{\partial u}\frac{\partial z}{\partial v}\right) \right.$$
$$\left. + \left(\frac{\partial \psi}{\partial x} - \frac{\partial \phi}{\partial y}\right)\left(\frac{\partial x}{\partial u}\frac{\partial y}{\partial v} - \frac{\partial y}{\partial u}\frac{\partial x}{\partial v}\right)\right\} du\, dv.$$

We can transform the expression on the right by collecting the terms involving ϕ, those in ψ, and those in χ. For the terms involving ϕ, for example, we obtain

$$-\frac{\partial \phi}{\partial y}\left(\frac{\partial x}{\partial u}\frac{\partial y}{\partial v} - \frac{\partial y}{\partial u}\frac{\partial x}{\partial v}\right) - \frac{\partial \phi}{\partial z}\left(\frac{\partial x}{\partial u}\frac{\partial z}{\partial v} - \frac{\partial z}{\partial u}\frac{\partial x}{\partial v}\right).$$

If to this we add the expression

$$-\frac{\partial \phi}{\partial x}\left(\frac{\partial x}{\partial u}\frac{\partial x}{\partial v} - \frac{\partial x}{\partial u}\frac{\partial x}{\partial v}\right),$$

which is identically zero, the terms involving ϕ in the integrand are

$$\frac{\partial x}{\partial v}\left(\frac{\partial \phi}{\partial x}\frac{\partial x}{\partial u} + \frac{\partial \phi}{\partial y}\frac{\partial y}{\partial u} + \frac{\partial \phi}{\partial z}\frac{\partial z}{\partial u}\right) - \frac{\partial x}{\partial u}\left(\frac{\partial \phi}{\partial x}\frac{\partial x}{\partial v} + \frac{\partial \phi}{\partial y}\frac{\partial y}{\partial v} + \frac{\partial \phi}{\partial z}\frac{\partial z}{\partial v}\right)$$
$$= \frac{\partial \phi}{\partial u}\frac{\partial x}{\partial v} - \frac{\partial \phi}{\partial v}\frac{\partial x}{\partial u}.$$

In the same way we obtain the two other terms

$$\frac{\partial \psi}{\partial u}\frac{\partial y}{\partial v} - \frac{\partial \psi}{\partial v}\frac{\partial y}{\partial u} \quad \text{and} \quad \frac{\partial \chi}{\partial u}\frac{\partial z}{\partial v} - \frac{\partial \chi}{\partial v}\frac{\partial z}{\partial u}$$

in the integrand. The double integral is therefore split up into the sum of the integrals of the three expressions

$$\frac{\partial(\phi, x)}{\partial(u, v)}, \quad \frac{\partial(\psi, y)}{\partial(u, v)}, \quad \frac{\partial(\chi, z)}{\partial(u, v)},$$

taken over the oriented region D, whose boundary curve K has an orientation corresponding to that of C. Now by Stokes's theorem for two dimensions (cf. p. 364) we have

$$\iint_D \left(\frac{\partial \phi}{\partial u}\frac{\partial x}{\partial v} - \frac{\partial \phi}{\partial v}\frac{\partial x}{\partial u}\right) du\, dv = \int_K \left(\phi \frac{\partial x}{\partial u} du + \phi \frac{\partial x}{\partial v} dv\right) = \int_C \phi \frac{dx}{ds} ds,$$

where the integrals are to be taken with corresponding orientations and the length of arc s on C increases in the direction in which the curve is positively described. If we add this formula to the two other corresponding ones, we obtain on the left the value of the surface integral and on the right the integral

$$\int_C \left(\phi \frac{dx}{ds} + \psi \frac{dy}{ds} + \chi \frac{dz}{ds}\right) ds.$$

The expression $\phi \frac{dx}{ds} + \psi \frac{dy}{ds} + \chi \frac{dz}{ds}$, however, is just the tangential component B_t of the vector $\boldsymbol{B}$ in the direction of the oriented boundary curve C, and we thus obtain Stokes's theorem

$$\iint_S (\operatorname{curl} \boldsymbol{B})_n\, dS = \int_C B_t\, ds,$$

or, written out in full,

$$\iint_S \left\{\left(\frac{\partial \chi}{\partial y} - \frac{\partial \psi}{\partial z}\right) dy\, dz + \left(\frac{\partial \phi}{\partial z} - \frac{\partial \chi}{\partial x}\right) dz\, dx + \left(\frac{\partial \psi}{\partial x} - \frac{\partial \phi}{\partial y}\right) dx\, dy\right\}$$
$$= \int_C (\phi\, dx + \psi\, dy + \chi\, dz).$$

This formula is true provided that the vector $\boldsymbol{A} = \operatorname{curl} \boldsymbol{B}$ is continuous in the region under consideration and that the surface S consists of one or more portions each of which can be continuously represented as above by parametric equations $x = x(u, v)$, $y = y(u, v)$, $z = z(u, v)$ with continuous first derivatives.

Stokes's theorem gives the answer to the question raised at the end of No. 1 of the preceding section (p. 389). We have seen that for a vector field whose divergence is identically zero the integral of the normal component over a surface bounded by a fixed curve C depends on the boundary curve C only and not on the particular nature of the surface. Since, as we shall prove in section 2 of the Appendix (p. 404), every vector field $\boldsymbol{A}$ whose divergence is identically zero has the form

$$\boldsymbol{A} = \operatorname{curl} \boldsymbol{B},$$

Stokes's theorem enables us to express the surface integral in a form which depends only on the boundary.

2. Interpretation of Stokes's Theorem.

The physical interpretation of Stokes's theorem in three dimensions is similar to that already given (p. 371) for Stokes's theorem in two dimensions.* Once again we interpret the vector field $\boldsymbol{B}$ as the velocity field of a steady flow of an incompressible fluid, and we call the integral $\int_C B_t\, ds$ taken round a closed curve C the *circulation* of the flow along this curve. Stokes's theorem states that the circulation round a curve is equal to the surface integral of the component of the curl in the direction of the positive normal to any surface bounded by the oriented curve, the orientation of the surface being given by that of the boundary curve. Suppose that we apply Stokes's theorem to a portion of a surface S with a continuously turning tangent plane. If we divide this surface integral by the area of the portion of surface and then perform a passage to the limit by letting the portion of surface and its boundary curve shrink to a point while remaining on the large surface S, on the left this process of space-differentiation gives us the component of the curl in the direction of the normal at that point of the surface to which the boundary curve C has shrunk. We therefore see that the component of the curl in the direction of the positive normal to the surface is to be regarded as the *specific circulation* or circulation-density of the flow in the surface at the corresponding point, where the sense of the circulation and the positive normal together form a right-handed screw.†

If we interpret the vector $\boldsymbol{B}$ as the field of a mechanical or electrical force, the line integral on the right-hand side of Stokes's theorem represents the work done by the field on a particle subject to the force when it is made to describe the curve C. By Stokes's theorem the expression for this work is transformed into an integral over the surface S bounded by the curve, the integrand being the normal component of the curl of the field of force.

* The student should note that in two dimensions Gauss's theorem and Stokes's theorem differ from one another formally by a sign only, while in three dimensions both the intuitive interpretation and the formal nature of the two theorems are essentially different.

† These considerations also show that the curl of a vector has a meaning independent of the co-ordinate system and therefore is itself a vector.

From Stokes's theorem we obtain a new proof for the main theorem on line integrals in space (cf. also p. 365, footnote). The chief question was, what must be the nature of the vector field $\boldsymbol{B}$ if the integral of the tangential component of the vector taken round an arbitrary closed curve is to vanish? Stokes's theorem yields a new proof of the fact that the vanishing * of this line integral is ensured if the curl of the vector field vanishes. The vanishing of the curl or, as we shall say, the *irrotational* nature of a vector field is therefore a sufficient condition—and, as we know from section 1 (p. 358), also a necessary one—that the line integral of the tangential component of the vector round any closed curve shall vanish. In this case the vector field $\boldsymbol{B}$ can itself, as we know from section 1 (p. 352), be represented as the gradient of a function $f(x, y, z)$:

$$\boldsymbol{B} = \operatorname{grad} f.$$

If the vector field $\boldsymbol{B}$ is not only irrotational but also source-free, that is, if its divergence vanishes, then the function f satisfies the equation

$$\operatorname{div} \operatorname{grad} f = 0,$$

or, in full,

$$\Delta f = \frac{\partial^2 f}{\partial x^2} + \frac{\partial^2 f}{\partial y^2} + \frac{\partial^2 f}{\partial z^2} = 0.$$

For the scalar quantity f, which as before we call the *potential* of the vector $\boldsymbol{B}$, we have Laplace's equation

$$\Delta f = 0,$$

which we have already met with (p. 93).

7. The Connexion between Differentiation and Integration for Several Variables

It is useful to reconsider, from a single point of view, the facts developed in this chapter.

In the case of one independent variable we regard the reciprocal relation between differentiation and integration as the

* Here, of course, we assume that a surface of the type described above and bounded by this curve exists. Since this may lead to difficulties or complications—for exampl- in the case of curves with multiple points—the proof of the theorem given in section 1 (p. 352) is preferable.

fundamental theorem of the differential and integral calculus (Vol. I, Chap. II, p. 117). For one independent variable this fundamental theorem is as follows: if $f(x)$ is a continuous function in the closed region $a \leqq x \leqq b$ and if $F(x)$ is a primitive of $f(x)$, then

$$\int_a^b f(x)\,dx = F(b) - F(a);$$

conversely, for every function $F(x)$ with a continuous derivative we can construct the corresponding function $f(x) = F'(x)$ in the above formula. In the present connexion the essential point is the first part of the fundamental theorem, that is, the transformation of an integral over a one-dimensional region into the expression $F(b) - F(a)$ depending only on the boundary points, which form, as we may say, a region of zero dimensions. In other words, if the integrand is given as the derivative of a function $F(x)$, the one-dimensional integral can be transformed by means of the function $F(x)$ into an expression depending on the boundary only.

The various integral theorems for regions in several dimensions now give us something analogous to the fundamental theorem for one independent variable. The point in question is always that of transforming an integral over a certain region lying in the region of the independent variables, no matter whether this region of integration is a curve, a surface, or a portion of space, into an expression that depends only on the boundary of the region. For example, Gauss's theorem in two dimensions is

$$\iint_R (a_x + b_y)\,dx\,dy = \int_{+C} (a\,dy - b\,dx).$$

This states that if the integrand of an integral $\iint_R f(x, y)\,dx\,dy$ over a closed region R is represented in the form

$$f(x, y) = a_x(x, y) + b_y(x, y),$$

then the double integral over the two-dimensional region can be transformed into an expression depending only on the one-dimensional boundary, namely, into a line integral round the boundary curve. Thus Gauss's theorem reduces the number of dimensions of the region of integration by 1. Instead of the boundary expression $F(b) - F(a)$ considered above, we have a line integral round the boundary of the plane region. Here, of course, we cannot speak of a primitive function F. The single primitive function is here in a sense represented by the vector field with components $a(x, y)$ and $b(x, y)$. On the other hand, the application of Gauss's theorem does require that the integrand of the double integral shall be expressed by means of the differentiation process, in fact, as the sum of a derivative with respect to x and a derivative with respect to y. The requirement that the integrand f shall be capable of being expressed in this way still allows a great deal of freedom in the choice of the primitive vector field (a, b), whereas for ordinary integrands the primitive function $F(x)$ is uniquely determined except for an arbitrary additive constant.*

For the case $n = 2$, besides Gauss's theorem and Stokes's theorem, which are essentially equivalent to one another, there is yet another generalization of the fundamental theorem, namely the main theorem on line integrals (p. 352). Within the two-dimensional region we have a closed one-dimensional bounded manifold, that is, a portion of curve with two end-points, and the problem is that of the reduction of this line integral to an expression depending only on the boundary. The main theorem on line integrals in section 1 (p. 352) states that this reduction is possible if, and only if, the integrand can be represented by means of a primitive function $U(x, y)$ in the form

$$\mathbf{t} \operatorname{grad} U,$$

where $\mathbf{t}$ is the tangential unit vector and the integration is with respect to the length of arc s. The value of the integral is then given by the equation

$$\int_{(\xi_0, \eta_0)}^{(\xi, \eta)} \mathbf{t} \operatorname{grad} U \, ds = U(\xi, \eta) - U(\xi_0, \eta_0),$$

which obviously corresponds to the state of affairs for $n = 1$.

* For a given integrand $f(x, y)$ there are many ways of finding a pair of functions $a(x, y)$ and $b(x, y)$ which satisfy the above equation. For example, we can take $b(x, y)$ as identically zero, or as equal to an arbitrary function, and then determine the corresponding function $a(x, y)$ in accordance with the equation $a_x = f - b_y$, choosing for $a(x, y)$ any indefinite integral of the function $f(x, y) - b_y(x, y)$ with respect to x, y acting as parameter. Every other vector field which arises by the addition of an arbitrary divergence-free field to the vector field found as above is likewise a primitive vector field.

The transformation of the line integral

$$\int_G (a\,dx + b\,dy)$$

into a boundary expression can therefore be carried out if, and only if, the vector $\boldsymbol{A}$ with the components a, b can be represented as the gradient of a potential. By comparing this with the ordinary fundamental theorem, we see that instead of expressing the integrand as the derivative we here express the integrand by means of a gradient and that the part played by the primitive function is taken by the potential of this gradient. An essential difference still remains between this case and the preceding one, however, since it is by no means true that the integrand of every line integral can be expressed as a gradient in this way; on the contrary, this depends on the condition of integrability $a_y = b_x$.

When there are three independent variables the conditions are very similar. By Gauss's theorem a triple integral over a bounded closed three-dimensional region is transformed into an integral over the closed boundary, which is a closed unbounded * two-dimensional region enclosed in three-dimensional space. The transformation is related to the expression of the integrand of the triple integral as the divergence of a vector field (a, b, c), and to a certain extent this vector field again plays the part of the primitive function.†

With regard to line integrals, the case of three independent variables is exactly like that of two independent variables and requires no further discussion.

In the case of three independent variables, the surface integral over a two-dimensional region, that is, a surface bounded by a space-curve, occupies a position between the line integral and the triple integral. Here the condition for the transformation of an integral taken over such a surface into an expression involving the boundary only is given by Stokes's theorem in section 6 (p. 393). The process of differentiation by means of which the integrand is constructed in Stokes's theorem amounts to the construction of the curl of a vector field, which here takes the place of the primitive function. Here again the situation

* That is, one having no boundary curve.

† Just as in the case of two independent variables, there are many different ways of constructing a primitive vector field corresponding to a given integrand.

resembles that in the case of the line integral. In order that the integrand of a surface integral

$$\iint_S (a\,dy\,dz + b\,dz\,dx + c\,dx\,dy)$$

may be expressible as the normal component of a curl the condition $a_x + b_y + c_z = 0$ must necessarily be fulfilled. Thus the transformation of the surface integral into a line integral is not always possible. We may remark that the necessary condition stated above is in fact a sufficient condition also.*

The situation is similar if there are more than three independent variables; we need not, however, discuss this here.

Examples

1. Evaluate the surface integral

$$\iint \frac{z}{p}\,dS$$

taken over the half of the ellipsoid $\frac{x^2}{a^2} + \frac{y^2}{b^2} + \frac{z^2}{c^2} = 1$ for which z is positive, where

$$\frac{1}{p} = \frac{lx}{a^2} + \frac{my}{b^2} + \frac{nz}{c^2},$$

l, m, n being the direction cosines of the outward-drawn normal.

2. Evaluate the surface integral

$$\iint H\,dS$$

taken over the sphere of radius unity with centre at the origin, where

$$H = a_1x^4 + a_2y^4 + a_3z^4 + 3a_4x^2y^2 + 3a_5y^2z^2 + 3a_6x^2z^2.$$

3*. Prove Gauss's theorem in n dimensions. That is, let B be a region in n-dimensional $x_1 \dots x_n$-space and let its boundary S be given by an equation

$$G(x_1, \dots, x_n) = 0$$

such that $G \leq 0$ in B. Let the functions $a_i(x_1, \dots, x_n)$, where $i = 1, \dots, n$, be continuously differentiable in B. Then

$$\iint \cdots \int_B \left(\frac{\partial a_1}{\partial x_1} + \dots + \frac{\partial a_n}{\partial x_n}\right) dx_1 \dots dx_n$$
$$= \int \cdots \int_S \left(a_1 \frac{\partial x_1}{\partial \nu} + \dots + a_n \frac{\partial x_n}{\partial \nu}\right) dS,$$

* For the proof of this see section 2 of the Appendix, p. 404.

where dS is the element of surface defined in Chap. IV, p. 301, and $\frac{\partial x_i}{\partial \nu}, \ldots$ are the derivatives of the co-ordinates with respect to the outward normal, that is,

$$\frac{\partial x_i}{\partial \nu} = \frac{G_{x_i}}{\sqrt{G_{x_i}^2 + \ldots + G_{x_n}^2}}.$$

Appendix to Chapter V

1. Remarks on Gauss's Theorem and Stokes's Theorem

In Chapter V we proved Gauss's theorem and Stokes's theorem by starting with multiple integrals and transforming them by simple integrations into boundary integrals. We can, however, arrive at the formal expressions of these theorems in the reverse way. The corresponding transformations, which in themselves are instructive, will be briefly discussed here.

For example, in order to obtain Stokes's theorem in the plane we consider two fixed points P and Q in the plane, joined by a

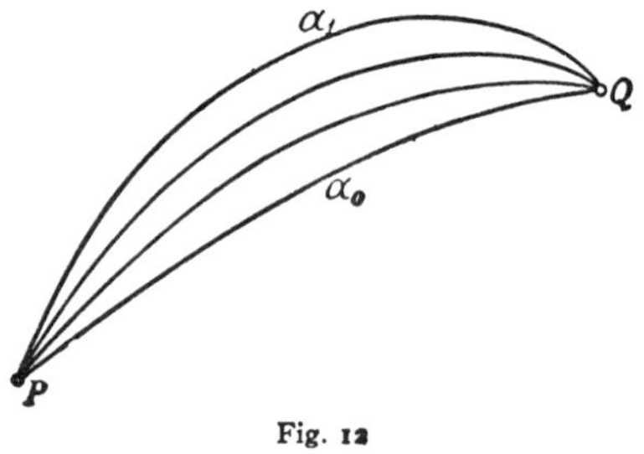

Fig. 12

curve C. This curve C, whose points are represented by means of a parameter t, is supposed to be deformed in such a way that during the deformation from its initial position to its final position it sweeps out the region R simply. We make this idea analytically precise in the following way: let the curve C, which depends on the parameter a, be given by the parametric equations

$$x = x(t, a), \; y = y(t, a); \quad t_0 \leqq t \leqq t_1,$$

where $x(t_0, a)$, $y(t_0, a)$ and $x(t_1, a)$, $y(t_1, a)$ are the co-ordinates of the two fixed points P and Q, which are independent of α. We suppose that when α describes an interval $\alpha_0 \leqq \alpha \leqq \alpha_1$ the curve describes a closed region R. We assume that the functions $x(t, \alpha)$, $y(t, \alpha)$ have continuous derivatives with respect to t and α and also continuous mixed second derivatives

$$\frac{\partial^2 x}{\partial \alpha \partial t} = x_{\alpha t}, \quad \frac{\partial^2 y}{\partial \alpha \partial t} = y_{\alpha t},$$

and, moreover, that everywhere in the region R except at the points P and Q the Jacobian $\frac{\partial(x, y)}{\partial(t, \alpha)}$ is different from zero, say positive. Then the region R, except for the points P and Q, is mapped in a one-to-one way on the rectangle $\alpha_0 \leqq \alpha \leqq \alpha_1$ and $t_0 \leqq t \leqq t_1$ of the αt-plane.

We assume that in the closed region R we are given the two functions $a(x, y)$ and $b(x, y)$ with continuous partial derivatives, and we consider the line integral

$$I(\alpha) = \int_{C_\alpha} (a(x, y)\,dx + b(x, y)\,dy) = \int_{t_0}^{t_1} (ax_t + by_t)\,dt,$$

taken along the curve C_α corresponding to the parameter α. Our object is to investigate how this integral $I(\alpha)$ depends on the variable α. For this purpose we form the derivative

$$\frac{dI(\alpha)}{d\alpha} = \int_{t_0}^{t_1} [(a_x x_\alpha + a_y y_\alpha)x_t + (b_x x_\alpha + b_y y_\alpha)y_t + ax_{\alpha t} + by_{\alpha t}]dt$$

according to the rules for differentiating an integral with respect to a parameter. On integration by parts we obtain

$$\int_{t_0}^{t_1} (ax_{\alpha t} + by_{\alpha t})\,dt = [ax_\alpha + by_\alpha]_{t_0}^{t_1} - \int_{t_0}^{t_1} (a_t x_\alpha + b_t y_\alpha)\,dt$$

$$= -\int_{t_0}^{t_1} [(a_x x_t + a_y y_t)x_\alpha + (b_x x_t + b_y y_t)y_\alpha]\,dt;$$

the last formula is obtained if we notice that the hypotheses assure that x_α and y_α vanish at $t = t_0$ and $t = t_1$. It follows that

$$\frac{dI(\alpha)}{d\alpha} = \int_{t_0}^{t_1} [a_y(y_\alpha x_t - y_t x_\alpha) + b_x(x_\alpha y_t - x_t y_\alpha)]\,dt,$$

that is,

$$\frac{dI(\alpha)}{d\alpha} = \int_{t_0}^{t_1} (a_y - b_x) \frac{\partial(x, y)}{\partial(t, \alpha)} dt.$$

If we integrate this last equation with respect to α between the limits α_0 and α_1, we obtain

$$I(\alpha_1) - I(\alpha_0) = \int_{\alpha_0}^{\alpha_1}\int_{t_0}^{t_1} (a_y - b_x) \frac{\partial(x, y)}{\partial(t, \alpha)} dt\, d\alpha;$$

or, if we introduce the independent variables x, y on the right instead of t, α,

$$I(\alpha_0) - I(\alpha_1) = \iint_R (b_x - a_y)\, dx\, dy.$$

On the left-hand side, however, we have simply the line integral $\int(a\,dx + b\,dy)$ taken round the boundary curve $C_{\alpha_0} - C_{\alpha_1}$ of the region R, and thus, subject to the assumptions made, we have obtained Stokes's theorem for the plane.

The reader may be left to deduce Stokes's theorem in three dimensions by the same method. Gauss's theorem in three dimensions may likewise be obtained by starting from a surface integral over a bounded surface and deforming this surface in such a way that it describes a region R of space.

It should, however, be pointed out that this way of deducing the integral theorems does not give exactly the same results as the proofs developed previously. In order to attain the same generality we must, e.g., show for Stokes's theorem in two dimensions that every region R of the type considered in § 2 (p. 362) in the plane can be covered by a family of curves C with the required properties of continuity and differentiability. Such a proof is possible, but it is so complicated that the previous method of proving Stokes's theorem remains preferable.

2. Representation of a Source-free Vector Field as a Curl

In view of the remarks at the end of Chap. V, section 6, No. 1 (p. 396), we shall investigate whether every source-free vector field, that is, every vector field $\boldsymbol{A}$ for which the expression div $\boldsymbol{A}$ vanishes everywhere in a closed region R of xyz-space, can be represented by means of a second vector $\boldsymbol{B}$ according to the formula

$$\boldsymbol{A} = \operatorname{curl} \boldsymbol{B}.$$

We shall show that this is actually the case. If $a(x, y, z)$, $b(x, y, z)$, $c(x, y, z)$ are the components of the vector $\boldsymbol{A}$, the problem is to find a vector $\boldsymbol{B}$ with components $u(x, y, z)$, $v(x, y, z)$, $w(x, y, z)$ such that the three equations

$$\begin{aligned} a &= w_y - v_z \\ b &= u_z - w_x \\ c &= v_x - u_y \end{aligned}$$

are satisfied in R. For the sake of simplicity we assume that the region R in which the vector $\boldsymbol{A}$ is defined and satisfies the condition $a_x + b_y + c_z = 0$ is a parallelepiped. We can then determine the vector $\boldsymbol{B}$ in many ways, e.g. in such a way that its third component $w(x, y, z)$ vanishes everywhere. If we make this assumption, we obtain the equations

$$\begin{aligned} a &= -v_z \\ b &= u_z \\ c &= v_x - u_y. \end{aligned}$$

The first equation is satisfied if we put

$$v = -\int_{z_0}^{z} a(x, y, \zeta)\, d\zeta,$$

where x and y act as parameters during the integration, and z_0 and subsequently y_0 are the z- and y-co-ordinates respectively of an arbitrary fixed point of R. To satisfy the second equation we put

$$u = \int_{z_0}^{z} b(x, y, \zeta)\, d\zeta + \alpha(x, y),$$

where $\alpha(x, y)$ is a function of x and y as yet undetermined. In virtue of the assumption $a_x + b_y = -c_z$ we can now satisfy the third equation also. We first arrive at the equation

$$c = v_x - u_y = -\int_{z_0}^{z} [a_x(x, y, \zeta) + b_y(x, y, \zeta)]\, d\zeta - \alpha_y(x, y),$$

and thus from $a_x + b_y = -c_z$ we obtain the further relation

$$c(x, y, z) = \int_{z_0}^{z} c_\zeta(x, y, \zeta)\, d\zeta - \alpha_y(x, y) = c(x, y, z) - c(x, y, z_0) - \alpha_y(x, y),$$

which we now use to determine the function $\alpha(x, y)$, putting

$$a_y = -c(x, y, z_0),$$

$$a = -\int_{y_0}^{y} c(x, \eta, z_0)\, d\eta.$$

The vector $\boldsymbol{B}$ defined by the functions

$$u = \int_{z_0}^{z} b(x, y, \zeta)\, d\zeta - \int_{y_0}^{y} c(x, \eta, z_0)\, d\eta,$$

$$v = -\int_{z_0}^{z} a(x, y, \zeta)\, d\zeta,$$

$$w = 0$$

is a solution of our problem. We at once arrive at the most general solution by writing down the three functions

$$U = u + \frac{\partial \Phi}{\partial x},$$

$$V = v + \frac{\partial \Phi}{\partial y},$$

$$W = w + \frac{\partial \Phi}{\partial z},$$

where $\Phi(x, y, z)$ is an arbitrary twice-continuously differentiable function. For we see at once that the vector $\boldsymbol{B}' = \boldsymbol{B} + \operatorname{grad} \Phi$ with components (U, V, W) satisfies our condition. Conversely, if $\boldsymbol{B}'$ is any vector which satisfies the condition $\operatorname{curl} \boldsymbol{B}' = \boldsymbol{A}$, we must have $\operatorname{curl} (\boldsymbol{B}' - \boldsymbol{B}) = 0$. Thus the vector $\boldsymbol{B}' - \boldsymbol{B}$ is irrotational, and by Chap. V, section 1 (p. 352) can be represented as the gradient of a function $\Phi(x, y, z)$, so that our statement is proved.

EXAMPLES

1. Let $f(x, y)$ be a continuous function with continuous first and second derivatives. Prove that if

$$f_{xx} f_{yy} - f_{xy}^2 \neq 0$$

the transformation

$$u = f_x(x, y), \quad v = f_y(x, y), \quad w = -z + x f_x(x, y) + y f_y(x, y)$$

has a unique inverse, which is of the form

$$x = g_u(u, v), \quad y = g_v(u, v), \quad z = -w + u g_u(u, v) + v g_v(u, v).$$

2. Represent the gravitational vector field

$$X = \frac{x}{\sqrt{(x^2+y^2+z^2)^3}}, \quad Y = \frac{y}{\sqrt{(x^2+y^2+z^2)^3}}, \quad Z = \frac{z}{\sqrt{(x^2+y^2+z^2)^3}}$$

as a curl.

Miscellaneous Examples

1. Let φ, a, and b be continuously differentiable functions of a parameter t, for $0 \leqq t \leqq 2\pi$, with $a(2\pi)=a(0)$, $b(2\pi)=b(0)$, $\varphi(2\pi)=\varphi(0)+2n\pi$ (n a rational integer), and let x, y be constants. Interpreting the equations

$$\xi = x\cos\varphi - y\sin\varphi + a, \quad \eta = x\sin\varphi + y\cos\varphi + b$$

as the parametric equations (with parameter t) of a closed plane curve Γ, prove that

$$\tfrac{1}{2}\int_\Gamma (\xi\, d\eta - \eta\, d\xi) = A(x^2+y^2) + Bx + Cy + D,$$

where

$$A = \tfrac{1}{2}\int_\Gamma d\varphi, \quad B = \int_\Gamma (a\cos\varphi + b\sin\varphi)\, d\varphi,$$

$$C = \int_\Gamma (-a\sin\varphi + b\cos\varphi)\, d\varphi, \quad D = \tfrac{1}{2}\int_\Gamma (a\, db - b\, da).$$

2*. Let a rigid plane P describe a closed motion with respect to a fixed plane Π with which it coincides. Every point M of P will describe a closed curve of Π bounding an area of algebraic value $S(M)$. Denote by $2n\pi$ (n a rational integer) the total rotation of P with respect to Π. Prove the following results:

(α) If $n \neq 0$, there is in P a point C such that for any other point M of P we have

$$S(M) = \pi n \overline{CM}^2 + S(C);$$

(β) If $n = 0$, then two cases may arise: (β_1) there is in P an oriented line Δ such that for every point M of P

$$S(M) = \lambda\, d(M),$$

where $d(M)$ is the distance of M from Δ and λ is a constant positive factor; or else (β_2) $S(M)$ has the same value for all the points M of the plane P. (Steiner's theorem.)

3*. A rigid line-segment AB describes in a plane Π one closed motion of a connecting-rod: B describes a closed counter-clockwise circular motion with centre C, while A describes a (closed) rectilinear motion on a line passing through C. Apply the results of the previous example to determine the area of the closed curve in Π described by a point M which is rigidly connected to the line-segment AB.

4. The end-points A and B of a rigid line-segment AB describe one full turn on a closed convex curve Γ. A point M on AB, where $AM = a$, $MB = b$, describes as a result of this motion a closed curve Γ'. Prove that the area between the curves Γ and Γ' is equal to πab. (Holditch's theorem.)

5*. Prove that if we apply to each element ds of a twisted, closed, and rigid curve Γ a force of magnitude ds/ρ in the direction of the principal normal vector (p. 86), the curve Γ remains in equilibrium; $1/\rho$ is the curvature of Γ at ds and is supposed to be finite and continuous at every point of Γ. (By the principles of the statics of a rigid body we have to prove that

$$\int_\Gamma \frac{\boldsymbol{n}}{\rho}\, ds = 0, \quad \int_\Gamma \frac{[\boldsymbol{x}\boldsymbol{n}]}{\rho}\, ds = 0,$$

where $\boldsymbol{n}$ denotes the unit principal normal vector of Γ at ds, and $\boldsymbol{x}$ is the position vector of ds.)

6. Prove that a closed rigid surface Σ remains in equilibrium under a uniform inward pressure on all its surface-elements. (If by $\boldsymbol{n}'$ we denote the inward-drawn unit vector normal to the surface-element $d\sigma$ and by $\boldsymbol{x}$ the position vector of $d\sigma$, the statement becomes equivalent to the vector equations

$$\iint_\Sigma \boldsymbol{n}' d\sigma = 0, \quad \iint_\Sigma [\boldsymbol{x}\boldsymbol{n}'] d\sigma = 0.)$$

7*. A rigid body of volume V bounded by the surface Σ is completely immersed in a fluid of specific gravity unity. Prove that the statical effect of the fluid pressure on the body is the same as that of a single force $\boldsymbol{f}$ of magnitude V, vertically upwards, applied at the centroid C of the volume V.

8*. Let p denote the distance from the centre of the ellipsoid Σ

$$x^2/a^2 + y^2/b^2 + z^2/c^2 = 1$$

to the tangent plane at the point $P(x, y, z)$, and dS the element of area at this point. Prove the relations

$$\text{(i)} \iint_\Sigma p\, dS = 4\pi abc, \quad \text{(ii)} \iint_\Sigma \frac{1}{p}\, dS = \frac{4\pi}{3abc}(b^2c^2 + c^2a^2 + a^2b^2).$$

9. An ordinary plane angle is measured by the length of the arc which its sides intercept on a unit circle with centre at the vertex. This idea can be extended to a *solid angle* bounded by a conical surface with vertex A as follows. The magnitude of the solid angle is by definition equal to the area which it intercepts on a unit sphere with centre A. Thus the measure of the solid angle of the domain $x \geqq 0$, $y \geqq 0$, $z \geqq 0$ is $4\pi/8 = \pi/2$. Now let Γ be a closed curve, Σ a surface bounded by Γ, and A a fixed point outside both Γ and Σ. An element of area dS at a point M of Σ defines an elementary cone with its vertex at A, and the solid

angle of this cone is readily found by an elementary argument to be

$$\frac{\cos\theta}{r^2}\,dS,$$

where $r = AM$ and θ is the angle between the vector MA and the normal to Σ at M. This elementary solid angle is positive or negative according as θ is acute or obtuse. Interpret the surface integral

$$\Omega = \iint_\Sigma \frac{\cos\theta}{r^2}\,dS$$

geometrically as a solid angle and show that

$$\Omega = \iint_\Sigma \frac{(a-x)\,dy\,dz + (b-y)\,dz\,dx + (c-z)\,dx\,dy}{[(a-x)^2+(b-y)^2+(c-z)^2]^{3/2}},$$

where (a, b, c) and (x, y, z) are the Cartesian co-ordinates of A and M respectively.

10. Prove, first directly and then by interpretation of the integral as a solid angle, that

$$\int_{-\infty}^{\infty}\int_{-\infty}^{\infty} \frac{dx\,dy}{(x^2+y^2+1)^{3/2}} = 2\pi.$$

11*. Prove that the solid angle which the whole surface of the hyperboloid of one sheet

$$x^2/a^2 + y^2/b^2 - z^2/c^2 = 1$$

subtends at its centre (0, 0, 0) is

$$8c\int_0^{\frac{\pi}{2}} \sqrt{\frac{b^2\cos^2\varphi + a^2\sin^2\varphi}{a^2b^2 + b^2c^2\cos^2\varphi + a^2c^2\sin^2\varphi}}\,d\varphi.$$

12. Show that the value of the integral

$$\Omega = \iint_\Sigma \frac{(a-x)\,dy\,dz + (b-y)\,dz\,dx + (c-z)\,dx\,dy}{[(a-x)^2+(b-y)^2+(c-z)^2]^{3/2}}$$

is independent of the choice of the surface Σ, provided its boundary Γ is kept fixed. By integrating over the outside of the surface, prove from this result that if Σ is a closed surface, then $\Omega = 4\pi$ or 0, according as $A(a, b, c)$ is within the volume bounded by Σ or outside this volume.

13*. Let the surface Σ be bounded by the closed curve Γ and consider the integral

$$\Omega(a, b, c) = \iint_\Sigma \frac{(a-x)\,dy\,dz + (b-y)\,dz\,dx + (c-z)\,dx\,dy}{r^3},$$

$$(r^2 = (a-x)^2 + (b-y)^2 + (c-z)^2),$$

as a function of a, b, c. Prove that the components of the gradient of Ω can be expressed as line-integrals as follows:

$$\frac{\partial \Omega}{\partial a} = \int_\Gamma \frac{(z-c)\,dy - (y-b)\,dz}{r^3}, \quad \frac{\partial \Omega}{\partial b} = \int_\Gamma \frac{(x-c)\,dz - (z-c)\,dx}{r^3},$$

$$\frac{\partial \Omega}{\partial c} = \int_\Gamma \frac{(y-b)\,dx - (x-a)\,dy}{r^3}.$$

(These formulæ, which have an important interpretation in electromagnetism, can be expressed by the following vector equation

$$\operatorname{grad} \Omega = -\int_\Gamma \frac{[\boldsymbol{x} \,.\, d\boldsymbol{x}]}{|\boldsymbol{x}|^3},$$

where $\boldsymbol{x}$ is the vector with components $(x-a)$, $(y-b)$, $(z-c)$.)

14 *. Verify that the expression

$$\frac{-4xy\,dx + 2(x^2 - y^2 - 1)\,dy}{(x^2 + y^2 - 1)^2 + 4y^2}$$

is the total differential of the angle which the segment $-1 \leqq x \leqq 1$, $y = 0$, subtends at the point (x, y). Using this fact, prove the following result by a geometrical argument:

Let Γ be an oriented closed curve in the xy-plane, not passing through either of the points $(-1, 0)$, $(1, 0)$. Let p be the number of times Γ crosses the line-segment $-1 < x < 1$, $y = 0$ from the upper half-plane $y > 0$ to the lower half-plane $y < 0$, and n the number of times Γ crosses this line-segment from $y < 0$ to $y > 0$. Then

$$\Theta = \int_\Gamma \frac{-4xy\,dx + (x^2 - y^2 - 1)\,dy}{(x^2 + y^2 - 1)^2 + 4y^2} = 2\pi(p - n).$$

Thus if Γ is the curve $r = 2\cos 2\theta$ $(0 \leqq \theta \leqq 2)$, in polar co-ordinates, $\Theta = 0$.

15 **. Consider the unit circle C

$$x' = \cos\varphi, \; y' = \sin\varphi, \; z' = 0 \qquad (0 \leqq \varphi \leqq 2\pi)$$

in the xy-plane. Denote by Ω the solid angle which the circular disc $x^2 + y^2 \leqq 1$, $z = 0$, subtends at the point $P = (x, y, z)$. Now let P describe an oriented closed curve Γ which does not meet the circle C. Let p be the number of times Γ crosses the circular disc $x^2 + y^2 < 1$, $z = 0$, from the upper half-space $z > 0$ to the lower half-space $z < 0$, and n the number of times Γ crosses this disc from $z < 0$ to $z > 0$. If P starts from a point P_0 on Γ with $\Omega = \Omega_0$, then P, describing Γ (while Ω varies continuously with P), will return to P_0 with a value $\Omega = \Omega_1$. Prove by a geometrical argument that

$$\Omega_1 - \Omega_0 = \int_\Gamma d\Omega = 4\pi(p - n).$$

Using the vector equation found above,

$$\operatorname{grad} \Omega = -\int_{\sigma} \frac{[\overline{PP'}\, dP']}{|\overline{PP'}|^3}$$

(example 13), prove that

$$\int_{\sigma}\int_{\Gamma} \frac{1}{|\overline{PP'}|^3} \begin{vmatrix} x'-x & dx & dx' \\ y'-y & dy & dy' \\ z'-z & dz & dz' \end{vmatrix} =$$

$$\int_{\Gamma}\int_{\sigma} \frac{(x'-x)(dy\,dz'-dz\,dy') + (y'-y)(dz\,dx'-dx\,dz') + (z'-z)(dx\,dy'-dy\,dx')}{[(x'-x)^2+(y'-y)^2+(z'-z)^2]^{3/2}}$$

$$= 4\pi(p-n).$$

(This repeated line-integral, which is due to Gauss, gives the number of times Γ is wound around C. It should be remarked that its vanishing is

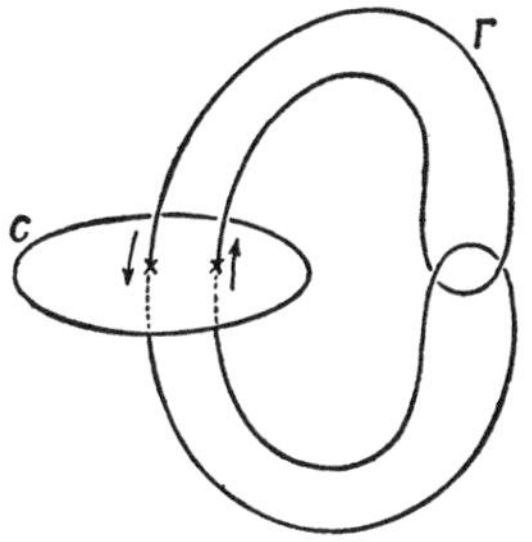

Fig. 13

necessary if the two curves Γ and C (thought of as being two strings) are to be separable, but not sufficient, as is shown by the example in fig. 13, where $p = n = 1$, yet Γ and C cannot be separated.)

CHAPTER VI

Differential Equations

We have already discussed special cases of differential equations in Vol. I, Chap. XI. We cannot attempt to develop the general theory in detail within the scope of this book. In this chapter, however, starting with further examples from mechanics, we shall give at least a sketch of the main principles of the subject.

1. The Differential Equations of the Motion of a Particle in Three Dimensions

1. The Equations of Motion.

In Vol. I, Chap. V, sections 4, 5 (p. 292), and Chap. XI (p. 502), we have already discussed the motion of a particle; we made the assumption, however, that this motion takes place along a pre-assigned fixed curve. We now drop this restriction and consider a mass m which we suppose concentrated at a point with co-ordinates (x, y, z). The position vector from the origin to the particle has components x, y, z and we denote it by $\boldsymbol{x}$. A motion of the particle will then be represented mathematically if we can find an expression for x, y, z or $\boldsymbol{x}$ as a function of the time t. If, as before, we denote differentiation with respect to the time t by a dot, then the vector $\dot{\boldsymbol{x}}$ with components $(\dot{x}, \dot{y}, \dot{z})$ and absolute value $v = \sqrt{(\dot{x}^2 + \dot{y}^2 + \dot{z}^2)}$ represents the velocity and the vector $\ddot{\boldsymbol{x}}$ with the components $\ddot{x}, \ddot{y}, \ddot{z}$ represents the acceleration of the particle.

We shall not deal with the foundations of mechanics, but take the following definitions and facts as our starting-point: we call the product of the acceleration vector $\ddot{\boldsymbol{x}}$ and m the *force* vector $\boldsymbol{f}$, and accordingly write

$$m\ddot{\boldsymbol{x}} = \boldsymbol{f}.$$

The components of this force vector, or, as we briefly say, of the force, will be denoted by

$$m\ddot{x} = X,$$
$$m\ddot{y} = Y,$$
$$m\ddot{z} = Z.$$

These three equations are known as Newton's *fundamental equations of mechanics*. From our preliminary point of view they represent nothing but a pure definition of the word force. It turns out, however, that in many cases this force vector can be determined without reference to the particular motion to be studied, a force field in space being previously known from physical assumptions. We can then regard the fundamental equations from quite a different point of view. They then represent conditions which must be satisfied by the acceleration in every particular motion if this motion takes place under the influence of the given field of force.

One example of such a field of force is the field of gravity. If we take gravity as acting in the direction of the negative z-axis, we know the components of the force to begin with. They are

$$X = 0, \quad Y = 0, \quad Z = -mg,$$

or, in vector notation,

$$\mathbf{f} = -mg \operatorname{grad} z,$$

where g is the constant acceleration due to gravity (cf. Vol. I, Chap. V, section 4, p. 294).

Another example is given by the field of force produced by a mass μ concentrated at the origin of the co-ordinate system and attracting according to Newton's law. If $r = \sqrt{(x^2 + y^2 + z^2)} = |\mathbf{x}|$ is the distance of the particle (x, y, z) with mass m from the origin, then in this case the field of force is given by the expression

$$\mathbf{f} = \mu m \gamma \operatorname{grad} \frac{1}{r}$$

(cf. p. 91), and Newton's fundamental equations are

$$\ddot{\mathbf{x}} = \mu\gamma \operatorname{grad} \frac{1}{r},$$

or, in components,

$$\ddot{x} = -\mu\gamma \frac{x}{r^3},$$

$$\ddot{y} = -\mu\gamma \frac{y}{r^3},$$

$$\ddot{z} = -\mu\gamma \frac{z}{r^3}.$$

In general, if f is a given field of force, with components $X(x, y, z)$, $Y(x, y, z)$, $Z(x, y, z)$ which are functions of position, the equations of motion

$$m\ddot{x} = f$$

or

$$m\ddot{x} = X,$$
$$m\ddot{y} = Y,$$
$$m\ddot{z} = Z$$

form a system of three differential equations for the three unknown functions $x(t)$, $y(t)$, $z(t)$. The fundamental problem of the mechanics of a particle is that of the determination of the actual path of the particle from the differential equations, when at the beginning of the motion, say at the time $t = 0$, the position of the particle (that is, the co-ordinates $x_0 = x(0)$, $y_0 = y(0)$, $z_0 = z(0)$) and the initial velocity (that is, the quantities $\dot{x}_0 = \dot{x}(0)$, $\dot{y}_0 = \dot{y}(0)$, $\dot{z}_0 = \dot{z}(0)$) are given. The problem of finding three functions which satisfy these initial conditions and also satisfy the three differential equations for all values of t is known as the problem of the *solution* or *integration* * of the system of differential equations.

2. The Principle of the Conservation of Energy.

Before we consider the integration of this system of differential equations in special cases, we shall state a number of general facts following from the equations of motion. The concept of the work done on the particle by the field of force during the motion was mentioned earlier (Chap. V, section 1, p. 350); we know that this work is given by the line integral

$$\int f\dot{x}dt = \int (X\,dx + Y\,dy + Z\,dz)$$

taken along the path described by the particle.

If the field of force can be represented as the gradient of a potential, say

$$f = \operatorname{grad}\Phi,$$

the work done during the motion is independent of the path

* This word is used because the solution of such differential equations may to a certain extent be regarded as a generalization of the process of ordinary integration.

and depends only on the initial and final points of the path (cf. Chap. V, section 1, p. 350). A field of force which can be represented as the gradient of a potential is called, following Helmholtz, a *conservative* * *field of force*. In such a field of force the equations of motion may be written in the vector form

$$m\ddot{\mathbf{x}} = -\operatorname{grad} U,$$

where instead of the potential Φ, which, it may be pointed out, is incompletely determined in that it contains an arbitrary additive constant, we introduce the *potential energy* $U = -\Phi$. In terms of the components the last equation becomes

$$m\ddot{x} = -U_x,$$
$$m\ddot{y} = -U_y,$$
$$m\ddot{z} = -U_z.$$

Although in general we cannot integrate this system of equations, we can deduce another equation from it in which the second derivatives do not occur and only the first derivatives of the functions $x(t)$, $y(t)$, $z(t)$ appear. If we use the vector notation, the argument may be carried out as follows. In the equation $m\ddot{\mathbf{x}} = -\operatorname{grad} U$, we form the scalar product of both sides and $\dot{\mathbf{x}}$. The left-hand side then becomes the derivative of the expression $\frac{1}{2}m\dot{\mathbf{x}}^2 = \frac{1}{2}mv^2$ with respect to t; the right-hand side is the derivative of the function $-U$ with respect to t (cf. p. 71), and by integration we therefore obtain

$$\tfrac{1}{2}m\dot{\mathbf{x}}^2 = -U + c,$$

where c is a constant, that is, a quantity independent of the time t. If we wish to avoid using vector analysis, then we may arrive at the same result by multiplying the three equations of motion by $\dot{x}$, $\dot{y}$, $\dot{z}$ respectively and adding; on the left-hand side we then have the derivative of the quantity

$$\tfrac{1}{2}m(\dot{x}^2 + \dot{y}^2 + \dot{z}^2)$$

with respect to t. The equation

$$\tfrac{1}{2}m(\dot{x}^2 + \dot{y}^2 + \dot{z}^2) + U = c$$

* "Conservative" in virtue of the theorem of the conservation of energy which we shall shortly deduce.

thus found is the mathematical expression of the *theorem of the conservation of energy*. We call the expression

$$T = \tfrac{1}{2}m(\dot{x}^2 + \dot{y}^2 + \dot{z}^2) = \tfrac{1}{2}mv^2$$

the *kinetic energy* (or *energy of motion*) of the moving particle, and the quantity U the *potential energy* (or *energy of position*) of the particle. Without going into the physical explanation of these concepts, we may mention that our equation has the following meaning:

In the case of motion in a conservative field of force the total energy, that is, the sum of the potential energy and the kinetic energy, remains constant.

The way in which this theorem can be used in the actual solution of the equations of motion will be shown in the examples in the next section.

3. Equilibrium. Stability.

The equations of motion, in conjunction with the assumption that $f = -\operatorname{grad} U$, i.e. that the field of force is conservative, now enable us to discuss the problem of equilibrium. We say that the particle is in equilibrium under the influence of the field of force if it remains at rest. In order that this may be the case its velocity and its acceleration must both be zero throughout the interval of time under consideration. The equations of motion therefore give the equations

$$\operatorname{grad} U = 0$$

or

$$U_x = 0, \quad U_y = 0, \quad U_z = 0$$

as the necessary conditions for equilibrium.

These same equations determine the points at which the potential energy U has a stationary value. It is particularly interesting to find that *a point at which the potential energy U has a proper minimum is a point of stable equilibrium.* By stability of equilibrium we mean that if we slightly disturb the state of equilibrium the whole resulting motion will differ only slightly from the state of rest.* More precisely, let R and ρ be any positive numbers.

* An example is given by a particle which rests under the influence of gravity at the lowest point of a spherical bowl which is concave upwards. On the other hand, a particle resting at the highest point of a spherical bowl which is concave downwards is in "unstable" equilibrium; the slightest disturbance results in a large change of position.

Corresponding to R and ρ we can find two positive numbers ϵ and δ so small that if the particle is moved a distance not more than ϵ from the position of equilibrium and started off with a velocity not greater than δ, then in the whole subsequent course of the motion the point never reaches a distance greater than R from the point of equilibrium and never has a velocity greater than ρ.

It is a remarkable fact that we can prove this statement about stability without integrating the equations of motion. In the proof we need only use the assumption that at the position of equilibrium in question the potential energy U has a proper minimum. For simplicity we assume that the position of equilibrium, the point where U has a minimum, is the origin; if not, we can make this point the origin by translation of axes. By definition the potential energy U involves an arbitrary additive constant; for the function U and the function $(U + \text{const.})$ give the same field of force, the constant disappearing in the process of differentiation. Thus without loss of generality we may take the value of the minimum $U(0, 0, 0)$ as zero.

About the origin we describe a sphere S_r with radius r; recalling the assumption that U is a minimum, we choose $r < R$ so small that everywhere in the interior and on the surface of this sphere, except at the origin, the inequality $U > 0$ is satisfied. The least value of U on the surface of the sphere we call a; by hypothesis, a is positive. It is therefore certain that the particle can never reach the surface of the sphere S_r as long as its potential energy remains less than a. Since U is continuous, we can find an ϵ, depending on a, so small that in the sphere S_ϵ with radius ϵ about the origin the value of U is at most $\frac{1}{2}a$. If we start the particle from a point of S_ϵ, and give it an initial velocity v_0 so small that for the initial kinetic energy we have

$$T_0 = \tfrac{1}{2}mv_0^2 < \tfrac{1}{2}a$$

(in other words, if $|v_0| < \sqrt{(a/m)}$), then by the law of the conservation of energy we always have

$$T + U = T_0 + U_0 < a.$$

Since T is always equal to or greater than zero, we shall always have U less than a, and therefore the particle can never reach a distance greater than r from the origin. Since U remains greater than or equal to zero, T remains less than a throughout the whole motion, and for the velocity we always have $v < \sqrt{(2a/m)}$. In virtue of the continuity of U, a tends to zero with r. We can therefore choose r so small that $\sqrt{(2a/m)} < \rho$ (that is, $a < \frac{1}{2}\rho^2 m$), so that the velocity is always less than ρ. Thus if the point starts inside S_ϵ with velocity v_0, and if $|v_0| < \sqrt{(a/m)}$, it always remains within the sphere S_r of radius $r < R$ and always has a velocity less than ρ.

2. Examples on the Mechanics of a Particle

1. Path of a Falling Body.

As a first example we shall consider the motion of a particle under the influence of gravity, taken as acting parallel to the negative z-axis. Newton's equations of motion take the form

$$m\ddot{x} = 0, \quad m\ddot{y} = 0, \quad m\ddot{z} = -mg;$$

that is,

$$\frac{d^2x}{dt^2} = 0, \quad \frac{d^2y}{dt^2} = 0, \quad \frac{d^2z}{dt^2} = -g.$$

From these equations by integration we find first the corresponding components of the velocity, and then the co-ordinates of the particle itself. We at once obtain

$$\frac{dx}{dt} = a_1, \quad \frac{dy}{dt} = b_1, \quad \frac{dz}{dt} = -gt + c_1,$$

where a_1, b_1, c_1 are constants; a second integration gives the equations

$$\begin{aligned} x &= a_1 t + a_2, \\ y &= b_1 t + b_2, \\ z &= -\tfrac{1}{2}gt^2 + c_1 t + c_2, \end{aligned}$$

where a_2, b_2, c_2 also represent constants. The meaning of the six constants of integration is found from the initial conditions. Without restricting the generality of the mechanical problem, we can choose the co-ordinates in such a way that at the time $t = 0$ the particle is at the origin. Accordingly, if we put $t = 0$ and at the same time $x = y = z = 0$ in the last equations, we at once obtain $a_2 = b_2 = c_2 = 0$. Moreover, we can assume without loss of generality that the initial velocity lies in the xz-plane, so that the component b_1 of the initial velocity has the value zero. With these assumptions the equation $y(t) = 0$ will hold for all values of t. The trajectory (that is, the path of the particle) therefore lies in a fixed plane, namely, the xz-plane. If we eliminate the time t from the remaining equations

$$x = a_1 t, \quad z = -\tfrac{1}{2}gt^2 + c_1 t,$$

we obtain the equation of the trajectory in the form

$$z = -\frac{g}{2a_1^2}x^2 + \frac{c_1}{a_1}x.$$

This curve is a parabola, with its axis parallel to the z-axis and its vertex upwards. The co-ordinates of the vertex, which correspond to the maximum of the function z, are found by equating the derivative of the right-hand side of our equation to zero. For the co-ordinates (x, z) of the vertex we thus obtain the values

$$x = \frac{a_1c_1}{g},$$

$$z = -\frac{g}{2a_1^2}\cdot\frac{a_1^2c_1^2}{g^2} + \frac{c_1}{a_1}\cdot\frac{a_1c_1}{g} = \frac{c_1^2}{2g}.$$

The time T at which the highest point of the path is reached is determined by the equation

$$T = \frac{x}{a_1} = \frac{c_1}{g}.$$

After twice this time, that is, $t = 2c_1/g$, the mass has reached the point with co-ordinates $x = 2a_1c_1/g$ and $z = 0$, and thus lies on the horizontal line $y = z = 0$ through the initial point.

2. Small Oscillations about a Position of Equilibrium.

In section 1, No. 3 (p. 416) we considered the question of the stability of equilibrium. The motion of a particle about a position of stable equilibrium, corresponding to a minimum of the potential energy, can be approximated to in a simple way. For the sake of brevity we restrict ourselves to a motion in the xy-plane and assume that there is no force acting in the direction of the z-axis. We imagine the potential energy in the neighbourhood of the origin (which we take at the minimum) expanded by Taylor's theorem in the form

$$U = U_0 + px + qy + \tfrac{1}{2}(ax^2 + 2bxy + cy^2) + \dots .$$

Here p, q and a, b, c denote the values of the derivatives U_x, U_y and U_{xx}, U_{xy}, U_{yy} respectively at the origin. In virtue of the assumption $U_0 = 0$, and since $U_x(0, 0) = 0$, $U_y(0, 0) = 0$, the constant term and the linear terms in this expansion disappear. We now assume that, corresponding to the fact that the origin is a minimum, the quadratic terms

$$Q(x, y) = \tfrac{1}{2}(ax^2 + 2bxy + cy^2)$$

form a positively definite quadratic form (p. 205), and that in a sufficiently small neighbourhood of the position of equilibrium the potential energy U can be replaced with sufficient accuracy by this quadratic form Q. With these assumptions the equations of motion take the form

$$m\ddot{\mathbf{x}} = -\operatorname{grad} Q$$

or

$$m\ddot{x} = -ax - by,$$
$$m\ddot{y} = -bx - cy.$$

These can easily be integrated completely if we first rotate the x- and y-axes through a suitably chosen angle. For if we consider the positively definite form $ax^2 + 2bxy + cy^2 = 2Q$, we know from elementary analytical geometry that by rotating the axes through a suitably chosen angle ϕ, that is, by making the substitution

$$x = \xi \cos\phi - \eta \sin\phi,$$
$$y = \xi \sin\phi + \eta \cos\phi,$$

this expression can be transformed into an expression of the form

$$\alpha\xi^2 + \beta\eta^2 = 2Q,$$

where ξ and η are the new rectangular co-ordinates and α and β are positive numbers.* In these new co-ordinates the equations of motion $m\ddot{\mathbf{x}} = -\operatorname{grad} Q$ transform into

$$m\ddot{\xi} = -\alpha\xi,$$
$$m\ddot{\eta} = -\beta\eta,$$

where ξ, η are the new components of the position vector $\mathbf{x}$. As in Vol. I, Chap. V, section 4 (pp. 296–7), both these equations can be integrated completely. We obtain

$$\xi = A_1 \sin\sqrt{\frac{\alpha}{m}}\,(t - c_1),$$

$$\eta = A_2 \sin\sqrt{\frac{\beta}{m}}\,(t - c_2),$$

where c_1, c_2, A_1, A_2 are constants of integration which enable us

* For the equation $Q = 1$ represents an ellipse, and by suitable choice of ϕ the term in xy can be removed.

to make the motion satisfy any arbitrarily assigned initial conditions.

The form of the solution shows that the motion about a position of stable equilibrium results from the superposition of simple harmonic oscillations in the two " principal directions ", the ξ-direction and the η-direction, the frequencies of these oscillations being given by $\sqrt{(a/m)}$ and $\sqrt{(\beta/m)}$. A general discussion of these oscillations, which we shall not carry out here, shows that the resultant motion may take a great variety of forms.

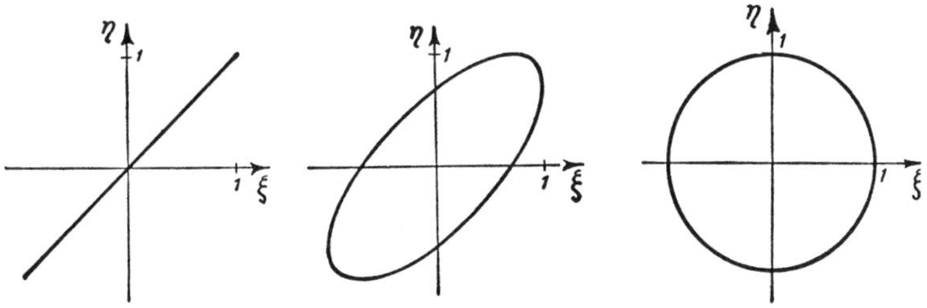

Figs. 1–3.—Oscillation diagrams

To give a few examples of these compound oscillations we first consider the motion represented by the equations

$$\xi = \sin(t + c),$$
$$\eta = \sin(t - c).$$

By eliminating the time t we obtain the equation

$$(\xi + \eta)^2 \sin^2 c + (\xi - \eta)^2 \cos^2 c = 4 \sin^2 c \cos^2 c,$$

which represents an ellipse. The two components of the oscillation have the same frequency 1 and the same amplitude 1, but a difference of phase $2c$. If this difference of phase successively takes all values between 0 and $\pi/4$, the corresponding ellipse passes from the degenerate straight-line case $\xi - \eta = 0$ to the circle $\xi^2 + \eta^2 = 1$, and the oscillation passes from the so-called linear oscillation to the circular (cf. figs. 1–3).

If as a second example we consider the motion represented by the equations

$$\xi = \sin t,$$
$$\eta = \sin 2(t - c),$$

where the frequencies are no longer equal, we obtain oscillation diagrams which are decidedly more complicated. In figs. 4, 5, and 6 these figures are given for the phase differences $c = 0$, $c = \pi/8$, and $c = \pi/4$ respectively. In the first two cases the particle moves continuously on a closed curve, but in the last case it swings backwards and forwards on an arc of the parabola $\eta = 2\xi^2 - 1$. The curves obtained by the superposition of

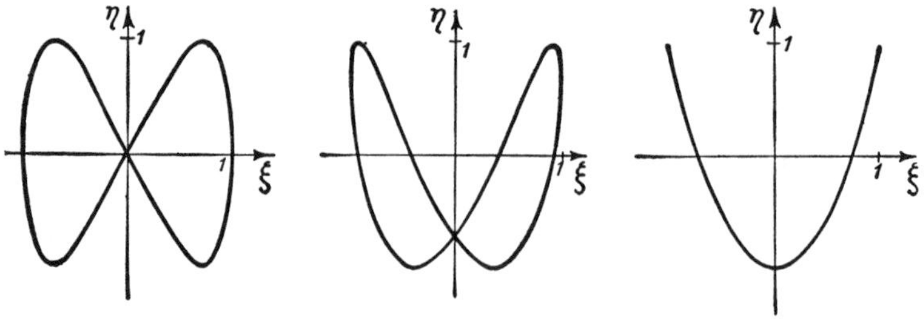

Figs. 4–6.—Oscillation diagrams

different simple harmonic oscillations in directions at right angles to one another are given the general name of *Lissajous figures.*

3. Planetary Motion.

In the examples discussed above the differential equations of the motion can immediately (or after a simple transformation) be written in such a way that each of the co-ordinates occurs in one differential equation only and can be determined by elementary integration. We shall now consider the most important case of a motion in which the equations of motion are no longer separable in this simple way, so that their integration involves a somewhat more difficult calculation. The problem in question is the *deduction of Kepler's laws of planetary motion from Newton's law of attraction.* We suppose that at the origin of the co-ordinate system there is a body of mass μ (e.g. the sun) whose gravitational field of force per unit mass is given by the vector

$$f = \gamma\mu \operatorname{grad} \frac{1}{r}.$$

What is the motion of a particle (a planet) under the influence of this field of force? The equations of motion are

$$\ddot{x} = -\gamma\mu \frac{x}{r^3},$$

$$\ddot{y} = -\gamma\mu \frac{y}{r^3},$$

$$\ddot{z} = -\gamma\mu \frac{z}{r^3}.$$

In order to integrate them we first state the theorem of conservation of energy for the motion in the form

$$\tfrac{1}{2}m(\dot{x}^2 + \dot{y}^2 + \dot{z}^2) - \frac{\gamma\mu m}{r} = C,$$

where C is constant throughout the motion and is determined by the initial conditions.

From the equations of motion we can now deduce other equations in which only the components of the velocity, not the acceleration, are present. If we multiply the first equation of motion by y, the second by x, and subtract, we obtain

$$\ddot{x}y - x\ddot{y} = 0, \quad \text{or} \quad \frac{d}{dt}(\dot{x}y - \dot{y}x) = 0,$$

whence by integration we have

$$x\dot{y} - y\dot{x} = c_1.$$

Similarly, from the remaining equations of motion we obtain *

$$y\dot{z} - z\dot{y} = c_2,$$
$$z\dot{x} - x\dot{z} = c_3.$$

These equations enable us to simplify our problem very considerably in a way which is highly plausible from the intuitive point of view. Without loss of generality we can choose the co-ordinate system in such a way that at the beginning of the motion, that is, at $t = 0$, the particle lies in the xy-plane and its velocity vector at that time also lies in that plane. Then

* We can also arrive at these three equations using vector notation, if we form the vector product of both sides of the equation of motion and the position vector $\boldsymbol{x}$. Since the force vector is in the same direction as the position vector, we obtain zero on the right, while the expression $[\boldsymbol{x}\ddot{\boldsymbol{x}}]$ on the left is the derivative of the vector $[\boldsymbol{x}\dot{\boldsymbol{x}}]$ with respect to the time. It therefore follows that this vector $[\boldsymbol{x}\dot{\boldsymbol{x}}] = \boldsymbol{c}$ has a value which is constant in time; this is exactly what is stated by the co-ordinate equations above.

As we see, this equation does not depend on our special problem, but holds in general for every motion in which the force has the same direction as the position vector.

The vector $[\boldsymbol{x}\dot{\boldsymbol{x}}]$ is called the *moment of velocity* and the vector $m[\boldsymbol{x}\dot{\boldsymbol{x}}]$ the *moment of momentum* of the motion. From the geometrical meaning of the vector product we easily obtain the following intuitive interpretation of the relation just given (cf. the subsequent discussions in the text). If we project the moving particle on to the co-ordinate planes and in each co-ordinate plane consider the area which the radius vector from the origin to the point of projection sweeps over in time t, this area is proportional to the time (theorem of areas).

$z(0) = 0$, and $\dot{z}(0) = 0$; and by substituting these values in the above equation and remembering that the right-hand sides are constants, we obtain

$$x\dot{y} - y\dot{x} = c_1 = h,$$
$$y\dot{z} - z\dot{y} = 0,$$
$$z\dot{x} - x\dot{z} = 0.$$

From these equations we conclude in the first place that the whole motion takes place in the plane $z = 0$. Since we naturally exclude the possibility of a collision between the sun and planet, we may assume that the three co-ordinates (x, y, z) do not vanish simultaneously, so that at the time $t = 0$ at which $z(0) = 0$ we have, say, $x(0) \neq 0$. Now from the last of the three equations above it follows that

$$\frac{d}{dt}\left(\frac{z}{x}\right) = -\frac{z\dot{x} - \dot{z}x}{x^2} = 0.$$

Therefore $z = ax$, where a is a constant. If we put $t = 0$ here, then from the equations $z(0) = 0$ and $x(0) \neq 0$ it follows that $a = 0$, so that z is always zero.

We may therefore base our problem of integration on the two differential equations

$$\tfrac{1}{2}m(\dot{x}^2 + \dot{y}^2) - \frac{\gamma\mu m}{r} = C,$$
$$x\dot{y} - y\dot{x} = h.$$

We next use the equations $x = r\cos\theta$, $y = r\sin\theta$ to transform the rectangular co-ordinates (x, y) into the polar co-ordinates (r, θ), which are now to be determined as functions of t. Since

$$\dot{x}^2 + \dot{y}^2 = \dot{r}^2 + r^2\dot{\theta}^2,$$

and

$$x\dot{y} - y\dot{x} = r^2\dot{\theta},$$

we have the two differential equations

$$\tfrac{1}{2}m(\dot{r}^2 + r^2\dot{\theta}^2) - \frac{\gamma\mu m}{r} = C,$$
$$r^2\dot{\theta} = h$$

for the polar co-ordinates r, θ. The first of these equations is the theorem of the conservation of energy, while the second expresses Kepler's law of areas. In fact (cf. Vol. I, Chap. V, section 2, pp. 273, 275) the expression $\frac{1}{2}r^2\dot{\theta}$ is the derivative with respect to the time of the area swept out in time t by the radius vector from the origin to the particle. This is found to be constant, or, as Kepler expressed it, *the radius vector describes equal areas in equal times.*

If the "area constant" h is zero, $\dot{\theta}$ must vanish, that is, θ must remain constant, so that the motion must take place on a straight line through the origin. We exclude this special case and expressly assume that $h \neq 0$.

In order to find the geometrical form of the orbit, we give up thinking of the motion as a function of the time * and consider the angle θ as a function of r, or r as a function of θ, and from our two equations we calculate the derivative $dr/d\theta$ as a function of r.

If we substitute the value $\dot{\theta} = h/r^2$ from the area equation in the energy equation and recall the equation

$$\dot{r} = \frac{dr}{dt} = \frac{dr}{d\theta}\dot{\theta},$$

we at once obtain the differential equation of the orbit in the form

$$\frac{m}{2}\left\{\frac{h^2}{r^4}\left(\frac{dr}{d\theta}\right)^2 + \frac{h^2}{r^2}\right\} - \frac{\gamma\mu m}{r} = C$$

or

$$\left(\frac{dr}{d\theta}\right)^2 = r^4\left(\frac{2C}{mh^2} + \frac{2\gamma\mu}{h^2}\frac{1}{r} - \frac{1}{r^2}\right).$$

To simplify the later calculations we make the substitution

$$r = \frac{1}{u}$$

* The course of the motion as a function of the time can be determined subsequently by means of the equation

$$\int_{\theta_0}^{\theta} r^2\,d\theta = h(t - t_0),$$

in which we suppose that r is known as a function of θ (cf. p. 428).

and introduce the following abbreviations:

$$\frac{1}{p} = \frac{\gamma\mu}{h^2},$$

$$\epsilon^2 = 1 + \frac{2Ch^2}{m\gamma^2\mu^2}.$$

The above differential equation then becomes

$$\left(\frac{du}{d\theta}\right)^2 = \frac{\epsilon^2}{p^2} - \left(u - \frac{1}{p}\right)^2,$$

and this can be integrated immediately. We have

$$\theta - \theta_0 = \int \frac{du}{\sqrt{(\epsilon^2/p^2 - (u - 1/p)^2)}},$$

or if for the moment we introduce $u - \frac{1}{p} = v$ as a new variable,

$$\theta - \theta_0 = \int \frac{dv}{\sqrt{(\epsilon^2/p^2 - v^2)}}.$$

For the integral (by Vol. I, Chap. IV, section 2, p. 213) we obtain the value $\arcsin \frac{vp}{\epsilon}$, and we thus obtain the equation of the orbit in the form

$$\frac{1}{r} - \frac{1}{p} = v = \frac{\epsilon}{p} \sin(\theta - \theta_0).$$

The angle θ_0 can be chosen arbitrarily, since it is immaterial from which fixed line the angle θ is measured. If we take $\theta_0 = \pi/2$, that is, if we let $v = 0$ correspond to the value $\theta = \pi/2$, we finally obtain the equation of the orbit in the form

$$r = \frac{p}{1 - \epsilon \cos\theta}.$$

We shall assume that the student already knows from analytical geometry that this is the equation in polar co-ordinates of a conic having one focus at the origin.

Our result therefore gives Kepler's law: *the planets move in conics with the sun at one focus.*

It is interesting to relate the constants of integration

$$p = \frac{h^2}{\gamma\mu}, \quad \epsilon^2 = 1 + \frac{2Ch^2}{m\gamma^2\mu^2}$$

to the initial motion. The quantity p is known as the semi-latus rectum or parameter of the conic; in the case of the ellipse and the hyperbola it is connected with the semi-axes a and b by the simple relation

$$p = \frac{b^2}{a}.$$

The square of the eccentricity, ϵ^2, determines the character of the conic; it is an ellipse, a parabola, or a hyperbola, according as ϵ^2 is less than, equal to, or greater than 1.

From the relation

$$\epsilon^2 = 1 + \frac{2Ch^2}{m\gamma^2\mu^2}$$

we see at once that the three different possibilities can also be stated in terms of the energy constant C; the orbit is an ellipse, a parabola, or a hyperbola, according as C is less than, equal to, or greater than zero.

If we suppose that the particle is brought at time $t = 0$ to the point x_0 in the field of force and is there started off with an initial velocity $\dot{x}_0$, then the relation

$$C = \tfrac{1}{2}mv_0^2 - \frac{\gamma\mu m}{r_0}$$

gives the surprising fact that the character of the orbit—ellipse, parabola, or hyperbola—does not depend on the direction of the initial velocity at all, but only on its absolute value v_0.

Kepler's third law is a simple consequence of the other two. It states that *in elliptic orbits the square of the period bears a constant ratio to the cube of the major semi-axis, the ratio depending on the field of force only and not on the particular planet.*

If we denote the period by T and the major semi-axis by a, we should then have

$$\frac{T^2}{a^3} = \text{const.},$$

where the constant on the right is independent of the particular problem and depends only on the magnitude of the attracting mass and on the gravitational constant γ.

To prove this we use the theorem of areas in the integrated form

$$\int_{\theta_0}^{\theta} r^2\,d\theta = h(t - t_0),$$

which defines the motion as a function of the time. If we take the integral over the interval from 0 to 2π, we obtain on the left-hand side twice the area of the orbital ellipse, that is, by previous results, $2\pi ab$, while on the right-hand side the time difference $t - t_0$ must be replaced by the period T. Therefore

$$2\pi ab = hT \quad \text{or} \quad 4\pi^2 a^2 b^2 = h^2 T^2.$$

We already know that h^2 is connected with the elements a and b of the orbit by the relation $h^2/\gamma\mu = p = b^2/a$. If we replace h^2 in the above equations by $\frac{b^2}{a}\gamma\mu$, it follows at once that

$$\frac{T^2}{a^3} = \frac{4\pi^2}{\gamma\mu},$$

which exactly expresses Kepler's third law.

EXAMPLES

1. Prove that as $t \to \infty$ the velocity $\sqrt{\dot{x}^2}$ of a planet tends to 0 if its orbit is a parabola and to a positive limit if it is a hyperbola.

2*. A planet is moving on an ellipse, and $\omega = \omega(t)$ denotes the angle PMP_s, where P is the position of the planet at the time t, P_s its position at the time t_s when it is nearest to the sun S, and M the centre of the ellipse. Prove that ω and t are connected by Kepler's equation

$$h(t - t_s) = ab(\omega - \varepsilon \sin\omega).$$

3. Prove that a body attracted towards a centre O by a force of magnitude mr moves on an ellipse with centre O.

4. Prove that the orbit of a body repelled by a force of magnitude $f(r)$, where f is a given function, from a centre O is given in polar co-ordinates (r, θ) by

$$\theta = \int^r \frac{dr}{r^2\sqrt{\left(\frac{2c}{h^2} + \frac{2}{h^2}\int^r f(r)\,dr - \frac{1}{r^2}\right)}}.$$

5. Prove that the equation of the orbit of a body repelled with a force $\frac{\mu}{r^3}$ from a centre O is

$$\frac{1}{r} = \begin{cases} \frac{2c}{h^2\varkappa}\cos(\varkappa\theta + \varepsilon) & \text{for} \quad \mu < h^2 \\ \frac{2c}{h^2\varkappa}\cosh(\varkappa\theta + \varepsilon) & \text{for} \quad \mu > h^2 \end{cases}$$

if

$$\varkappa = \sqrt{\left(\left| 1 - \frac{\mu}{h^2} \right|\right)} \text{ and } \varepsilon \text{ is a constant of integration.}$$

3. Further Examples of Differential Equations

Before discussing the foundations of the theory of differential equations, which we shall do in the next section, we shall here consider some further examples of problems involving differential equations, also arising in part from mechanics.

1. The General Linear Differential Equation of the First Order.

In Vol. I, Chap. III, section 7 (pp. 178, 182) we have already integrated the equation $y' + ay + b = 0$ completely in the case where a and b are constants. We can, however, also completely integrate this "linear differential equation of the first order" *

$$y' + ay + b = 0$$

for the unknown function $y(x)$ in the general case where a and b are any continuous functions of x. The solution is obtained by means of the exponential function and ordinary integration (which, however, cannot in general be performed in terms of elementary functions).

We first suppose that $b = 0$. Then the differential equation can be put in the form

$$-a = \frac{y'}{y} = \frac{d}{dx}\log|y|,$$

provided that $y \neq 0$. From this it follows that

$$\log|y| = -\int a(x)\,dx,$$

* The word "linear" expresses the fact that the unknown function and its derivatives are only linearly involved in the differential equation. A differential equation is said to be "of the first order" when it contains first derivatives only and no higher derivatives.

and finally, if for brevity we denote any indefinite integral of the function $a(x)$ by $A(x)$,

$$y = ce^{-A(x)},$$

where c is an arbitrary constant of integration. This formula gives a solution even when $c = 0$, namely $y = 0$.

If now $b(x)$ is not equal to zero, we attempt to find a solution of the form

$$y = u(x)e^{-A(x)},$$

where $u(x)$ must be suitably determined.*

Since $A'(x) = a(x)$,

$$y' = u'(x)e^{-A(x)} - u(x)a(x)e^{-A(x)},$$

and for the unknown function $u(x)$ we therefore have the differential equation

$$u'(x)e^{-A(x)} = -b,$$

from which it follows that

$$u(x) = -\int b(x)e^{A(x)}\,dx.$$

The expression

$$y(x) = -e^{-A(x)}\int b(x)e^{A(x)}\,dx,$$

where

$$A(x) = \int a(x)\,dx$$

therefore gives a solution of the differential equation. This solution is formed from known functions by means of the exponential function and of ordinary processes of integration only. Since the function $u(x)$ involves an arbitrary additive constant, we see that the expression

$$y(x) = e^{-A(x)}\left(c - \int b(x)e^{A(x)}\,dx\right),$$

where

$$A(x) = \int a(x)\,dx,$$

gives a solution which still contains an arbitrary constant of integration c. This solution really contains only *one* arbitrary constant, although $A(x)$ also involves an additive constant. For if we replace $A(x)$ by $A(x) + c_1$, the solution becomes one of similar type obtained from the original solution by replacing c by ce^{-c_1}.

* This device is known as "variation of the parameter" (see also p. 445).

For example, in the case of the differential equation

$$y' + xy + x = 0$$

we have

$$A(x) = \int x\,dx = \tfrac{1}{2}x^2, \quad \int e^{A(x)}b(x)\,dx = \int xe^{x^2/2}\,dx = e^{x^2/2},$$

and hence the solution

$$y = e^{-x^2/2}(c - e^{x^2/2}) = ce^{-x^2/2} - 1,$$

as we may verify by differentiation.

2. Separation of the Variables.

The idea which underlies the above solution is that of separation of the variables. If a differential equation is of the form

$$y' = -\frac{\alpha(x)}{\beta(y)},$$

where α depends on x only and β on y only, it may also be expressed symbolically by

$$\alpha\,dx + \beta\,dy = 0$$

or

$$\alpha\,dx = -\beta\,dy,$$

in which the variables x and y are separated. Introducing the two indefinite integrals

$$A = \int \alpha\,dx, \quad B = -\int \beta\,dy,$$

which are obtained by ordinary quadratures, we at once obtain

$$\frac{d}{dx}(A - B) = \alpha + \beta y' = 0,$$

that is,

$$A - B = c,$$

where c is an arbitrary constant of integration. This equation may now be imagined as solved for y, and the required solution is thus obtained by quadrature.

Another example in which the same idea is applied is the so-called *homogeneous* differential equation

$$y' = f\left(\frac{x}{y}\right).$$

If we take $z = y/x$, so that $y' = xz' + z$, the differential equation becomes

$$xz' + z = f(z),$$

or

$$z' = \frac{f(z) - z}{x},$$

an equation for z in which y does not appear explicitly. Hence

$$\int \frac{dz}{f(z) - z} = \int \frac{dx}{x} + c = c + \log |x|,$$

where c is an arbitrary constant of integration. Using this equation to express z as a function of x, we obtain the required solution.

Examples.—From $y' = \dfrac{y}{x}$ we at once have

$$\frac{dy}{y} = \frac{dx}{x},$$

the solution of which is

$$\log \left| \frac{y}{x} \right| = c.$$

Again, the equation

$$y' = \frac{y^2}{x^2}$$

gives

$$\int \frac{dz}{z^2 - z} = \log \frac{z-1}{z} = c + \log |x|;$$

hence

$$y = \frac{x}{1 - kx},$$

where k is a constant.

EXAMPLES

1. Integrate the following equations by separation of the variables:

   (a) $(1 + y^2)x\,dx + (1 + x^2)\,dy = 0.$

   (b) $ye^{2x}\,dx - (1 + e^{2x})\,dy = 0.$

2. Solve the following homogeneous equations:

   (a) $y^2\,dx + x(x - y)\,dy = 0.$

   (b) $xy\,dx + (x^2 + y^2)\,dy = 0.$

   (c) $x^2 - y^2 + 2xyy' = 0.$

   (d) $(x + y)\,dx + (y - x)\,dy = 0.$

   (e) $(x^2 + xy)y' = x\sqrt{(x^2 - y^2)} + xy + y^2.$

3. Show that a differential equation of the form

$$y' = \varphi\left(\frac{ax + by + c}{a_1x + b_1y + c_1}\right) \qquad (a, a_1, \ldots \text{ constant})$$

can be reduced to a homogeneous equation as follows. If $ab_1 - a_1b \neq 0$, we take a new unknown function and a new independent variable

$$\eta = ax + by + c, \quad \xi = a_1x + b_1y + c_1.$$

If $ab_1 - a_1b = 0$, we need only change the unknown function by putting

$$\eta = ax + by$$

to reduce the equation to a new equation in which the variables are separated.

4. Apply the method of the previous example to

(a) $(2x + 4y + 3)y' = 2y + x + 1.$

(b) $(3y - 7x + 3)y' = 3y - 7x + 7.$

5. Integrate the following linear differential equations of the first order:

(a) $y' + y\cos x = \cos x \sin x.$ (b) $y' - \dfrac{ny}{x + 1} = e^x(x + 1)^n.$

(c) $x(x - 1)y' + (1 - 2x)y + x^2 = 0.$ (d) $y' - \dfrac{2}{x}y = x^4.$

(e) $(1 + x^2)y' + xy = \dfrac{1}{1 + x^2}.$

6. Integrate the equation

$$y' + y^2 = \frac{1}{x^2}.$$

3. Determination of the Solution by Boundary Values. The Loaded Cable and the Loaded Beam.

In the problems of mechanics and the other examples previously discussed, we selected from the whole family of functions satisfying the differential equation a particular one by means of so-called initial conditions, that is, we chose the constants of integration in such a way that the solution and in certain cases also its derivatives up to the $(n - 1)$-th order assume pre-assigned values at a definite point. In many applications we are concerned neither with finding the general solution nor with solving definite initial-value problems, but instead with solving a so-called *boundary-value problem*. In a boundary-value problem we are required to find a solution which must satisfy pre-assigned conditions at *several* points and which must be considered in the intervals between those points. Here we shall discuss a few typical examples without going into the general theory of such boundary-value problems.

Ex. 1.—*The Differential Equation of a Loaded Cable.*

In a vertical xy-plane—in which the y-axis is vertical—we suppose that a cable whose (constant) horizontal component of tension is S is stretched from the origin to the point $x = a$, $y = b$ (cf. fig. 7). The cable is acted on by a load whose density per unit length of horizontal projection is given by a sectionally continuous function $p(x)$. Then the sag $y(x)$ of the cable, that is, the y-co-ordinate, is given by the differential equation

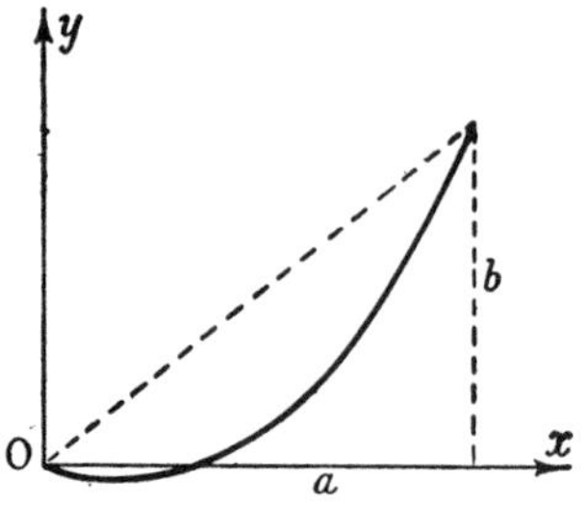

Fig. 7.—Loaded cable

$$y''(x) = g(x), \quad \text{where} \quad g(x) = \frac{p}{S}.$$

The shape of the cable will then be given by that solution $y(x)$ of the differential equation which satisfies the conditions $y(0) = 0$, $y(a) = b$. The solution of this boundary-value problem can be written down at once, since the general solution of the homogeneous equation $y'' = 0$ is the linear function $c_0 + c_1x$, and the solution of the non-homogeneous equation which, with its first derivative, vanishes at the origin is given by the integral $\int_0^x g(\xi)(x - \xi)\,d\xi$ (see below, pp. 441–8). In the general solution

$$y(x) = c_0 + c_1x + \int_0^x g(\xi)(x - \xi)\,d\xi$$

the condition $y(0) = 0$ at once gives $c_0 = 0$, and then the condition $y(a) = b$ gives the equation

$$b = c_1a + \int_0^a g(\xi)(a - \xi)\,d\xi$$

for the determination of c_1.

In practice, besides this very simple form of boundary-value problem a more complicated case occurs, in which the cable is subject not only to the continuously-distributed load but also to concentrated loads, that is, loads which are concentrated at a definite point of the cable, say at the point $x = x_0$. Such concentrated loads we shall consider as ideal limiting cases arising as $\varepsilon \to 0$ from a loading $p(x)$ which acts only in the interval $x_0 - \varepsilon$ to $x_0 + \varepsilon$ and for which

$$\int_{x_0-\epsilon}^{x_0+\epsilon} p(x)\,dx = P,$$

that is, the total loading remains constant during the passage to the limit $\varepsilon \to 0$; the number P is then called the concentrated load acting at the point x_0. By integrating both sides of the differential equation $y'' = p(x)/S$ over the interval from $x - \varepsilon$ to $x + \varepsilon$ before making the passage to the limit $\varepsilon \to 0$, we see that the equation $y'(x_0 + \varepsilon) - y'(x_0 - \varepsilon) = P/S$ holds. If we now perform the passage to the limit $\varepsilon \to 0$, we obtain the result that a concentrated load P acting at the point x_0 corresponds to a jump of the derivative $y'(x)$ by an amount P/S at the point x_0.

The following example suffices to show how the occurrence of a concentrated load modifies the boundary-value problem. We suppose that the cable is stretched between the points $x = 0$, $y = 0$ and $x = 1$, $y = 1$ and that the only load is a concentrated load of magnitude P acting at the mid-point $x = \frac{1}{2}$. According to the above discussion, this physical problem corresponds to the following mathematical problem: we have to find a continuous function $y(x)$ which satisfies the differential equation $y'' = 0$ everywhere in the interval $0 \leqq x \leqq 1$, except at the point $x_0 = \frac{1}{2}$, which takes the values $y(0) = 0$, $y(1) = 1$ on the boundary, and whose derivative has a jump of the amount P/S at the point x_0. In order to find this solution, we express it in the following way:

$$y(x) = ax + b \quad \text{for} \quad 0 \leqq x \leqq \tfrac{1}{2}$$

and

$$y(x) = c(1 - x) + d \quad \text{for} \quad \tfrac{1}{2} \leqq x \leqq 1.$$

The condition $y(0) = 0$, $y(1) = 1$ gives $b = 0$, $d = 1$. From the condition that both parts of the function shall give the same value at the point $x = \frac{1}{2}$ we find that

$$\tfrac{1}{2}a = \tfrac{1}{2}c + 1.$$

Finally, the requirement that the derivative y' shall increase by the amount P/S on passing the point $\frac{1}{2}$ gives the condition

$$-c - a = \frac{P}{S}.$$

We therefore have the constants

$$a = 1 - \frac{P}{2S}, \quad b = 0, \quad c = -1 - \frac{P}{2S}, \quad d = 1,$$

and our solution is thus determined. Moreover, it is easy to show that no other solution with the same properties exists.

*Ex. 2.—The Loaded Beam.**

The situation in the case of loaded beams is very similar (cf. fig. 8). Let us suppose that in its position of rest the beam coincides with the x-axis between the abscissæ $x = 0$ and $x = a$. Then it is found that the sag $y(x)$ due to a force acting vertically in the y-direction is given by the linear differential equation of the fourth order

$$y^{iv} = \varphi(x),$$

* For the theory of loaded beams cf. e.g. Morley, *Theory of Structures* (Longmans, Green & Co., 1927).

where the right-hand side $\varphi(x)$ is $p(x)/EI$, $p(x)$ being the density of loading, E the modulus of elasticity of the material of the beam (E is the stress

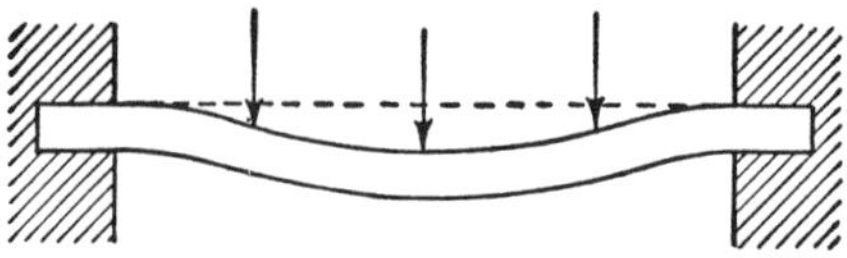

Fig. 8.—Loaded beam

divided by the elongation), and I the moment of inertia of the cross-section of the beam about a horizontal line through the centre of mass of the cross-section.

The general solution of this differential equation can at once be expressed (p. 446) in the form

$$y(x) = c_0 + c_1x + c_2x^2 + c_3x^3 + \int_0^x \varphi(\xi)\,\frac{(x-\xi)^3}{3!}\,d\xi,$$

where c_0, c_1, c_2, c_3 are arbitrary constants of integration. The real problem, however, is not that of finding this general solution, but that of finding a particular solution, i.e. of determining the constants of integration in such a way that certain definite boundary conditions are satisfied. If e.g. the beam is clamped at the ends, the boundary conditions

$$y(0) = 0, \quad y(a) = 0, \quad y'(0) = 0, \quad y'(a) = 0$$

hold. It then follows at once that $c_0 = c_1 = 0$, and the constants c_2 and c_3 are to be determined from the equations

$$c_2a^2 + c_3a^3 + \int_0^a \varphi(\xi)\,\frac{(a-\xi)^3}{3!}\,d\xi = 0,$$

$$2c_2a + 3c_3a^2 + \int_0^a \varphi(\xi)\,\frac{(a-\xi)^2}{2!}\,d\xi = 0.$$

With beams the occurrence of concentrated loads is again of particular interest. As before, we shall think of the concentrated load acting at the point $x = x_0$ as arising from a loading $p(x)$, distributed continuously over the interval $x_0 - \varepsilon$, to $x_0 + \varepsilon$, for which $\int_{x_0-\varepsilon}^{x_0+\varepsilon} p(\xi)\,d\xi = P$; we again let ε approach zero and at the same time let $p(x)$ increase in such a way that the value of P remains constant during the passage to the limit $\varepsilon \to 0$. P is then the value of the concentrated load at $x = x_0$. Just as in the example above, we integrate both sides of the differential equation over the interval from $x - \varepsilon$ to $x + \varepsilon$ and then perform the passage to the limit $\varepsilon \to 0$. It is found that the third derivative of the solution $y(x)$ must have a jump at the point $x = x_0$, this jump amounting to

$$y'''(x_0 + 0) - y'''(x_0 - 0) = \frac{P}{EI}.$$

Here $y(x_0 + 0)$ means the limit of $y(x_0 + h)$ as h tends to zero through positive values, $y(x_0 - 0)$ being the corresponding limit from the left.

Thus the following mathematical problem arises: we attempt to find a solution of $y^{iv} = 0$ which, together with its first and second derivatives, is continuous, for which $y(0) = y(1) = y'(0) = y'(1) = 0$, and whose third derivative has a jump of the amount P/EI at the point $x = x_0$ and elsewhere is continuous.

If the beam is fixed at a point $x = x_0$ (cf. fig. 9), i.e. if at this point the

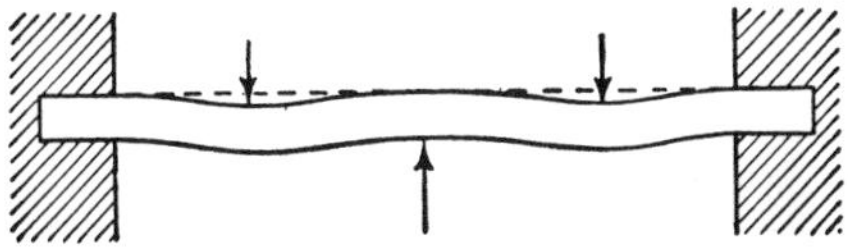

Fig. 9.—Sag of beam supported in the middle

sag has the fixed pre-assigned value $y = 0$, we can think of the fixation as being effected by means of a concentrated load acting at that point. By the mechanical principle that action is equal to reaction the value of this concentrated load will be equal to the force which the fixed beam exerts on its support. The magnitude P of this force is then given at once by the formula

$$P = EI\{y'''(x_0 + 0) - y'''(x_0 - 0)\},$$

where $y(x)$ satisfies the differential equation $y^{iv} = P/EI$ everywhere in the interval $0 \leqq x \leqq 1$ except at the point $x = x_0$ and in addition also satisfies the conditions $y(0) = y(1) = y'(0) = y'(1) = 0$, $y(x_0) = 0$, and y, y', and y'' are also continuous at $x = x_0$.

In order to illustrate these ideas we consider a beam extending from the point $x = 0$ to the point $x = 1$, clamped at its end-points $x = 0$ and $x = 1$, carrying a uniform load of density $p(x) = 1$, and supported at the point $x = \frac{1}{2}$ (cf. fig. 9). For the sake of simplicity we assume that $EI = 1$, so that the beam satisfies the differential equation

$$y^{iv} = 1$$

everywhere, except at the point $x = \frac{1}{2}$.

As the formula shows, the general solution of the differential equation is a polynomial of the fourth degree in x, the coefficient of x^4 being $1/4!$. The solution will be expressed by a polynomial of this type in each of the two half-intervals. For the first half-interval we write the polynomial in the form

$$y = b_0 + b_1x + b_2x^2 + b_3x^3 + \frac{1}{4!}x^4,$$

in the second half-interval, in the form

$$y = c_0 + c_1(x - 1) + c_2(x - 1)^2 + c_3(x - 1)^3 + \frac{1}{4!}(x - 1)^4.$$

Since the beam is clamped at the ends $x = 0$ and $x = 1$, it follows that

$$y(0) = y(1) = y'(0) = y'(1) = 0,$$

whence we obtain $b_0 = b_1 = c_0 = c_1 = 0$. In addition, $y(x)$, $y'(x)$, $y''(x)$ must be continuous at the point $x = \frac{1}{2}$; that is, the values of $y(\frac{1}{2})$, $y'(\frac{1}{2})$, $y''(\frac{1}{2})$ calculated from the two polynomials must be the same, and the value of $y(\frac{1}{2})$ must be zero. This gives

$$\frac{1}{4}b_2 + \frac{1}{8}b_3 + \frac{1}{384} = \frac{1}{4}c_2 - \frac{1}{8}c_3 + \frac{1}{384} = 0,$$

$$b_2 + \frac{3}{4}b_3 + \frac{1}{48} = -c_2 + \frac{3}{4}c_3 - \frac{1}{48},$$

$$2b_2 + 3b_3 = 2c_2 - 3c_3.$$

From this we obtain the following values for b_2, b_3, c_2, c_3:

$$b_2 = c_2 = \frac{1}{96}; \quad b_3 = -c_3 = -\frac{1}{24},$$

and the force which must act on the beam at the point $x = \frac{1}{2}$ in order that no sag may occur at that point is given by

$$y'''\left(\frac{1}{2} + 0\right) - y'''\left(\frac{1}{2} - 0\right) = \left(6c_3 - \frac{1}{2}\right) - \left(6b_3 + \frac{1}{2}\right) = -\frac{1}{2}.$$

4. Linear Differential Equations

1. Principle of Superposition. General Solutions.

Many of the examples previously discussed belong to the general class of linear differential equations. A differential equation in the unknown function $u(x)$ is said to be linear of the n-th order if it has the form

$$u^{(n)}(x) + a_1 u^{(n-1)}(x) + \dots + a_n u(x) = \phi(x),$$

where $a_1, a_2, a_3, \dots, a_n$ are given functions of the independent variable x, as is also the right-hand side $\phi(x)$. The expression on the left-hand side we shall denote by the abbreviation $L[u]$ (" linear differential expression of the n-th order ").

If $\phi(x)$ is identically zero in the interval under consideration, we say that the equation is *homogeneous*; otherwise, we say that it is *non-homogeneous*. We see at once (as in the special case of the linear differential equation of the second order with constant coefficients, discussed in Vol. I, p. 510) that the following *principle of superposition* holds: if u_1, u_2 are any two solutions of the homogeneous equation, every linear combination of them, $u = c_1 u_1 + c_2 u_2$, where the coefficients c_1, c_2 are constants, is also a solution.

If we know a single solution $v(x)$ of the non-homogeneous equation $L[u] = \phi(x)$, we can obtain other such solutions by adding to $v(x)$ any solution of the homogeneous equation. Conversely, any two solutions of the non-homogeneous equation differ only by a solution of the homogeneous equation.

For $n = 2$ and constant coefficients a_1, a_2 we proved in Vol. I, Chap. XI (p. 508) that every solution of the homogeneous equation can be expressed in terms of two suitably chosen solutions u_1, u_2 in the form $c_1u_1 + c_2u_2$. An analogous theorem holds for any homogeneous differential equation with arbitrary continuous coefficients.

To begin with, we explain what we mean by saying that a system of functions are linearly dependent or linearly independent, by means of the following definition: n functions $\phi_1(x)$, $\phi_2(x), \ldots, \phi_n(x)$ are *linearly dependent* if n constants $c_1, \ldots, c_n$ exist, which do not all vanish and which satisfy the equation

$$c_1\phi_1(x) + c_2\phi_2(x) + \ldots + c_n\phi_n(x) = 0$$

identically, that is, for all values of x in the interval under consideration. Then if $c_n \neq 0$, say, $\phi_n(x)$ may be expressed in the form

$$\phi_n(x) = a_1\phi_1(x) + \ldots + a_{n-1}\phi_{n-1}(x),$$

and ϕ_n is said to be *linearly dependent* on the other functions. If no linear relation of the form

$$c_1\phi_1(x) + c_2\phi_2(x) + \ldots + c_n\phi_n(x)$$

exists, the n functions $\phi_i(x)$ are said to be *linearly independent.*

Ex. 1.—The functions $1, x, x^2, \ldots, x^{n-1}$ are linearly independent. Otherwise, constants $c_0, c_1, \ldots, c_{n-1}$ would have to exist such that the polynomial

$$c_0 + c_1x + \ldots + c_{n-1}x^{n-1}$$

vanishes for all values of x in a certain interval. This, however, is impossible unless all the coefficients of the polynomial are zero.

Ex. 2. — The functions e^{a_ix} are linearly independent, provided $a_1 < a_2 < \ldots < a_n$.

Proof.—We assume that this statement has been proved true for $(n-1)$ such exponential functions. Then if

$$c_1e^{a_1x} + c_2e^{a_2x} + \ldots + c_ne^{a_nx} = 0$$

is an identity in x, we divide by $e^{a_n x}$ and, putting $a_i - a_n = b_i$, obtain

$$c_1 e^{b_1 x} + c_2 e^{b_2 x} + \ldots + c_{n-1} e^{b_{n-1} x} + c_n = 0.$$

If we differentiate this equation with respect to x, the constant c_n disappears and we have an equation which implies that the $(n - 1)$ functions $e^{b_1 x}, e^{b_2 x}, \ldots, e^{b_{n-1} x}$ are linearly dependent, from which it follows that $e^{a_1 x}, e^{a_2 x}, \ldots, e^{a_{n-1} x}$ are linearly dependent, contrary to our original assumption. Hence there cannot be a linear relation between the n original functions either.

Ex. 3.—The functions $\sin x$, $\sin 2x$, $\sin 3x, \ldots,$ $\sin nx$ are linearly independent in the interval $0 \leqq x \leqq \pi$. We leave the reader to prove this, using the fact that $\int_{-\pi}^{+\pi} \sin mx \sin nx\, dx = \begin{cases} 0 \text{ if } m \neq n, \\ \pi \text{ if } m = n. \end{cases}$ (Cf. Vol. I, p. 217.)

If we assume that the functions $\phi_i(x)$ have continuous derivatives up to the $(n - 1)$-th order, we have the following theorem:

The necessary and sufficient condition that the system of functions $\phi_i(x)$ *shall be linearly dependent is that the equation*

$$W = \begin{vmatrix} \phi_1(x) & \phi_2(x) & \ldots & \phi_n(x) \\ \phi_1'(x) & \phi_2'(x) & \ldots & \phi_n'(x) \\ \cdot & \cdot & \cdot & \cdot \\ \phi_1^{(n-1)}(x) & \phi_2^{(n-1)}(x) & \ldots & \phi_n^{(n-1)}(x) \end{vmatrix} = 0$$

shall be an identity in x. *In addition the* n *determinants formed from* (n − 1) *of the functions must not vanish simultaneously at any point.* The function W is called the *Wronskian* of the system of functions.*

That the condition is *necessary* follows immediately: if we assume that

$$\Sigma c_i \phi_i(x) = 0,$$

successive differentiation gives the further equations

$$\Sigma c_i \phi_i'(x) = 0,$$

$$\cdot \quad \cdot \quad \cdot \quad \cdot \quad \cdot \quad \cdot \quad \cdot$$

$$\Sigma c_i \phi_i^{(n-1)}(x) = 0.$$

These, however, form a homogeneous system of n equations, which are satisfied by the n coefficients $c_1, \ldots, c_n$; hence W, the determinant of the system of equations, must vanish.

That the condition is *sufficient*, that is, that if $W = 0$ the functions are linearly dependent, may be proved in various ways.

* In this proof and the following one a knowledge of the elements of the theory of determinants is assumed.

One proof is as follows. From the vanishing of W we may deduce that the system of equations

$$\begin{aligned} c_1\phi_1 \quad &+ \dots + c_n\phi_n &= 0 \\ c_1\phi_1' \quad &+ \dots + c_n\phi_n' &= 0 \\ \dots \quad &\dots \quad \dots & \dots \\ c_1\phi_1^{(n-1)} &+ \dots + c_n\phi_n^{(n-1)} &= 0 \end{aligned}$$

possesses a solution $c_1, c_2, \dots, c_n$ which is not trivial, where c_i may still be a function of x. Here we may assume without loss of generality that $c_n = 1$. Further, we may assume that V, the Wronskian of the $(n-1)$ functions $\phi_1, \phi_2, \dots, \phi_{n-1}$, is not zero, for we may suppose that our theorem has already been proved for $(n-1)$ functions; then $V = 0$ implies the existence of a linear relation between $\phi_1, \phi_2, \dots, \phi_{n-1}$, and hence between $\phi_1, \phi_2, \phi_3, \dots, \phi_n$. By differentiating * the first equation with respect to x and combining the result with the second, we obtain

$$c_1'\phi_1 + c_2'\phi_2 + \dots + c_{n-1}'\phi_{n-1} = 0;$$

similarly, by differentiating the second equation and combining the result with the third, we obtain

$$c_1'\phi_1' + c_2'\phi_2' + \dots + c_{n-1}'\phi_{n-1}' = 0,$$

and so on, up to

$$c_1'\phi_1^{(n-2)} + c_2'\phi_2^{(n-2)} + \dots + c_{n-1}'\phi_{n-1}^{(n-2)} = 0.$$

Since V, the determinant of these equations, is assumed not to vanish, it follows that $c_1', c_2', \dots, c_{n-1}'$ are zero; that is, $c_1, c_2, \dots, c_{n-1}$ are constants. Hence the equation

$$\sum_1^n c_i\phi_i(x) = 0$$

does actually express a linear relation, as was asserted.

We now state the fundamental theorem on linear differential equations:

Every homogeneous linear differential equation

$$L[u] = a_0(x)u^{(n)}(x) + a_1(x)u^{n-1}(x) + \dots + a_n u(x) = 0$$

* It is easy to see that the coefficients c_i are continuously differentiable functions of x; for, if the determinant V is not zero, they can be expressed rationally in terms of the functions ϕ_i and their derivatives.

possesses systems of n *linearly independent solutions* $u_1, u_2, \ldots, u_n$. *By superposing these fundamental solutions every other solution* u *may be expressed * as a linear expression with constant coefficients* $c_1, \ldots, c_n$:

$$u = \sum_{i=1}^{n} c_i u_i.$$

In particular, a system of fundamental solutions can be determined by the following conditions. At a prescribed point, say $x = \xi$, u_1 is to have the value 1 and all the derivatives of u_1 up to the $(n-1)$-th order are to vanish; u_i, where $i > 1$, and all the derivatives of u_i up to the $(n-1)$-th order, except the i-th, are to vanish, while the i-th derivative is to have the value 1.

The existence of a system of fundamental solutions follows from the existence theorem proved in the next section (p. 450). It follows from Wronski's condition, which we have just proved, that a linear relation must exist between any further solution u and $u_1, \ldots, u_n$; for from the equations

$$\sum_{l=0}^{n} a_l u^{(n-l)} = 0$$

$$\sum_{l=0}^{n} a_l u_i^{(n-l)} = 0 \qquad (i = 1, \ldots, n)$$

it follows that the Wronskian of the $(n+1)$ functions $u, u_1, u_2, \ldots, u_n$ must vanish, so that $u, u_1, u_2, \ldots, u_n$ are linearly dependent. Since $u_1, \ldots, u_n$ are independent, u depends linearly on $u_1, \ldots, u_n$.

2. **Homogeneous Differential Equations of the Second Order.**

We shall consider differential equations of the second order in more detail, as they have very important applications.

Let the differential equation be

$$L[u] = au'' + bu' + cu = 0.$$

If $u_1(x)$, $u_2(x)$ are a system of fundamental solutions, $W = u_1 u_2' - u_2 u_1'$ is its Wronskian, and $W' = u_1 u_2'' - u_2 u_1''$. Since

$$L[u_1] = 0 \quad \text{and} \quad L[u_2] = 0,$$

* Two different systems of fundamental solutions $u_1, \ldots, u_n$; $v_1, \ldots, v_n$ can be transformed into one another by a linear transformation

$$v_i = \sum_{k=1}^{n} c_{ik} u_k,$$

where the coefficients c_{ki} are constants and form a matrix whose determinant does not vanish.

it follows that

$$u_1L[u_2] - u_2L[u_1] = aW' + bW = 0.$$

Hence by integration

$$k + \log | W | = -\int \frac{b}{a}\, dx,$$

or

$$W = ce^{-\int \frac{b}{a} dx},$$

where c is a constant. This formula is used a great deal in the more detailed theory of differential equations of the second order.

Another property worth mentioning is that a linear homogeneous differential equation of the second order can always be transformed into an equation of the first order, known as *Riccati's differential equation.* Riccati's equation is of the form

$$v' + v^2 + qv + r = 0,$$

where v is a function of x; or, in a slightly more general form,

$$v' + pv^2 + qv + r = 0,$$

which is obtained from the first form by putting $v = z/p$. The linear equation is transformed into Riccati's equation by putting $u' = uz$, so that $u'' = u'z + uz' = uz^2 + uz'$, and we have

$$az' + az^2 + bz + c = 0.$$

A third remark: if we know *one* solution $v(x)$ of our linear homogeneous differential equation of the second order, the problem is reduced to that of solving a differential equation of the first order, and can be carried out by quadratures. In fact, if we assume that $L[v] = 0$ and put $u = zv$, where $z(x)$ is the new function which we are seeking, we obtain the differential equation

$$az''v + 2az'v' + bz'v + zL[v] = avz'' + (2av' + bv)z' = 0$$

for z. This, however, is a linear homogeneous differential equation for the unknown function $z' = w$; its solution is given on p. 429. From w we then obtain the factor z, and hence the solution u, by a further quadrature.

Example.—The linear equation of the second order

$$y'' - 2\frac{y'}{x} + 2\frac{y}{x^2} = 0$$

is equivalent to Riccati's equation

$$z' + z^2 - \frac{2}{x} z + \frac{2}{x^2} = 0,$$

where $z = y'/y$. The original equation has $y = x$ as a particular solution; hence it may be reduced to the equation of the first order

$$v''x = 0,$$

where $v = y/x$. That is, $v = ax + b$. Hence the general integral of the original equation is given by

$$y = ax^2 + bx.$$

We would expressly emphasize that exactly the same method can be used to reduce a linear differential equation of the n-th order to one of the $(n - 1)$-th order, when one solution of the first equation is known.

EXAMPLES

1. Prove that if $a_1, \ldots, a_k$ are different numbers and $P_1(x), \ldots, P_k(x)$ are arbitrary polynomials (not identically zero), then the functions $\varphi_1(x) = P_1(x)e^{a_1 x}, \ldots, \varphi_k(x) = P_k(x)e^{a_k x}$ are linearly independent.

2. Show that the so-called Bernoulli's equation

$$y' + a(x)y = b(x)y^n \qquad (n \neq 1)$$

reduces to a linear differential equation for the new unknown function $z = y^{1-n}$. Use this to solve the equations

(a) $xy' + y = y^2 \log x$.
(b) $xy^2(xy' + y) = a^2$.
(c) $(1 - x^2)y' - xy = axy^2$.

3. Show that Riccati's differential equation

$$y' + P(x)y^2 + Q(x)y + R(x) = 0$$

can be transformed into a linear differential equation if we know a particular integral $y_1 = y_1(x)$. (Introduce the new unknown function $u = 1/(y - y_1)$. Use this to solve the equation

$$y' - x^2y^2 + x^4 - 1 = 0$$

which possesses the particular integral $y_1 = x$.

4. Find the integrals which are common to the two differential equations

(a) $y' = y^2 + 2x - x^4$. (b) $y' = -y^2 - y + 2x + x^2 + x^4$.

5*. Integrate the differential equation

$$y' = y^2 + 2x - x^4$$

in terms of definite integrals, using the particular integral found in Ex. 4. Draw a rough graph of the integral curves of the equation throughout the xy-plane.

6*. Let y_1, y_2, y_3, y_4 be four solutions of Riccati's equation (cf. Ex. 3). Prove that the expression

$$\frac{y_1 - y_3}{y_1 - y_4} \div \frac{y_2 - y_3}{y_2 - y_4}$$

is a constant.

7. Show that if two solutions, $y_1(x)$ and $y_2(x)$, of Riccati's equation are known, then the general solution is given by

$$y - y_1 = c(y - y_2)e^{\int P(y_2 - y_1)\,dx},$$

where c is an arbitrary constant.

Hence find the general solution of

$$y' - y \tan x = y^2 \cos x - \frac{1}{\cos x},$$

which has solutions of the form $a \cos^n x$.

8. Prove that the equations

(a) $(1 - x)y'' + xy' - y = 0,$

(b) $2x(2x - 1)y'' - (4x^2 + 1)y' + y(2x + 1) = 0$

have a common solution. Find it, and hence integrate both equations completely.

3. The Non-homogeneous Differential Equation. Method of Variation of Parameters.

To solve the non-homogeneous differential equation

$$L[u] = a_0 u^{(n)} + \dots + a_n u = \phi(x)$$

it is sufficient, by what we have said on p. 439, to find a single solution. This may be done as follows. By proper choice of the constants $c_1, c_2, \dots, c_n$, we first determine a solution of the homogeneous equation $L[u] = 0$ in such a way that the equations

$$u(\xi) = 0, \quad u'(\xi) = 0, \quad \dots, \quad u^{(n-2)}(\xi) = 0, \quad u^{(n-1)}(\xi) = 1$$

are satisfied. This solution, which depends on the parameter ξ, we denote by $u(x, \xi)$. The function $u(x, \xi)$ is a continuous function of ξ for fixed values of x, and so are its first n derivatives with respect to x. As an example, for the differential equation $u'' + k^2 u = 0$ the solution $u(x, \xi)$ has the form $\sin k(x - \xi)/k$, and this fulfils the conditions stated above. We now assert that the formula

$$v(x) = \int_0^x \phi(\xi)u(x, \xi)\,d\xi$$

gives a solution of $L[u] = \phi$ which, together with its first $n-1$ derivatives, vanishes at the point* $x = 0$. To verify this statement we differentiate the function $v(x)$ repeatedly with respect to x by the rule for the differentiation of an integral with respect to a parameter (cf. Chap. IV, section 1, p. 220), and recall the relations

$$u(x, x) = 0, \quad u'(x, x) = 0, \quad \ldots, \quad u^{(n-2)}(x, x) = 0, \quad u^{(n-1)}(x, x) = 1$$

(where e.g. $u'(x, x) = \partial u(x, \xi)/\partial x$ for $\xi = x$).

We thus obtain

$$v'(x) = \phi(\xi)u(x, \xi)\Big|_{\xi=x} + \int_0^x \phi(\xi)u'(x, \xi)\,d\xi = \int_0^x \phi(\xi)u'(x, \xi)\,d\xi,$$

$$v''(x) = \phi(\xi)u'(x, \xi)\Big|_{\xi=x} + \int_0^x \phi(\xi)u''(x, \xi)\,d\xi = \int_0^x \phi(\xi)u''(x,\xi)\,d\xi,$$

. . . .

$$v^{(n-1)}(x) = \phi(\xi)u^{(n-2)}(x, \xi)\Big|_{\xi=x} + \int_0^x \phi(\xi)u^{(n-1)}(x, \xi)\,d\xi$$

$$= \int_0^x \phi(\xi)u^{(n-1)}(x, \xi)\,d\xi,$$

$$v^{(n)}(x) = \phi(\xi)u^{(n-1)}(x, \xi)\Big|_{\xi=x} + \int_0^x \phi(\xi)u^{(n)}(x, \xi)\,d\xi$$

$$= \phi(x) + \int_0^x \phi(\xi)u^{(n)}(x, \xi)\,d\xi.$$

Since $L[u(x, \xi)] = 0$, this establishes the equation $L[v] = \phi(x)$ and shows that the initial conditions $v(0) = 0$, $v'(0) = 0, \ldots,$ $v^{(n-1)}(0) = 0$ are satisfied.

The same solution can also be obtained by the following

* The physical meaning of this process is this. If $x = t$ denotes the time and u the co-ordinate of a point moving on a straight line subject to a force $\phi(x)$, the effect of this force may be thought of as arising from the superposition of the small effects of small impulses. The above solution $u(x, \xi)$ then corresponds to an impulse of amount 1 at time ξ, and our solution gives the effect of impulses of amount $\phi(\xi)$ during the time between 0 and x. We cannot go further into the details here.

apparently different method. We seek to find a solution u of the non-homogeneous equation in the form of a linear combination

$$u = \Sigma\gamma_i(x)u_i(x),$$

but now we must allow the coefficients γ_i to be functions of x. On these functions we impose the following conditions:

$$\begin{aligned}
\gamma_1'u_1 \quad &+ \gamma_2'u_2 \quad + \dots + \gamma_n'u_n \quad = 0\\
\gamma_1'u_1' \quad &+ \gamma_2'u_2' \quad + \dots + \gamma_n'u_n' \quad = 0\\
&\cdots\cdots\cdots\cdots\\
\gamma_1'u_1^{(n-2)} &+ \gamma_2'u_2^{(n-2)} + \dots + \gamma_n'u_n^{(n-2)} = 0.
\end{aligned}$$

From these it follows that the derivatives of u are given by the following formulæ:

$$\begin{aligned}
u' \quad &= \Sigma\gamma_iu_i'\\
u'' \quad &= \Sigma\gamma_iu_i''\\
&\cdots\cdots\\
u^{(n-1)} &= \Sigma\gamma_iu_i^{(n-1)}\\
u^{(n)} \quad &= \Sigma\gamma_i'u_i^{(n-1)} + \Sigma\gamma_iu_i^{(n)}.
\end{aligned}$$

Substituting these expressions in the differential equation and remembering that $L[u] = \phi$, we have

$$\Sigma\gamma_i'u_i^{(n-1)} = \phi(x).$$

For the coefficients γ_i' we obtain a linear system of equations, whose determinant is W, the Wronskian of the system of fundamental solutions u_i, and therefore does not vanish. Thus the coefficients γ_i' are determined, and hence by quadratures the coefficients γ_i. As the whole argument can be reversed, a solution of the equation has actually been found, and in fact *all* solutions, in virtue of the integration constants concealed in the coefficients γ_i.

We leave it to the reader to show that the two methods are really identical, by expressing $u(x, \xi)$, the solution of the homogeneous equation defined above, in the form

$$u(x, \xi) = \Sigma a_i(\xi)u_i(x).$$

The latter method is known as *variation of parameters*, because here the solution appears as a linear combination of functions with variable coefficients, whereas in the case of the homogeneous equation these coefficients were constants.

Example.—We consider the equation

$$u'' - 2\frac{u'}{x} + 2\frac{u}{x} = xe^x.$$

By p. 443, a system of fundamental solutions of the corresponding homogeneous equation

$$u'' - 2\frac{u'}{x} + 2\frac{u}{x^2} = 0$$

is given by $u_1 = x$, $u_2 = x^2$. Hence if we seek solutions of the form

$$u = \gamma_1 x + \gamma_2 x^2,$$

we have the conditions

$$\gamma_1' x + \gamma_2' x^2 = 0,$$
$$\gamma_1' + 2\gamma_2' x = xe^x$$

for γ_1 and γ_2. That is,

$$\gamma_1' = -xe^x, \quad \gamma_2' = e^x.$$

Hence the general solution of the original non-homogeneous equation is

$$u = xe^x + c_1 x + c_2 x^2.$$

4. Forced Vibrations.

As an application we shall give a brief account of a method for dealing with forced vibrations, in which the right-hand side of the differential equation need no longer be a periodic function, as in the cases considered in Vol. I, Chap. XI, section 3 (p. 510), but may instead be an arbitrary continuous function $f(t)$. For the sake of simplicity we restrict ourselves to the case where there is no friction and take $m = 1$ (or, what amounts to the same thing, divide through by m). We accordingly write the differential equation in the form

$$\ddot{x}(t) + \kappa^2 x(t) = \phi(t),$$

where the quantity κ^2 is what we previously called k, and the external force is denoted by ϕ instead of f.

According to p. 446, the function

$$F(t) = \frac{1}{\kappa}\int_0^t \phi(\lambda)\sin\kappa(t - \lambda)\,d\lambda$$

is a solution of the differential equation $\ddot{x} + \kappa^2 x = \phi(t)$, and satisfies the initial conditions

$$F(0) = 0, \quad \dot{F}(0) = 0.$$

For the general solution of the differential equation we thus obtain, just as before, the function

$$x(t) = \frac{1}{\kappa}\int_0^t \phi(\lambda)\sin\kappa(t-\lambda)\,d\lambda + c_1 \sin\kappa t + c_2 \cos\kappa t,$$

where c_1 and c_2 are arbitrary constants of integration.

If, in particular, the function on the right-hand side of the differential equation is a purely periodic function of the form $\sin\omega t$ or $\cos\omega t$, a simple calculation shows that we again obtain the results of Vol. I, Chap. XI, section 3.

Examples

1*. Prove that the linear homogeneous equation

$$L(y) = y^{(n)} + c_1 y^{(n-1)} + \dots + c_{n-1}y' + c_n = 0$$

with *constant* coefficients c has a system of fundamental solutions of the form $x^\mu e^{a_k x}$, where the a_k's are the roots of the polynomial

$$f(z) = z^n + c_1 z^{n-1} + \dots + c_n.$$

2. Integrate the following equations:

(*a*) $y''' - y = 0$. (*b*) $y''' - 4y'' + 5y' - 2y = 0$.

(*c*) $y''' - 3y'' + 3y' - y = 0$. (*d*) $y^{iv} - 3y'' + 2y = 0$.

(*e*) $x^2y'' + xy' - y = 0$.

3. Let

$$a_0 y + a_1 y' + \dots + a_n y^{(n)} = P(x)$$

be a linear non-homogeneous differential equation of the n-th order with constant coefficients, and let $P(x)$ be a polynomial. Let $a_0 \neq 0$ and consider the formal identity

$$\frac{1}{a_0 + a_1 t + \dots + a_n t^n} = b_0 + b_1 t + b_2 t^2 + \dots .$$

Prove that

$$y = b_0 P(x) + b_1 P'(x) + b_2 P''(x) + \dots$$

is a particular integral of the differential equation.

If $a_0 = 0$, but $a_1 \neq 0$, then the expansion

$$\frac{1}{a_1 t + a_2 t^2 + \dots + a_n t^n} = bt^{-1} + b_0 + b_1 t + b_2 t^2 + \dots$$

is possible. Prove that now

$$y = b\int P(x)\,dx + b_0P(x) + b_1P'(x) + b_2P''(x) + \dots$$

is a particular integral of the differential equation.

4. Apply the method of Ex. 3 to find particular integrals of

$$(a)\quad y'' + y = 3x^2 - 5x; \qquad (b)\quad y'' + y' = (1 + x)^2.$$

5. A particular integral of the equation

$$a_0y + a_1y' + \dots + a_ny^{(n)} = e^{kx}P(x),$$

where k, a_0, a_1, ... are real constants and $P(x)$ is a polynomial, can be found by introducing a new unknown function $z = z(x)$ given by

$$y = ze^{kx},$$

and applying the method of Ex. 3 to the equation in z.

Use this method to find particular integrals of

$$(a)\quad y'' + 4y' + 3y = 3e^x. \qquad (b)\quad y'' - 2y' + y = xe^x.$$

6. Integrate the equation

$$y'' - 5y' + 6y = e^x(x^2 - 3)$$

completely.

5. General Remarks on Differential Equations

Although a complete theory of differential equations would extend far beyond the compass of this book, we shall here sketch at least the elements of a general method for their treatment.

1. Differential Equations of the First Order and their Geometrical Interpretation.

We begin by considering a differential equation of the first order, that is, an equation in which the first derivative of the function $y(x)$, but no higher derivative, occurs in addition to x and $y(x)$. The general expression for a differential equation of this type is

$$F(x, y, y') = 0,$$

where we assume that the function F is a continuously differentiable function of its three arguments x, y, y'. We now attempt to visualize the geometrical meaning of this equation. In the points of a plane region with rectangular co-ordinates (x, y), this equation prescribes a condition for the direction of the tangent to any curve $y(x)$ which passes through this point and satisfies the differential equation. We assume that in a certain region R of a plane, say in a rectangle, the differential equation $F(x, y, y') = 0$ can be solved uniquely for y', and thus expressed in the form

$$y' = f(x, y),$$

where the function $f(x, y)$ is a continuously differentiable function of x and y. Then to each point (x, y) of R this differential equation $y' = f(x, y)$ assigns a "direction of advance". The differential equation is therefore represented geometrically by a *field of directions*; and the problem of solving the differential equation geometrically consists in the finding of those curves which belong to this field of directions, that is, whose tangents at every point have the direction pre-assigned by the equation $y' = f(x, y)$. We call these curves the *integral curves of the differential equation.*

It is now intuitively plausible that through each point (x, y) of R there passes just one integral curve of the differential equation $y' = f(x, y)$. These facts are stated more precisely in the following fundamental existence theorem:

If in the differential equation $y' = f(x, y)$ *the function* f *is continuous and has a continuous derivative with respect to* y *in a region* R, *then through each point* (x_0, y_0) *of* R *there passes one, and only one, integral curve, that is, there exists one, and only one, solution* $y(x)$ *of the differential equation for which* $y(x_0) = y_0$.

We shall return to the proof of this theorem in sub-section 4 (p. 459). Here we confine ourselves to the consideration of some examples.

For the differential equation

$$y' = -\frac{x}{y},$$

which we consider in the region $y < 0$, say, the direction of the field of directions is readily seen to be perpendicular to the vector from the origin to the point (x, y). From this we infer by geometry that the circular arcs about the origin must be the integral curves of the differential equation. This result is very easily verified analytically. For from the equation of these circles,

$$y = \sqrt{(c^2 - x^2)},$$

it follows at once that

$$y' = -\frac{x}{\sqrt{(c^2 - x^2)}},$$

which shows that these circles satisfy the differential equation.

At each point the field of directions of the differential equation

$$y' = \frac{y}{x}$$

obviously has the direction of the line joining that point to the origin. Thus the lines through the origin belong to this field of directions and are therefore integral curves. As a matter of fact, we see at once that the function $y = cx$ satisfies the differential equation * for any arbitrary constant c.

In the same way we can verify analytically that the differential equations

$$y' = \frac{x}{y} \qquad (y \neq 0)$$

and

$$y' = -\frac{y}{x} \qquad (x \neq 0)$$

are satisfied by the respective families of hyperbolas

$$y = \sqrt{(c + x^2)},$$

$$y = \frac{c}{x},$$

where c is the parameter specifying the particular curve of the family.

Our fundamental theorem shows in general that differential equations of the first order are satisfied by a one-parameter family of functions, that is, by functions of x which depend not only on x but also on a parameter c (for example, on $c = y_0 = y(0)$); as we say, the solutions depend on an arbitrary constant of integration. The ordinary integration of a function $f(x)$ is merely a special case of the solution of this differential equation, namely, the special case in which $f(x, y)$ does not involve y. All the directions of the field of directions are then determined by the x-co-ordinate alone, and we see at once that the integral curves are obtained from one another by translation in the direction of the y-axis. Analytically this corresponds to the familiar fact that in indefinite integration, that is, in the solution of the differential

* At the origin the field of directions is no longer uniquely defined; this is connected with the fact that an infinite number of integral curves pass through this "singular point" of the differential equation.

equation $y' = f(x)$, the function y involves an arbitrary additive constant c.

The geometrical interpretation of the differential equation now enables us to carry out an *approximate graphical integration*, that is, a graphical construction of the integral curves, in much the same way as in the special case of the indefinite integration of

10

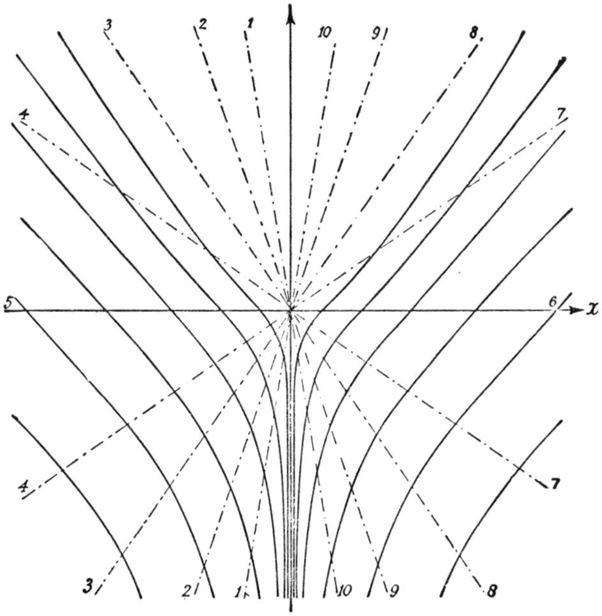

11

Fig. 10.—Directions of the integral curves on the isoclines in fig. 11

Fig. 11.—Solutions of $y' = \sqrt{(x^2 + y^2)}/x$ by the isoclinal method

a function of x (Vol. I, pp. 119–21). We have only to think of the integral curve as replaced by a polygon in which each side has the direction assigned by the field of directions for its initial point (or for any other one of its points). Such a polygon can be constructed by starting from an arbitrary point in R. The smaller we take the length of the sides of the polygon, the greater the accuracy with which the sides of the polygon will agree with the field of directions of the differential equation, not only at their initial points but throughout their whole length. Without going into the proof, we here state the fact that by successively diminishing the length of side a polygon constructed in this way may actually be made to approach closer and closer to the integral curve through the initial point. For this process $f(x, y)$ need not be given explicitly; it need only be given graphically.

Such a graphical integration is frequently carried out in practice by the so-called *isoclinal* method. The field of directions is represented by joining points with the same direction by curves (*isoclines*), that is, by sketching the family of curves $f(x, y) = c = \text{const.}$ To every value c of this constant there then corresponds a definite direction which can, for example, be sketched in an auxiliary figure. An integral curve must then cut every isocline in the corresponding direction obtained from the auxiliary figure, and the construction of the integral curves is therefore easily carried out by drawing parallels.

Fig. 11 shows the graphical integration * of $y' = \dfrac{\sqrt{(x^2 + y^2)}}{x}$. Here the isoclines are half-lines through the origin. The corresponding directions y' agree with the correspondingly-numbered directions in the auxiliary fig. 10.

2. The Differential Equation of a Family of Curves. Singular Solutions. Orthogonal Trajectories.

The existence theorem shows that a family of curves corresponds to every differential equation. This suggests the question whether this statement is reversible. In other words, does every one-parameter family of curves $\phi(x, y, c) = 0$ or $y = g(x, c)$ have a corresponding differential equation

$$F(x, y, y') = 0$$

which is satisfied by all the curves of the family, and how can we find this differential equation? Here the essential point is that c, the parameter of the family of curves, does not occur in the differential equation, so that the differential equation is in a sense a representation of the family of curves *not* involving a

* This differential equation can be integrated explicitly by introducing polar co-ordinates, but the result of this explicit integration is by no means so clear and easy to discuss.

parameter. In fact, it is easy to find such a differential equation. Differentiating the equation

$$\phi(x, y, c) = 0$$

with respect to x, we have

$$\phi_x + \phi_y y' = 0.$$

If ϕ_y is not identically zero, and if we eliminate the parameter c between this equation and the equation $\phi = 0$, the result is the desired differential equation. This elimination is always possible for a region of the plane in which the equation $\phi = 0$ can be solved for the parameter c in terms of x and y. We then have only to substitute the expression $c = c(x, y)$ thus found in the expressions for ϕ_x and ϕ_y in order to obtain a differential equation for the family of curves.

As a first example we consider the family of concentric circles $x^2 + y^2 - c^2 = 0$, from which, by differentiation with respect to x, we obtain the differential equation

$$x + yy' = 0,$$

in agreement with p. 451.

Another example is the family $(x - c)^2 + y^2 = 1$ of circles with unit radius and centre on the x-axis. By differentiation with respect to x we obtain

$$(x - c) + yy' = 0,$$

and on eliminating c we obtain the differential equation

$$1 - y^2 = y^2 y'^2 \quad \text{or} \quad y^2(1 + y'^2) = 1.$$

The family $y = (x - c)^2$ of parabolas touching the x-axis likewise leads by way of the equation $y' = 2(x - c)$ to the required differential equation

$$y'^2 = 4y.$$

In the last two examples we see that the corresponding differential equations are satisfied not only by the curves of the family, but in the first case by the lines $y = 1$ and $y = -1$ also, in the second case by the x-axis $y = 0$ also. These facts, which can at once be verified analytically, follow without calculation from the geometrical meaning of the differential equation. For these lines are the envelopes of the corresponding family of curves, and since the envelopes at each point touch a curve of the family, they must at that point have the direction prescribed by the field of directions. Therefore every envelope of a family of integral curves must itself satisfy the differential equation. Solutions of the differential equation which are found by forming the envelope of a one-parameter family of integral curves are called *singular solutions.**

If to each point P of a region R which is simply covered by a one-parameter family of curves $\Phi(x, y) = c =$ const. we assign the direction of the tangent of the curve passing through P, we obtain a field of directions defined by the differential equation $y' = -\dfrac{\Phi_x}{\Phi_y}$ (see above). If, on the other hand, to each point P we assign the direction of the normal to the curve passing through it, the resulting field of directions is defined by the differential equation

$$y' = \frac{\Phi_y}{\Phi_x}.$$

The solutions of this differential equation are called the *orthogonal trajectories* of the original family of curves $\Phi(x, y) = c$. The curves $\Phi = c$ and their orthogonal trajectories intersect everywhere at right angles. Hence if a family of curves is given by the differential equation $y' = f(x, y)$, we can find the differential equation of the orthogonal trajectories without integrating the

* It is remarkable that we can find singular solutions of a differential equation $F(x, y, y') = 0$ without integrating the differential equation, that is, without having the one-parameter family of ordinary solutions to start from. For we recall that by our fundamental theorem the solution of the differential equation is *uniquely* determined in the neighbourhood of a point (x, y) when in this neighbourhood the differential equation can be written in the form $y' = f(x, y)$, where $f(x, y)$ is a continuously differentiable function. It follows that at the points through which both a member of the family and also a singular solution pass, such an expression must be impossible. In the neighbourhood of this point (x, y) the differential equation $F(x, y, y')$ cannot have a solution in the above form. The theorem on implicit functions in Chap. III, section 1 (p. 117), however, states that such a solution is possible if $F_{y'} \neq 0$ at the place in question. We thus find that a *necessary* (but by no means a *sufficient*) condition for a point of a singular solution is that the equation

$$\frac{\partial}{\partial y'} F(x, y, y') = 0$$

is satisfied. If we eliminate y' between this equation and the given differential equation, we obtain an equation between x and y which the singular solution must satisfy (if it exists). The examples above confirm this rule. Thus from the differential equation $y^2(1 + y'^2) = 1$ we obtain the equation $y^2 y' = 0$ by differentiating with respect to y'. From these two relations we have $y^2 = 1$, or $y = \pm 1$, which are the singular solutions found above.

given differential equation, for the equation of the orthogonal trajectories is

$$y' = -\frac{1}{f(x, y)}.$$

In the examples discussed above, from the differential equation satisfied by the circles $\sqrt{(x^2 + y^2)} = c$ we find that the differential equation of the orthogonal trajectories is $y' = y/x$. The orthogonal trajectories are therefore straight lines through the origin (see p. 452).

If $p > 0$, the family of confocal parabolas (cf. Chap. III, p. 137) $y^2 - 2p(x + p/2) = 0$ satisfies the differential equation

$$y' = \frac{1}{y}\{-x + \sqrt{(x^2 + y^2)}\}.$$

Hence the differential equation of the orthogonal trajectories of this family is

$$y' = \frac{-1}{\{-x + \sqrt{(x^2 + y^2)}\}/y} = \frac{1}{y}\{-x - \sqrt{(x^2 + y^2)}\}.$$

The solutions of this differential equation are the parabolas

$$y^2 - 2p(x + p/2) = 0,$$

where $p < 0$, which are parabolas confocal with one another and with the curves of the first family.

3. The Integrating Factor. (Euler's Multiplier.)

If we write the differential equation $y' = f(x, y)$ in the form

$$dy - f(x, y)\,dx = 0,$$

where dx and dy are the differentials of the independent and dependent variables respectively (for the idea of the differential see Chap. II, p. 66), and multiply by an arbitrary non-vanishing factor $b(x, y)$, we arrive at an equivalent differential equation of the form

$$a(x, y)\,dx + b(x, y)\,dy = 0.$$

The problem of the general solution of the differential equation consists in finding a function $y(x)$ such that this differential equation for the differentials dx and dy is satisfied identically in x.

In one case such a solution can be given immediately; namely, when the expression $a\,dx + b\,dy$ is the total differential of a function $F(x, y)$, that is, if a function $F(x, y)$ exists for which

$a = \partial F/\partial x$ and $b = \partial F/\partial y$. The differential equation then becomes

$$dF = 0.$$

This is solved if we put

$$F(x, y) = c,$$

where c is an arbitrary constant of integration c, and from this equation we calculate y as a function of x and of the constant of integration c.

According to Chap. V (p. 354), a necessary and sufficient condition that $a\,dx + b\,dy$ may be the total differential of a function F is that the condition of integrability $\partial a/\partial y = \partial b/\partial x$ is satisfied. If this condition is satisfied, the line integral of the expression $a\,dx + b\,dy$ is independent of the path and for a fixed initial point P_0 represents a function $F(x, y)$ of the end-point P with co-ordinates (x, y), and this function F gives us the above solution.

In general, the coefficients a and b of a differential equation $a\,dx + b\,dy = 0$ do not satisfy the condition of integrability. This is true e.g. for the differential equation $dx + \frac{y}{x}\,dy = 0$. We can then attempt to multiply the differential equation by a factor $\mu(x, y)$ which is chosen in such a way that after the multiplication the coefficients do satisfy the condition of integrability, so that the differential equation can be solved by evaluating a line integral along a particular path, that is, by a simple integration. In our example $\mu(x, y) = x$ is such a factor. It leads to the differential equation $x\,dx + y\,dy = 0$, the left-hand side of which is the differential of the function $\frac{1}{2}(x^2 + y^2)$. Thus in agreement with the previous result on p. 451 the solutions of the differential equation are the circles $x^2 + y^2 = 2c$.

In general, such a factor $\mu(x, y)$, which we call an integrating factor or multiplier of the differential equation, is determined by the condition that

$$\frac{\partial}{\partial y}(\mu a) = \frac{\partial}{\partial x}(\mu b) \quad \text{or} \quad a\mu_y - b\mu_x + (a_y - b_x)\mu = 0.$$

The still unknown integrating factor $\mu(x, y)$ is therefore itself determined by an equation involving derivatives, and in fact partial derivatives with respect to x and y. Thus the finding of an integrating factor is not in theory any simpler than the original problem. Nevertheless, in many cases such a factor is easily found by trial and error, as in the above example. The integrating factor, however, is chiefly of theoretical interest, and we shall not discuss it further here.

4. Theorem of the Existence and Uniqueness of the Solution.

We now prove the theorem of the existence and uniqueness of the solution of the differential equation $y' = f(x, y)$ which we stated on p. 451. Without loss of generality we can assume that for the solution $y(x)$ in question we have $y(0) = 0$, for otherwise we could introduce $y - y_0 = \eta$ and $x - x_0 = \xi$ as new variables and should then obtain a new differential equation, $d\eta/d\xi = f(\xi + x_0, \eta + y_0)$, of the same type, to which we could apply our argument.

In the proof we may confine ourselves to a sufficiently small neighbourhood of the point $x = 0$. If we have proved the existence and uniqueness of the solution for such an interval about the point $x = 0$, we can then prove the existence and uniqueness for a neighbourhood of one of its end-points, and so on.

We first convince ourselves that there cannot be *more* than one solution of the differential equation satisfying the initial conditions. For if there were two solutions $y_1(x)$ and $y_2(x)$, for the difference $d(x) = y_1 - y_2$ we should have

$$d'(x) = f(x, y_1(x)) - f(x, y_2(x)).$$

By the mean value theorem the right-hand side of this equation can be put in the form $(y_1 - y_2)f_y(x, \bar{y}) = d(x)f_y(x, \bar{y})$, where $\bar{y}$ is a value intermediate between y_1 and y_2. In a neighbourhood $|x| \leqq a$ of the origin y_1 and y_2 are continuous functions of x which vanish at $x = 0$. Let b be an upper bound of the absolute values of the two functions in this neighbourhood, so that $|\bar{y}| \leqq b$ whenever $|x| \leqq a$. Moreover, by M we shall mean a bound of $|f_y|$ in the region $|x| \leqq a$, $|y| \leqq b$. Finally, let D be the greatest value of $|d(x)|$ in the interval $|x| \leqq a$. We suppose that this value is assumed at $x = \xi$. Then

$$|d'(x)| = |d(x)f_y(x, \bar{y})| \leqq DM,$$

and therefore

$$D = |d(\xi)| = \left| \int_0^\xi d'(x)\, dx \right| \leqq |\xi|\, DM \leqq aDM$$

We can choose a so small that $aM < 1$, for if $|f_y(x, y)|$ is less than M in a region $|x| \leqq a$, $|y| \leqq b$, it continues to be less than M in every region obtained by reducing a. But if $aM < 1$, from $D \leqq aMD$ it follows that $D = 0$. That is: in such an interval $|x| \leqq a$ we have * $y_1(x) = y_2(x)$.

By a similar integral estimate we arrive at a proof of the existence of the solution. We construct the solution by a method which is also important in applications, in particular, in the numerical solution of differential equations. This is the process of *iteration* or *successive approximation*. Here we obtain the solution as the limit function of a sequence of approximate solutions $y_0(x)$, $y_1(x)$, $y_2(x)$, ... As a first approximation $y_0(x)$ we take $y_0(x) = 0$. Using the differential equation, we take

$$y_1(x) = \int_0^x f(\xi, 0)\, d\xi$$

as the second approximation: from this we obtain the next approximation $y_2(x)$,

$$y_2(x) = \int_0^x f(\xi, y_1(\xi))\, d\xi,$$

and in general the $(n + 1)$-th approximation is obtained from the n-th by the equation

$$y_n(x) = \int_0^x f(\xi, y_{n-1}(\xi))\, d\xi.$$

If in an interval $|x| \leqq a$ these approximating functions converge uniformly to a limit function $y(x)$, we can at once perform the passage to the limit under the integral sign, and for the limit function we obtain the equation

$$y(x) = \int_0^x f(\xi, y(\xi))\, d\xi,$$

from which it follows by differentiation that $y' = f(x, y)$, so that y is actually the required solution.

We carry out the proof of convergence for a sufficiently small interval $|x| \leqq a$ by means of the following estimate. We put $y_{n+1}(x) - y_n(x) = d_n(x)$ and by D_n denote the maximum of $|d_n(x)|$ in the interval $|x| \leqq a$.

* The root idea of this proof is the fact that for bounded integrands integration gives a quantity which vanishes to the same order as the interval of integration, as that interval tends to zero.

From the equation

$$d'_n(x) = y'_{n+1} - y'_n = f(x, y_n) - f(x, y_{n-1})$$

the mean value theorem gives

$$d'_n(x) = d_{n-1}(x) f_y(x, \bar{y}_{n-1}(x)),$$

where $\bar{y}_{n-1}$ is a value intermediate between y_n and y_{n-1}. Let the inequalities $|f_y(x, y)| \leqq M$, $|f(x, y)| \leqq M_1$ hold in the rectangular region $|x| \leqq a$, $|y| \leqq b$. If we assume that for the function y_n the relation $|y_n| \leqq b$ holds in the interval $|x| \leqq a$, then by the definition of y_{n+1} we have

$$|y_{n+1}(x)| = \left| \int_0^x f(\xi, y_n(\xi))\, d\xi \right| \leqq |x| M_1 \leqq aM_1.$$

We shall therefore choose the bound a for x so small that $aM_1 \leqq b$. Then in the interval $|x| \leqq a$ we shall certainly have $|y_{n+1}(x)| \leqq b$. Since for $y_0(x) = 0$ it is obvious that $|y_0| \leqq b$, in the interval $|x| \leqq a$ we have $|y_n(x)| \leqq b$ for every n. Hence in the equation

$$d_{n+1}(x) = \int_0^x f_y(\xi, \bar{y}_n(\xi))\, d_n(\xi)\, d\xi$$

we may estimate the integral on the right by using $|f_y| \leqq M$, and for the maximum D_{n+1} of $|d_{n+1}(x)|$ in the interval $|x| \leqq a$ we thus at once obtain

$$D_{n+1} \leqq aMD_n.$$

We now take a so small that $aM \leqq q < 1$, where q is a fixed proper fraction, say $q = \frac{3}{4}$. Then $D_{n+1} \leqq qD_n \leqq q^n D_0$.

Let us now consider the series

$$d_0(x) + d_1(x) + d_2(x) + \dots + d_{n-1}(x) + \dots .$$

The n-th partial sum of this series is $y_n(x)$. The absolute value of the n-th term is not greater than the number $D_0 q^{n-1}$ when $|x| \leqq a$. Our series is therefore dominated by a convergent geometric series with constant terms. Hence (cf. Vol. I, p. 392) it converges uniformly in the interval $|x| \leqq a$ to a limit function $y(x)$, and thus we see that an interval $|x| \leqq a$ exists in which the differential equation has a unique solution.

All that now remains to be shown is that this solution can

be extended step by step until it reaches the boundary of the (closed bounded) region R in which we assume $f(x, y)$ to be defined. The proof so far shows that if the solution has been extended to a certain point, it can be continued onward over an x-interval of length a, where a, however, depends on the co-ordinates (x, y) of the end-point of the portion already constructed. It might be imagined that this advance a diminishes from step to step so rapidly that the solution cannot be extended by more than a small amount, no matter how many steps are made. This, as we shall show, is not the case.

Suppose that R' is a closed bounded region entirely within R. Then we can find a b so small that for every point (x_0, y_0) in R' the whole square $x_0 - b \leqq x \leqq x_0 + b$, $y_0 - b \leqq y \leqq y_0 + b$ lies in R. If by M and M_1 we denote the upper bounds of $|f_y(x, y)|$ and $|f(x, y)|$ in the region R, then we find that in the preceding proof all the conditions imposed on a are certainly satisfied if we take a to be, say, the smallest of the numbers b, $1/2M$, and b/M_1. This no longer depends on (x_0, y_0); hence at each step we can advance by an amount a which is a constant. Thus we can proceed step by step until we reach the boundary of R'. Since R' can be chosen as any closed region in R, we see that the solution can be extended to the boundary of R.

5. Systems of Differential Equations and Differential Equations of Higher Order.

Many of the above arguments extend to systems of differential equations of the first order with as many unknown functions of x as there are equations. As an example of sufficient generality we shall here consider a system of two differential equations for two functions $y(x)$ and $z(x)$,

$$y' = f(x, y, z),$$
$$z' = g(x, y, z).$$

We again assume that the functions f and g are continuously differentiable. This system of differential equations can be interpreted by a field of directions in xyz-space. To the point (x, y, z) of space a direction is assigned whose direction cosines are in the ratio $dx : dy : dz = 1 : f : g$. The problem of integrating the differential equation again consists, geometrically speaking, in finding curves in space which belong to this field of directions.

As in the case of a single differential equation, we again have the fundamental theorem that through every point of a region R in which the above functions are continuously differentiable there passes one, and only one, integral curve of the system of differential equations. The region R is covered by a two-parameter family of curves in space. These give the solutions of the system of differential equations as two functions $y(x)$ and $z(x)$ which both depend on the independent variable x and also on two arbitrary parameters c_1 and c_2, the constants of integration.

Systems of differential equations of the first order are particularly important in that equations of higher order, that is, differential equations in which derivatives higher than the first occur, can always be reduced to such systems.

For example, the differential equation of the second order

$$y'' = h(x, y, y')$$

can be written as a system of two differential equations of the first order. We have only to take the first derivative of y with respect to x as a new unknown function z and then write down the system of differential equations

$$y' = z,$$
$$z' = h(x, y, z).$$

This is exactly equivalent to the given differential equation of the second order, in the sense that every solution of the one problem is at the same time a solution of the other.

The reader may use this as a starting-point for the discussion of the linear differential equation of the second order, and thus prove the fundamental existence theorem for linear differential equations.

Here we cannot enter into further discussion of these questions, and for illustrations of these general remarks we shall merely refer to the differential equations of the second order which we have dealt with above (cf. pp. 442, 448).

6. Integration by the Method of Undetermined Coefficients.

In conclusion, we mention yet another general device which can frequently be applied to the integration of differential equations. This is the method of integration by power series. We assume that in the differential equation

$$y' = f(x, y)$$

the function $f(x, y)$ can be expanded as a power series in the variables x and y and accordingly possesses derivatives of any order with respect to x and y. We can then attempt to find the solutions of the differential equation in the form of a power series

$$y = c_0 + c_1 x + c_2 x^2 + \dots$$

and to determine the coefficients of this power series by means of the differential equation.* To do this we may e.g. proceed by forming the differentiated series

$$y' = c_1 + 2c_2 x + 3c_3 x^2 + \dots,$$

replacing y in the power series for $f(x, y)$ by its expression as a power series, and then equating the coefficients of each power of x on the right and on the left (*method of undetermined coefficients*). Then if $c_0 = c$ is given any arbitrary value, we can attempt to determine the coefficients

$$c_1, c_2, c_3, c_4, \dots$$

successively.

The following process, however, is often simpler and more elegant. We assume that we are seeking to find that solution of the differential equation for which $y(0) = 0$, that is, for which the integral curve passes through the origin. Then $c_0 = c = 0$. If we recall that by Taylor's theorem the coefficients of the power series are given by the expressions

$$c_\nu = \frac{1}{\nu!} y^{(\nu)}(0),$$

we can calculate them easily. In the first place, $c_1 = y'(0) = f(0, 0)$. To obtain the second coefficient c_2 we differentiate both sides of the differential equation with respect to x and obtain

$$y''(x) = f_x + f_y y'.$$

If we here substitute $x = 0$ and the already known values $y(0) = 0$ and $y'(0) = f(0, 0)$, we obtain the value $y''(0) = 2c_2$. In the same way we can continue the process and determine the other coefficients $c_3, c_4, \dots$, one after the other.

It can be shown that this process always gives a solution if

* The first few terms of the series then form a polynomial of approximation to the solution. To a certain extent, therefore, the method is the analytical counterpart of the approximate graphical integration mentioned on p. 453.

the power series for $f(x, y)$ converges absolutely in the interior of a circle about $x = 0, y = 0$. We shall not give the proof here.

Examples

1. Verify that the left-hand sides of the following differential equations are total differentials, and integrate the equations:

$$(a)\quad (3x^2 + 6xy^2)\,dx + (6x^2y + 4y^3)\,dy = 0.$$

$$(b)\quad \frac{x\,dx + y\,dy}{\sqrt{1 + x^2 + y^2}} + \frac{y\,dx - x\,dy}{x^2 + y^2} = 0.$$

2. Show how to solve the equation $Mdx + Ndy = 0$, where M and N are homogeneous functions of the same degree.

3. Integrate the equation

$$(xy^2 - y^3)\,dx + (1 - xy^2)\,dy = 0,$$

which has an integrating factor independent of x.

4. Integrate the equation

$$2y^3\,dx + (3xy^2 - 1)\,dy = 0,$$

and from its general integral state an integrating factor.

5. Let

$$f(x, y, c) = 0$$

be a family of plane curves. By eliminating the constant c between this and the equation

$$\frac{\partial f}{\partial x} + \frac{\partial f}{\partial y}\,y' = 0,$$

we get the differential equation

$$F(x, y, y') = 0$$

of the family of curves (cf. p. 455). Now let $\varphi(p)$ be a given function of p; a curve C satisfying the differential equation

$$F(x, y, \varphi(y')) = 0$$

is called a *trajectory* of the family of curves $f(x, y, c) = 0$. The second and third equations show that

$$y' = \varphi(Y')$$

is the relation between the slope Y' of C at any given point, and the slope y' of the curve $f(x, y, c) = 0$ passing through this point. The most important case is $\varphi(p) = -1/p$, leading to the equation

$$F\left(x, y, -\frac{1}{y'}\right) = 0,$$

which is the differential equation of the *orthogonal trajectories* of the family of curves (cf. p. 456).

Use this method to find the orthogonal trajectories of the following families of curves:

(a) $x^2 + y^2 + cy - 1 = 0.$ (b) $y = cx^2.$

(c) $\dfrac{x^2}{a^2 + c} + \dfrac{y^2}{b^2 + c} = 1$ $(a > b > 0, \; -b^2 < c < \infty).$

(d) $y = \cos x + c.$ (e) $(x - c)^2 + y^2 = a^2.$

In each case draw the graphs of the two orthogonal families of curves.

6. For the family of lines $y = cx$ find the two families of trajectories in which (a) the slope of the trajectory is twice as large as the slope of the line; (b) the slope of the trajectory is equal and of opposite sign to the slope of the line.

7. Differential equations of the type

$$y = xp + \psi(p), \quad p = y'$$

were first investigated by Clairaut. Differentiating, we get

$$[x + \psi'(p)]\frac{dp}{dx} = 0,$$

which gives $p = c = \text{const.}$, so that

$$y = xc + \psi(c)$$

is the general integral of the differential equation; it represents a family of straight lines. Another solution is

$$x = -\psi'(p),$$

which together with

$$y = -p\psi'(p) + \psi(p)$$

gives a parametric representation of the so-called *singular integral.* Note that the curve given by the last two equations is the envelope of the family of lines.

Use this method to find the singular solutions of the equations

$$(a) \quad y = xp - \frac{p^2}{4}.$$

$$(b) \quad y = xp + e^p.$$

8. Find the differential equation of the tangents to the catenary

$$y = a \cosh \frac{x}{a}.$$

9. Lagrange investigated the most general differential equation which is linear in both x and y, namely,

$$y = x\varphi(p) + \psi(p).$$

Differentiating, we get

$$p = \varphi(p) + [x\varphi'(p) + \psi'(p)]\frac{dp}{dx},$$

which is equivalent to the linear differential equation

$$\frac{dx}{dp} + \frac{\varphi'(p)}{\varphi(p) - p}x + \frac{\psi'(p)}{\varphi(p) - p} = 0,$$

provided $\varphi(p) - p \neq 0$ and p is not constant. Integrating and using the first equation, we get a parametric representation of the general integral. From the second equation we see that the equations $\varphi(p) - p = 0$, $p = \text{const.}$ lead to a certain number of singular solutions representing straight lines.

The solutions can be interpreted geometrically as follows. Consider the Clairaut equation

$$y = xp + \psi[\varphi^{-1}(p)],$$

where $\varphi^{-1}(p)$ is the inverse function of $\varphi(p)$, i.e. $\varphi^{-1}(\varphi(p)) \equiv p$. From this we see that the solutions of the differential equation are a family of trajectories of the family of straight lines

$$y = xc + \psi[\varphi^{-1}(c)],$$

or

$$y = x\varphi(c) + \psi(c) \qquad (c = \text{const.}).$$

Thus e.g.

$$y = -\frac{x}{p} + \psi(p)$$

is the differential equation of the involutes (orthogonal trajectories of the tangents) of the curve which represents the singular integral of the Clairaut equation

$$y = xp + \psi\left(-\frac{1}{p}\right).$$

Use this method to integrate the equation

$$y = x(p + a) - \tfrac{1}{2}(p + a)^2.$$

10. Express, when possible, the integrals of the following differential equations by elementary functions:

(a) $\left(\frac{dy}{dx}\right)^2 = 1 - y^2$; (b) $\left(\frac{dy}{dx}\right)^2 = \frac{1}{1 - y^2}$;

(c) $\left(\frac{dy}{dx}\right)^2 = \frac{2a - y}{y}$; (d) $\left(\frac{dy}{dx}\right)^2 = \frac{1 - y^2}{1 + y^2}$.

In each case draw a graph of the family of integral curves, and detect the singular solutions, if any, from the figures.

11. A differential equation of the form

$$f(y, y', y'') = 0$$

(note that x does not occur explicitly) may be reduced to an equation of the first order as follows. Choose y as the independent variable and $p = y'$ as the unknown function. Then

$$y' = p, \quad y'' = \frac{dp}{dx} = \frac{dp}{dy}\frac{dy}{dx} = p'p,$$

and the differential equation becomes $f(y, p, pp') = 0$.

Use this method to solve the following problem:

At a variable point M of a plane curve Γ draw the normal to Γ; mark on this normal the point N where the normal meets the x-axis and C, the centre of curvature of Γ at M. Find the curves such that

$$MN \,.\, MC = \text{const.} = k.$$

Discuss the various possible cases for $k > 0$ and $k < 0$, and draw the graphs.

12*. Find the differential equation of the third order satisfied by all circles

$$x^2 + y^2 + 2ax + 2by + c = 0.$$

13. Integrate the homogeneous equation

$$\left(xy' - y\right)^2 = \left(x^2 - y^2\right)\left(\text{arc sin}\,\frac{y}{x}\right)^2$$

and find the singular solutions.

14*. Solve the differential equation

$$y'' + \frac{1}{x}y' + y = 0,$$

with $y(0) = 1$, $y'(0) = 0$, by means of a power series. Prove that this function is identical with the Bessel function $J_0(x)$ defined in Ex. 4, p. 223.

6. The Potential of Attracting Charges

Differential equations for functions of a single independent variable, such as we have discussed above, are usually called *ordinary* differential equations, to indicate that they involve only the "ordinary" derivatives of functions of one independent variable. In many branches of analysis and its applications, however, an important part is played by *partial* differential equations for functions of several variables, that is, equations between the variables and the *partial* derivatives of the unknown function. Here we shall touch upon some typical cases of partial differential equations, and shall begin by considering the theory of attractions.

We have already considered the fields of force produced by masses according to Newton's law of attraction, and we have represented them as the gradient of a potential Φ (cf. Chap. IV, p. 283 *et seq.*). In this section we shall study the potential in somewhat greater detail.*

1. Potentials of Mass Distributions.

As an extension of the cases considered previously we now take μ as a positive or negative mass or charge. Negative masses do not enter into the ordinary Newtonian law of attraction, but they do occur in the theory of electricity, where mass is replaced by electric charge and we distinguish between positive and negative electricity; Coulomb's law of attracting charges has the same form as the law of attraction of mechanical masses. If a charge μ is concentrated at a single point of space with co-ordinates (ξ, η, ζ), we call the expression μ/r, where

$$r = \sqrt{\{(x-\xi)^2 + (y-\eta)^2 + (z-\zeta)^2\}},$$

the potential † of this mass at the point (x, y, z). By adding up a number of such potentials for different "sources" or "poles" (ξ_i, η_i, ζ_i) we obtain as before (cf. p. 283) the potential of a system of particles

$$\Phi = \sum_i \frac{\mu_i}{r_i}.$$

The corresponding fields of force are given by the expression $f = \gamma \operatorname{grad} \Phi$, where γ is a constant independent of the masses and of their positions.

If the masses, instead of being concentrated at single points or "sources", are distributed with density $\mu(\xi, \eta, \zeta)$ over a definite portion R of $\xi\eta\zeta$-space, we have already taken the potential of this mass-distribution to be

$$\Phi = \iiint \frac{\mu}{r}\, d\xi\, d\eta\, d\zeta.$$

* An extensive literature is devoted to this important branch of analysis; see, e.g., Kellogg's *Foundations of Potential Theory* (Springer, Berlin, 1929).

† We could call this *a* potential of the mass. Any function obtained by adding an arbitrary constant to this could equally well be called a potential of the mass, since it would give the same field of force.

If the masses are distributed over a surface S with surface-density μ, then the surface integral

$$\iint \frac{\mu(u, v)}{r} d\sigma$$

taken over the surface S with surface element $d\sigma$ represents the potential of this surface, if the surface is given parametrically (p. 159 *et seq.*) by u, v as parameters.

For the potential of a mass distributed along a curve we likewise obtain an expression of the form

$$\int \frac{\mu(s)}{r} ds,$$

where s is the length of arc on this curve, $\mu(s)$ the linear density of the mass, and r the distance of the point (x, y, z) from the point S of the curve.

For every such potential the surfaces

$$\Phi = \text{const.}$$

represent the *equipotential surfaces* or *level surfaces.**

As an example of the potential of a line-distribution we take this case: a mass of constant linear density μ is distributed along the segment $-l \leqq z \leqq +l$ of the z-axis. We consider a point P with co-ordinates (x, y) in the plane $z = 0$; if for brevity we introduce $\rho = \sqrt{(x^2 + y^2)}$, the distance of the point P from the origin, we obtain the potential in the form

$$\Phi(x, y) = \mu \int_{-l}^{+l} \frac{dz}{\sqrt{(\rho^2 + z^2)}} + C.$$

Here we have added a constant C to the integral, which does not affect the field of force derived from the potential. The indefinite integral on the right can be evaluated as in Vol. I, p. 213, and we obtain

$$\int \frac{dz}{\sqrt{(\rho^2 + z^2)}} = \text{ar sinh} \frac{z}{\rho} = \log \frac{z + \sqrt{(z^2 + \rho^2)}}{\rho},$$

* Curves which at every point have the direction of the force vector are called *lines of force*. The lines of force are therefore curves which everywhere intersect the level surfaces at right angles. We thus see that the families of lines of force corresponding to potentials generated by a single pole or by a finite number of poles run out from these poles as if from a source. In the case of a single pole, for example, the lines of force are simply the straight lines passing through the pole.

so that the potential in the xy-plane is given by

$$\Phi(x, y) = 2\mu \log \frac{l + \sqrt{(l^2 + \rho^2)}}{\rho} + C.$$

To obtain the potential of a line extending to infinity in both directions we give the value $-2\mu \log 2l$ to the constant * C and thus obtain

$$\Phi(x, y) = 2\mu \log \frac{l + \sqrt{(l^2 + \rho^2)}}{2l} - 2\mu \log \rho.$$

If we now let the length l increase without limit, that is, if we let the length of the line tend to infinity, the expression $\{l + \sqrt{(l^2 + \rho^2)}\}/2l$ tends to unity, and for the limiting value of $\Phi(x, y)$ we obtain the expression

$$\Phi(x, y) = -2\mu \log \rho.$$

We thus see that apart from the factor -2μ *the expression*

$$\log \rho = \log \sqrt{(x^2 + y^2)}$$

is the *potential of a straight line perpendicular to the* xy-*plane over which a mass is distributed uniformly.*

In addition to the distributions previously considered, potential theory also deals with so-called *double layers*, which we obtain in the following way. We suppose that at the point (ξ, η, ζ) a charge M is concentrated and at the point $(\xi + h, \eta, \zeta)$ a charge $-M$ is concentrated. The potential of this pair of charges is given by

$$\Phi = \frac{M}{\sqrt{(x - \xi)^2 + (y - \eta)^2 + (z - \zeta)^2}} - \frac{M}{\sqrt{(x - \xi - h)^2 + (y - \eta)^2 + (z - \zeta)^2}}.$$

If we let h, the distance between the two poles, tend to zero and at the same time let the charge M increase without limit in such a way that M is always equal to $-\mu/h$, where μ is a constant, Φ in the limit tends to the expression

$$\mu \frac{\partial}{\partial \xi}\left(\frac{1}{r}\right).$$

We call this expression the *potential of a dipole* or *doublet* with

* We make this choice in order that in the passage to the limit $l \to \infty$ the potential Φ shall remain finite.

its axis in the ξ-direction and with "moment" μ. Physically it represents the potential of a pair of equal and opposite charges lying very close to one another. In the same way we can express the potential of a dipole in the form

$$\frac{\partial}{\partial \nu}\left(\frac{1}{r}\right),$$

where $\partial/\partial\nu$ denotes differentiation in an arbitrary direction $\boldsymbol{\nu}$, that of the axis of the dipole.

If we imagine dipoles distributed over a surface S with moment-density μ, and if we assume that at each point the axis of the dipole is normal to the surface, we obtain an expression of the form

$$\iint_S \mu(\xi, \eta, \zeta) \frac{\partial}{\partial \nu}\left(\frac{1}{r}\right) d\sigma,$$

where $\partial/\partial\nu$ denotes differentiation in the direction of the positive normal to the surface (we can, as before, choose either direction of the normal as positive), r is the distance of the point of the surface (ξ, η, ζ) from the point (x, y, z), and the point (ξ, η, ζ) ranges over the surface. This *potential of a double layer* can be thought of as arising in the following way. On each side of the surface and at a distance h we construct surfaces, and we give one of these surfaces a surface-density $\mu/2h$, the other a surface-density $-\mu/2h$. At an external point these two layers together create a potential which tends to the expression above as $h \to 0$. We shall assume that in all our expressions the point (x, y, z) considered is at a point in space at which no charge is present, so that the integrands and their derivatives with respect to x, y, z are continuous.

2. The Differential Equation of the Potential.

In virtue of these hypotheses we can obtain a relation which all our potential expressions satisfy, namely, the differential equation

$$\Phi_{xx} + \Phi_{yy} + \Phi_{zz} = 0,$$

or in abbreviated form

$$\Delta\Phi = 0,$$

which is known as Laplace's equation. As we have already

(Vol. I, p. 470) verified by simple calculation, this equation is satisfied by the expression $1/r$. It therefore holds also for all the other expressions formed from it by summation or integration, since we can perform the differentiations with respect to x, y, z under the integral sign. This differential equation is also satisfied by the potential of a double layer, for in virtue of the reversibility of the order of differentiation * we find that for the potential of a single dipole the equation

$$\Delta \frac{\partial}{\partial \nu}\left(\frac{1}{r}\right) = \frac{\partial}{\partial \nu}\Delta \frac{1}{r} = 0$$

holds.

Laplace's equation is also satisfied by the expression $\log\sqrt{(x^2 + y^2)}$ obtained for the potential of a vertical line, as we can readily verify (cf. also Chap. II, p. 76). Since this no longer depends on the variable z, it in fact satisfies the simpler Laplace's equation in two dimensions,

$$\Phi_{xx} + \Phi_{yy} = 0.$$

The study of these and related partial differential equations forms one of the most important branches of analysis. We may, however, point out that potential theory is not by any means chiefly directed to the search for general solutions of the equation $\Delta\Phi = 0$, but rather to the question of the existence and to the investigation of those solutions which satisfy pre-assigned conditions. Thus a central problem of the theory is the "boundary-value problem", in which we have to find a solution Φ of $\Delta\Phi = 0$ which together with its derivatives up to the second order is continuous in a region R, and which has pre-assigned continuous values on the boundary of R.

3. Uniform Double Layers.

We cannot enter here into a more detailed study of *potential functions*, that is, of functions which satisfy Laplace's equation $\Delta u = 0$. In this subject Gauss's theorem and Green's theorem

* It must be noted that the differentiation $\partial/\partial\nu$ refers to the variables (ξ, η, ζ) and the expression Δ to the variables (x, y, z). Moreover, the function $1/r$, considered as a function of the six variables $(x, y, z;\ \xi, \eta, \zeta)$, is symmetrical in the two sets of variables, and therefore satisfies the differential equation

$$\Delta\Phi = \Phi_{\xi\xi} + \Phi_{\eta\eta} + \Phi_{\zeta\zeta} = 0$$

with respect to the variables (ξ, η, ζ) also.

(Chap. V, pp. 388, 390) are among the chief tools employed. It will be sufficient to show by some examples how such investigations are carried out.

We shall first consider the potential of a double layer with constant moment-density $\mu = 1$, that is, an integral of the form

$$V = \iint_S \frac{\partial}{\partial \nu}\left(\frac{1}{r}\right) d\sigma.$$

This integral has a simple geometrical meaning. Let us assume that each point of the surface carrying the double layer can be seen from the point P with co-ordinates (x, y, z), that is, that it can be joined to this point P by a straight line which meets the surface nowhere else. The surface S, together with the rays joining its boundary to the point P, forms a conical region R of space. We now state that *the potential of the uniform double layer, except perhaps for sign, is equal to the solid angle which the boundary of the surface* S *subtends at the point* P. By this solid angle we mean the area of that portion of the spherical surface of unit radius about the point P as centre which is cut out of the spherical surface by the rays going from P to the boundary of S. We give this solid angle the positive sign when the rays pass through the surface S in the same direction as the positive normal ν, otherwise we give it the negative sign (cf. Ex. 9, p. 408).

To prove this we recall that the function $u = 1/r$, when considered not only as a function of (x, y, z) but also as a function of (ξ, η, ζ), still satisfies the differential equation

$$\Delta u = u_{\xi\xi} + u_{\eta\eta} + u_{\zeta\zeta} = 0.$$

We fix the point P with co-ordinates (x, y, z), and denote the rectangular co-ordinates in the conical region R by (ξ, η, ζ), and by a small sphere of radius ρ about the point P we cut off the vertex from R; the residual region we call R_ρ. To the function $u = 1/r$, considered as a function of (ξ, η, ζ) in the region R_ρ, we now apply Green's theorem (Chap. V, p. 390) in the form

$$\iiint_{R_\rho} \Delta u \, d\xi \, d\eta \, d\zeta = \iint_{S'} \frac{\partial u}{\partial n} d\sigma,$$

where S' is the boundary surface of R_ρ and $\partial/\partial n$ denotes differentiation in the direction of the outward normal. Since $\Delta u = 0$, the value of the left-hand side is zero.* If we have chosen the positive normal direction ν on S so as to coincide with the outward normal n, the surface integral on the right-hand side consists of three parts: (1) the surface integral

$$\iint_S \frac{\partial}{\partial n}\left(\frac{1}{r}\right) d\sigma = \iint_S \frac{\partial}{\partial \nu}\left(\frac{1}{r}\right) d\sigma$$

over the surface S, which is the expression V considered above (p. 474); (2) an integral over the lateral surface formed by the linear rays; (3) an integral over a portion Γ_ρ of the surface of the small sphere of radius ρ. The second part is zero, since there the normal direction n is perpendicular to the radius, and therefore is tangential to the sphere $r =$ const. For the inner sphere with radius ρ the symbol $\partial/\partial n$ is equivalent to $-\partial/\partial \rho$, since the outward direction of the normal points in the direction of diminishing values of r. We thus obtain the equation

$$V - \iint_{\Gamma_\rho} \frac{\partial}{\partial \rho}\left(\frac{1}{\rho}\right) d\sigma = 0$$

or

$$V = -\frac{1}{\rho^2} \iint_{\Gamma_\rho} d\sigma,$$

where on the right we have to integrate over the portion Γ_ρ of the small spherical surface which belongs to the boundary of R_ρ. If we now write the surface element on the sphere with radius ρ in the form $d\sigma = \rho^2 d\omega$, where $d\omega$ is the surface element on the unit sphere, we at once obtain

$$V = -\iint d\omega.$$

The integral on the right is to be taken over the portion of the spherical surface of unit radius lying in the cone of rays, and we see at once that the right-hand side has the geometrical meaning stated above; it is the apparent magnitude, except for sign, if the normal direction on S is chosen so that it points

* From this form of Green's theorem it follows in general that the surface integral $\iint \frac{\partial u}{\partial n} d\sigma$ taken over a closed surface must always vanish when the function u satisfies Laplace's equation $\Delta u = 0$ everywhere in the interior of the surface.

outwards * from the conical region R. Otherwise the positive sign is to be taken.

If the surface S is not in the simple position relative to P described above, but instead is intersected several times by some of the rays through P, we have only to divide the surface into a number of portions of the simpler kind in order to see that the statement still holds good. *The potential of the uniform double layer (of moment* 1) *on a bounded surface is therefore, except perhaps for sign, equal to the "apparent" magnitude which the boundary has when looked at from the point* (x, y, z).

For a *closed surface* we see by subdividing it into two bounded portions that our expression is equal to zero if the point P is outside, and equal to -4π if it is inside.

A similar argument shows in the case of two independent variables that the integral

$$\int_{\sigma} \frac{\partial}{\partial \nu} (\log r) \, ds$$

along the curve C, except possibly for sign, is equal to the angle which this curve subtends at the point P with the co-ordinates (x, y).

This result, like the corresponding result in space, can also be explained geometrically as follows. Let the point Q with the co-ordinates (ξ, η) lie on the curve C. Then the derivative of $\log r$ at the point Q in the direction of the normal to the curve is given by the equation

$$\frac{\partial}{\partial \nu} (\log r) = \frac{\partial}{\partial r} (\log r) \cos(\nu, r) = \frac{1}{r} \cos(\nu, r),$$

where the symbol (ν, r) denotes the angle between this normal and the direction of the radius vector r. On the other hand, when written in polar co-ordinates (r, θ) the element of arc ds of the curve has the form

$$ds = \frac{r \, d\theta}{\cos(\nu, r)}$$

* The negative sign is explained by the fact that with this choice of the normal direction the negative charge lies "next" the point P.

(cf. Vol. I, pp. 266 and 280), so that the integral is transformed as follows:

$$\int \frac{\partial}{\partial \nu}(\log r)\,ds = \int \frac{1}{r}\cos(\nu, r)\,\frac{r\,d\theta}{\cos(\nu, r)} = \int d\theta.$$

The integral on the right, however, is the analytical expression of our statement.

4. The Theorem of Mean Value.

As a second application of Green's transformation we prove the following theorem: every potential function, that is, every function u which in a certain region R satisfies the differential equation $\Delta u = 0$, has the following mean value property:

The value of the potential function at the centre P *of an arbitrary sphere of radius* r *lying completely in the region* R *is equal to the mean value of the function* u *on the surface* $\mathrm{S_r}$ *of the sphere; that is,*

$$u(x, y, z) = \frac{1}{4\pi r^2}\iint_{S_r} \bar{u}\,d\sigma,$$

where u(x, y, z) *is the value at the centre* P *and* ū *the value on the surface* $\mathrm{S_r}$ *of the sphere of radius* r.

To prove this we proceed as follows: let S_ρ be a concentric sphere inside S_r with radius $0 < \rho \leqq r$. Since $\Delta u = 0$ everywhere in the interior of S_ρ, by the footnote on p. 475 we have

$$\iint_{S_\rho} \frac{\partial u}{\partial n}\,d\sigma = 0,$$

where $\partial u/\partial n$ is the derivative of u in the direction of the outward normal to S_ρ. If (ξ, η, ζ) are current co-ordinates and if with the point (x, y, z) as pole we introduce polar co-ordinates by the equations

$$\xi - x = \rho\cos\phi\sin\theta, \quad \eta - y = \rho\sin\phi\sin\theta, \quad \zeta - z = \rho\cos\theta,$$

the above equation becomes

$$\iint_{S_\rho} \frac{\partial u(\rho, \theta, \phi)}{\partial \rho}\,d\sigma = 0.$$

Since the surface element $d\sigma$ of the sphere S_ρ is equal to $\rho^2 d\bar{\sigma}$, where $d\bar{\sigma}$ is the element of surface of the sphere S of unit radius (cf. Chap. IV, p. 274), we find that if $\rho > 0$

$$\iint_S \frac{\partial u}{\partial \rho}\, d\bar{\sigma} = 0,$$

where the region of integration no longer depends on ρ. Consequently

$$\int_0^r d\rho \iint_S \frac{\partial u}{\partial \rho}\, d\bar{\sigma} = 0,$$

and on interchanging the order of integration and performing the integration with respect to ρ we have

$$\iint_S \{u(r,\, \theta,\, \phi) - u(0,\, \theta,\, \phi)\}\, d\bar{\sigma} = 0.$$

Since $u(0, \theta, \phi) = u(x, y, z)$ is independent of θ and ϕ,

$$\iint_S u(r,\, \theta,\, \phi)\, d\sigma = u(x,\, y,\, z) \iint_S d\bar{\sigma} = 4\pi\, u(x,\, y,\, z).$$

As

$$\iint_S u(r,\, \theta,\, \phi)\, d\bar{\sigma} = \frac{1}{r^2} \iint_{S_r} u(r,\, \theta,\, \phi)\, d\sigma,$$

where the integral on the right is to be taken over the surface of S_r, the mean value property of u is proved.

In exactly the same way, for functions u of two variables which satisfy Laplace's equation $u_{xx} + u_{yy} = 0$ we have the corresponding *mean value property of the circle* expressed by the formula

$$2\pi r\, u(x,\, y) = \int_{S_r} \bar{u}\, ds,$$

where $\bar{u}$ denotes the value of the potential function on a circle S_r with radius r about the point (x, y) and ds is the element of arc of this circle.

5. Boundary-value Problem for the Circle. Poisson's Integral.

As an example of a boundary-value problem we shall now discuss Laplace's equation in two independent variables x, y for the case of a circular boundary. Within the circular region $x^2 + y^2 \leq R^2$ we introduce polar co-ordinates (r, θ). We wish to find a function $u(x, y)$ which is continuous within the circle and on the boundary, possesses continuous derivatives of the first and second order within the region, satisfies Laplace's equation $\Delta u = 0$, and has prescribed values $u(R, \theta) = f(\theta)$ on the boundary. Here we assume that $f(\theta)$ is a continuous periodic function of θ with sectionally continuous first derivatives.

The solution of this problem, in terms of polar co-ordinates, is given by the so-called Poisson's integral

$$u = \frac{R^2 - r^2}{2\pi} \int_0^{2\pi} \frac{f(a)}{R^2 - 2Rr\cos(\theta - a) + r^2}\, da.$$

To prove this, we begin by constructing as many solutions of Laplace's equations as we please in the following way. We transform Laplace's equation to polar co-ordinates, obtaining

$$\Delta u = \frac{1}{r}(ru_r)_r + \frac{1}{r^2} u_{\theta\theta} = 0,$$

and seek to find solutions which can be expressed in the form $u = \phi(r)\psi(\theta)$, that is, as a product of a function of r and a function of θ. If we substitute this expression for u in Laplace's equation, the equation becomes

$$r\frac{(r\phi'(r))_r}{\phi(r)} = -\frac{\psi''(\theta)}{\psi(\theta)}.$$

As the left-hand side does not involve θ and the right-hand side does not involve r, the two sides must each be independent of both variables, that is, must be equal to the same constant k. For $\psi(\theta)$ we accordingly have the differential equation $\psi'' + k\psi = 0$.

Since the function u and hence also $\psi(\theta)$ must be periodic with period 2π, it follows that the constant k is equal to n^2, where n is an integer. Hence

$$\psi(\theta) = a\cos n\theta + b\sin n\theta,$$

where a and b are arbitrary constants.

The differential equation for $\phi(r)$,

$$r^2\phi''(r) + r\phi'(r) - n^2\phi(r) = 0,$$

is a linear differential equation and, as we can immediately verify, the functions r^n and r^{-n} are independent solutions. Since the second solution becomes infinite at the origin, while u is to be continuous there, we are left with the first solution $\phi = r^n$, and the solutions of Laplace's equation are

$$r^n(a \cos n\theta + b \sin n\theta).$$

We now use the fact that by linear combination of such solutions according to the principle of superposition (cf. section 4, p. 438) we can obtain other solutions

$$\tfrac{1}{2}a_0 + \Sigma r^n(a_n \cos n\theta + b_n \sin n\theta).$$

Even an infinite series of this form will be a solution, provided that the series converges uniformly and can be differentiated term by term twice in the interior of the circle.

If we now imagine the prescribed boundary function $f(\theta)$ expanded in a Fourier series

$$f(\theta) = \tfrac{1}{2}a_0 + \sum_{n=1}^{\infty}(a_n \cos n\theta + b_n \sin n\theta),$$

this series, regarded as a series in θ, certainly converges absolutely and uniformly (cf. Vol. I, Chap. IX, p. 451). Hence the series

$$u(r, \theta) = \tfrac{1}{2}a_0 + \sum_{n=1}^{\infty} \frac{r^n}{R^n}(a_n \cos n\theta + b_n \sin n\theta)$$

a fortiori converges uniformly and absolutely in the interior of the circle. This series, however, can be differentiated term by term, provided $r < R$, because the resulting series again converge uniformly (cf. the account of power series in Vol. I, Chap. VI, p. 399). This function is accordingly a potential function; it has the prescribed value on the boundary, and hence is a solution of our boundary-value problem.

We can reduce this solution to the integral form given above by introducing the integrals for the Fourier coefficients,

$$a_n = \frac{1}{\pi}\int_0^{2\pi} f(\alpha) \cos n\alpha \, d\alpha, \quad b_n = \frac{1}{\pi}\int_0^{2\pi} f(\alpha) \sin n\alpha \, d\alpha.$$

Since the convergence is uniform, we can interchange integration and summation, and obtain

$$u(r, \theta) = \frac{1}{\pi}\int_0^{2\pi} f(\alpha)\left\{\frac{1}{2} + \sum_{n=1}^{\infty} \frac{r^n}{R^n} \cos n(\theta - \alpha)\right\} d\alpha.$$

Poisson's integral formula is therefore proved, provided that we can establish the relation

$$\frac{1}{2} + \sum_{n=1}^{\infty} \frac{r^n}{R^n} \cos n\tau = \frac{1}{2} \frac{R^2 - r^2}{R^2 - 2Rr\cos\tau + r^2}.$$

But this can be proved by the method used in Vol. I, Chap. IX, p. 436; we leave the proof to the reader.

Examples

1. By applying inversion to Poisson's formula, find a potential function $u(x, y)$ which is bounded in the region *outside* the unit circle and assumes given values $f(\theta)$ on its boundary (the so-called *outer* boundary-value problem).

2*. Find (*a*) the equipotential surfaces and (*b*) the lines of force for the potential of the segment $x = y = 0$, $-l \leqq z \leqq +l$, of constant linear density μ.

3*. Prove that if the values of a harmonic $u(x, y, z)$ and of its normal derivative $\partial u/\partial n$ are given on a closed surface S, then the value of u at any interior point is given by the expression

$$u(x, y, z) = \frac{1}{4\pi}\iint_S \left(\frac{1}{r}\frac{\partial u}{\partial n} - u\frac{\partial(1/r)}{\partial n}\right) d\sigma,$$

where r is the distance from the point (x, y, z) to the variable point of integration. (Apply Green's theorem to the functions u and $1/r$.)

7. Further Examples of Partial Differential Equations

We shall now briefly discuss a few partial differential equations which are of frequent occurrence.

1. The Wave Equation in One Dimension.

The phenomena of wave propagation, e.g. of light or sound are governed by the so-called *wave equation*. We begin by considering the simple idealized case of a so-called " one-dimensional wave ". Such a wave depends on some property u, for example, the pressure, the change of position of a particle, or the intensity of an electric field; and u depends not only on the co-ordinate of position x (we take the direction of propagation as the x-axis) but also on the time t.

The function $u(x, t)$ then satisfies a partial differential equation of the form

$$u_{xx} = \frac{1}{a^2} u_{tt},$$

where a is a constant depending on the physical nature of the medium. We can express solutions of this equation in the form

$$u = f(x - at),$$

where $f(\xi)$ is an arbitrary function of ξ, about which we assume only that it has continuous derivatives of the first and second order. If we put $\xi = x - at$, we see at once that our differential equation is actually satisfied, for

$$u_{xx} = f''(\xi), \quad u_{tt} = a^2 f''(\xi).$$

In the same way, using another arbitrary function $g(\xi)$, we obtain a solution of the form

$$u = g(x + at).$$

Both these solutions represent wave motions which are propagated with the velocity a along the x-axis; the first represents a wave travelling in the positive x-direction, the second a wave travelling in the negative x-direction. For let u have the value $u(x_1, t_1)$ at any point x_1 at time t_1; then u has the same value at time t at the point $x = x_1 + a(t - t_1)$. For then $x - at = x_1 - at_1$, so that $f(x - at) = f(x_1 - at_1)$. In the same way, we can see that the function $g(x + at)$ represents a wave travelling in the negative x-direction with velocity a.

We shall now solve the following initial-value problem for this wave equation. From all possible solutions of the differential equation we wish to select those for which the initial state (at $t = 0$) is given by two prescribed functions $u(x, 0) = \phi(x)$ and $u_t(x, 0) = \psi(x)$. To solve this problem, we have merely to write

$$u = f(x - at) + g(x + at)$$

and determine the functions f and g from the two equations

$$\phi(x) = f(x) + g(x)$$

$$\frac{1}{a}\psi(x) = -f'(x) + g'(x).$$

The second equation gives

$$c + \frac{1}{a}\int_0^x \psi(\tau)\, d\tau = -f(x) + g(x),$$

where c is an arbitrary constant of integration. From this we readily obtain the required solution in the form

$$u(x, t) = \frac{\phi(x + at) + \phi(x - at)}{2} + \frac{1}{2a} \int_{x-at}^{x+at} \psi(\tau)\, d\tau.$$

The reader should prove for himself, by introducing new variables $\xi = x - at$, $\eta = x + at$ instead of x and t, that no solutions of the differential equation exist other than those given.

2. The Wave Equation in Three-dimensional Space.

In the wave equation for space of three dimensions the function u depends on four independent variables, namely, the three space co-ordinates x, y, z and the time t. The wave equation is then

$$u_{xx} + u_{yy} + u_{zz} = \frac{1}{a^2} u_{tt},$$

or, more briefly,

$$\Delta u = \frac{1}{a^2} u_{tt}.$$

Here again we can easily find solutions which represent the propagation of a plane wave in the physical sense.

In fact, any function $f(\xi)$, provided we assume that it is twice continuously differentiable, gives us a solution of the differential equation, if we make ξ a linear expression of the form

$$\xi = \alpha x + \beta y + \gamma z \pm at,$$

whose coefficients satisfy the relation

$$\alpha^2 + \beta^2 + \gamma^2 = 1.$$

For since

$$\Delta u = (\alpha^2 + \beta^2 + \gamma^2) f''(\xi) = f''(\xi)$$

and

$$u_{tt} = a^2 f''(\xi),$$

we see that $u = f(\alpha x + \beta y + \gamma z \pm at)$ is really a solution of the equation

$$\Delta u = \frac{1}{a^2} u_{tt}.$$

If q is the distance of the point (x, y, z) from the plane

$\alpha x + \beta y + \gamma z = 0$, we know by analytical geometry (cf. Chap. I, p. 9) that

$$q = \alpha x + \beta y + \gamma z.$$

Hence, in the first place, we see from the expression

$$u = f(q \pm at)$$

that at all points of a plane at a distance q from the plane $\alpha x + \beta y + \gamma z = 0$ and parallel to it the property which is being propagated (represented by u) has the same value at a given moment. The property is propagated in space in such a way that planes parallel to $\alpha x + \beta y + \gamma z = 0$ are always surfaces on which the property is constant; the velocity of propagation is a in the direction perpendicular to the planes.

In theoretical physics a propagated phenomenon of this kind is referred to as a *plane wave*.

A case of particular importance is that in which the property varies periodically with the time. If the frequency of the vibration is ω, a phenomenon of this kind may be represented by

$$u = e^{ik(\alpha x + \beta y + \gamma z)} e^{i\omega t},$$

where k, as usual, denotes the reciprocal of the wave-length: $k = \frac{1}{\lambda} = \frac{\omega}{a}$.

In the case of the wave equation with four independent variables we can find other solutions, which represent a *spherical* wave spreading out from a given point, say the origin. A spherical wave is defined by the statement that the property is the same at a given instant at every point of a sphere with its centre at the origin, that is, that u has the same value at every point of the sphere. To find solutions which satisfy this condition, we transform Δu to polar co-ordinates (r, θ, ϕ), and then we have merely to assume that u depends on r and t only and not on θ and ϕ. If we accordingly equate the derivatives of u with respect to θ and ϕ to zero (cf. p. 391), the differential equation becomes

$$u_{rr} + \frac{2}{r} u_r = \frac{1}{a^2} u_{tt}$$

or

$$(ru)_{rr} = \frac{1}{a^2} (ru)_{tt}.$$

If for the moment we replace ru by w, w is a solution of the equation

$$w_{rr} = \frac{1}{a^2} w_{tt},$$

which we have already discussed, and hence must be expressible in the form

$$w = f(r - at) + g(r + at).$$

Then

$$u = \frac{1}{r} \{f(r - at) + g(r + at)\}.$$

The reader should now verify for himself directly that a function of this type is actually a solution of the differential equation

$$\Delta u = \frac{1}{a^2} u_{tt}.$$

Physically the function $u = \frac{1}{r} f(r - at)$ represents a wave which is propagated outwards into space from a centre with velocity a.

3. Maxwell's Equations in Free Space.

As a concluding example we shall discuss the system of equations, known as *Maxwell's equations*, which form the foundations of electrodynamics. We shall not, however, attempt to approach the equations from the physical point of view, but shall merely consider them as illustrating the various mathematical concepts developed above.

The electromagnetic condition in free space is determined by two vectors, an electric vector $\boldsymbol{E}$ with components E_1, E_2, E_3, and a magnetic vector $\boldsymbol{H}$ with components H_1, H_2, H_3. These vectors satisfy Maxwell's equations:

$$\operatorname{curl} \boldsymbol{E} + \frac{1}{c} \frac{\partial \boldsymbol{H}}{\partial t} = 0,$$

$$\operatorname{curl} \boldsymbol{H} - \frac{1}{c} \frac{\partial \boldsymbol{E}}{\partial t} = 0,$$

where c is the velocity of light in free space. Expressed in terms of the components of the vectors, the equations are:

$$\frac{\partial E_3}{\partial y} - \frac{\partial E_2}{\partial z} + \frac{1}{c}\frac{\partial H_1}{\partial t} = 0,$$

$$\frac{\partial E_1}{\partial z} - \frac{\partial E_3}{\partial x} + \frac{1}{c}\frac{\partial H_2}{\partial t} = 0,$$

$$\frac{\partial E_2}{\partial x} - \frac{\partial E_1}{\partial y} + \frac{1}{c}\frac{\partial H_3}{\partial t} = 0,$$

and

$$\frac{\partial H_3}{\partial y} - \frac{\partial H_2}{\partial z} - \frac{1}{c}\frac{\partial E_1}{\partial t} = 0,$$

$$\frac{\partial H_1}{\partial z} - \frac{\partial H_3}{\partial x} - \frac{1}{c}\frac{\partial E_2}{\partial t} = 0,$$

$$\frac{\partial H_2}{\partial x} - \frac{\partial H_1}{\partial y} - \frac{1}{c}\frac{\partial E_3}{\partial t} = 0.$$

For the components as functions of position and time we thus have a system of six partial differential equations of the first order, that is, of equations involving the first partial derivatives of the components with respect to the space co-ordinates and to the time.

We shall now deduce some distinctive consequences of Maxwell's equations. If we form the " divergence " of both equations, and remember that $\operatorname{div}\operatorname{curl}\boldsymbol{A} = 0$ and that the order of differentiation with respect to the time and formation of the divergence is interchangeable, we obtain

$$\operatorname{div}\boldsymbol{E} = \text{const.},$$

$$\operatorname{div}\boldsymbol{H} = \text{const.};$$

that is, the two " divergences " are independent of the time.

If we assume that the constants are initially zero, then they remain zero for all time.

We now consider any closed surface S lying in the field and take the volume integrals

$$\iiint \operatorname{div}\boldsymbol{E}\,d\tau$$

and

$$\iiint \operatorname{div}\boldsymbol{H}\,d\tau$$

throughout the volume enclosed by it. If we apply Gauss's theorem (Chap. V, p. 388) to these integrals, they become integrals of the normal components E_n, H_n over the surface S. That is, the equations

$$\operatorname{div} \boldsymbol{E} = 0,$$
$$\operatorname{div} \boldsymbol{H} = 0$$

give

$$\iint_S E_n \, d\sigma = 0,$$
$$\iint_S H_n \, d\sigma = 0.$$

In electrical theory surface integrals

$$\iint_S E_n \, d\sigma \quad \text{or} \quad \iint H_n \, d\sigma$$

are called the electric or magnetic flux across the surface S, and our result may accordingly be stated as follows:

The electric flux and the magnetic flux across a closed surface, subject to the assumptions we have made above, are zero.

We obtain a further deduction from Maxwell's equations if we consider a portion of surface S bounded by the curve Γ and lying in the surface.

If we denote the components of a vector normal to the surface A by the suffix n, it immediately follows from Maxwell's equations that

$$(\operatorname{curl} \boldsymbol{E})_n = -\frac{1}{c}\frac{\partial H_n}{\partial t},$$
$$(\operatorname{curl} \boldsymbol{H})_n = +\frac{1}{c}\frac{\partial E_n}{\partial t}.$$

If we integrate these equations over the surface with surface element $d\sigma$, we can transform the left-hand sides into line integrals taken round the boundary Γ by Stokes's theorem (cf. Chap. V, p. 395). If we do this, and carry out the differentiation with respect to t outside the integral sign, we obtain the equations

$$\int_\Gamma E_s \, ds = -\frac{1}{c}\frac{d}{dt}\iint_S H_n \, d\sigma,$$
$$\int_\Gamma H_s \, ds = +\frac{1}{c}\frac{d}{dt}\iint_S E_n \, d\sigma,$$

where the symbols E_s and H_s under the integral signs on the left are the tangential components of the electric and magnetic vectors in the direction of increasing arc, and the sense of description of the curve Γ in conjunction with the direction of the normal n forms a right-handed screw.

The facts expressed by these equations may be expressed in words as follows: The line integral of the electric or the magnetic force round an element of surface is proportional to the rate of change of the electric or magnetic flux across the element of surface, the constant of proportionality being $-1/c$ or $+1/c$.

Finally, we shall establish the connexion between Maxwell's equations and the wave equation. We find, in fact, that each of the vectors $\boldsymbol{E}$ and $\boldsymbol{H}$, that is, each component of the vectors, satisfies the wave equation

$$\Delta u = \frac{1}{c^2} u_{tt}.$$

For we can eliminate the vector $\boldsymbol{H}$, say, from the two equations, by differentiating the second equation with respect to the time and substituting for $\frac{\partial \boldsymbol{H}}{\partial t}$ from the first equation.

It then follows that

$$c \operatorname{curl} \operatorname{curl} \boldsymbol{E} + \frac{1}{c} \frac{\partial^2 \boldsymbol{E}}{\partial t^2} = 0.$$

If we now use the vector relation *

$$\operatorname{curl} \operatorname{curl} \boldsymbol{A} = -\Delta \boldsymbol{A} + \operatorname{grad} \operatorname{div} \boldsymbol{A},$$

and remember that

$$\operatorname{div} \boldsymbol{E} = 0,$$

we at once have

$$\Delta \boldsymbol{E} = \frac{1}{c^2} \frac{\partial^2 \boldsymbol{E}}{\partial t^2}.$$

In the same way we can show that the vector $\boldsymbol{H}$ satisfies the same equation:

$$\Delta \boldsymbol{H} = \frac{1}{c^2} \frac{\partial^2 \boldsymbol{H}}{\partial t^2}.$$

* This vector relation follows immediately from the expressions in terms of co-ordinates.

Examples

1. Integrate the following partial differential equations:

$$(a)\quad u_{xy} = 0;$$

$$(b)\quad u_{xyz} = 0;$$

$$(c)\quad u_{xy} = a(x, y).$$

2*. Solve the equation

$$u_{xx} + 5u_{xy} + 6u_{yy} = e^{x+y}$$

by reducing it to one of the form of Ex. 1(*c*).

3. Find the partial differential equation satisfied by the two-parameter family of spheres

$$z^2 = 1 - (x - a)^2 - (y - b)^2.$$

4*. Let $u(x, t)$ denote a solution of the wave equation

$$u_{xx} = \frac{1}{a^2} u_{tt}, \quad (a > 0)$$

which is twice continuously differentiable. Let $\varphi(x)$ be a given function which is twice continuously differentiable and such that

$$\varphi(0) = \varphi'(0) = \varphi''(0) = 0.$$

Find the solution u for $x \geqq 0$ and $t \geqq 0$ which is determined by the boundary conditions

$$u(x, 0) = u_t(x, 0) = 0 \text{ for } x \geqq 0,$$

$$u(0, t) = \varphi(t) \text{ for } t \geqq 0.$$

5. Find a solution of the equation

$$u_{xy} = u,$$

for which $u(x, 0) = u(0, y) = 1$, in the form of a power series.

6. (*a*) Find particular solutions of the equation

$$u_x^2 + u_y^2 = 1$$

of the form $u = f(x) + g(y)$.

(*b*) Find particular solutions of the equation

$$u_x u_y = 1$$

of the forms $u = f(x) + g(y)$ and $u = f(x)g(y)$.

7*. Prove that if

$$z = u(x, y, a, b)$$

is a solution, depending on two parameters a, b, of the partial differential equation of the first order,

$$F(x, y, z, z_x, z_y) = 0,$$

then the envelope of every one-parameter family of solutions chosen from $z = u(x, y, a, b)$ is again a solution.

8. Use this result to obtain other solutions of equation 6(b) by putting $b = ka$ in $u = ax + \frac{1}{a}y + b$ (where k is a constant).

CHAPTER VII

Calculus of Variations

1. Introduction

1. Statement of the Problem.

In the theory of ordinary maxima and minima of a differentiable function $f(x_1, \ldots, x_n)$ of n independent variables, the necessary condition (p. 184) for the occurrence of an extreme value in a certain region of the independent variables is

$$df = 0 \quad \text{or} \quad \operatorname{grad} f = 0 \quad \text{or} \quad f_{x_i} = 0 \quad (i = 1, \ldots, n).$$

These equations express the *stationary character* of the function f at the point in question. The question whether these stationary points are actually maximum or minimum points can only be decided after further investigation. In contrast to the *equations* given above, the corresponding sufficient conditions take the form of *inequalities*.

The calculus of variations is likewise concerned with the problem of extreme values (stationary values). Here, however, we have to deal with a completely new situation. For now the functions which are to have an extreme value no longer depend on one independent variable or a finite number of independent variables within a certain region, but are so-called functions of functions. That is, to determine them we require a knowledge of the behaviour of one or more functions or curves (or surfaces, as the case may be), the so-called " argument functions ".

General attention was first drawn to problems of this type in 1696 by John Bernoulli's statement of the *brachistochrone problem*.

In a vertical xy-plane a point $A(x_0, y_0)$ is to be joined to a point $B(x_1, y_1)$, such that $x_1 > x_0$, $y_1 > y_0$, by a smooth curve $y = u(x)$ in such a way that the time taken by a particle sliding without friction from A to B along the curve under gravity (which is taken as acting in the direction of the positive y-axis) is as short as possible.

The mathematical expression of the problem is based on the physical assumption that in such a curve $y = \phi(x)$ the velocity ds/dt (s being the length of arc of the curve) is proportional to $\sqrt{2g(y - y_0)}$, the square root of the height of fall. The time taken in the fall of the particle is therefore given by

$$T = \int_{x_0}^{x_1} \frac{dt}{ds}\frac{ds}{dx}\,dx = \frac{1}{\sqrt{2g}}\int_{x_0}^{x_1} \frac{\sqrt{(1 + y'^2)}}{\sqrt{(y - y_0)}}\,dx$$

(cf. Vol. I, pp. 299–301). If we drop the unimportant factor $\sqrt{2g}$ and take $y_0 = 0$ (which we can do without loss of generality), we have the following problem:

Among all continuously differentiable functions $y = \phi(x)$, $y \geqq 0$, for which $\phi(x_0) = 0$, $\phi(x_1) = y_1$, to find that for which the integral

$$I\{\phi\} = \int_{x_0}^{x_1} \sqrt{\left(\frac{1 + y'^2}{y}\right)}\,dx$$

has the least possible value.

On p. 505 we shall obtain the result, which was very surprising to Bernoulli's contemporaries, that the curve $y = \phi(x)$ must be a *cycloid*. Here we wish to emphasize that Bernoulli's problem and the elementary problems of maxima and minima are absolutely different. The expression $I\{\phi\}$ depends on the whole behaviour of the function ϕ. It cannot be determined by stating the values of a finite number of independent variables, that is, it cannot be regarded as a function in the ordinary sense. We indicate its character of "function of a function $\phi(x)$" by means of curly brackets.

The following is another problem of a similar nature:

Two points $A(x_0, y_0)$ and $B(x_1, y_1)$, where $x_1 > x_0$, $y_0 > 0$, $y_1 > 0$, are to be joined by a curve $y = u(x)$ lying above the x-axis, in such a way that the *area of the surface of revolution* formed when the curve is rotated about the x-axis is *as small as possible.*

Using the expression given on p. 274 for the area of a surface of revolution and dropping the unimportant factor 2π, we have the following mathematical statement of the problem:

Among all continuously differentiable functions $y = \phi(x)$ for which $\phi(x_0) = y_0$, $\phi(x_1) = y_1$, $\phi(x) > 0$, to find that for which the integral

$$I\{\phi\} = \int_{x_0}^{x_1} y\sqrt{(1 + y'^2)}\,dx \qquad (y = \phi(x))$$

has the least possible value. It will be found that the solution is a *catenary*.

The elementary geometrical problem of finding the shortest curve joining two points A and B in the plane belongs in theory to the same category. Analytically, in fact, the problem is that of finding two functions $x(t)$, $y(t)$ of a parameter t in an interval $t_0 \leqq t \leqq t_1$, for which the values $x(t_0) = x_0$, $x(t_1) = x_1$ and $y(t_0) = y_0$, $y(t_1) = y_1$ are prescribed, and for which the integral

$$\int_{t_0}^{t_1} \sqrt{(\dot{x}^2 + \dot{y}^2)}\,dt \qquad \left(\dot{x} = \frac{dx}{dt}, \dot{y} = \frac{dy}{dt}\right)$$

has the least possible value. The solution is of course a straight line.

On the other hand, the corresponding problem of finding the *geodesics on a given surface* $G(x, y, z) = 0$, that is, of joining two points on the surface with co-ordinates (x_0, y_0, z_0) and (x_1, y_1, z_1) by the shortest possible line lying in the surface, unlike the problem of the shortest distance between two points in a plane, is not a trivial one. In analytical language this problem is as follows:

Among all triads of functions $x(t)$, $y(t)$, $z(t)$ of the parameter t which make the equation

$$G(x, y, z) = 0,$$

an identity in t, and for which $x(t_0) = x_0$, $y(t_0) = y_0$, $z(t_0) = z_0$; $x(t_1) = x_1$, $y(t_1) = y_1$, $z(t_1) = z_1$, to find that for which the integral

$$\int_{t_0}^{t_1} \sqrt{(\dot{x}^2 + \dot{y}^2 + \dot{z}^2)}\,dt$$

has the least possible value.

The *isoperimetric problem* of finding a closed curve of given length enclosing the largest possible area, already discussed on p. 214, also belongs to the same category. We have proved above that the solution is a circle.*

* The proof given there applied only to convex curves; the following remark, however, enables us to extend the result immediately to any curve:

The general statement of the simplest type of problems of the kind dealt with here is as follows:

We are given a function $F(x, \phi, \phi')$ of three arguments, which in the region of the arguments considered is continuous and has continuous derivatives of the first and second orders. If in this function F we replace ϕ by a function $y = \phi(x)$ and ϕ' by the derivative $y' = \phi'(x)$, F becomes a function of x, and an integral of the form

$$I\{\phi\} = \int_{x_0}^{x_1} F(x, y, y')\,dx$$

becomes a definite number depending on the behaviour of the function $y = \phi(x)$, i.e. it is a "function of the function $\phi(x)$". The fundamental problem of the calculus of variations is now as follows:

Among all the functions which are defined and continuous and possess continuous first and second derivatives in the interval $x_0 \leqq x \leqq x_1$, and for which the boundary values $y_0 = \phi(x_0)$ and $y_1 = \phi(x_1)$ are prescribed, to find that for which the integral $I\{\phi\}$ has the least possible value (or the greatest possible value).

In discussing this problem the absolutely essential point is the nature of the "conditions of admission" imposed on the functions $\phi(x)$. The problem merely requires that when $\phi(x)$ is substituted F shall be a sectionally continuous function of x, and this is assured if the derivative $\phi'(x)$ is sectionally continuous. But we have made the conditions of admission more stringent by requiring that the first derivatives, and even the second derivatives, of the functions $\phi(x)$ shall be continuous. The field in which the maximum or minimum is to be sought is of course thereby restricted. It will, however, be found that this restriction does not, in fact, affect the solution, i.e. that the function which is most favourable when the wider field is available will always

We consider the "convex envelope" K of a curve C (cf. Ex. 2, p. 100), i.e. the convex curve of least area enclosing the interior of C. This curve K consists of convex arcs of C and rectilinear portions of tangents to C, which touch C at two points and bridge over concave parts of C by straight lines. It is evident that the area of K exceeds that of C, provided C is not convex, and, on the other hand, that the perimeter of K is less than that of C. If we now make K expand uniformly, so that it always retains the same shape, until the resulting curve K' has the prescribed perimeter, K' will be a curve of the same perimeter as C, but enclosing a greater area. Hence in the isoperimetric problem we may from the outset confine ourselves to convex curves, in order to obtain the maximum area.

be found in the more restricted field of functions with continuous first and second derivatives.

Problems of this type occur very frequently in geometry and physics. Here we mention only one example. The fundamental principle of geometrical optics can be formulated as a variation problem of this type. If we consider a ray of light in the xy-plane and assume that the velocity of light is a given function $v(x, y, y')$ of the point (x, y) and of the direction y' ($y = \phi(x)$ being the equation of the light-path and $y' = \phi'(x)$ the corresponding derivative), then *Fermat's principle of least time* is as follows:

The path of a ray of light between two given points A, B is such that the time taken by the light in traversing it is less than the time which light would take to traverse any other path from A to B.

In other words, if t is the time and s the length of arc of any curve $y = \phi(x)$ joining the points A and B, the time which light would take to traverse the portion of curve between A and B is given by the integral

$$I\{\phi\} = \int_{x_0}^{x_1} \frac{dt}{ds}\frac{ds}{dx}\,dx = \int_{x_0}^{x_1} \frac{\sqrt{(1+y'^2)}}{v(x, y, y')}\,dx.$$

To determine the actual path of the light we accordingly require to solve the problem of finding a function $y = \phi(x)$ for which this integral has the least possible value.

We see that the optical problem in this form is actually equivalent to the general problem stated above if we relate the two functions F and v to one another by putting $F = \dfrac{\sqrt{(1+y'^2)}}{v}$. In most optical cases the velocity of light v is independent of the direction and is merely a function of position, $v(x, y)$.

2. Necessary Conditions for Extreme Values.

Our object is to find necessary conditions that a function $u = \phi(x)$ may give a maximum or minimum, or, to use a general term, an extreme value, of the above integral $I\{\phi\}$. Here we proceed by a method quite analogous to that used in the elementary problem of finding the extreme values of a function of one or more variables. We assume that $y = \phi = u(x)$ is the solution. Then we have to express the fact that (for a minimum)

I must *increase* when u is replaced by another admissible function ϕ. Here, moreover, as we are merely concerned with obtaining necessary conditions, we may confine ourselves to the consideration of functions ϕ which approximate to u, i.e. functions for which the absolute value of the difference $\phi - u$ remains between prescribed bounds.

We think of the function u as a member of a one-parameter family with parameter ϵ, constructed as follows. We take any function $\eta(x)$ which vanishes on the boundary of the interval, i.e. for which $\eta(x_0) = 0$, $\eta(x_1) = 0$, and which has continuous first and second derivatives everywhere in the closed interval. We then form the family of functions

$$\phi(x, \epsilon) = u(x) + \epsilon\eta(x).$$

The expression $\epsilon\eta(x) = \delta u$ is called the *variation of the function* u. (Since $\eta(x) = \partial\phi/\partial\epsilon$, the symbol δ denotes the differential obtained when ϵ is regarded as the independent variable and x as a parameter.) Then, if we regard the function u as well as the function η as fixed,

$$I\{u + \epsilon\eta\} = \Phi(\epsilon) = \int_{x_0}^{x_1} F(x, u + \epsilon\eta, u' + \epsilon\eta')\,dx$$

is a function of ϵ; and the postulate that u shall give a minimum of $I\{\phi\}$ implies that the function above shall possess a minimum for $\epsilon = 0$, so that as necessary conditions we have the equation

$$\Phi'(0) = 0$$

and further the inequality

$$\Phi''(0) \geqq 0.$$

In the same way, if we were seeking a maximum, we should have the same equation $\Phi'(0) = 0$ and the inequality $\Phi''(0) \leqq 0$ as necessary conditions. The condition $\Phi'(0) = 0$ must be satisfied for every function η which satisfies the above conditions but is otherwise arbitrary.

Putting aside the question of discrimination between maxima and minima, we say that if a function u satisfies the equation $\Phi'(0) = 0$, for all functions η, the integral I is *stationary* for $\phi = u$. If, as before, we use the symbol δ to denote differentiation with respect to ϵ, we may also say that the equation

$$\delta I = \epsilon\Phi'(0) = 0,$$

when satisfied by a function $\phi = u$ and an arbitrary η, expresses the stationary character of I. The expression

$$\epsilon\Phi'(0) = \epsilon\left\{\frac{d}{d\epsilon}\int_{x_0}^{x_1} F(x, u + \epsilon\eta, u' + \epsilon\eta')\,dx\bigg|_{\epsilon=0}\right\}$$

is called the *variation*, or more accurately the *first variation*,* of the integral. *Stationary character of an integral and vanishing of the first variation, therefore, mean exactly the same thing.*

Stationary character is *necessary* for the occurrence of maxima or minima, but, as in the case of ordinary maxima or minima, it is not a *sufficient* condition for the occurrence of either of these possibilities. Here we cannot go into the problem of sufficient conditions in more detail, and in what follows we confine ourselves to the problem of stationary character.

Our main object is to transform the condition $\Phi'(0) = 0$ for the stationary character of the integral in such a way that it becomes a condition for u only and no longer contains the arbitrary function η.

Examples

1. In connexion with the brachistochrone problem (see pp. 491, 492), calculate the time of fall when the points A and B are joined by a straight line.

2. Let the velocity of a particle with polar co-ordinates (r, θ, φ) moving in three-dimensional space be $v = 1/f(r)$. What time does the particle take to describe the portion of a curve given by a parameter σ (the co-ordinates of a point on the curve being $r(\sigma)$, $\theta(\sigma)$, $\varphi(\sigma)$) between the points A and B?

2. Euler's Differential Equation in the Simplest Case

1. Deduction of Euler's Differential Equation

The fundamental criterion of the calculus of variations is as follows:

The necessary and sufficient condition that the integral

$$I\{\phi\} = \int_{x_0}^{x_1} F(x, \phi, \phi')\,dx$$

* From this comes the use of the term *calculus of variations*, which is meant to indicate that in this subject we are concerned with the behaviour of functions of a function when this independent function or "argument function" is made to vary by altering a parameter ϵ.

shall be stationary when $\phi = \mathrm{u}$ *is that* $\phi = \mathrm{u}$ *shall be an admissible function satisfying Euler's differential equation**

$$L[u] = F_u - \frac{d}{dx} F_{u'} = 0,$$

or, in full,

$$F_{u'u'} u'' + F_{uu'} u' + F_{xu'} - F_u = 0.$$

To prove this we note that we can differentiate the expression

$$\Phi(\epsilon) = \int_{x_0}^{x_1} F(x, u + \epsilon\eta, u' + \epsilon\eta')\, dx$$

with respect to ϵ under the integral sign (cf. Chap. IV, § 1, p. 218), provided that the differentiation gives rise to a continuous function, or at least a sectionally continuous function of x, under the integral sign. In this case, on putting $u + \epsilon\eta = y$ and differentiating, we obtain under the integral sign the expression $\eta F_y + \eta' F_{y'}$, which, owing to the assumptions made about f, u, and η, satisfies the conditions just stated. Hence we immediately obtain

$$\Phi'(0) = \int_{x_0}^{x_1} (\eta F_u + \eta' F_{u'})\, dx = 0 \qquad (F(x, u, u')).$$

For subsequent purposes (see the next page), we note that in the formation of this equation we have used nothing beyond the continuity of the functions u and η and the sectional continuity of their first derivatives. In this equation the arbitrary function appears under the integral sign in a twofold form, namely, as η and η'. We can, however, immediately get rid of η' by integration by parts; we have

$$\int_{x_0}^{x_1} \eta' F_{u'}\, dx = \eta F_{u'} \Big|_{x_0}^{x_1} - \int_{x_0}^{x_1} \eta \left(\frac{d}{dx} F_{u'}\right) dx = -\int_{x_0}^{x_1} \eta \left(\frac{d}{dx} F_{u'}\right) dx,$$

for by hypothesis $\eta(x_0)$ and $\eta(x_1)$ vanish. In this integration by parts we have to assume that the expression $\frac{d}{dx} F_{u'}$ can be formed, but this assumption certainly holds good, for we began by assuming continuity of the second derivatives. Hence, if we write

$$\mathrm{L}[u] = F_u - \frac{d}{dx} F_{u'}$$

* The terms *principal equation, characteristic equation* are also used.

for brevity, we have the equation

$$\int_{x_0}^{x_1} \eta L[u]\, dx = 0.$$

Now this equation must be satisfied for every function η which satisfies our conditions but is otherwise arbitrary. We thus conclude that

$$L[u] = 0,$$

in virtue of the following

Lemma I.—If a function $C(x)$ which is continuous in the interval under consideration satisfies the relation

$$\int_{x_0}^{x_1} \eta(x)C(x)\, dx = 0,$$

where $\eta(x)$ is any function such that $\eta(x_0) = \eta(x_1) = 0$ and $\eta''(x)$ is continuous, then $C(x) = 0$ for every value of x in the interval. The proof of this lemma will be postponed to the next sub-section (p. 501).

We could, however, obtain our condition in a different way,* by getting rid of the term in η in the equation

$$\int_{x_0}^{x_1} (\eta F_u + \eta' F_{u'})\, dx = 0$$

by integration by parts. For if we write $F_{u'} = A$, $F_u = b = B'$ for brevity and remember the boundary condition for η, on integrating by parts we obtain

$$\int_{x_0}^{x_1} \eta F_u\, dx = \int_{x_0}^{x_1} \eta B'\, dx = -\int_{x_0}^{x_1} \eta' B\, dx.$$

If we put $\zeta = \eta'$, we have the condition

$$\int_{x_0}^{x_1} \zeta(A - B)\, dx = 0.$$

In this method we need not make any further assumptions about the second derivatives of η and u. On the contrary, it is sufficient to assume that φ (or u) and η are continuous and have sectionally continuous first derivatives. Now our equation must hold, not, it is true, for any arbitrary (sectionally continuous) function ζ, but only for those functions ζ which are derivatives of a function $\eta(x)$ satisfying our conditions. If,

* The first method is due to Lagrange, the second to P. Du Bois Reymond.

however, $\zeta(x)$ is any given sectionally continuous function satisfying the relation

$$\int_{x_0}^{x_1} \zeta(x)\,dx = 0$$

and otherwise arbitrary, we can put

$$\eta = \int_{x_0}^{x} \zeta(t)\,dt;$$

we have then constructed an admissible η, for $\eta' = \zeta$ and $\eta(x_0) = \eta(x_1) = 0$. We thus obtain the following result:

A necessary condition that the integral should be stationary is

$$\int_{x_0}^{x_1} \zeta(A - B)\,dx = 0,$$

where ζ is an arbitrary sectionally continuous function merely satisfying the condition

$$\int_{x_0}^{x_1} \zeta dx = 0.$$

We now require the help of

Lemma II.—If a sectionally continuous function $S(x)$ satisfies the condition

$$\int_{x_0}^{x_1} \zeta S\,dx = 0,$$

for all functions $\zeta(x)$ which are sectionally continuous in the interval and for which

$$\int_{x_0}^{x_1} \zeta\,dx = 0,$$

then $S(x)$ is a constant c.

This lemma will also be proved in the next sub-section (p. 501). If meanwhile we assume its truth, it follows, if we substitute the above expressions for A and B, that

$$\int_{x_0}^{x} F_u\,dx + c = F_{u'}.$$

The left-hand side regarded as an indefinite integral may be differentiated with respect to x and has F_u as its derivative; the same is therefore true of the right-hand side. Hence the expression $\frac{d}{dx} F_{u'}$ for the supposed solution u exists, and the equation

$$F_u = \frac{d}{dx} F_{u'}$$

holds.

Thus Euler's equation still remains the necessary condition for an extreme value, or the condition that the integral should be stationary, when the class of admissible functions $\varphi(x)$ is extended from the outset by requiring only sectional continuity of the first derivative of $\varphi(x)$.

Euler's equation is an *ordinary differential equation of the second order*. Its solutions are called the *extremals* of the minimum problem. To solve the minimum problem we have, among all the extremals, to find one which satisfies the prescribed boundary conditions. If "Legendre's condition"

$$F_{u'u'} \neq 0$$

is satisfied for $\phi = u(x)$, the differential equation can be brought into the "regular" form $u'' = f(x, u, u')$, where the right-hand side is a known expression involving x, u, u'.

2. Proofs of the Lemmas.

We have now to prove the two lemmas used above.

To prove Lemma I, we assume that at some point, say $x = \xi$, $C(x)$ is not zero, say positive. Then in virtue of the continuity of $C(x)$ we can certainly mark off a sub-interval

$$\xi - a \leqq x \leqq \xi + a$$

within the complete interval in such a way that $C(x)$ remains positive everywhere in the sub-interval. We now define η as given in this sub-interval by

$$\eta(x) = (x - \xi + a)^4(x - \xi - a)^4 = \{(x - \xi)^2 - a^2\}^4$$

and elsewhere as zero. This function η certainly fulfils all the prescribed conditions; $\eta(x)C(x)$ is positive inside the sub-interval, and zero outside it. The integral $\int_{x_0}^{x_1} \eta C\,dx$ therefore cannot be zero.* Since this contradicts our hypothesis, $C(\xi)$ cannot be positive. For the same reasons, $C(\xi)$ cannot be negative. Hence $C(\xi)$ must vanish for all values of ξ within the interval, as was stated in the lemma.

To prove Lemma II, we note that our assumption about $\zeta(x)$ immediately leads to the relations

$$\int_{x_0}^{x_1} \zeta(x)\,dx = 0 \text{ and } \int_{x_0}^{x_1} \zeta(x)\{S(x) - c\}\,dx = 0,$$

* The integral of a continuous non-negative function is positive, except when the integrand vanishes everywhere; this follows immediately from the definition of integral.

where c is an arbitrary constant. We now choose c in such a way that $S(x) - c$ is an admissible function $\zeta(x)$, that is, we determine c by the equation

$$0 = \int_{x_0}^{x_1} \zeta\, dx = \int_{x_0}^{x_1} \{S(x) - c\}\, dx = \int_{x_0}^{x_1} S(x)\, dx - c(x_1 - x_0).$$

Substituting this value of c in the above equation and taking $\zeta = S(x) - c$, we at once have

$$\int_{x_0}^{x_1} \{S(x) - c\}^2\, dx = 0.$$

Since by hypothesis the integrand is continuous, or at least sectionally continuous, it follows that

$$S(x) - c = 0$$

is an identity in x, as was stated in the lemma.

3. Solution of Euler's Differential Equation. Examples.

To find the solutions u of the minimum problem we have (p. 497) to find a particular solution of Euler's differential equation for the interval $x_0 \leqq x \leqq x_1$ which assumes the prescribed boundary values y_0 and y_1 at the end-points. As the complete integral of Euler's differential equation of the second order contains two constants of integration, it is generally possible to make these two constants fit the boundary conditions, the latter giving two equations which the constants of integration must satisfy.

In general it is not possible to solve Euler's differential equation explicitly in terms of elementary functions or quadratures. In the general case we have to be content to establish the fact that the variation problem does reduce to a problem in differential equations. On the other hand, in important special cases and, in fact, in most of the classical examples, the equation can be solved by means of quadratures.

The first case is that in which F does not contain the derivative $y' = \phi'$ explicitly: $F = F(\phi, x)$. Here Euler's differential equation is simply $F_u(u, x) = 0$; that is, it is no longer a differential equation at all but forms an implicit definition of the solution $y = u(x)$. Here of course there is no question of integration constants or the possibility of satisfying boundary conditions.

The second important special case is that in which F does not contain the function $y = \phi(x)$ explicitly: $F = F(y', x)$. Here

Euler's differential equation is $\frac{d}{dx}(F_{u'}) = 0$, which at once gives

$$F_{u'} = c,$$

where c is an arbitrary constant of integration. We may use this equation to express u' as a function, $f(x, c)$, of x and c, and we then have the equation

$$u' = f(x, c),$$

from which by a simple integration (quadrature) we obtain

$$u = \int_0^x f(\xi, c)\, d\xi + a,$$

that is, u is expressed as a function of x and c, together with an additional arbitrary constant of integration a. In this case, therefore, Euler's differential equation can be completely solved by quadrature.

The third case, which is the most important in examples and applications, is that in which F does not contain the independent variable x explicitly: $F = F(y, y')$. In this case we have the following important theorem:

If the independent variable x *does not occur explicitly in the variation problem, then*

$$E = F(u, u') - u'F_{u'}(u, u') = c$$

is an integral of Euler's differential equation. That is, if we substitute in this expression a solution u(x) *of Euler's differential equation for* F, *the expression becomes a constant independent of* x.

The truth of this statement follows at once if we form the derivative dE/dx. We have

$$\frac{dE}{dx} = F_u u' + F_{u'} u'' - u'' F_{u'} - u'^2 F_{uu'} - u'u'' F_{u'u'},$$

or

$$\frac{dE}{dx} = u'L[u] = 0;$$

hence for every solution u of Euler's differential equation we have $E = c$, where c is a constant.

If we think of u' as calculated from the equation $E = c$, say $u' = f(u, c)$, a simple quadrature applied to the equation

$$\frac{dx}{du} = \frac{1}{f(u, c)}$$

gives $x = g(u, c) + a$ (where a is another constant of integration), i.e. x is expressed as a function of u, c, and a, and by solving for u we obtain the function $u(x, c, a)$. Hence the general solution of Euler's differential equation, depending on two arbitrary constants of integration, is obtained by a quadrature.

We shall now use these methods to discuss a number of examples.

1. *General Note.*—There is a general class of examples in which F is of the form $F = g(y)\sqrt{(1 + y'^2)}$, where $g(y)$ is a function depending explicitly on y only. For the extremals $y = u$ our last rule at once gives

$$g(u)\sqrt{(1 + u'^2)} - g(u)u'^2/\sqrt{(1 + u'^2)} = c$$

or

$$\frac{g(u)}{\sqrt{(1 + u'^2)}} = c,$$

$$u'^2 = \frac{\{g(u)\}^2}{c^2} - 1,$$

$$\frac{dx}{du} = \frac{1}{\sqrt{\{g(u)\}^2/c^2 - 1}},$$

and on integrating we have the equation

$$x - b = \int \frac{du}{\sqrt{\{g(u)\}^2/c^2 - 1}},$$

where b is another constant of integration. By evaluating the integral on the right and imagining the equation solved for u, we obtain u as a function of x and of the two constants of integration c and b.

2. *The Surface of Revolution of Least Area.*—In this case $g = y$. The integral given above becomes

$$x - b = \int \frac{du}{\sqrt{u^2/c^2 - 1}} = c \operatorname{ar\,cosh} \frac{u}{c};$$

hence the result is

$$y = u = c \cosh \frac{x - b}{c}.$$

That is, the solution of the problem of finding a curve which on rotation gives a surface of revolution with stationary area is a *catenary*.

A necessary condition for the occurrence of such a stationary curve is that the two given points A and B can be joined by a catenary for which $y > 0$. The question whether the catenary really represents a minimum cannot be discussed here.

3. *The Brachistochrone.*—Another example is obtained by taking $g = 1/\sqrt{y}$. This is the problem of the brachistochrone. By means of the substitutions $1/c^2 = k$, $u = k\tau$, $\tau = \sin^2\theta/2$ the integral

$$\int \frac{du}{\sqrt{1/(uc^2) - 1}}$$

is immediately transformed into

$$x - b = k\int\sqrt{\left(\frac{\tau}{1-\tau}\right)}\,d\tau = \tfrac{1}{2}k\int(1 - \cos\theta)\,d\theta,$$

whence

$$x - b = \tfrac{1}{2}k(\theta - \sin\theta),$$

$$y = u = \tfrac{1}{2}k(1 - \cos\theta).$$

The brachistochrone is accordingly (cf. Vol. I, p. 261) a *common cycloid* with its cusps on the x-axis.

EXAMPLES

Find the extremals for the following integrands:

1. $F = \sqrt{y(1 + y'^2)}$. 2. $F = \sqrt{1 + y'^2}/y$. 3. $F = y\sqrt{1 - y'^2}$.

4. Find the extremals for the integrand $F = x^n y'^2$, and prove that if $n \geqq 1$ two points lying on opposite sides of the y-axis cannot be joined by an extremal.

5. Find the extremals for the integrand $y^n y'^m$, where n and m are even integers.

6. Find the extremals for the integrand $F = ay'^2 + 2byy' + cy^2$, where a, b, c are given continuously differentiable functions of x. Prove that Euler's differential equation is a linear differential equation of the second order. Why is it that when b is constant this constant does not enter into the differential equation at all?

7. Show that the extremals for the integrand $F = e^x\sqrt{1 + y'^2}$ are given by the equations $\sin(y - b) = e^{-(x-a)}$ and $y = b$, where a, b are constants. Discuss the form of these curves, and investigate how the two points A and B must be situated if they can be joined by an extremal arc of the form $y = f(x)$.

8. For the case where F does not contain the derivative y', deduce Euler's condition $F_y = 0$ by an elementary method.

9. Find a function giving the absolute minimum of

$$I(y) = \int_0^1 y'^2\,dx$$

with the boundary conditions

$$(a) \quad y(0) = y(1) = 0;$$

$$(b) \quad y(0) = 0, \quad y(1) = 1.$$

4. Identical Vanishing of Euler's Expression.

Euler's differential equation for $F(x, y, y')$ may degenerate into an identity which tells us nothing, i.e. into a relation which is satisfied by every admissible function $y = \phi(x)$. In other words, the corresponding integral may be stationary for any admissible function $y = \phi(x)$. If this degenerate case is to occur, Euler's expression $F_y - F_{xy'} - F_{yy'}y' - F_{y'y'}y''$ must vanish at every point x of the interval, no matter what function $y = \phi(x)$ is substituted in it. We can, however, always find a curve for which $y = \phi$, $y' = \phi'$, and $y'' = \phi''$ have arbitrary prescribed values at a definite point. Euler's expression must therefore vanish for every quartet of numbers x, y, y', y''. We conclude that the coefficient of y'', i.e. $F_{y'y'}$, must vanish identically. F must therefore be a linear function of y', say $F = ay' + b$, where a and b are functions of x and y only. If we substitute this in the remaining part of the differential equation,

$$F_{yy'}y' + F_{xy'} - F_y = 0,$$

it follows at once that

$$a_y y' + a_x - a_y y' - b_y,$$

or

$$a_x - b_y,$$

must vanish identically in x and y. In other words, Euler's expression vanishes identically if, and only if, the integral is of the form

$$I = \int \{a(x, y)y' + b(x, y)\}\, dx = \int a\, dy + b\, dx,$$

where a and b satisfy the condition of integrability which we have already met with in Chap. V, § 1 (p. 353), that is, where $a\,dy + b\,dx$ is a perfect differential.

3. Generalizations

1. Integrals with More than one Argument Function.

The problem of finding the extreme values (stationary values) of an integral can be extended to the case where this integral depends not on a single argument function but on a number of argument functions $\phi_1(x), \phi_2(x), \ldots, \phi_n(x)$. The typical problem of this type may be formulated as follows:

Let $F(x, \phi_1, \ldots, \phi_n, \phi_1', \ldots, \phi_n')$ be a function of the $(2n+1)$ arguments $x, \phi_1, \ldots, \phi_n'$, which is continuous and has continuous derivatives up to and including the second order in the interval under consideration. If we replace $y_i = \phi_i$ by a function of x with continuous first and second derivatives, and ϕ_i' by its derivative, F becomes a function of the single variable x, and the integral

$$I\{\phi_1, \ldots, \phi_n\} = \int_{x_0}^{x_1} F(x, \phi_1, \ldots, \phi_n, \phi_1', \ldots, \phi_n')\,dx$$

over a given interval $x_0 \leqq x \leqq x_1$ has a definite value determined by the choice of these functions.

In the comparison we regard all functions $\phi_i(x)$ as admissible which satisfy the above continuity conditions and for which the boundary values $\phi_i(x_0)$ and $\phi_i(x_1)$ have prescribed fixed values. In other words, we consider the curves $y_i = \phi_i(x)$ joining two given points A and B in $(n+1)$-dimensional space in which the co-ordinates are $y_1, y_2, \ldots, y_n, x$. The variation problem now requires us to find, among all these systems of functions $\phi_i(x)$, one $(y_i = \phi_i(x) = u_i(x))$ for which the above integral $I\{\phi_1, \ldots, \phi_n\}$ has an extreme value (a maximum or a minimum).

Here again we cannot discuss the actual nature of the extreme value, but shall confine ourselves to inquiring for what systems of argument functions $\phi_i(x) = u_i(x)$ the integral is stationary.

We define the concept of stationary value in exactly the same way as we did in § 1 (p. 496). We include the system of functions $u_i(x)$ in a one-parameter family of functions depending on the parameter ϵ, in the following way. Let $\eta_1(x), \ldots, \eta_n(x)$ be n arbitrarily chosen functions which vanish for $x = x_0$ and $x = x_1$, are continuous in the interval, and possess continuous first and second derivatives there. Then we consider the family of functions $y_i = \phi_i(x) = u_i(x) + \epsilon\eta_i(x)$.

The term $\epsilon\eta_i(x) = \delta u_i$ is called the *variation* of the function u_i. If we substitute the expressions for ϕ_i in $I\{\phi_1, \ldots, \phi_n\}$, this integral is transformed into

$$\Phi(\epsilon) = \int_{x_0}^{x_1} F(x, u_1 + \epsilon\eta_1, \ldots, u_n + \epsilon\eta_n, u_1' + \epsilon\eta_1', \ldots, u_n' + \epsilon\eta_n')\,dx,$$

which is a function of the parameter ϵ. A necessary condition that there may be an extreme value for $\phi_i = u_i$, i.e. for $\epsilon = 0$, is

$$\Phi'(0) = 0.$$

Just as in § 1, p. 496, we say that if the equation $\Phi'(0) = 0$ holds, or, as we may also say, if the equation

$$\delta I = \epsilon\Phi'(0) = 0$$

holds, no matter how the functions η_i are chosen subject to the conditions stated above, the integral I has a stationary value for $\phi_i = u_i$. In other words, stationary character of the integral for a fixed system of functions $u_i(x)$ and vanishing of the first variation δI mean the same thing.

We have still the problem of setting up conditions for the stationary character of the integral which no longer contain the arbitrary variations η_i. To do this we do not require any new ideas, but proceed as follows. If we take $\eta_2, \eta_3, \ldots, \eta_n$ as identically zero, i.e. if we do not let the functions $u_2, \ldots, u_n$ vary, and thus consider the first function $\phi_1(x)$ as alone variable, the condition $\Phi'(0) = 0$, by § 2, p. 498, is equivalent to Euler's differential equation

$$F_{u_1} - \frac{d}{dx} F_{u_1'} = 0.$$

As we can pick out any one of the functions $u_i(x)$ in the same way, we obtain the following result:

A necessary and sufficient condition that the integral $I\{u_1, u_2, \ldots, u_n\}$ *may be stationary is that the* n *functions* $u_i(x)$ *shall satisfy the system of Euler's equations*

$$F_{u_i} - \frac{d}{dx} F_{u_i'} = 0 \qquad (i = 1, 2, \ldots, n).$$

This is a system of differential equations of the second order, n in number, for the n functions $u_i(x)$. All solutions of this system of differential equations are said to be *extremals* of the variation problem. Thus the problem of finding stationary values of the integral reduces to the problem of solving these differential equations and adapting the general solution to the given boundary conditions.*

2. Examples.

The possibility of giving a general solution of the system of Euler's differential equations is even more remote than in the case in § 2. It is only in very special cases that we can find all the extremals explicitly. Here the following theorem, analogous to the particular case on p. 503, is often useful:

If the function F *does not contain the independent variable* x *explicitly,* $F = F(\phi_1, \ldots, \phi_n, \phi_1', \ldots, \phi_n')$, *then the expression*

$$E = F(u_1, \ldots, u_n, u_1', \ldots, u_n') - \sum_{i=1}^{n} u_i' F_{u_i'}$$

is an integral of Euler's system of differential equations. That is, if we consider a system of solutions $u_i(x)$ of Euler's system of differential equations, we have for this solution

$$E = F - \Sigma u_i' F_{u_i'} = \text{const.} = c,$$

where, of course, the value of this constant depends upon the system of solutions which is substituted.

The proof follows the same lines as in § 2 (p. 503); we differentiate the left-hand side of our expression with respect to x and, using Euler's differential equations, verify that the result is zero.

A trivial example is the problem of finding the shortest distance between

* Using Lemma II (§ 2, p. 500), we can prove that these differential equations must hold, under the general assumption that the admissible functions need only have sectionally continuous first derivatives. For the beginner who wishes to concentrate on the essential mechanism of the subject, however, it is more convenient to include continuity of the second derivatives in the conditions of admissibility of the functions $\phi_i(x)$. We can then work out the expressions $\frac{d}{dx} F_{u_i'}$ and write them in the more explicit form

$$\sum_{k=1}^{n} F_{u_k' u_i'} u_k'' + \sum_{k=1}^{n} F_{u_k u_i'} u_k' + F_{x u_i'}.$$

two points in three-dimensional space. Here we have to determine two functions $y = y(x)$, $z = z(x)$ such that the integral

$$\int_{x_0}^{x_1} \sqrt{(1 + y'^2 + z'^2)}\, dx$$

has the least possible value, the values of $y(x)$ and $z(x)$ at the end-points of the interval being prescribed. Euler's differential equations give

$$\frac{d}{dx}\frac{y'}{\sqrt{(1 + y'^2 + z'^2)}} = \frac{d}{dx}\frac{z'}{\sqrt{(1 + y'^2 + z'^2)}} = 0,$$

whence it follows at once that the derivatives $y'(x)$ and $z'(x)$ are constant; hence the extremals must be straight lines.

Somewhat less trivial is the problem of the *brachistochrone in three dimensions.* (Gravity is again taken as acting along the positive y-axis.) Here we have to determine $y = y(x)$, $z = z(x)$ in such a way that the integral

$$\int_{x_0}^{x_1} \sqrt{\left(\frac{1 + y'^2 + z'^2}{y}\right)} dx = \int_{x_0}^{x_1} F(y, y', z')\, dx$$

is stationary. Euler's differential equations give

$$\frac{z'}{\sqrt{y}}\frac{1}{\sqrt{(1 + y'^2 + z'^2)}} = a,$$

$$F - y'F_{y'} - z'F_{z'} = \frac{1}{\sqrt{y}}\frac{1}{\sqrt{(1 + y'^2 + z'^2)}} = b,$$

where a and b are constants. By division it follows that $z' = a/b = k$ is likewise constant. The curve for which the integral is stationary must therefore lie in a plane $z = kx + h$. From the further equation

$$\frac{1}{\sqrt{y}}\frac{1}{\sqrt{(1 + k^2 + y'^2)}} = b$$

there follows the fact, obvious from § 2 (p. 505), that this curve must again be a cycloid.

EXAMPLE

Write down the differential equations for the path of a ray of light in three dimensions in the case where (polar co-ordinates r, θ, φ being used) the velocity of light is a function of r (cf. § 1, Ex. 2, p. 497). Show that the rays are plane curves.

3. Hamilton's Principle. Lagrange's Equations.

Euler's system of differential equations has a very important bearing on many branches of applied mathematics, especially dynamics. For the motion of a mechanical system consisting of a finite number of heavy particles can be expressed by the condition that a certain expression, the so-called Hamilton's integral, is stationary. Here we shall briefly explain this connexion.

A mechanical system has n degrees of freedom if its position is determined by n independent co-ordinates $q_1, q_2, \ldots, q_n$. If, for example, the system consists of a single particle, $n = 3$, since for q_1, q_2, q_3 we can take the three rectangular co-ordinates or the three polar co-ordinates. Again, if the system consists of two particles which are held at unit distance apart by a rigid connexion—assumed to have no mass—then $n = 5$, since for the co-ordinates q_i we can take the three rectangular co-ordinates of one particle and two other co-ordinates determining the direction of the line joining the two particles.

A dynamical system can be described with sufficient generality by means of two functions, the *kinetic energy* and the *potential energy*. If we think of the system as moving in any way, the co-ordinates q_i will be functions $q_i(t)$ of the time t, the "components of velocity" being $\dot{q}_i = dq_i/dt$. Then associated with the dynamical system there is a function which we call the kinetic energy and which is of the form

$$T(q_1, \ldots, q_n, \dot{q}_1, \ldots, \dot{q}_n) = \sum_{i,k=1}^{n} a_{ik}\dot{q}_i\dot{q}_k \qquad (a_{ik} = a_{ki}).$$

The kinetic energy, therefore, is a homogeneous quadratic expression in the components of velocity, the coefficients a_{ik} being taken as known functions, *not* depending explicitly on the time, of the co-ordinates $q_1, \ldots, q_n$ themselves.*

In addition to the kinetic energy, the dynamical system is supposed to be characterized by another function, the potential energy $U(q_1, \ldots, q_n)$, which depends on the co-ordinates of position q_i only and not on the velocities or the time.†

Now Hamilton's principle is as follows: the actual motion of

* We obtain this expression for the kinetic energy T by thinking of the individual rectangular co-ordinates of the particles of the system as expressed as functions of the co-ordinates $q_1, \ldots, q_n$. Then the rectangular velocity components of the individual particles can be expressed as linear homogeneous functions of the $\dot{q}_i$'s, and finally the elementary expression for the kinetic energy is formed, namely, half the sum of the products of the individual masses and the squares of the corresponding velocities.

† As is shown in dynamical textbooks, this potential energy determines the external forces acting on the system. In bringing the system from one position into another mechanical work is done; this is equal to the difference between the corresponding values of U and does not depend on the path by which the transference from one position to another takes place.

a dynamical system in the interval of time $t_0 \leqq t \leqq t_1$ from a given initial position to a given final position is such that for this motion the integral

$$H\{q_1, \ldots, q_n\} = \int_{t_0}^{t_1} (T - U)\, dt$$

is stationary, if in the comparison we include all continuous functions $q_i(t)$ which have continuous derivatives up to and including the second order and which for $t = t_0$ and $t = t_1$ have the prescribed boundary values.

This principle of Hamilton's is a fundamental principle of dynamics. The advantage of it is that it forms a brief summary of the laws of dynamics. When applied to Hamilton's principle, the general theory of this chapter gives *Lagrange's equations,*

$$\frac{d}{dt}\frac{\partial T}{\partial \dot{q}_i} - \frac{\partial T}{\partial q_i} = -\frac{\partial U}{\partial q_i}, \qquad (i = 1, 2, \ldots, n)$$

which are the fundamental equations of higher dynamics.

Here we shall merely make one noteworthy deduction, namely, the law of the *conservation of energy.*

Since the integrand in Hamilton's integral does not depend explicitly on the independent variable t, the solution $q_i(t)$ of the differential equations of dynamics must be such as to make the expression

$$E = T - U - \Sigma \dot{q}_i \frac{\partial (T - U)}{\partial \dot{q}_i}$$

constant. Since U does not depend on the $\dot{q}_i$'s and T is a homogeneous quadratic function in them (cf. p. 109),

$$\Sigma \dot{q}_i \frac{\partial (T - U)}{\partial \dot{q}_i} = \Sigma \dot{q}_i \frac{\partial T}{\partial \dot{q}_i} = 2T.$$

Hence

$$T + U = \text{const.};$$

that is, during the motion the sum of the kinetic energy and the potential energy does not vary with the time.

4. Integrals Involving Higher Derivatives.

Methods analogous to those used in the examples discussed previously can be used to attack the problem of the extreme values of integrals in which the integrand F not only contains the required function $y = \phi$ and its derivative ϕ', but also involves higher derivatives, e.g. the second. For example, suppose we wish to find the extreme values of an integral of the form

$$I\{\phi\} = \int_{x_0}^{x_1} F(x, \phi, \phi', \phi'')\, dx,$$

where in the comparison those functions $y = \phi(x)$ are admissible which, together with their first derivatives, have prescribed values at the end-points of the interval, and which also have continuous derivatives up to and including the fourth order.

To find necessary conditions for an extreme value we again assume that $y = u(x)$ is the desired function. We then include it in a family of functions $y = \phi(x) = u(x) + \epsilon\eta(x)$, where ϵ is an arbitrary parameter and $\eta(x)$ an arbitrarily-chosen function with continuous derivatives up to and including the fourth order, which together with its derivatives vanishes at the end-points. The integral then takes the form $\Phi(\epsilon)$, and the necessary condition

$$\Phi'(0) = 0$$

must be satisfied for all these functions $\eta(x)$. Proceeding in a way analogous to that in § 2 (p. 498), we differentiate under the integral sign and thus obtain the above condition in the form

$$\int_{x_0}^{x_1} (\eta F_u + \eta' F_{u'} + \eta'' F_{u''})\, dx = 0,$$

which must be satisfied if u is substituted for $\phi(x)$. Integrating once by parts we reduce the term in $\eta'(x)$ to one in η, and integrating twice by parts we reduce the term in $\eta''(x)$ to one in η; taking the boundary conditions into account, we easily obtain

$$\int_{x_0}^{x_1} \eta \left(F_u - \frac{d}{dx} F_{u'} + \frac{d^2}{dx^2} F_{u''} \right) dx = 0.$$

Hence the necessary condition for an extreme value, i.e. that the integral may be stationary, is Euler's differential equation

$$L[u] = F_u - \frac{d}{dx} F_{u'} + \frac{d^2}{dx^2} F_{u''} = 0.$$

The reader can verify for himself that this is a differential equation of the fourth order.

EXAMPLE

Consider

$$I = \int_{x_0}^{x_1} (\varphi''^2 - 2f\varphi)\, dx,$$

where $f(x)$ is a given function. Here Euler's differential equation is

$$u^{iv} - f(x) = 0.$$

5. **Several Independent Variables.**

The general method for finding necessary conditions for an extreme value can equally well be applied when the integral is no longer a simple integral but a multiple integral. Let D be a given region bounded by a sectionally smooth curve Γ in the xy-plane. Let $F(x, y, \phi, \phi_x, \phi_y)$ be a function which is continuous and twice continuously differentiable with respect to all five of its arguments. If in F we substitute for ϕ a function $\phi(x, y)$, which has continuous derivatives up to and including the second order in the region D and has prescribed boundary values on Γ, and if we replace ϕ_x and ϕ_y by the partial derivatives of ϕ, F becomes a function of x and y, and the integral

$$I\{\phi\} = \iint_D F(x, y, \phi, \phi_x, \phi_y)\, dx\, dy$$

has a value depending on the choice of ϕ. The problem is that of finding a function $\phi = u(x, y)$ for which this value is an extreme value.

To find necessary conditions we again use the old method. We choose a function $\eta(x, y)$ which vanishes on the boundary Γ, has continuous derivatives up to and including the second order, and is otherwise arbitrary; we assume that u is the required function and then substitute $\phi = u + \epsilon\eta$ in the integral, where ϵ is an arbitrary parameter. The integral again becomes a function $\Phi(\epsilon)$ and a necessary condition for an extreme value is

$$\Phi'(0) = 0.$$

As before, this condition takes the form

$$\iint_D (\eta F_u + \eta_x F_{u_x} + \eta_y F_{u_y})\, dx\, dy = 0.$$

To get rid of the terms in η_x and η_y under the integral sign we regard the double integral as a repeated integral, and integrate one term by parts with respect to x and the other with respect to y. Since η vanishes on Γ, the boundary values on Γ fall out, and we have

$$\iint \eta \left\{ F_u - \frac{\partial}{\partial x} F_{u_x} - \frac{\partial}{\partial y} F_{u_y} \right\} dx\, dy = 0.$$

Lemma I of § 2 (p. 499) can be extended at once to more dimensions than one, and we immediately obtain *Euler's partial differential equation of the second order,*

$$F_u - \frac{\partial}{\partial x} F_{u_x} - \frac{\partial}{\partial y} F_{u_y} = 0.$$

EXAMPLES

1. $F = \varphi_x^2 + \varphi_y^2$. If we omit the factor 2, Euler's differential equation becomes

$$\Delta u = u_{xx} + u_{yy} = 0.$$

That is, Laplace's equation has been obtained from a variation problem.

2. *Minimal Surfaces. Plateau's Problem.*—To find a surface $z = f(x, y)$ over the region D, which passes through a prescribed curve in space whose projection is Γ, and whose area

$$\iint_D \sqrt{(1 + \varphi_x^2 + \varphi_y^2)}\, dx\, dy$$

is a minimum.

Here Euler's differential equation is

$$\frac{\partial}{\partial x} \frac{u_x}{\sqrt{(1 + u_x^2 + u_y^2)}} + \frac{\partial}{\partial y} \frac{u_y}{\sqrt{(1 + u_x^2 + u_y^2)}} = 0,$$

or, in expanded form,

$$u_{xx}(1 + u_y^2) - 2u_{xy}u_x u_y + u_{yy}(1 + u_x^2) = 0.$$

This is the celebrated differential equation of minimal surfaces, which we cannot discuss further here.

6. Problems Involving Subsidiary Conditions. Euler's Multiplier.

In discussing the theory of ordinary extreme values of functions of several variables in Chapter III, § 6 (p. 191) we considered the case where these variables are subject to certain subsidiary conditions. In this case the method of undetermined multipliers led to a particularly clear expression for the conditions that the function may have a stationary value. An analogous method is of even greater importance in the calculus of variations. Here we shall briefly discuss the simplest cases only.

(a) *Ordinary Subsidiary Conditions.*—As a typical case we consider that of finding a curve $x = x(t)$, $y = y(t)$, $z = z(t)$ $(t_0 \leqq t \leqq t_1)$ in three-dimensional space, expressed in terms of the parameter t, subject to the subsidiary condition that the curve shall lie on a given surface $G(x, y, z) = 0$ and shall pass through two given points A and B on that surface. What we have to do, then, is to make an integral of the form

$$\int_{t_0}^{t_1} F(x, y, z, \dot{x}, \dot{y}, \dot{z})\, dt$$

stationary by suitable choice of the functions $x(t)$, $y(t)$, $z(t)$, subject to the subsidiary condition $G(x, y, z) = 0$ and the usual boundary conditions and continuity conditions.

This problem can be immediately reduced to the cases discussed in sub-section 1 (p. 507). We assume that $x(t)$, $y(t)$, $z(t)$ are the required functions. We assume further that on the portion of surface on which the required curve is to lie z can be expressed in the form $z = g(x, y)$. This is certainly the case if G_z differs from zero on this portion of the surface. If we assume that on the surface in question the three equations $G_x = 0$, $G_y = 0$, $G_z = 0$ are not simultaneously true and confine ourselves to a sufficiently small portion of surface, we can suppose without loss of generality that $G_z \neq 0$. If we then substitute $z = g(x, y)$ and $\dot{z} = g_x\dot{x} + g_y\dot{y}$ under the integral sign, the problem becomes one in which $x(t)$ and $y(t)$ are functions independent of one another. Thus we can immediately apply the result of sub-section 1 (p. 508) and write down the conditions that the integral I may be stationary, by applying the aforesaid result to the integrand

$$F(x, y, g(x, y), \dot{x}, \dot{y}, \dot{x}g_x + \dot{y}g_y) = H(x, y, \dot{x}, \dot{y}).$$

We then have the two equations

$$\frac{d}{dt} H_{\dot{x}} - H_x = \frac{d}{dt} F_{\dot{x}} - F_x + \frac{d}{dt}(F_{\dot{z}} g_x) - F_z g_x - F_{\dot{z}} \frac{\partial \dot{z}}{\partial x} = 0,$$

$$\frac{d}{dt} H_{\dot{y}} - H_y = \frac{d}{dt} F_{\dot{y}} - F_y + \frac{d}{dt}(F_{\dot{z}} g_y) - F_z g_y - F_{\dot{z}} \frac{\partial \dot{z}}{\partial y} = 0$$

But
$$\frac{d}{dt} g_x = \frac{\partial \dot{z}}{\partial x}, \quad \frac{d}{dt} g_y = \frac{\partial \dot{z}}{\partial y},$$

as we see at once on differentiation. Hence we have

$$\frac{d}{dt} F_{\dot{x}} - F_x + g_x\left(\frac{d}{dt} F_{\dot{z}} - F_z\right) = 0,$$

$$\frac{d}{dt} F_{\dot{y}} - F_y + g_y\left(\frac{d}{dt} F_{\dot{z}} - F_z\right) = 0.$$

If for brevity we write

$$\frac{d}{dt} F_{\dot{z}} - F_z = \lambda G_z, \quad . \; . \; . \; . \; . \; . \quad \text{(A)}$$

that is, if we introduce a multiplier $\lambda(t)$, and use the facts that $g_x = -G_x/G_z$, $g_y = -G_y/G_z$, we obtain the two further equations

$$\frac{d}{dt} F_{\dot{x}} - F_x = \lambda G_x, \quad . \; . \; . \; . \; . \; . \quad \text{(A)}$$

$$\frac{d}{dt} F_{\dot{y}} - F_y = \lambda G_y. \quad . \; . \; . \; . \; . \; . \quad \text{(A)}$$

We thus have the following condition that the integral may be stationary:

If we assume that G_x, G_y, G_z do not all vanish simultaneously on the surface $G = 0$, the necessary condition for an extreme value is the existence of a multiplier $\lambda(t)$ such that the three equations (A) given above are simultaneously satisfied in addition to the subsidiary condition $G(x, y, z) = 0$. That is, we have four symmetrical equations determining the functions $x(t)$, $y(t)$, $z(t)$ and the multiplier λ.

The most important special case of this is the problem of finding the shortest line joining two points A and B on a given surface $G = 0$, on which it is assumed that the gradient grad G does not vanish. Here

$$F = \sqrt{(\dot{x}^2 + \dot{y}^2 + \dot{z}^2)},$$

and Euler's differential equations are

$$\frac{d}{dt} \frac{\dot{x}}{\sqrt{(\dot{x}^2 + \dot{y}^2 + \dot{z}^2)}} = \lambda G_x,$$

$$\frac{d}{dt}\frac{\dot{y}}{\sqrt{(\dot{x}^2+\dot{y}^2+\dot{z}^2)}}=\lambda G_y,$$

$$\frac{d}{dt}\frac{\dot{z}}{\sqrt{(\dot{x}^2+\dot{y}^2+\dot{z}^2)}}=\lambda G_z.$$

These equations are invariant with respect to the introduction of a new parameter t. That is, as the reader may easily verify for himself, they retain the same form if t is replaced by any other parameter $\tau=\tau(t)$, provided that the transformation is one-to-one, reversible, and continuously differentiable. If we take the arc as the new parameter, in other words, if we assume that after the introduction of the new parameter $\dot{x}^2+\dot{y}^2+\dot{z}^2=1$, our differential equations take the form

$$\frac{d^2x}{ds^2}=\lambda G_x,\quad \frac{d^2y}{ds^2}=\lambda G_y,\quad \frac{d^2z}{ds^2}=\lambda G_z.$$

The geometrical meaning of these differential equations is that the osculating planes * of the extremals of our problem are orthogonal to the surface $G=0$. We call these curves *geodesics* of the surface. The shortest distance between two points on a surface, then, is necessarily given by an arc of a geodesic.

EXAMPLE

Show that the same geodesics are also obtained as the paths of a particle which is constrained to move on the given surface $G=0$, subject to no external forces. (In this case the potential energy U vanishes and the reader may apply Hamilton's principle (p. 512).)

(*b*) *Other Types of Subsidiary Conditions*.—In the problem discussed above we were able to eliminate the subsidiary condition by solving the equation determining the subsidiary condition and thus reducing the problem directly to the type discussed previously. With other kinds of subsidiary conditions which frequently occur, however, it is not possible to do this. The most important case of this type is the case of "isoperimetric" subsidiary conditions. The following is a typical example.

With the previous boundary conditions and continuity conditions, the integral

* I.e. the planes containing the vectors $(\dot{x}, \dot{y}, \dot{z})$ and $(\ddot{x}, \ddot{y}, \ddot{z})$ (cf. Ex. 1, 2, 4, pp. 93–4).

$$I\{\phi\} = \int_{x_0}^{x_1} F(x, \phi, \phi')\, dx$$

is to be made stationary, the argument function $\phi(x)$ being subject to the further subsidiary condition

$$H\{\phi\} = \int_{x_0}^{x_1} G(x, \phi, \phi')\, dx = \text{a given constant } c.$$

A particular case of this ($F = \phi$, $G = \sqrt{(1 + \phi'^2)}$) is the classical isoperimetric problem.

This type of problem cannot be attacked by our previous method of forming the " varied " function $\phi = u + \epsilon\eta$ by means of an arbitrary function $\eta(x)$ vanishing on the boundary only. For in general these functions do not satisfy the subsidiary condition in a neighbourhood of $\epsilon = 0$, except at $\epsilon = 0$. We can attain the desired result, however, by a method similar to that used in the original problem, by introducing, instead of one function η and one parameter ϵ, two functions $\eta_1(x)$ and $\eta_2(x)$, which vanish on the boundary, and two parameters ϵ_1 and ϵ_2. Assuming that $\phi = u$ is the required function, we then form the varied function

$$\phi = u + \epsilon_1\eta_1 + \epsilon_2\eta_2.$$

If we introduce this function into the two integrals, we obtain the following as a necessary condition for an extreme value or stationary character of the integral

$$I = \int_{x_0}^{x_1} F(x, u + \epsilon_1\eta_1 + \epsilon_2\eta_2, u' + \epsilon_1\eta_1' + \epsilon_2\eta_2')\, dx = \Phi(\epsilon_1, \epsilon_2),$$

subject to the subsidiary condition

$$H = \int_{x_0}^{x_1} G(x, u + \epsilon_1\eta_1 + \epsilon_2\eta_2, u' + \epsilon_1\eta_1' + \epsilon_2\eta_2')\, dx = \Psi(\epsilon_1, \epsilon_2) = c:$$

the function $\Phi(\epsilon_1, \epsilon_2)$ is to be stationary for $\epsilon_1 = 0$, $\epsilon_2 = 0$, where ϵ_1, ϵ_2 satisfy the subsidiary condition

$$\Psi(\epsilon_1, \epsilon_2) = c.$$

A simple discussion, based on the previous results for ordinary extreme values with subsidiary conditions, and in other respects following the same lines as the account given in § 2 (p. 498), then leads to this result:

Stationary character of the integral is equivalent to the existence of a constant multiplier λ *such that the equation* $\mathrm{H} = \mathrm{c}$ *and Euler's differential equation*

$$\frac{d}{dx}(F_{u'} + \lambda G_{u'}) - (F_u + \lambda G_u) = 0$$

are satisfied. An exception to this can only occur if the function u *satisfies the equation*

$$\frac{d}{dx} G_{u'} - G_u = 0.$$

The details of the proof may be left to the reader, who may consult the literature on this subject.*

EXAMPLES

1. Use the method of Euler's multiplier to prove that the solution of the classical isoperimetric problem is a circle.

2. A thread of uniform density and given length is stretched between two points A and B. If gravity acts in the direction of the negative y-axis, the equilibrium position of the thread is that in which the centre of gravity has the lowest possible position. It is accordingly a question of making an integral of the form $\int_{x_0}^{x_1} y\sqrt{(1 + y'^2)}\,dx$ a minimum, subject to the subsidiary condition that $\int_{x_0}^{x_1} \sqrt{(1 + y'^2)}\,dx$ has a given constant value. Show that the thread will hang in a catenary.

MISCELLANEOUS EXAMPLES VII

1. Show that the geodesics on a cylinder are helices.

2. Find Euler's equations in the following cases:

(a) $F = \sqrt{(1 + y'^2)} + y g(x)$,

(b) $F = \dfrac{y''^2}{(1 + y'^2)^3} + y g(x)$,

(c) $F = y''^2 - y'^2 + y^2$,

(d) $F = \sqrt[4]{(1 + y'^2)}$.

3. If there are two independent variables, find Euler's equations in the following cases:

* E.g. O. Bolza, *Lectures on the Calculus of Variations* (University of Chicago Press, 1904); G. A. Bliss, *The Calculus of Variations* (Open Court Publishing Company, Chicago, 1925).

(a) $F = a\varphi_x^2 + 2b\varphi_x\varphi_y + c\varphi_y^2 + \varphi^2 d$,

(b) $F = (\varphi_{xx} + \varphi_{yy})^2 = (\Delta\varphi)^2$,

(c) $F = (\Delta\varphi)^2 + (\varphi_{xx}\varphi_{yy} - \varphi^2_{xy})$.

4. Find Euler's equations for the isoperimetric problem in which

$$\int_{x_0}^{x_1} (au'^2 + 2buu' + cu^2)\,dx$$

is to be stationary subject to the condition

$$\int_{x_0}^{x_1} u^2\,dx = 1.$$

5. Let $f(x)$ be a given function. The integral

$$I(\varphi) = \int_0^1 f(x)\varphi(x)\,dx$$

is to be made a maximum subject to the integral condition

$$H(\varphi) = \int_0^1 \varphi^2\,dx = K^2$$

(where K is a given constant).

(a) Find the solution $u(x)$ from Euler's equation;

(b) Prove by applying Schwarz's inequality that the solution found in (a) gives the absolute maximum for I.

CHAPTER VIII

Functions of a Complex Variable

In Chap. VIII, § 7 (p. 410) of Vol. I we touched on the theory of functions of a complex variable and saw that this theory throws new light on the structure of functions of a real variable. Here we shall give a brief but more systematic account of the elements of that theory.

1. Introduction

1. Limits and Infinite Series with Complex Terms.

We start from the elementary concept of a complex number $z = x + iy$ (cf. Vol. I, p. 73) formed from the imaginary unit i and any two real numbers x, y. We operate with these complex numbers just as we do with ordinary numbers, with the additional rule that i^2 may always be replaced by -1. We represent x, the real part, and y, the imaginary part of z, by rectangular co-ordinates in an xy-plane or a "complex z-plane". The number $\bar{z} = x - iy$ is called the complex number *conjugate* to z. If we introduce polar co-ordinates (r, θ) by means of the relations $x = r \cos\theta$, $y = r \sin\theta$, θ is called the *argument* (or *amplitude*) of the complex number and $r = \sqrt{(x^2 + y^2)} = \sqrt{z\bar{z}} = |z|$ its *absolute value* (or *modulus*).

We can immediately establish the so-called "triangle inequality" satisfied by the complex numbers z_1, z_2, and $z_1 + z_2$,

$$|z_1 + z_2| \leqq |z_1| + |z_2|,$$

and the further inequality

$$|u_1| - |u_2| \leqq |u_1 - u_2|,$$

which follows immediately from it, if we put $z_1 = u_1 - u_2$, $z_2 = u_2$

The "triangle inequality" may be interpreted geometrically as follows: we can represent the complex numbers z_1, z_2 by vectors in the xy-plane with components x_1, y_1 and x_2, y_2 respectively. The vector which represents the sum $z_1 + z_2$ is then simply obtained by vector addition of the two first vectors. The lengths of the sides of the triangle so formed are $|z_1|$, $|z_2|$, $|z_1 + z_2|$. Thus the "triangle inequality" merely expresses the fact that any one side of a triangle is less than the sum of the other two.

The essentially new concept which we now have to consider is that of the *limit of a sequence of complex numbers.* We state the following definition: a sequence of complex numbers z_n tends to a limit z provided $|z_n - z|$ tends to zero. This of course means that the real part and the imaginary part of $z_n - z$ both tend to zero. Cauchy's test applies: the necessary and sufficient condition for the existence of a limit z of a sequence z_n is $\lim_{\substack{n \to \infty \\ m \to \infty}} |z_n - z_m| = 0$.

A particularly important class of limits arises from *infinite series with complex terms.* We say that the infinite series with complex terms,

$$\sum_{\nu=0}^{\infty} c_\nu,$$

converges and has the sum S, if the sequence of partial sums

$$S_n = \sum_{\nu=0}^{n} c_\nu$$

tends to the limit S. If the real series with non-negative terms,

$$\sum_{\nu=0}^{\infty} |c_\nu|,$$

converges, it follows, just as in Chap. VIII of Vol. I (p. 369), that the original series with complex terms also converges. The latter series is then said to be *absolutely convergent.*

If the terms c_ν of the series, instead of being constants, depend on (x, y), the co-ordinates of a point varying in a region R, the concept of *uniform convergence* acquires a meaning. The series is said to be uniformly convergent in R if for an arbitrarily small prescribed positive ϵ a fixed bound N can be found, depending on ϵ only, such that for every $n \geqq N$ the relation $|S_n - S| < \epsilon$ holds, no matter where the point $z = x + iy$ lies in the region R. *Uniform convergence of a sequence* of complex functions $S_n(z)$ depending on the point z of R may of course be defined in exactly the same way. All these relations and definitions and the associated proofs correspond exactly to those with which we are already familiar from the theory of real variables.

The simplest example of a convergent series is the geometric series

$$1 + z + z^2 + z^3 + \ldots .$$

Just as in the case of the real variable, we have

$$S_n = \frac{1 - z^{n+1}}{1 - z},$$

$$1 + z + z^2 + \ldots = \frac{1}{1 - z} \text{ for } |z| < 1;$$

we see that the geometric series converges absolutely provided $|z| < 1$, and also that the convergence is uniform provided $|z| \leqq q$, where q is any fixed positive number between 0 and 1. In other words, *the geometric series converges absolutely for all values of* z *within the unit circle and converges uniformly in every closed circle concentric with the unit circle and with a radius less than unity.*

For the investigation of convergence the *principle of comparison* is again available: if $|c_\nu| \leqq p_\nu$, where p_ν is real and non-negative, and if the infinite series $\sum_{\nu=0}^{\infty} p_\nu$ converges, then the complex series Σc_ν converges absolutely.

If the p_ν's are constants, while the c_ν's depend on a point z varying in R, the series Σc_ν converges *uniformly* in the region in question. The proofs are word for word the same as the corresponding proofs for the real variable (Vol. I, Chap. VIII, p. 392) and therefore need not be repeated here.

If M is an arbitrary positive constant and q a positive number between 0 and 1, the infinite series with the positive terms $p_\nu = Mq^\nu$ or $\nu M q^{\nu-1}$ or $\frac{M}{\nu + 1} q^{\nu+1}$ also converge, as we know from Vol. I, Chap. VIII, p. 401. We shall immediately make use of these expressions for purposes of comparison.

2. **Power Series.**

The most important infinite series with complex terms are power series, in which c_ν is of the form $c_\nu = a_\nu z^\nu$; that is, a power series may be expressed in the form

$$P(z) = \sum_{\nu=0}^{\infty} a_\nu z^\nu,$$

or, somewhat more generally, in the form

$$\sum_{\nu=0}^{\infty} a_\nu (z - z_0)^\nu.$$

where z_0 is a fixed point. As this form can, however, always be reduced to the preceding one by the substitution $z' = z - z_0$, we need only consider the case where $z_0 = 0$.

The main theorem on power series is word for word the same as the corresponding theorem for real power series in Chap. VIII of Vol. I (p. 399). *If the power series converges for* $z = \xi$, *it converges absolutely for every value of* z *such that* $|z| < |\xi|$. *Further, if* q *is a positive number less than* 1, *the series converges uniformly within the circle* $|z| \leqq q|\xi|$.

We can at once proceed to the following further theorem: *The two series*

$$D(z) = \sum_{\nu=1}^{\infty} \nu a_\nu z^{\nu-1}$$

$$I(z) = \sum_{\nu=0}^{\infty} \frac{a_\nu}{\nu+1} z^{\nu+1}$$

also converge absolutely and uniformly if $|z| \leqq q|\xi|$.

The proof follows exactly as before. Since the series $P(z)$ converges for $z = \xi$, it follows that the n-th term, $a_n \xi^n$, tends to zero as n increases. Hence a positive constant M certainly exists such that the inequality $|a_n \xi^n| < M$ holds for all values of n. If now $|z| = q|\xi|$, where $0 < q < 1$, we have

$$|a_n z^n| < Mq^n, \quad |na_n z^{n-1}| < \frac{M}{|\xi|} nq^{n-1}, \quad \left|\frac{a_n}{n+1} z^{n+1}\right| < \frac{M|\xi|}{n+1} q^{n+1}.$$

We thus obtain comparison series which, as we have seen already (p. 524), converge absolutely. Our theorem is thus proved.

In the case of a power series there are two possibilities: either it converges for all values of z, or there are values $z = \eta$ for which it diverges. Then by the theorem above the series must diverge for all values of z for which $|z| > \eta$ (cf. Vol. I, p. 400), and, just as in the case of real power series, there is a *radius of convergence* ρ such that the series converges when $|z| < \rho$ and diverges when $|z| > \rho$. The same applies to the two series $D(z)$ and $I(z)$, the value of ρ being the same as for the original series. The circle $|z| = \rho$ is called the *circle of convergence* of the power series. No general statements can be made about the convergence or divergence of the series on the circumference of the circle itself, i.e. for $|z| = \rho$.

3. Differentiation and Integration of Power Series.

It is natural to call an expression of the form

$$f(z) = a_0 + a_1 z + a_2 z^2 + \ldots + a_n z^n$$

with fixed (complex) coefficients a_ν a *function* of z, and more particularly a *polynomial of the* n-th *degree in* z. In the same way, a convergent power series

$$P(z) = \sum_{\nu=0}^{\infty} a_\nu z^\nu$$

is regarded as a function of the complex variable z in the interior of its circle of convergence. In that region it is the limit to which the polynomial

$$P_n(z) = \sum_{\nu=0}^{n} a_\nu z^\nu$$

tends as n tends to infinity.

A polynomial $f(z)$ may be differentiated with respect to the independent variable z in exactly the same way as for the real variable. In the first place we notice that the algebraic identity

$$\frac{z_1^n - z^n}{z_1 - z} = z_1^{n-1} + z_1^{n-2} z + \ldots + z^{n-1}$$

holds. If we now let z_1 tend to z*, we immediately have

$$\frac{d}{dz} z^n = \lim_{z_1 \to z} \frac{z_1^n - z^n}{z_1 - z} = n z^{n-1}.$$

* The concept of a limit for a continuous complex variable ($z_1 \to z$) can be introduced in exactly the same way as for the real variable.

In the same way we immediately have

$$P_n'(z) = \frac{d}{dz} P_n(z) = \lim_{z_1 \to z} \frac{P_n(z_1) - P_n(z)}{z_1 - z} = \sum_{\nu=1}^{n} \nu a_\nu z^{\nu-1} = D_n(z).$$

We naturally call the expression $P_n'(z)$ the *derivative* of the complex polynomial $P_n(z)$.

We now have the following theorem, which is fundamental in the theory of power series:

A convergent power series

$$P(z) = \sum_{\nu=0}^{\infty} a_\nu z^\nu$$

may be differentiated term by term in the interior of its circle of convergence. That is, the limit

$$P'(z) = \lim_{z_1 \to z} \frac{P(z_1) - P(z)}{z_1 - z}$$

exists, and

$$P'(z) = \sum_{\nu=1}^{\infty} \nu a_\nu z^{\nu-1} = \lim_{n \to \infty} P_n'(z) = \lim_{n \to \infty} D_n(z) = D(z).$$

From this theorem it is at once clear that the power series

$$I(z) = \sum_{\nu=0}^{\infty} \frac{a_\nu}{\nu + 1} z^{\nu+1}$$

may be regarded as the *indefinite integral* of the first power series, i.e. that $I'(z) = P(z)$.

The term-by-term differentiability of the power series is proved in the following way:

From p. 526 we know that the relation $D(z) = \lim_{n \to \infty} D_n(z)$ holds within the circle of convergence. We have to prove that the absolute value of the difference quotient $\frac{P(z_1) - P(z)}{z_1 - z}$ differs from $D(z)$ by less than a prescribed positive number ϵ, if only we take z_1 sufficiently close to z within the circle of convergence. For this purpose we form the difference quotient

$$D(z_1, z) = \frac{P(z_1) - P(z)}{z_1 - z} = \frac{P_n(z_1) - P_n(z)}{z_1 - z} + \sum_{\nu=n+1}^{\infty} a_\nu \lambda_\nu,$$

where for brevity we write

$$\lambda_\nu = \frac{z_1^\nu - z^\nu}{z_1 - z} = z_1^{\nu-1} + z_1^{\nu-2}z + \ldots + z^{\nu-1}.$$

If we keep to the notation used on p. 525, and if $|z| < q|\xi|$ and also $|z_1| < q|\xi|$, then it is certain that

$$|\lambda_\nu| \leqq \nu q^{\nu-1}|\xi|^{\nu-1}.$$

Hence

$$|R_n| = \left|\sum_{\nu=n+1}^{\infty} a_\nu\lambda_\nu\right| \leqq \sum_{\nu=n+1}^{\infty} |a_\nu|\,\nu q^{\nu-1}|\xi|^{\nu-1} \leqq \frac{M}{|\xi|}\sum_{\nu=n+1}^{\infty} \nu q^{\nu-1}.$$

Owing to the convergence of the series of positive terms $\Sigma\nu q^{\nu-1}$, the expression $|R_n|$ can therefore be made as small as we please, provided we make n sufficiently large. We choose n so large that this expression is less than $\epsilon/3$, and also so large—increasing n further if necessary—that $|D(z) - D_n(z)| < \epsilon/3$. We now choose z_1 so close to z that the absolute value of $\dfrac{P_n(z_1) - P_n(z)}{z_1 - z}$ also differs from $D_n(z)$ by less than $\epsilon/3$. Then

$$|D(z_1, z) - D(z)| \leqq \left|\frac{P_n(z_1) - P_n(z)}{z_1 - z} - D_n(z)\right| + |D_n(z) - D(z)| + R_n$$
$$< \frac{\epsilon}{3} + \frac{\epsilon}{3} + \frac{\epsilon}{3} = \epsilon,$$

and this inequality expresses the fact asserted.

Since the derivative of the function is again a power series with the same radius of convergence, we can differentiate again and repeat the process as often as we like. That is, *a power series can be differentiated as often as we please in the interior of its circle of convergence.*

Power series are the Taylor series of the functions P(z) *which they represent:* that is, *the coefficients* a_ν *may be expressed by the formula*

$$a_\nu = \frac{1}{\nu!}P^{(\nu)}(0).$$

The proof is word for word the same as for the real variable (cf. Vol. I, p. 404).

4. Examples of Power Series.

As we mentioned in Chap. VIII, § 7 (p. 413) of Vol. I, the power series for the elementary functions can immediately be extended to the complex variable; in other words, we can regard the power series for the elementary functions as complex power series and extend the definitions of these functions to the complex realm in this way. For example, the series

$$\sum_{\nu=0}^{\infty} \frac{z^\nu}{\nu!}, \quad \sum_{\nu=0}^{\infty} (-1)^\nu \frac{z^{2\nu}}{(2\nu)!}, \quad \sum_{\nu=0}^{\infty} \frac{(-1)^\nu z^{2\nu+1}}{(2\nu+1)!}, \quad \sum_{\nu=0}^{\infty} \frac{z^{2\nu}}{(2\nu)!}, \quad \sum_{\nu=0}^{\infty} \frac{z^{2\nu+1}}{(2\nu+1)!}$$

converge for all values of z. (This follows at once from comparison tests.) The functions represented by these power series are again denoted respectively by the symbols e^z, $\cos z$, $\sin z$, $\cosh z$, $\sinh z$, just as in the real case. The relations

$$\cos z + i \sin z = e^{iz},$$

$$\cosh z = \cos iz, \quad i \sinh z = \sin iz$$

now follow immediately from the power series. Again, by differentiating term by term we obtain the relation

$$\frac{d}{dz} e^z = e^z.$$

As examples of power series with a finite radius of convergence, other than the geometric series, we consider the series

$$\log(1+z) = \sum_{\nu=1}^{\infty} (-1)^{\nu+1} \frac{z^\nu}{\nu}$$

$$\text{arc tan}\, z = \sum_{\nu=0}^{\infty} (-1)^\nu \frac{z^{2\nu+1}}{2\nu+1} = \frac{1}{2i}\{\log(1+iz) - \log(1-iz)\},$$

whose sums we again denote by the symbols log, arc tan. Here the radius of convergence is again 1. Differentiating term by term, we have

$$\frac{d \log(1+z)}{dz} = \frac{1}{1+z}, \quad \frac{d}{dz}(\text{arc tan}\, z) = \frac{1}{1+z^2}.$$

Examples

1. For which points $z = x + iy$ is

$$\left| \frac{z-1}{z+1} \right| \leq 1?$$

2. Prove that if $\Sigma a_n z^n$ is *absolutely* convergent for $z = \zeta$, then it is uniformly convergent for every z such that $|z| \leq |\zeta|$.

3. Using the power series for $\cos z$ and $\sin z$, show that

$$\cos^2 z + \sin^2 z = 1.$$

4*. For what values of z is

$$\sum_{\nu=1}^{\infty} \frac{z^\nu}{1 - z^\nu}$$

convergent?

2. Foundations of the Theory of Functions of a Complex Variable

1. The Postulate of Differentiability.

As we have seen above, all functions which are represented by power series possess a derivative and an indefinite integral. This fact may be made the starting-point for the general theory of functions of a complex variable. The object of such a theory is to extend the differential and integral calculus to functions of a complex variable. In particular, it is important that the con cept of function should be generalized for complex independent variables in such a way that the function is differentiable in the complex region.

We could, of course, confine ourselves from the very beginning to the consideration of functions which are represented by power series and thus satisfy the postulate of differentiability. There are, however, two objections to this procedure. In the first place, we cannot tell *a priori* whether the postulate of the differentiability of a complex function does necessarily imply that the function can be expanded in a power series. (In the case of the real variable we saw that functions even exist which possess derivatives of any order and yet cannot be expanded in a power series (cf. Vol. I, p. 335).) In the second place, we learn even from the case of the simple function $1/(1 - z)$, whose power series, the geometric series, converges in the unit circle only, that even for simple functional expressions the power series does not represent the whole behaviour of the function, which in this particular case we already know in other ways.

These difficulties can, it is true, be avoided by a method due to Weierstrass, and the theory of functions of a complex variable can actually be developed on the basis of the theory of power series. It is desirable, however, to emphasize another point of view, which is due to Cauchy and Riemann. In their method, functions are characterized not by *explicit expressions* but by simple *properties*. More precisely, the postulate that a function shall be differentiable, and not that it shall be capable of being represented by a power series, is to be used to mark out the region in which a function is defined.

We could start *a priori* from the following general concept of a complex function $\zeta = f(z)$ of the complex variable z. If R is a region of the z-plane and if with every point $z = x + iy$ in R we associate a complex number $\zeta = u + iv$ by means of any relation, ζ is said to be a complex function of z in R. This definition, therefore, would merely express the fact that every pair of real numbers x, y, such that the point (x, y) lies in R, has a corresponding pair of real numbers u, v; i.e. that u and v are any two real functions $u(x, y)$ and $v(x, y)$, defined in R, of the two real variables x and y.

This concept of function, however, would be much too wide. We limit it in the first place by the condition that $u(x, y)$ and $v(x, y)$ must be continuous functions in R with continuous first derivatives u_x, u_y, v_x, v_y. Further, we insist that our expression $u + iv = \zeta = f(z) = f(x + iy)$ shall be differentiable in R with respect to the complex independent variable z; that is, the limit

$$\lim_{z_1 \to z} \frac{f(z_1) - f(z)}{z_1 - z} = \lim_{h \to 0} \frac{f(z + h) - f(z)}{h} = f'(z)$$

shall exist for all values of z in R. This limit is then called the *derivative* of $f(z)$.

In order that the function may be differentiable it is by no means sufficient that u and v should possess continuous derivatives with respect to x and y. Our postulate of differentiability implies far more than differentiability in the real region, for $h = r + is$ can tend to zero through both real values $(s = 0)$ and purely imaginary values $(r = 0)$ or in any other way, and the *same* limit $f'(z)$ must result in all cases, if the function is to be differentiable.

If, for example, we put $u = x$, $v = 0$, that is, $f(z) = f(x + iy) = x$, we should have a correspondence in which $u(x, y)$ and $v(x, y)$ are continuously differentiable. For the derivative, however, by putting $h = r$ we obtain

$$\lim_{r \to 0} \frac{f(z + r) - f(z)}{r} = \lim_{r \to 0} \frac{x + r - x}{r} = 1,$$

whereas if we put $h = is$ we have

$$\lim_{s \to 0} \frac{f(z + is) - f(z)}{is} = \lim_{s \to 0} \frac{0}{is} = 0,$$

that is, we obtain two entirely different limits. For $\zeta = u + iy = x + 2iy$ we similarly obtain different limits for the difference quotient as h tends to zero in different ways.

Thus in order to ensure the differentiability of $f(z)$ we have to impose yet another restriction. This fundamental fact in the theory of functions of a complex variable is expressed by the following theorem:

If $\zeta = u(x, y) + i v(x, y) = f(z) = f(x + iy)$, *where* $u(x, y)$ *and* $v(x, y)$ *are continuously differentiable, the necessary and sufficient conditions that the function* $f(z)$ *shall be differentiable in the complex region are*

$$u_x = v_y, \quad u_y = -v_x,$$

the so-called Cauchy-Riemann differential equations.

In every region R *where* u *and* v *satisfy these conditions* f(z) *is said to be an analytic * function of the complex variable* z, *and the derivative of* f(z) *is given by*

$$f'(z) = u_x + iv_x = v_y - iu_y = \frac{1}{i}(u_y + iv_y).$$

We shall first show that the Cauchy-Riemann differential equations form a *necessary* condition. If we accordingly assume that $f'(z)$ exists, we must obtain the limit $f'(z)$ by taking h equal to a real quantity r. That is,

$$\begin{aligned} f'(z) &= \lim_{r \to 0} \frac{u(x + r, y) - u(x, y)}{r} + i\,\frac{v(x + r, y) - v(x, y)}{r} \\ &= u_x + iv_x. \end{aligned}$$

In the same way, we must obtain $f'(z)$ if we take h to be a pure imaginary is, that is, we must have

$$\begin{aligned} f'(z) &= \lim_{s \to 0} \frac{u(x, y + s) - u(x, y)}{is} + i\,\frac{v(x, y + s) - v(x, y)}{is} \\ &= \frac{1}{i}(u_y + iv_y). \end{aligned}$$

Hence

$$u_x + iv_x = \frac{1}{i}(u_y + iv_y).$$

* The term *regular* is also used.

By equating real and imaginary parts we at once obtain the Cauchy-Riemann equations.

These equations, however, also form a *sufficient* condition for the differentiability of the function $f(z)$. To prove this, we form the difference quotient

$$\frac{f(z+h)-f(z)}{h}$$

$$=\frac{u(x+r,\,y+s)-u(x,\,y)+i\{v(x+r,\,y+s)-v(x,\,y)\}}{r+is}$$

$$=\frac{ru_x+su_y+irv_x+isv_y+\epsilon_1\,|\,h\,|+\epsilon_2\,|\,h\,|}{r+is},$$

where ϵ_1 and ϵ_2 are two real quantities which tend to zero with $|\,h\,|=\sqrt{(r^2+s^2)}$. If now the Cauchy-Riemann equations hold, the above expression immediately becomes

$$u_x+iv_x+\epsilon_1\frac{|\,h\,|}{r+is}+\epsilon_2\frac{|\,h\,|}{r+is}.$$

We see at once that as $h\to 0$ this expression tends to the limit u_x+iv_x, and that independently of the way in which the passage to the limit $h\to 0$ is carried out.

We now use the Cauchy-Riemann equations, or the property of differentiability which is equivalent to them, as the definition of an analytic function, on which we shall base our deduction of all the properties of such functions.

2. The Simplest Operations of the Differential Calculus.

All polynomials, and all power series in the interior of their circle of convergence, are analytic functions, by § 1 (p. 527). We see at once that the operations which lead to the elementary rules of the differential calculus can be carried out in exactly the same way as for the real variable. In particular, the following rules hold: the sum, the difference, the product, and (provided the denominator does not vanish) the quotient of analytic functions can be differentiated according to the elementary rules of the calculus, and hence are again analytic functions. Further, an analytic function of an analytic function can be differentiated according to the chain rule and therefore is itself an analytic function.

We also note the following theorem: *if the derivative of an analytic function* $\zeta = f(z)$ *vanishes everywhere in a region* R, *the function is a constant.*

Proof.—We have $u_x - iu_y = 0$ everywhere in R. Hence $u_x = 0$, $u_y = 0$, and in virtue of the Cauchy-Riemann equations $v_x = 0$, $v_y = 0$; that is, u and v are constants; hence ζ is a constant.

Application to the Exponential Function.—We use this theorem to define the exponential function, which we have already defined by means of the power series $e^z = \sum_{\nu=0}^{\infty} z^\nu/\nu!$, by means of its differential property, in the complex region also:

If a complex function f(z) *satisfies the differential equation*

$$f'(z) = f(z),$$

then $f(z) = ce^z$, *where* c *is a constant.*

Proof.—As we see at once by differentiating the power series (which converges everywhere) term by term, the exponential function certainly satisfies the condition. If $g(z)$ is another function for which $g'(z) = g(z)$, it immediately follows that $f(z)g'(z) - g(z)f'(z) = 0$ everywhere in R. We are entitled to assume that $g(z)$ is not zero at any point, as otherwise our relation would be satisfied at that point by $f(z) = 0$, or $c = 0$, which gives $f(z) = 0$ everywhere. Then the equation $(fg' - f'g)/g^2 = 0$ means that the derivative of the quotient f/g vanishes, i.e. that f/g is constant, which is what we asserted.

From this follows the *functional equation* of the exponential function,

$$e^z e^{z_1} = e^{z+z_1}.$$

(On the basis of the power series definition this functional equation is by no means a trivial assertion.) We obtain it by considering the function $g(z) = e^{z+z_1}$, where z_1 is fixed. By the chain rule, $g(z)$ satisfies the differential equation $g'(z) = g(z)$. Hence by the above theorem $g(z) = ce^z$. To determine c we put $z = 0$ and bear in mind that according to the power series definition $e^0 = 1$. Thus we at once have $g(0) = e^{z_1} = c$, and the functional equation follows.

In § 3 (p. 542) we shall develop a more satisfactory method for discussing the exponential function independently of the power series. Here we merely mention that in particular for $z = x$, $z_1 = iy$

$$e^{x+iy} = e^x e^{iy} = e^x(\cos y + i \sin y).$$

It follows further that the exponential function can never vanish, for if e^{z_1} vanished, then $e^z = e^{z_1} e^{z-z_1}$ would vanish for all values of z, which is certainly not the case.

Making use of the facts that $\cos 2\pi = 1$ and $\sin 2\pi = 0$, we immediately have

$$e^{2\pi i} = 1.$$

The exponential function therefore satisfies the equation

$$e^z = e^{z+2\pi i};$$

that is, it is *periodic* with period $2\pi i$.

EXAMPLE

Prove that the product and the quotient of analytic functions and the function of an analytic function are again analytic, using not the property of differentiability but the Cauchy-Riemann differential equations.

3. Conformal Representation. Inverse Functions.

By means of the functions $u(x, y)$ and $v(x, y)$ the points of the z-plane or xy-plane are made to correspond to points of the ζ-plane or uv-plane. Thus we have a transformation or mapping (Chap. III, § 3, p. 133) of regions of the xy-plane on to regions of the uv-plane. The Jacobian of the transformation is

$$D = \frac{\partial(u, v)}{\partial(x, y)} = u_x v_y - u_y v_x = u_x^2 + v_x^2 = |f'(z)|^2.$$

The Jacobian is therefore different from zero, and is in fact positive, wherever $f'(z) \neq 0$. If we assume that $f'(z) \neq 0$, our previous results (Chap. III, § 3, p. 152) show that a neighbourhood of the point z_0 in the z-plane, if sufficiently small, is mapped uniquely, reversibly, and continuously on a region of the ζ-plane in the neighbourhood of the point $\zeta_0 = f(z_0)$. This mapping is *conformal*, i.e. angles are unchanged by it. For, as we have seen in Chap. III, p. 166, the Cauchy-Riemann equations are the necessary and sufficient conditions that the transformation may be conformal, not only the magnitude but also the sign of angles being preserved. We thus have the following result:

Conformality of the transformation given by u(x, y) *and* v(x, y) *and analytic character of the function* f(z) = u + iv *mean exactly the same thing, provided we avoid points* z_0 *for which* $f'(z_0) = 0$.

The reader should study the examples of conformal representation discussed in Chap. III, § 3, p. 136, and prove that all these transformations can be expressed by analytic functions of simple form.

Since in the case of a unique reversible conformal representation of a neighbourhood of z_0 on a neighbourhood of ζ_0 the reverse transformation is also conformal, it follows that $z = x + iy$ may also be regarded as an analytic function $\phi(\zeta)$ of $\zeta = u + iv$. This function is called the *inverse* of $\zeta = f(z)$.

Instead of using our geometrical argument, we can at once establish the analytic character of this inverse by calculating the derivatives of $x(u, v)$, $y(u, v)$ as on p. 143. We have

$$x_u = \frac{v_y}{D}, \quad x_v = -\frac{u_y}{D}, \quad y_u = -\frac{v_x}{D}, \quad y_v = \frac{u_x}{D},$$

and we see that the Cauchy-Riemann equations $x_u = y_v$, $x_v = -y_u$ are satisfied by the inverse function. As we can at once verify, the derivative of the inverse $z = \phi(\zeta)$ of the function $\zeta = f(z)$ is given by the formula

$$\frac{dz}{d\zeta}\frac{d\zeta}{dz} = 1.$$

EXAMPLES

1. Find where the following functions are continuous:

$$(a)\ \bar{z}; \qquad (b)\ |z|; \qquad (c)\ \frac{z + \bar{z}}{1 + |z|}; \qquad (d)\ \frac{z^2 + \bar{z}^2}{|z|^2}.$$

2. Which of the functions in Ex. 1 are also differentiable?

3*. Prove that a substitution of the form

$$\zeta = \frac{\alpha z + \bar{\beta}}{\beta z + \bar{\alpha}},$$

where α and β are any complex numbers satisfying the relation

$$\alpha\bar{\alpha} - \beta\bar{\beta} = 1,$$

transforms the circumference of the unit circle into itself and the interior of the circle into itself. Prove also that if

$$\beta\bar{\beta} - \alpha\bar{\alpha} = 1,$$

the interior is transformed into the exterior.

4. Prove that in the transformation $\zeta = \frac{1}{2}(z + 1/z)$ the circles with centres at the origin and the straight lines through the origin of the z-plane are respectively transformed into confocal ellipses and hyperbolas in the ζ-plane (cf. Ex. 5, p. 158).

5. Prove that a substitution $\zeta = \dfrac{\alpha z + \beta}{\gamma z + \delta}$ leaves the cross ratio $\dfrac{z_1 - z_3}{z_2 - z_3} \Big/ \dfrac{z_1 - z_4}{z_2 - z_4}$ of four points z_1, z_2, z_3, z_4 unaltered.

6*. Prove that any circle may be transformed by a substitution of the form $\zeta = \dfrac{\alpha z + \beta}{\gamma z + \delta}$ into the upper half-plane bounded by the real axis. (Use Ex. 1, p. 529.)

7. Prove the following property of the general linear transformation

$$\zeta = \frac{az + b}{cz + d},$$

where a, b, c, d are constants and $ad - bc \neq 0$:

All circles and straight lines in the z-plane are transformed by this relation into all straight lines and circles in the ζ-plane.

If the z-plane and the ζ-plane are imagined to coincide, the points z for which $\zeta = z$ are called *fixed points*. In general there are two different fixed points. Show that in this case the family of circles through the two fixed points and the family of circles orthogonal to them transform into themselves.

8. The inverse of the power function $\zeta = z^n$ is unique in the neighbourhood of every point z_0, provided $z_0 \neq 0$, for then the derivative nz^{n-1} does not vanish. The point $z_0 = 0$, where the derivative vanishes, however, forms an exception; hence the many-valuedness of the function $\sqrt[n]{\zeta}$. We shall discuss these relations more closely in § 6, p. 563.

3. The Integration of Analytic Functions

1. Definition of the Integral.

The central fact of the differential and integral calculus of functions of a real variable is expressed in the theorem that the integral of a function (the upper limit being undetermined) may be regarded as the primitive function or " indefinite integral " of the original function (Vol. I, p. 109). A corresponding relation forms the nucleus of the theory of analytic functions of a complex variable.

We begin by extending the definition of the definite integral of a given function $f(z)$. Here it is convenient to use $t = r + is$ instead of the independent variable z, as we shall use t to denote the variable of integration. Let the function $f(t)$ be analytic in a region R, and let $t = t_0$ and $t = z$ be two points in this region, joined by an oriented curve C which is piecewise smooth and lies wholly within R (fig. 1). We then subdivide the curve C into n portions by means of the successive points $t_0, t_1, \ldots, t_n = z$ and form the sum

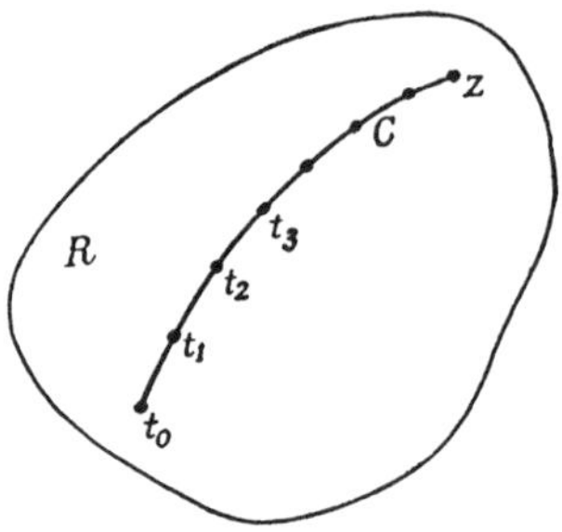

Fig. 1

$$S_n = \sum_{\nu=1}^{n} f(t_\nu')(t_\nu - t_{\nu-1}),$$

where t_ν' denotes any point lying on C between $t_{\nu-1}$ and t_ν. If we now make the subdivision finer and finer by letting the number of points increase without limit in such a way that the greatest of the intervals $| t_\nu - t_{\nu-1} |$ tends to zero, S_n tends to a limit which is independent of the choice of the particular intermediate point t_ν' and of the points t_ν.

This can be proved directly by a method analogous to that used to prove the corresponding theorem of the existence of the definite integral for real variables. For our purpose, however, it is more convenient to reduce the theorem to what we already know about real curvilinear integrals (cf. Chap. V, § 1, p. 344), as follows. We put $f(t) = u(r, s) + iv(r, s)$, $t_\nu = r_\nu + is_\nu$, $t_\nu' = r_\nu' + is_\nu'$, $\Delta t_\nu = t_\nu - t_{\nu-1} = \Delta r_\nu + i\Delta s_\nu$. Then we have

$$S_n = \sum_{\nu=1}^{n} u(r_\nu', s_\nu')\Delta r_\nu - v(r_\nu', s_\nu')\Delta s_\nu$$

$$+ i\left\{\sum_{\nu=1}^{n} v(r_\nu', s_\nu')\Delta r_\nu + u(r_\nu', s_\nu')\Delta s_\nu\right\}.$$

As n increases the sums on the right-hand side tend to the real curvilinear integrals $\int_C (u\,dx - v\,dy)$ and $i\int_C (v\,dx + u\,dy)$ respectively, and hence, as we asserted, S_n tends to a limit. We call this limit the definite integral of the function $f(t)$ along the curve C from t_0 to z, and write it

$$\int_{t_0}^{z} f(t)\,dt \quad \text{or} \quad \int_C f(t)\,dt.$$

Thus $$\int_C f(t)\,dt = \int_C (u\,dx - v\,dy) + i\int_C (v\,dx + u\,dy).$$

The definition of this definite integral (cf. Chap. V, § 1, p. 349) at once gives the following important estimate: *if* $|\,f(t)\,| \leqq M$ *on the path of integration, where* M *is a constant and* L *is the length of the path of integration, then*

$$\left|\int_C f(t)\,dt\right| \leqq ML.$$

In addition we may point out that operations with complex integrals (in particular, combination of different paths of integration) satisfy all the rules stated in this connexion for curvilinear integrals in Chap. V, § 1, p. 347–9.

2. Cauchy's Theorem.

The essential fact of the theory of functions of a complex variable is that the integral between t_0 and z is largely independent of the choice of the path of integration C. In fact, we have Cauchy's theorem:

If the function f(t) *is analytic in a simply-connected region* R, *the integral*

$$\int_{t_0}^{z} f(t)\,dt = \int_C f(t)\,dt$$

is independent of the particular choice of the path of integration C *joining* t_0 *and* z *in* R; *the integral is an analytic function* F(z) *such that*

$$\frac{d}{dz}F(z) = \frac{d}{dz}\left(\int_{t_0}^{z} f(t)\,dt\right) = f(z).$$

F(z) *is accordingly a primitive function or indefinite integral of* f(z).

Cauchy's theorem may also be expressed as follows:

If subject to the above assumptions we take the integral of f(t) *round a closed curve lying in a simply-connected region, the integral has the value zero.*

The proof that the integral is independent of the path follows immediately from the main theorem on curvilinear integrals (cf. Chap. V, § 1, p. 353); for both $u\,dx - v\,dy$, the integrand in the real part, and $v\,dx + u\,dy$, the integrand in the imaginary part, satisfy the condition of integrability, in virtue of the Cauchy-

Riemann equations (p. 532). Thus the integral is a function of x, y or $x + iy = z$, $F(z) = U(x, y) + iV(x, y)$, and from our previous results for curvilinear integrals we have the relations

$$U_x = u,\quad U_y = -v,\quad V_x = v,\quad V_y = u,$$

that is,

$$U_x = V_y,\quad U_y = -V_x,\quad U_x + iV_x = u + iv,$$

which shows that $F(z)$ is actually an analytic function in R with the derivative $F'(z) = f(z)$.

The assumption that the region is *simply-connected* is *essential* for the validity of Cauchy's theorem.

For example, we may consider the function $1/t$, which is analytic everywhere in the t-plane except at the origin. We are, however, not entitled to conclude from Cauchy's theorem that the integral of $1/t$, taken round a closed curve enclosing the origin, vanishes. For this curve cannot be enclosed in a simply-connected region in which the function is analytic. The simple connectivity of the region is destroyed by the exceptional point $t = 0$. If we take the integral e.g. round a circle K given by $|t| = r$ or $t = re^{i\theta}$ in the positive sense, and make θ the variable of integration ($dt = rie^{i\theta}d\theta$), we have

$$\int_K \frac{dt}{t} = \int_0^{2\pi} \frac{rie^{i\theta}}{re^{i\theta}}\, d\theta = 2\pi i;$$

that is, the value of the integral is not zero but $2\pi i$.

We can, however, extend Cauchy's theorem to multiply-connected regions as follows:

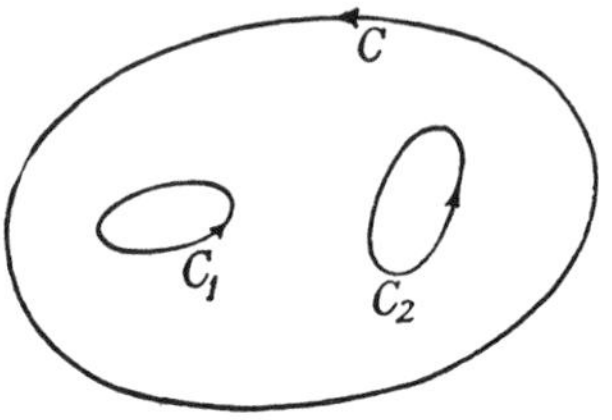

Fig. 2.— $\int_C = \int_{C_1} + \int_{C_2}$

If a multiply-connected region R *is bounded by a finite number of sectionally smooth closed curves* $C_1, C_2, \ldots$, *and if* f(z) *is analytic in the interior of this region and also on its boundary,* then the sum of the integrals of the function along all the boundary curves is zero, provided that all the boundaries are described in the same sense relative to the interior of the region* R, *i.e. that the region* R *is always on the same side, say the left-hand side, of the curve as it is described.*

The proof follows at once, on the model of the corresponding

* A function is said to be analytic on a curve if it is analytic throughout a neighbourhood, no matter how small, of this curve.

proofs for curvilinear integrals; we cut up the region R into a finite number of simply-connected regions (figs. 2, 3), apply Cauchy's theorem to these regions separately, and add the results.

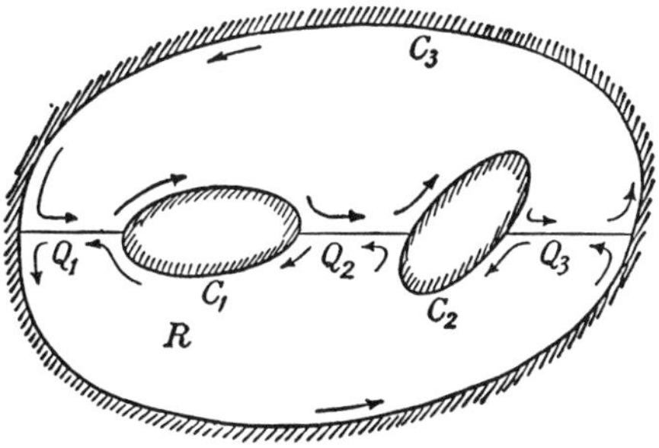

Fig. 3.—A multiply-connected region R subdivided by Q_1, Q_2, .. into simply-connected regions

We can express this theorem in a somewhat different way:

If the region R *is formed from the interior of a closed curve* C *by cutting out of this interior the interiors of further curves* C_1, C_2, . . . , *then*

$$\int_C f(t)\,dt = \sum_\nu \int_{C_\nu} f(t)\,dt,$$

where the integrals round the external boundary C *and the internal boundaries are to be taken in the same sense.*

3. Applications. The Logarithm, the Exponential Function, and the General Power Function.

We can now use Cauchy's theorem as the basis for a satisfactory theory of the logarithm, the exponential function, and hence of the other elementary functions, following a procedure similar to that adopted for the real variable (Vol. I, Chap. III, § 6, p. 167).

We begin by defining the logarithm as the integral of the function $1/t$. At first we limit the path of integration by making it lie in a simply-connected region, making a cut along the negative real x-axis, that is, permitting no path of integration which crosses the negative real axis. More precisely: if we put $t = |t| (\cos\theta + i\sin\theta)$, we limit θ by the inequality $-\pi < \theta \leqq \pi$. In the t-plane, after the cut has been made, we join the point

$t = 1$ to an arbitrary point z by any curve C, and we can then use Cauchy's theorem to integrate the function $1/t$ between these two points, independently of the path. The result is an analytic function, which we call $\log z$:

$$\zeta = \log z = \int_1^z \frac{dt}{t} = f(z).$$

The logarithm has the property that

$$\frac{d}{dz}(\log z) = \frac{1}{z}.$$

As this derivative does not vanish anywhere, we can form the

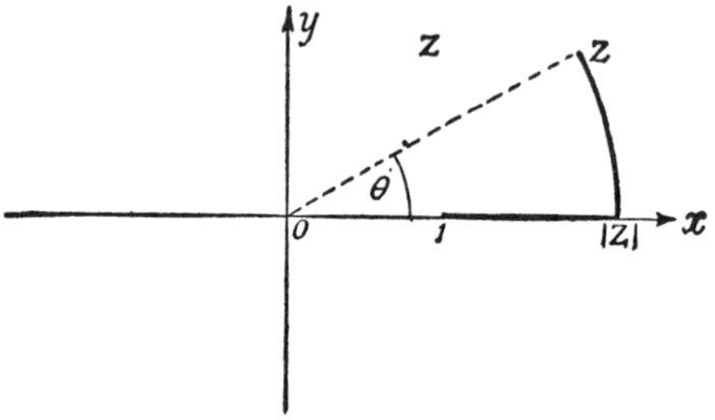

Fig. 4.— $\log z = \log |z| + i\theta$

inverse function of the logarithm, $z = g(\zeta)$. We have $g(0) = 1$, and by the formula for the derivative of the inverse

$$g'(\zeta) = 1/f'(z) = z = g(\zeta).$$

By § 2, p. 536, the inverse is thus determined uniquely and is identical with the exponential function defined previously: $g(\zeta) = e^{\zeta}$.

The function $f(z) = \log z$ is uniquely determined, except for an additive constant, by its differentiation property $f'(z) = 1/z$. For if there were another function $g(z)$ with this property, their difference would have the derivative zero and would therefore be constant. Since the function $g(z) = f(az) = \log(az)$ satisfies the condition $g'(z) = af'(az) = a/az = 1/z$, by the chain rule, we have $\log(az) = g(z) = c + \log z$, where c is a constant independent of z. Its value is determined by putting $z = 1$, i.e. $\log z = 0$, and we thus have $\log(a) = c$. This gives the *addition theorem for the logarithm*,

$$\log(az) = \log a + \log z.$$

The integral

$$\log z = \int_1^z \frac{dt}{t}$$

is easily evaluated explicitly by taking the straight line joining the points $t = 1$ and $t = |z|$ together with the circular arc $|t| = |z|$ as the path of integration. We have

$$\log z = \log|z| + i\theta,$$

where θ is the argument of the complex number z (fig. 4).

The value obtained in this way for the logarithm of any complex number z, whose argument lies in the interval $-\pi < \theta \leqq \pi$,

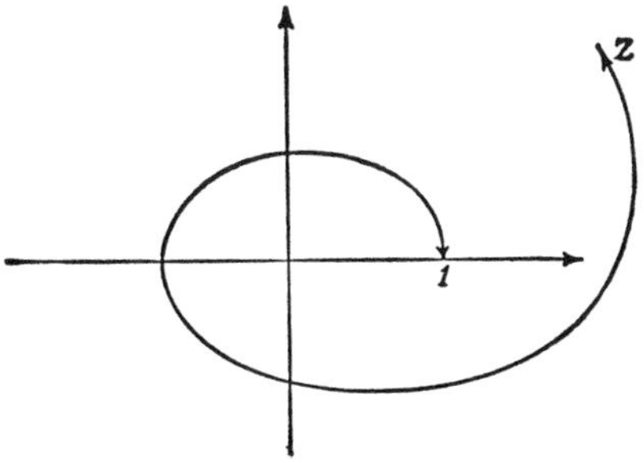

Fig. 5.— $\log z = \log|z| + i\theta + 2\pi i$

is often called the *principal value* of the logarithm. This terminology is justified by the fact that other values of the logarithm can be obtained by removing the condition that the negative real axis must not be crossed. We can then join the point 1 to the point z by a point which encloses the origin $t = 0$. On this curve the argument of t will increase up to a value which is greater or less than the argument previously assigned to z by 2π. We then have the value

$$\log z = \log|z| + i\theta + 2\pi i$$

for the integral (fig. 5). In the same way, by making the curve travel round the origin in one direction or the other any integral number of times n, we obtain the value

$$\log z = \log|z| + i\theta + 2n\pi i.$$

This expresses the *many-valuedness of the logarithm.*

In the case of the exponential function this many-valuedness is exhibited in the equation $e^{2\pi i} = 1$. For the same value of z corresponds to all the different values $\zeta = \log z$, which differ only by multiples of $2\pi i$. In the inverse of the logarithm, i.e. the exponential function, the addition or subtraction of $2\pi i$ to or from the argument must not alter the value of the function: $\phi(\zeta + 2\pi i) = \phi(\zeta)$, or $e^{\zeta+2\pi i} = e^{\zeta}$. If $\zeta = 0$, we have the equation $e^{2\pi i} = 1$.

If we now introduce the trigonometric functions $\sin z$ and $\cos z$ by means of the equation

$$e^{iz} = \cos z + i \sin z,$$

which we now may take as their definition, we see at once that these functions have the period 2π. Thus we have deduced the periodic character of the trigonometric functions without reference to their elementary geometrical definitions.

Now that we have introduced the logarithm and the exponential function it is easy to introduce the general power functions a^z and z^α, where a and α are constants (cf. the corresponding discussion for the real variable in Vol. I (Chap. III, § 6, p. 173)). We define a^z by the relation

$$a^z = e^{z \log a},$$

where the principal value of $\log a$ is to be taken. In the same way we define z^α by the relation

$$z^\alpha = e^{\alpha \log z}.$$

While the function a^z is defined uniquely if we use the principal value of $\log a$ in the definition, the many-valuedness of the function z^α goes deeper. Taking the many-valuedness of $\log z$ into account, we see that along with any one value of z^α we also have all the other values which are obtained by multiplying one value by $e^{2n\pi i\alpha}$, where n is any positive or negative integer. If α is rational, say $\alpha = p/q$, where p and q are integers prime to one another, among these multipliers there are only a finite number of different values (whose q-th power must be unity). If, however, a is irrational, we obtain an infinite number of different multipliers. The many-valuedness of the function z^a will be discussed in greater detail in § 6 (p. 563).

As we see from the chain rule, these functions satisfy the differentiation formulæ

$$\frac{d(a^z)}{dz} = a^z \log a, \quad \frac{d(z^a)}{dz} = az^{a-1}.$$

Examples

1. (The gamma function.) Prove that the integral

$$\Gamma(z) = \int_0^\infty t^{z-1} e^{-t}\, dt$$

(where the principal value of t^{z-1} is taken), extended over all real values of the variable of integration t, is an analytic function of the parameter $z = x + iy$, if $x > 0$. (Show directly that the expression $\Gamma(z)$ can be differentiated with respect to z.) Prove that the gamma function thus defined for the complex variable satisfies the functional equation $\Gamma(z + 1) = z\Gamma(z)$.

2*. (Riemann's zeta function.) Taking the principal value of n^z, form the infinite series

$$\sum_{n=1}^{\infty} \frac{1}{n^z} = \zeta(z).$$

Prove that this series converges if $x > 1$ and represents a differentiable function ($\zeta(z)$ is called Riemann's *zeta function*). The proof can be carried out directly by a method like that for power series (cf. Vol. I, p. 382).

4. Cauchy's Formula and its Applications

1. Cauchy's Formula.

Cauchy's theorem for multiply-connected regions leads to a fundamental formula, again due to Cauchy, which expresses the value of an analytic function $f(z)$ at any point $z = a$ in the interior of a closed region R, throughout which the function is analytic, by means of the values which the function takes on the boundary C.

We assume that the function $f(z)$ is analytic in the simply-connected region R and on its boundary C. Then the function

$$g(z) = \frac{f(z)}{z - a}$$

is analytic everywhere in the region R, the boundary C included, except at the point $z = a$. Out of the region R we cut a circle of small radius ρ about the point $z = a$, lying entirely within R (fig. 6), and then apply Cauchy's theorem (p. 541) to the function $g(z)$. If K denotes the circumference of the circle described in the positive sense and C the boundary of R described in the positive sense, Cauchy's theorem states that

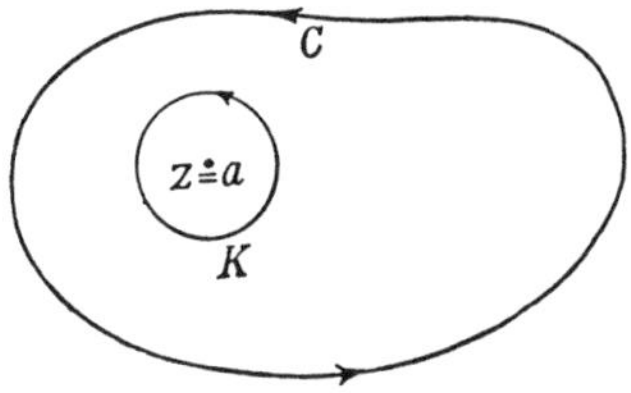

Fig. 6

$$\int_C g(z)\,dz = \int_K g(z)\,dz.$$

On the circle K we have $z = a + \rho e^{i\theta}$, where the angle θ determines the position of the point on the circumference. On the circle, therefore, $dz = \rho i e^{i\theta}\,d\theta$, and hence

$$\int_K g(z)\,dz = i\int_0^{2\pi} f(a + \rho e^{i\theta})\,d\theta.$$

Since $f(z)$ is continuous at the point a, we have, provided ρ is sufficiently small,

$$f(a + \rho e^{i\theta}) = f(a) + \eta,$$

where $|\eta|$ is less than an arbitrary prescribed positive quantity ϵ. Hence

$$\left|\int_0^{2\pi} f(a + \rho e^{i\theta})\,d\theta - \int_0^{2\pi} f(a)\,d\theta\right| = \left|\int_0^{2\pi} \eta\,d\theta\right| \leqq 2\pi\epsilon,$$

and therefore

$$\int_0^{2\pi} f(a + \rho e^{i\theta})\,d\theta = 2\pi f(a) + \kappa,$$

where $|\kappa| \leqq 2\pi\epsilon$. Thus if ρ is sufficiently small

$$\int_C g(z)\,dz = 2\pi i f(a) + \kappa i,$$

where $|\kappa i| < \epsilon$.

If we make ϵ tend to zero (by making ρ tend to zero), the right-hand side of the equation tends to $2\pi i f(a)$, while the value of the left-hand side, namely, $\int_C g(z)\,dz$, is unaltered.

We thus obtain Cauchy's *fundamental integral formula*

$$f(a) = \frac{1}{2\pi i}\int_C \frac{f(z)}{z-a}\,dz.$$

If we now revert to the use of t as variable of integration and then replace a by z, the formula takes the form

$$f(z) = \frac{1}{2\pi i}\int_C \frac{f(t)}{t-z}\,dt.$$

This formula expresses the values of a function in the interior of a closed region in which the function is analytic by means of the values which the function takes on the boundary of the region.

If in particular C is a circle $t = z + re^{i\theta}$ with centre z, that is, if $dt = ire^{i\theta}d\theta$, then

$$f(z) = \frac{1}{2\pi}\int_0^{2\pi} f(z + re^{i\theta})\,d\theta.$$

In words: *the value of a function at the centre of a circle is equal to the mean of its values on the circumference, provided that the closed area of the circle is a region in which the function is analytic.*

2. Expansion of Analytic Functions in Power Series.

Cauchy's formula has a number of important theoretical applications, the chief of which is the proof of the fact that *every analytic function can be expanded in a power series*, which thus connects the present theory with that given in § 1 (p. 527). More precisely, we have the following theorem: *if the function* f(z) *is analytic in the interior and on the boundary of a circle* $|z - z_0| \leqq R$, *it can be expanded as a power series in* $z - z_0$ *which converges in the interior of that circle.*

In proving this we can take $z_0 = 0$ without loss of generality. (Otherwise we should merely have to introduce a new independent variable z' by means of the transformation $z - z_0 = z'$.) We now apply Cauchy's integral formula to the circle C, $|z| = R$, and write the integrand (using the geometric series) in the form

$$\frac{f(t)}{t-z} = \frac{f(t)}{t}\,\frac{1}{1-z/t} = \frac{f(t)}{t}\left(1 + \frac{z}{t} + \dots \frac{z^n}{t^n}\right) + \frac{f(t)}{t}\left(\frac{z}{t}\right)^{n+1}\frac{1}{1-z/t}.$$

Since z is a point in the interior of the circle, $|z/t| = q$ is a positive number less than unity, and for $r_n = \frac{1}{t}\frac{z^{n+1}}{t^{n+1}}\frac{1}{1-z/t}$, the remainder of the geometric series, we obviously have the estimate

$$| r_n | \leq \frac{1}{R} q^{n+1} \frac{1}{1-q}.$$

Introducing our expressions into Cauchy's formula and integrating term by term, we obtain

$$f(z) = c_0 + c_1 z + \dots + c_n z^n + R_n,$$

where

$$c_\nu = \frac{1}{2\pi i}\int_C \frac{f(t)}{t^{\nu+1}}\,dt,$$

$$R_n = \frac{1}{2\pi i}\int_C f(t) r_n\,dt.$$

If M is an upper bound of the values of $|f(t)|$ on the circumference of the circle, our estimation formula for complex integrals (cf. § 3, p. 539) immediately gives

$$| R_n | \leq \frac{1}{2\pi R}\frac{q^{n+1}}{1-q} 2\pi R M = \frac{q^{n+1}}{1-q} M$$

for the remainder. Since q is a proper fraction this remainder tends to zero as n increases, and for $f(z)$ we obtain the power series

$$f(z) = \sum_{\nu=0}^{\infty} c_\nu z^\nu,$$

where

$$c_\nu = \frac{1}{2\pi i}\int_C \frac{f(t)}{t^{\nu+1}}\,dt.$$

Our assertion is thus proved.

This theorem has important results. To begin with, we know from § 1 (p. 528) that every power series can be differentiated as often as we please in the interior of its circle of convergence. Since every analytic function can be represented by a power series, it follows that *the derivative of a function* in the interior of a region where the function is analytic is also differentiable, i.e. is again *an analytic function.* In other words, *the operation of differentiation does not lead us out of the class of analytic functions.* As we already know that the same is true for the operation of integration, we see that *differentiation and integration of analytic functions can be carried out without any restrictions*. This is an agreeable state of affairs, which does not exist in the case of real functions.

Since, as we saw in § 1, p. 528, every power series is the Taylor series of the function which it represents, it now follows in general that every analytic function can be expanded in the neighbourhood of a point $z = z_0$ in a region R where the function is analytic in a Taylor series

$$f(z) = f(z_0) + \sum_{\nu=1}^{\infty} \frac{f^{(\nu)}(z_0)}{\nu!} (z - z_0)^\nu;$$

the coefficients c_ν above are accordingly given by the formulæ

$$c_\nu = \frac{f^{(\nu)}(0)}{\nu!} = \frac{1}{2\pi i} \int_C \frac{f(t)}{t^{\nu+1}} dt.$$

From our result we may also deduce an important fact about the radius of convergence of a power series. The Taylor series of a function $f(z)$ in the neighbourhood of a point $z = z_0$ certainly converges in the interior of the largest circle whose interior lies wholly within the region where the function is defined and is analytic.

In virtue of the theorems on differentiation and integration which we have now established as valid for the complex variable also, all the elementary functions which we expanded in Taylor series for the real variable have exactly the same Taylor series for the complex variable. For most of these functions we have already seen that this is true.

Here we may point out that e.g. the binomial series

$$(1 + z)^\alpha = \sum_{\nu=0}^{\infty} \binom{\alpha}{\nu} z^\nu$$

is also valid for the complex variable if $|z| < 1$, provided that

$$(1 + z)^\alpha = e^{\alpha \log(1+z)}$$

is formed from the principal value of $\log(1 + z)$.

The fact that the radius of convergence of this series is equal to unity follows from what we have just said, together with the remark that the function $(1 + z)^\alpha$ is no longer analytic at the point $z = -1$. For if it were, all the derivatives must exist there, which is certainly not the case. The circle with radius 1 with the point $z = 0$ as centre is therefore the largest circle in the interior of which the function is still analytic.

As we have already pointed out in Chap. VIII of Vol. I (p. 414), the behaviour of power series as regards convergence only becomes completely intelligible in the light of the fact which we have just proved about the radius of convergence.

For example, the failure of the geometric series representing $1/(1 + z^2)$ to converge on the unit circle is a simple consequence of the fact that the function is no longer analytic for $z = +i$ and $z = -i$. We also see now that the power series

$$\frac{z}{e^z - 1} = \Sigma \frac{B_\nu z^\nu}{\nu!},$$

which defines Bernoulli's numbers (cf. Vol. I, Chap. VIII, Appendix, p. 422), must have the circle $|z| = 2\pi$ as its circle of convergence, for the denominator of the function vanishes for $z = 2\pi i$ but (apart from the origin) at no point interior to the circle $|z| \leqq 2\pi$.

Example

Prove, without using the theory of power series directly, that the derivative of an analytic function is differentiable, by successive differentiation under the integral sign in Cauchy's formula and justification of the validity of this process.

3. The Theory of Functions and Potential Theory.

From the fact that analytic functions may be differentiated as often as we please it also follows that the functions $u(x, y)$ and $v(x, y)$ have continuous derivatives of any order. We may therefore differentiate the Cauchy-Riemann equations. If we differentiate the first equation with respect to x and the second with respect to y and add, we have

$$\Delta u = u_{xx} + u_{yy} = 0;$$

in the same way, the imaginary part v satisfies the same equation

$$\Delta v = v_{xx} + v_{yy} = 0.$$

In other words, *the real part and the imaginary part of an analytic function are potential functions.*

If two potential functions u, v satisfy the Cauchy-Riemann equations, v is said to be conjugate to u, and $-u$ conjugate to v.

We accordingly find that the theory of functions of a complex variable and potential theory in two dimensions are essentially equivalent to one another.

EXAMPLE

Show that for every potential function u it is possible to construct a conjugate function v and to determine it uniquely apart from an additive constant.

4. The Converse of Cauchy's Theorem.

As a further deduction we have the *converse of Cauchy's theorem*:

If the continuous function $\zeta = u + iv = f(z)$ is such that its integral round every closed curve C in its region of definition R vanishes, then $f(z)$ is an analytic function in R.

To prove this we note that in any case, by § 3, p. 539, the integral $\int_{t_0}^{z} f(t)\,dt$ taken along any path joining a fixed point t_0 and a variable point z is a differentiable function $F(z)$, where $F'(z) = f(z)$. $F(z)$ is therefore analytic, and by our result above so is its derivative $F'(z) = f(z)$.

This converse of Cauchy's theorem shows that the postulate of differentiability could have been replaced by the postulate of integrability. The equivalence of these two postulates is a very characteristic feature of the theory of functions of a complex variable.

5. Zeros, Poles, and Residues of an Analytic Function.

If the function $f(z)$ vanishes at the point $z = z_0$, the constant term in the Taylor series of the function in powers of $z - z_0$,

$$f(z) = f(z_0) + (z - z_0)f'(z_0) + \dots,$$

vanishes, and possibly further terms of the series vanish in addition. A factor $(z - z_0)^n$ may then be taken out of the power series and we may write

$$f(z) = (z - z_0)^n g(z),$$

where $g(z_0) \neq 0$. A point z_0 for which this occurs is said to be a *zero of the function* f(z) *of the* n-*th order*.

The reciprocal $1/f(z) = q(z)$ of an analytic function, as we saw above, is also analytic, except at the points where $f(z)$ vanishes. If z_0 is a zero of $f(z)$ of the n-th order, the function $q(z)$ can be represented in the neighbourhood of the point z_0 in the form

$$q(z) = \frac{1}{(z - z_0)^n} \frac{1}{g(z)} = \frac{1}{(z - z_0)^n} h(z),$$

where $h(z)$ is analytic in the neighbourhood of $z = z_0$. At the point $z = z_0$ the function $q(z)$ ceases to be analytic. We call this point a *singularity* (*singular point*), in this particular case a *pole* of the function $q(z)$ *of the n-th order*. If we think of the function $h(z)$ as expanded in powers of $(z - z_0)$ and then divided by $(z - z_0)^n$ term by term, in the neighbourhood of the pole we obtain an expansion of the form

$$q(z) = c_{-n}(z - z_0)^{-n} + \ldots + c_{-1}(z - z_0)^{-1} + c_0 + c_1(z - z_0) + \ldots,$$

where the coefficients of the powers of $(z - z_0)$ are denoted by $c_{-n}, \ldots, c_{-1}, c_0, c_1, \ldots$.

If we are dealing with a pole of the first order, i.e. if $n = 1$, we obtain the coefficient c_{-1} immediately from the relation

$$c_{-1} = \lim_{z \to z_0} (z - z_0) q(z).$$

Since

$$\frac{1}{q(z)(z - z_0)} = \frac{f(z)}{z - z_0} = \frac{f(z) - f(z_0)}{z - z_0},$$

we have

$$c_{-1} = \frac{1}{f'(z_0)}.$$

In the same way, if $q(z) = r(z)/\phi(z)$, and $\phi(z)$ has a zero of the first order at $z = z_0$, while $r(z_0) \neq 0$, we have

$$c_{-1} = \frac{r(z_0)}{\phi'(z_0)}.$$

If a function is defined and analytic everywhere in the neighbourhood of a point z_0, but not at the point itself, its integral round a complete circle enclosing the point z_0 will in general not be zero. By Cauchy's theorem, however, the integral is independent of the radius of this circle and in general has the same value for all closed curves C which form the boundary of a sufficiently small region enclosing the point z_0. The value of the integral taken round the point in the positive sense is called the *residue* at the point.

If the singularity is a pole of the n-th order and if we integrate the expansion of the function, the integral of the series with positive indices is zero, as this power series is still analytic at the point z_0.

When integrated the term $c_{-1}(z-z_0)^{-1}$ gives the value $2\pi i c_{-1}$, while the terms with higher negative indices give zero, for the indefinite integral of $(z - z_0)^{-\nu}$ for $\nu > 1$ is $(z - z_0)^{-\nu+1}/(1 - \nu)$, as in the real case, so that the integral round a closed curve vanishes.

The residue of a function at a pole is therefore $2\pi i c_{-1}$.

In the next section we shall become acquainted with the usefulness of this idea as expressed by the following theorem:

Theorem of Residues. If the function f(z) *is analytic in the interior of a region* R *and on its boundary* C, *except at a finite number of poles, the integral of the function taken round* C *in the positive sense is equal to the sum of the residues of the function at the poles enclosed by the boundary* C.

The proof follows at once from the statements above.

EXAMPLES

1*. Show that the function

$$f(z) - \frac{1}{2\pi i}\int \frac{f(\zeta)}{\zeta - z}\frac{z^n}{\zeta^n}\,d\zeta,$$

where the integral is taken round a simple contour enclosing the points $\zeta = 0$ and $\zeta = z$, is a polynomial $g(z)$ of degree $n - 1$ such that

$$g^{(m)}(0) = f^{(m)}(0) \quad \text{for} \quad m = 0, 1, \ldots, n - 1.$$

2. Let $f(z)$ be analytic for $|z| \leq \rho$. If M is the maximum of $|f(z)|$ on the circle $|z| = \rho$, then the coefficients of the power series for f,

$$f(z) = \sum_{\nu=0}^{\infty} a_\nu z^\nu,$$

satisfy the inequality

$$|a_\nu| \leq \frac{M}{\rho^\nu}.$$

3*. Prove that if a region is bounded by a single closed curve C, and if $f(z)$ is analytic in the interior of C and on C and does not vanish on C, then

$$\frac{1}{2\pi i}\int_C \frac{f'(z)}{f(z)}\,dz$$

is the number of zeros of f in the interior of C.

4. (*a*) Two polynomials $P(z)$ and $Q(z)$ are such that at every point on a certain closed contour C

$$|Q(z)| < |P(z)|.$$

Prove that the equations $P(z) = 0$ and $P(z) + Q(z) = 0$ have the same numbers of roots within C. (Consider the family of functions $P(z) + \theta Q(z)$, where the parameter θ varies from 0 to 1.)

(*b*) Prove that all the roots of the equation

$$z^5 + az + 1 = 0$$

lie within the circle $|z| = r$ if

$$|a| < r^4 - \frac{1}{r}.$$

5. If $f(z) = 0$ has one simple root a within a closed curve C, prove that this root is given by

$$a = \frac{1}{2\pi i}\int_C z\frac{f'(z)}{f(z)}\,dz.$$

5. Applications to Complex Integration (Contour Integration)

Cauchy's theorem and the theorem of residues frequently enable us to evaluate real definite integrals by regarding these as integrals along the real axis of a complex plane and then simplifying the argument by suitable modification of the path of integration. In this way we sometimes obtain surprisingly elegant evaluations of apparently complicated definite integrals, without necessarily being able to calculate the corresponding indefinite integrals. We shall discuss some typical examples.

1. Proof of the Formula

$$\int_0^\infty \frac{\sin x}{x}\,dx = \frac{\pi}{2}.$$

Here we give the following instructive proof of this important formula, which we have already discussed by other methods (Vol. I, pp. 251, 418, 450; Vol. II, p. 315).

We integrate the function e^{iz}/z in the complex z-plane along the path C shown in fig. 7, which consists of a semicircle H_R of radius R, a semicircle H_r of radius r, both having their centres at the origin, and the two symmetrical intervals I_1 and I_2 of the real axis. Since the function e^{iz}/z is regular in the circular ring enclosed by these boundaries, the value of the integral in question is zero. Combining the integrals along I_1 and I_2, we have

$$\int_{H_R} \frac{e^{iz}}{z}\,dz + \int_{H_r} \frac{e^{iz}}{z}\,dz + 2i\int_r^R \frac{\sin x}{x}\,dx = 0.$$

We now let R tend to infinity. Then the integral along the semi-circle H_R tends to zero. For if we put $z = R(\cos\theta + i\sin\theta) = Re^{i\theta}$

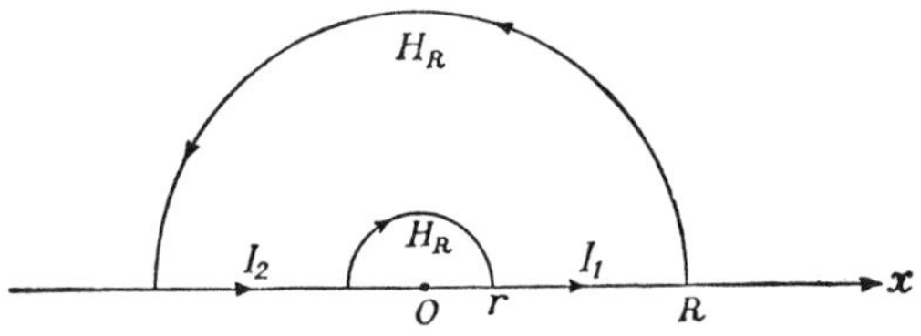

Fig. 7

for points on the semicircle, we have $e^{iz} = e^{iR\cos\theta}e^{-R\sin\theta}$, and the integral becomes $i\int_0^{\pi} e^{iR\cos\theta}e^{-R\sin\theta}\,d\theta$. The absolute value of the factor $e^{iR\cos\theta}$ is 1, while the absolute value of the factor $e^{-R\sin\theta}$ is less than 1 and, moreover, tends uniformly to zero as R tends to infinity, in every interval $\epsilon \leqq \theta \leqq \pi - \epsilon$. Hence it follows at once that the integral along H_R tends to zero as $R \to \infty$. As the reader can easily prove for himself, the integral along the semicircle H_r tends to $-\pi i$ as $r \to 0$. The integral along the two symmetrical intervals I_1, I_2 of the real axis tends to $2i\int_0^{\infty} \frac{\sin x}{x}\,dx$ as $R \to \infty$ and $r \to 0$. Combining these statements, we immediately obtain the relation given above.

2. **Proof of the Formula**

$$\int_0^{\infty} \cos ax\, e^{-x^2}\,dx = \tfrac{1}{2}\sqrt{\pi}\,e^{-\frac{1}{4}a^2}.$$

We have already proved this formula in Chap. IV (Ex. 4a, p. 318), but we shall now obtain it by means of Cauchy's theorem.

We integrate the expression e^{-z^2} along a rectangle $ABB'A'$ (fig. 8), in which the length of the vertical sides AA', BB' is $a/2$, and that of the horizontal sides AB, $A'B'$ is $2R$. This integral has the value zero, by Cauchy's theorem. On the vertical sides we have $|e^{-z^2}| = |e^{-(x^2-y^2)}e^{-2ixy}| = e^{-R^2}e^{y^2} < e^{-R^2}e^{\frac{1}{4}a^2}$, and this expression tends uniformly to zero as R tends to infinity. Thus the portions of the whole integral which arise from the vertical sides tend to zero, and if we carry out the passage to the limit

$R \to \infty$ and note that on $A'B'$ $dz = d(x + \frac{1}{2}ia) = dx$, we may express the result of Cauchy's theorem as follows:

$$\int_{-\infty}^{\infty} e^{-(x+\frac{1}{2}ia)^2} dx = \int_{-\infty}^{\infty} e^{-x^2} dx.$$

That is, we can displace the path of integration of the infinite integral parallel to itself. By our previous result * (p. 262) the

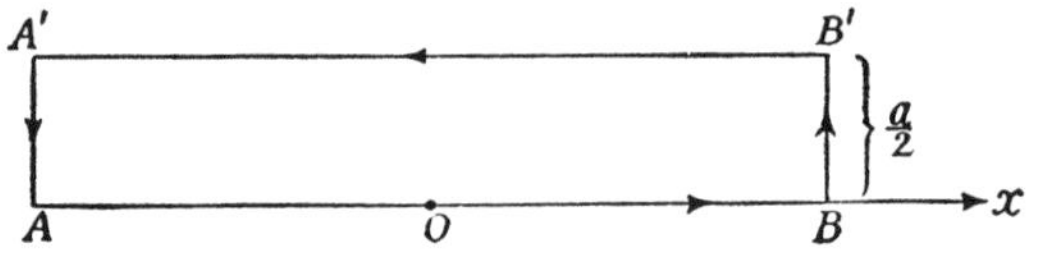

Fig. 8

value of the integral on the right is $\sqrt{\pi}$. The integral on the left immediately becomes

$$e^{\frac{1}{4}a^2}\int_{-\infty}^{\infty} e^{-x^2}(\cos ax - i \sin ax)\, dx = 2e^{\frac{1}{4}a^2}\int_{0}^{\infty} \cos ax\, e^{-x^2} dx,$$

if we remember that $\sin ax$ is an odd function and $\cos ax$ an even function. This proves the formula.

3. Application of the Theorem of Residues to the Integration of Rational Functions.

If in the rational function

$$Q(z) = \frac{a_0 + a_1 z + \ldots + a_m z^m}{b_0 + b_1 z + \ldots + b_n z^n}$$

the denominator has no real zeros and its degree exceeds that of the numerator by at least two, the integral

$$I = \int_{-\infty}^{\infty} Q(x)\, dx$$

can be evaluated in the following way.

We begin by taking the integral along a contour consisting of the boundary of a semicircle H of radius R (on which $z = Re^{i\theta}$, $0 \leqq \theta \leqq \pi$), where R is chosen so large that no pole of $Q(z)$ lies on or outside the circumference of the circle, and the real axis from $-R$ to $+R$. Then on the one hand the integral is equal

* Cf. also sub-section 6, p. 561.

to the sum of the residues of $Q(z)$ within the semicircle, while on the other hand it is equal to the integral

$$I_R = \int_{-R}^{R} Q(x)\,dx$$

plus the integral along the semicircle H. By our assumptions, a fixed positive constant M exists such that for sufficiently large values of R we have *

$$|Q(z)| < \frac{M}{R^2}.$$

The length of the circumference of the semicircle is πR. By our estimation formula on p. 539, the integral along the semicircle is therefore less in absolute value than $\pi R \dfrac{M}{R^2} = \dfrac{\pi M}{R}$, and hence tends to zero as $R \to \infty$. This means that *the integral*

$$I = \int_{-\infty}^{\infty} Q(x)\,dx$$

is equal to the sum of the residues of Q(z) *in the upper half-plane.*

We now apply this principle to some interesting special cases.

We begin by taking

$$Q(z) = \frac{1}{az^2 + bz + c} = \frac{1}{f(z)},$$

where the coefficients a, b, c are real and satisfy the conditions $a > 0$, $b^2 - 4ac < 0$. Then the function $Q(z)$ has only one simple pole

$$z = z_1 = \frac{1}{2a}\{-b + i\sqrt{(4ac - b^2)}\},$$

where the square root is to be taken positive, in the upper half-plane. By the general rule (p. 553), therefore, the residue is $2\pi i \dfrac{1}{f'(z_1)}$. Since

$$f'(z_1) = 2az_1 + b = i\sqrt{(4ac - b^2)},$$

we have

$$\int_{-\infty}^{\infty} \frac{1}{ax^2 + bx + c}\,dx = \frac{2\pi}{\sqrt{(4ac - b^2)}}.$$

* This follows immediately from the fact that $Q(z) = \dfrac{1}{z^2} R(z)$, where $R(z)$ tends to zero as $z \to \infty$ (when $n > m + 2$) or to a_m/b_n (when $n = m + 2$).

As a second example we shall prove the formula (cf. Vol. I, p. 234)

$$\int_{-\infty}^{+\infty} \frac{dx}{1+x^4} = \tfrac{1}{2}\pi\sqrt{2}.$$

Here again we can immediately apply our general principle. In the upper half-plane the function $1/(1+z^4) = 1/f(z)$ has the two poles $z_1 = \varepsilon = e^{\frac{1}{4}\pi i}$, $z_2 = -\varepsilon^{-1}$ (the two fourth roots of -1 which have a positive imaginary part). The sum of the residues is

$$2\pi i \left\{\frac{1}{f'(z_1)} + \frac{1}{f'(z_2)}\right\} = 2\pi i \frac{1}{4}\left(\frac{1}{z_1{}^3} + \frac{1}{z_2{}^3}\right) = \frac{\pi i}{2}(\varepsilon^{-3} - \varepsilon^3),$$

$$= -\pi i \,.\, i \sin\frac{3\pi}{4} = \pi \sin\frac{\pi}{4} = \tfrac{1}{2}\pi\sqrt{2},$$

as was asserted.

EXAMPLES

1. Prove the formula

$$\int_{-\infty}^{\infty} \frac{x^2}{1+x^4}\, dx = \tfrac{1}{2}\pi\sqrt{2}$$

in the same way as above.

2. Prove that in general if n and m are positive integers and $n > m$,

$$\int_{-\infty}^{\infty} \frac{x^{2m}}{1+x^{2n}}\, dx = \frac{\pi}{n} \sin\left(\frac{2m+1}{2n}\pi\right).$$

The following proof of the formula

$$\int_{-\infty}^{\infty} \frac{dx}{(1+x^2)^{n+1}} = \frac{\pi}{4^n}\frac{(2n)!}{(n!)^2}$$

exemplifies the case where the residue at a pole of higher order has to be calculated.

If we replace x by z, the denominator of the integrand is of the form $(z+i)^{n+1}(z-i)^{n+1}$, and the integrand accordingly has a pole of the $(n+1)$-th order at the point $z = +i$. To find the residue at that point we write

$$\frac{1}{(z^2+1)^{n+1}} = \frac{1}{f(z)} = \frac{1}{(z-i)^{n+1}} \frac{1}{(2i+z-i)^{n+1}}$$

$$= \frac{1}{(z-i)^{n+1}} \frac{1}{(2i)^{n+1}} \left(1 + \frac{z-i}{2i}\right)^{-n-1}.$$

If we expand the last factor by the binomial theorem, the term in $(z-i)^n$ has the coefficient

$$\frac{1}{(2i)^n}\binom{-n-1}{n} = \frac{1}{(2i)^n}(-1)^n \frac{(n+1)\ldots 2n}{1.2\ldots n} = \frac{i^n}{2^n}\frac{(2n)!}{(n!)^2}.$$

The coefficient c_{-1} in the series for the integrand in the neighbourhood of the point $z = i$ is therefore equal to $\frac{1}{2^{2n+1}}\frac{1}{i}\frac{(2n)!}{(n!)^2}$. The residue $2\pi i c_{-1}$ is therefore $\frac{\pi}{2^{2n}}\frac{(2n)!}{(n!)^2}$, which proves the formula.

As a further exercise the reader may prove for himself by the theory of residues that

$$\int_{-\infty}^{\infty} \frac{x \sin x}{x^2 + c^2}\, dx = \tfrac{1}{2}\pi e^{-|c|}$$

(replacing $\sin x$ by e^{ix}).

Example

Let $f(x)$ be a polynomial of degree n with the simple roots $\alpha_1, \alpha_2, \ldots, \alpha_n$. Prove that

$$\sum_{\nu=1}^{n} \frac{\alpha_\nu{}^k}{f'(\alpha_\nu)} = 0 \quad (k = 0, 1, \ldots, n-2).$$

(Consider $\int \frac{z^k}{f(z)}\, dz$ round a closed curve enclosing all the α_ν's.)

4. The Theorem of Residues and Linear Differential Equations with Constant Coefficients.

If

$$a_0 + a_1 z + a_2 z^2 + \ldots + a_n z^n = P(z)$$

is a polynomial of the n-th degree, and t a real parameter, we think of the integral

$$u(t) = \int_C \frac{e^{tz} f(z)}{P(z)}\, dz,$$

taken along any closed path C in the z-plane, which does not pass through any of the zeros of $P(z)$, as a function $u(t)$ of the parameter t. Let $f(z)$ be a constant or any polynomial in z, of a degree which we shall assume to be less than n. By the rules for differentiation under the integral sign, which hold unaltered for the complex region, we can differentiate the expression $u(t)$ once or repeatedly with respect to t. This differentiation with respect to t under the integral sign is equivalent to multiplication of the integrand by $z, z^2, z^3, \ldots$, as the case may be. If we now form the differential expression $L[u] = a_0u + a_1u' + a_2u'' + \ldots + a_nu^{(n)}$, or, in symbolic notation, $P(D)u$, where D denotes the symbol of differentiation $D = d/dt$, we have

$$P(D)u = L[u] = \int_C e^{tz} f(z)\, dz.$$

By Cauchy's theorem the value of the complex integral on the right is zero; i.e. the function $u(t)$ is a solution of the differential equation $L[u] = 0$. If $f(z)$ is any polynomial of the $(n-1)$-th degree, this solution contains n arbitrary constants. We may accordingly expect to get in this way the most general solution of the linear differential equation with constant coefficients, $L[u] = 0$.

In fact we do obtain the solutions in the form which we already know (cf. Chap. VI, § 4, p. 449), on evaluating the integral by the theory of residues, with the assumption that the curve C encloses all the zeros $z_1, z_2, \ldots, z_n$ of the denominator $P(z) = a_n(z - z_0)(z - z_1) \ldots (z - z_n)$. If we assume to begin with that all these zeros are simple zeros, they are simple poles of the integrand, and the residue at the point a_ν is $2\pi i \dfrac{f(z_\nu)}{P'(z_\nu)} e^{tz_\nu}$. By suitable choice of the polynomial $f(z)$ the expressions $f(z_\nu)/P'(z_\nu)$ can be made arbitrary constants; we accordingly obtain the solution in the form

$$u(t) = \sum_{\nu=1}^{n} c_\nu e^{z_\nu t},$$

in agreement with our previous results.

If a zero z_ν of the polynomial $P(z)$ is multiple, say r-fold, so that the corresponding pole of the integrand is of the r-th order, the residue at the point z_ν must be determined by imagining the numerator $e^{tz} f(z) = e^{tz_\nu} e^{t(z-z_\nu)} f(z)$ also expanded in powers of $z - z_\nu$. We leave it to the reader to show that the residue at the point z_ν gives the solutions $te^{tz_\nu}, \ldots, t^{r-1}e^{tz_\nu}$ as well as the solution e^{tz_ν}.

6. Proof of the Formula

$$\int_{-\infty}^{\infty} e^{-x^2}\,dx = \sqrt{\pi}.$$

In evaluating the integral on p. 555 we took over this formula as known from the theory of real variables. It is, however, possible to obtain the result by complex integration, using the theory of residues. As this proof is very instructive, we shall give it here, although from our elementary point of view its starting-point may appear artificial. We begin with a complex integral which arises in other branches of mathematics (e.g. the theory of numbers).

We use the symbol $/\frac{1}{2}$ to denote the straight line $z = \frac{1}{2} + \rho e^{i\pi/4}$ $(-\infty < \rho < \infty)$ in the z-plane, that is, a straight line making an angle of 45° with the x-axis and cutting it at the point $\frac{1}{2}$. The symbol $/-\frac{1}{2}$ or $/0$ will bear a similar meaning. Let u be a real parameter. We then consider the integral

$$f(u) = \int_{/\frac{1}{2}} \frac{e^{\pi i z^2 + 2\pi i u z}}{e^{2\pi i z} - 1}\,dz.$$

This integral is to be regarded as an improper integral, that is, we integrate in the first place between the limits $\rho = -R$, $\rho = R$, and then let R tend to infinity. The reader may verify that this integral exists by means of an argument following the pattern of similar arguments for real integrals. Then

$$\begin{aligned} f(u+1) - f(u) &= \int_{/\frac{1}{2}} \frac{e^{\pi i z^2}}{e^{2\pi i z} - 1}\, e^{2\pi i u z}(e^{2\pi i z} - 1)\,dz \\ &= \int_{/\frac{1}{2}} e^{\pi i z^2 + 2\pi i u z}\,dz \\ &= e^{-\pi i u^2} \int_{/\frac{1}{2}} e^{\pi i (z+u)^2}\,dz. \end{aligned}$$

As the integrand on the right is regular everywhere, we can use Cauchy's theorem to displace the path of integration parallel to itself to any extent, as on p. 556, writing, for example,

$$f(u+1) - f(u) = e^{-\pi i u^2} \int_{/0} e^{\pi i z^2}\,dz = e^{-\pi i u^2} I,$$

where $z = \rho e^{i\pi/4}$ on the path of integration and hence

$$I = e^{i\pi/4}\int_{-\infty}^{\infty} e^{-\pi\rho^2}\,d\rho.$$

That is, if we substitute $\sqrt{\pi}\rho = t$, we have

$$I = e^{i\pi/4}\frac{1}{\sqrt{\pi}}\int_{-\infty}^{\infty} e^{-t^2}\,dt.$$

Again, if we put $z = \lambda + 1$ and take λ as the new variable of integration, we obtain the expression

$$f(u) = -\int_{/-\frac{1}{2}} \frac{e^{\pi i\lambda^2 + 2\pi i\lambda u}}{e^{2\pi i\lambda} - 1}\, e^{2\pi iu} e^{2\pi i\lambda}\,d\lambda,$$

using the facts that $e^{2\pi i} = 1$, $e^{\pi i} = -1$, or

$$-e^{-2\pi iu} f(u) = \int_{/-\frac{1}{2}} e^{\pi i\lambda^2 + 2\pi i\lambda u}\,d\lambda + \int_{/-\frac{1}{2}} \frac{e^{\pi i\lambda^2 + 2\pi i\lambda u}}{e^{2\pi i\lambda} - 1}\,d\lambda.$$

By the above result, as we can again displace the path of integration parallel to itself, the first integral on the right is equal to $e^{-\pi iu^2}I$. If we replace the second integral by the integral obtained for $f(u)$ by displacing the path of integration through an interval 1 to the right, we have to note that the pole $\lambda = 0$ of the integrand lies between the two paths of integration.

We now apply the theorem of residues—the fact that the path of integration $/-\frac{1}{2}$ and $/\frac{1}{2}$ extends to infinity gives us no trouble, in virtue of the analogous discussion on p. 556—prove that the residue of the integrand at the point $\lambda = 0$ has the value 1, and then at once obtain the result

$$-f(u)e^{-2\pi iu} = e^{-\pi iu^2}I + f(u) - 1$$

from our equation. Here neither I nor the function $f(u)$ is explicitly known. If, however, we put $u = \frac{1}{2}$, $f(u)$ disappears from the equation, and we are left with

$$e^{-\pi i/4}I = 1.$$

But since

$$I = e^{\pi i/4}\frac{1}{\sqrt{\pi}}\int_{-\infty}^{\infty} e^{-t^2}\,dt,$$

the real integral formula follows at once.

6. Many-valued Functions and Analytic Extension

In defining functions both real and complex we have hitherto always adopted the point of view that for each value of the independent variable the value of the function must be *unique*. Even Cauchy's theorem, for example, is based on the assumption that the function can be defined uniquely in the region under consideration. All the same, many-valuedness often arises of necessity in the actual construction of functions, e.g. in finding the inverse of a unique function such as the n-th power. In the real case we separated different *one-valued branches* of the inverse function in inversion processes such as $\sqrt{z}$ or $\sqrt[n]{z}$. We shall see, however, that in the complex case this separation is no longer possible, for the various one-valued branches are now interconnected.

We must be content here with a very simple discussion based on typical examples.

For instance, we shall consider the inverse $\zeta = \sqrt{z}$ of the function $z = \zeta^2$. To one value of z there correspond the two possible solutions ζ and $-\zeta$ of the equation $z = \zeta^2$. These two branches of the function are connected in the following way. Let $z = re^{i\theta}$. If we then put $\zeta = \sqrt{r}\, e^{i\theta/2} = f(z)$, $\zeta = f(z)$ is certainly analytic in every simply-connected region R excluding the origin (where $f(z)$ is no longer differentiable). In such a region ζ is uniquely defined, by our previous statement. If, however, we let the point z move round the origin on a concentric circle K, say in the positive direction, $\zeta = \sqrt{r}\, e^{i\theta/2}$ will vary continuously; the angle θ, however, will not return to its original value, but will be increased by 2π. Hence in this continuous extension when we come back to the point z we no longer have the initial value $\zeta = \sqrt{r}\, e^{i\theta/2}$, but the value $\sqrt{r}\, e^{i\theta/2} e^{2\pi i/2} = -\zeta$. We say that when it is continuously extended on the closed curve K the function $f(z)$ is not unique.

The function $\sqrt[n]{z}$, where n is an integer, exhibits exactly the same behaviour. Here every revolution multiplies the value of the function by the n-th root of unity, namely $\varepsilon = e^{2\pi i/n}$, and the function only returns to its original value after n revolutions.

In the case of the function $\log z$ we saw (p. 543) that there is a similar many-valuedness, in that in travelling once continuously round the origin in the positive sense the value of $\log z$ is increased by $2\pi i$.

Again, the function z^a is multiplied by $e^{2\pi i a}$ per revolution.

All these functions, although in the first instance uniquely defined in a region R, are found to be many-valued when we extend them continuously (as analytic functions) and return to the starting-point by a certain closed path. This phenomenon of many-valuedness and the associated general theory of analytic extension cannot be investigated in greater detail within the limits of this book. We would merely point out that the uniqueness of the values of a function can theoretically be ensured by drawing certain lines in the z-plane which the path traced by z is not allowed to cross, or, as we say, by making cuts along certain lines. These cuts are so arranged that closed paths in the plane which lead to many-valuedness are no longer possible.

For example, the function $\log z$ is made one-valued by cutting the z-plane along the negative real axis. The same applies to the function $\sqrt{z}$. The function $\sqrt{(1-z^2)}$ becomes one-valued if we make a cut along the real axis between -1 and $+1$.

Once the plane has been cut in this way, Cauchy's theorem can at once be applied to these functions.

We now give a simple example showing how Cauchy's theorem is applied in a case where many-valued functions arise, by proving the formula

$$I = \int_{-1}^{+1} \frac{1}{(x-k)\sqrt{(1-x^2)}}\,dx = \frac{2\pi}{\sqrt{(k^2-1)}},$$

where k is a constant which does not lie on the real axis between -1 and $+1$.

We begin by noting that the function $\dfrac{1}{(z-k)\sqrt{(1-z^2)}}$ is one-valued in the z-plane provided we make a cut along the real axis from -1 to $+1$. If in the complex plane we approach this cut S first from above and then from below, we obtain equal and opposite values for the square root $\sqrt{(1-z^2)}$, say positive from above and negative from below. We now take the complex integral

$$\int_C \frac{dz}{(z-k)\sqrt{(1-z^2)}}$$

along a path C as indicated in fig. 9. By Cauchy's theorem we can make this path contract round the cut without altering the value of the integral. The integral is therefore equal to the limiting value obtained when this contraction is made, which is obviously equal to $2I$. On the other hand, if we take the integral of the same integrand along the circumference of a circle K with radius R and centre the origin, this integral, by our previous

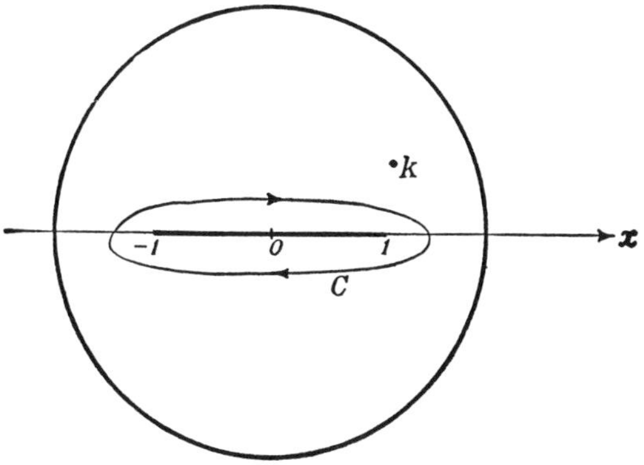

Fig. 9

investigations, tends to zero* as R increases. By the theorem of residues, however, the sum of the integrals along C and K is equal to the residue of the integrand at the enclosed pole $z = k$; hence $2I$ is equal to the residue in question. This residue is

$$2\pi i \lim_{z \to k} (z - k) \frac{1}{\sqrt{(1 - z^2)}} \frac{1}{(z - k)} = \frac{2\pi}{\sqrt{(k^2-1)}},$$

which proves our statement.

Example of Analytic Extension. The Gamma Function.—In conclusion we give yet another example showing how an analytic function, originally defined in a part of the plane only, can be extended beyond the original region of definition. We shall extend the gamma function, which was defined for $x > 1$ by the equation

$$\Gamma(z) = \int_0^\infty t^{z-1} e^{-t}\, dt,$$

analytically for $x \leq 1$ also. We can do this e.g. by means of the functional equation $\Gamma(z) = \frac{1}{z}\Gamma(z + 1)$, using this equation to define $\Gamma(z - 1)$ when $\Gamma(z)$ is known. By means of this equation we can imagine $\Gamma(z)$ as extended first in the parallel strip

* In fact, its value is actually zero, since by Cauchy's theorem it is independent of the radius R, provided that the circle encloses the pole $z = k$.

$-1 < x \leqq 0$ and subsequently extended to the next parallel strip $-2 < x \leqq -1$, and so on.

We can, however, adopt another method, of greater theoretical interest, for extending the gamma function. We consider the path C in the t-plane indicated in fig. 10, which surrounds the

Fig. 10.—Loop-integral for the gamma function

positive real axis of the t-plane and approaches this axis asymptotically on either side. We easily see from Cauchy's theorem that the value of the "loop-integral",*

$$\int_C t^{z-1}e^{-t}\,dt,$$

is unaltered when the loop is made to contract into the x-axis. The integrand $t^{z-1}e^{-t}$ then tends to different values as we approach the x-axis from above and below, the values differing by the factor $e^{2\pi iz}$. For $x > 0$ we thus obtain the formula

$$(1 - e^{2\pi iz})\Gamma(z) = \int_C t^{z-1}e^{-t}\,dt.$$

This formula is deduced subject to the assumption that x, the real part of z, is positive. We see now, however, that the loop-integral has a meaning, no matter what the complex number z is, since it avoids the origin $t = 0$. This loop-integral therefore represents a function which is defined throughout the z-plane. We then define this function by stating that it is equal to $(1 - e^{2\pi iz})\Gamma(z)$ throughout the z-plane. The gamma function has thus been analytically extended to the whole of the z-plane, except the points $x < 0$ for which the factor $(1 - e^{2\pi iz})$ vanishes, that is, except the points $z = 0$, $z = -1$, $z = -2$, and so on.

For more detailed and more extensive investigations the reader must be referred to the literature of the theory of functions.†

* This is again an improper integral, which arises by a passage to a limit from an integral along a finite portion of C. The reader may satisfy himself that it exists, by an argument similar to those previously employed.

† E.g. MacRobert, *Functions of a Complex Variable* (Macmillan); Whittaker and Watson, *Modern Analysis* (Cambridge University Press); Watson, *Complex Integration and Cauchy's Theorem* (Cambridge Tracts, No. 15).

Miscellaneous Examples VIII

1. Write down the condition that three points z_1, z_2, z_3 may lie in a straight line.

2*. Write down the condition that four points z_1, z_2, z_3, z_4 may lie on a circle.

3*. Let A, B, C, D in the z-plane be four points in order on the circumference of a circle, with co-ordinates z_1, z_2, z_3, z_4. Using these complex co-ordinates, show that $AB \,.\, CD + BC \,.\, AD = AC \,.\, BD$.

4. Prove that the equation $\cos z = c$ can be solved for all values of c.

5. For which values of c has the equation $\tan z = c$ no solution?

6. For which values of z is (a) $\cos z$, (b) $\sin z$ real?

7. Find the radius of convergence of the power series $\Sigma a_n z^n$, where

(a) $a_n = \dfrac{1}{n^s}$, s being a complex number with a positive real part;

(b) $a_n = n^n$;

(c) $a_n = \log n$.

8. Prove the formula

$$e^z = \lim_{n \to \infty} \left(1 + \frac{z}{n}\right)^n,$$

where z is complex.

9. Evaluate the integrals

$$(a) \int_0^\infty \frac{\cos x}{1 + x^4}\,dx, \qquad (b) \int_0^\infty \frac{x^2 \cos x}{1 + x^4}\,dx, \qquad (c) \int_0^\infty \frac{\cos x}{q^2 + x^2}\,dx,$$

$$(d) \int_0^\infty \frac{x^{\alpha-1}}{(x+1)(x+2)}\,dx \text{ for } 1 < \alpha < 2$$

by complex integration.

10. Find the poles and residues of the functions

$$\frac{1}{\sin z}, \quad \frac{1}{\cos z}, \quad \Gamma(z), \quad \cot z = \frac{\cos z}{\sin z}.$$

11*. Find the limiting value of the integral

$$\int_{C_n} \frac{\cot \pi t}{t - z}\,dt$$

as $n \to \infty$, where the path of integration is a square C_n with its sides parallel to the axes at a distance $n \pm \frac{1}{2}$ from the origin. Hence, using the theorem of residues, obtain the expression for $\cot \pi z$ in partial fractions.

12*. Using the equation

$$\log(1 + z) = \int_0^z \frac{dt}{1 + t},$$

show that the power series for $\log(1+z)$ converges everywhere on the unit circle $|z| = 1$, except at the point $z = -1$. By equating the imaginary part of the series to the imaginary part of $\log(1 + e^{i\theta})$, establish the truth of the Fourier series (cf. Vol. I, p. 440)

$$\tfrac{1}{2}\theta = \sin\theta - \tfrac{1}{2}\sin 2\theta + \tfrac{1}{3}\sin 3\theta - \ldots \quad (-\pi < \theta < \pi).$$

13*. (*a*) Prove that the series

$$f(z) = f(x + iy) = \sum_{\nu=1}^{\infty} \frac{(-1)^{\nu+1}}{\nu^z}$$

converges for $x > 0$.

(*b*) Prove that this series provides an extension of the zeta function (defined in Ex. 2, p. 545) to values of z such that $0 < x \leqq 1$, by means of the formula

$$f(z) = (1 - 2^{1-z})\zeta(z),$$

which is valid for $x > 1$.

(*c*) Prove that the zeta function has a pole of residue 1 at $z = 1$.

SUPPLEMENT

Real Numbers and the Concept of Limit

In Vol. I, Chapter I, it was taken for granted that the real numbers form an aggregate within which the ordinary operations of arithmetic may be performed as with the rational numbers. We shall investigate this assumption more closely here. We take the arithmetical operations on the rational numbers as given. Our object is then to make an abstract analytical extension of the class of rational numbers which shall yield the wider class of real numbers, and to do this without relying on intuition in our proofs. We must frame our definitions in such a way that, as a logical consequence of them, the ordinary rules of arithmetic apply to all real numbers just as they do to rational numbers.

The introduction of irrational numbers will be undertaken in close conjunction with a thorough consideration of the concept of limit, in which we shall repeat in a revised form the discussion of Vol. I, Chapter I, Appendix (p. 58 *et seq.*).*

1. Definition of the Real Numbers by means of Nests of Intervals.

The irrational numbers and, in general, the real numbers were defined in Vol. I, Chapter I, § 1, p. 8, by means of decimals, the rational numbers being represented by terminating or recurring decimals. By such a decimal, say $\alpha = 0{\cdot}a_1a_2a_3 \ldots$, we mean that the number represented, called α, lies between the rational number $\alpha_n = 0{\cdot}a_1 \ldots a_n$ and the rational number $\alpha_n + 10^{-n}$. The number α is thus determined by means of a sequence or *nest* of progressively smaller and smaller intervals, each inside the previous one, the n-th interval being of length 10^{-n}.

* The only difference in the point of view will be that here we shall start with the logical abstract concept of real numbers, while on the former occasion the properties of real numbers were taken for granted.

For our present purpose it would be inconvenient to restrict ourselves to special nests of intervals where the length of the n-th interval is 10^{-n}. We begin with the following general definition.

By a rational interval $(a \mid b)$ we mean the aggregate of all the rational numbers x which satisfy the inequalities $a \leqq x \leqq b$, where $a < b$ and a and b are rational numbers. The number $(b - a)$ is called the length of the interval. We say that the interval $(c \mid d)$ is contained in the interval $(a \mid b)$ if $a \leqq c < d \leqq b$. An infinite sequence of rational intervals $(a_1 \mid b_1)$, $(a_2 \mid b_2)$, ... is called a *nest of intervals* if every interval $(a_n \mid b_n)$ contains the next in order, $(a_{n+1} \mid b_{n+1})$, and the lengths $b_n - a_n$ tend to zero. That is, given any positive number ϵ, however small (the number ϵ must, of course, be rational, since no other numbers have as yet been introduced), there is a number $N(\epsilon)$ such that the lengths $b_n - a_n$ are less than ϵ for all suffixes n which exceed N.

From the intuitive meaning of a nest of intervals, and remembering in particular how we may pick out any point on the number axis by means of a nest of intervals, as on p. 9 of Vol. I, we arrive at the idea that we may *define* an arbitrary real number by a nest of intervals. This is to be taken as meaning the following: the real number is given by an unending process of approximation which is determined by the nest of intervals. The nest whose general member is $(a_n \mid b_n)$ gives us, with regard to the number a to be defined, the fact that this real number lies between a_1 and b_1; again, it lies between a_2 and b_2, between a_3 and b_3, and so on. The nest of intervals will thus give us two rational numbers, as near together as we please, between which the real number lies.

The essential step is now that we abandon the notion of obtaining an *objective* definition of the irrational numbers. We give up the attempt to characterize the irrational numbers as given mathematical entities with specific properties. We do not say that an irrational number *is* such and such a mathematical object; instead, we are content with the process of approximation which gives the nest of intervals and regard each such process as defining a real number. If there is a rational number a contained in all the intervals $(a_n \mid b_n)$, the real number defined by the nest of intervals $(a_n \mid b_n)$ is said to be identical with a. By this assumption the rational numbers become real numbers also.

The words *irrational number* or, more generally, *real number* may thus be regarded merely as a brief way of referring to a nest of intervals.*

This is what is meant by the statement that an irrational number is given or defined by a nest of intervals. In practice it comes to this, that every operation with real numbers is an operation with nests of intervals. This offers the possibility of making calculations with real numbers depend logically on operations with rational numbers.

It is necessary to lay down a procedure for defining addition, multiplication, &c., of real numbers by nests of intervals. Here the rules must be framed in such a way that the ordinary laws of calculation still apply. Moreover, we must ensure that the rules of calculation with rational numbers are not contradicted.

We shall begin by showing that our definition implies an ordering of the real numbers by magnitude. This in itself provides a sufficient groundwork for the axiomatic foundation of the concept of limit and a more thorough understanding of it. When this has been achieved, we shall return to the question of the rules of calculation with real numbers.

2. The Real Numbers in Order of Magnitude.

Let two numbers a and γ be given by nests of intervals $(a_n | b_n) = i_n$ and $(c_n | d_n) = j_n$. The following three cases may occur.

(1) From a certain stage $n = n_0$ onward every interval j_n lies to the right of the interval i_n; that is, for $n = n_0$, and of course for every $n > n_0$, we have $b_n < c_n$. We then say that γ is greater than a, or $\gamma > a$.

(2) If, on the other hand, from a certain n_0 onward i_n lies to the right of j_n, then we say that $a > \gamma$. In this case for $n \geqq n_0$ we have always $d_n < a_n$.

(3) Neither of the above situations arises. We then say that the two nests of intervals i_n and j_n define the same number: $a = \gamma$. Thus two nests of intervals define the same number if, and only if, the intervals i_n and j_n always overlap; that is, if

* Some process of this kind is often essential in giving a precise formulation to mathematical concepts. For instance, in projective geometry, when points at infinity are introduced these points are not treated as definite mathematical entities in themselves; we merely say that a point at infinity is given by a pencil of parallel lines.

both $a_n \leqq d_n$ and $b_n \geqq c_n$; or if the two intervals i_n and j_n have rational points in common for every n. A special consequence of this definition is that if, of two nests of intervals, one is obtained from the other by the omission of a finite or infinite number of constituent intervals, the two nests define the same real number.

All these rules giving the magnitude relations between two real numbers can be understood immediately from the point of view of the intuitive meaning of nests of intervals.

A few simple facts about inequalities between real numbers will now be noticed. They will be of use in what follows.

We first make the following observation. The relation $\alpha \leqq \gamma$ can be inferred from the two defining nests of intervals $(a_n | b_n)$ for α and $(c_n | d_n)$ for γ if we note that from a certain $n = n_0$ onwards the inequality $a_n \leqq d_n$ holds.*

In just the same way we see that the condition $c_n \leqq b_n$ for all large values of n is equivalent to $\alpha \geqq \gamma$.

We see at once from the above that if α is a real number determined by the nest of intervals $(a_n | b_n)$, then $a_n \leqq \alpha \leqq b_n$. This fact justifies our rule, for it shows that any real number is actually contained in every interval of the nest which defines it.

If α and β are two real numbers and $\alpha < \beta$, then by the interval $(\alpha | \beta)$ is meant the aggregate of all real numbers ξ such that $\alpha \leqq \xi \leqq \beta$. We call the interval a rational interval if its "end-points" α and β are rational numbers. We say that the real number ξ lies in the *interior* of the interval if the signs of equality are absent, so that $\alpha < \xi < \beta$. We describe $(\alpha | \beta)$ as a *neighbourhood* of the real number γ if γ lies in the interior of $(\alpha | \beta)$.

Every interval has rational numbers r in its interior.

For let $(a_n | b_n)$ and $(c_n | d_n)$ be nests of intervals defining the numbers α and β. Since $\alpha < \beta$, there is a number n_0 such that $b_{n_0} < c_{n_0}$. Thus $\alpha \leqq b_{n_0} < c_{n_0} \leqq \beta$. We see that $r = (b_{n_0} + c_{n_0})/2$ is a number with the required property.

From this we obtain the following statement: if $(\alpha | \beta)$ is a neighbourhood of γ, then $(\alpha | \beta)$ contains a rational neighbour-

* For if $\alpha = \gamma$ this inequality is satisfied, as can be seen from the definition of equality, and if $\alpha < \gamma$ then from some number onwards we have $b_n < c_n$, so that *a fortiori* $a_n < d_n$. Conversely, if from some number onwards $a_n \leqq d_n$, then either $b_n \geqq c_n$ for all such values of n, and then $\alpha = \gamma$ by definition, or else, for some value of n, $b_n < c_n$, which gives $\alpha < \gamma$.

hood $(a|b)$ of γ. It is only necessary to choose two rational numbers a and b such that $\alpha < a < \gamma < b < \beta$. It is also easy to see that if $\alpha < \beta$, then rational neighbourhoods $(a|b)$ of α and $(c|d)$ of β can be found such that $b < c$; in other words, the two neighbourhoods have no points in common.

We shall not deal with the fundamental rules of calculation until we come to sub-section 8, p. 580. Our next step is to resume the analysis of the concept of limit with the help of the ideas just explained.

3. The Principle of the Point of Accumulation.*

The determination of real numbers by nests of intervals forms the essential basis of the proof of the principle of the point of accumulation, which is due to Weierstrass. A few remarks on the concept of the point of accumulation will first be made.

Let M be an infinite set of real numbers in which it is permissible for the same number to occur more than once, and indeed an infinity of times. (For example, 1, 1, 1, . . . is such a set.) *If ξ is a number such that every neighbourhood of ξ contains an infinity of numbers belonging to* M, *then ξ is called a point of accumulation of the set* M. The name of course recalls the geometrical connexion between numbers and points. Since every neighbourhood of ξ contains a rational neighbourhood, it is sufficient to formulate the above requirement in terms of rational neighbourhoods only.

An infinite set of numbers need not necessarily have a point of accumulation. The set of integers provides an example. A point of accumulation of a set need not itself be a member of the set. For example, the set $1, \frac{1}{2}, \frac{1}{3}, \ldots, 1/n, \ldots$ has 0 as a point of accumulation, but the definition of the set shows that 0 is not one of its members. A set which contains all its points of accumulation is said to be *closed*. The set of all numbers x such that $0 < x < 25$ is not closed, since the points of accumulation 0 and 25 do not belong to it. On the other hand, $0 \leq x \leq 25$ defines a closed set. A set $a \leq x \leq b$ is called a *closed interval*.

A set may have an infinity of points of accumulation. For example, every real number is a point of accumulation of the

* The above discussion is essentially a repetition of the text in Vol. I, Chap. I, p. 58. The same is true of the next three sub-sections.

set of rational numbers. For if a is any real number, which may be thought of as given by a nest of intervals $(a_n | b_n)$, then every neighbourhood of a contains an infinity of intervals $(a_n | b_n)$, and hence of rational numbers a_n, b_n.

The principle of the point of accumulation, which will now be proved, runs as follows:

Every bounded infinite set of real numbers, that is, every infinite set of real numbers lying in a definite interval, possesses at least one point of accumulation.

To prove this we have to construct a nest of intervals defining a real number which has the property of a point of accumulation of the set.

We first observe that it is legitimate to assume that the given set is contained in a rational interval; for if this were not the case we could replace the given interval by a larger interval with rational end-points. We now divide this rational interval into two equal sub-intervals. At least one of these contains an infinite number of points of the set. For if this is not the case the original interval contains only a finite number of points of the set, and the hypothesis is contradicted. We take the sub-interval containing an infinite number of points of the set, or, if such occur in both, we take one or other of the sub-intervals, and divide it into two equal sub-intervals. Just as before, at least one of these sub-intervals contains an infinity of points of the set. Either this one, or one of the two containing an infinity of points of the set, is now sub-divided, and so on. In this way a nest of intervals $(a_n | b_n)$ is constructed; for each interval taken is contained in the previous one and the length of the nth interval is one 2^n-th part of the length of the original interval. This nest of intervals defines a real number ξ. It will be shown that ξ is a point of accumulation of the set.

Consider any rational neighbourhood $(r | s)$ of ξ, so that $r < \xi < s$. Then from a certain number n_1 onward we must have $r < a_{n_1}$ and from another (possibly different) number n_2 onward $b_{n_2} < s$. In any case, if $n > n_1$ and also $n > n_2$, then $(a_n | b_n)$ is contained in $(r | s)$. The construction of our nest $(a_n | b_n)$ shows that each interval of the nest contains an infinity of points of the set, and therefore the arbitrary rational neighbourhood $(r | s)$ of ξ also contains an infinity of points of the set. But this asserts precisely the fact that ξ is a point of accumulation of the set.

4. Upper and Lower Points of Accumulation. Upper and Lower Limits.

In the construction which has just led us to a point of accumulation of a bounded infinite set, we might have made the restriction that the second interval (that with the larger numbers as its end-points) should always be chosen whenever it contained an infinity of points of the set. If this were done, the nest of intervals obtained would define a perfectly definite point of accumulation β of the set. This number β is the greatest of the numbers corresponding to points of accumulation of the set.

This follows at once from the remark that there can only be a finite number of points of the set in any interval to the " right " of each interval $(a_n \mid b_n)$ of the nest described above.

If γ is an arbitrary number greater than β and if n is sufficiently large, the number b_n is less than γ. Only a finite number of members of the set can be greater than b_n. Thus γ cannot be a point of accumulation, so that β is in fact the greatest number corresponding to a point of accumulation. It is called the *upper limit* ($\overline{\lim}$) of the set.

If in the construction we agree to choose the first interval of the two (that with the smaller numbers as end-points) whenever it contains an infinity of points of the set, we arrive in the same way at the *lower limit* ($\underline{\lim}$) of the set.

The upper limit β and the lower limit α need not belong to the set. For example, in the case of the set of numbers $a_{2n} = 1/n$, $a_{2n-1} = (n-1)/n$, we have $\alpha = 0$, $\beta = 1$, but the numbers 0 and 1 are not members of the given set.

In the example just given the set contains no number greater than 1. We say that in this case 1, besides being the upper limit, is the *upper bound* G of the set. The general definition is as follows: *the number* G *is called the upper bound of a set of numbers if the set contains no number greater than* G, *and if every number less than* G *is exceeded by at least one number belonging to the set.*

It is important to notice the distinction between the upper limit and the upper bound of a set. Take, for example, the set of numbers $1, \frac{1}{2}, \frac{1}{3}, \ldots$. The upper bound is 1 and the upper limit is 0, the number 0 giving the only point of accumulation of the set.

We shall now show that *every set of numbers which is bounded above has an upper bound.* A set of numbers is said to be bounded above if there is a number M such that all members of the set are smaller than M. We first note that *if the set contains a greatest member* G, *then* G *is the upper bound of the set.* But a set which is bounded above need not have a greatest member, as is seen from the example $(n - 1)/n$, $(n = 1, 2, \ldots)$. We now assert that *if the set has no greatest member, its upper limit is also its upper bound.*

For suppose the set contains a number $x > \beta$. We consider all members of the set which are not less than x. There can only be a finite number of these, for otherwise the interval $(x|M)$ would contain an infinity of members of the set and thus at least one point of accumulation, contrary to the assumption that β is the upper limit. Among the finite number of members of the set which are not less than x there would be a greatest one, and this would at the same time be the greatest member of the whole set. Thus we should be thrown back on the case already dealt with. It follows that if the set contains no greatest member, then no member of the set exceeds the upper limit. The number β also fulfils the second condition that it should be the upper bound. For suppose that y is any number less than β; then the interval $(y|M)$ is a neighbourhood of β. But since β is a point of accumulation the neighbourhood contains an infinity of points of the set, all greater than y.

The *lower bound* g of a set of numbers is correspondingly defined as that number which is not greater than any member of the set, and which has the property that every number greater than g is also greater than at least one member of the set. Every set which is bounded below has a lower bound, which is either the least member of the set or else the lower limit of the set.

5. Convergent Sequences.

We consider sequences of numbers $a_1, a_2, \ldots$, always assuming that they are bounded. The principle of the point of accumulation shows that the set of numbers $a_1, a_2, \ldots$ has at least one point of accumulation. A sequence of numbers is called *convergent* if it has only one point of accumulation a. This number a is then called the *limit* of the sequence, and we write

$$\lim_{n \to \infty} a_n = a.$$

The following definition is clearly equivalent to the one just given.

A sequence of numbers $a_1, a_2, \ldots$ has the limit a if, and only if, every neighbourhood of a contains all the members a_n of the sequence, with the possible exception of a finite number of members.

For if the bounded sequence a_n has only one point of accumulation a, then only a finite number of members can lie outside any neighbourhood of a; otherwise there would be some other point of accumulation. Conversely, if *all* neighbourhoods of a contain all the numbers a_n, with only a finite number of exceptions, then the sequence a_n is certainly bounded. It can only possess the one point of accumulation a. For if a' were another, we could choose quite separate neighbourhoods of a and a', and in each of these there would be an infinity of numbers belonging to the sequence. This would contradict the hypothesis that only a finite number of members of the sequence lie outside any neighbourhood of a.

A sequence which does not possess a limit should not be regarded as anything abnormal. On the contrary, the existence of a limit is in a sense exceptional. For example, the sequence whose members are $a_{2n} = 1/n$, $a_{2n-1} = (n-1)/n$, $n = 1, 2, \ldots$ has two points of accumulation, namely 0 and 1.

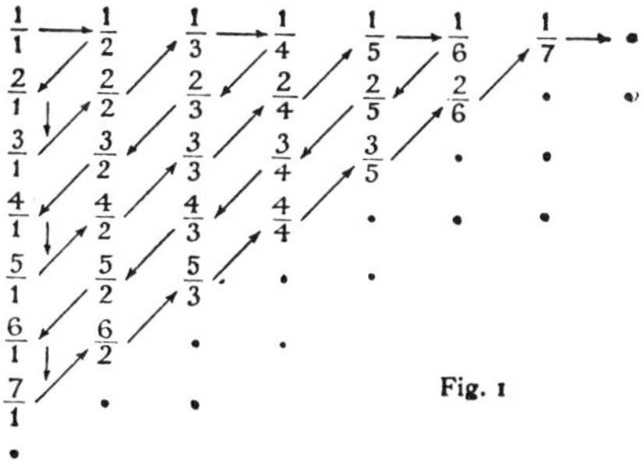

Fig. 1

The aggregate of the positive rational numbers can be regarded as a sequence of numbers, though we must first entirely dislocate the ordering by magnitude. The simplest way to arrive at such a sequence is to order the members by means of the array in fig. 1. The line drawn in the figure shows the order in which the numbers should be taken, any number which has already appeared in the sequence being disregarded. As has already been mentioned, the set of all rational numbers has every real number as a point of accumulation.

The concept of convergence enables us to make a very useful deduction from the principle of the point of accumulation. *If* M *is a given bounded infinite set of numbers with ξ as point of accumulation, then* M *contains an infinite sequence $a_1, a_2, \ldots$ of numbers converging to the limit ξ.*

To prove this we assume that ξ is given by a nest of intervals $(a_n \mid b_n)$ where $a_n < \xi < b_n$. Since ξ is a point of accumulation, $(a_1 \mid b_1)$ contains an infinity of points of M. We choose one of these and call it α_1. Again, $(a_2 \mid b_2)$ contains points of M. We choose one of these and call it α_2, and so on. The resulting sequence $\alpha_1, \alpha_2, \ldots$ is bounded and can have no point of accumulation other than ξ. It therefore converges to the limit ξ.

We now call attention to the two following theorems on convergent sequences, which, though simple, are important in what follows.

If the sequence $\alpha_1, \alpha_2, \ldots$ *converges to the limit* α, *then every infinite sub-sequence converges to* α. For instance, $\alpha_1, \alpha_3, \alpha_5, \ldots$ converges to α.

This follows immediately from the observation that any point of accumulation of a sub-sequence must be a point of accumulation of the original sequence. An infinite sub-sequence must have at least one point of accumulation, and this can only be α.

If $\alpha_1, \alpha_2, \ldots$ *and* $\beta_1, \beta_2, \ldots$ *are two sequences with the same limit* γ, *then the mixed sequence* $\alpha_1, \beta_1, \alpha_2, \beta_2, \alpha_3, \ldots$ *converges to* γ.

Any neighbourhood of γ contains all the numbers α_n and all the numbers β_n, with the possible exception of a finite number of members of each sequence. It therefore contains all members of the mixed sequence, except possibly for a finite number of these.

6. **Bounded Monotonic Sequences.**

A sequence of numbers $\alpha_1, \alpha_2, \ldots$ is said to be monotonic if *either*

$$\alpha_n \leqq \alpha_{n+1}$$

for all values of n *or*

$$\alpha_n \geqq \alpha_{n+1}$$

for all values of n. In the first case we say that the sequence is monotonic non-decreasing, and in the second that it is monotonic non-increasing.

We now prove the important statement that *every bounded monotonic sequence is convergent.* We may restrict ourselves to the proof for the non-decreasing sequence. The other case is exactly similar.

Since every bounded sequence has at least one point of

accumulation, we need only show that our monotonic sequence cannot possess more than one. Suppose, then, that there are two such, a and a', say, and that $a < a'$. About a and a' we construct two quite separate neighbourhoods U_a and $U_{a'}$. Each must contain an infinity of members a_n of the sequence. Take one of the members contained in $U_{a'}$, say a_r. Now let a_s be the first member beyond a_r which lies in U_a. There must be such a member, since U_a contains an infinity of members. Now all the members in U_a are smaller than any in $U_{a'}$. It follows that $a_r > a_s$, which contradicts the hypothesis that the sequence is non-decreasing.

We may add the following remark: if $a_1, a_2, \ldots$ is non-decreasing and bounded, then $\lim_{n\to\infty} a_n \geqq a_N$ for every N. For only a finite number of members a_n, that is to say $a_1, a_2, \ldots, a_{N-1}$, can be less than a_N. Therefore the limit is not less than a_N. In the same way, we see that the limit of a non-increasing sequence is not greater than any member of the sequence.

7. Cauchy's Convergence Test for Sequences of Rational Numbers.

Before we can lay the foundations of calculation with real numbers we need a convergence test which is not restricted to sequences of rational numbers; but we cannot formulate this until we have defined subtraction for real numbers. We shall therefore prove the convergence test for rational numbers here, and return to the general case in sub-section 9, p. 585.

The test in question is as follows:

A sequence of rational numbers $a_1, a_2, \ldots$ *is convergent if, and only if, corresponding to every positive number* ϵ, *however small, we can find a number* $N(\epsilon)$ *such that for every* $n > N$ *and* $m > N$

$$|a_n - a_m| < \epsilon.$$

We shall first show that if this inequality is satisfied for all sufficiently large numbers m and n, then the sequence is convergent.* The boundedness of the sequence is proved as follows. We take the special value $\epsilon = 1$. Then for a sufficiently large value of n and all sufficiently large values of m

$$|a_n - a_m| < 1.$$

* Attention must be drawn to the fact that the elements of the sequence $a_1, a_2, \ldots$ are assumed to be rational, but that this is not the case with the limit a

With a finite number of possible exceptions, then, all the numbers a_m lie in the interval $(a_n - 1 | a_n + 1)$. Thus a properly chosen interval will contain all the numbers a_m without exception. The principle of the point of accumulation shows that the sequence has at least one point of accumulation. We have still to show that there cannot be more than one. Suppose there are two, α and α'. About α and α' we could construct quite separate neighbourhoods $(c | d)$ and $(c' | d')$ so that

$$c < \alpha < d < c' < \alpha' < d',$$

where we assume, as we may without restriction of generality, that $\alpha < \alpha'$. Since α and α' are assumed to be points of accumulation, $(c | d)$ contains an infinite number of points a_n and $(c' | d')$ contains an infinite number of points a_m. Thus, in particular, for an infinite set of values of n and m we have

$$a_m - a_n \geqq c' - d > 0.$$

But this contradicts the hypothesis, which shows that for all sufficiently large values of m and n

$$| a_m - a_n | < c' - d.$$

Hence the sequence has one, and only one, point of accumulation.

We next show that if the sequence $a_1, a_2, \ldots$ converges to α, then for every $\epsilon > 0$ and for all sufficiently large values of n and m

$$| a_n - a_m | < \epsilon.$$

We take a neighbourhood $(c | d)$ of α, whose length $(d - c)$ is less than or equal to ϵ. If N is suitably chosen, then whenever n exceeds N, a_n lies in $(c | d)$. Thus if $n > N$ and $m > N$, both a_n and a_m lie in $(c | d)$. From this it follows that

$$| a_n - a_m | < d - c \leqq \epsilon.$$

8. **Calculation with Real Numbers.**

So far our work has given us the definition of real numbers by means of nests of intervals, and their ordering by magnitude. The theorem last proved provides a simple means of defining the rules of arithmetical calculation with real numbers.

Let a real number α be given by a nest of intervals $(a_n | b_n)$. Since the intervals form a nest, the numbers a_n form a monotonic non-decreasing sequence and the numbers b_n a monotonic non-increasing sequence. These sequences are bounded; for we may note that every a_n is less than or equal to b_1, and every b_n greater than or equal to a_1. The sequences therefore converge. In both cases, moreover, the limit is the real number α. For every neighbourhood of α contains all the intervals $(a_n | b_n)$, except possibly for a finite number of these, and thus the neighbourhood contains all but a finite number of members of the a_n and the b_n sequences. We may therefore say that *every real number can be exhibited as the limit of sequences of rational numbers.*

If now we wish to define any operation of arithmetic for two real numbers α and β, we choose two sequences a_n and b_n of rational numbers with the limits α and β respectively. We perform the operation on the pairs of numbers a_n and b_n and thus obtain a new sequence. When we have proved that this sequence has a limit, we shall say, by way of definition, that it is the result of the operation on the two real numbers α and β.

Let α and β be two arbitrary real numbers and let $\lim_{n\to\infty} a_n = \alpha$ and $\lim_{n\to\infty} b_n = \beta$. We consider the sequences $a_n + b_n$, $a_n - b_n$, $a_n b_n$, and $1/a_n$. If we can prove that these sequences converge, we can set up the *definitions*

$$\alpha + \beta = \lim_{n\to\infty} (a_n + b_n),$$

$$\alpha - \beta = \lim_{n\to\infty} (a_n - b_n),$$

$$\alpha\beta = \lim_{n\to\infty} (a_n b_n),$$

$$\frac{1}{\alpha} = \lim_{n\to\infty} \left(\frac{1}{a_n}\right).$$

The convergence of these sequences will be proved by means of Cauchy's convergence test.

It follows from the convergence of $a_1, a_2, \ldots$ that if ϵ is a given positive number and n and m are sufficiently large, say $n > N_1$ and $m > N_1$, then

$$| a_n - a_m | < \epsilon/2,$$

and, from the convergence of $b_1, b_2, \ldots$, that if n and m are sufficiently large, say $n > N_2$ and $m > N_2$, then

$$|b_n - b_m| < \epsilon/2.$$

If $N(\epsilon)$ denotes the larger of the two numbers N_1 and N_2, then if $n > N(\epsilon)$, $m > N(\epsilon)$,

$$|(a_n + b_n) - (a_m + b_m)| \leqq |a_n - a_m| + |b_n - b_m| < \frac{\epsilon}{2} + \frac{\epsilon}{2} = \epsilon$$

and

$$|(a_n - b_n) - (a_m - b_m)| \leqq |a_n - a_m| + |b_n - b_m| < \frac{\epsilon}{2} + \frac{\epsilon}{2} = \epsilon$$

By Cauchy's test both the sequences $a_n + b_n$ and $a_n - b_n$ converge.

To prove that $a_n b_n$ converges, we must first notice that the numbers a_n and b_n form bounded sets. There are therefore two positive rational numbers A and B such that for all values of n

$$|a_n| \leqq A, \quad |b_n| \leqq B.$$

Now

$$\begin{aligned} |a_n b_n - a_m b_m| &= |a_n(b_n - b_m) + b_m(a_n - a_m)| \\ &\leqq |a_n|\,|b_n - b_m| + |b_m|\,|a_n - a_m| \\ &\leqq A\,|b_n - b_m| + B\,|a_n - a_m|. \end{aligned}$$

Since the sequences $a_1, a_2, \ldots$ and $b_1, b_2, \ldots$ converge, we can find numbers N_1 and N_2 corresponding to any given $\epsilon > 0$, such that

$$|a_n - a_m| < \epsilon/2B \quad \text{when} \quad n > N_1 \text{ and } m > N_1$$

and

$$|b_n - b_m| < \epsilon/2A \quad \text{when} \quad n > N_2 \text{ and } m > N_2.$$

Thus if n and m are both greater than the larger of the two numbers N_1 and N_2, the above inequalities hold simultaneously. We have therefore

$$|a_n b_n - a_m b_m| < A\frac{\epsilon}{2A} + B\frac{\epsilon}{2B} = \epsilon.$$

Cauchy's test shows at once that the sequence $a_n b_n$ is convergent.

We now suppose that $a \neq 0$ and $\lim_{n\to\infty} a_n = a$. We have to show that $1/a_n$ converges. It is first necessary to show that if n is sufficiently large, $|a_n|$ is greater than a positive number p independent of n. We take a rational neighbourhood of a which does not contain 0. This is possible, since $a \neq 0$. From a suitable $n = n_0$ onward all the members of the sequence $a_1, a_2, \ldots$ lie in this neighbourhood. This shows that, for $n > n_0$, $|a_n| \geqq p$, where p is the absolute value corresponding to the end-point of the interval that is nearer to 0. The convergence of the sequence $\frac{1}{a_1}, \frac{1}{a_2}, \ldots$ is not affected by the omission of the first n_0 members, and we may therefore now assume that for all values of n

$$|a_n| \geqq p > 0.$$

We observe that

$$\left|\frac{1}{a_n} - \frac{1}{a_m}\right| = \frac{|a_m - a_n|}{|a_n a_m|} = \frac{|a_m - a_n|}{|a_n|\,|a_m|} \leqq \frac{|a_m - a_n|}{p^2}.$$

Let $\epsilon > 0$ be given. If N is suitably chosen, then, since $a_1, a_2, \ldots$ converges, $n > N$ and $m > N$ give

$$|a_m - a_n| < \epsilon p^2,$$

so that

$$\left|\frac{1}{a_n} - \frac{1}{a_m}\right| < \frac{\epsilon p^2}{p^2} = \epsilon.$$

This proves the convergence of $1/a_n$, provided that $a \neq 0$.

It is obvious that any real number may be exhibited as the limit of more than one sequence of rational numbers. It might be thought that the definitions given above do not define the arithmetical operations uniquely. For instance, suppose that $\lim\limits_{n\to\infty} a_n = a$ and $\lim\limits_{n\to\infty} b_n = \beta$ give one representation of the numbers a and β and $\lim\limits_{n\to\infty} a_n' = a$ and $\lim\limits_{n\to\infty} b_n' = \beta$ another. Then possibly the two sequences $a_n + b_n$ and $a_n' + b_n'$ might have different limits. (We have proved that they do have limits.) We shall now prove that this difficulty does not arise. It will be shown that if

$$\lim_{n\to\infty} a_n = \lim_{n\to\infty} a_n' \quad \text{and} \quad \lim_{n\to\infty} b_n = \lim_{n\to\infty} b_n',$$

then

$$\lim_{n\to\infty} (a_n \pm b_n) = \lim_{n\to\infty} (a_n' \pm b_n'),$$

$$\lim_{n\to\infty} (a_n b_n) = \lim_{n\to\infty} (a_n' b_n'),$$

and, if $\lim_{n\to\infty} a_n = \lim a_n' \neq 0$,

$$\lim_{n\to\infty} \frac{1}{a_n} = \lim_{n\to\infty} \frac{1}{a_n'}.$$

The proof is very simple. It has already been shown that if $\lim_{n\to\infty} a_n = \lim_{n\to\infty} a_n' = a$, then the mixed sequence $a_1, a_1', a_2, a_2', \ldots$ has the limit a. In the same way, we see that $b_1, b_1', b_2, b_2', \ldots$ converges to $\beta = \lim_{n\to\infty} b_n = \lim_{n\to\infty} b_n'$. From this and the above theorems we find that the mixed sequences $a_1 \pm b_1, a_1' \pm b_1', \ldots,$ and $a_1b_1, a_1'b_1', \ldots$ and, if $a \neq 0$, $\frac{1}{a_1}, \frac{1}{a_1'}, \ldots$ are convergent.

It has already been proved that every sub-sequence of a given convergent sequence converges to the same limit. From this it follows that the sequences

$$a_1 \pm b_1, a_2 \pm b_2, \ldots \quad \text{and} \quad a_1' \pm b_1', a_2' \pm b_2', \ldots,$$

which are sub-sequences of a convergent sequence, must converge to the same limit. In the same way,

$$a_1b_1, a_2b_2, \ldots \quad \text{and} \quad a_1'b_1', a_2'b_2', \ldots$$

have the same limit, and the same is true of

$$\frac{1}{a_1}, \frac{1}{a_2}, \ldots \quad \text{and} \quad \frac{1}{a_1'}, \frac{1}{a_2'}, \ldots.$$

The results just obtained allow us to settle another important question which is connected with our definitions of the operations of arithmetic.

The class of real numbers contains the rational numbers. In the course of our definitions of operations on the real numbers we have thus incidentally defined these operations for the rational numbers. But we began by taking the operations on rational numbers as known. We must therefore verify that the new definitions do not give rise to any contradiction in the case of the rational numbers. What we have to show is that if $\lim_{n\to\infty} a_n = a$ and $\lim_{n\to\infty} b_n = b$ are rational numbers, then

$$\lim_{n\to\infty} (a_n \pm b_n) = a \pm b,$$

$$\lim_{n\to\infty} (a_nb_n) = ab$$

and, if $a \neq 0$,

$$\lim_{n\to\infty} \frac{1}{a_n} = \frac{1}{a}.$$

It should first be noticed that a rational number a is the limit of the rational sequence $a, a, \ldots$. For the two sequences $a_1', a_2', \ldots$ and $b_1', b_2', \ldots$ we may take the special sequences $a, a, \ldots$ and $b, b, \ldots$. The above theorems then yield

$$\lim_{n\to\infty} (a_n \pm b_n) = \lim_{n\to\infty} (a \pm b) = a \pm b,$$

$$\lim_{n\to\infty} (a_n b_n) = \lim_{n\to\infty} (ab) = ab,$$

$$\lim_{n\to\infty} \frac{1}{a_n} = \lim_{n\to\infty} \frac{1}{a} = \frac{1}{a},$$

which is the required result.

It need hardly be mentioned that, as a result of our definitions, all the rules of calculation that hold for rational numbers also hold for all real numbers. We have only to apply the rules to the rational numbers forming the sequences. Let us, for instance, prove the distributive law, $\alpha(\beta + \gamma) = \alpha\beta + \alpha\gamma$.

Let $\alpha = \lim_{n\to\infty} a_n$, $\beta = \lim_{n\to\infty} b_n$, $\gamma = \lim_{n\to\infty} c_n$. Then the left-hand side of the equality to be proved is $\lim_{n\to\infty} \{a_n(b_n + c_n)\}$, and the right-hand side is $\lim_{n\to\infty} (a_n b_n + a_n c_n)$. But since the distributive law is true for the rational numbers, the two sequences are the same, and this must also be true of their limits.

9. The General Form of Cauchy's Convergence Test.

We return to Cauchy's convergence test, which we have already proved for rational sequences on p. 579. Now that the operations of arithmetic, in particular subtraction, have been established for real numbers, we can formulate the convergence test quite generally for real numbers. *The sequence $a_1, a_2, \ldots$ is convergent if, and only if, for any given $\epsilon > 0$, we can find an suffix $N(\epsilon)$ such that whenever m and n are both greater than $N(\epsilon)$*

$$| a_n - a_m | < \epsilon.$$

The proof is exactly like that given on p. 579, and need not be repeated.

The following point is of great theoretical importance. Cauchy's convergence test contains in its enunciation a means of estimation of error. For if we are given the sequence and know the number $N(\epsilon)$, we can state at once that the limit of the sequence lies between the numbers $a_n + \epsilon$ and $a_n - \epsilon$ whenever $n > N(\epsilon)$.

In this respect Cauchy's test differs from the test for monotonic sequences. The latter proves the existence of the limit, but it gives no means of estimating the limit. Thus in proofs of convergence which depend on this test any estimation of the limit (and theoretically it is always necessary to give one) must depend on separate and extraneous considerations.

MISCELLANEOUS EXAMPLES

1. Two vectors x, y (or three vectors x, y, z) are said to be *linearly independent* if a linear relation

$$ax + by = 0 \quad (\text{or } ax + by + cz = 0)$$

is possible only when $a = b = 0$ (or $a = b = c = 0$). They are said to be *linearly dependent* if such relations exist without all the coefficients vanishing. Prove the following statements:

(*a*) Three vectors x, y, z such that any two of them are orthogonal to one another are linearly independent.

(*b*) The vectors x, y (or x, y, z) are linearly independent if, and only if,

$$[xy] \neq 0$$

$$\left(\text{or } x[yz] = \begin{vmatrix} x_1 & x_2 & x_3 \\ y_1 & y_2 & y_3 \\ z_1 & z_2 & z_3 \end{vmatrix} \neq 0\right).$$

(*c*) If two vectors x, y in a plane are linearly independent, then any vector v in their plane may be written in the form $v = ax + by$. Similarly, if x, y, z are linearly independent, then any vector v may be written in the form $v = ax + by + cz$.

2. We know already that if x, y, z are three vectors,

$$x[yz] = y[zx] = z[xy] = \begin{vmatrix} x_1 & x_2 & x_3 \\ y_1 & y_2 & y_3 \\ z_1 & z_2 & z_3 \end{vmatrix}$$

(the common scalar value of these expressions may be conveniently denoted by (x, y, z)). Prove the further vectorial identities

(*a*) $(x, y, z)(x', y', z') = \begin{vmatrix} xx' & xy' & xz' \\ yx' & yy' & yz' \\ zx' & zy' & zz' \end{vmatrix}.$

(*b*) $[xy][x'y'] = (xx')(yy') - (xy')(yx')$ (cf. Ex. 5, p. 19).

(*c*) $[x[yz]] = (xz)y - (xy)z$.

(*d*) $\big([x[yz]], [y[zx]], [z[xy]]\big) = 0$.

Use the last result to deduce that if a plane is drawn through each of three concurrent straight lines perpendicular to the plane of the other two, the three planes thus obtained meet in a straight line.

3. Let Ox, Oy be a system of rectangular axes in a plane. Let Ox', Oy' be a second such system and let the angle xOx' be φ. Prove that the passage from one system of co-ordinates to the other is given by the formulæ

$$x = x' \cos\varphi - y' \sin\varphi, \quad x' = x\cos\varphi + y\sin\varphi,$$
$$y = x' \sin\varphi + y' \cos\varphi, \quad y' = -x\sin\varphi + y\cos\varphi.$$

4. From the result of Ex. 3 deduce the addition formulæ

$$\cos(\varphi + \psi) = \cos\varphi\cos\psi - \sin\varphi\sin\psi, \quad \sin(\varphi + \psi) = \sin\varphi\cos\psi + \cos\varphi\sin\psi.$$

5. Let Ox, Oy, Oz and Ox', Oy', Oz' be two co-ordinate systems, both having the same orientation, the cosines of the various angles being indicated by the following scheme:

	x'	y'	z'
x	α_1	β_1	γ_1
y	α_2	β_2	γ_2
z	α_3	β_3	γ_3

In Ex. 1, p. 12, and Ex. 9, p. 38, the relations

$$\alpha_1^2 + \beta_1^2 + \gamma_1^2 = 1, \quad \alpha_2\alpha_3 + \beta_2\beta_3 + \gamma_2\gamma_3 = 0,$$
$$\alpha_2^2 + \beta_2^2 + \gamma_2^2 = 1, \quad \alpha_3\alpha_1 + \beta_3\beta_1 + \gamma_3\gamma_1 = 0,$$
$$\alpha_3^2 + \beta_3^2 + \gamma_3^2 = 1, \quad \alpha_1\alpha_2 + \beta_1\beta_2 + \gamma_1\gamma_2 = 0,$$

$$\Delta = \begin{vmatrix} \alpha_1 & \beta_1 & \gamma_1 \\ \alpha_2 & \beta_2 & \gamma_2 \\ \alpha_3 & \beta_3 & \gamma_3 \end{vmatrix} = 1$$

were proved. A three-rowed determinant Δ whose elements satisfy these relations is said to be *orthogonal*.

Prove (*a*) that to any orthogonal determinant Δ equal to $+1$ there correspond two co-ordinate systems Ox, Oy, Oz and Ox', Oy', Oz' with the same orientation, such that the cosines of the angles between the various co-ordinate axes are given by the elements of Δ.

(*b*) That for any orthogonal determinant the relations

$$\alpha_1^2 + \alpha_2^2 + \alpha_3^2 = 1, \quad \beta_1\gamma_1 + \beta_2\gamma_2 + \beta_3\gamma_3 = 0$$
$$\beta_1^2 + \beta_2^2 + \beta_3^2 = 1, \quad \gamma_1\alpha_1 + \gamma_2\alpha_2 + \gamma_3\alpha_3 = 0$$
$$\gamma_1^2 + \gamma_2^2 + \gamma_3^2 = 1, \quad \alpha_1\beta_1 + \alpha_2\beta_2 + \alpha_3\beta_3 = 0$$

are also satisfied.

6*. Let Ox, Oy, Oz and Ox', Oy', Oz' be two co-ordinate systems as in Ex. 5. Assume that Oz and Oz' do not coincide; let the angle zOz' be θ $(0 < \theta < \pi)$. Draw the half-line Ox_1 at right angles to both Oz and Oz', and such that the system Ox_1, Oz, Oz' has the same orientation as Ox, Oy, Oz. Then Ox_1 is the line of intersection of the planes Oxy and $Ox'y'$. Let the angle xOx_1 be φ and the angle x_1Ox' be ψ and let them be measured in the usual positive sense in their respective planes, Oxy and $Ox'y'$.

Prove that the passage from Ox, Oy, Oz to Ox', Oy', Oz' is given by the scheme

	x'	y'	z'
x	$\cos\varphi\cos\psi$ $-\sin\varphi\sin\psi\cos\theta$	$-\cos\varphi\sin\psi$ $-\sin\varphi\cos\psi\cos\theta$	$\sin\varphi\sin\theta$
y	$\sin\varphi\cos\psi$ $+\cos\varphi\sin\psi\cos\theta$	$-\sin\varphi\sin\psi$ $+\cos\varphi\cos\psi\cos\theta$	$-\cos\varphi\sin\theta$
z	$\sin\psi\sin\theta$	$\cos\psi\sin\theta$	$\cos\theta$.

(Note that this result holds also for $\theta = 0$ or π, when φ and ψ become indeterminate with $\varphi + \psi = \angle xOx'$ or $\varphi - \psi = \angle xOx'$ respectively. The angles φ, ψ, θ are the so-called *Eulerian angles*, and our result, together with Example 5, shows that the most general orthogonal determinant Δ of value $+1$ may be expressed " parametrically " by means of the *three* variables φ, ψ, θ, subject to the inequalities

$$0 \leqq \theta \leqq \pi, \quad 0 \leqq \varphi < 2\pi, \quad 0 \leqq \psi < 2\pi.)$$

7. Let ABC be a spherical triangle of sides a, b, c and angles A, B, C on the " unit sphere " (i.e. the sphere of radius unity). From Ex. 6 deduce the " cosine theorem "

$$\cos a = \cos b \cos c + \sin b \sin c \cos A.$$

8. Find the angle φ between the plane

$$Ax + By + Cz + D = 0$$

and the line

$$x = x_0 + \alpha t, \quad y = y_0 + \beta t, \quad z = z_0 + \gamma t.$$

9. Solve the equations

$$\begin{aligned} 2x - 3y + 4z &= 4 \\ 4x - 9y + 16z &= 10 \\ 8x - 27y + 64z &= 34. \end{aligned}$$

10. Prove the identity

$$(a^2 + b^2)(c^2 + d^2) = (ac + bd)^2 + (bc - ad)^2$$

by forming the product of the determinants

$$\begin{vmatrix} a & b \\ -b & a \end{vmatrix} \text{ and } \begin{vmatrix} c & d \\ -d & c \end{vmatrix}.$$

11*. Prove that the value of the determinant

$$D = \begin{vmatrix} \cos(\theta + \alpha) & \cos(\theta + \beta) & \cos(\theta + \gamma) \\ \sin(\theta + \alpha) & \sin(\theta + \beta) & \sin(\theta + \gamma) \\ \sin(\beta - \gamma) & \sin(\gamma - \alpha) & \sin(\alpha - \beta) \end{vmatrix}$$

is independent of θ.

12. If $A = x^2 + y^2 + z^2$, $B = xy + yz + zx$, show that

$$D = \begin{vmatrix} B & A & B \\ B & B & A \\ A & B & B \end{vmatrix} = (x^3 + y^3 + z^3 - 3xyz)^2.$$

13. Show that

$$\Delta = \begin{vmatrix} t_1 + x & a + x & a + x & a + x \\ b + x & t_2 + x & a + x & a + x \\ b + x & b + x & t_3 + x & a + x \\ b + x & b + x & b + x & t_4 + x \end{vmatrix}$$

is of the form $A + Bx$, where A and B are independent of x. Hence by giving particular values to x, prove that

$$A = \frac{af(b) - bf(a)}{a - b}, \quad B = \frac{f(b) - f(a)}{a - b},$$

where

$$f(t) = (t_1 - t)(t_2 - t)(t_3 - t)(t_4 - t).$$

14*. Prove that if u and v are functions of x and $v = 1/u$, then $v''' = D/u^4$, where D is the determinant

$$D = \begin{vmatrix} u''' & 3u'' & 3u' \\ u'' & 2u' & u \\ u' & u & 0 \end{vmatrix}.$$

15. (*a*) Show that a function u of the form $u(x, y) = f(x)\,g(y)$ satisfies the partial differential equation

$$uu_{xy} - u_x u_y = 0.$$

(*b*)* Prove the converse statement.

16. Prove that

$$u(x, y, z) = \frac{f(t + r)}{r} + \frac{g(t - r)}{r} \quad (r^2 = x^2 + y^2 + z^2)$$

satisfies the equation

$$\Delta u = u_{tt}.$$

17. Show that a function u satisfies the equation

$$u_{xx}u_{yy} - u^2_{xy} = 0$$

if its first derivatives satisfy a relation of the form

$$F(u_x, u_y) = 0.$$

18*. Prove that a surface $u = f(x, y)$ generated by straight lines meeting the u-axis or, what comes to the same thing, a surface cut in straight lines by vertical planes $\frac{y}{x} = c$, satisfies the equation

$$x^2 u_{xx} + 2xy\, u_x u_y + y^2 u_{yy} = 0.$$

19. Find a $\delta = \delta(\varepsilon, x, y)$ for the continuous functions (cf. pp. 44–5)

$$(a)\ f(x, y) = \sqrt{(1 + x^2 + 2y^2)},$$
$$(b)\ f(x, y) = \sqrt{(1 + e^{xy})}.$$

20. Show that the functions

$$f(x, y) = \frac{x^4y^4}{(x^2 + y^4)^3}, \quad g(x, y) = \frac{x^2}{x^2 + y^2 - x}$$

tend to zero if (x, y) approaches the origin along any straight line, but that f and g are discontinuous at the origin.

21. Let C be a smooth curve with a continuously turning tangent. Let d denote the shortest distance between two points on the curve and l the length of arc between the two points. Prove that $d - l = o(d)$ when d is small.

22. Evaluate

$$S = \sum_{a=0}^{\infty} \sum_{b=0}^{\infty} \frac{(a + b)!}{a!\ b!} \frac{a}{x^a y^b}$$

if $\dfrac{1}{x} + \dfrac{1}{y} < 1$, $x > 0$, $y > 0$.

23. Show by using Euler's relation (p. 109) that a homogeneous function $S_n(x, y, z)$ of degree n which satisfies Laplace's equation $\Delta S_n = 0$ also satisfies the relation

$$\Delta(r^{2m}S_n) = 2m(2n + 2m + 1)r^{2m-2}S_n,$$

where

$$r^2 = x^2 + y^2 + z^2.$$

24. Prove that the curvature of the curve $x = x(t)$ (t being an arbitrary parameter) is given by

$$k = \pm \frac{(x'^2x''^2 - (x'x'')^2)^{\frac{1}{2}}}{(x'^2)^{\frac{3}{2}}}.$$

25*. Let a twisted curve C be defined by $x = x(s)$, $y = y(s)$, $z = as$, s being the length of arc of the plane curve $x = x(s)$, $y = y(s)$. Prove that the osculating plane of the curve at a point P (cf. Ex. 1, p. 93) contains the normal to the cylinder $x = x(s)$, $y = y(s)$ at P. Show that the curvature and torsion of C are respectively given by

$$k = \frac{x'y'' - x''y'}{1 + a}, \quad \tau = \frac{a(x'y'' - x''y')}{1 + a^2}.$$

(A curve of this kind is called a *circular helix*.)

26. Find the equation of the osculating plane (cf. Ex. 1, p. 93) at the point θ of the curve $x = \cos\theta$, $y = \sin\theta$, $z = f(\theta)$. Show that if $f(\theta) = \dfrac{1}{A}\cosh A\theta$, each osculating plane touches a sphere whose centre is the origin and whose radius is $\sqrt{(1 + 1/A^2)}$.

27. A curve is drawn on the cylinder $x^2 + y^2 = a^2$, such that the angle between the z-axis and the tangent at any point P of the curve is equal to the angle between the y-axis and the tangent plane at P to the cylinder. Prove that the co-ordinates of any point P of the curve can be expressed in terms of a parameter θ by the equations

$$x = a\cos\theta, \quad y = a\sin\theta, \quad z = c \pm a\log\sin\theta,$$

and that the curvature of the curve is $(1/a)\sin\theta(1 + \sin^2\theta)^{\frac{1}{2}}$.

28. (*a*) Prove that the equation of the plane passing through the three points t_1, t_2, t_3 on the curve

$$x = \tfrac{1}{3}at^3, \quad y = \tfrac{1}{2}bt^2, \quad z = ct$$

is

$$\frac{3x}{a} - 2(t_1 + t_2 + t_3)\frac{y}{b} + (t_2t_3 + t_3t_1 + t_1t_2)\frac{z}{c} - t_1t_2t_3 = 0.$$

(*b*) Show that the point of intersection of the osculating planes at t_1, t_2, t_3 lies in this plane.

29. Let a, b, c, A, B, C, be the sides and angles of a triangle of area s, and let R be the radius of its circumscribed circle. Show that

$$ds = R(\cos A\, da + \cos B\, db + \cos C\, dc).$$

30. Consider a fixed point A in space and a variable point P whose motion is given as a function of the time. Denoting by $\dot{P}$ the velocity vector of P and by $\boldsymbol{a}$ a unit vector in the direction from P to A, show that

$$\frac{d}{dt}(AP) = -\boldsymbol{a}\dot{P}.$$

31. Let A, B, C be three fixed points and let the components of the velocity vector $\dot{P}$ of a moving point P in the directions PA, PB, PC be u, v, w. Let $\boldsymbol{a}, \boldsymbol{b}, \boldsymbol{c}$ be unit vectors in the directions PA, PB, PC. Prove that

$$\frac{d\boldsymbol{a}}{dt} = \dot{\boldsymbol{a}} = \left(\frac{\cos APB}{PA}v + \frac{\cos APC}{PA}w\right)\boldsymbol{a} - \frac{v}{PA}\boldsymbol{b} - \frac{w}{PA}\boldsymbol{c}.$$

32. Prove that the acceleration vector $\ddot{P}$ of the point P is

$$\ddot{P} = \alpha\boldsymbol{a} + \beta\boldsymbol{b} + \gamma\boldsymbol{c},$$

where

$$\alpha = \dot{u} + uv\left(\frac{\cos APB}{PA} - \frac{1}{PB}\right) + uw\left(\frac{\cos APC}{PA} - \frac{1}{PC}\right),$$

with two similar expressions for β and γ.

33. Find the envelope of a variable circle in a plane which passes through a fixed point O, and whose centre describes a given conic with centre O.

34. If Γ is a plane curve and O a point in its plane, the locus Γ' of the orthogonal projections of O on a variable tangent of Γ is called the *pedal curve* of Γ with respect to the point O. Prove that if the point M describes the curve Γ, the pedal curve Γ' is the envelope of the variable circle with the radius vector OM as diameter.

35. What is the envelope of the variable sphere with the radius vector OM (cf. Ex. 34) as diameter?

36. What are the envelopes of the variable circles and spheres of Ex. 34, 35, if Γ is a circle and O a point on its circumference?

37. MM' is a variable chord of an ellipse parallel to the minor axis. Find the envelope of the variable circle with MM' as diameter.

38. A plane moves so as to touch the parabolas

$$z = 0,\ y^2 = 4x \text{ and } y = 0,\ z^2 = 4x.$$

Show that its envelope consists of two parabolic cylinders.

39.† Generalize the investigation of § 1 of the Appendix to Chap. III (p. 204) to functions of n variables, proving the following results. Let $f(x_1, \ldots, x_n)$ be three times continuously differentiable in the neighbourhood of a *stationary* point $x_1 = x_1^0, \ldots, x_n = x_n^0$, that is, a point where $f_{x_1} = f_{x_2} = \ldots = f_{x_n} = 0$. Consider the second total differential of f at the point x^0, $d^2f^0 = \sum_{i,k=1}^{n} f^0_{x_i x_k} dx_i\, dx_k$; this is a quadratic form in the variables $dx_1, \ldots, dx_n$. If this quadratic form is *non-degenerate*, that is, if

$$D = \begin{vmatrix} f^0_{x_1x_1} & \cdots & f^0_{x_1x_n} \\ \vdots & & \vdots \\ f^0_{x_nx_1} & \cdots & f^0_{x_nx_n} \end{vmatrix} \neq 0,$$

then d^2f^0 may be (1) *positively definite*, (2) *negatively definite* or (3) *indefinite*. Prove that these possible cases correspond respectively to the following properties of f at the point (x^0): (1) f has a minimum, (2) f has a maximum, (3) f has neither a minimum nor a maximum.

40. Consider the function of two variables $f = (y - x^2)(y - 2x^2)$, which is stationary at the origin O ($x = y = 0$). Prove (1) that along any straight line through O, f has a minimum at O, (2) that f, considered as a function of (x, y), has neither a minimum nor a maximum at O.

41. Let $P_1P_2P_3$ be a plane triangle with all three angles less than 120°. Prove by the criterion of Ex. 39 that at the point P interior to $P_1P_2P_3$ such that $\angle P_2PP_3 = \angle P_3PP_1 = \angle P_1PP_2 = 120°$, the sum $PP_1 + PP_2 + PP_3$ is actually a minimum (cf. Ex. 4, p. 187).

42. Where does the minimum of the sum $PP_1 + PP_2 + PP_3$ occur if in the triangle of Ex. 41 the angle $P_2P_1P_3$ is greater than or equal to 120°?

† For Ex. 39, 41, 43, and 44, the reader is assumed to be familiar with the elements of the theory of quadratic forms.

43. To investigate stationary points of $f = f(x_1, \ldots, x_n)$, where the variables satisfy the relations

$$\varphi_1(x_1, \ldots, x_n) = 0, \ldots, \varphi_m(x_1, \ldots, x_n) = 0 \quad (m < n), \quad . \quad (1)$$

we may assume that we have found numerical values for the variables and the multipliers λ_μ, such that $F = f + \lambda_1\varphi_1 + \ldots + \lambda_m\varphi_m$ satisfies the equations

$$\partial F/\partial x_1 = 0, \ldots, \partial F/\partial x_n = 0, \quad . \quad . \quad . \quad . \quad . \quad (2)$$

and such that the Jacobian of $\varphi_1, \ldots, \varphi_m$ with respect to the variables $x_1, \ldots, x_m$ is not zero. To apply the criterion of Ex. 39 we may proceed as follows. Regarding $x_{m+1}, \ldots, x_n$ as independent variables, by differentiating (1) we can obtain the first and second differentials of $x_1 \ldots, x_m$ as functions of $x_{m+1}, \ldots, x_n$, and finally introduce these values into

$$d^2f = \sum_{i,k=1}^{n} f_{x_ix_k} dx_i dx_k + f_{x_1} d^2x_1 + \ldots + f_{x_m} d^2x_m. \quad . \quad . \quad (3)$$

Prove the following second rule, not involving the computation of the second differentials $d^2x_1, \ldots, d^2x_m$. Regarding $x_1, \ldots, x_n$ as independent variables, consider

$$d^2F = \Sigma F_{x_ix_k} dx_i dx_k = d^2f + \lambda_1 d^2\varphi_1 + \ldots + \lambda_m d^2\varphi_m;$$

compute $dx_1, \ldots, dx_m$ from the equations

$$d\varphi_\mu = \varphi_{\mu x_1} dx_1 + \ldots + \varphi_{\mu x_n} dx_n = 0 \quad (\mu = 1, \ldots, m)$$

and introduce these values into d^2F, thus obtaining a quadratic form δ^2F in the variables $dx_{m+1}, \ldots, dx_n$. If this quadratic form is non-degenerate, then f has respectively a minimum, a maximum, or neither of these, according as δ^2F is positively definite, negatively definite, or indefinite.

44. In the problem of finding the maximum of $f = x_1x_2 \ldots x_n$ subject to the condition $\varphi = x_1 + x_2 + \ldots + x_n - a = 0$ $(a > 0)$, the rule of undetermined multipliers gives a stationary value of f at the point $x_1 = x_2 = \ldots = x_n = a/n$. Apply the rule of Ex. 43, instead of the consideration of the absolute maximum, to show that f has a maximum value at this point.

45. Apply the criterion of Ex. 43 to prove that among all triangles of constant perimeter the equilateral triangle has the largest area (cf. Ex. 2, p. 200).

46. The curve $x^3 + y^3 - 3axy = 0$ has a double point at the origin. What are its tangents there?

47. Draw a graph of the curve $(y - x^2)^2 - x^5 = 0$, and show that it has a cusp at the origin. What is the peculiarity of this cusp as compared with the cusp of the curve $x^2 - y^3 = 0$?

48. (*a*) Prove that if all the symbols denote positive quantities the stationary value of $lx + my + nz$ subject to the condition $x^p + y^p + z^p = c^p$ is

$$c(l^q + m^q + n^q)^{1/q},$$

where $q = p/(p - 1)$.

(*b*) Show that the value is a maximum or minimum according as $p \gtrless 1$.

49. Find the values of x, y which make

$$2x^3 + (x - y)^2 - 6y$$

stationary.

50. Prove that if Σ is a closed convex curve and ABC is the circumscribed triangle of least area, then the points of contact of Σ with the sides of the triangle are the centres of the sides.

51. Show that each of the curves

$$(x \cos\alpha - y \sin\alpha - b)^3 = c(x \sin\alpha + y \cos\alpha)^2,$$

where α is variable, has a cusp, and that all the cusps lie on a circle.

52. If $C = f(a, b)$ is a true maximum or minimum of $f(x, y)$ subject to the condition $\varphi(x, y) = C'$, show that in general $C' = \varphi(a, b)$ is a true maximum or minimum of $\varphi(x, y)$ subject to the condition $f(x, y) = C$.

53. A circle of radius a rolls on a fixed straight line, carrying a tangent fixed relatively to the circle. Taking axes at the point of contact where the moving tangent coincides with the fixed line, show that the envelope of the tangent is given by

$$x = a(\theta + \cos\theta \sin\theta - \sin\theta)$$
$$y = a(\cos^2\theta - \cos\theta).$$

54. If the co-ordinates (x, y, z) of a point on a sphere are given by the equations (cf. p. 160)

$$x = a \sin\theta \cos\varphi, \quad y = a \sin\theta \sin\varphi, \quad z = a \cos\theta$$

show that the two curves of the systems $\theta + \varphi = \alpha$, $\theta - \varphi = \beta$, which pass through any point (θ, φ) cut one another at the angle $\arccos \{(1 - \sin^2\theta)/(1 + \sin^2\theta)\}$ (cf. p. 164).

Show that the radius of curvature of either curve is equal to

$$a(1 + \sin^2\theta)^{\frac{3}{2}}/(5 + 3 \sin^2\theta)^{\frac{1}{2}}$$

(cf. Ex. 24).

55. If

$$f(x) = \int_0^{\pi/2} \log(1 - x^2 \cos^2\theta)\, d\theta,$$

prove that $f(x)$ is finite if $x^2 \leqq 1$, and that if $x^2 < 1$

$$\frac{d}{dx} f(x) = \int_0^{\pi/2} \frac{d}{dx} \{\log(1 - x^2 \cos^2\theta)\}\, d\theta,$$

and hence evaluate the integral.

56. Show that the area Σ of the right conoid

$$x = r\cos\theta, \quad y = r\sin\theta, \quad z = f(\theta),$$

included between two planes through the axis of z and the cylinder with generating lines parallel to this axis and cross-section $r = f'(\theta)$, and the area of its orthogonal projection on $z = 0$ are in the ratio

$$[\sqrt{2} + \log(1 + \sqrt{2})] : 1.$$

57. Assuming that the earth is a sphere of radius R for which the density at a distance r from the centre is of the form

$$\rho = A - Br^2$$

and the density at the surface is $2\frac{1}{2}$ times the density of water, while the mean density is $5\frac{1}{2}$ times that of water, show that the attraction at an internal point is equal to

$$\frac{1}{11} g \frac{r}{R}\left(20 - 9\frac{r^2}{R^2}\right),$$

where g is the value of gravity at the surface.

58. Let (x_1, y_1), (x_2, y_2), (x_3, y_3) be the vertices of a triangle of area A (the order of the suffixes giving the positive orientation). Prove that the moment of inertia of the triangle with respect to the x-axis is given by

$$\frac{A}{6}(y_1^2 + y_2^2 + y_3^2 + y_1y_2 + y_2y_3 + y_3y_1).$$

59. A hemisphere of radius a and of uniform density ρ is placed with its centre at the origin, so as to lie entirely on the positive side of the xy-plane. Show that its potential at the point $(0, 0, z)$ is

$$\frac{2\pi\rho}{3z}\left[(a^2 + z^2)^{\frac{3}{2}} - a^3 + \frac{3}{2}a^2z\right] - \frac{4}{3}\pi\rho z^2 \text{ if } 0 < z < a$$

and

$$\frac{2\pi\rho}{z}\left[(a^2 + z^2)^{\frac{3}{2}} + a^3 - \frac{3}{2}a^2z\right] - \frac{2}{3}\pi\rho z^2 \text{ if } z > a.$$

60. Sketch the curve

$$x = \frac{1 - \lambda^2}{1 + \lambda^2}, \quad y = \frac{\lambda - \lambda^3}{1 + \lambda^2}, \quad -1 \leqq \lambda \leqq +1$$

and calculate the area included by the curve.

61. Prove that the attraction at either pole of a uniform spheroid with density ρ and semi-axes a, a, c is equal to

$$2\pi\rho\int_0^{2c} r(1 - \cos\theta)\,dr,$$

where

$$r = 2a^2c\cos\theta/(a^2\cos^2\theta + c^2\sin^2\theta).$$

62. In the integral

$$I = \int_2^4 dx \int_{4/x}^{(20-4x)/(8-x)} (y-4)/dy$$

change the order of integration and evaluate the integral.

63. (*a*) By transforming to polar co-ordinates, show that the value of the integral

$$K = \int_0^{a \sin \beta} \left\{ \int_{y \cot \beta}^{\sqrt{a^2 - y^2}} \log(x^2 + y^2)\, dx \right\} dy \qquad \left(0 < \beta < \frac{\pi}{2}\right)$$

is $a^2 \beta (\log a - \frac{1}{2})$.

(*b*) Change the order of integration in the original integral.

64. Find the volume V cut off from the right cone

$$x^2 + y^2 = (h - z)^2 \tan^2 \alpha, \quad 0 \leqq z \leqq h,$$

by the right cylinder whose base is the curve

$$(h \tan \alpha - r)^3 = h^3 \tan^3 \alpha \sin^4 \theta \cos^2 \theta,$$

where z, r, θ are cylindrical polar co-ordinates, the volume being outside the cylinder and inside the cone.

65. Show that for the hyperbolic paraboloid $z = xy$ the value of

$$I = \iint \frac{z_{xx} z_{yy} - z_{xy}^2}{(1 + z_x^2 + z_y^2)^{\frac{3}{2}}}\, dS$$

taken over the surface bounded by the generators through the origin and the point (ξ, η, ζ) is

$$-\text{arc} \tan [\xi^2 \eta^2 / (\eta^2 \zeta^2 + \zeta^2 \xi^2 + \xi^2 \eta^2)^{\frac{1}{2}}].$$

66. Prove that for $-1 < a < 1$ and $-\frac{\pi}{2} < \text{arc} \sin a < \frac{\pi}{2}$

$$K(a) = \int_0^{\pi} \frac{\log(1 + a \cos x)}{\cos x}\, dx = \pi \text{ arc} \sin a.$$

67. Show that the area in the positive quadrant bounded by the curves $x^3 = a^2 y$, $x^3 = b^2 y$, $y^2 = cx$, $y^2 = dx$ is

$$\tfrac{5}{12}(a^{\frac{8}{5}} - b^{\frac{8}{5}})(c^{\frac{6}{5}} - d^{\frac{6}{5}}).$$

68*. Let Γ be a closed curve in space on which a definite sense of description of the curve has been assigned. Prove that there is a vector $\boldsymbol{a}$ with the following characteristic property: for any unit vector $\boldsymbol{n}$, the scalar product $\boldsymbol{an}$ is equal to the algebraic value of the area enclosed by the orthogonal projection of Γ on the plane Π orthogonal to $\boldsymbol{n}$. (Note that $\boldsymbol{n}$ gives the orientation of Π, and Γ gives the orientation of its projection on Π.) In particular, the projection of Γ on any plane parallel to $\boldsymbol{a}$ has the algebraic area zero. (The vector $\boldsymbol{a}$ may be called the *area vector* of Γ.)

69. Prove that in a central orbit the attraction p per unit mass is given by

$$p = \frac{h^2}{q^3}\frac{dq}{dr},$$

where q is the distance of the tangent of the orbit from the pole and h the area constant (p. 425).

Hence prove that the cardioid $r = a(1 + \cos\theta)$ can be described under an attraction to the pole equal to μr^{-4} per unit mass.

70*. Let there be n fixed particles in a plane, all attracting with a central force of magnitude $\frac{1}{r}$. Prove that there are not more than $n - 1$ positions of equilibrium for a particle in the field.

Calculate these positions for the case of four attracting particles with co-ordinates (a, b), $(-a, b)$, $(a, -b)$, $(-a, -b)$, where $a > b > 0$.

71*. A particle of unit mass moves under the action of two forces, of which the first is always towards the origin, and is equal to λ^2 times the distance of the particle from that point, while the second is always at right angles to the path of the particle, and is equal to 2μ times its velocity. Prove that if the particle is projected from the origin along the axis of x with velocity u, its co-ordinates at any subsequent time t are

$$x = \frac{u}{\sqrt{(\lambda^2 + \mu^2)}} \sin\sqrt{(\lambda^2 + \mu^2)}t \cos\mu t,$$

$$y = \frac{u}{\sqrt{(\lambda^2 + \mu^2)}} \sin\sqrt{(\lambda^2 + \mu^2)}t \sin\mu t.$$

72. (*a*) If u, v are two independent solutions of the equation

$$f(x)y''' - f'(x)y'' + \varphi(x)y' + \lambda(x)y = 0,$$

prove that the complete solution is $Au + Bv + Cw$, where

$$w = u\int\frac{vf(x)\,dx}{(uv' - u'v)^2} - v\int\frac{uf(x)\,dx}{(uv' - u'v)^2}$$

and A, B, C are arbitrary constants.

(*b*) Solve the equation

$$x^2(x^2 + 5)y''' - x(7x^2 + 25)y'' + (22x^2 + 40)y' - 30xy = 0,$$

which has solutions of the form x^n.

73. The tangent at a point P of a curve cuts the axis of y at a point T below the origin O and the curve is such that $OP = n\,.\,OT$. Prove that its polar equation is of the form

$$r = a\frac{(1 + \sin\theta)^n}{\cos^{n+1}\theta}.$$

74. Determine the solutions of the equation

$$\frac{\partial^2 z}{\partial t^2} = a^2 \frac{\partial^2 z}{\partial x^2}$$

which are also solutions of

$$\left(\frac{\partial z}{\partial t}\right)^2 = a^2 \left(\frac{\partial z}{\partial x}\right)^2.$$

75. Prove that if K is a homogeneous function of x, y, z the equation

$$\frac{\partial}{\partial x}\left(K \frac{\partial u}{\partial x}\right) + \frac{\partial}{\partial y}\left(K \frac{\partial u}{\partial y}\right) + \frac{\partial}{\partial z}\left(K \frac{\partial u}{\partial z}\right) = 0$$

has a solution which is a power of $(x^2 + y^2 + z^2)$.

76. (*a*) Apply Cauchy's theorem to the integral

$$\int \left(z + \frac{1}{z}\right)^m z^{n-1}\, dz \qquad (n > m > 0)$$

taken along a path consisting of the positive quadrant of the unit circle $|z| = 1$ and the parts of the axes between the origin and this circle, a small circular detour being made round $z = 0$; and hence deduce that

$$\int_0^{\pi/2} \cos^m \theta \cos n\theta \, d\theta = \frac{\sin \frac{(n-m)\pi}{2}}{2^{m+1}} \frac{\Gamma(m+1)\,\Gamma\left(\frac{n-m}{2}\right)}{\Gamma\left(\frac{n+m}{2}+1\right)}.$$

(*b*) Prove that if $n = m$ the value of the latter integral is $\pi/2^{m+1}$.

(In the complex integral the integrand may be taken as real on the positive half of the axis).

77. Prove that if f is analytic $\frac{d^n}{dx^n} f(\sqrt{x})$ is equal to the result obtained by putting y and a each equal to $\sqrt{x}$ in the expression for

$$2 \frac{\partial^n}{\partial y^n} \frac{y f(y)}{(y+a)^{n+1}}.$$

78. Show that if x and y are real

$$|\sinh(x + iy)| \geq A(x),$$

where $A(x)$ is independent of y and tends to ∞ as $x \to \pm\infty$.

By integrating $\frac{1}{(z-w)\sinh z}$ round a suitable sequence of contours, show that

$$\frac{1}{\sinh w} = \frac{1}{w} + 2w \sum_1^{\infty} \frac{(-1)^n}{w^2 + \pi^2 n^2}.$$

SUMMARY OF IMPORTANT THEOREMS AND FORMULÆ

1. Differentiation

Chain Rule for Functions of Several Variables.

If $u = f(\xi, \eta, \zeta, \dots)$, where $\xi = \xi(x, y)$, $\eta = \eta(x, y)$, . .

$$
\begin{aligned}
u_x &= f_\xi \xi_x + f_\eta \eta_x + f_\zeta \zeta_x + \dots, \\
u_{xx} &= f_{\xi\xi}\xi_x^2 + f_{\eta\eta}\eta_x^2 + f_{\zeta\zeta}\zeta_x^2 + \dots \\
&\quad + 2f_{\xi\eta}\xi_x\eta_x + 2f_{\xi\zeta}\xi_x\zeta_x + \dots \\
&\quad + \dots\dots\dots\dots\dots \\
&\quad + f_\xi \xi_{xx} + f_\eta \eta_{xx} + f_\zeta \zeta_{xx} + \dots,
\end{aligned}
$$

with corresponding formulæ for u_{xy} and u_{yy} (p. 73).

Implicit Functions. If $F(x, y) = 0$,

$$\frac{dy}{dx} = -\frac{F_x}{F_y},$$

$$\frac{d^2y}{dx^2} = -\frac{F_{xx}F_y^2 - 2F_{xy}F_xF_y + F_{yy}F_x^2}{F_y^3} \quad \text{(p. 115–16)}$$

Jacobians. If $\xi = \phi(x, y)$, $\eta = \psi(x, y)$,

$$\frac{\partial x}{\partial \xi} = \frac{\psi_y}{D}, \quad \frac{\partial x}{\partial \eta} = -\frac{\phi_y}{D}, \quad \frac{\partial y}{\partial \xi} = -\frac{\psi_x}{D}, \quad \frac{\partial y}{\partial \eta} = \frac{\phi_x}{D},$$

where

$$D = \frac{\partial(\xi, \eta)}{\partial(x, y)} = \begin{vmatrix} \phi_x & \phi_y \\ \psi_x & \psi_y \end{vmatrix} = \phi_x\psi_y - \phi_y\psi_x$$

(Jacobian or functional determinant) (p. 143).

Rules for Jacobians.

(1)
$$\frac{\partial(x, y)}{\partial(\xi, \eta)} = \frac{1}{\dfrac{\partial(\xi, \eta)}{\partial(x, y)}} \qquad \text{(p. 144).}$$

(2) If $u = u(\xi, \eta)$, $v = v(\xi, \eta)$ and $\xi = \xi(x, y)$, $\eta = \eta(x, y)$, then

$$\frac{\partial(u, v)}{\partial(x, y)} = \frac{\partial(u, v)}{\partial(\xi, \eta)} \frac{\partial(\xi, \eta)}{\partial(x, y)} \qquad \text{(p. 147).}$$

2. Convergence of Double Sequences

Convergence Test for Double Sequences (pp. 102–3). The sequence a_{nm} converges, or, in symbols,

$$\lim_{\substack{n \to \infty \\ m \to \infty}} a_{nm} = a,$$

if, and only if, for every positive ϵ there is an N such that

$$| a_{nm} - a_{n'm'} | < \epsilon$$

when $n > N$, $m > N$, $n' > N$, $m' > N$. Then

$$\lim_{\substack{n \to \infty \\ m \to \infty}} a_{nm} = \lim_{n \to \infty} \left(\lim_{m \to \infty} a_{nm} \right) = \lim_{m \to \infty} \left(\lim_{n \to \infty} a_{nm} \right),$$

provided that $\lim_{m \to \infty} a_{nm}$ and $\lim_{n \to \infty} a_{nm}$ respectively exist.

3. Uniform Convergence and Interchange of Infinite Operations

Dini's Theorem. If a series of *positive* continuous functions converges to a continuous limit function in a closed region, it converges *uniformly* to that limit (p. 106).

Interchange of Differentiation and Integration. (Differentiation of an integral with respect to a parameter.)

$$\frac{d}{dx}\int_a^b f(x, y)\,dy = \int_a^b f_x(x, y)\,dy,$$

provided that $f(x, y)$ and $f_x(x, y)$ are continuous in the interval under consideration (p. 218).

Interchange of Differentiation and Integration in Improper Integrals.

$$\frac{d}{dx}\int_0^\infty f(x, y)\,dy = \int_0^\infty f_x(x, y)\,dy,$$

provided that $f_x(x, y)$ is continuous in the interval under consideration and the integrals $\int_0^\infty f(x, y)\,dy$ and $\int_0^\infty f_x(x, y)\,dy$ converge uniformly in that interval (p. 312).

Interchange of Two Integrations. If $f(x, y)$ is continuous and a, b, α, β are constants,

$$\int_a^b dx \int_\alpha^\beta f(x, y)\,dy = \int_\alpha^\beta dy \int_a^b f(x, y)\,dx \qquad \text{(p. 239).}$$

The order of integration may also be reversed when the limits are not constants, provided that both integrations are performed over the *whole* of the region concerned and corresponding new limits are introduced (p. 242).

Interchange of Two Integrations in Improper Integrals.

$$\int_\alpha^\beta dx \int_0^\infty f(x, y)\,dy = \int_0^\infty dy \int_\alpha^\beta f(x, y)\,dx,$$

provided that the integral $\int_0^\infty f(x, y)\,dy$ converges uniformly in the interval $\alpha \leqq x \leqq \beta$ (p. 310).

4. Special Definite Integrals

$$\int_0^\infty e^{-x^2}\,dx = \tfrac{1}{2}\sqrt{\pi} \qquad \text{(pp. 262, 561; see also Vol. I, p. 496).}$$

$$\int_0^\infty \frac{\sin x}{x}\,dx = \tfrac{1}{2}\pi \qquad \text{(pp. 315, 554; see also Vol. I, pp. 251–3, 418, 450).}$$

Fresnel's Integrals:

$$\int_{-\infty}^{\infty} \sin(\tau^2)\,d\tau = \int_{-\infty}^{+\infty} \cos(\tau^2)\,d\tau = \sqrt{\frac{\pi}{2}} \quad \text{(p. 317).}$$

Fourier's Integral Theorem: If $f(x)$ is sectionally smooth and $\int_{-\infty}^{+\infty} |f(x)|\,dx$ converges and if $f(x+0)+f(x-0)=2f(x)$, then

$$f(x) = \frac{1}{\sqrt{2\pi}} \int_{-\infty}^{+\infty} g(\tau) e^{i\tau x}\,d\tau,$$

where
$$g(\tau) = \frac{1}{\sqrt{2\pi}} \int_{-\infty}^{+\infty} f(t) e^{-it\tau}\,dt \qquad \text{(p. 319).}$$

The Gamma Function (pp. 325–38). If $x > 0$, the gamma function $\Gamma(x)$ is defined by the equation

$$\Gamma(x) = \int_0^{\infty} e^{-t} t^{x-1}\,dt = 2\int_0^{\infty} e^{-t^2} t^{2x-1}\,dt.$$

It satisfies the functional equation

$$\Gamma(x+1) = x\Gamma(x);$$

hence if x is a positive integer n,

$$\Gamma(n) = (n-1)!$$

For all values of x other than $0, -1, -2, \ldots$ it may be expressed by the formulæ

$$\Gamma(x) = \lim_{n\to\infty} \frac{(n-1)!}{x(x+1)\ldots(x+n-1)} n^x = \frac{1}{x} \prod_{\nu=1}^{\infty} \frac{(1+1/\nu)^x}{1+x/\nu},$$

$$\frac{1}{\Gamma(x)} = x e^{\gamma x} \prod_{\nu=1}^{\infty} \left(1+\frac{x}{\nu}\right) e^{-x/\nu},$$

where $\gamma = \lim_{n\to\infty} \left(\sum_{\nu=1}^{n} \frac{1}{\nu} - \log n \right)$ is Euler's constant. Further, for every integer $m \geqq 2$,

$$\sum_{\nu=0}^{\infty} \frac{1}{(x+\nu)^m} = \frac{(-1)^m}{(m-1)!} \frac{d^m}{dx^m} \log \Gamma(x).$$

Again,

$$\Gamma(x)\Gamma(1-x) = \frac{\pi}{\sin \pi x}$$

(" extension theorem "). Hence, in particular,

$$\Gamma(\tfrac{1}{2}) = 2\int_0^\infty e^{-t^2}\,dt = \sqrt{\pi}.$$

The *beta function* $B(x, y)$ is defined as follows for positive values of x and y:

$$B(x, y) = \int_0^1 t^{x-1}(1-t)^{y-1}\,dt = \int_{-\frac{1}{2}}^{+\frac{1}{2}} (\tfrac{1}{2}+t)^{x-1}(\tfrac{1}{2}-t)^{y-1}\,dt$$

$$= 2\int_0^{\pi/2} \sin^{2x-1}\phi \cos^{2y-1}\phi\, d\phi.$$

The beta and gamma functions are connected by the relation

$$B(x, y) = \frac{\Gamma(x)\Gamma(y)}{\Gamma(x+y)} \qquad \text{(p. 337).}$$

For any complex z,

$$(1 - e^{2\pi i z})\Gamma(z) = \int t^{z-1}\, e^{-t} dt,$$

where C denotes a path which surrounds the positive real axis and approaches it asymptotically on either side (p. 566).

5. Mean Value Theorems

Mean Value Theorem for Functions of Two Variables (p. 80).

$$f(x+h, y+k) - f(x, y) = hf_x(x+\theta h, y+\theta k)$$
$$+ kf_y(x+\theta h, y+\theta k), \quad 0 < \theta < 1.$$

Taylor's Theorem for Functions of Two Variables (p. 80).

$$f(x+h, y+k) - f(x, y) = hf_x + kf_y$$
$$+ \frac{1}{2!}\{h^2 f_{xx} + 2hk f_{xy} + k^2 f_{yy}\}$$
$$+ \dots$$
$$+ \frac{1}{n!}\left\{h^n f_{x^n} + \binom{n}{1} h^{n-1}k f_{x^{n-1}y} + \dots + k^n f_{y^n}\right\}$$
$$+ R_n,$$

where the remainder R_n (in the symbolical notation of p. 79) is given by

$$R_n = \frac{1}{(n+1)!}\{hf_x(x+\theta h, y+\theta k)+kf_y(x+\theta h, y+\theta k)\}^{(n+1)}, \quad 0<\theta<1.$$

If as n increases this remainder tends to zero, we have the infinite *Taylor series*

$$f(x+h, y+k)$$
$$= f(x, y) + \frac{1}{1!}\{hf_x + kf_y\} + \frac{1}{2!}\{h^2 f_{xx} + 2hkf_{xy} + k^2 f_{yy}\}$$
$$+ \dots + \frac{1}{n!}\left\{h^n f_{x^n} + \binom{n}{1} h^{n-1}k f_{x^{n-1}y} + \dots + k^n f_{y^n}\right\} + \dots .$$

Mean Value Theorems for Multiple Integrals (p. 232).

$$\iint_R f(x, y)\,dS = \mu \Delta R,$$

where ΔR is the area of R and μ a value intermediate between the maximum and the minimum of $f(x, y)$ in R.

Similarly, if $p(x, y) \geqq 0$,

$$\iint_R f(x, y)p(x, y)\,dS = \mu \iint_R p(x, y)\,dS.$$

6. Vectors

For the definition of a vector see p. 3.

Let $\boldsymbol{v}$ be a vector in three dimensions with the components v_1, v_2, v_3.

Length of a Vector.

$$|\boldsymbol{v}| = \sqrt{(v_1^2 + v_2^2 + v_3^2)}.$$

Addition of Vectors.

$$\boldsymbol{z} = \boldsymbol{u} + \boldsymbol{v}$$

means the vector which has the components

$$z_1 = u_1 + v_1, \quad z_2 = u_2 + v_2, \quad z_3 = u_3 + v_3 \text{ (p. 5).}$$

Scalar Product (inner product)

$$\boldsymbol{uv} = |\boldsymbol{u}|\,|\boldsymbol{v}|\cos\delta$$
$$= u_1v_1 + u_2v_2 + u_3v_3,$$

where δ is the angle between $\boldsymbol{u}$ and $\boldsymbol{v}$ (p. 7).

Vector Product (outer product)

$$\boldsymbol{z} = [\boldsymbol{uv}]$$

means the vector which has the components

$$z_1 = \begin{vmatrix} u_2 & u_3 \\ v_2 & v_3 \end{vmatrix}, \quad z_2 = \begin{vmatrix} u_3 & u_1 \\ v_3 & v_1 \end{vmatrix}, \quad z_3 = \begin{vmatrix} u_1 & u_2 \\ v_1 & v_2 \end{vmatrix} \qquad \text{(p. 17).}$$

Differentiation.

$$\frac{d(\boldsymbol{u}+\boldsymbol{v})}{dt} = \frac{d\boldsymbol{u}}{dt} + \frac{d\boldsymbol{v}}{dt};$$

$$\frac{d(\boldsymbol{uv})}{dt} = \frac{d\boldsymbol{u}}{dt}\boldsymbol{v} + \boldsymbol{u}\frac{d\boldsymbol{v}}{dt};$$

$$\frac{d[\boldsymbol{uv}]}{dt} = \left[\frac{d\boldsymbol{u}}{dt}\boldsymbol{v}\right] + \left[\boldsymbol{u}\frac{d\boldsymbol{v}}{dt}\right] \qquad \text{(p. 85).}$$

If the co-ordinate axes are rotated, the vector components are transformed in the same way as x, y, z, the components of the position vector (p. 84).

By the *derivative of the function* f(x, y) *in the direction of the unit vector* $\boldsymbol{n}$ whose components are $\cos\alpha$, $\sin\alpha$, we mean the limit

$$\lim_{\rho\to 0} \frac{f(x+\rho\cos\alpha,\ y+\rho\sin\alpha) - f(x, y)}{\rho} = \frac{\partial f(x, y)}{\partial n}.$$

Hence

$$\frac{\partial}{\partial n} = \cos\alpha\frac{\partial}{\partial x} + \sin\alpha\frac{\partial}{\partial y}.$$

In particular,

$$\frac{\partial x}{\partial n} = \cos\alpha, \quad \frac{\partial y}{\partial n} = \sin\alpha,$$

and hence in general

$$\frac{\partial f}{\partial n} = \frac{\partial f}{\partial x}\frac{\partial x}{\partial n} + \frac{\partial f}{\partial y}\frac{\partial y}{\partial n}.$$

In the same way, in three dimensions the derivative in the direction of the vector $\boldsymbol{n}$ whose components are $\cos\alpha$, $\cos\beta$, $\cos\gamma$ is

$$\frac{\partial f}{\partial n} = \cos\alpha\,\frac{\partial f}{\partial x} + \cos\beta\,\frac{\partial f}{\partial y} + \cos\gamma\,\frac{\partial f}{\partial z}$$

$$= \frac{\partial f}{\partial x}\frac{\partial x}{\partial n} + \frac{\partial f}{\partial y}\frac{\partial y}{\partial n} + \frac{\partial f}{\partial z}\frac{\partial z}{\partial n} \qquad \text{(p. 64).}$$

The Differential Operations.

With every scalar function $f(x_1, x_2, x_3)$ there is associated a vector grad f with the components $f_{x_1}, f_{x_2}, f_{x_3}$ (p. 89). The derivative of f in the direction of the unit vector $\boldsymbol{n}$ is $\boldsymbol{n}$ grad f.

With every vector field $\boldsymbol{u}(x_1, x_2, x_3)$ there is associated a vector curl $\boldsymbol{u}$ with the components

$$\frac{\partial u_3}{\partial x_2} - \frac{\partial u_2}{\partial x_3},\quad \frac{\partial u_1}{\partial x_3} - \frac{\partial u_3}{\partial x_1},\quad \frac{\partial u_2}{\partial x_1} - \frac{\partial u_1}{\partial x_2}, \qquad \text{(p. 92)}$$

and a scalar function

$$\operatorname{div} \boldsymbol{u} = \frac{\partial u_1}{\partial x_1} + \frac{\partial u_2}{\partial x_2} + \frac{\partial u_3}{\partial x_3} \qquad \text{(p. 91).}$$

Using the symbolic vector ∇ (*nabla*) with "components" $\frac{\partial}{\partial x_1}, \frac{\partial}{\partial x_2}, \frac{\partial}{\partial x_3}$, we have

$$\operatorname{grad} f = \nabla f,\ \operatorname{curl} \boldsymbol{u} = [\nabla \boldsymbol{u}],\ \operatorname{div} \boldsymbol{u} = \nabla \boldsymbol{u} \qquad \text{(p. 92).}$$

Further,

$$\operatorname{curl} \operatorname{grad} f = 0, \quad \operatorname{div} \operatorname{curl} \boldsymbol{u} = 0,$$

$$\operatorname{div} \operatorname{grad} f = \Delta f = \frac{\partial^2 f}{\partial x_1^2} + \frac{\partial^2 f}{\partial x_2^2} + \frac{\partial^2 f}{\partial x_3^2}.$$

7. Multiple Integrals

For the definition of a multiple integral see p. 224.

The rules for the addition of integrands and combination of regions of integration are the same as for ordinary integrals (p. 231).

Transformation of a Multiple Integral. If the oriented region R of the xy-plane is mapped on a correspondingly-oriented region R' of the uv-plane by means of a reversible one-to-one transformation whose Jacobian

$$D = \frac{\partial(x, y)}{\partial(u, v)}$$

does not vanish anywhere, then

$$\iint_R f(x, y)\,dx\,dy = \iint_{R'} f(x, y)\,D\,du\,dv \quad \text{(pp. 253, 377).}$$

An analogous formula holds for any number of dimensions (p. 254).

In particular, transformation to *polar co-ordinates*

$$x = r\cos\theta, \quad y = r\sin\theta$$

or

$$x = r\cos\phi\sin\theta, \quad y = r\sin\phi\sin\theta, \quad z = r\cos\theta$$

gives the formulæ

$$\iint_R f(x, y)\,dx\,dy = \iint_{R'} f(r\cos\theta, r\sin\theta)\,r\,dr\,d\theta,$$

(Vol. I, p. 494).

$$\iiint_R f(x, y, z)\,dx\,dy\,dz = \iiint_{R'} f(x, y, z)\,r^2\sin\theta\,dr\,d\theta\,d\phi \quad \text{(p. 254).}$$

Reduction of a Multiple Integral to Ordinary Integrals (p. 243). Let $\alpha \leqq y \leqq \beta$ in R, and for every y let $a = a(y) \leqq x \leqq b(y) = b$; then

$$\iint_R f(x, y)\,dx\,dy = \int_\alpha^\beta dy \int_a^b f(x, y)\,dx.$$

8. Integral Theorems of Gauss, Green, and Stokes

For the definition of a curvilinear integral (line integral), see pp. 344 *et seq.*

1. Two Dimensions.

If the region R is simply connected, the line integral

$$\int_C (a\,dx + b\,dy) = \int_C \mathbf{A}\,d\mathbf{x}$$

is independent of the path C joining two points in R if, and only if, the condition of integrability

$$a_y = b_x$$

holds at every point of R. In this case, if the initial point is fixed, the integral is a function $U(\xi, \eta)$ of the end-point, such that the vector $\boldsymbol{A}$ with components a, b satisfies the relation

$$\boldsymbol{A} = \operatorname{grad} U \qquad \text{(p. 352).}$$

Gauss's Theorem. Let R be a simply-connected region and C its boundary. Then

$$\iint_R \{f_x(x, y) + g_y(x, y)\}\,dx\,dy = \int_{+C} \{f(x, y)\,dy - g(x, y)\,dx\}, \qquad \text{(p. 360)}$$

or, in vector notation,

$$\iint_R \operatorname{div} \boldsymbol{A}\,dx\,dy = \int_C \boldsymbol{A}\boldsymbol{n}\,ds = \int_C A_n\,ds, \qquad \text{(p. 364)}$$

where $\boldsymbol{n}$ is the unit vector in the direction of the outward-drawn normal, A_n the normal component of the vector $\boldsymbol{A}$ with components f, g, and ds the element of arc of the boundary curve.

Green's Theorem (p. 366).

$$\iint_R (u_x v_x + u_y v_y)\,dx\,dy = -\iint_R u\,\Delta v\,dx\,dy + \int_{+C} (-u v_y\,dx + u v_x\,dy)$$

$$= -\iint_R v\,\Delta u\,dx\,dy + \int_{+C} v\frac{\partial u}{\partial n}\,ds,$$

$$\iint_R (u\,\Delta v - v\,\Delta u)\,dx\,dy = \int_{+C} \{(v u_y - u v_y)\,dx - (v u_x - u v_x)\,dy\}$$

$$= \int_{+C} \left(u\frac{\partial v}{\partial n} - v\frac{\partial u}{\partial n}\right) ds.$$

In vector notation the first form of the theorem is

$$\iint_R (\operatorname{grad} u \operatorname{grad} v)\,dx\,dy = -\iint_R v \operatorname{div} \operatorname{grad} u\,dx\,dy + \int_C v\frac{\partial u}{\partial n}\,ds,$$

where

$$\Delta u = \operatorname{div} \operatorname{grad} u = u_{xx} + u_{yy},$$

and $\partial/\partial n$ denotes differentiation in the direction of the outward-drawn normal.

2. Three Dimensions.

The necessary and sufficient condition that the line integral

$$\int_C (a\,dx + b\,dy + c\,dz) = \int_C \boldsymbol{A}\,d\boldsymbol{x}$$

shall be independent of the path C joining two points in a simply-connected region R is

$$\operatorname{curl} \boldsymbol{A} = 0,$$

or, in full,

$$a_y = b_x, \quad b_z = c_y, \quad c_x = a_z \qquad \text{(p. 358)}.$$

Surface Integral (p. 381). This is given by

$$\iint_S \{a(x, y, z)\,dy\,dz + b(x, y, z)\,dz\,dx + c(x, y, z)\,dx\,dy\}$$

or

$$\iint_B \left\{a(x, y, z)\frac{\partial(y, z)}{\partial(u, v)} + b(x, y, z)\frac{\partial(z, x)}{\partial(u, v)} + c(x, y, z)\frac{\partial(x, y)}{\partial(u, v)}\right\} du\,dv,$$

if $x = x(u, v)$, $y = y(u, v)$, $z = z(u, v)$ and the oriented region B in the uv-plane corresponds to the surface S.

Gauss's Theorem. Let $\boldsymbol{n}$ be the unit vector in the direction of the outward-drawn normal and A_n the normal component of the vector $\boldsymbol{A}$ with components a, b, c; further, let $\partial/\partial n$ denote differentiation in the direction of the outward-drawn normal. Then

$$\iiint_R (a_x + b_y + c_z)\,dx\,dy\,dz = \iint_S \left(a\frac{\partial x}{\partial n} + b\frac{\partial y}{\partial n} + c\frac{\partial z}{\partial n}\right) dS, \qquad \text{(p. 386)}$$

or, in vector notation,

$$\iiint_R \operatorname{div} \boldsymbol{A}\,dx\,dy\,dz = \iint_S \boldsymbol{A}\boldsymbol{n}\,dS = \iint_S A_n\,dS, \quad \text{(p. 388)}$$

the integrals on the right being taken over the closed surface S bounding the region R.

Green's Theorem (p. 390).

$$\iiint_R (u_x v_x + u_y v_y + u_z v_z)\,dx\,dy\,dz$$

$$= -\iiint_R u\Delta v\,dx\,dy\,dz + \iint_S u\frac{\partial v}{\partial n}\,dS,$$

$$\iiint_R (u\Delta v - v\Delta u)\,dx\,dy\,dz = \iint_S \left(u\frac{\partial v}{\partial n} - v\frac{\partial u}{\partial n}\right) dS,$$

where $\partial/\partial n$ and S have the same meanings as before and

$$\Delta u = u_{xx} + u_{yy} + u_{zz}.$$

Stokes's Theorem (p. 393). Let the oriented surface S be bounded by the correspondingly-oriented curve C. Then

$$\iint_S \left\{ \left(\frac{\partial \chi}{\partial y} - \frac{\partial \psi}{\partial z} \right) dy\,dz + \left(\frac{\partial \phi}{\partial z} - \frac{\partial \chi}{\partial x} \right) dz\,dx + \left(\frac{\partial \psi}{\partial x} - \frac{\partial \phi}{\partial y} \right) dx\,dy \right\}$$

$$= \int_C (\phi\,dx + \psi\,dy + \chi\,dz).$$

In vector notation: let A_t be the tangential component of the vector $\boldsymbol{A} = (\phi, \psi, \chi)$ in the direction in which the curve C is described, $(\text{curl}\,\boldsymbol{A})_n$ the component of curl $\boldsymbol{A}$ in the direction of the outward-drawn normal, and ds the element of arc on C measured in the direction in which the curve is described: then

$$\iint_S (\text{curl}\,\boldsymbol{A})_n\,dS = \int_C A_t\,ds.$$

9. Maxima and Minima

The following rules hold only for maxima and minima in the *interior* of the region under consideration.

Free Maxima and Minima of a Function of Two Variables.

The *necessary* conditions for an extreme value of the function $u = f(x, y)$ are

$$f_x = 0, \quad f_y = 0 \qquad \text{(p. 184).}$$

If these conditions are satisfied and if

$$f_{xx} f_{yy} - f_{xy}^2 > 0,$$

there is an extreme value at the point in question. It is a maximum or a minimum according as f_{xx} (and hence also f_{yy}) is negative or positive. If

$$f_{xx} f_{yy} - f_{xy}^2 < 0,$$

the point is a saddle point (p. 207).

Maxima and Minima subject to Subsidiary Conditions (Method of Undetermined Multipliers) (pp. 188–99).

If in the function $u = f(x_1, \ldots, x_n)$ the n variables are connected by the m subsidiary conditions ($m < n$)

$$\phi_1(x_1, \ldots, x_n) = 0, \quad \ldots, \quad \phi_m(x_1, \ldots, x_n) = 0,$$

we introduce m multipliers $\lambda_1, \ldots, \lambda_m$ and form the function

$$F = f + \lambda_1\phi_1 + \lambda_2\phi_2 + \ldots + \lambda_m\phi_m.$$

Then the m conditions and the n additional equations

$$\frac{\partial F}{\partial x_1} = 0, \quad \ldots, \quad \frac{\partial F}{\partial x_n} = 0$$

give $(m + n)$ *necessary* conditions for the extreme points.

10. Curves * and Surfaces

In what follows (ξ, η), or (ξ, η, ζ), are current co-ordinates.

1. Plane Curves.

Equation of the curve:

$$(a)\ y = f(x), \quad (b)\ F(x, y) = 0, \quad (c)\ x = \phi(t),\ y = \psi(t).$$

Equation of the tangent at the point (x, y) (Vol. I, p. 263; Vol. II, p. 122):

$$(a)\ \eta - y = (\xi - x)f'(x), \quad (b)\ (\xi - x)F_x + (\eta - y)F_y = 0,$$
$$(c)\ \{\xi - \phi(t)\}\psi'(t) - \{\eta - \psi(t)\}\phi'(t) = 0.$$

Equation of the normal at the point (x, y) (Vol. I, p. 263; Vol. II, p. 123):

$$(a)\ \xi - x + (\eta - y)f'(x) = 0, \quad (b)\ (\xi - x)F_y - (\eta - y)F_x = 0,$$
$$(c)\ \{\xi - \phi(t)\}\phi'(t) + \{\eta - \psi(t)\}\psi'(t) = 0.$$

Curvature (Vol. I, p. 281; Vol. II, p. 125):

$$(a)\ k = \frac{y''}{(1 + y'^2)^{\frac{3}{2}}}, \quad (b)\ k = -\frac{F_{xx}F_y^2 - 2F_{xy}F_xF_y + F_{yy}F_x^2}{(F_x^2 + F_y^2)^{\frac{3}{2}}},$$
$$(c)\ k = \frac{\dot\phi\ddot\psi - \ddot\phi\dot\psi}{(\dot\phi^2 + \dot\psi^2)^{\frac{3}{2}}}.$$

* Some formulæ discussed in Vol. I have been repeated here for convenience of reference.

Radius of curvature (Vol. I, p. 282; Vol. II, p. 126):

$$\rho = \frac{1}{|k|}.$$

Evolute (locus of centre of curvature) (Vol. I, pp. 283, 307--311):

$$(a)\ \xi = x - y'\frac{1+y'^2}{y''}, \quad \eta = y + \frac{1+y'^2}{y''},$$

$$(b)\ \xi = x + F_x \frac{F_x^2 + F_y^2}{F_{xx}F_y^2 - 2F_{xy}F_xF_y + F_{yy}F_x^2},$$

$$\eta = y + F_y \frac{F_x^2 + F_y^2}{F_{xx}F_y^2 - 2F_{xy}F_xF_y + F_{yy}F_x^2}.$$

$$(c)\ \xi = \phi - \dot{\psi}\frac{\dot{\phi}^2 + \dot{\psi}^2}{\dot{\phi}\ddot{\psi} - \ddot{\phi}\dot{\psi}}, \quad \eta = \psi + \dot{\phi}\frac{\dot{\phi}^2 + \dot{\psi}^2}{\dot{\phi}\ddot{\psi} - \ddot{\phi}\dot{\psi}}.$$

Involute (Vol. I, p. 309):

$$\xi = x + (a - s)\dot{x}, \quad \eta = y + (a - s)\dot{y},$$

where a is an arbitrary constant and s the length of arc measured from a given point (s being the parameter).

Point of inflection (Vol. I, pp. 159, 266; Vol. II, p. 125):
The necessary condition for a point of inflection is

$$(a)\ y'' = 0, \quad (b)\ F_{xx}F_y^2 - 2F_{xy}F_xF_y + F_{yy}F_x^2 = 0,$$
$$(c)\ \dot{x}\ddot{y} - \ddot{x}\dot{y} = 0.$$

Angle between two curves (Vol. I, p. 264; Vol. II, p. 126):

$$(b)\ \cos\omega = \frac{F_xG_x + F_yG_y}{\sqrt{(F_x^2 + F_y^2)}\sqrt{(G_x^2 + G_y^2)}},$$

$$(c)\ \cos\omega = \frac{\dot{x}\dot{x}_1 + \dot{y}\dot{y}_1}{\sqrt{(\dot{x}^2 + \dot{y}^2)}\sqrt{(\dot{x}_1^2 + \dot{y}_1^2)}}.$$

In particular, the curves are orthogonal if

$$(b)\ F_xG_x + F_yG_y = 0, \quad (c)\ \dot{x}\dot{x}_1 + \dot{y}\dot{y}_1 = 0;$$

the curves touch if

$$(b)\ F_xG_y - F_yG_x = 0, \quad (c)\ \dot{x}\dot{y}_1 - \dot{x}_1\dot{y} = 0.$$

Two curves $y = f(x)$, $y = g(x)$ have contact of order n at a point x, if

$$f(x) = g(x),\ f'(x) = g'(x),\ \ldots,\ f^{(n)}(x) = g^{(n)}(x),$$
$$f^{(n+1)}(x) \neq g^{(n+1)}(x)$$

(Vol. I, pp. 331–3).

2. **Curves in Space.**

Equation of the curve:

$$x = \phi(t), \quad y = \psi(t), \quad z = \chi(t).$$

Direction cosines of the tangent (p. 86):

$$\frac{\dot{x}}{\sqrt{(\dot{x}^2 + \dot{y}^2 + \dot{z}^2)}}, \quad \frac{\dot{y}}{\sqrt{(\dot{x}^2 + \dot{y}^2 + \dot{z}^2)}}, \quad \frac{\dot{z}}{\sqrt{(\dot{x}^2 + \dot{y}^2 + \dot{z}^2)}}.$$

Curvature (p. 86):

$$|k| = \frac{1}{\rho} = \sqrt{\left\{\left(\frac{d^2x}{ds^2}\right)^2 + \left(\frac{d^2y}{ds^2}\right)^2 + \left(\frac{d^2z}{ds^2}\right)^2\right\}},$$

where ds is the element of arc.

3. **Surfaces.**

Equation of the surface:

$$(a)\ z = f(x, y), \quad (b)\ F(x, y, z) = 0,$$
$$(c)\ x = \phi(u, v),\ y = \psi(u, v),\ z = \chi(u, v).$$

Equation of the tangent plane (p. 130):

$$(a)\ \zeta - z = (\xi - x)f_x + (\eta - y)f_y,$$
$$(b)\ (\xi - x)F_x + (\eta - y)F_y + (\zeta - z)F_z = 0,$$
$$(c)\ (\xi - x)(\psi_u\chi_v - \psi_v\chi_u) + (\eta - y)(\chi_u\phi_v - \chi_v\phi_u) + (\zeta - z)(\phi_u\psi_v - \phi_v\psi_u) = 0.$$

Direction cosines of the normal (Vol. II, pp. 130, 163):

$$(a)\ \cos\alpha = -\frac{f_x}{\sqrt{(1 + f_x^2 + f_y^2)}}, \quad \cos\beta = -\frac{f_y}{\sqrt{(1 + f_x^2 + f_y^2)}},$$
$$\cos\gamma = +\frac{1}{\sqrt{(1 + f_x^2 + f_y^2)}};$$

$$(b)\ \cos\alpha = \frac{F_x}{\sqrt{(F_x^2 + F_y^2 + F_z^2)}},\quad \cos\beta = \frac{F_y}{\sqrt{(F_x^2 + F_y^2 + F_z^2)}},$$

$$\cos\gamma = \frac{F_z}{\sqrt{(F_x^2 + F_y^2 + F_z^2)}};$$

$$(c)\ \cos\alpha = \frac{A}{\sqrt{(A^2 + B^2 + C^2)}},\quad \cos\beta = \frac{B}{\sqrt{(A^2 + B^2 + C^2)}},$$

$$\cos\gamma = \frac{C}{\sqrt{(A^2 + B^2 + C^2)}},$$

where

$$A = \psi_u\chi_v - \psi_v\chi_u,\quad B = \chi_u\phi_v - \chi_v\phi_u,\quad C = \phi_u\psi_v - \phi_v\psi_u.$$

Angle between two surfaces (Vol. II, p. 130):

$$\cos\omega = \cos\alpha_1 \cos\alpha_2 + \cos\beta_1 \cos\beta_2 + \cos\gamma_1 \cos\gamma_2;$$

in particular, the condition that the surfaces are orthogonal is

$$\cos\alpha_1 \cos\alpha_2 + \cos\beta_1 \cos\beta_2 + \cos\gamma_1 \cos\gamma_2 = 0.$$

4. **Envelopes** (Vol. II, pp. 171–83).

To obtain the envelope of the family of plane curves

$$f(x, y, c) = 0$$

or of the family of surfaces

$$f(x, y, z, c) = 0,$$

we calculate the "discriminant" by eliminating c from the equations

$$f = 0,\quad f_c = 0.$$

The discriminant contains the envelope and also the geometrical locus of the singular points.

If the family of curves is given by the parametric equations $x = \phi(t, c)$, $y = \psi(t, c)$, the discriminant is obtained by eliminating c and t from the equations

$$x = \phi(t, c),\quad y = \psi(t, c),\quad \frac{\partial\phi}{\partial t}\frac{\partial\psi}{\partial c} - \frac{\partial\phi}{\partial c}\frac{\partial\psi}{\partial t} = 0 \quad \text{(p. 174).}$$

The envelope of a two-parameter family of surfaces

$$f(x, y, z, c_1, c_2) = 0$$

is contained in the equation obtained by eliminating the two parameters c_1, c_2 from the equations

$$f = 0, \quad f_{c_1} = 0, \quad f_{c_2} = 0.$$

11. Length of Arc, Area, Volume

Length of Arc (Vol. I, pp. 276–80). Let a plane curve be given by the equations

$$(a)\ y = f(x), \quad (b)\ F(x, y) = 0, \quad (c)\ x = \phi(t),\ y = \psi(t),$$
$$(d)\ \text{(polar co-ordinates)}\ r = r(\theta).$$

The length of arc is

$$(a)\ s = \int_{x_0}^{x_1} \sqrt{(1 + y'^2)}\, dx, \qquad (c)\ s = \int_{t_0}^{t_1} \sqrt{(\dot{x}^2 + \dot{y}^2)}\, dt,$$

$$(b)\ s = \int_{x_0}^{x_1} \frac{1}{F_y} \sqrt{(F_x{}^2 + F_y{}^2)}\, dx, \qquad (d)\ s = \int_{\theta_0}^{\theta_1} \sqrt{(r^2 + r'^2)}\, d\theta$$

The length of arc of the three-dimensional curve

$$x = \phi(t), \quad y = \psi(t), \quad z = \chi(t)$$

is

$$s = \int_{t_0}^{t_1} \sqrt{(\dot{x}^2 + \dot{y}^2 + \dot{z}^2)}\, dt \qquad \text{(p. 86)}.$$

Area of Plane Surface. The area bounded by the curve

$$r = r(\theta)$$

and two radii vectores θ_0, θ_1, where r, θ are polar co-ordinates, is given by

$$\tfrac{1}{2} \int_{\theta_0}^{\theta_1} r^2\, d\theta \qquad \text{(Vol. I, p. 275)}.$$

The area enclosed by the curve

$$y = f(x),$$

the two ordinates $x = x_0$, $x = x_1$, and the x-axis, is

$$\int_{x_0}^{x_1} y\, dx \qquad \text{(Vol. I, p. 80)}.$$

Let R be a positively-oriented plane surface and C its boundary (for the orientation and sign of an area cf. Chap. V, section 4, p. 375). Then the area of the surface is

$$\iint_R dx\,dy = -\int_C y\,dx = \int_C x\,dy = \tfrac{1}{2}\int_C (x\,dy - y\,dx)$$

(pp. 347, 375–6).

Area of Curved Surface (pp. 268–74). Let the equation of the surface be

$$(a)\ z = f(x, y), \quad (b)\ F(x, y, z) = 0,$$
$$(c)\ x = \phi(u, v),\ y = \psi(u, v),\ z = \chi(u, v).$$

In case (c) let E, F, G be the so-called fundamental quantities of the surface, i.e. let

$$\left.\begin{aligned} E &= \phi_u^2 + \psi_u^2 + \chi_u^2, \\ F &= \phi_u\phi_v + \psi_u\psi_v + \chi_u\chi_v, \\ G &= \phi_v^2 + \psi_v^2 + \chi_v^2. \end{aligned}\right\} \qquad \text{(pp. 162, 273).}$$

Then

$$EG - F^2 = (\phi_u\psi_v - \phi_v\psi_u)^2 + (\psi_u\chi_v - \psi_v\chi_u)^2 + (\chi_u\phi_v - \chi_v\phi_u)^2$$

(p. 273).

The length of arc of the curve

$$u = u(t), \quad v = v(t)$$

drawn on the surface is then

$$s = \int_{t_0}^{t_1} \sqrt{(E\dot{u}^2 + 2F\dot{u}\dot{v} + G\dot{v}^2)}\,dt \qquad \text{(p. 162).}$$

The area of the curved surface lying vertically above the region R in the xy-plane is

$$A = \iint d\sigma:$$

$$(a)\ A = \iint_R \sqrt{(1 + f_x^2 + f_y^2)}\,dx\,dy,$$

$$(b)\ A = \iint_R \frac{1}{F_z}\sqrt{(F_x^2 + F_y^2 + F_z^2)}\,dx\,dy,$$

$$(c)\ A = \iint_B \sqrt{(EG - F^2)}\,du\,dv,$$

the last integral being taken over the region B of the uv-plane which corresponds to the region R.

The area of the *surface of revolution*

$$x = u\cos v, \quad y = u\sin v, \quad z = \phi(u),$$

which is produced when the curve

$$z = \phi(x)$$

is rotated about the z-axis, is

$$A = 2\pi\int_{u_0}^{u_1} u\sqrt{\{1+\phi'^2(u)\}}\,du = 2\pi\int_{s_0}^{s_1} u\,ds,$$

where s is the arc of the meridian curve $z = \phi(x)$ (Vol. II, p. 274; cf. also Vol. I, p. 285).

The surface ω_n of the unit sphere in n dimensions,

$$x_1^2 + x_2^2 + \ldots + x_n^2 = 1,$$

is given by

$$\omega_n = \frac{2(\sqrt{\pi})^n}{\Gamma(n/2)}.$$

Volumes. The volume bounded below by the region R and above by the surface S with the equation

$$z = f(x, y)$$

is given by

$$V = \iint_R f(x, y)\,dx\,dy \qquad \text{(p. 225).}$$

(for the sign see Chap. V, section 4, p. 380).

If the surface S is *closed* and forms the whole boundary of the region V, the volume of this region is given by

$$V = \iiint_V dx\,dy\,dz = -\iint_S z\,dx\,dy = -\iint_S x\,dy\,dz = -\iint_S y\,dz\,dx \qquad \text{(p. 387).}$$

In *polar co-ordinates* the same volume is given by

$$V = \iiint_B r^2\sin\theta\,dr\,d\theta\,d\phi,$$

where B is the region of $r\theta\phi$-space corresponding to the region V (p. 254).

The volume of the *surface of revolution*

$$x = u\cos v, \; y = u\sin v, \; z = \phi(u),$$

which is produced when the curve

$$z = \phi(x)$$

is rotated about the z-axis, is

$$V = \pi \int_{z_0}^{z_1} u^2 \, dz$$

(Vol. II, p. 267; cf. also Vol. I, p. 285).

The volume v_n of the unit sphere in n dimensions,

$$x_1^2 + x_2^2 + \ldots + x_n^2 = 1,$$

is given by

$$v_n = \frac{(\sqrt{\pi})^n}{\Gamma\{(n+2)/2\}} \qquad \text{(p. 304).}$$

The volume swept out by a moving plane area P of area A is

$$V = \int_{t_0}^{t_1} A(t) \frac{dn}{dt} \, dt,$$

where dn/dt is the component of the velocity of the mean centre of P perpendicular to the plane of P (p. 295).

12. Calculus of Variations

The necessary and sufficient condition that the integral

$$I(u) = \int_{x_0}^{x_1} F(x, u, u') dx$$

shall be stationary is *Euler's equation*

$$F_u - \frac{d}{dx} F_{u'} = 0,$$

or

$$F_{u'u'} u'' + F_{uu'} u' + F_{xu'} - F_u = 0 \qquad \text{(p. 498).}$$

If F involves several functions $u_1(x)$, $u_2(x)$, . . . , $u_n(x)$ and their derivatives, then a necessary and sufficient condition that the integral

$$I(u) = \int_{x_0}^{x_1} F(x, u_1, \ldots, u_n, u'_1, \ldots, u'_n)$$

shall be stationary is that $u_1, \ldots, u_n$ satisfy the system of Euler's equations

$$F_{u_i} - \frac{d}{dx} F_{u_i'} = 0 \quad (i = 1, \ldots, n), \qquad \text{(p. 508).}$$

If F depends on x, $u(x)$, $u'(x)$, $u''(x)$, Euler's equation is

$$F_u - \frac{d}{dx} F_{u'} + \frac{d^2}{dx^2} F_{u''} = 0 \qquad \text{(p. 513).}$$

If $\int F(x, y, z, \dot{x}, \dot{y}, \dot{z})dt$ is to be made stationary subject to the subsidiary condition $G(x, y, z) = 0$, then a necessary condition is

$$\frac{d}{dt} F_{\dot{x}} - F_x = \lambda G_x,$$

$$\frac{d}{dt} F_{\dot{y}} - F_y = \lambda G_y,$$

$$\frac{d}{dt} F_{\dot{z}} - F_z = \lambda G_z,$$

where λ denotes Lagrange's multiplier (p. 517).

13. Analytic Functions

For definition, see p. 532.

The necessary and sufficient condition that

$$f(z) = f(x + iy) = u(x, y) + iv(x, y)$$

shall be analytic in a region R is that in R the *Cauchy-Riemann differential equations*

$$u_x = v_y, \quad u_y = -v_x$$

hold (p. 532).

Cauchy's theorem: If $f(t)$ is analytic in a simply-connected region R, then

$$\int_C f(t)\, dt = 0$$

if C is a closed curve in the interior of R (p. 539).

Cauchy's formula: Under the same condition as Cauchy's theorem the formula

$$f(z) = \frac{1}{2\pi i} \int_C \frac{f(t)}{t - z}\, dt$$

holds if z is a point in the interior of C.

If $f(z)$ is analytic in the interior and on the boundary of a circle $|z - z_0| \leqq R$, it can be expanded in a power series

$$f(z) = f(z_0) + \sum_{\nu=1}^{\infty} c_\nu (z - z_0)^\nu$$

which converges in the interior of the circle. Here

$$c_\nu = \frac{f^{(\nu)}(z_0)}{\nu!} = \frac{1}{2\pi i}\int_C \frac{f(t)}{(t - z_0)^{\nu+1}}\,dt \qquad \text{(pp. 547–9).}$$

ANSWERS AND HINTS

CHAPTER I

§ 1, p. 12.

3. Let the vectors joining O to the points P, Q, R, S be denoted by $\boldsymbol{p}$, $\boldsymbol{q}$, $\boldsymbol{r}$, $\boldsymbol{s}$. Then the vector from O to the centre of mass of the triangle PQR is given by $\frac{1}{3}(\boldsymbol{p} + \boldsymbol{q} + \boldsymbol{r})$, and (cf. Ex. 2) the vector joining O to the centre of mass of the tetrahedron by $\frac{3}{4}\,\frac{1}{3}(\boldsymbol{p} + \boldsymbol{q} + \boldsymbol{r}) + \frac{1}{4}\boldsymbol{s} = \frac{1}{4}(\boldsymbol{p} + \boldsymbol{q} + \boldsymbol{r} + \boldsymbol{s})$; this expression is independent of the order in which the vertices are taken.

4. A, A', . . . , C' are the final points of the vectors $\frac{1}{2}(\boldsymbol{p} + \boldsymbol{q})$, $\frac{1}{2}(\boldsymbol{r} + \boldsymbol{s})$, . . . , $\frac{1}{2}(\boldsymbol{q} + \boldsymbol{r})$, and the three lines AA', BB', CC' all have the same mid point, the final point of the vector $\frac{1}{4}(\boldsymbol{p} + \boldsymbol{q} + \boldsymbol{r} + \boldsymbol{s})$, which is the centre of mass of the tetrahedron.

§ 2, p. 18.

1. The distance is given by the length of the vector product of a unit vector lying along l and any vector joining P to a point A (b, d, f) in l:

$$\frac{1}{\sqrt{(a^2+b^2+c^2)}}\sqrt{\left\{\begin{vmatrix} x_0-b & y_0-d \\ a & c \end{vmatrix}^2 + \begin{vmatrix} y_0-d & z_0-f \\ c & e \end{vmatrix}^2 + \begin{vmatrix} z_0-f & x_0-b \\ e & a \end{vmatrix}^2\right\}}.$$

2. The shortest distance h between l and l', two straight lines in space is perpendicular to both l and l', i.e. is parallel to the vector product of two arbitrary vectors lying along l and l' respectively. Also, the shortest distance between l and l' is obtained by projecting a line joining any two points on l and l' on to the line h:

$$\frac{1}{\sqrt{\{(ac' - a'c)^2 + (ae' - a'e)^2 + (ce' - c'e)^2\}}}\begin{vmatrix} a & c & e \\ a' & c' & e' \\ b - b' & d - d' & f - f' \end{vmatrix}.$$

3. The left-hand side may be interpreted as the volume of a tetrahedron.

4. The length of the vector product of the vectors $(\omega\alpha, \omega\beta, \omega\gamma)$ and (x, y, z): $\omega\sqrt{\{(\beta z - \gamma y)^2 + (\gamma x - \alpha z)^2 + (\alpha y - \beta x)^2\}}$.

6. It is sufficient to prove the statement for the case where the origin is inside the polygon, as the sum of the determinants is unaltered by translation of the co-ordinate system. If the origin is inside the polygon, all the determinants have the same sign and give the areas of the triangles OP_1P_2, OP_2P_3, ..., OP_nP_1.

§ 3, p. 26.

2. If we write $-d = d \times -1$, $-e = e \times -1$, $-f = f \times -1$, the three equations may be regarded as three homogeneous equations in $x, y, -1$; the necessary condition for the existence of a solution is therefore $D = 0$. If $D = 0$ and e.g. $\begin{vmatrix} a_1 & a_2 \\ b_1 & b_2 \end{vmatrix} \neq 0$, then the third equation is a consequence of the first two, and the first two equations in x and y have a solution, as their determinant does not vanish.

3. The lines intersect if the three equations

$$\begin{aligned} a_1t + b_1 &= c_1\tau + d_1 \\ a_2t + b_2 &= c_2\tau + d_2 \\ a_3t + b_3 &= c_3\tau + d_3 \end{aligned}$$

for t, τ have a solution (cf. Ex. 2). The condition is

$$\begin{vmatrix} a_1 & c_1 & d_1 - b_1 \\ a_2 & c_2 & d_2 - b_2 \\ a_3 & c_3 & d_3 - b_3 \end{vmatrix} = 0.$$

5. Subtract the last row of the determinant from the first three.

§ 4, p. 37.

1. (a) 0, (b) 2, (c) 12, (d) $(x - y)(y - z)(z - x)(x + y + z)$.

2. $a + c = 2b$.

3. (a) Introduce the three vectors $\mathbf{x} = (a, b, c)$, $\mathbf{y} = (a', b', c')$, $\mathbf{z} = (a'', b'', c'')$. Then $D = [\mathbf{x}\mathbf{y}]\mathbf{z}$. Now for any two vectors $\mathbf{a}$ and $\mathbf{b}$ we have

$$|[\mathbf{ab}]| \leqq |\mathbf{a}|\,|\mathbf{b}| \text{ and } |\mathbf{ab}| \leqq |\mathbf{a}|\,|\mathbf{b}|. \quad \text{Hence } D \leqq |\mathbf{x}|\,|\mathbf{y}|\,|\mathbf{z}|.$$

(b) If, and only if, the vectors represented by the columns of D are mutually orthogonal.

4. $ab + cd = 0$, $a^2 + c^2 = b^2 + d^2 = 1$.

6. It is sufficient to show that there is one point (x_0, y_0, z_0) which remains on the same ray through the origin, i.e. that there are four quantities x_0, y_0, z_0, λ (the first three of which do not all vanish) such that the equations

$$\begin{aligned} \lambda x_0 &= ax_0 + by_0 + cz_0 \\ \lambda y_0 &= dx_0 + ey_0 + fz_0 \\ \lambda z_0 &= gx_0 + hy_0 + kz_0 \end{aligned}$$

are satisfied. Now we have only to choose λ so that the determinant of these three homogeneous equations in x_0, y_0, z_0 vanishes; this gives an equation of the third degree in λ, which can always be satisfied, as an equation of the third degree always has a real root.

7. $x' = \frac{1}{2}(1 + \cos\varphi)x - \frac{1}{2}\sqrt{2}\sin\varphi \,.\, y - \frac{1}{2}(1 - \cos\varphi)z.$
$y' = \frac{1}{2}\sqrt{2}\sin\varphi \,.\, x + \cos\varphi \,.\, y + \frac{1}{2}\sqrt{2}\sin\varphi \,.\, z.$
$z' = \frac{1}{2}(\cos\varphi - 1)x - \frac{1}{2}\sqrt{2}\sin\varphi \,.\, y + \frac{1}{2}(\cos\varphi + 1)z.$

9. By Ex. 1, p. 12, and the rule for the multiplication of determinants, the square of the determinant is equal to $+1$.

CHAPTER II

§§ 1, 2, p. 49.

2. $\frac{1}{2}(n + 1)(n + 2)$. **4.** $\begin{vmatrix} a & b \\ b & c \end{vmatrix} > 0.$

5. (*a*) No. (*b*) No. (*c*) No. (*d*) No. (*e*) Yes. (*f*) No. (*g*) Yes. (*h*) No.

§ 3, p. 58.

1. Cf. Ex. 2, p. 49: $\frac{1}{2}(n + 1)(n + 2)$.

3. $6x + 2(a + e + k)$.

§§ 4, 5, p. 77.

2. We may take the origin at the vertex of the cone; its equation is then of the form $u = \varphi\left(\frac{y}{x}\right)x$.

4. (*a*) $g_{rr} + \frac{2}{r} g_r$. **5.** $g_{rr} + \frac{n-1}{r} g_r$.

6. Cf. p. 391.

§ 6, p. 81.

1. xy.

2. Use Taylor's theorem, expressing $f(2h, e^{-1/2h})$ and $f(0, 0)$ in terms of f and its first and second derivatives in $(h, e^{-1/h})$; add and divide by h^2.

4. (*a*) $\sum_{n=0}^{\infty} \sum_{m=0}^{\infty} \binom{m+n}{n} x^m y^n$: $|x| + |y| < 1$. (*b*) $\sum_{n=0}^{\infty} \sum_{m=0}^{\infty} \frac{x^m y^n}{m!\,n!}$: all values of x and y.

§ 7, p. 93.

1. Use Taylor's theorem to express the co-ordinates of a point on the curve in terms of f, g, h and their first and second derivatives in t_0; then apply Ex. 3, p. 19:

$$\begin{vmatrix} x - f(t_0) & f'(t_0) & f''(t_0) \\ y - g(t_0) & g'(t_0) & g''(t_0) \\ z - h(t_0) & h'(t_0) & h''(t_0) \end{vmatrix} = 0.$$

3. If $\mathfrak{y}$ is the centre of the sphere, the expression $A = \sqrt{\{\mathfrak{x}(s) - \mathfrak{y}\}^2}$ must be as stationary as possible, that is, $\dot{A}$, $\ddot{A}$, $\dddot{A}$ must vanish (the dots denoting differentiation with respect to s). Using the relations $\dot{x}^2 = 1$, $\dot{x}\ddot{x} = 0$, we obtain the equations $(y - x)\dot{x} = 0$, $(y - x)\ddot{x} = 1$, $(y - x)\dddot{x} = 0$. Hence we have $y - x = \dfrac{[\dot{x}\ddot{x}]}{[\dot{x}\ddot{x}]\dddot{x}}$.

5. Cf. Ex. 3 and also Ex. 5, p. 19.

7. From the definitions of ξ_1, ξ_2, ξ_3 we have $\xi_1 = \dot{x}$, $\dot{x}^2 = 1$, $\xi_2 = \ddot{x}/k$, $\xi_3 = [\xi_1\xi_2]$, $\pm\sqrt{\dot{\xi}_3^2} = 1/\tau$. Obviously $\dot{\xi}_1 = k\xi_2$. To determine $\dot{\xi}_2$, $\dot{\xi}_3$, we calculate their components with respect to a rectangular coordinate system $O\xi_1$, $O\xi_2$, $O\xi_3$. From the relations

$$\xi_2^2 = 1,\ \xi_3^2 = 1,\ \xi_1\xi_2 = \xi_2\xi_3 = \xi_3\xi_1 = 0$$

we obtain by differentiation

$$\dot{\xi}_3\xi_1 = -\dot{\xi}_1\xi_3 = 0,\ \dot{\xi}_3\xi_3 = 0;$$

hence $\dot{\xi}_3$ is perpendicular both to ξ_1 and to ξ_3, and therefore

$$\dot{\xi}_3 = \pm\sqrt{(\dot{\xi}_3^2)}\xi_2 = \pm\,\xi_2/\tau.$$

We define the sign of τ so as to give $\dot{\xi} = -\,\xi_2/\tau$. This implies that τ is positive or negative according as the screw defined by the motion of the osculating plane in the direction of increasing s is right-handed or left handed. To prove the second formula, note that

$$\dot{\xi}_2\xi_1 = -\dot{\xi}_1\xi_2 = -k,\ \dot{\xi}_2\xi_2 = 0,\ \dot{\xi}_2\xi_3 = -\dot{\xi}_3\xi_2 = 1/\tau.$$

8. Use Ex. 6 and Ex. 3: (*a*) $\dot{k}\xi_2 - k^2\xi_1 + \frac{k}{\tau}\xi_3$, (*b*) $\frac{\dot{k}}{k^2\tau}\xi_3 + \frac{\xi_2}{\tau}$.

9. $1/|\tau| = \sqrt{\dot{\xi}_3^2} = 0$; hence ξ_3 is a constant vector η, say; $\dot{x}\eta = \xi_1\eta = \xi_1\xi_3 = 0$, so that $x\eta = \text{const.}$, where η is a fixed vector. That is, the curve lies in a fixed plane.

10. (*b*) If the curve is given by $x = f(t)$, $y = g(t)$, $z = h(t)$, the surface has the parametric equations

$$x = f(t) + sf'(t)$$
$$y = g(t) + sg'(t)$$
$$z = h(t) + sh'(t);$$

then express $\dfrac{\partial^2 z}{\partial x^2}, \dfrac{\partial^2 z}{\partial x \partial y}, \dfrac{\partial^2 z}{\partial y^2}$ in terms of the derivatives with respect to t and s.

Appendix, § 1, p. 100.

1. (*a*) As R is closed, there is a point B in R whose distance from A is less than that of any other point in R. Let n be the normal to AB at B. Then no point C in R lies on the same side of n as A; for otherwise not only B and C, but the whole segment BC, would belong to R, and on this segment there would be points nearer to A than B is. Hence the parallel to n through A cannot meet R.

(*b*) There is a sequence of points P_1, P_2, . . . , not in R, converging to P. Let $l_1, l_2, \ldots$ be straight lines passing through $P_1, P_2, \ldots$ respectively and dividing the plane into two half-planes, one of which contains no point of R (cf. (*a*)). From these straight lines we can choose a sub-sequence for which the directions also converge. The limiting straight line is then a line of support through P.

(*c*) If A were not in R, then by the proof of (*a*) a line of support n separating A from R would exist.

(*d*) Let G be the centre of mass of R and g any line of support, which we take as x-axis. Then the y-co-ordinates of all points in R have the same sign. By the definition of the centre of mass (cf. Vol. I, p. 284), the y-co-ordinate of G also has this sign; that is, G and R are on the same side of the arbitrary line of support. Now apply (*c*).

(*f*) The curvature is equal to $d\varphi/ds$, where φ is the angle which the tangent makes with the x-axis, s the length of arc; φ is a continuous function of s. Hence φ increases monotonically from $\varphi(0)$ to $\varphi(0) + 2\pi$; that is, φ cannot have the same value for two different points of the curve.

If the curve were cut at three points s_0, s_1, s_2 by a straight line l ($ax + by = c$), then the function

$$F(s) = ax(s) + by(s) - c$$

would have three zeros; in this case $F'(s)$ would also have at least three zeros, i.e. there would be three tangents parallel to l. In addition, two of these would certainly have the same sense, i.e. they would have the same value of φ, which contradicts the statement above.

2. (*a*) The set consisting of the points which lie in all convex regions containing S has the properties (1), (2), (3).

(*b*) If P is in E, there can be no straight line l separating P from S; for otherwise one could take e.g. a large square Q with one side on l and containing S; Q would then be a convex region containing S but not P.

If P is not in E, there is at least one convex region Q containing S but not P; then (cf. Ex. 1(*a*)) there is a straight line separating Q from P, and therefore, as Q contains S, also separating S from P.

(*c*) Cf. Ex. 1(*d*).

Appendix, § 2 (p. 107).

1. (*a*) No. (*b*) No. (*c*) Yes (cf. Vol. I, p. 436).

CHAPTER III

§ 1, p. 122.

2. (*a*) $-\frac{5}{4}$; (*b*) $-\frac{\pi}{2}$; (*c*) -1; (*d*) -1.

3. (*a*) $-\frac{21}{32}$; (*b*) π; (*c*) 2; (*d*) $-\frac{19}{3}$.

4. Max. value $+6$, min. value -6.

5. $\partial z/\partial x = -1$, $\partial z/\partial y = -1$.

§ 2, p. 131.

1. (*a*) $5x + 7y - 21z + 9 = 0$; (*b*) $20x + 13y + 3z = 36$;
(*c*) $x - y - z + \pi/6 = 0$.

2. 1.

3. Use the fact that the tangents at the origin are given by $y = 0$ and $ax + by = 0$. $2c/a$, $2(a^3g - a^2bf + ab^2e - b^3c)/a(a^2 + b^2)^{3/2}$.

4. Write equation in form $0 = F = f(\sqrt{x^2 + y^2}, \text{arc tan } y/x)$:

$$k = \frac{2r'^2 - rr'' + r^2}{(r'^2 + r^2)^{3/2}}, \text{ where } r' = \frac{df}{d\theta}, \ r'' = \frac{d^2f}{d\theta^2}.$$

6. $x(y + z) = ay$.

8. (*a*) Double point.
(*b*) Two branches touching one another.
(*c*) Corner.
(*d*) Cusp.
(*e*) Cusp.

9. Differentiate the equation $F = 0$ twice with respect to x and use the fact that $F_y = 0$.

$$\varphi = \text{arc tan } 2\sqrt{F_{xy}^2 - F_{xx}F_{yy}}/(F_{xx} + F_{yy}); \ (a)\ \pi/2; \ (b)\ \pi/2.$$

10. $a = 1$, $b = -\frac{1}{2}$.

12. The circles K, K', K'' may be denoted by the equations

$$\begin{aligned} K \ &= x^2 + y^2 + ax + by + c = 0, \\ K' \ &= x^2 + y^2 + a'x + b'y + c' = 0, \\ K'' &= x^2 + y^2 + a''x + b''y + c'' = 0. \end{aligned}$$

Then any circle passing through A and B is given by $K' + \lambda K'' = 0$. The conditions that the circle K should be orthogonal to K' and K'' are $aa' + bb' - 2(c + c') = 0$, $aa'' + bb'' - 2(c + c'') = 0$. From these conditions the corresponding relation expressing the orthogonality of K and $K' + \lambda K''$ readily follows.

13. $z_x = \dfrac{yz - x^2}{z^2 - xy}$, $\quad z_y = \dfrac{zx - y^2}{z^2 - xy}$.

§ 3, p. 157.

2. (*c*) $\dfrac{\partial(\xi, \eta)}{\partial(x, y)} = \dfrac{-1}{(x^2 + y^2)^2}$.

3. Take O as the origin and invert; then the curvilinear triangle is transformed into an ordinary triangle with the same angles.

5. (*b*) If we denote the left-hand side of the equation defining t_1 and t_2 by $F(x, y, t)$, two curves $t_1 =$ const. and $t_2 =$ const. are given implicitly by the equations $F(x, y, t_1) = 1$ and $F(x, y, t_2) = 1$ respectively. The condition that these should be orthogonal is therefore

$$0 = F_x(x, y, t_1)F_x(x, y, t_2) + F_y(x, y, t_1)F_y(x, y, t_2)$$
$$= \frac{4x^2}{(a-t_1)(a-t_2)} + \frac{4y^2}{(b-t_1)(b-t_2)};$$

but this relation is an immediate consequence of $F(x, y, t_1) - F(x, y, t_2) = 0$.

(c) The coefficients of the quadratic equation defining t_1 and t_2 are respectively equal to t_1, t_2, and $-(t_1 + t_2)$. We thus obtain two linear equations in x^2 and y^2, whence

$$x = \pm\sqrt{\frac{(a-t_1)(a-t_2)}{a-b}}, \quad y = \pm\sqrt{\frac{(b-t_1)(b-t_2)}{b-a}}.$$

(d) $$\frac{\partial(t_1, t_2)}{\partial(x, y)} = \frac{4xy(a-b)}{\sqrt{\{(a+b)^2 - 2(a-b)(x^2-y^2) + (x^2+y^2)^2\}}}.$$

(e) $$\frac{f_1'g_1'}{(a-t_1)(b-t_1)} = \frac{f_2'g_2'}{(a-t_2)(b-t_2)}.$$

6. (a) Let $F(t)$ be the left-hand side of the equation defining t. F is a continuous function of t in $-\infty < t < c$, for which $F(-\infty) = 0$, $F(c-0) = +\infty$; hence $F = 1$ at one point at least of that interval. Similar conclusions apply to the other intervals.

(b) Cf. Ex. 5(b).

(c) Cf. Ex. 5(c): $x = \pm\sqrt{\dfrac{(a-t_1)(a-t_2)(a-t_3)}{(a-b)(a-c)}}$,

with similar formulæ for y and z.

7. (b) Let $x = r\cos\theta$, $y = r\sin\theta$. Then the straight line $\theta = \text{const.}$ is transformed into the conic $t_1 = \frac{1}{2} - \cos^2\theta$ and the circle $r = \text{const.}$ into the conic $t_2 = -\frac{1}{4}\left(r^2 + \frac{1}{r^2}\right)$.

9. $$\frac{\partial(u, x^2+y^2)}{\partial(x, y)} = 2\begin{vmatrix} u_x & u_y \\ x & y \end{vmatrix} = 0.$$

§ 4, p. 167.

1. (b) A circle on the sphere is given by a linear equation in x, y, z.

(d) $ds^2 = 4\dfrac{du^2 + dv^2}{(u^2+v^2+1)^2}$.

2. (a) $ds^2 = \sin^2 v\, du^2 + dv^2$;

(b) $ds^2 = \cosh^2 v\, du^2 + (1 + 2\sinh^2 v)dv^2$;

(c) $ds^2 = (1 + f'^2)dz^2 + f^2 d\theta^2$;

(d) $$ds^2 = \frac{(t_1-t_2)(t_1-t_3)}{4(a-t_1)(b-t_1)(c-t_1)}\,dt_1^2 + \frac{(t_2-t_1)(t_2-t_3)}{4(a-t_2)(b-t_2)(c-t_2)}\,dt_2^2.$$

3. $EG - F^2 = \begin{vmatrix} y_u & z_u \\ y_v & z_v \end{vmatrix}^2 + \begin{vmatrix} z_u & x_u \\ z_v & x_v \end{vmatrix}^2 + \begin{vmatrix} x_u & y_u \\ z_t & y_t \end{vmatrix}^2$; use the transformation formula for Jacobians.

4. Introduce co-ordinates x, y, z such that P becomes the origin, the tangent plane at P the xy-plane, and t the x-axis. The equation of S then takes the form $z = f(x, y)$, where $f(0, 0) = f_x(0, 0) = f_y(0, 0) = 0$. A plane Σ through t is given by the equation $z = \alpha y$. We now introduce $r = \sqrt{(y^2 + z^2)}$ and x as co-ordinates in Σ; then the intersection of Σ and S is given implicitly by the equation

$$\frac{r\alpha}{\sqrt{(1+\alpha^2)}} = f\left\{x, \frac{r}{\sqrt{(1+\alpha^2)}}\right\}.$$

The curvature of the curve of intersection at the point $x = 0, r = 0$ is therefore (cf. p. 125) given by

$$k = f_{xx}\frac{\sqrt{(1+\alpha^2)}}{\alpha}.$$

Thus the centre of curvature of this section has the co-ordinates

$$x = 0,\ y = \frac{1}{k\sqrt{(1+\alpha^2)}} = \frac{\alpha}{f_{xx}(1+\alpha^2)},\ z = \frac{\alpha}{k\sqrt{(1+\alpha^2)}} = \frac{\alpha^2}{f_{xx}(1+\alpha^2)};$$

that is, it lies on the circle

$$f_{xx}(y^2 + z^2) - z = 0.$$

5. Take the tangent plane at P as the xy-plane. Then the equation of S may be taken to be $z = f(x, y)$. A normal plane is given by the equation $x = \alpha y$. Take $r = \sqrt{(x^2 + y^2)}$ and z as co-ordinates in the plane; then the curve of intersection is given by

$$z = f\left\{\frac{ar}{\sqrt{(1+\alpha^2)}}, \frac{r}{\sqrt{(1+\alpha^2)}}\right\},$$

and its curvature at $r = 0$ by

$$k = f_{xx}(0,0)\frac{\alpha^2}{1+\alpha^2} + 2f_{xy}(0,0)\frac{\alpha}{1+\alpha^2} + f_{yy}(0,0)\frac{1}{1+\alpha^2};$$

the final point of the vector of length $1/\sqrt{k}$ along the line t then has the co-ordinates

$$x = \frac{\alpha}{\sqrt{(1+\alpha^2)}}\frac{1}{\sqrt{k}},\quad y = \frac{1}{\sqrt{(1+\alpha^2)}}\frac{1}{\sqrt{k}},\quad z = 0;$$

that is, it lies on the conic

$$x^2 f_{xx} + 2xy f_{xy} + y^2 f_{yy} = 1.$$

6. (*a*) By differentiating the two equations with respect to a parameter t of the curve, we obtain

$$xx' + yy' + zz' = 0, \quad axx' + byy' + czz' = 0.$$

From these relations we can find the ratio $x' : y' : z'$, i.e. the direction of the tangent. If (ξ, η, ζ) are current co-ordinates, the equations of the tangent are

$$(\xi - x) : (\eta - y) : (\zeta - z) = \frac{c-b}{x} : \frac{a-c}{y} : \frac{b-a}{z}.$$

(*b*) By differentiating the equations of the curve a second time and using the result of (*a*), we obtain

$$xx'' + yy'' + zz'' = -(x'^2 + y'^2 + z'^2) = \lambda \left\{ \frac{(c-b)^2}{x^2} + \frac{(a-c)^2}{y^2} + \frac{(b-a)^2}{z^2} \right\}$$

and

$$axx'' + byy'' + czz'' = \lambda \left\{ \frac{a(c-b)^2}{x^2} + \frac{b(a-c)^2}{y^2} + \frac{c(b-a)^2}{x^2} \right\},$$

where λ is a factor of proportionality. Eliminating λ, we have

$$(xx'' + yy'' + zz'') \left\{ \frac{a(c-b)^2}{x^2} + \frac{b(a-c)^2}{y^2} + \frac{c(b-a)^2}{z^2} \right\}$$

$$= (axx'' + byy'' + czz'') \left\{ \frac{(c-b)^2}{x^2} + \frac{(a-c)^2}{y^2} + \frac{(b-a)^2}{z^2} \right\}.$$

This linear equation in x'', y'', z'' remains valid if we substitute x', y', z' for x'', y'', z''. Hence it is still satisfied if we replace x'', y'', z'' by some linear combination $\lambda x' + \mu x''$, $\lambda y' + \mu y''$, $\lambda z' + \mu z''$ respectively. Now if (ξ, η, ζ) is in the osculating plane, $\xi - x$, $\eta - y$, $\zeta - z$ are just such a linear combination (cf. Ex. 6, p. 94).

The equation of the osculating plane is hence found to be

$$\frac{ax^3}{c-b}(\xi - x) + \frac{by^3}{(a-c)}(\eta - y) + \frac{cz^3}{b-a}(\xi - z) = 0.$$

§ 5, p. 182.

1. Let $P(x, y, z)$ be a point on the tube-surface Σ, and let S be the sphere of the family which has the point P in common with Σ. Then S and Σ have the same tangent plane at P, i.e. the same values of x, y, z, z_x, z_y at that point. It is therefore sufficient to prove that the relation is true for any sphere of unit radius which has its centre in the xy-plane, i.e. for $u(x, y) = \sqrt{\{1 - (x-a)^2 + (y-b)^2\}}$.

2. (*a*) $\sqrt{x} + \sqrt{y} + \sqrt{z} = 1$; (*b*) $x^{\frac{2}{3}} + y^{\frac{2}{3}} + z^{\frac{2}{3}} = 1$.

4. We may introduce t as parameter on the curve, so that the latter is given by $x = x(t)$, $y = y(t)$, $z = z(t)$ and the tangent at the point with parameter t lies in the two planes corresponding to t; this gives the relations

$$ax' + by' + cz' = 0, \quad dx' + ey' + fz' = 0.$$

By differentiating the equations of the straight lines with respect to t, we thus obtain

$$a'x + b'y + c'z = 0, \quad d'x + e'y + f'z = 0.$$

With the relation

$$ax + by + cz = dx + ey + fz$$

we then have three homogeneous equations in x, y, z, and the determinant must vanish.

5. For the envelope we have the two equations

$$x \cos t + y \sin t + z = t$$
$$-x \sin t + y \cos t = 1.$$

These two equations give a family of straight lines with parameter t; if a curve having these lines as tangents exists, it must also satisfy the equations obtained by differentiating once again.

(*a*) $r \sin\{z + \sqrt{(r^2 - 1)} - \theta\} + 1 = 0$; (*b*) the curve is given by $z = \theta - \pi/2$, $r = 1$.

7. Use inversion. Since S_1, S_2, S_3 pass through the origin, they are transformed into planes; we have then merely to find the envelope of the spheres touching three planes, i.e. a certain circular cone, which we reinvert:

$$(x^2 + y^2 + z^2)^2 - 2(x^2 + y^2 + z^2)(x + y + z)$$
$$- 3(x^2 + y^2 + z^2 - 2xy - 2xz - 2yz) = 0.$$

8. (*b*) $\xi^2(1 - a^2) + \eta^2(1 - b^2) - 2ab\xi\eta + 2a\xi + 2b\eta = 1$;
(*c*) $a^2\xi^2 + b^2\eta^2 = 1$.

§ 6, p. 202.

1. $(4 + \sqrt{5})/\sqrt{2}$, $(4 - \sqrt{5})/\sqrt{2}$.

2. $a/20$, $a/10$, $a/10$.

3. Maxima for $x = 0$, $y = \pm 1$, minimum for $x = y = 0$.

4. The maximum value is the same as for the expression $ax^2 + 2bxy + cy^2$ subject to the subsidiary condition $ex^2 + 2fxy + gy^2 = 1$.

5. Cf. Ex. 4. (*a*) $\dfrac{14}{3} + \dfrac{2\sqrt{67}}{3}$;

(*b*) the function has an *improper maximum* (p. 184) equal to 1·95, when $y/x = 0{\cdot}64$.

6. Saddle points: $y = 0$, $x = \pi/3, 7\pi/3, 13\pi/3, \ldots$.
Minima: $y = 0$, $z = 5\pi/3, 11\pi/3, 17\pi/3, \ldots$.

7. The ellipse obviously touches the circle, i.e. the two equations must give a double root in x; hence the condition for contact is $a^2(b^2 - 1) = b^4$: $a = 3/\sqrt{2}$, $b = \sqrt{(3/2)}$.

8. Introduce the angles between a, b and c, d as variables: the cyclic quadrilateral.

9. $(-1/\sqrt{14}, \ -2/\sqrt{14}, \ -3/\sqrt{14})$.

10. Cf. the similar proof for triangles on p. 187. A minimum point O does exist. First show that if O is not one of the vertices, then it can only be the point of intersection of the diagonals. Use the fact that the final points of four unit vectors whose vector sum is zero form a rectangle. Then prove that the sum of the distances from the vertices is less for the point of intersection of the diagonals than for any of the vertices.

11. $A = a^2/x$, $B = b^2/y$, $C = c^2/z$, together with the subsidiary condition

$$\frac{x^2}{a^2} + \frac{y^2}{b^2} + \frac{z^2}{c^2} = 1:$$

$$(a)\ x = \frac{a^{\frac{4}{3}}}{\sqrt{(a^{\frac{2}{3}} + b^{\frac{2}{3}} + c^{\frac{2}{3}})}}, \text{ \&c.}; \quad (b)\ x = \frac{a^{\frac{3}{2}}}{\sqrt{(a + b + c)}}, \text{ \&c.}$$

12. The vertices are given by $x = \pm a/\sqrt{3}$, $y = \pm b/\sqrt{3}$, $z = \pm c/\sqrt{3}$.

13. The vertices are given by $x = a^2/\sqrt{(a^2 + b^2)}$, $y = b^2/\sqrt{(a^2 + b^2)}$.

14. $x = 1$, $y = 1$.

15. The greatest axis is given by the maximum of $\sqrt{(x^2 + y^2 + z^2)}$, with the subsidiary condition that (x, y, z) lies on the ellipsoid. Hence we have the three equations

$$\frac{x}{\sqrt{(x^2 + y^2 + z^2)}} = \frac{x}{l} = \lambda(ax + dy + ez), \text{ \&c.}$$

Multiplying these by (x, y, z) respectively and adding, we have $\lambda = \sqrt{(x^2 + y^2 + z^2)} = l$. On the other hand, we may regard the equations as three linear homogeneous equations in x, y, z, whose determinant must vanish.

Appendix, p. 208.

1. $$f(x) + f(y) + f(z) = 3f(a) + \{(x - a) + (y - a) + (z - a)\}f'(a) + \tfrac{1}{2}\rho^2\{f''(a) + \varepsilon\},$$

where $\rho^2 = (x - a)^2 + (y - a)^2 + (z - a)^2$. On the other hand, the subsidiary condition gives

$$(x-a)+(y-a)+(z-a)=\rho^2\left(-\frac{\phi''(a)}{2\phi'(a)}+\varepsilon\right)$$
$$-\frac{\phi'(a)}{\phi(a)}\{(x-a)(y-a)+(x-a)(z-a)+(y-a)(z-a)\}$$
$$=\left\{-\frac{\phi''(a)}{2\phi'(a)}+\frac{\phi'(a)}{2\phi(a)}+\varepsilon\right\}\rho^2,$$

where $\lim\limits_{x,y,z\to a}\varepsilon=0$.

CHAPTER IV

§ 1, p. 222.

1. $F=0$ for $y>0$.

2. Use the relation

$$\frac{1}{z}(f_x\cos\varphi+f_y\sin\varphi)=f_{xx}\sin^2\varphi-2f_{xy}\sin\varphi\cos\varphi+f_{yy}\cos^2\varphi$$
$$+\frac{1}{z}\frac{d}{d\varphi}(f_x\sin\varphi-f_y\cos\varphi).$$

3. Integrate u_{xx} by parts twice (special precautions necessary in the case where $p<5/2$).

4. Integrate J_0' by parts.

§ 2, 3, p. 247.

1. $\pi/24$. **2.** 0. **3.** 0.

4. $\pi/8$ if region of integration is restricted by the condition $z>0$; otherwise zero.

5. 1/50400. **6.** $\pi(2-\frac{3}{2}\log 3)$.

7. Introduce polar co-ordinates and integrate first with respect to φ and θ: $\pi(2+\frac{3}{2}\log 3)$.

8. $4\log(1+\sqrt{2})$.

9. Divide up the interval of integration into the segments $-1\leqq x\leqq-\sqrt[4]{h}$, $-\sqrt[4]{h}\leqq x\leqq\sqrt[4]{h}$, $\sqrt[4]{h}\leqq x\leqq 1$, and find the limits of the integrals along each segment.

§ 4, p. 255.

1. Apply the substitution $x+y=\xi$, $x-y=\eta$: $\frac{1}{4}\left(e-\frac{1}{e}\right)$.

2. Introduce polar co-ordinates: (a) $\frac{\pi}{4}-\frac{1}{2}$, (b) $\frac{\sqrt{3}}{2}\arctan\frac{1}{2}$.

3. Substitute $x=a\xi$, $y=b\eta$, $z=c\zeta$: $\frac{1}{8}a^2b^2c^2$.

7. Introduce new rectangular co-ordinates (ξ', η', ζ') such that $\xi' = (x\xi + y\eta + z\zeta)/r$. Then $d\xi d\eta d\zeta = d\xi' d\eta' d\zeta'$ (cf. Ex. 9, p. 38), and

$$I = \iiint \cos(r\xi')\, d\xi' d\eta' d\zeta'$$

throughout the sphere $\xi'^2 + \eta'^2 + \zeta'^2 \leq 1$. Hence, if we perform the integrations with respect to η' and ζ',

$$I = \pi \int_{-1}^{+1} \cos(r\xi')(1 - \xi'^2)\, d\xi'.$$

Answer, $\dfrac{4\pi}{r^2}\left(\dfrac{\sin r}{r} - \cos r\right)$, where $r^2 = x^2 + y^2 + z^2$.

8. Substitute $\xi' = (x\xi + y\eta)/r$, $\eta' = -(y\xi + x\eta)/r$, and integrate with respect to η'.

§ 6, p. 275.

1. Apply Guldin's rule, using the fact that the centre of an ellipse is also its centre of mass: $2\pi^2 ab$.

2. $\pi abh^2/2$.

3. Substitute $x = a\xi$, $y = b\eta$, $z = c\zeta$:

$$\frac{1}{3}\pi abc\left(1 - \frac{p}{\sqrt{(a^2l^2 + b^2m^2 + c^2n^2)}}\right)^2\left(2 + \frac{p}{\sqrt{(a^2l^2 + b^2m^2 + c^2n^2)}}\right).$$

4. (*a*) Compare corresponding elements of area.

(*b*) $a^2 \int_0^{2\pi} \{1 - \cos f(\varphi)\}\, d\varphi$; (*c*) $2\pi(1 - \frac{1}{2}\sqrt{2})a^2$.

5. $2\pi a^2\left\{1 + (1 - e^2)\dfrac{\operatorname{ar\,tanh} e}{e}\right\}$, where $2a$ is the major axis.

6. Volume $= \frac{1}{3}\pi cp^2$, surface area $= \pi(a + b)p$, where a, b, c are the sides of the triangle and p the perpendicular from C to AB.

7. From the differential equation (cf. Ex. 1, p. 182) satisfied by the tube-surface $u = u(x, y)$ we have

$$A = 2\iint \sqrt{(1 + u_x^2 + u_y^2)}\, dx\, dy = 2\iint \frac{dx\, dy}{u}.$$

If we introduce as parameters the length of arc s on L and the distance t along the normal to L, then (cf. Ex. 22, Vol. I, p. 291, and Ex. 3, Vol. II, p. 182)

$$A = 2\int_L ds \int_{-1}^{+1} \frac{(1 + kt)}{\sqrt{(1 - t^2)}}\, dt = 2\pi \int_L ds,$$

where k denotes the curvature of L.

8. Integrate first with respect to x and y: (*a*) $16r^3/9$; (*b*) $8r^3$.

9. $\frac{\pi}{2}\left\{R\sqrt{(R^2+h^2)} - r\sqrt{(r^2+h^2)} + h^2 \log \frac{R+\sqrt{(R^2+h^2)}}{r+\sqrt{(r^2+h^2)}}\right\}.$

10. Introduce polar co-ordinates: $\pi^2/2$.

§ 7, p. 285.

1. On the axis of the cone, two-thirds of the way from the vertex to the centre of the base.

2. $x = 2x_0/3$, $y = z = 0$.

3. Cf. Ex. 7, p. 275, and Ex. 1, p. 182.

4. (a) $\pi h(R^4 - R'^4)$; (b) $2\pi h(R^2 - R'^2)\{\frac{1}{4}(R^2 + R'^2) + \frac{1}{3}h^2\}$.

5. For example, $A + B - C = 2\iiint \mu z^2\, dx\, dy\, dz$, which is positive.

6. Substitute $x = a\xi$, $y = b\eta$, $z = c\zeta$; use the expressions for the moments of inertia given in the text and the properties of symmetry of the ellipsoid:

$$(a)\ \frac{4}{15}\pi abc(a^2+b^2);\quad (b)\ \frac{4}{15}\pi abc\{(1-\alpha^2)a^2 + (1-\beta^2)b^2 + (1-\gamma^2)c^2\}.$$

7. The distance of the point (x, y, z) from the plane $ux + vy + wz = -1$ is given by

$$\frac{ux+vy+wz+1}{\sqrt{(u^2+v^2+w^2)}}.$$

The moment of inertia of the ellipsoid with respect to this plane is therefore given by

$$\frac{Au^2+Bv^2+Cw^2+V}{u^2+v^2+w^2},$$

where A, B, C denote the moments of inertia with respect to the co-ordinate planes and V is the volume of the ellipsoid, i.e. $A = 4a^3bc/15$, $B = 4ab^3c/15$, $C = 4abc^3/15$, and $V = 4abc/3$. We have now to find the envelope of the planes for which this expression is equal to h. The envelope is given by the equations

$$(A-h)u = \lambda x,\ (B-h)v = \lambda y,\ (C-h)w = \lambda z,$$

where λ denotes a common multiplier, which from the expression for the moment of inertia and the equation of the plane is found to be V. By squaring the three equations we obtain the equation of the envelope, namely,

$$\frac{x^2}{h-A} + \frac{y^2}{h-B} + \frac{z^2}{h-C} = \frac{1}{V}.$$

9. $a^2(x-\xi)^2 + b^2(y-\eta)^2 + c^2(z-\zeta)^2$
$= \{a^2+b^2+c^2+5(\xi^2+\eta^2+\zeta^2)\}\{(x-\xi)^2+(y-\eta)^2+(z-\zeta)^2\}.$

10. $(\frac{1}{3}, 0, 0)$. **11.** $x = \frac{5a}{16}\frac{2a^2 + b^2 + c^2}{a^2 + b^2 + c^2}$.

13. $\frac{2\pi a^2 b}{\sqrt{(b^2 - a^2)}} \log\left\{\frac{b}{a} + \sqrt{\left(\frac{b^2}{a^2} - 1\right)}\right\}$.

14. Integrate first with respect to x and y:

$$2\pi \int_a^b \sqrt{z^2 + \{f(z)\}^2}\, dz - \pi \,|\, b^2 \mp a^2 \,|,$$

where the upper or lower sign is to be taken according as the origin is inside the body or not.

Appendix, § 2, p. 298.

1. S consists of unit circles orthogonal to C and having their centres on C (cf. p. 295).

Appendix, § 3, p. 307.

1. Substitute $x_1 = a_1\xi_1, \ldots, x_n = a_n\xi_n$: $\frac{(\sqrt{\pi})^n}{\Gamma\left(\frac{n+2}{2}\right)} a_1 a_2 \ldots a_n$.

2. By p. 301,

$$I = \int \cdots \int \frac{f(x_1) + f(-x_1)}{\sqrt{(1 - x_2^2 - \ldots - x_n^2)}}\, dx_2 \ldots dx_n$$

taken throughout the interior of the $(n - 1)$-dimensional unit sphere in $x_2 \ldots x_n$-space. Introducing polar co-ordinates, we obtain

$$I = \int_0^1 dr \int_{S(r)} \frac{f(\sqrt{1 - r^2}) + f(-\sqrt{1 - r^2})}{\sqrt{1 - r^2}}\, d\sigma,$$

where $S(r)$ denotes the sphere of radius r and centre O in $x_2 \ldots x_n$-space. As the integrand depends on r only,

$$I = \omega_{n-1} \int_0^1 \frac{f(\sqrt{1 - r^2}) + f(-\sqrt{1 - r^2})}{\sqrt{1 - r^2}}\, r^{n-2}\, dr.$$

Putting $y = \sqrt{1 - r^2}$, we have

$$I = \omega_{n-1} \int_{-1}^{+1} f(y)(1 - y^2)^{\frac{n-3}{2}}\, dy.$$

Appendix, § 4, p. 317.

1. Put $I_n(a) = \int_0^\infty x^n e^{-ax^2} dx$; then $I_n(a) = -I''_{n-2}(a)$, where dashes denote differentiation with respect to a. Alternatively, integrate by parts. $\frac{1}{2}\left(\frac{n-1}{2}\right)!$ when n is odd, $\sqrt{\pi}\,\frac{1 \cdot 3 \cdot \cdots (n-1)}{2^{(n+2)/2}}$ when n is even.

2. Substitute $\xi = \alpha x + \beta y$, $\eta = \gamma x + \delta y$, where α, β, γ, δ are chosen so that

$$\xi^2 + \eta^2 = ax^2 + 2bxy + cy^2.$$

Then $(\alpha\delta - \beta\gamma)^2 = ac - b^2$, and the integral is transformed into

$$\frac{1}{\sqrt{(ac-b^2)}}\int_{-\infty}^{\infty}\int_{-\infty}^{\infty} e^{-(\xi^2+\eta^2)}\,d\xi\,d\eta.$$

$ac - b^2 = \pi^2$, $a > 0$.

3. Make the same substitution as in Ex. 2 and evaluate the resulting integrals, (*a*) using the result of Ex. 1, (*b*) introducing polar co-ordinates.

$$(a)\ \frac{\pi(aC + cA - 2bB)}{(ac-b^2)^{\frac{3}{2}}};\quad (b)\ \frac{2\pi}{(ac-b^2)^{\frac{1}{2}}}.$$

4. (*a*) Forming $K'(a)$, where the dash denotes differentiation with respect to a, and integrating by parts twice (taking xe^{-ax^2} as one factor), we have $K'(a) = -K(a)/2a + K(a)/4a^2$, i.e.

$$K(a) = Ca^{-\frac{1}{2}}e^{-\frac{1}{4a}},$$

where C is given by $C = \lim_{a\to\infty}\sqrt{a}\,K(a) = \lim_{a\to\infty}\int_0^{\infty} e^{-t^2}\cos\frac{t}{\sqrt{a}}\,dt = \frac{1}{2}\sqrt{\pi}$.

$K(a) = \frac{1}{2}\sqrt{\frac{\pi}{a}}\,e^{-\frac{1}{4a}}$.

(*b*) Integrate the formula $\frac{t}{1+t^2} = \int_0^{\infty} e^{-tx}\cos x\,dx$ with respect to t from a to b.

$\frac{1}{2}\log\frac{1+a^2}{1+b^2}$.

(*c*) Substituting $x = 1/t$ in the expression for $I'(a)$, prove that $I' = -2I$, i.e.

$$I = Ce^{-2a},$$

where $C = \lim_{a\to 0} I = \int_0^{\infty} e^{-x^2}\,dx$.

$\frac{1}{2}\sqrt{\pi}e^{-2a}$.

(*d*) Substitute the integral expression for J_0 and change the order of integration. Use the formula $2\sin ax\cos bxt = \sin(a+bt)x + \sin(a-bt)x$; cf. the expression for $\int_0^{\infty}\frac{\sin xy}{y}\,dy$ on pp. 307–8.

$\pi/2$ when $a > b$, $\arcsin a/b$ when $a < b$.

6. There exists an $\varepsilon > 0$ such that for every A there is an $A' > A$ such that

$$\left| \int_{A'}^{\infty} f(x, y)\, dy \right| \geqq \varepsilon$$

for some value of x.

Appendix, § 6, p. 338.

1. Substitute $x^m = a^m \xi$, $y^m = b^m \eta$.

3. Integrate first with respect to y and z:

$$x = \tfrac{3}{4} a \frac{\Gamma(2n)\Gamma(3n)}{\Gamma(n)\Gamma(4n)}.$$

4. $2R^4 B(\tfrac{3}{2}, \tfrac{11}{2}) = \dfrac{21\pi}{2^9} R^4.$

5. Show that

$$G_{2n}(2x) = \tfrac{1}{2} 2^{2x} G_n(x) G_n(x + \tfrac{1}{2}) \frac{(2n)!\, \sqrt{n}}{2^{2n}(n!)^2};$$

then let $n \to \infty$ and apply Wallis's formula (Vol. I, p. 225).

CHAPTER V

§ 1, p. 359.

1. $e^{\xi} \sin \eta$.

2. Let $u = \dfrac{e^x}{x^2 + y^2}(x \cos y + y \sin y)$,

$$v = \frac{-e^x}{x^2 + y^2}(-x \sin y + y \cos y),$$

and let u_1 and v_1 be defined by the equations

$$u = \frac{x}{x^2 + y^2} + u_1, \quad -v = -\frac{y}{x^2 + y^2} + v_1.$$

Then u_1 and v_1 are twice continuously differentiable (and that at the origin also), and $(u_1)_y = (v_1)_x$. Hence $\int_C u_1 dx + v_1 dy = 0$ and

$$\int_C u\, dy - v\, dx = \int \frac{x}{x^2 + y^2} dy - \frac{y}{x^2 + y^2} dx = 2\pi,$$

by the footnote on p. 359.

§ 5, p. 392.

1. (*a*) Cf. Ex. **3**, p. **37**.

(*c*) Let R be an arbitrary region and v an arbitrary function vanishing on the boundary of R. Then by Green's first formula

$$\iiint_R (u_{x_1}v_{x_1} + u_{x_2}v_{x_2} + u_{x_3}v_{x_3})\,dx_1\,dx_2\,dx_3$$
$$= -\iiint_R v\,\Delta u\,dx_1\,dx_2\,dx_3 = -\iiint_R v\,\Delta u\,\sqrt{e_1e_2e_3}\,dp_1\,dp_2\,dp_3.$$

Now

$$u_{x_i} = u_{p_1}\frac{\partial p_1}{\partial x_i} + u_{p_2}\frac{\partial p_2}{\partial x_i} + u_{p_3}\frac{\partial p_3}{\partial x_i}$$
$$= u_{p_1}\frac{a_{i1}}{e_1} + u_{p_2}\frac{a_{i2}}{e_2} + u_{p_3}\frac{a_{i3}}{e_3}$$

and

$$v_{x_i} = v_{p_1}\frac{a_{i1}}{e_1} + v_{p_2}\frac{a_{i2}}{e_2} + v_{p_3}\frac{a_{i3}}{e_3}.$$

Hence

$$\iiint_R (u_{x_1}v_{x_1} + u_{x_2}v_{x_2} + u_{x_3}v_{x_3})\,dx_1\,dx_2\,dx_3$$
$$= \iiint\left(\frac{1}{e_1}u_{p_1}v_{p_1} + \frac{1}{e_2}u_{p_2}v_{p_2} + \frac{1}{e_3}u_{p_3}v_{p_3}\right)dx_1\,dx_2\,dx_3$$
$$= \iiint\left(\sqrt{\frac{e_2e_3}{e_1}}\,u_{p_1}v_{p_1} + \sqrt{\frac{e_3e_1}{e_2}}\,u_{p_2}v_{p_2} + \sqrt{\frac{e_1e_2}{e_3}}\,u_{p_3}v_{p_3}\right)dp_1\,dp_2\,dp_3$$
$$= \iiint(U_1v_{p_1} + U_2v_{p_2} + U_3v_{p_3})\,dp_1\,dp_2\,dp_3,$$

where we write $U_i = \dfrac{\sqrt{e_1e_2e_3}}{e_i}\,u_{p_i}$.

Applying Gauss's theorem to the vector $(U_1v,\ U_2v,\ U_3v)$, we obtain

$$-\iiint\left(\frac{\partial U_1}{\partial p_1} + \frac{\partial U_2}{\partial p_2} + \frac{\partial U_3}{\partial p_3}\right)v\,dp_1\,dp_2\,dp_3.$$

Thus for an arbitrary v vanishing on the boundary of R we have

$$\iiint v\,\Delta u\,\sqrt{e_1e_2e_3}\,dp_1\,dp_2\,dp_3$$
$$= \iiint v\left(\frac{\partial U_1}{\partial p_1} + \frac{\partial U_2}{\partial p_2} + \frac{\partial U_3}{\partial p_3}\right)dp_1\,dp_2\,dp_3.$$

and hence (cf. lemma I, p. 499).

$$\Delta u = \left(\frac{\partial U_1}{\partial p_1} + \frac{\partial U_2}{\partial p_2} + \frac{\partial U_3}{\partial p_3}\right) \frac{1}{\sqrt{e_1 e_2 e_3}}$$

$$= \frac{1}{\sqrt{e_1 e_2 e_3}} \left[\frac{\partial}{\partial p_1}\left(\sqrt{\frac{e_2 e_3}{e_1}} \frac{\partial u}{\partial p_1}\right) + \frac{\partial}{\partial p_2}\left(\sqrt{\frac{e_3 e_1}{e_2}} \frac{\partial u}{\partial p_2}\right) + \frac{\partial}{\partial p_3}\left(\sqrt{\frac{e_1 e_2}{e_3}} \frac{\partial u}{\partial p_3}\right)\right].$$

(*d*) Use Ex. 6(*c*), p. 158.

$$\tfrac{1}{4}(t_2 - t_1)(t_3 - t_1)(t_3 - t_2)\,\Delta u = (t_3 - t_2)\sqrt{\varphi(t_1)}\, \frac{\partial}{\partial t_1}\left(\sqrt{\varphi(t_1)}\, \frac{\partial u}{\partial t_1}\right)$$

$$+ (t_3 - t_1)\sqrt{-\varphi(t_2)}\, \frac{\partial}{\partial t_2}\left(\sqrt{-\varphi(t_2)}\, \frac{\partial u}{\partial t_2}\right) + (t_2 - t_1)\sqrt{\varphi(t_3)}\, \frac{\partial}{\partial t_3}\left(\sqrt{\varphi(t_3)}\, \frac{\partial u}{\partial t_3}\right),$$

where $\varphi(x) = (a - x)(b - x)(c - x)$.

§ 7, p. 401.

1. $$\iint \frac{z}{p}\, dS = \left(\frac{1}{a^2} + \frac{1}{b^2} + \frac{2}{c^2}\right) \iiint z\, dx\, dy\, dz,$$

where the volume integral is to be extended throughout the upper half of the ellipsoid. (The base of this half-ellipsoid contributes nothing to the surface integral): $\frac{\pi}{4}\left(\frac{1}{a^2} + \frac{1}{b^2} + \frac{2}{c^2}\right) abc^2$.

2. Since H is a homogeneous function of the fourth degree, we have

$$4 \iint H\, dS = \iint (x H_x + y H_y + z H_z)\, dS$$

$$= \iint \frac{\partial H}{\partial n}\, dS = \iiint \Delta H\, dx\, dy\, dz$$

$$= 6 \iiint [x^2(2a_1 + a_4 + a_6) + y^2(2a_2 + a_4 + a_5) + z^2(2a_3 + a_5 + a_6)]\, dx\, dy\, dz.$$

$$\frac{4\pi}{5}(a_1 + a_2 + a_3 + a_4 + a_5 + a_6).$$

Appendix, § 2, p. 406.

1. The two equations $u = f_x$, $v = f_y$ can be solved for x and y, since $\frac{\partial(u, v)}{\partial(x, y)} \neq 0$. Let $x = \sigma(u, v)$, $y = \tau(u, v)$; since $u_y = v_x$, we have (cf. p. 143) $x_v = y_u$, $\sigma_v = \tau_u$. Hence a function g exists such that $x = g_u(u, v)$, $y = g_v(u, v)$.

2. $$u = \frac{yz}{(x^2 + y^2)\sqrt{(x^2 + y^2 + z^2)}}, \quad v = \frac{-xz}{(x^2 + y^2)\sqrt{(x^2 + y^2 + z^2)}},$$

$w = 0$.

Miscellaneous Examples V, p. 407.

2. If (ξ, η) and (x, y) are rectangular co-ordinates in Π and P respectively, then the motion of the point $M(x, y)$ can be described by the equations $\xi = x\cos\varphi - y\sin\varphi + a$, $\eta = x\sin\varphi + y\cos\varphi + b$ (i.e. by a rotation and a translation). Then

$$S(M) = A(x^2 + y^2) + Bx + Cy + D.$$

(α) If $A = n\pi \neq 0$, we have $S(M) = n\pi[(x - x_0)^2 + (y - y_0)^2] + S(C)$, where C is the point $x = x_0 = -B/2n\pi$, $y = y_0 = -C/2n\pi$, hence A, B, C, D have the values in Ex. 1. (β_1) If $A = n\pi = 0$, but $B^2 + C^2 > 0$, then

$$S_M = \sqrt{B^2 + C^2}\,\frac{Bx + Cy + D}{\sqrt{B^2 + C^2}} = \lambda d(M),$$

where $\lambda = \sqrt{B^2 + C^2}$ and Δ is the line $Bx + Cy + D = 0$. (β_2) If $A = B = C = 0$, we have $S(M) = D =$ constant.

3. For the motion of the plane P rigidly attached to the connecting-rod AB we have $n = 0$, $S(A) = 0$, $S(B) = \pi\overline{CB}^2 = \pi\gamma^2$. Hence Δ passes through A, and by symmetry Δ is perpendicular to AB at A. Hence $S(M) = \pi\gamma^2 l^{-1} d(M)$, where $l = \overline{AB}$.

4. For the motion of the plane P rigidly attached to the chord AB we have $n = 1$, $S(A) = S(B) = S =$ area of Γ. The point C of Steiner's theorem is therefore equidistant from A and B and $S(A) = \pi\overline{CA}^2 + S(C)$, $S(M) = \pi\overline{CM}^2 + S(C)$, hence $S(A) - S(M) =$ area of $\Gamma -$ area of $\Gamma' = \pi(\overline{CA}^2 - \overline{CM}^2) = \pi ab$.

5. If l is the length of Γ, the Frenet formulæ (p. 94) give

$$\int \frac{\boldsymbol{n}}{\rho}\,ds = \int \frac{\boldsymbol{\xi}_2}{\rho}\,ds = \int \dot{\boldsymbol{\xi}}_1 ds = \int \frac{d^2\boldsymbol{x}}{ds^2}\,ds = 0;$$

$$\int \frac{[\boldsymbol{x}\boldsymbol{n}]}{\rho}\,ds = \int [\boldsymbol{x}\dot{\boldsymbol{\xi}}_1]ds = [\boldsymbol{x}\boldsymbol{\xi}_1]\Big|_0^l - \int [\dot{\boldsymbol{x}}\boldsymbol{\xi}_1]ds$$

$$= -\int [\boldsymbol{\xi}_1\boldsymbol{\xi}_1]ds = 0 \quad \text{(cf. p. 85).}$$

6. Let $\boldsymbol{n}' = (\alpha, \beta, \gamma)$, $\boldsymbol{x} = (x, y, z)$. If in Gauss's formula

$$\iint (a\alpha + b\beta + c\gamma)d\sigma = -\iiint \left(\frac{\partial a}{\partial x} + \frac{\partial b}{\partial y} + \frac{\partial c}{\partial z}\right) dx\,dy\,dz$$

we substitute $a = 1$, $b = c = 0$, and $a = 0$, $b = -z$, $c = y$, we get

$$\iint \alpha\,d\sigma = 0 \quad \text{and} \quad \iint (y\gamma - z\beta)d\sigma = 0 \text{ respectively.}$$

7. Take rectangular co-ordinates (x, y, z) such that $z = 0$ is the free horizontal surface of the fluid and Oz points downwards. The pressure on $d\sigma$ is $\boldsymbol{n}z\,d\sigma$, where z is the depth of $d\sigma$. By repeated applications of

Gauss's formula in three dimensions, with obvious choices of the functions a, b, c, we find for the components of the resultant of the fluid pressure

$$\iint \alpha z\,d\sigma = 0,\quad \iint \beta z\,d\sigma = 0,\quad \iint \gamma z\,d\sigma = -\iiint dx\,dy\,dz = -V.$$

For the components of the resultant moment with respect to the origin O we find, again by Gauss's formula,

$$\iint (yz\gamma - z^2\beta)\,d\sigma = \iiint y\,dx\,dy\,dz = Vy_0,\quad \iint (z^2\alpha - xz\gamma)\,d\sigma =$$

$$-\iiint x\,dx\,dy\,dz = -Vx_0,\quad \iint (xz\beta - yz\alpha)\,d\sigma = 0,$$

(x_0, y_0, z_0 are the co-ordinates of the centroid C).

Now we note that the components of the force f are 0, 0, $-V$, and the components of its moment with respect to O are Vy_0, $-Vx_0$, 0.

8. From the parametric equations

$$x = a\cos u \cos v,\quad y = b \sin u \cos v,\quad z = c \sin v$$

$$\left(0 \leqq u < 2\pi,\quad -\frac{\pi}{2} \leqq v < \frac{\pi}{2}\right)$$

of the ellipsoid we readily obtain the formulæ

$$p\,dS = abc \cos v\, du\,dv,\quad dS/p = D^2\,du\,dv/(abc\cos v),$$

where

$$D^2 = b^2c^2 \cos^2 u \cos^2 v + a^2c^2 \sin^2 u \cos^2 v + a^2b^2 \sin^2 v \cos^2 v.$$

10. The integral represents the flat solid angle which the plane $z=0$ subtends at the point $M = (0, 0, 1)$. For a direct analytical proof use plane polar co-ordinates.

12. Verify the identity

$$\frac{\partial}{\partial x}\left(\frac{a-x}{\gamma^3}\right)+\frac{\partial}{\partial y}\left(\frac{b-y}{\gamma^3}\right)+\frac{\partial}{\partial z}\left(\frac{c-z}{\gamma^3}\right)=0,\quad \gamma^2=(x-a)^2+(y-b)^2+(z-c)^2,$$

for all points (x, y, z) different from (a, b, c). From Gauss's formula in three dimensions we conclude (i) that $\Omega = 0$ if Σ is a closed surface such that $A = (a, b, c)$ is outside the volume bounded by Σ; (ii) that if A is within Σ, the value of the integral is independent of the shape of Σ. Taking for Σ a sphere with centre A, we easily see that $\Omega = 4\pi$.

13. The integral

$$\frac{\partial\Omega}{\partial a} = \iint_\Sigma \frac{\partial}{\partial a}\left(\frac{a-x}{\gamma^3}\right)dy\,dz + \frac{\partial}{\partial a}\left(\frac{b-x}{\gamma^3}\right)dz\,dx + \frac{\partial}{\partial a}\left(\frac{c-z}{\gamma^3}\right)dx\,dy$$

is independent of Σ and depends only on the boundary Γ of Σ, for the identity given in the answer to Ex. 12 implies that

$$\frac{\partial}{\partial x}\left[\frac{\partial}{\partial a}\left(\frac{a-x}{\gamma^3}\right)\right]+\frac{\partial}{\partial y}\left[\frac{\partial}{\partial a}\left(\frac{b-y}{\gamma^3}\right)\right]+\frac{\partial}{\partial z}\left[\frac{\partial}{\partial a}\left(\frac{c-z}{\gamma^3}\right)\right]=0.$$

By Stokes's theorem and the discussion of Chap. V, Appendix, § 2 (pp. 393, 404), the surface-integral expression for $\partial\Omega/\partial a$ may be expressed as a line integral $\int u\,dx + v\,dy + w\,dz$ along Γ. Verify that the functions

$$u = 0, \quad v = \frac{z-c}{\gamma^3}, \quad w = -\frac{y-b}{\gamma^3}$$

satisfy the identities

$$\frac{\partial w}{\partial y} - \frac{\partial v}{\partial z} = \frac{\partial}{\partial a}\left(\frac{a-x}{\gamma^3}\right), \quad \frac{\partial u}{\partial z} - \frac{\partial w}{\partial x} = \frac{\partial}{\partial a}\left(\frac{b-y}{\gamma^3}\right), \quad \frac{\partial v}{\partial x} - \frac{\partial u}{\partial y} = \frac{\partial}{\partial a}\left(\frac{c-z}{\gamma^3}\right).$$

14. Note the following facts: (1) the value of the line-integral Θ remains unchanged if Γ is deformed in such a way that Γ never sweeps over any of the points $(-1, 0)$ or $(1, 0)$ during its deformation; (2) $\Theta = 2\pi$ if Γ is a small circle around $(1, 0)$ oriented counter-clockwise; (3) $\Theta = 2\pi$ if Γ is a small circle around $(-1, 0)$ oriented clockwise.

15. Think of C as being a rigid circle made of wire and of Γ as being a string. Now deform the string Γ to a new position Γ' lying entirely within the plane $y = 0$. The numbers p and n are not changed during this deformation, and the first formula now follows directly if Ex. 14 is applied to the curve Γ' within the plane $y = 0$ and the line-segment $-1 < x < 1$, $y = 0$, $z = 0$ of this plane. The factor 4π (instead of 2π, as in the previous example) is due to the fact that the solid angle Ω increases by 4π along a closed path for which $p = 1$, $n = 0$.

One way of carrying out the above deformation of Γ into Γ' analytically is as follows. Assume that Γ does not meet the z-axis and let

$$x = \gamma(t)\cos\varphi(t), \quad y = \gamma(t)\sin\varphi(t), \quad z = z(t) \ (0 \leqq t \leqq 2\pi)$$

be the parametric equations of Γ. Consider now the family of curves

$$\Gamma(\tau): x = \gamma(t)\cos[\tau\varphi(t)], \quad y = \gamma(t)\sin[\tau\varphi(t)], \quad z = z(t)$$

depending on the parameter τ which decreases from $\tau = 1$ to $\tau = 0$. Note that $\Gamma(1) = \Gamma$ and that $\Gamma' = \Gamma(0)$ is a closed curve which lies in the plane $y = 0$. Note also that (for a fixed value of z) each point P of $\Gamma(\tau)$ rotates about the z-axis as τ varies; hence the solid angle Ω which C subtends at P does not vary with τ. This implies that $\Omega_1 - \Omega_0$ will have the same value for $\Gamma(0)$ as for $\Gamma(1) = \Gamma$. To prove the second formula, note that

$$\Omega_1 - \Omega_0 = \int_\Gamma d\Omega = \int_\Gamma \operatorname{grad}\Omega \,.\, dP = -\int_\Gamma dP \,.\int_C \frac{[\overline{PP'}\,dP']}{|\overline{PP'}|^3}$$

$$= -\int_\Gamma\int_C \frac{dP\,.\,[\overline{PP'}v\,dP']}{|\overline{PP'}|^3} = \int_\Gamma\int_C \frac{\overline{PP'}\,.\,[dP\,.\,dP']}{|\overline{PP'}|^3}.$$

CHAPTER VI

§ 2, p. 428.

1. Use the theorem of the conservation of energy, and prove that $r \to \infty$ as $t \to \infty$.

2. If (ξ, η) are the co-ordinates with respect to the axes of the ellipse, then

$$\xi = a \cos\omega = x + \varepsilon a$$
$$\eta = b \sin\omega = y$$

give the equation of the ellipse; and by the law of areas

$$h(t - t_s) = \int_0^\omega \left(x \frac{\partial y}{\partial \omega} - y \frac{\partial x}{\partial \omega}\right) d\omega$$

$$= ab \int_0^\omega (1 - \varepsilon \cos\omega)\, d\omega.$$

[Note that the question ought to read: ". . . the angle $P'MP_s$, where P' is the point on the auxiliary circle corresponding to P, the position of the planet . . .".]

3, 4. Use the theorem of the conservation of energy and the law of areas.

§ 3, p. 432.

1. (*a*) $y = \tan \log c/\sqrt{(1 + x^2)}$. (*b*) $y = c\sqrt{(1 + e^{2x})}$.

2. (*a*) $y = ce^{y/x}$. (*b*) $y^2(2x^2 + y^2) = c^2$.

(*c*) $x^2 - 2cx + y^2 = 0$ (circles).

(*d*) $\arctan(y/x) + c = \log\sqrt{(x^2 + y^2)}$, or, in polar co-ordinates, $r = e^{\phi + c}$ (logarithmic spirals).

(*e*) $c + \log|x| = \arcsin(y/x) - \dfrac{1}{x}\sqrt{(x^2 - y^2)}$.

3. If $ab_1 - a_1 b \neq 0$, we have

$$\frac{d\eta}{d\xi} = \frac{a + by'}{a_1 + b_1 y'} = \frac{a + b\varphi(\eta/\xi)}{a_1 + b_1\varphi(\eta/\xi)},$$

which is a homogeneous equation.

If $ab_1 - a_1 b = 0$ or $a_1/a = b_1/b = k$, then

$$\frac{d\eta}{dx} = a + b\frac{dy}{dx} = a + b\varphi\left(\frac{\eta + c}{k\eta + c_1}\right),$$

and the variables are separated.

4. (*a*) $4x + 8y + 5 = ce^{4x - 8y}$.

(*b*) $x = c - \frac{1}{3}(3y - 7x) - \frac{2}{9}\log(3y - 7x)$.

5. (*a*) $y = ce^{-\sin x} + \sin x - 1.$ (*b*) $y = (x + 1)^n(e^x + c).$
(*c*) $y = cx(x - 1) + x.$ (*d*) $y = \frac{1}{3}x^5 + cx^2.$
(*e*) $y = \frac{c}{\sqrt{(1 + x^2)}} - \frac{1}{(1 + x^2)(x + \sqrt{1 + x^2})}.$

6. Introduce $1/y$ as new unknown function; the equation then becomes homogeneous:

$$\frac{1}{x}\,\frac{1 - cx^{\sqrt{5}}}{cx^{\sqrt{5}}\,(\frac{1}{2} - \frac{1}{2}\sqrt{5}) - \frac{1}{2} - \frac{1}{2}\sqrt{5}}.$$

§ 4, p. 444.

1. Use induction. Suppose that a linear relation $c_1\varphi_1 + \ldots + c_k\varphi_k = 0$ holds. Divide by $e^{a_k x}$ and differentiate $(n_k + 1)$ times, if $P_k(x)$ is of degree n_k. The degree of the coefficients of the other $e^{a_i x}$'s is unchanged, so that they remain different from zero.

2. Multiply both sides of the equation by $(1 - n)y^{-n}$.

$$(a)\quad y^{-1} = cx + \log x + 1; \qquad (b)\quad y^3 = cx^{-3} + \frac{3a^2}{2x};$$

$$(c)\quad (y^{-1} + a)^2 = c(x^2 - 1).$$

3. If we put $y = y_1 + u^{-1}$, the equation reduces to the linear equation $u' - (2Py_1 + Q)u = P$.

$$y = x - \frac{e^{\frac{1}{2}x^4}}{c + \int_0^x x^2 e^{\frac{1}{2}x^4}\,dx}.$$

4. By equating the right-hand sides of (*a*) and (*b*), we obtain the common integral $y = x^2$.

5. $$y = x^2 - \frac{e^{\frac{2}{3}x^3}}{c + \int_{-\infty}^x e^{\frac{2}{3}x^3}\,dx} \quad (= f(x, c)).$$

To draw the graphs of the corresponding family of curves, first plot the two branches of the curve

$$y^2 + 2x - x^4 = 0 \qquad (y = \pm\sqrt{(x^3 - 2)x}),$$

which divides the plane into two regions where $y' < 0$ and one region where $y' > 0$. The two infinite branches of this curve are asymptotic to the two parabolas $y = \pm x^2$. Show that all the integral curves are asymptotic to these parabolas by proving the two relations

$$f(x, c) = -x^2 + o(1) \quad \text{as} \quad x \to +\infty \; (-\infty < c < \infty)$$

and

$$f(x, c) = x^2 + o(1) \quad \text{as} \quad x \to -\infty \; (c \neq 0),$$

where $o(1)$ denotes a function which tends to zero.

6. Put

$$y_1 - y_3 = a, \quad y_1 - y_4 = b, \quad y_2 - y_3 = c, \quad y_2 - y_4 = d.$$

Then

$$a' + Pa(y_1 + y_3) + Qa = 0,$$

so that

$$P(y_1 + y_3) = -Q - \frac{a'}{a},$$

$$P(y_1 - y_3) = aP,$$

or

$$2Py_1 = aP - Q - \frac{a'}{a}.$$

Similarly,

$$2Py_1 = bP - Q - \frac{b'}{b}.$$

Hence

$$\frac{d\log(a/b)}{dx} = P(a - b) = -P(y_3 - y_4),$$

and similarly,

$$\frac{d\log(c/d)}{dx} = -P(y_3 - y_4);$$

by subtraction,

$$\log\left(\frac{a}{b} \div \frac{c}{d}\right) = \text{const.}$$

7. Cf. the relation

$$\frac{d\log(a/b)}{dx} = P(y_4 - y_3),$$

in the proof of the preceding example.

Particular solutions of the special equation are $y_1 = \frac{1}{\cos x}$ and $y_2 = -\frac{1}{\cos x}$: $y = \frac{1 + ce^{2x}}{(1 - ce^{2x})\cos x}$.

8. The common solution e^x of (*a*) and (*b*) is obtained by eliminating y'' from the two equations.

(*a*) $c_1e^x + c_2x$;

(*b*) $c_1e^x + c_2\sqrt{x}$.

§ 4, p. 449.

1. From the fundamental theorem of algebra it follows that $f(z)$ may be written $f(z) = (z - a_1)^{\mu_1}(z - a_2)^{\mu_2} \ldots (z - a_k)^{\mu_k}$ (cf. Vol. I, p. 230), where the μ_ν's are positive integers such that $\mu_1 + \ldots + \mu_k = n$; and

$$f(a_\nu) = f'(a_\nu) = \ldots = f^{(\mu_\nu - 1)}(a_\nu) = 0.$$

Now

$$L(e^{\lambda x}) = f(\lambda)e^{\lambda x}.$$

On differentiating this relation $(\mu_\nu - 1)$ times and putting $\lambda = a_\nu$ in the result, we get (cf. Leibnitz's rule, Vol. I, p. 202)

$$L(e^{a_\nu x}) = f(a_\nu)e^{a_\nu x} = 0$$

$$L(xe^{a_\nu x}) = [f'(a_\nu) + xf(a_\nu)]e^{a_\nu x} = 0$$

$$L(x^2e^{a_\nu x}) = [f''(a_\nu) + 2xf'(a_\nu) + x^2f(a_\nu)]e^{a_\nu x} = 0$$

$$\cdots\cdots\cdots\cdots\cdots\cdots\cdots\cdots$$

$$L(x^{\mu_\nu - 1}e^{a_\nu x}) = \left[\binom{\mu_\nu - 1}{0} f^{(\mu_\nu - 1)}(a_\nu) + \binom{\mu_\nu - 1}{1} f^{(\mu_\nu - 2)}(a_\nu)x + \ldots + \binom{\mu_\nu - 1}{\mu_\nu - 1} f(a_\nu)x^{\mu_\nu - 1}\right] e^{a_\nu x} = 0.$$

So we have n particular solutions

$$e^{a_1 x},\ xe^{a_1 x},\ \ldots,\ x^{\mu_1 - 1}e^{a_1 x}$$
$$e^{a_2 x},\ xe^{a_2 x},\ \ldots,\ x^{\mu_2 - 1}e^{a_2 x}$$
$$\cdots\cdots\cdots\cdots$$
$$e^{a_k x},\ xe^{a_k x},\ \ldots,\ x^{\mu_k - 1}e^{a_k x},$$

which are linearly independent, by Ex. 1, p. 444.

2. (*a*) $y = c_1e^x + c_2e^{-\frac{1}{2}x}\cos\frac{\sqrt{3}x}{2} + c_3e^{-\frac{1}{2}x}\sin\frac{\sqrt{3}x}{2}.$

(*b*) $y = c_1e^x + c_2xe^x + c_3e^{2x}.$

(*c*) $y = c_1e^x + c_2xe^x + c_3x^2e^x.$

(*d*) $y = c_1e^x + c_2e^{-x} + c_3e^{\sqrt{2}x} + c_4e^{-\sqrt{2}x}.$

(*e*) Substitute $x = e^t$:

$$y = c_1x + c_2/x.$$

3. On substituting in the differential equation, we get

$$(a_0b_0 - 1)P(x) + (a_0b_1 + a_1b_0)P'(x) + (a_0b_2 + a_1b_1 + a_2b_0)P''(x) + \ldots = 0,$$

and this is an identity if $a_0b_0 = 1$, $a_0b_1 + a_1b_0 = 0$, $\ldots$, from the expansion. The second case reduces to the first if we substitute y' for y.

4. (*a*) $\frac{1}{1+t^2} = 1 - t^2 + t^4 - \ldots$; hence $y = P(x) - P''(x) = 3x^2 - 5x - 6.$

(*b*) $\frac{1}{t + t^2} = \frac{1}{t} - 1 + t - t^2 + \ldots$; hence

$$y = \int P(x)\,dx - P(x) + P'(x) - P''(x) = -\tfrac{2}{3} + x + \tfrac{1}{3}x^3.$$

5. (*a*) $y = \frac{3}{8}e^x$, (*b*) $y = \frac{1}{6}x^3e^x$.

6. $y = e^x\left(\frac{x^2}{2} + \frac{3}{2}x + \frac{7}{4}\right) + c_1e^{3x} + c_2e^{2x}$.

§ 5, p. 465.

1. (*a*) Use the fact that the curvilinear integral

$$\int (3x^2 + 6xy^2)\,dx + (6x^2y + 4y^3)\,dy$$

is independent of the path. Integrating between (0, 0) and (x, y) along the broken line $(0, 0) \ldots (x, 0) \ldots (x, y)$, we get

$$\int_{0,\,0}^{x,\,y} (3x^2 + 6xy^2)dx + (6x^2y + 4y^3)dy = x^3 + 3x^2y^2 + y^4 = c.$$

(*b*) By inspection we find the general integral

$$\sqrt{(1 + x^2 + y^2)} - \operatorname{arc\,tan}(y/x) = c.$$

2. Here dy/dx is a function of y/x alone.

3. $x^2y - 2xy^2 - 2cy - 2 = 0$ (integrating factor $\mu = 1/y^2$).

4. The equation is linear in x and its general integral is $(xy^2 + 1)^2 = cy$. The identity

$$d\left(\frac{(xy^2 + 1)^2}{y}\right) = \frac{xy^2 + 1}{y^2}\,[2y^3dx + (3xy^2 - 1)dy]$$

displays an integrating factor of the equation.

5. (*a*) $x^2 + y^2 + cx + 1 = 0$ $(-\infty < c < \infty)$ and the line $x = 0$.

(*b*) $x^2 + 2y^2 = c^2$.

(*c*) The differential equation of this family of confocal conics (cf. p. 158) is found to be

$$y'^2 + \frac{x^2 - y^2 - a^2 + b^2}{xy}\,y' - 1 = 0,$$

which is unaltered if y' is replaced by $-1/y'$; the family of ellipses $(-b^2 < c < \infty)$ is orthogonal to the family of hyperbolas $(-a^2 < c < -b^2)$.

(*d*) $y = \log|\tan(x/2)| + c$ and the vertical lines $x = k\pi$ (k an integer).

(*e*) The family of curves (tractrix)

$$x - c = \pm(\sqrt{(a^2 - y^2)} - a \operatorname{ar\,cosh}(a/y))$$

and the same family reflected in the x-axis.

6. (*a*) The family of parabolas $y = cx^2$.

(*b*) The family of hyperbolas $xy = c$.

7. (*a*) $y = x^2$, (*b*) $y = -x + x\log(-x)$ $(0 > x > -\infty)$.

8. $y = xp + a\sqrt{1 + p^2} - ap \operatorname{ar\,sinh} p$.

9. $x = ce^{-p/a} + \frac{1}{2}p$

$$y = c(p + a)e^{-p/a} + \tfrac{1}{2}p(p + a) - \tfrac{1}{4}(p + a)^2.$$

Note that for $c = 0$ this gives the parabola $y = x^2 - \dfrac{a^2}{4}$. What is the geometrical meaning of this result?

10. (*a*) $y = \sin(x + c)$, singular solutions $y = \pm 1$.

(*b*) $x = \pm \frac{1}{2}(\arcsin y + y\sqrt{1 - y^2}) + c$.

(*c*) $x = \mp \left(\sqrt{(2a - y)y} - 2a \arctan\sqrt{\dfrac{y}{2a - y}}\right) + c$,

which is a family of cycloids and can be expressed in the parametric form $x = c + a(\varphi - \sin\varphi)$, $y = a(1 - \cos\varphi)$. Singular solution $y = 2a$.

(*d*) $x = \pm\displaystyle\int_0^y \sqrt{\frac{1 + y^2}{1 - y^2}}\, dy + c \quad (-1 \leqq y \leqq 1)$; singular solutions $y = \pm 1$. (The reader should prove that these curves are not sine curves.)

11. $MN = y\sqrt{1 + y'^2}$, $MC = -\dfrac{(1 + y'^2)^{\frac{3}{2}}}{y''}$, and the differential equation is

$$(1 + y'^2)^2 y + ky'' = 0.$$

By the general method this is easily reduced to

$$\left(\frac{dy}{dx}\right)^2 = \frac{k + c - y^2}{y^2 - c} \quad (c \text{ an arbitrary constant}).$$

The various cases, all of importance in the differential geometry of surfaces,* are as follows:

(1) $k = \kappa^2 (> 0)$, $c = -\gamma^2 (< 0, \gamma^2 < \kappa^2)$. The curve is everywhere smooth, and oscillates, alternately touching the lines $y = \pm\sqrt{\kappa^2 - \gamma^2}$. It looks like a sine curve, but is not one.

(2) $k = \kappa^2$, $c = 0$. The curve is a circle of radius κ with centre on the x-axis.

(3) $k = \kappa^2$, $c = \gamma^2 (> 0)$. The curve consists of a sequence of identical arcs, joined by cusps lying on the line $y = \gamma$, and all touched by $y = \sqrt{\kappa^2 + \gamma^2}$. It looks like a cycloid, but is not one.

(4) $k = -\kappa^2 (< 0)$, $c = \gamma^2 > \kappa^2$. The curve consists of a sequence of identical arcs upside-down, with their cusps on $y = \gamma$ and touched by $y = \sqrt{\gamma^2 - \kappa^2}$.

(5) $k = -\kappa^2$, $c = \gamma^2 = \kappa^2$. The curve is a tractrix.

(6) $k = -\kappa^2$, $c = \gamma^2 < \kappa^2$. The curve has an infinity of cusps, perpendicular to the lines $y = \gamma$ and $y = -\gamma$ alternatively.

12. Eliminate b from the equations obtained by differentiating the equation of the circle twice and thrice: $(1 + y'^2)y''' - 3y'y''^2 = 0$.

* See Eisenhart, *Differential Geometry*, pp. 270–4 (Princeton Press).

13. $y = x \sin ax$; singular solutions $y = x$ and $y = -x$.

14. If $y(x) = \Sigma c_\nu x^\nu$, then

$$c_{\nu+2} = -\frac{c_\nu}{(\nu+2)^2} \quad \text{and} \quad c_0 = 1,\ c_1 = 0;$$

$$y(x) = \sum_{\nu=0}^{\infty} \frac{(-1)^\nu}{2^{2\nu}\nu!^2} x^{2\nu}.$$

If we substitute the power series for $\cos xt$ in the expression for $J_0(x)$ in Ex. 4, p. 223, and interchange summation and integration (why is this permissible?), we get

$$J_0(x) = \frac{1}{\pi} \sum_{\nu=0}^{\infty} \frac{x^{2\nu}}{(2\nu)!} (-1)^\nu \int_{-1}^{+1} \frac{t^{2\nu}}{\sqrt{(1-t^2)}}\, dt;$$

the value of $\int_{-1}^{+1} \frac{t^{2\nu}}{\sqrt{(1-t^2)}}\, dt$ is $\frac{(2\nu)!\,\pi}{\nu!^2 2^{2\nu}}$, as is found by putting $t = \sin\tau$ and referring to Vol. I, p. 223. The power series for $y(x)$ and $J_0(x)$ are therefore identical.

§ 6, p. 481.

1. Poisson's formula gives a potential function $u(r, \theta)$ inside the unit circle, with boundary values $f(\theta)$. Now $u\left(\frac{1}{r}, \theta\right)$ is also a potential function (cf. Vol. I, p. 479, Ex. 3) with the same boundary values, and it is bounded in the region outside the unit circle; thus the expression

$$\frac{r^2-1}{2\pi} \int_0^{2\pi} f(\alpha) \frac{d\alpha}{1 - 2r\cos(\theta-\alpha) + r^2}$$

is a solution of the problem.

2. The potential is

$$\mu \log \frac{z + l + \sqrt{(z+l)^2 + x^2 + y^2}}{z - l + \sqrt{(z-l)^2 + x^2 + y^2}}.$$

Since on the ellipsoid $z = l\alpha \cos\varphi$, $\sqrt{x^2+y^2} = l\sqrt{\alpha^2-1}\, \sin\varphi$, the potential is

$$\mu \log \frac{\alpha+1}{\alpha-1},$$

the confocal ellipsoids

$$\frac{z^2}{l^2\alpha^2} + \frac{x^2+y^2}{l^2(\alpha^2-1)} = 1 \qquad (1 \leq \alpha \leq \infty)$$

are equipotential surfaces. The lines of force are the orthogonal trajectories and hence (cf. Ex. 5c, p. 466) are the confocal hyperbolas given by the same equation when $0 \leq \alpha \leq 1$ and the ratio of x to y is constant.

3. Let Σ be a sphere of radius ρ and centre (x, y, z), lying inside S. Since $\Delta\left(\frac{1}{r}\right) = 0$ and $\Delta u = 0$ in the region bounded by Σ and S, by Green's theorem (cf. p. 390) we have

$$0 = \iint_S \left(\frac{1}{r}\frac{\partial u}{\partial n} - u\frac{\partial(1/r)}{\partial n}\right) d\sigma - \iint_\Sigma \left(\frac{1}{r}\frac{\partial u}{\partial n} - u\frac{\partial(1/r)}{\partial n}\right) d\sigma,$$

where in the first integral n is the outward normal to S and in the second the outward normal to Σ. Now on the sphere Σ we have $\frac{\partial(1/r)}{\partial n} = \frac{\partial(1/r)}{\partial r} = -\frac{1}{\rho^2}$, $r = \text{const.} = \rho$; therefore

$$\iint_\Sigma \frac{1}{r}\frac{\partial u}{\partial n}\, d\sigma = \frac{1}{\rho}\iint_\Sigma \frac{\partial u}{\partial n}\, d\sigma = 0,$$

since u is a harmonic function (cf. p. 475); in addition,

$$-\frac{1}{4\pi}\iint_\Sigma u\,\frac{\partial(1/r)}{\partial n}\, d\sigma = \frac{1}{4\pi\rho^2}\iint_\Sigma u\, d\sigma,$$

and as $\rho \to 0$ this expression obviously tends to $u(x, y, z)$, for it is the mean value of u on Σ.

§ 7, p. 489.

1. (*a*) $u = f(x) + g(y)$ (f and g are arbitrary functions).

(*b*) $u = f(x, y) + g(x, z) + h(y, z)$ (f, g, h are arbitrary functions).

(*c*) The most general solution is obtained from a particular solution by adding the general solution of the homogeneous equation $u_{xy} = 0$.

$u = \int_0^x d\xi \int_0^y a(\xi, \eta)\, d\eta + f(x) + g(y)$, where f and g are arbitrary.

2. Apply the linear transformation

$$x = \xi + \eta$$
$$y = 3\xi + 2\eta.$$
$$u = f(y - 2x) + g(3x - y) + \frac{1}{12}e^{x+y}.$$

3. $z^2(z_x^2 + z_y^2 + 1) = 1$.

4. $u(x, t) = f(x - at) + g(x + at)$; then for $x \geqq 0$

$$0 = u(x, 0) = f(x) + g(x)$$
$$0 = u_t(x, 0) = -af'(x) + ag'(x);$$

by differentiating the first equation and comparing with the second, we have

$$f'(x) = 0, \quad g'(x) = 0,$$

or

$$f(x) = \text{const.} = c, \quad g(x) = -c \quad \text{for} \quad x \geqq 0.$$

For $t \geqq 0$, moreover,

$$\varphi(t) = u(0, t) = f(-at) + g(at) = f(-at) - c,$$

that is, $f(\xi) = c + \varphi\left(\frac{\xi}{-a}\right)$ if $\xi < 0$. As $x + at \geqq 0$ always, and hence $g(x + at) = -c$, it follows that

$$u(x, t) = \begin{cases} 0 \text{ for } x - at \geqq 0 \\ \varphi\left(\dfrac{x - at}{-a}\right) \text{ for } x - at \leqq 0 \end{cases}$$

if both x and t are non-negative.

5. If $u(x, y) = \Sigma a_{\nu\mu} x^\nu y^\mu$, then

$$a_{\nu+1, \mu+1} = \frac{a_{\nu\mu}}{(\nu + 1)(\mu + 1)};$$

in addition,

$$a_{\nu 0} = a_{0\nu} = 0 \quad \text{for} \quad \nu \geqq 1 \quad \text{and} \quad a_{00} = 1.$$

Hence

$$u(x, y) = \sum_{\nu=0}^{\infty} \frac{x^\nu y^\nu}{\nu!^2} = J_0(2i\sqrt{xy}),$$

where J_0 is the Bessel function of Ex. 4, p. 223.

6. (*a*) From the differential equation we get

$$(f'(x))^2 + (g'(y))^2 = 1$$

or

$$(f'(x))^2 = 1 - (g'(y)^2).$$

As the left-hand side does not depend on y, nor the right-hand side on x, both sides are equal to a constant (which has to be positive or zero), say c^2; that is,

$$(f'(x))^2 = c^2, \quad 1 - (g'(y))^2 = c^2.$$

Hence

$$u = cx + \sqrt{(1 - c^2)}\, y + b$$

is a solution, where c and b are arbitrary and $c^2 \leqq 1$.

(*b*) $u = f(x) + g(y)$ gives

$$f'(x) = \frac{1}{g'(y)} = \text{const.} = a,$$

so that

$$u = ax + \frac{1}{a}y + b$$

(where a and b are constants).

If $u = f(x)g(y)$, then

$$\frac{d}{dx}(f(x))^2 = 4\Big/\frac{d}{dy}(g(y))^2 = \text{const.} = 2c;$$

so in this case

$$u = \sqrt{\left\{(2cx + a)\left(\frac{2}{c}y + b\right)\right\}},$$

where a, b, c are arbitrary constants.

7. A one-parameter family is obtained from the two-parameter family of solutions $z = u(x, y, a, b)$ by making a and b depend in some way on a parameter t:

$$a = f(t), \quad b = g(t),$$
$$z = u(x, y, f(t), g(t)).$$

The envelope of this one-parameter family is obtained by finding t from the equation

$$0 = z_t = u_a f' + u_b g',$$

and substituting this expression for t in $z = u(x, y, f(t), g(t))$. The result is again a solution of $F(x, y, z, z_x, z_y) = 0$, as

$$z = u(x, y, a, b)$$
$$z_x = u_x + u_t t_x = u_x(x, y, a, b)$$
$$z_y = u_y + u_t t_y = u_y(x, y, a, b)$$

and $z = u(x, y, a, b)$ satisfies the equation $F(x, y, z, z_x, z_y) = 0$.

8. $$u = x\sqrt{\frac{y}{x+k}} + y\sqrt{\frac{x+k}{y}} + k\sqrt{\frac{y}{x+k}}.$$

CHAPTER VII

§ 1, p. 497.

1. $\dfrac{2}{\sqrt{2g}}\sqrt{\dfrac{(x_1 - x_0)^2 + (y_1 - y_0)^2}{y_1 - y_0}}$.

2. $T = \displaystyle\int_{\sigma_0}^{\sigma_1} f(r)\sqrt{\dot{r}^2 + r^2\dot{\theta}^2 + r^2\sin^2\theta\,\dot{\varphi}^2}\,d\sigma.$

§ 2, p. 505.

1. Parabolas $y = c^2 + \dfrac{x^2}{4c^2}$.

2. Circle with centre on x-axis.

3. $y = c \sin \dfrac{x-a}{c}$.

4. $y = \dfrac{a}{x^{n-1}} + b$ for $n > 1$ and $y = a \log x + b$ for $n = 1$.

5. $y = a(x - b)^{n/(n+m)}$ if $n + m \neq 0$; $y = ae^{bx}$ if $n = -m$.

6. $ay'' + a'y' + (b' - c)y = 0$

$$\left(\text{for } b = \text{const.}, \int_{x_0}^{x_1} byy'\,dx = \frac{b}{2}(y_2^2 - y_1^2)\right.$$

only depends on the end-points of the curve $y = y(x)$).

7. $y_1 - y_0 < \dfrac{\pi}{2}$.

8 Consider $F(x, y)$ for fixed x as a function of y; let this function of y have a minimum for $y = \bar{y}$. Then $F(x, y) \geqq F(x, \bar{y})$ for a certain neighbourhood of $\bar{y}$ and $F_y(x, \bar{y}) = 0$. $\bar{y}$ will depend on the parameter x; i.e. $\bar{y} = \bar{y}(x)$. Then for any neighbouring function y we have

$$\int_{x_0}^{x_1} F(x, y(x))\,dx \geqq \int_{x_0}^{x_1} F(x, \bar{y}(x))\,dx,$$

where $\bar{y}(x)$ satisfies the equation $F_y(x, \bar{y}(x)) = 0$.

9. (*a*) $y = 0$.

(*b*) Use Schwarz's inequality. For any admissible x,

$$1 = y(1) - y(0) = \int_0^1 y'\,dx \leqq \sqrt{\left(\int_0^1 1^2\,dx\right)}\sqrt{\left(\int_0^1 y'^2\,dx\right)} = \sqrt{I},$$

and the equality sign holds for $y = x$.

§ 3, p. 510.

1. If $v = 1/f(r)$, then T is given by Ex. 2, p. 497:

$$F = f(r)\sqrt{(\dot{r}^2 + r^2\dot{\theta}^2 + r^2 \sin^2\theta\, \dot{\varphi}^2)}.$$

Euler's equation for the variable φ gives

$$F_{\dot{\varphi}} = \frac{\dot{\varphi} f^2 r^2 \sin^2\theta}{F} = \text{const.} = C$$

along a ray. Now let the polar co-ordinates be chosen in such a way that the plane $\varphi = 0$ passes through the initial point and the end-point; since $\varphi = 0$ at both these points we have $\dot{\varphi} = 0$ for some intermediate point, by the mean value theorem, that is, $C = 0$; but then $\dot{\varphi} = 0$ for the whole ray, i.e. $\varphi \equiv 0$. Hence the whole ray must lie in the plane $\varphi = 0$.

§ 3, p. 518.

The law of conservation of energy gives

$$T + U = T = \frac{1}{2}\left(\frac{ds}{dt}\right)^2 = \text{const.} = \frac{1}{2} C^2;$$

hence $\frac{ds}{dt} = \text{const.} = C = \text{initial velocity.}$

Then Hamilton's principle asserts the stationary character of

$$\int_{t_0}^{t_1} (T - U)\,dt = \int_{t_0}^{t_1} T\,dt = \frac{1}{2} C^2 \int_{t_0}^{t_1} dt = \frac{1}{2} C \int_{s_0}^{s_1} ds;$$

stationary character of Hamilton's integral implies that the length of path is stationary.

Miscellaneous Examples VII, p. 520.

1. From the differential equations for geodesics (p. 518) we find that for a cylinder, i.e. if G does not depend on z, $\frac{dz}{dt} = \text{const.}$; hence the geodesics on a cylinder make a constant angle with the xy-plane.

2. (*a*) $g(x) - \frac{y''}{\sqrt{(1 + y'^2)^3}} = 0.$

(*b*) $g(x) - \frac{6y''(y''^2 + 4y'y''')}{(1 + y'^2)^4} + \frac{2y^{iv}}{(1 + y'^2)^3} + \frac{48y'^2y''^3}{(1 + y'^2)^5} = 0.$

(*c*) $y + y'' + y^{iv}.$

(*d*) $(2 - y'^2)y'' = 0.$

3. (*a*) $\varphi d = (a_x + b_y)\varphi_x + (b_x + c_y)\varphi_y + a\varphi_{xx} + 2b\varphi_{xy} + c\varphi_{yy}.$

(*b*) $\Delta^2\varphi = 0.$

(*c*) $\Delta^2\varphi = 0.$

4. $\frac{au'' + a'u' + u(b' - c)}{u} = \lambda = \text{const.}$

5. (*a*) Euler's equation gives

$$f + 2\lambda u = 0;$$

from this equation and $\int_0^1 \varphi^2\,dx = K^2$, we have

$$\lambda = \pm \frac{\sqrt{(\int_0^1 f^2\,dx)}}{2K}, \quad u = \frac{\pm Kf}{\sqrt{(\int_0^1 f^2\,dx)}};$$

(*b*) For any continuous admissible φ we have

$$I = \int_0^1 f\varphi\,dx \leq \sqrt{\left(\int_0^1 f^2\,dx\right)} \sqrt{\left(\int_0^1 \varphi^2\,dx\right)} = K \sqrt{\left(\int_0^1 f^2\,dx\right)},$$

the equality sign holding for $\varphi = u$.

CHAPTER VIII

§ 1, p. 529.

1. For $x \geqq 0$.

2. Use the principle of comparison.

3. The coefficient of z^n in the expansion of $\cos^2 z + \sin^2 z$ for $n > 0$ is

$$(-1)^{\frac{n}{2}} \sum_{\nu=0}^{n} \frac{(-1)^\nu}{\nu!(n-\nu)!} = \frac{(-1)^{\frac{n}{2}}}{n!} \sum_{\nu=0}^{n} (-1)^\nu \binom{n}{\nu} = 0$$

(cf. Vol. I, p. 28, Ex. 2(*b*)).

4. The series is convergent if, and only if, $|z| < 1$. For if $|z| = \theta < 1$, then

$$\left|\frac{z^\nu}{1-z^\nu}\right| \leqq \frac{\theta^\nu}{1-\theta^\nu} \leqq \frac{1}{1-\theta}\theta^\nu$$

and we may compare with the geometric series. If $|z| > 1$, then $\dfrac{z^\nu}{1-z^\nu}$ tends to -1 as ν increases, whereas in a convergent series the terms must tend to 0. If $|z| = 1$, either there are terms in the series which are not defined, or at least its terms are not bounded, since z^ν may approach 1 as closely as we please.

§ 2, p. 535.

Let $f(z) = u + iv$, $g(z) = u' + iv'$, $fg = p + iq$, where $p = uu' - vv'$, $q = uv' + vu'$. Assume that u, v and also u', v' satisfy the Cauchy-Riemann equations, and prove that the same is true of p, q.

§ 2, p. 536.

1. The functions (*a*), (*b*), (*c*) are continuous everywhere; (*d*) is discontinuous at $z = 0$.

2. None.

3. $|\zeta|^2 = \zeta\bar{\zeta} = \dfrac{\alpha\bar{\alpha}z\bar{z} + \beta\bar{\beta} + (\alpha\beta z + \bar{\alpha}\bar{\beta}\bar{z})}{\beta\bar{\beta}z\bar{z} + \alpha\bar{\alpha} + (\alpha\beta z + \bar{\alpha}\bar{\beta}\bar{z})}$.

Now for $\alpha\bar{\alpha} - \beta\bar{\beta} = 1$ the difference between the numerator and the denominator is

$$z\bar{z} - 1,$$

so that the numerator is greater than the denominator for $|z| > 1$, and smaller for $|z| < 1$. If $\beta\bar{\beta} - \alpha\bar{\alpha} = 1$, the converse is the case.

4. If $z = re^{i\phi}$, $\zeta = \xi + i\eta$, then

$$\xi = \frac{1}{2}\left(r + \frac{1}{r}\right)\cos\varphi$$

$$\eta = \frac{1}{2}\left(r - \frac{1}{r}\right)\sin\varphi.$$

If $r = \text{const.} = c$, then

$$\frac{\xi^2}{\frac{1}{4}\left(c + \frac{1}{c}\right)^2} + \frac{\eta^2}{\frac{1}{4}\left(c - \frac{1}{c}\right)^2} = 1;$$

if $\varphi = \text{const.} = c$, then

$$\frac{\xi^2}{\cos^2 c} + \frac{\eta^2}{\cos^2 c - 1} = 1$$

(cf. Ex. 5, p. 158).

6. First transform, by putting $\zeta = az + b$, into the unit circle; then apply the transformation $\zeta' = i\dfrac{1+\zeta}{1-\zeta}$.

7. The equation of a circle or straight line in the ζ-plane is of the form

$$\alpha\zeta\bar{\zeta} + \beta\zeta + \bar{\beta}\bar{\zeta} + \gamma = 0,$$

where α and γ are real; if we here substitute the expression for ζ, we get an equation of the same form for z.

For a fixed point $\zeta = z$ we have the quadratic equation

$$cz^2 + dz - az - b = 0,$$

which has in general two different solutions. A circle through the fixed points is, as we have just shown, transformed into a circle and must again pass through the fixed points; the family of orthogonal circles transforms into itself, because circles become circles and the transformation is conformal.

§ 3, p. 545.

2. The series is absolutely convergent, by Vol. I, p. 382.

§ 3, p. 551.

1. By the Cauchy-Riemann equations the partial derivatives v_x and v_y of v are given; a function v with these derivatives does exist, since the condition of integrability $u_{xx} + u_{yy} = 0$ is satisfied (cf. p. 353); v is uniquely determined, apart from an additive constant c, and is given by the curvilinear integral

$$v(x, y) = \int_{(x_0, y_0)}^{(x, y)} (v_y\,dy + v_x\,dx) + c.$$

It also follows from the Cauchy-Riemann equations that v is a potential function.

§ 4, p. 553.

1. It is easily seen that

$$h(z) = \frac{1}{2\pi i} \int \frac{f(\zeta)}{\zeta - z} \frac{z^n}{\zeta^n} d\zeta$$

is an analytic function of z. By differentiating under the integral sign and using Leibnitz's rule (cf. Vol. I, p. 202), we find that $h^{(\mu)}(z)$ is

$$\frac{1}{2\pi i} \sum_{\nu=0}^{\mu} \binom{\mu}{\nu} \nu!\, n\,.\,(n-1)\ldots(n-\mu+\nu+1) \int \frac{f(\zeta)}{(\zeta - z)^{\nu+1}} \frac{z^{n-\mu+\nu}}{\zeta^n} d\zeta$$

$$= \frac{\mu!}{2\pi i} \sum_{\nu=0}^{\mu} \binom{n}{\mu-\nu} \int \frac{f(\zeta)}{(\zeta - z)^{\nu+1}} \frac{z^{n-\mu+\nu}}{\zeta^n} d\zeta.$$

Only the terms with $\mu - \nu \leqq n$ differ from zero, as otherwise $\binom{n}{\mu-\nu}$ vanishes. On the other hand, a term with $\mu - \nu < n$ vanishes for $z = 0$; if $\mu < n$, there are no other terms, so that $h^{(\mu)}(0) = 0$. If $\mu \geqq n$, there remains only the term with $\mu - \nu = n$, so that

$$h^{(\mu)}(0) = \frac{\mu!}{2\pi i} \int \frac{f(\zeta)}{(\zeta - z)^{\mu+1}} d\zeta = f^{(\mu)}(0).$$

2. $| a_\nu | = \left| \frac{1}{2\pi i} \int_C \frac{f(t)}{t^{\nu+1}} dt \right| \leqq \frac{1}{2\pi} \frac{M}{\rho^{\nu+1}} 2\pi\rho$, where C is the circle of radius ρ about the origin.

3. $\int_C \frac{f'(z)}{f(z)} dz$ is equal to the sum of the residues of $\frac{f'}{f}$ in the interior of C. Now if f has a zero of order n at $z = z_0$,

$$f(z) = (z - z_0)^n \varphi(z),$$

where $\varphi(z_0) \neq 0$;

$$\frac{f'(z)}{f(z)} = \frac{n\varphi(z) + (z - z_0)\varphi'(z)}{(z - z_0)\varphi(z)},$$

so the residue of $\frac{f'(z)}{f(z)}$ at $z = z_0$ is $2\pi i n$.

4. (*a*) The number of roots of the equation $P(z) + \theta Q(z) = 0$, by Ex. 3, is

$$\frac{1}{2\pi i} \int_C \frac{P'(z) + \theta Q'(z)}{P(z) + \theta Q(z)} dz.$$

The denominator differs from zero for every θ for which $0 \leqq \theta \leqq 1$ at any point of C; the whole integral is therefore a continuous function of θ.

As its value is always an integer, it is constant, and hence the same for $\theta = 0$ and $\theta = 1$.

(*b*) If $|a| < r^4 - \frac{1}{r}$, then $r > 1$; so the equation $z^5 + 1 = 0$ has five roots inside the circle $|z| = r$; if we put $P(z) = z^5 + 1$, $Q(z) = az$ we have on the circle $|z| = r$,

$$|Q(z)| = |a|r < r^5 - 1 < |z^5 + 1| = |P(z)|.$$

5. Cf. the proof of Ex. 3.

§ 5, 3, p. 559.

1. The left-hand side of the formula is the sum of the residues of the function $z^k/f(z)$, and is therefore equal to $\frac{1}{2\pi i}\int \frac{z^k}{f(z)}dz$ round a circle enclosing all the roots α_ν. But this integral tends to zero as the radius of the circle tends to infinity (the centre remaining fixed).

Miscellaneous Examples VIII (p. 567).

1. $\frac{z_1 - z_2}{z_2 - z_3}$ must be real.

2. $\Delta = \frac{z_1 - z_3}{z_2 - z_3} \Big/ \frac{z_1 - z_4}{z_2 - z_4}$ must be real. For if C is the circle through z_1, z_2, z_3, we may transform C by a linear transformation $\zeta = \frac{\alpha z + \beta}{\gamma z + \delta}$ into the real axis (cf. Ex. 6, p. 537); by Ex. 5, p. 537, Δ is unchanged; then a necessary condition that the image of z_4 shall lie on the same circle as the images of z_1, z_2, z_3 is that it is real, which is equivalent to Δ being real.

3. The equality to be proved is

$$\sqrt{|z_1 - z_2|\,|z_3 - z_4|} + \sqrt{|z_2 - z_3|\,|z_1 - z_4|} = \sqrt{|z_1 - z_3|\,|z_2 - z_4|}$$

or

$$1 + \sqrt{\left|\frac{(z_1 - z_2)(z_3 - z_4)}{(z_2 - z_3)(z_1 - z_4)}\right|} = \sqrt{\left|\frac{(z_1 - z_3)(z_2 - z_4)}{(z_2 - z_3)(z_1 - z_4)}\right|}.$$

Now the expressions under the square roots are invariant in a linear transformation (cf. Ex. 5, 6, p. 537). If by a suitable linear transformation we transform the circle into the real axis, we have only to prove the relation $AB \cdot CD + BC \cdot AD = AC \cdot BD$ for four points on a straight line, where it is trivial.

4. $\zeta = e^{iz}$ takes every value except $\zeta = 0$, as is easily seen from the relation $e^{iz} = e^{-y}(\cos x + i \sin x)$. Now we have to choose ζ so that

$$c = \cos z = \frac{1}{2}\left(\zeta + \frac{1}{\zeta}\right);$$

this quadratic equation always has a solution

$$\zeta = c \pm \sqrt{c^2 - 1},$$

and this solution is not zero, so that a corresponding z exists.

5. Cf. Ex. 4. If $\zeta = e^{iz}$, then

$$\tan z = \frac{1}{i} \frac{\zeta - \frac{1}{\zeta}}{\zeta + \frac{1}{\zeta}} = c$$

or

$$\zeta = \sqrt{\frac{1 + ic}{1 - ic}};$$

there is a finite $\zeta \neq 0$ only when $c \neq \pm i$; hence $\tan z = c$ only has a solution if c is neither $+i$ nor $-i$.

6. If $z = x + iy$, $\cos z$ is real if $x = \pi n$ or $y = 0$, and $\sin z = 0$ if $x = \pi n + \frac{\pi}{2}$ or $y = 0$ (where n is an integer).

7. (*a*) $r = 1$ (for $|z| > 1$ the individual terms tend to ∞, for $|z| < 1$ compare with the geometric series).

(*b*) $r = 0$. (*c*) $r = 1$.

8. Cf. Vol. I, p. 175.

9. (*a*) Integrate $\frac{e^{iz}}{1 + z^4}$ over upper semicircle:

$$\frac{\pi\sqrt{2}}{4} e^{-\frac{\sqrt{2}}{2}} \left(\sin \frac{\sqrt{2}}{2} + \cos \frac{\sqrt{2}}{2}\right).$$

(*b*) Integrate $\frac{z^2 e^{iz}}{1 + z^4}$ over upper semicircle:

$$\frac{\pi\sqrt{2}}{4} e^{-\frac{\sqrt{2}}{2}} \left(\cos \frac{\sqrt{2}}{2} - \sin \frac{\sqrt{2}}{2}\right).$$

(*c*) Integrate $\frac{e^{iz}}{q^2 + z^2}$ over upper semicircle: $\frac{\pi}{2q} e^{-q}$.

(*d*) Integrate $\frac{x^{\alpha - 1}}{(x + 1)(x + 2)}$ over a region bounded by a large circle about the origin and slit along the positive real axis: $\frac{\pi(2^{\alpha-1} - 1)}{\sin \pi\alpha}$.

10. (*a*) $+2\pi i$ at $z = 2n\pi$, $-2\pi i$ at $z = (2n + 1)\pi$.

(*b*) $+2\pi i$ at $z = 2n\pi + \frac{3\pi}{2}$, $-2\pi i$ at $z = 2n\pi + \frac{\pi}{2}$.

(c) Use the functional equation $\Gamma(z) = \Gamma(z+\nu+1)/z(z+1)\dots(z+\nu)$:

$$\frac{(-1)^n}{n!}\,2\pi i \text{ at } z = -n.$$

(d) $2\pi i$ at $z = n\pi i$.

11. Write $\dfrac{\cot \pi t}{t-z} = \dfrac{\cot \pi t}{t} + \dfrac{z \cot \pi t}{t(t-z)}$; $\cot \pi t$ is bounded on the squares C_n, and the integrals of $\dfrac{\cot \pi t}{t}$ over opposite sides of the square almost cancel one another; hence

$$\lim_{n\to\infty} \int_{C_n} \frac{\cot \pi t}{t-z}\,dt = \lim_{n\to\infty} \int_{C_n} \frac{z \cot \pi t}{t(t-z)}\,dt = 0$$

If we put together residues of opposite poles, the sum of the residues converges and we obtain $\cot \pi x = \dfrac{2x}{\pi}\left(\dfrac{1}{2x^2} + \dfrac{1}{x^2-1^2} + \dfrac{1}{x^2-2^2} + \dots\right)$ (cf. Vol. I, p. 444).

12.
$$\frac{1}{1+t} = 1 - t + t^2 - + \dots \pm t^{n-1} + (-1)^n \frac{t^n}{1+t}.$$

Hence

$$\log(1+z) = z - \frac{z^2}{2} + \frac{z^3}{3} - \dots \pm \frac{z^n}{n} + R_n,$$

where

$$R_n = (-1)^n \int_0^z \frac{t^n}{1+t}\,dt.$$

If we take $z = e^{i\theta}$ and the straight line from 0 to $e^{i\theta}$ as path of integration, we have, for $e^{i\theta} \neq -1$,

$$|R_n| = \left| \int_0^1 \frac{t^n}{1+e^{i\theta}t}\,dt \right| \leqq \frac{1}{m} \int_C^1 t^n\,dt = \frac{1}{m(n+1)},$$

where m denotes the minimum of $1 + e^{i\theta}$ for $0 \leqq t \leqq 1$. Hence if $z = e^{i\theta} \neq -1$, R_n tends to 0.

13. (a)
$$f(z) = \sum_{\nu=1}^{\infty} \left(\frac{1}{(2\nu-1)^z} - \frac{1}{(2\nu)^z} \right);$$

now

$$\frac{1}{(2\nu-1)^z} - \frac{1}{(2\nu)^z} = z \int_{2\nu-1}^{2\nu} \frac{1}{y^{z+1}}\,dy$$
$$\leqq \frac{|z|}{|(2\nu-1)^{z+1}|} = \frac{|z|}{(2\nu-1)^{1+x}},$$

and the series $\sum_\nu \dfrac{1}{(2\nu-1)^{1+x}}$ is absolutely convergent for $x > 0$.

(*b*) $(1 - 2^{1-z})\zeta(z)$

$$= 1 + \frac{1}{2^z} + \frac{1}{3^z} + \frac{1}{4^z} + \ldots - \frac{2}{2^z} - \frac{2}{4^z} - \frac{2}{6^z} - \ldots$$

$$= 1 - \frac{1}{2^z} + \frac{1}{3^z} - \frac{1}{4^z} + \ldots = f(z).$$

(*c*) $$\lim_{z\to 1} (z - 1)\zeta(z) = f(1) \, . \lim_{z\to 1} \frac{z-1}{1 - 2^{1-z}} = \frac{f(1)}{g'(1)} = 1,$$

where

$$g(z) = 1 - 2^{1-z}.$$

MISCELLANEOUS EXAMPLES

1. (*a*) If there were a linear relation $ax + by + cz = 0$, where e.g. $a \neq 0$, then by scalar multiplication of this relation by x we should get $axx + byx + czx = ax^2 = 0$; hence $a = 0$, since $x^2 \neq 0$.

(*b*) The relation $ax + by + cz = 0$ is equivalent to the system of linear equations for a, b, c,

$$ax_1 + by_1 + cz_1 = 0$$
$$ax_2 + by_2 + cz_2 = 0$$
$$ax_3 + by_3 + cz_3 = 0.$$

These equations have the unique solution $a = b = c = 0$, unless the determinant vanishes.

(*c*) The vector equation $v = ax + by + cz$ corresponds to three ordinary linear equations for a, b, c which certainly have a solution, since, by (*b*), the determinant is not zero.

2. Take a co-ordinate system Ox, Oy, Oz. Then (*a*) reduces to the multiplication theorem for determinants; (*b*) reduces to the identity,

$$\begin{vmatrix} x_1x_1' + x_2x_2' + x_3x_3' & x_1y_1' + x_2y_2' + x_3y_3' \\ y_1x_1' + y_2x_2' + y_3x_3' & y_1y_1' + y_2y_2' + y_3y_3' \end{vmatrix} = \begin{vmatrix} x_2x_3 \\ y_2y_3 \end{vmatrix} \times \begin{vmatrix} x_2'x_3' \\ y_2'y_3' \end{vmatrix}$$
$$+ \begin{vmatrix} x_3x_1 \\ y_3y_1 \end{vmatrix} \times \begin{vmatrix} x_3'x_1' \\ y_3'y_1' \end{vmatrix} + \begin{vmatrix} x_1x_2 \\ y_1y_2 \end{vmatrix} \times \begin{vmatrix} x_1'x_2' \\ y_1'y_2' \end{vmatrix},$$

which is easily verified by splitting up the left-hand determinant into a sum of nine determinants; (*c*) may be verified by calculating the components of x, y, z; (*d*) is an immediate consequence of (*c*) and 1(*b*), since by (*c*)

$$[x[yz]] + [y[zx]] + [z[xy]] = 0.$$

Finally, if x, y, z are vectors lying respectively in the three concurrent straight lines, then the plane through x which is perpendicular to y and z passes through x and $[yz]$, i.e. its normal has the direction of $[x[yz]]$; the three normals lie in one plane, hence the planes pass through one line.

4. A rotation of $Ox'y'$ through the angle ψ leads to a new system $Ox''y''$. A direct passage from Oxy to $Ox''y''$ gives the desired result.

5. (*a*) In the co-ordinate system Ox, Oy, Oz take the vectors $(\alpha_1, \beta_1, \gamma_1)$, $(\alpha_2, \beta_2, \gamma_2)$, $(\alpha_3, \beta_3, \gamma_3)$. If the determinant is orthogonal, the vectors will form a new orthogonal co-ordinate system Ox', Oy', Oz'.

(*b*) The passage from the system Ox', Oy', Oz' to the system Ox, Oy, Oz is given by the determinant

$$\begin{vmatrix} \alpha_1 & \alpha_2 & \alpha_3 \\ \beta_1 & \beta_2 & \beta_3 \\ \gamma_1 & \gamma_2 & \gamma_3 \end{vmatrix},$$

which again must be orthogonal.

6. Pass from Ox, Oy, Oz to Ox', Oy', Oz' by the following three rotations: (1) Rotate Ox, Oy, Oz through the angle φ about Oz, so as to form the new system Ox_1, Oy_1, Oz_1 ($Oz = Oz_1$). (2) Rotate Ox_1, Oy_1, Oz_1 through θ about Ox_1, obtaining Ox_2, Oy_2, Oz_2 ($Ox_1 = Ox_2$, $Oz_2 = Oz'$). (3) Rotate Ox_2, Oy_2, Oz_2 through ψ about Oz_2, obtaining Ox', Oy', Oz'. In each of these steps the change of variables is to be performed according to Ex. 3. Finally, eliminate the intermediate variables x_1, y_1, z_1, x_2, y_2, z_2; this is best done by multiplying, in the correct order, the three determinants corresponding to the above rotations.

7. Note that $\cos xOx' = \cos\varphi\cos\psi - \sin\varphi\sin\psi\cos\theta$.

8. If a is a unit vector in the direction of the normal to the plane and b a unit vector lying in the straight line, then $\frac{\pi}{2} - \varphi$ is the angle between a and b. It follows that

$$\varphi = \text{arc sin}\frac{A\alpha + B\beta + C\gamma}{\sqrt{(A^2 + B^2 + C^2)(\alpha^2 + \beta^2 + \gamma^2)}}.$$

9. $x = 3$, $y = 2$, $z = 1$.

11. $$D = \begin{vmatrix} \cos\theta & -\sin\theta & 0 \\ \sin\theta & \cos\theta & 0 \\ 0 & 0 & 1 \end{vmatrix} \times \begin{vmatrix} \cos\alpha & \cos\beta & \cos\gamma \\ \sin\alpha & \sin\beta & \sin\gamma \\ \sin(\beta - \gamma) & \sin(\gamma - \alpha) & \sin(\alpha - \beta) \end{vmatrix};$$

the first factor is equal to unity.

12. Adding the third and second column to the first, dropping the factor $A + 2B$, and subtracting the first row from the second and third row, we have $D = (A + 2B)(B - A)^2$

$$= \{(x + y + z)(x^2 + y^2 + z^2 - xy - xz - yz)\}^2.$$

13. In order to see that the determinant represents a linear function, subtract the first column from the other columns. By substituting $x = -a$ or $x = -b$ in Δ, we get A and B.

14. As $uv = 1$, Leibnitz's rule (cf. Vol. I, p. 202) gives

$$\begin{aligned} u'v + uv' &= 0 \\ u''v + 2u'v' + uv'' &= 0 \\ u'''v + 3u''v' + 3u'v'' &= -uv'''. \end{aligned}$$

These equations, considered as linear equations for v, v', v'', have the determinant $-D$. If we solve the equations for v by the rule given on p. 25, we have

$$v = -\frac{1}{D}\begin{vmatrix} 0 & u & 0 \\ 0 & 2u' & u \\ -uv''' & 3u'' & 3u' \end{vmatrix} = u^3v'''/D,$$

i.e. $v''' = Dv/u^3 = D/u^4$.

15. (*b*) Put $z = \log u$; for z we then have the equation $z_{xy} = 0$, i.e. z_x does not depend on y. Let $z_x = \varphi(x)$; then

$$z = \int^x \varphi(x)\,dx + \psi(y).$$

If we put

$$e^{\int^x \phi(x)\,dx} = f(x), \quad e^{\psi(y)} = g(y),$$

then

$$u = e^z = f(x) \times g(y).$$

17. Differentiate $F(u_x, u_y) = 0$ with respect to x and y.

18. u is of the form

$$u = xf\left(\frac{x}{y}\right) + g\left(\frac{x}{y}\right).$$

19. (*a*) $|f(x+h, y+k) - f(x, y)|$

$$= \left|\frac{2hx + 4ky + h^2 + 2k^2}{\sqrt{\{1 + (x+h)^2 + 2(y+k)^2\}} + \sqrt{\{1 + x^2 + 2y^2\}}}\right|$$

$$\leqq |2hx + 4ky + h^2 + 2k^2|$$
$$\leqq 2\,|h^2 + k^2 + 2hx + 2ky|$$
$$\leqq 2(h^2 + k^2 + 2\sqrt{(h^2 + k^2)}\sqrt{(x^2 + y^2)})$$
$$\leqq 2\sqrt{(h^2 + k^2)}\,\{1 + 2\sqrt{x^2 + y^2}\}$$

if we assume that $h^2 + k^2 < 1$. Thus e.g.

$$|f(x+h, y+k) - f(x, y)| < \varepsilon$$

for

$$\sqrt{(h^2 + k^2)} \leqq \frac{\varepsilon}{2 + 4\sqrt{(x^2 + y^2)}}.$$

20. Let $x = at$, $y = bt$; then

$$\lim_{\to 0} f(x, y) = \lim_{t \to 0} a^4b^4\frac{t^2}{(a^2 + b^4t^2)^3} = 0,$$

$$\lim_{t \to 0} g(x, y) = \lim_{t \to 0}\frac{a^2t^2}{a^2t^2 + b^2t^2 - at} = 0,$$

but if (x, y) approaches the origin along the parabola $y^2 = x$, $f(x, y) = \frac{1}{8}$, $g(x, y) = 1$.

21. Let C be given by the equations $x = x(t)$, $y = y(t)$, where $x(t)$ and

$y(t)$ have continuous derivatives. Let two points on C correspond to t_1 and t_2. Applying the mean value theorem, we get

$$l = \int_{t_1}^{t_2} \sqrt{\dot{x}^2 + \dot{y}^2}\, dt = (t_2 - t_1)\sqrt{\dot{x}(\tau_1)^2 + \dot{y}(\tau_1)^2},$$

$$d = \sqrt{[x(t_2) - x(t_1)]^2 + [y(t_2) - y(t_1)]^2}$$
$$= (t_2 - t_1)\sqrt{\dot{x}(\tau_2)^2 + \dot{y}(\tau_3)^2},$$

where τ_1, τ_2, τ_3 lie between t_1 and t_2;

$$d - l = o(t_2 - t_1),$$

since

$$\sqrt{\dot{x}(\tau_1)^2 + \dot{y}(\tau_1)^2} - \sqrt{\dot{x}(\tau_2)^2 + \dot{y}(\tau_3)^2} \to 0 \text{ as } t_2 - t_1 \to 0.$$

22. As the series has positive terms, it is sufficient to prove its convergence and find its sum for any order of the terms. Put $a + b = n$; then

$$S = \sum_{n=0}^{\infty} \sum_{a=0}^{n} \binom{n}{a} \frac{a}{x^a y^{n-a}} = \sum_{n=0}^{\infty} \frac{1}{y^n} \sum_{a=0}^{n} \binom{n}{a} a \left(\frac{y}{x}\right)^a$$
$$= \sum_{n=0}^{\infty} \frac{1}{y^n} \frac{y}{x} n \left(1 + \frac{y}{x}\right)^{n-1},$$

since the relation

$$\sum_{a=0}^{n} \binom{n}{a} a z^a = nz(1 + z)^{n-1}$$

holds. (This may be proved by differentiating the identity

$$\sum_{a=0}^{n} \binom{n}{a} z^a = (1 + z)^n).$$

Thus

$$S = \frac{1}{x} \sum_{n=0}^{\infty} n \left(\frac{1}{y} + \frac{1}{x}\right)^{n-1}$$
$$= \frac{1}{x} \frac{1}{\left(1 - \frac{1}{x} - \frac{1}{y}\right)^2}.$$

24. If dots denote differentiation with respect to the length of arc s, we have

$$k = \sqrt{\ddot{x}^2}.$$

Now

$$\ddot{x} = x''\dot{t}^2 + x'\ddot{t},$$

$$\dot{t} = \frac{1}{\sqrt{x'^2}}, \quad \ddot{t} = \frac{d}{dt}(\dot{t})\frac{dt}{ds} = -\frac{(x'x'')}{(x'^2)^2};$$

hence

$$\ddot{x} = \frac{x''}{x'^2} - \frac{(x'x'')}{(x'^2)^2} x',$$

$$\ddot{x}^2 = \frac{x''^2 x'^2 - (x'x'')^2}{(x'^2)^3}.$$

25. By definition

$$x'^2 + y'^2 = 1;$$

hence, by differentiating, we get

$$x'x'' + y'y'' = 0. \quad \ldots\ldots\ldots \quad (a)$$

If we put

$$x'y'' - x''y' = \gamma,$$

we have

$$x'' = -\gamma y', \quad y'' = \gamma x',$$

and hence

$$x''^2 + y''^2 = \gamma^2.$$

Now as $z' = a$, $z'' = 0$, the osculating plane (cf. Ex. 1, p. 93) is given by

$$-ay''(\xi - x) + (\eta - y)ax'' + (\zeta - z)\gamma = 0; \quad \ldots \quad (b)$$

it obviously contains the normal of the cylinder, given by

$$(\xi - x)x' + (\eta - y)y' = 0, \ \zeta = z.$$

By (a) the curvature (cf. Ex. 24) is given by

$$k^2 = \frac{(x'^2 + y'^2 + a^2)(x''^2 + y''^2)}{(x'^2 + y'^2 + a^2)^3} = \frac{\gamma^2}{(1 + a^2)^2}.$$

By (b) the binormal vector (cf. Ex. 7, p. 94) is given by

$$\left(\frac{-ay''}{\sqrt{a^2(x''^2 + y''^2) + \gamma^2}}, \quad \frac{ax''}{\sqrt{a^2(x''^2 + y''^2) + \gamma^2}}, \quad \frac{\gamma}{\sqrt{a^2(x''^2 + y''^2) + \gamma^2}}\right),$$

or

$$\left(\frac{-ay''}{\gamma\sqrt{1 + a^2}}, \quad \frac{ax''}{\gamma\sqrt{1 + a^2}}, \quad \frac{1}{\sqrt{1 + a^2}}\right),$$

or

$$\left(\frac{-ax'}{\sqrt{1 + a^2}}, \quad \frac{-ay'}{\sqrt{1 + a^2}}, \quad \frac{1}{\sqrt{1 + a^2}}\right).$$

Since τ is the length of the derivative of this vector with respect to the length of arc (the element of which is $(1 + a^2)ds$), we have

$$\tau^2 = \frac{a^2}{(1 + a^2)^2}(x''^2 + y''^2) = \frac{\gamma^2 a^2}{(1 + a^2)^2}.$$

26. Cf. Ex. 1, p. 93. The equation of the osculating plane is

$$f + f'' = \xi(f'' \cos\theta + f' \sin\theta) + \eta(f'' \sin\theta - f' \cos\theta) + \zeta;$$

the distance of the plane from the origin is

$$(f + f'')/\sqrt{(1 + f'^2 + f''^2)},$$

which reduces to $\sqrt{(1 + 1/A^2)}$ in the special case.

27. Cf. Ex. 24.

28. (*a*) According to Ex. 3, p. 19, the plane is given by

$$\begin{vmatrix} \frac{1}{3}at_1^3 - x & \frac{1}{2}bt_1^2 - y & ct_1 - z \\ \frac{1}{3}at_2^3 - x & \frac{1}{2}bt_2^2 - y & ct_2 - z \\ \frac{1}{3}at_3^3 - x & \frac{1}{2}bt_3^2 - y & ct_3 - z \end{vmatrix} = 0.$$

(*b*) By Ex. 1, p. 93, the osculating plane at the point t is the limit of the plane through three points which tend towards the same point, and is therefore given by

$$t^3 - \frac{3t^2z}{c} + \frac{6ty}{b} - \frac{3x}{a} = 0.$$

At the point of intersection (x, y, z) of the osculating planes at t_1, t_2, t_3 this equation must be satisfied for $t = t_1$ and $t = t_2$ and $t = t_3$. Hence t_1, t_2, t_3 are the three roots of the equation above. Therefore

$$\frac{3z}{c} = t_1 + t_2 + t_3,$$

$$\frac{6y}{b} = t_1t_2 + t_1t_3 + t_2t_3,$$

$$\frac{3x}{a} = t_1t_2t_3.$$

These expressions for x, y, z satisfy the equation of the plane in (*a*).

29. If b, c are kept fixed and a alone varies, we have $s = \frac{1}{2}bc \sin A$, $ds = \frac{1}{2}bc \cos A\,dA$. From $a^2 = b^2 + c^2 - 2bc \cos A$, we have by differentiation $a\,da = bc \sin A\,dA$; hence

$$ds = \frac{a}{2 \sin A} \cos A\,da = R \cos A\,da.$$

30. Denote the components of the vector AP by x, y, z. Then

$$AP = \sqrt{(x^2 + y^2 + z^2)}, \quad d(AP) = \frac{x\,dx + y\,dy + z\,dz}{\sqrt{(x^2 + y^2 + z^2)}} = -\boldsymbol{a} \cdot dP.$$

31. Using a self-evident notation, we have

$$P = A - PA.\boldsymbol{a},\ \dot{P} = -PA.\dot{\boldsymbol{a}} - \frac{d}{dt}(PA).\boldsymbol{a}, \text{ or } PA.\dot{\boldsymbol{a}} = -\frac{d}{dt}(PA).\boldsymbol{a} - \dot{P}.$$

By Ex. 30,

$$\frac{d}{dt}(PA) = -\boldsymbol{a}\dot{P} = -\boldsymbol{a}(\boldsymbol{a}u + \boldsymbol{b}v + \boldsymbol{c}w) = -u - (\boldsymbol{ab})v - (\boldsymbol{ac})w.$$

Now

$$\begin{aligned} PA\dot{\boldsymbol{a}} &= \boldsymbol{a}u + (\boldsymbol{ab})v\boldsymbol{a} + (\boldsymbol{ac})w\boldsymbol{a} - \boldsymbol{a}u - \boldsymbol{b}v - \boldsymbol{c}w \\ &= [(\boldsymbol{ab})v + (\boldsymbol{ac})w]\boldsymbol{a} - v\boldsymbol{b} - w\boldsymbol{c}. \end{aligned}$$

32. $\dot{P} = \boldsymbol{a}u + \boldsymbol{b}v + \boldsymbol{c}w$, hence $\ddot{P} = \dot{\boldsymbol{a}}u + \dot{\boldsymbol{b}}v + \dot{\boldsymbol{c}}w + \boldsymbol{a}\dot{u} + \boldsymbol{b}\dot{v} + \boldsymbol{c}\dot{w}$. Introducing the expression for $\dot{\boldsymbol{a}}$ from the previous example and the similar expressions for $\dot{\boldsymbol{b}}$ and $\dot{\boldsymbol{c}}$, we get the required expression for $\ddot{P}$.

33. If $(x^2/a^2) \pm (y^2/b^2) = 1$ is the equation of the conic, then $(x^2 + y^2)^2 = 4(a^2x^2 \pm b^2y^2)$ is the equation of the envelope. Note that if the conic is a rectangular hyperbola this envelope is an ordinary lemniscate $(x^2 + y^2)^2 = 4a^2(x^2 - y^2)$.

35. If P describes the pedal curve Γ' of Γ, construct on OP as diameter a circle in the plane perpendicular to the plane of Γ; the envelope is the surface generated by this variable circle.

37. An ellipse.

38. A plane touching the parabolas has an equation of the form

$$-c^2x + cy + cz = 1 \text{ or } -c^2x + cy - cz = 1.$$

The corresponding envelopes are

$$(y + z)^2 = 4x \text{ and } (y - z)^2 = 4x.$$

39. The proof resembles that for $n = 2$ (Appendix to Chap. III, § 1, p. 204). A positively definite quadratic form $\Sigma a_{ik}x_ix_k$ can be brought by a suitable transformation $x_i = \sum_{k=1}^{n} c_{ik}y_k$ $(i = 1, \ldots, n)$ with a non-vanishing determinant into the form $\Sigma a_{ik}x_ix_k = y_1^2 + y_2^2 + \ldots + y_n^2 > m(x_1^2 + \ldots + x_n^2)$, where m is a suitable positive constant. For the applications it is important to remember that a necessary and sufficient condition that a form $\Phi = \Sigma a_{ik}x_ix_k$ shall be positively definite is that its principal first minors of order 1, 2, . . . , n, as indicated below,

$$\begin{vmatrix} a_{11} & a_{12} & a_{13} & \cdots & a_{1n} \\ a_{21} & a_{22} & a_{23} & & \vdots \\ a_{31} & a_{32} & a_{33} & & \vdots \\ \vdots & & & & \vdots \\ a_{n1} & \cdots & \cdots & \cdots & a_{nn} \end{vmatrix}$$

shall all be positive. Φ is negatively definite if $-\Phi$ is positively definite.

40. Sketch the curve $f = 0$ and investigate the sign of f throughout the plane.

41. If $P_i = (x_i, y_i)$, $r_i = PP_i$, we have

$$d^2f = \sum_{1}^{3} d^2r_i = \sum_{i=1}^{3} r_i^{-3}[(y - y_i)dx - (x - x_i)dy]^2,$$

which is positively definite.

42. At the point P_1. Note that the function $f = r_1 + r_2 + r_3$ is continuous in the whole plane, but not differentiable at the points P_1, P_2, P_3, where it has *conical points* (like the function $z = \sqrt{(x - x_1)^2 + (y - y_1)^2}$, which geometrically represents a circular cone). Investigate the derivative of f at P_1 in all directions round this point.

43. According to the first rule we have to compute d^2f from (3), with

$dx_1, \ldots, dx_m, d^2x_1, \ldots, d^2x_m$ substituted from (1). Note that (1) implies that

$$d^2\varphi_\mu = \Sigma\varphi_{\mu x_i x_k} dx_i dx_k + \varphi_{\mu x_1} d^2x_1 + \ldots + \varphi_{\mu x_m} d^2x_m = 0 \; (\mu = 1, \ldots, m);$$

if this is multiplied by λ_μ and added to (3) for all values of μ, we have $d^2f = d^2F = \Sigma F_{x_i x_k} dx_i dx_k$, because $d^2x_1, \ldots, d^2x_m$ drop out on account of the relations (2).

44. For $F = f + \lambda\varphi$ (disregarding a positive factor) we get

$$d^2F = \sum_{i,k}^{1,n} dx_i dx_k, \text{ with } d\varphi = dx_1 + \ldots + dx_n = 0.$$

Eliminating dx_n, we have to show that the quadratic form

$$-d^2F = (dx_1 + \ldots + dx_{n-1})^2 - \sum_{i,k}^{1,n-1} dx_i dx_k = \sum^{1,n-1} dx_i^2 + \sum_{i,k}^{1,n-1} dx_i dx_k$$

is positively definite.

46. The co-ordinate axes.

47. $y = x^2(1 \pm x^{\frac{1}{2}})$. The two branches of the curve forming the cusp at the origin lie on the same side of their common tangent.

48. (a) If we put $f = lx + my + nz$, $\varphi = x^p + y^p + z^p - c^p$, $F = f - \lambda\varphi$, then the conditions for stationary values are

$$l = \lambda p x^{p-1}, \quad m = \lambda p y^{p-1}, \quad n = \lambda p z^{p-1}. \quad . \quad . \quad (A)$$

Multiplying these equations by x, y, z respectively and adding, we have

$$lx + my + nz = \lambda p c^p. \quad . \quad . \quad . \quad . \quad . \quad . \quad . \quad (B)$$

Calculating x, y, z from (A) and substituting in $\varphi = 0$, we get

$$\lambda p = (l^q + m^q + n^q)^{1/q} c^{1-p}.$$

Substitution of this expression for λp in (B) gives the stationary value.

(b) Cf. Ex. 43. Here we have

$$d^2F = -\lambda p(p-1)(x^{p-2}dx^2 + y^{p-2}dy^2 + z^{p-2}dz^2);$$

as $\lambda p > 0$, this quadratic form is positively or negatively definite according as $p \lesseqgtr 1$.

49. Minimum for $x = 1$, $y = 4$, saddle point for $x = -1$, $y = 2$.

50. Let AB touch Σ at P. Let $A'B'$ be another tangent, and let the new point of contact be P'. Then if $d\varphi$ is the angle between AB and $A'B'$, and we neglect terms of the second order, the difference of the areas of $A'B'C$ and ABC is

$$\Delta S = \frac{d\varphi}{2}(\overline{AP}^2 - \overline{BP}^2).$$

For the triangle of least area $\Delta S = 0$, that is, $AP = BP$.

51. Apply the transformation

$$x' = x\cos\alpha - y\sin\alpha, \quad y' = x\sin\alpha + y\cos\alpha.$$

52. Let S denote the curve $f(x, y) = C$ and S' the curve $\varphi(x, y) = C'$. S and S' have a point of contact in (a, b). In general, $f(x, y) - C$ is positive on one side of S and negative on the other side in some neighbourhood; similarly with $\varphi(x, y) - C'$ and S'. If e.g. $f(a, b)$ is a maximum of f, then $f(x, y) - C \leqq 0$ on S', i.e. S' is wholly on one side of S; then S is also on one side of S'. That is, $\varphi(x, y) - C'$ has a constant sign on S and as it is equal to zero at (a, b), it has either a maximum or a minimum there.

53. The equation of the generating tangent is

$$x\sin\theta + y\cos\theta = a(\theta\sin\theta + \cos\theta - 1).$$

55. $$\frac{d}{dx}f(x) = \frac{2}{x}\int_0^{\frac{\pi}{2}}\left(1 - \frac{1}{1 - x^2\cos^2\theta}\right)d\theta;$$

since

$$\int\frac{1}{1 - x^2\cos^2\theta}\,d\theta = \frac{1}{\sqrt{1 - x^2}}\arctan\left(\frac{\tan\theta}{\sqrt{1 - x^2}}\right),$$

we have

$$\frac{d}{dx}f(x) = \frac{\pi}{x}\left(1 - \frac{1}{\sqrt{1 - x^2}}\right),$$

and therefore

$$f(x) = \pi\log(1 + \sqrt{1 - x^2}) - \pi\log 2.$$

56. According to p. 273,

$$\Sigma = \iint\sqrt{EG - F^2}\,dr\,d\theta$$

$$= \int_{\theta_1}^{\theta_2}d\theta\int_0^{f'(\theta)}\sqrt{r^2 + f'^2}\,dr$$

$$= [\sqrt{2} + \log(1 + \sqrt{2})]\int_{\theta_1}^{\theta_2}\tfrac{1}{2}f'^2 d\theta \quad \text{(cf. Vol. I, p. 215)},$$

which is $[\sqrt{2} + \log(1 + \sqrt{2})]$ times the area of the projection

$$\theta_1 \leqq \theta \leqq \theta_2, \quad O \leqq r \leqq f'(\theta).$$

57. As $A - BR^2 = \frac{5}{2}$, $A - \frac{3}{5}BR^2 = \frac{11}{2}$, we have $A = 10$, $B = \frac{15}{2}/R^2$. The attraction at an internal point is equal to the attraction of the total mass of the points inside of the sphere of radius r concentrated at the centre of the sphere.

58. By translation we can ensure that the triangle lies in the upper half-plane. Then its moment of inertia is equal to

$$\varphi(x_1y_1,\ x_2y_2) + \varphi(x_2y_2,\ x_3y_3) + \varphi(x_3y_3,\ x_1y_1),$$

where $\varphi(x_1y_1,\ x_2y_2)$ denotes the moment of inertia of the quadrilateral with vertices $(x_1,\ 0)$, $(x_1,\ y_1)$, $(x_2,\ y_2)$, $(x_2,\ 0)$ multiplied by the sign of $(x_1 - x_2)$. Then show that

$$\varphi(x_1y_1,\ x_2y_2) = \tfrac{1}{12}\,(x_1 - x_2)(y_1^3 + y_1^2y_2 + y_1y_2^2 + y_2^3).$$

60. $2 - \dfrac{\pi}{2}$.

61. Introduce polar co-ordinates with the pole as origin.

62. $I = \displaystyle\int_1^2 (y-4)\,dy \int_{(8y-20)/(y-4)}^{4/y} dx = 12 - 16\log 2.$

63. (*a*) $K = \displaystyle\int_0^\beta d\theta \int_0^a r \log r^2\,dr.$

(*b*) $K = \displaystyle\int_0^a \left\{\int_0^{\phi(x)} \log(x^2+y^2)\,dy\right\} dx,$

where

$$\varphi(x) = x\tan\beta \text{ for } 0 \leqq x \leqq a\cos\beta;\ \sqrt{a^2 - x^2} \text{ for } a\cos\beta \leqq x \leqq a.$$

64. $V = \frac{7}{72}\,\pi h^3 \tan^2\alpha.$

For $V = \displaystyle\iint z\,d\sigma = \iint \left(h - \frac{r}{\tan\alpha}\right) d\sigma$, where the integral is to be extended over the region

$$h\tan\alpha(1 - \sin^{4/3}\theta\cos^{2/3}\theta) \leqq r \leqq h\tan\alpha.$$

That is, $V = \displaystyle\int_0^{2\pi} d\theta \int_{h\tan\alpha(1-\sin^{4/3}\theta\cos^{2/3}\theta)}^{h\tan\alpha} \left(h - \frac{r}{\tan\alpha}\right) r\,dr$

$$= h^3\tan^2\alpha \int_0^{2\pi} \tfrac{1}{2}\sin^{8/3}\theta\cos^{4/3}\theta\,d\theta - h^3\tan^2\alpha \int_0^{2\pi} \tfrac{1}{3}\sin^4\theta\cos^2\theta\,d\theta;$$

if we substitute $\sin^2\theta = y$ in the first integral, it becomes

$$\int_0^1 y^{5/6}(1-y)^{1/6}\,dy = B(\tfrac{11}{6}, \tfrac{7}{6}) = \frac{\Gamma(\frac{11}{6})\,\Gamma(\frac{7}{6})}{\Gamma(3)}$$

$$= \tfrac{5}{72}\,\Gamma(\tfrac{1}{6})\,\Gamma(\tfrac{5}{6}) = \tfrac{5}{36}\,\pi,$$

where we have made use of the extension theorem for the gamma function (cf. pp. 335 and 337).

65. The generators are the lines of the surface given by $x =$ const., or by $y =$ const. Thus, as $dS = (1 + z_x^2 + z_y^2)^{1/2} dx\,dy$,

$$I = -\int_0^\eta dy \int_0^\xi dx \frac{1}{(1 + x^2 + y^2)^{3/2}} = -\int_0^\eta dy \frac{\xi}{(1 + y^2)(1 + \xi^2 + y^2)^{1/2}}$$

$$= -\text{arc tan} \frac{\xi\eta}{(1 + \xi^2 + \eta^2)^{1/2}}.$$

66. $$\frac{d}{da} K(a) = \int_0^\pi \frac{d}{da} \left(\frac{\log(1 + a \cos x)}{\cos x}\right) dx = \int_0^\pi \frac{dx}{1 + a \cos x} = \frac{\pi}{\sqrt{1 - a^2}};$$

thus $K(a) = \pi$ arc sin a + const.; the constant is determined from the condition $K(0) = 0$.

67. Introduce new variables u and v by the equations $u = x^3/y$, $v = y^2/x$. The area then becomes

$$\int_{a^3}^{b^3} \int_c^d \frac{\partial(x, y)}{\partial(u, v)} du\,dv = \tfrac{1}{5} \int_{a^3}^{b^3} u^{-2/5}\,du \int_c^d v^{-1/5}\,dv$$

(cf. p. 253).

68. Take a co-ordinate system Ox_1, Ox_2, Ox_3, and denote the position vector of a variable point on Γ by $\boldsymbol{x}$. Then $\boldsymbol{a} = \frac{1}{2}\int_\Gamma \boldsymbol{x} \times d\boldsymbol{x}$ has the required properties, for $a_{x_3} = \frac{1}{2}\int_\Gamma (x_1\,dx_2 - x_2\,dx_1)$ is the area of the projection of Γ on the plane Ox_1x_2.

69. The motion takes place in a plane, since p is a central force proved for the case $p = 1/r^2$ on pp. 423–4). Hence

$$\ddot{x} = -\frac{x}{r} p,$$

$$\ddot{y} = -\frac{y}{r} p.$$

It follows that

$$x\dot{y} - \dot{x}y = \text{const.} = h,$$

$$\ddot{x}\dot{x} + \ddot{y}\dot{y} = \frac{-x\dot{x} - y\dot{y}}{r} p = -\dot{r}p.$$

Hence

$$\tfrac{1}{2} \frac{d}{dt} (\dot{x}^2 + \dot{y}^2) = -\dot{r}p.$$

The distance of the tangent from the origin is

$$q = \frac{|x\dot{y} - \dot{x}y|}{\sqrt{\dot{x}^2 + \dot{y}^2}} = \frac{h}{\sqrt{\dot{x}^2 + \dot{y}^2}};$$

therefore

$$\tfrac{1}{2}\frac{d}{dt}\frac{h^2}{q^2} = -p\frac{dr}{dt}$$

or

$$\tfrac{1}{2}\frac{d}{dr}\frac{h^2}{q^2} = -p$$

which proves the first statement. For the cardioid we have $q = r^2/\sqrt{2ar}$

70. Let $(x_1, y_1), \ldots, (x_n, y_n)$ be the attracting particles. Then the resultant force at a point (x, y) has the components

$$X = \sum_\nu \frac{x - x_\nu}{\sqrt{\{(x - x_\nu)^2 + (y - y_\nu)^2\}}}, \quad Y = \sum_\nu \frac{y - y_\nu}{\sqrt{\{(x - x_\nu)^2 + (y - y_\nu)^2\}}}.$$

If we introduce the complex quantities $z_1 = x_1 + iy_1, \ldots, z_n = x_n + iy_n$, $z = x + iy$, $Z = X + iY$, we have

$$Z = \sum_\nu \frac{1}{\bar{z} - \bar{z}_\nu} = \overline{\frac{f'(z)}{f(z)}},$$

where $f(z)$ denotes the polynomial $(z - z_1) \ldots (z - z_n)$ and $\bar{z}$ the complex quantity conjugate to z. The positions of equilibrium correspond to $Z = 0$, i.e. to the zeros of the polynomial $f'(z)$, of which there are $n - 1$ at most.

Positions of equilibrium in the particular case: $(0, 0)$, $(\sqrt{(a^2 - b^2)}, 0)$, $(-\sqrt{(a^2 - b^2)}, 0)$.

71. By definition

$$\begin{aligned} \ddot{x} &= -\lambda^2 x - 2\mu\dot{y} \\ \ddot{y} &= -\lambda^2 y + 2\mu\dot{x}. \end{aligned} \qquad \ldots\ldots\ldots \quad (A)$$

Or differentiating the two equations twice and combining them we get an equation involving x only,

$$\ddddot{x} + (2\lambda^2 + 4\mu^2)\ddot{x} + \lambda^4 x = 0,$$

and a corresponding equation involving y only,

$$\ddddot{y} + (2\lambda^2 + 4\mu^2)\ddot{y} + \lambda^4 y = 0.$$

Thus x and y are linear combinations of $e^{\pm i(\mu \pm \sqrt{\lambda^2+\mu^2}\,t)}$ (cf. Ex. 1, p. 444), or of $\cos(\mu + \sqrt{\lambda^2 + \mu^2})t$, $\cos(\mu - \sqrt{\lambda^2 + \mu^2})t$, $\sin(\mu + \sqrt{\lambda^2 + \mu^2})t$, $\sin(\mu - \sqrt{\lambda^2 + \mu^2})t$, with constant coefficients a, b, c, d, and a', b', c', d'. From (A) it follows that $a' = -c$, $b' = -d$, $c' = a$, $d' = b$. Using the initial conditions $x(0) = y(0) = \dot{y}(0) = 0$, $\dot{x}(0) = u$, we obtain the result given.

72. (*b*) The equation becomes of the form treated in (*a*) if we multiply it by x^3. It has the particular solutions $u = x^3$ and $v = x^5$; hence, by (*a*), a third solution is given by $w = 1 + x^2$; the general solution is then

$$A(1 + x^2) + Bx^3 + Cx^5.$$

73. The curve satisfies the differential equation

$$n\left(x \frac{dy}{dx} - y\right) = r,$$

or in polar co-ordinates r, θ, with θ as independent variable,

$$\frac{nr^2}{\cos\theta \dfrac{dr}{d\theta} - r\sin\theta} = r;$$

that is,

$$\frac{d\log r}{d\theta} = \frac{n}{\cos\theta} + \tan\theta,$$

whence

$$r = a \frac{\left[\tan\left(\dfrac{\theta}{2} + \dfrac{\pi}{4}\right)\right]^n}{\cos\theta} = a \frac{(1 + \sin\theta)^n}{\cos^{n+1}\theta}$$

(cf. Vol. I, pp. 214–5).

74. According to p. 482, a solution of the first equation is of the form

$$z = f(x + at) + g(x - at).$$

On substituting this expression in the second equation we have

$$f'g' = 0,$$

i.e. either $f =$ const. or $g =$ const. Hence $z = f(x + at)$ or $z = f(x - at)$ is the most general solution of both equations.

75. Put $u = (x^2 + y^2 + z^2)^{n/2}$ and let K be of degree h. Then

$$\Delta u = u_{xx} + u_{yy} + u_{zz} = n(n + 1)(x^2 + y^2 + z^2)^{\frac{n-2}{2}},$$

$$x\frac{\partial K}{\partial x} + y\frac{\partial K}{\partial y} + z\frac{\partial K}{\partial z} = hK \qquad \text{(cf. p. 109).}$$

Hence $u = (x^2 + y^2 + z^2)^{-\frac{1+h}{2}}$ is a solution.

76. (*a*) The value of the integral round the small circular detour tends to zero as the circle becomes smaller. If we put $z = e^{i\theta}$ on the unit circle and $z = x$, $z = iy$ respectively on the axes, Cauchy's theorem gives

$$0 = \int_0^1 \left(x + \frac{1}{x}\right)^m x^{n-1}\,dx + i\int_0^{\frac{\pi}{2}} (e^{i\theta} + e^{-i\theta})^m e^{i\theta n}\,d\theta$$

$$-i\int_0^1 \left(iy + \frac{1}{iy}\right)^m (iy)^{n-1}\,dy$$

$$= \int_0^1 \left(x + \frac{1}{x}\right)^m x^{n-1}\,dx + i\,.\,2^m \int_0^{\frac{\pi}{2}} \cos^m\theta e^{in\theta}\,d\theta$$

$$-e^{\frac{i\pi(n-m)}{2}} \int_0^1 \left(-y + \frac{1}{y}\right)^m y^{n-1}\,dy;$$

by equating the imaginary parts of this equation, we get

$$2^m \int_0^{\frac{\pi}{2}} \cos^m\theta \cos n\theta\,d\theta = \sin\frac{\pi(n-m)}{2} \int_0^1 \left(-y + \frac{1}{y}\right)^m y^{n-1}\,dy$$

$$= \tfrac{1}{2}\sin\frac{\pi(n-m)}{2} \int_0^1 (1-\eta)^m \eta^{(n-m-2)/2}\,d\eta$$

$$= \tfrac{1}{2}\sin\frac{\pi}{2}(n-m)\,B\left(m+1, \frac{n-m}{2}\right) \qquad \text{(cf. p. 337).}$$

(*b*) Use the relation

$$\sin\frac{(n-m)\pi}{2}\,\Gamma\left(\frac{n-m}{2}\right) = \frac{\pi}{\Gamma\left(1 - \frac{n-m}{2}\right)} \qquad \text{(cf. p. 335).}$$

77. If $x \neq 0$ and if C' is a contour in the region in which f is regular, and contains y but not 0, then, by p. 549,

$$\frac{d^n}{dy^n}\frac{yf(y)}{(y+a)^{n+1}} = \frac{n!}{2\pi i}\int_{C'} \frac{tf(t)}{(t+a)^{n+1}(t-y)^{n+1}}\,dt.$$

If we put $a = y = \sqrt{x}$ the latter integral becomes

$$\frac{n!}{2\pi i}\int_{C'} \frac{tf(t)}{(t^2 - x)^{n+1}}\,dt.$$

If we then substitute $t^2 = \tau$, the integral becomes

$$\frac{n!}{4\pi i}\int_{C} \frac{f(\sqrt{\tau})}{(\tau - x)^{n+1}}\,d\tau,$$

where C is a contour containing x but not 0; this integral is equal to

$$\tfrac{1}{2}\frac{d^n}{dx^n} f(\sqrt{x}).$$

78. $$|\sinh(x+iy)|^2 = \left(\frac{e^{x+iy}-e^{-x-iy}}{2}\right)\left(\frac{e^{x-iy}-e^{-x+iy}}{2}\right)$$

$$= \tfrac{1}{2}(\cosh 2x - \cos 2y)$$

$$\geqq \tfrac{1}{2}(\cosh 2x - 1).$$

Integrate along the boundary of a square with sides $x = \pm\pi(n+\frac{1}{2})$ and $y = \pm\pi(n+\frac{1}{2})$, where n is an integer. As $n \to \infty$ the integral tends to zero; hence the sum of the residues tends to zero.

INDEX